Advances in Friction Stir Welding and Processing

Related titles:

Friction Stir Welding
(ISBN 978-1-84569-450-0)

Self-Piercing Riveting
(ISBN 978-1-84569-535-4)

Advances in Laser Materials Processing technology
(ISBN 978-1-84569-476-6)

Advances in Friction Stir Welding and Processing

Mohammad Kazem Besharati Givi and
Parviz Asadi

ELSEVIER

AMSTERDAM • BOSTON • CAMBRIDGE • HEIDELBERG • LONDON
NEW YORK • OXFORD • PARIS • SAN DIEGO
SAN FRANCISCO • SINGAPORE • SYDNEY • TOKYO
Woodhead Publishing is an imprint of Elsevier

WP
WOODHEAD
PUBLISHING

Woodhead Publishing is an imprint of Elsevier
80 High Street, Sawston, Cambridge, CB22 3HJ, UK
225 Wyman Street, Waltham, MA 02451, USA
Langford Lane, Kidlington, OX5 1GB, UK

British Library Cataloguing-in-Publication Data
A catalogue record for this book is available from the British Library

Library of Congress Control Number: 2014946706

ISBN 978-0-85709-454-4 (print)
ISBN 978-0-85709-455-1 (online)

For information on all Woodhead Publishing publications visit our website at http://store.elsevier.com/

Typeset by RefineCatch Limited, Bungay, Suffolk

To Imam Mohammad bin Ali who was just as described in Torah,
Splitter of Science and to his son Imam Jafar bin Mohammad,
who taught science and wisdom to thousands, one being Geber

Contents

List of figures and tables

Figures

Tables

List of abbreviations

AA	aluminium alloy
ABS (ch 14)	butadiene styrene
AMC	aluminium matrix composite
ANN	artificial neural network
ANOVA	analysis of variance
AS	advancing side
ASTM	American Society for Testing and Materials
b.c.c.	body centred cubic
BEI	backscattered electron image
BM	base metal
CDRX	continuous dynamic recrystallization
CeO_2	cerium oxide
CFD	computational fluid dynamics
CNC	Computer Numerical Control
COF	coefficient of friction
CSL	coincidence site lattice
CTOD	crack tip open displacement
Cu	copper
1-D	one-dimensional
2-D	two-dimensional
3-D	three-dimensional
DDRX	discontinuous dynamic recrystallization
DOE	design of experiment
DSC	differential scanning calorimetry
DSS	duplex stainless steel
DXZ	dynamically recrystallized zone
EA	evolutionary algorithm
EADS	European Aeronautic Defense and Space company
EBSD	electron back scatter diffraction
EBW	electron beam welding
ECAE	equal channel angular extrusion
EDAX (ch 10)	energy dispersive spectroscopy

EDM	electrical discharge machine
EDS (ch. 6)	energy dispersive spectroscopy
EIS	electrochemical impedance spectroscopy
ESA – ES TEC	European Space Agency – European Space Research and Technology Centre
FC300	flake graphite cast iron
FC700	nodular graphite cast iron
f.c.c.	face centred cubic
FE	field emission
FEM	finite element method
FGM	functionally gradient material
FRW	friction welding
FS	friction stir
FSKW	friction stir knead welding
FSLLW	friction stir lap linear welding
FSLW	friction stir lap welding
FSP	friction stir processing
FSSW	friction stir spot welding
FSW	friction stir welding
FVM	finite volume method
Fx	advancing force
Fz	vertical downward force
FZ	fusion zone
GA	genetic algorithm
GDRX	geometrical dynamic recrystallization
GMA-P	pulse gas metal arc welding
GMAW	gas metal arc welding
GMDH	group method of data handling
GONNS	genetically optimized neural network systems
GTAW	gas tungsten arc welding
HABs	high-angle boundaries
HAZ	heat affected zone
HCl	hydrogen chloride
hcp	hexagonal close packed
HDPE	high-density polyethylene
HPDC	high-pressure die cast
HT	heat treated
HV (ch 9)	Hardness Vickers
IF	interstitial-free
iHDT	incremental hole drilling technique
IM	ingot metallurgy

INCO	International Nickel Company
I-STIR PDS	Intelligent Stir Welding for Industry and Research Process Development System
JAA-FAA	Joint Aviation Authorities – Federal aviation administration
LABs	low-angle boundaries
LAMIG	laser-MIG hybrid welding
LBW	laser beam welding
LDPE	low-density polyethylene
LFW	liner friction welding
LHT	left-hand thread
LM	laser melted
LOC	lack of consolidation
MA	mechanical alloying
MCP	mechanochemical processing
Mg	magnesium
MIG	metal inert gas
MMC	metal-matrix composite
MOEA	multi-objective evolutionary algorithm
MOGA	multi-objective genetic algorithm
MOOP	multi-objective optimization problem
MRE	Meal, Ready-to-Eat
NAB	NiAl bronze
NACE	National Association of Corrosion Engineers
ND	normal direction
NPGA	niched-Pareto genetic algorithm
NSGA	non-dominated sorting genetic algorithm
NVH	noise-vibration harshness
NZ	nugget zone
OA	orthogonal array
ODS	oxide dispersion-strengthened
OM	optical microscopy
PAES	Pareto archived evolution strategy
PAW	plasma arc welding
PCBN	polycrystalline cubic boron nitride
PET	polyethyleneterepthalate
P-FSW	plasma-assisted friction stir welding
PFZs	precipitate-free zones
PHI	Pseudo Heat Index
PID	Proportional, Integral and Derivative
PSN	particle stimulated nucleation

PVD	physical vapour deposition
QFN	Quad Flat Non-lead
RBFNN	radial basis function neural network
RD	reference direction
RE	rare-earth
rpm	revolutions per minute
RPT	retractable pin tool
RS	retreating side
RSM	response surface method
SCC	stress corrosion cracking
SCE	saturated calomel electrode
SCT	surface crack tension
SD	shear direction
SEM	scanning electron microscope
SFSW	submerged friction stir welding
SH	straight hexagon
SHTA	solution heat treatment and ageing
SMAW	submerged arc welding
S/N	signal-to-noise
SNZ	stir nugget zone
SO	straight octagon
SPD	severe plastic deformation
SPEA	strength Pareto evolutionary algorithm
SPN	shear plane normal
SRV	static recovery
SRX	static recrystallization
SS	stainless steel
SSq	straight square
SSFSW	stationary shoulder friction stir welding
SSRT	strain rate tensile test
SUS	stochastic universal selection
SZ	stir zone
TD	traverse direction
TEM	transmission electron microscope
Tg	glass transition
Ti	titanium
TIG	tungsten inert gas
T_m	melt transition
TMAZ	thermo-mechanically affected zone
TO	tapered octagon
TRC	transfer ratio of the conveyor

TRS	tool rotation speed
TS	tapered square
TWB	tailor welded blanks
TWI	The Welding Institute
UTS	ultimate tensile strength
VEBW	vacuum electron beam welding
VEGA	vector evaluated genetic algorithm
WC	tungsten carbide
WD	welding direction
XRD	X-ray diffraction
YS	yield strength

Acknowledgements

It gives us a great pleasure to acknowledge all our colleagues contributing to this book.

We sincerely thank the students and academic members at the University of Tehran for their encouragement during this endeavour. Special thanks must go to Nader Parvin, Professor at the Amirkabir University of Technology, Iran, for his helpful guidance.

We would like to show our gratitude to all at Woodhead Publishing Limited for their efforts made to give us the opportunity to prepare the book on time.

Lastly we are thankful to the readers and look forward for their constructive suggestions, which will enable us to improve the next edition.

Mohammad Kazem Besharati Givi
Parviz Asadi
School of Mechanical Engineering, University of Tehran

Preface

In the Name of Allah the Compassionate, the Merciful

Friction stir welding and processing are the most state-of-the-art techniques in solid state welding and processing among the joining and processing methods used in the manufacturing of industrial products.

Lack of standards, handbooks and other types of documents necessitate collection of all the experimental and analytical procedures, results and ideas as much as possible, in order to draw a clear prospect for the future of these new and non-traditional approaches.

Advances in friction stir welding and processing is intended to give the students, researchers and persons engaged in the field, using these newly developed solid state welding and surface improving techniques, the most up-to-date information available. The book contains relatively broad comprehensive research methods and the results found at the time writing the materials, and is presented in 15 chapters that will give the readers a relatively clear roadmap.

Mohammad Kazem Besharati Givi
Parviz Asadi
School of Mechanical Engineering, University of Tehran

About the authors

Mohammad Kazem Besharati Givi is an Associate Professor in the Faculty of Mechanical Engineering, University of Tehran. He has over 40 years teaching experience, having taught courses including machine elements, material science and engineering, metal forming, metallurgy in manufacturing, innovation management and technology innovation systems. He has supervised more than 65 post-graduate students, won the National Book of the Year award in 2006, and has published more than 80 ISI articles in fields of manufacturing methods (focusing on friction-welding and processing methods), metal forming, deep drawing extrusion and advanced processing of materials. Besharati has an MSc in Mechanical Engineering from the University of Tabriz, and a PhD awarded by the University of Leeds for his investigation into the effects of deep drawing on the orientation of sheet steels. He has fulfilled numerous industrial R&D projects, particularly in automotive engineering and industrial machinery, some leading to solving severe problems in catastrophic accidents, with others entailing industrial independence from abroad.

Parviz Asadi is a member of Iran's National Elites Foundation (INEF) and has a PhD in Mechanical Engineering from the University of Tehran. He has focused on simulation, analyzing and producing the metal matrix composites with nano-particles during his post-graduate course, also dealing with friction stir welding and processing of various similar and dissimilar materials. His seven-year research studies in the fields of friction stir welding/processing and materials science are published in more than 25 ISI papers, also including several patents. He has succeeded in achieving numerous scientific awards for his scientific activities, such as first top student researcher. Asadi is currently a lecturer at the University of Tehran, and also referee of several scientific journals. Along with his scientific activities, Asadi has taken part in some industrial projects for designing and manufacturing the mechanisms and apparatus for the aerospace, oil and gas industries.

Donya Ahmadkhaniha is a PhD student at the University of Tehran. She is working on microstructure modification of Mg alloys by friction stir processing for biodegradable materials applications. In addition, she has been involved in other research on in situ formation of Ai-Ni composite via friction stir processing. Donya received her BSc degree in Materials and Science Engineering from the University of Tehran in 2006. She also received her MSc from Amirkabir University of Technology (Polytechnic) in 2009 and she ranked first. Her MSc research project was on plasma nitriding of steels. She has already published a number of papers both on friction stir processing and plasma nitriding.

Mostafa Akbari has a BSc in Mechanical Engineering from the University of Tehran and an MSc in Automotive Engineering from the Iran University of Science and Technology (IUST) and is now a PhD student at IUST. He has seven years' experience in friction stir welding, optimization and finite element methods. Akbari has worked on the development and/or implementation of over 30 friction stir welding projects, including optimizing FSW parameters, simulation of FSW using finite element methods, investigation of mechanical and microstructural properties of friction stir welded joints and has published over 20 ISI papers in these areas.

Esther Titilayo Akinlabi is currently a Senior Lecturer at the Department of Mechanical Engineering Science, University of Johannesburg. She has both industrial and academic experience of about 12 years. She completed her Doctorate in Mechanical Engineering in the field of Friction Stir Welding at the Nelson Mandela Metropolitan University, Port Elizabeth, South Africa. Her doctorate research work focused mainly on dissimilar joining using the friction stir welding process specifically as applied to joining aluminium and copper. Her other research areas include manufacturing processes, surface modification techniques, laser material processing and materials characterization. She has more than 70 publications to her credit. She is registered with the Engineering Council of South Africa and the Council for Regulation of Engineering in Nigeria. She is also a member of the South African Institute of Mechanical Engineers and a member of the Nigerian Society of Engineers.

Stephen Akinwale Akinlabi is currently a staff member at the School of Mechanical, Industrial and Aeronautical Engineering, University of the Witwatersrand. An emerging professional academic and researcher with interests in the area of manufacturing processes, laser material processing and its applications, material science and characterization, welding processes and friction stir processing, and surface modifications techniques. He is a professional mechanical engineer with over 12 years' industrial work experience both upstream and downstream of the oil and gas industry, delivering timely and quality results in the areas of Project Management and Control, and Maintenance Engineering Systems. Furthermore, Akinlabi has over five years' academic experience in conducting research, teaching and tutoring both undergraduate and graduate students. He is also a registered member of relevant professional institutions such as the Nigerian Society of Engineers (MNSE), Council for Regulation of Engineering in Nigeria (MCOREN), South Africa Institution of Mechanical Engineering (MSAIMechE) and the Engineering Council of South Africa (ECSA).

Ali Amini is a member of Iran's National Elites Foundation (INEF). In 2013, he received his BSc in Mechanical Engineering from the University of Tehran, and also was placed second in Iran's National Olympiad of Mechanical Engineering, winning the Silver Medal. Being a distinguished student, he was awarded a Master's Degree scholarship and a Research Assistant position in Fluids Mechanical Engineering at the same university. Since 2011, he has contributed to industrial issues throughout his co-operations in projects concerned with power plant systems and optimization. He has a strong research background in biomechanics, energy systems and materials science.

Harpreet Singh Arora is a Post-Doctorate Research Associate in the Department of Material Science and Engineering at University of North Texas. His research interests include severe plastic deformation of materials, friction stir processing and welding, structure-property relationships, degradation of materials, and bulk metallic glasses. He has worked on improving surface degradation of materials by wear, erosion and corrosion through structural modifications using friction stir processing. He has published 19 papers in reputed international journals. He has a Bachelor's degree in Mechanical Engineering from Punjab Technical University, Jalandhar, and a PhD in Material Processing from the Indian Institute of Technology, Ropar.

Swarup Bag is currently an Assistant Professor at the Department of Mechanical Engineering, Indian Institute of Technology Guwahati, Assam, India. He has more than three years teaching experience, two years industrial experience and eight years of research experience. He did his PhD in Mechanical Engineering at the Indian Institute of Technology Bombay (IIT Bombay), Mumbai, India. He has 20 months post-doctoral experience from CEMEF-MinesParisTech, Sophia Antipolis, France. His research interest mainly focuses on materials and manufacturing technology of welding and joining. His research area includes finite element modelling of manufacturing processes, laser material processing, micro-joining, recrystallization and optimization in manufacturing processes. Bag has more than 32 publications in peer-reviewed international journals and conferences.

Mohsen Barmouz is a researcher at the University of Tarbiat Modares. He has over five years experience in research and experiments within university projects. He started working on friction stir processing in 2009 and has worked on many projects on fabrication of micro- and nano-composites by adding different types of fillers such as SiC, WC and CNT. In 2011, he began work on fabrication of polymeric nano-composites via friction stir processing and has successfully published in this field. Barmouz has a Master's degree from the University of Tehran in Manufacturing Engineering and is studying for his PhD at the University of Tarbiat Modares.

Daniel Filipe Oliveira Braga graduated in 2012 from the Faculty of Engineering at the University of Porto (FEUP) with a Master's degree in Mechanical Engineering. He then worked as a research scholar in the Institute of Mechanical Engineering and Industrial Management (INEGI), more specifically in the Laboratory of Optics and Experimental Mechanics (LOME). Between 2013 and 2014, Braga did an internship at EADS-IW (now Airbus Group Innovations) in Ottobrunn, Germany. In 2014, he became a PhD student in Leaders for Technical Industries in FEUP within the MIT-Portugal programme. Braga was also awarded a PhD scholarship by the Portuguese National Science and Technology Foundation in 2014.

B.K. Dhindaw graduated from the Indian Institute of Technology Kharagpur, India and earned BTech (Hons) and PhD degrees. He has more than four decades of teaching and research experience, mostly at the Indian Institute of Technology, Kharagpur and various universities and research institutions in USA, Sweden, Japan, UK, Canada, Germany and Malaysia. His research interests include general physical metallurgy, solidification processing, composites and friction stir processing. He has guided 12 PhDs and published more than 160 papers in peer-reviewed journals, 5 books and monographs. He has received certificates of recognition from NASA, USA and best metallurgist award from the Government of India. He has been reviewer for several high impact factor journals and is on the editorial board of the *International Journal of Cast Metals Research*.

Harpreet Singh Grewal is a research scholar at the School of Mechanical, Materials and Energy Engineering of Indian Institute of Technology, Ropar, India. His research interests include tribology and surface modification of materials. He is a life member of the Tribological Society of India and has published over 21 research articles in peer-reviewed international journals. He received his Bachelor's degree in Mechanical Engineering and Master's degree in Machine Design from the Punjab Technical University, India. He is doing his PhD in Mechanical Engineering at IIT Ropar. Currently he is working as a research associate in the Department of BioMicrosystems of the Korea Institute of Science and Technology, Seoul.

Hamid Hanifian has an MSc degree in Mechanical Engineering/Manufacturing from the Tehran University of Science and Research, Islamic Azad University and is a faculty member of the Islamic Azad University, Shahindezh branch. He is a specialist in the field of metallurgy and welding and has published two ISI papers as a result of his four years research in these fields. He has five years of teaching experience and has carried out industrial and research projects.

Akbar Heidarzadeh has an MSc degree in material engineering (casting field) from the University of Tehran, Iran, and is now a PhD candidate in Materials Science (welding field) at the Sahand University of Technology, Iran. He is an academic staff member of the Azarbaijan Shahid Madani University, Iran. His research experience is focused on production of Al-Mg_2Si metal matrix *in situ* composites, friction stir welding (of nonferrous metals such as pure copper, copper alloys, aluminium alloys and magnesium alloys), and using response surface methodology (RSM) for optimizing the friction stir welding parameters. He has published several ISI papers in these fields.

Virgínia Infante has an MSc in 1998 and a PhD in 2002 in Mechanical Engineering from the Instituto Superior Técnico (IST), University of Lisbon, Portugal. Since 1995, she has been working on mechanical behaviour of materials and has been involved in several national and international research projects. Currently, Infante is Assistant Professor in the Mechanical Engineering Department at IST. She has special expertise in experimental testing and has published over 30 journals and conference papers in the field of fatigue. She is a member of the editorial board of the *Engineering Failure Analysis Journal*.

Hamid Karimi-Nemch is a mechanical engineer focused on manufacturing processes. He has an MSc in Mechanical Engineering-Manufacturing and graduated from Iran University of Science and Technology. He has been working as a lecturer in Islamic Azad University and Payame Noor University since 2010. Furthermore, he has experience of research on high strain rate metal forming processes such as electromagnetic forming and electrohydraulic forming, resulting in publications in these fields. He has also gained much experience in working in the oil, gas and petrochemical industries.

Kamel Kazemi-Choobi has seven years' experience in welding and joining of materials and shape memory materials (SMAs). He is the author of a book, 12 scientific publications and 2 patents. His main research interests include smart materials and actuators, welding and forming of materials. He has served as a researcher at research centres for advanced materials and mineral processing in the Sahand University of Technology, where he conducted research in SMAs and friction stir welding and processing (FSP/FSW). He recently became a member of the technical staff at D&T, where he is project manager for investigation of alloying advantages in Pb-Zn alloys.

Seyyed Tohid Khandani has BSc and MSc degrees in Mechanical Engineering from the University of Tabriz, and is a PhD student at the School of Mechanical Engineering, College of Engineering, University of Tehran. He has several years of experience in research fields consisting of metal forming, severe plastic deformation, metal matrix composites, solid state welding and processing, powder, etc. As a result, he has several published research papers on composite fabrication by FSP methods in both indexed journals and international conference proceedings. Also, he has two years working experience as a technical consultant for Iranian Science and Technology Park.

Ghazaleh Khodabandehlouie is a researcher. She has a Bachelor's degree in computer engineering and a Master's degree in Industrial Engineering from the Islamic Azad University of Iran. Her research experience is focused on optimization techniques as well as artificial neural networks (ANNs) and has studied classification and clustering methods. Along the same lines, she proposed a new initialization method for k-means algorithm in clustering problems. Also she is a member of the Robotics Research Centre in the NAJ Robotics Company, where she is applying composition of machine learning algorithms and clustering methods in order to process images.

Vladimir N. Malyshev graduated in 1976 from the Ufa Petroleum Institute (UPI), majoring in 'Machinery and Equipment for Oil and Gas Fields'. In 1984 he post-graduated at the Gubkin Moscow Petroleum Institute and defended his thesis 'Enhancement of the wear resistance of friction knots by microarc oxidation' and in 1999, his doctoral thesis on 'Hardening of friction surfaces by microarc oxidation', at the Gubkin Russian State University of Oil and Gas. He has worked as an engineer, as a Junior Scientific Collaborator in UPI (1976–1980), Senior Engineer of Lab for Special Materials in the Kharkiv branch of VNIPImorneftegaz (1984–1986) and Assistant Professor of Brest Polytechnic Institute (1986–1994). Since 2000, he has been Associate Professor, then from 2004 Professor of Chair for 'Tribology and Repair Technologies of Oil and Gas Equipment' at the Gubkin Russian State University of Oil and Gas. His main research areas are materials science, tribology, wear resistant and protective ceramic coatings on metals, and synergetics. Malyshev is the author of more than 190 scientific papers, including 2 textbooks, a monograph, 18 copyright certificates and 5 patents of the Russian Federation, more than 140 articles and abstracts and 8 learning-methodical developments. He was awarded with Sign 'Inventor of the USSR' in 1982 and Sign of Laureate of All-Russian Contest 'Engineer of Year' in 2012.

Pedro M.G.P. Moreira is an expert in Structural Integrity and Mechanical Testing. He received his PhD in 2008 in Mechanical Engineering from the University of Porto. He started his research activities in 2001 and since 2008 he has been a member of INEGI. His research interests mainly focus on fracture mechanics, fatigue, structural integrity and experimental mechanics. He has published 1 book, 3 chapters in books, more than 45 papers in scientific journals, and around 100 publications/presentations in international scientific conferences. These publications are the result of his participation in several projects, mostly focused on the development and analysis of aero-structures. Since 2002, he has lectured in several Mechanical Engineering courses in university institutions, and has been supervisor or co-supervisor of several MSc and PhD students. He has been involved as a member in the organization of around 20 scientific conferences and a coordinator or researcher in several international (EC and ESA) and national projects.

Sundeep Mukherjee is an Associate Professor in the Department of Material Science and Engineering at University of North Texas, USA. Prior to joining the University of North Texas in 2013, he worked as Senior Process Engineer at Intel Corporation for nearly six years and briefly as a Post-Doctoral Research Associate at Yale University, USA. His research interests include bulk metallic glasses, thin films, semiconductor devices, electrochemistry and energy conversion. He has published more than 25 papers in reputed international journals and given many invited talks. He has a Bachelor's degree in Material Engineering from the Indian Institute of Technology, Kharagpur, and a Master's and PhD degree in Material Science from the California Institute of Technology, USA.

Sukhomayl Pal is an Assistant Professor at the Department of Mechanical Engineering, IIT Guwahati, India. He has over four years of teaching and eight years of research experiences. His research interests include monitoring of welding processes, monitoring of machining processes, friction stir welding, modelling and optimization of manufacturing processes using soft computing and artificial neural networks. He has published more than 25 peer-reviewed papers in reputed international journals and conferences. He is also involved as Investigator and Consultant in several projects funded by the DST and North Eastern Electrical Power Corporation Ltd. Pal received his Bachelor's degree in Production Engineering from Jadavpur University, his Master's degree in Production Engineering from Bengal Engineering and Science University, and his PhD in Mechanical Engineering from IIT Kharagpur, India. He was a Post-doctoral Fellow at the University of Pretoria, South Africa.

Peyman Salami is a teacher assistant at the Islamic Azad University. He has four years' experience in consulting and research and has specialized in friction stir processing since 2010. He completed a number of projects in metal forming and tool designing. He has presented articles in national and international conferences and published an article in an international journal. Salami has a Bachelor's degree in Manufacturing Engineering from Tabriz University and a Master's degree in Manufacturing Engineering from the K.N. Toosi University of Technology.

Peyman Shahi received his Bachelor of Engineering degree from the Amirkabir University of Technology. He also obtained his MSc and PhD degrees in 2010 and 2014, respectively, in Mechanical Engineering from Tarbiat Modares University. In addition, he was a visiting researcher at the University of Ontario Institute of Technology (UOIT) during his PhD period. His main research areas involve the processing and characterization of polymer and composites with thermoplastic matrices, as he has published several papers in ISI journals and conference proceedings. He also has industrial experience in fabrication technology, engineering design, materials and has acted as a consultant to various industrial companies within Iran, since 2006. Furthermore, he has four years teaching experience in several Iranian universities.

Ana Silva is a Metallurgical and Materials Engineer with a Master's degree from the Faculty of Engineering of the University of Porto (FEUP) in 2011. He was an early researcher scholar at the Department of Materials and Metallurgic Engineering (DEMM, FEUP) in 2010 and 2011 and then began working as a research scholar in the Laboratory of Optics and Experimental Mechanics (LOME) at the Institute of Mechanical Engineering and Industrial Management (INEGI). He is currently working in the residual stress measurements and friction stir welding (FSW) process fields, among others. Silva began a traineeship in Fraunhofer IWS, Dresden, Germany, concerning bobbin tool friction stir welding (BTFSW) at the end of 2013.

Harpreet Singh is currently working as an Associate Professor in Mechanical Engineering at the Indian Institute of Technology Ropar, Punjab. After obtaining a BE and ME in Mechanical Engineering, he did his PhD at the Indian Institute of Technology, Roorkee, India in 2005. His areas of interest include surface engineering-degradation of materials, high temperature corrosion and its prevention, slurry erosion of hydraulic turbines and its control, biomedical coatings, and friction stir processing/welding. He has authored a monograph and 2 book chapters, and contributed more than 150 research papers in various international/national journals and conferences. He has produced six PhDs. He is a recipient of many Indian research-related awards and has been involved in several sponsored research projects. He has visited USA, UK, Netherlands, Canada, South Korea and Germany, and delivered several invited talks.

Catarina Vidal is a PhD student in Mechanical Engineering at Instituto Superior Técnico, University of Lisbon, Portugal. She has an MSc degree in Aerospace Engineering from the Technical University of Lisbon. She is also author of several publications, with particular expertise in fatigue testing and friction stir welding and channelling.

Deepak Kumar Yaduwanshi is currently a Research Scholar in the Department of Mechanical Engineering at the Indian Institute of Technology Guwahati, Assam, India. He completed his diploma in Mechanical Engineering from Sardar Vallabh Bhai Polytechnic College, Bhopal, Madhya Pradesh, in 2002. He graduated in Mechanical Engineering from Bhopal Institute of Technology and Science, Bhopal, Madhya Pradesh, in 2007. He worked as a post-graduate in Production Engineering at the National Institute of Technology, Rourkela, Orissa, India in 2010. His research interest includes thermo-mechanical analysis of hybrid friction stir welding process.

Parisa Zolghadr is a Master of Science student in the Faculty of Mechanical Engineering at Payame Noor University (PNU) Tehran, Iran and has a BSc in Mechanical Engineering from the University of Applied Science, Qazvin Science and Technologies Branch. She is also author of several conference papers with particular expertise in welding. Currently her research focuses on the Active Tungsten Inert Gas (ATIG) welding of aluminium alloys.

Contributor contact details

Chapter 1

Mohammad Kazem Besharati Givi and Parviz Asadi*
School of Mechanical Engineering
College of Engineering
University of Tehran
Iran

e-mail: p.asadi85@gmail.com

Chapter 2

S. Bag*, D. Yaduwanshi and S. Pal
Department of Mechanical Engineering
Indian Institute of Technology Guwahati (IIT Guwahati)
Guwahati 781039
Assam, India

e-mail: swarupbag@iitg.ernet.in

Chapter 3

Akbar Heidarzadeh*
Faculty of Engineering
Azarbaijan Shahid Madani University
Tabriz, Iran

e-mail: ak.hz62@gmail.com

Kamel Kazemi-Choobi
Research Center for Advanced Materials and Mineral Processing
Faculty of Materials Engineering
Sahand University of Technology
Tabriz, Iran

Hamid Hanifian
Department of Mechanical Engineering
College of Mechanical Engineering
Miandoab Branch
Islamic Azad University
Miandoab, Iran

Parviz Asadi
School of Mechanical Engineering
College of Engineering
University of Tehran
Iran

Chapter 4

D.F.O. Braga*, A.C.F. da Silva and P.M.G.P. Moreira
Institute of Mechanical Engineering and Industrial Management (INEGI)
University of Porto (UP)
Campus da FEUP Rua Dr Roberto Frias
400, 4200-465 Porto, Portugal

e-mail: dbraga@inegi.up.pt

Chapter 5

Virgínia Infante* and Catarina Vidal
Institute of Materials and Surfaces Science and Engineering (ICEMS)
Instituto Superior Técnico
University of Lisbon
Av. Rovisco Pais
1, 1049-001 Lisbon, Portugal

e-mail: virginia@dem.ist.utl.pt

Chapter 6

E.T. Akinlabi*
Department of Mechanical Engineering Science
University of Johannesburg
P.O Box 524
Auckland Park Kingsway Campus
Johannesburg, South Africa

e-mail: etakinlabi@uj.ac.za

S.A. Akinlabi
School of Mechanical, Industrial and Aeronautical Engineering
University of the Witwatersrand
South West Campus
Private Bag 3
Wits 2050
Johannesburg, South Africa

Chapter 7

H.S. Arora and S. Mukherjee
Department of Materials Science and Engineering
University of North Texas
Denton
Texas 76203, USA

H.S. Grewal and H. Singh
School of Mechanical, Materials and Energy Engineering
Indian Institute of Technology Ropar
Rupnagar
Punjab-140001, India

B.K. Dhindaw*
Mechanical Engineering,
Faculty of Engineering
Christ University
Bangalore-560060
India

e-mail: dhindaw@metal.iitkgp.ernet.in

Chapter 8

Vladimir N. Malyshev
Tribology and Repair technologies of oil and gas equipment
Gubkin Russian State University of Oil and Gas

e-mail: vmal@inbox.ru

Chapter 9

Donya Ahmadkhaniha
School of Metallurgy and Materials
College of Engineering
University of Tehran
Iran

Parviz Asadi*
School of Mechanical Engineering
College of Engineering
University of Tehran
Iran

e-mail: p.asadi85@gmail.com

Chapter 10

Peyman Salami*
School of Mechanical Engineering
College of Engineering
University of Khaje nasireldin toosi
Tehran, Iran

e-mail: salami.peyman@gmail.com

Tohid Khandani, Parviz Asadi and Mohammad Kazem Besharati Givi
School of Mechanical Engineering
College of Engineering
University of Tehran
Iran

Chapter 11

Peyman Shahi* and Mohsen Barmouz
College of Engineering
Tarbiat Modares University
Tehran, Iran

e-mail: peyman.shahi@gmail.com

Parviz Asadi
School of Mechanical Engineering
College of Engineering
University of Tehran
Iran

Chapter 12

Parviz Asadi
School of Mechanical Engineering
College of Engineering
University of Tehran
Iran

Mostafa Akbari*
School of Automotive Engineering
Iran University of Science and Technology
Tehran, Iran

e-mail: mr.mostafaakbari@yahoo.com

Hamid Karimi-Nemch
School of Mechanical Engineering
Iran University of Science and Technology
Tehran, Iran

Chapter 13

Mostafa Akbari*
School of Automotive Engineering
Iran University of Science and Technology
Tehran, Iran

e-mail: mr.mostafaakbari@yahoo.com

Parviz Asadi and Mohammad Kazem Besharati Givi
School of Mechanical Engineering
College of Engineering
University of Tehran
Iran

Ghazaleh Khodabandehlouie
Faculty of Industrial and Mechanical Engineering
Qazvin Branch
Islamic Azad University
Qazvin, Iran

Chapter 14

Mohsen Barmouz* and Peyman Shahi
College of Engineering
Tarbiat Modares University
Tehran, Iran

e-mail: mbarmouz@ut.ac.ir

Parviz Asadi
School of Mechanical Engineering
College of Engineering
University of Tehran
Iran

Chapter 15

Ali Amini and Parviz Asadi*
School of Mechanical Engineering
College of Engineering
University of Tehran
Iran

e-mail: p.asadi85@gmail.com

Parisa Zolghadr
Faculty of Mechanical Engineering
Payame Noor University (PNU)
Tehran, Iran

General introduction

M. K. Besharati Givi and P. Asadi, University of Tehran, Iran

DOI: 10.1533/9780857094551.1

Abstract: This first chapter of the book gives a general introduction to the chronological development of different welding methods in fusion and solid-state categories. Then the different solid-state friction welding (FW) methods are explained, with the emphasis on friction stir welding and its required tools and resulting microstructure. Advantages and disadvantages of FSW in comparison with fusion welds are described and friction stir processing as a surface modification technique is discussed. Finally the outline of the book, with the content of the different chapters is briefly introduced.

Key words: welding development, friction stir welding, friction stir processes, solid-state welding.

1.1 Historical development of welding

Welding, a fabrication or sculptural process that joins materials, has a history as far back as the Bronze Age, where gold circular boxes estimated to be formed by pressure welding some 2000 years ago, have been recovered. The people in the Eastern Mediterranean area, especially the Egyptians, learned to weld pieces of iron together 3000 years ago. They left behind many tools and weapons, and also items of iron and bronze that exhibit forging, and forge welding operations have been recovered from the pyramids of Egypt. During the Middle Ages, many items of iron were produced that have been welded by hammering. One of the largest welds from this period is the Iron Pillar of Delhi in India, which was established nearly 1700 years ago. It was not until the 19th century that professional welding, as known today, was invented.

Sir Humphrey Davies from England discovered acetylene in the mid-1800s and also managed to produce an arc between two carbon electrodes by using a battery. Then the electric generator was invented and arc lighting became popular in the mid-1800s. During the late 1800s and early 1900s, gas welding and cutting, and arc welding with the carbon arc and metal arc, were invented. The resistance welding process became a practical jointing process.

In 1881, De Meritens in France used the heat of an arc to join lead plates for storage batteries. His Russian pupil Nikolai Benardos was granted a patent for welding along with a Russian fellow, Stanislaus Olszewski, in the late 1800s, thus starting the era of carbon arc welding. Bernardos' endeavours were limited to carbon arc welding, though he was able to weld both iron and lead. Although Benardos was not successful with metallic electrodes, his attempts made carbon arc welding popular during the late 1890s and early 1900s.

In 1892, C.L. Coffin of Detroit earned the first US patent for an arc welding process by use of a metal electrode, making this the first record of the metal melted from the electrode actually carried across the arc to deposit filler metal in the joint to form the weld. At about the same time, N.G. Slavianoff, a Russian, presented the same idea of transferring metal across an arc, but to cast metal in a mould.

Around 1900, Strohmeyer in England invented a coated metal electrode, with a thin wash coating of clay or lime, to provide a more stable arc. In 1907, Oscar Kjellberg from Sweden worked on a covered or coated electrode, which took him 7 to 8 years to produce stick electrodes by dipping short lengths of bare iron wire into thick mixtures of carbonates and silicates, and allowing the coating to dry.

Meanwhile, the resistance welding processes, including spot welding, seam welding, flash butt welding and projection welding were developed. In 1903, Goldschmidt invented thermite welding, which was used to weld train track rails.

Gas welding and cutting was also promoted during this time period by the production of oxygen and later the liquefying of air, accompanied by applying a torch (a blow pipe). Before 1900, hydrogen and coal gas were used with oxygen. When a torch suitable for use with low-pressure acetylene was developed, oxyacetylene gas welding and cutting processes were launched.

Generally, during the first two decades of the 1900s, oxyacetylene welding and cutting processes along with carbon arc welding and metal arc welding processes with lightly covered electrodes were used for repair and maintenance work.

With the start of World War I, demands for metal material production drastically increased, hugely focusing on welding as a servicing process. To meet these demands, many firms in the US and Europe rose to the challenge to produce all types of welding machines and electrodes. In 1919, when the war was over, the American Welding Society was founded by 20 members, from the Wartime Welding Committee of the Emergency Fleet Corporation, led by Comfort Avery Adams. This organization was founded as a non-profit making organization with the aim of advancing welding and allied processes.

In 1920, automatic welding was introduced by the General Electric Company. It used bare electrode wire, which was operated by direct current and utilized arc voltage as the basis of the regulator for the wire feed rate. It was used to build up worn motor shafts and worn crane wheels. Furthermore, automobile industries utilized it in production of rear axle housings.

Several grades of electrodes were developed during the 1920s. Mild steel or low carbon steel, usually with 0.20% or less carbon content, was used to weld practically all types of rolled steel. Higher carbon electrodes and alloy steel electrodes also became available. Copper alloy rods were used for carbon arc welding and brazing.

In addition, during this decade, some research works were conducted to protect the arc and welding zone, which concluded with the development of gas-shielded welding techniques. It was clearly understood that the diffusion of oxygen and nitrogen in the molten weld metal resulted in brittle and sometimes porous welds. Alexander and Langmuir used hydrogen as a shield for the welding atmosphere and used two electrodes starting with carbon electrodes, later changing to tungsten electrodes.

During welding, the shielding hydrogen was atomized under the arc, and then blown out of the arc, generating a very hot flame and releasing the heat. This arc produces half again as much heat as an oxyacetylene flame. This process, named the atomic hydrogen welding process, never became very popular; however, it was utilized for special applications and somehow for welding of tool steels. The covered electrode became the backbone of the welding industry, although it was utilized manually. Concurrently, many attempts were made to improve automatic welding by using bare wire, such as using a covering on the electrodes, which were then coiled, and at the welding point the coating was milled away to introduce the welding current to the core wire.

Stud welding was developed in 1930 at the New York Navy Yard, specifically for attaching wooden decking over a metal surface, in which the process welded studs, screws, etc., to the base metal using a special gun that controlled the arc automatically. Addition of fluxing elements into the stud improved the weld properties. Stud welding attracted the attention of many in the shipbuilding and construction industries.

Also in the 1930s, the submerged arc welding (SMAW) process became very popular, being developed by the National Tube Company for a new pipe mill at McKeesport, Pennsylvania. This 'under powder' or smothered arc welding process was designed to make longitudinal seams in the pipe. It is still one of the most productive and popular welding processes used up to the present day.

In 1890, Coffin patented a welding method in a non-oxidizing gas atmosphere, leading to what is known as gas tungsten arc welding (GTAW or tungsten inert gas, TIG). Hobart scrutinized GTAW in the late 1920s by using helium for shielding, and later Devers used argon. Again it was improved and patented by Meredith in 1941, and named Heliarc® welding, which was later licensed to Linde Air Products.

The Air Reduction Company sponsored the Battelle Memorial Institute to develop the gas shielded metal arc welding process in 1948. This used the gas-shielded arc similar to the gas tungsten arc, but replaced the tungsten electrode with a continuously fed electrode wire.

Utilization of consumable electrodes for welding in an atmosphere of CO_2 gas was first announced in 1953 by Lyubavskii and Novoshilov, which became very popular because of using equipment developed for inert gas metal arc (GMAW or metal inert gas, MIG) welding, which could now be used for economically welding steels. Also, because the CO_2 arc is a hot arc and a high current is required by the larger electrode wires, the process has become more acceptable for welding, entailing small-diameter electrode wires and refined power supplies.

In the 1960s, a special electrode wire was developed, which was described as an inside-outside electrode. This tubular wire, with the fluxing agents on the inside, could be used with the same equipment as the GMAW process.

In the late 1950s, R. Gage introduced the plasma arc welding (PAW) process, which is similar to GTAW. PAW uses a tight arc (through an orifice), which has a higher temperature than the tungsten arc. It became popular for nonferrous metals as a cutting process. It is also used for metal spraying and cutting, and for spraying both wires and powders. PAW is popular for low-current welding and is becoming increasingly popular in higher-current applications.

In the late 1940s, the Zeiss Company and the French Atomic Energy Commission developed the electron beam welding process, which uses a focused beam of electrons as a heat source in a vacuum chamber. This process is witnessing an increasing popularity since recent developments have allowed it to be taken out of the vacuum chamber, which was its major disadvantage. More and more applications are being found for this process, which promises a growth in popularity in the future.

Friction welding, to be discussed in more detail, was developed in the Soviet Union. This method uses high rotational speeds and high pressure to supply friction heat. Then, it gained approval in Great Britain and the USA. This process, also called inertia welding, has found many uses and is still increasing in popularity.

Laser beam welding (LBW) is a more recently developed welding process. The laser was first developed at the Bell Telephone Laboratories as a communications device; however, owing to its severely concentrated energy, it is also a powerful heat source. It has been used for cutting metals and other materials and now can be used for welding of common metals. Short pulses of energy were the early problems faced by use of the laser. However, today continuous pulse type equipment is available. Although the equipment is still massive and very expensive, with the passage of time its cost and size should decrease.

1.2 Fusion and solid-state welding techniques

There are many other types of welding processes and without doubt more welding processes will be developed and as the need arises, which will be accommodated in metal-working requirements. In general, two common groups of welding processes are fusion welding and solid-state welding, in which all the various welding techniques can be summarized.

In fusion welding processes, a chemical bonding of metal occurs in the molten phase and may require a filler material such as a consumable electrode or a filler wire roll. MIG, TIG, LBW and electron beam welding (EBW) can be considered as the most general examples of fusion welding. The fusion welding processes face many disadvantages, because in all these techniques the metal melts and then solidifies to form the joint. Therefore, the final weld has a casting structure that leads to a decrease in mechanical properties of the weld such as ultimate tensile strength, fatigue strength and increase in the ductility. As an example, the cross-section of a titanium alloy plate welded using LBW and laser-MIG hybrid welding (LAMIG) is illustrated in Figure 1.1.

The disadvantages of casting structure in fusion welding also include porosity, oxidation, micro-segregation, hot cracking and other macro- and micro-structural defects in the joint. As an example, the formation of cavity defects in the weld pool of vacuum electron beam welding (VEBW) is shown in Figure 1.2. In addition, the fusion welding process has limitations in the joining together of dissimilar metals and alloys, because

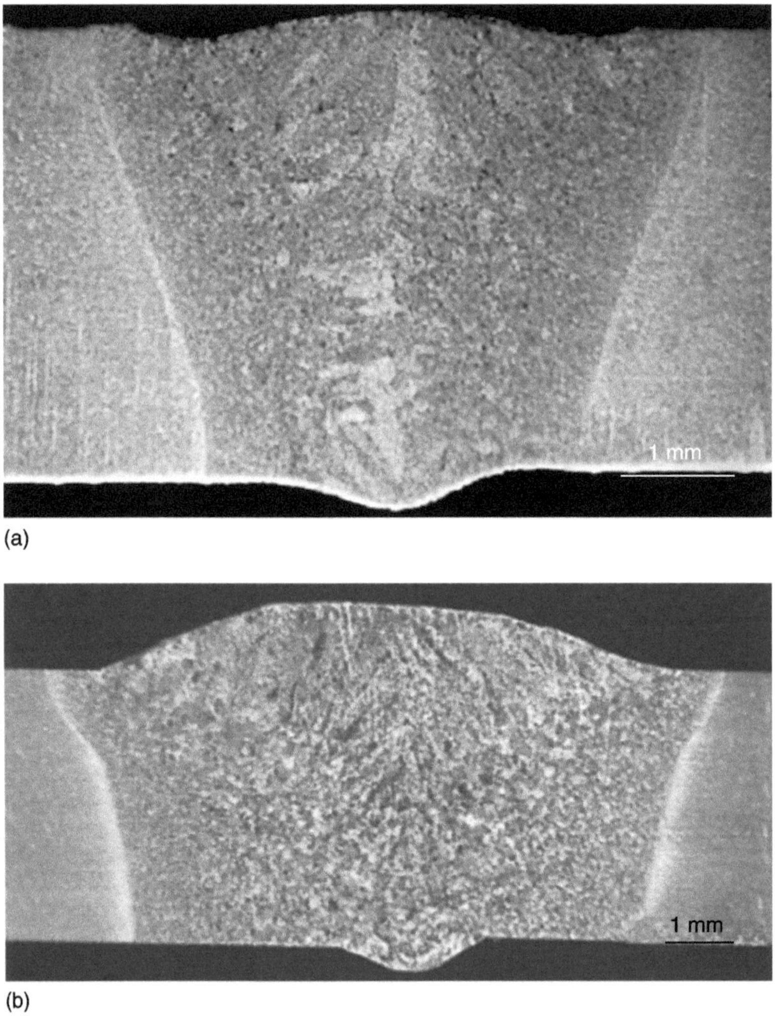

Figure 1.1 (a) LBW; and (b) LAMIG welded joint cross-sections [2]

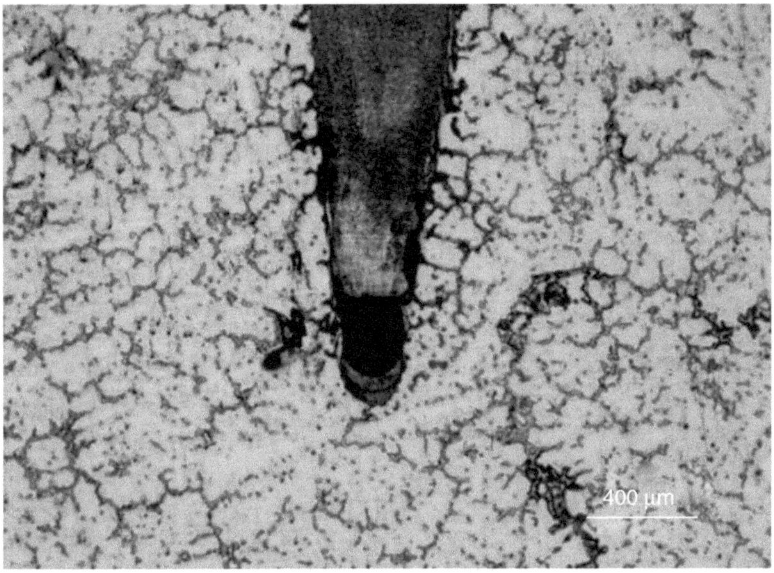

(a)

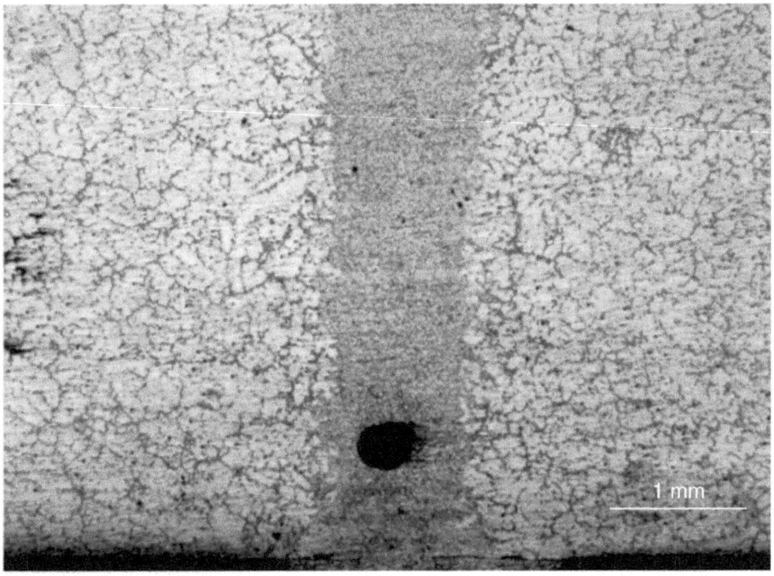

(b)

Figure 1.2 Porosity defects in VEBW welds: (a) in non-penetrated weld; and (b) in full penetrated weld of different thermal conductivity and expansion coefficients of the various metals. Furthermore, because of the high heat-input in fusion welding processes, the heat affected zone (HAZ) is wide and causes a decrease in mechanical properties of the joints. In order to eliminate these defects in fusion-welded joints, solid-state welding processes were introduced.

In solid-state welding processes, the joints are produced at temperatures below the melting temperatures of the base metal (BM). Therefore, no filler material is needed in such techniques. Furthermore, the process does not need any inert ambience to emerge the oxide phases, because the metal does not reach its melting temperature. Some examples of solid-state welding are friction welding, explosion welding, forge welding, hot pressure welding and ultrasonic welding. In these processes, there are fewer joint defects because metal melting and solidification do not occur, and casting structures are avoided. Therefore, the jointed metals retain their original properties. The HAZ is very narrow compared to fusion welding techniques, because of lower heat input conditions. Moreover, dissimilar metals can be welded together, because the thermal expansion coefficients and the thermal conductivity coefficients are less important in these methods than in fusion welding techniques.

1.3 Conventional friction welding

Friction welding (FRW) is a class of solid-state welding processes, in which heat is generated by mechanical friction between a moving component and a stationary one, and at the same time a lateral force called 'upset' is applied to the parts, in order to plastically displace and fuse the material. Since no melting occurs, FRW is not technically a welding process in the traditional sense, but is considered as a forging technique. However, due to the similarities between this method and traditional welding, the term has been commonly used. This method of welding is widely used with metals and also thermoplastics in aviation, automotive and other similar applications. FRW includes different types, which are summarized below.

1.3.1 Spin welding

The spin weld method consists of two chucks for holding the two to-be-welded parts, one of which is fixed and the other rotates. Before the process begins, one of the workpieces is attached to the rotating chuck along with a flywheel of a given weight. The piece is then spun up to a high rotational speed to provide the required energy to be stored in the flywheel. Once the desired rotational speed is reached, the motor is removed and the pieces are forced together under pressure. The force is kept on the pieces even after the spinning stops, so that it allows the weld to set. This technique is also known as inertia, rotational or inertial friction welding (Figure 1.3).

1.3.2 Linear friction welding

Linear friction welding (LFW) is similar to spin welding, except that the moving chuck oscillates laterally instead of spinning, which requires the pieces to be kept continuously under pressure, and also requires the parts to have high shear strengths. This welding method needs more complex machinery than spin welding, but in this way parts of any shape can be joined, as opposed to parts with only circular sections as in spin welding. Another advantage is that in this method the quality of joint is better than that obtained using the rotating technique. Two samples welded by the LFW technique are demonstrated in Figure 1.4.

1.3.3 Friction surfacing

The friction surfacing process is derived from FRW, the difference being that a coating material is applied to a substrate. A rod composed of the coating material, called a mechtrode, is rotated under pressure, generating a plasticized layer in the rod interface with the substrate. By moving the substrate across the face of the rotating rod, a plasticized layer between 0.2 and 2.5 mm thick is formed, depending on mechtrode diameter and coating material.

1.4 Friction stir welding

Friction stir welding (FSW) is an upgraded version of the conventional friction welding process. It was invented in 1991 at the The Welding Institute (TWI). It is considered the latest development and the most important one in metal joining during the past two decades. At first, FSW gained significant attention as a solid-state joining process of aluminium alloys, but now its application is extended to relatively harder metals and also to plastics.

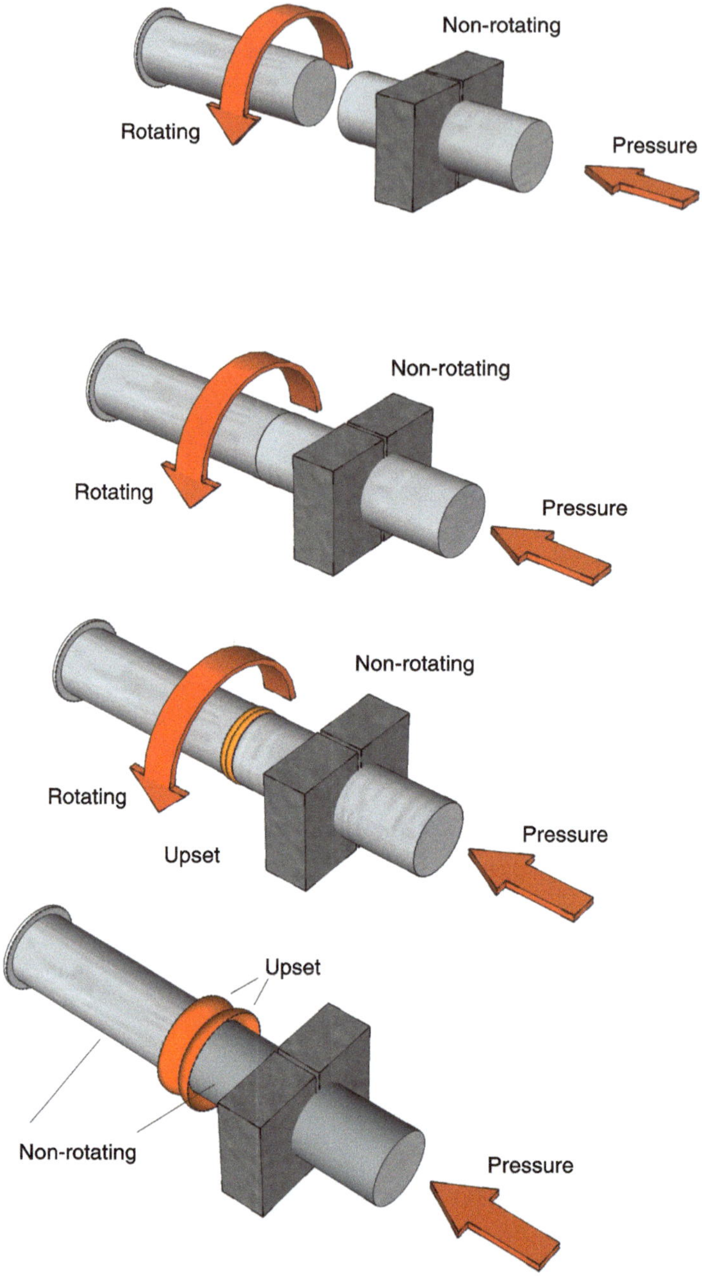

Figure 1.3 Conventional friction welding (rotational welding) [5]

Figure 1.4 **Linear friction welding [6]**

In the FSW technique, a tool with a shoulder and a profiled pin is rotated and slowly plunged into the joint line between two rigidly clamped substrates on a backing plate support. In Figure 1.5, a cylindrical tool pin profile, a schematic illustration of FSW, and a designed fixture for clamping the plates, are presented. The tool shoulder is in firm contact to the top surface of the workpiece with the load applied. The heat needed for softening the material is generated by friction and plastic deformation. The frictional heat is generated between the welding tool (shoulder and pin) and the workpieces. The tool shoulder experiences larger amounts of heat than the pin surface. However, the deformation or stirring is generated by rotation of the tool pin, thus leading to generation of additional heat.

The plasticized material around the pin experiences severe plastic deformation and a material flow from the front of the tool to the trailing edge, where it is forged into the joint. Consequently, this results in a solid-state bond between two plates. The velocities of two symmetrical points on the advancing side (AS) and retreating side (RS) are not the same, due to the combination of rotational and traverse motions of the

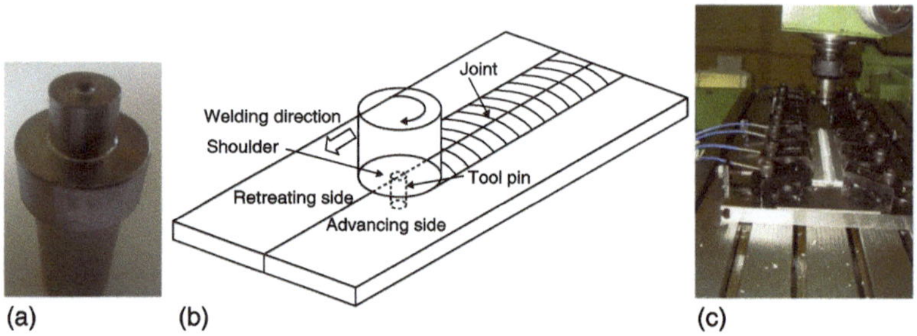

Figure 1.5 **(a) A simple FSW tool; (b) schematic of FSW process; and (c) set-up of FSW showing the designed fixture**

tool. This actually leads to asymmetry in heat transfer and material flow of the two sides of the weld. In fact, in the AS, the direction of tool rotational speed and traverse speed are the same, which is different in the RS.

1.4.1 Microstructure

Generally, in an FSW process, the joint material can be separated into four visually distinct microstructural zones according to how they are affected. These zones are distinguished in Figure 1.6 as:

a) unaffected parent material;
b) heat affected zone (HAZ);
c) thermo-mechanically affected zone (TMAZ); and
d) weld nugget (or stir zone (SZ)).

In the HAZ, properties and microstructures are affected by the heat generated by the welding process, while there is no mechanical deformation. This zone retains the same grain structure as the parent materials. The TMAZ shows characteristics suggesting that it has undergone plastic deformation, but recrystallization does not occur, due to insufficient temperature and deformation strain. In the weld nugget, or dynamically recrystallized zone (DXZ), intense plastic deformation and frictional heating during the FSW process result in a recrystallized fine-grained microstructure. The interface between the recrystallized nugget zone and the parent metal is relatively diffused and smooth on the RS of the tool, while it is quite sharp on the AS.

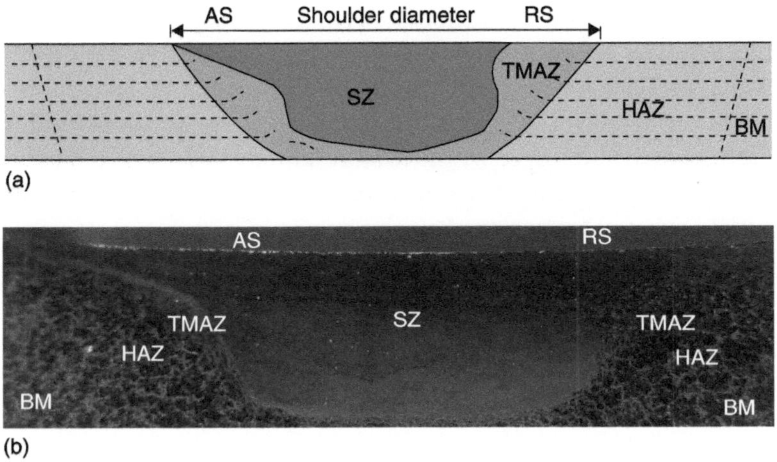

Figure 1.6 **(a) Schematic of microstructural zones in FSW in aluminium; (b) micrograph showing various microstructural zones [7]. AS: advancing side; RS: retreating side; SZ: stir zone; TMAZ: thermo-mechanically affected zone; HAZ: heat affected zone; BM: base material**

1.4.2 Advantages and disadvantages of friction stir welding

The solid-state nature of FSW leads to several advantages over fusion welding methods, since any probable defects associated with cooling from the liquid phase are immediately avoided. Problems such as porosity, solute redistribution and solidification cracking do not occur during FSW. In general, FSW has been found to be a method that produces low concentration of defects and is very tolerant to variations in parameters and materials. Nevertheless, it is associated with a number of unique defects. Insufficient weld temperatures, resulting from low rotational speeds or high traverse speeds, make the weld material unable to accommodate the extensive deformation during the welding process. This may result in long tunnel defects running along the weld line, which may be produced on the surface or subsurface.

Low temperatures may also limit the forging action of the tool and reduce the continuity of the bond between the materials from each side of the weld. The light contact between the materials has given a rise to the name 'kissing-bond' for such cases. This defect is extremely worrying, particularly as it is very difficult to detect using non-destructive methods, such as X-ray or ultrasonic testing. If the pin is not long enough or the tool rises out the plate, then the interface at the bottom of the weld may not be disrupted and forged by the tool, resulting in a lack-of-penetration defect. This is essentially a notch in the material, which can become a potential source of fatigue cracks.

FSW as a relatively new welding process has many significant advantages in comparison to fusion processes:

- energy efficient;
- environmentally friendly;
- good mechanical properties in the as-welded condition;
- joins high-strength, similar or dissimilar, metallic alloys that are difficult to weld by conventional fusion methods;
- keeps original properties of the material, despite the joining process;
- improved safety due to the absence of toxic fumes or the spatter of molten material;
- no consumables, as conventional steel tools can weld over 1000 m of aluminium, and no filler or gas shield is required;
- easily automated on simple milling machines, resulting in lower set-up costs and less training;
- can operate in all positions (horizontal, vertical, etc.), as there is no weld pool;
- generally good weld appearance and minimal thickness under/over-matching, thus reducing the need for expensive machining after welding
- welding preparation not usually required;
- reduced distortion in parent components.

Besides the advantages, FSW involves some drawbacks:

- exit hole left when tool is withdrawn;
- large down forces are required, with heavy-duty clamping necessary to hold the plates;
- less flexible than manual and arc processes, i.e. the difficulties associated with thickness variations and non-linear welds;

- slower travel rates than fusion welding techniques for hard metals, although this may be compensated for if more than one pass of fusion welding is required;
- critical tolerances.

1.5 Friction stir processing

Surface properties, such as wear resistance, can determine the service life of components in many industrial applications. Components that are produced by most metals and their alloys exhibit poor mechanical and tribological characteristics. Hence, there is a strong drive to develop new metal-based materials with improved mechanical and tribological properties.

Recently, much attention has been paid to friction stir processing (FSP), developed based on the principle of FSW, and is known as the surface modification technique. Similar to FSW, a rotating tool with a pin (often shorter than the thickness of the sheet) and a shoulder, is inserted into a single piece of material and travels as the tool in FSW does. This results in significant microstructural changes in the processed zone, due to the occurrence of intense plastic deformation. FSP has been proved to be an effective way to refine the microstructure of a material, and thereby improve its mechanical properties. Also, it can be used to produce surface composite layers by addition of ceramic powders during the process (Figure 1.7). The

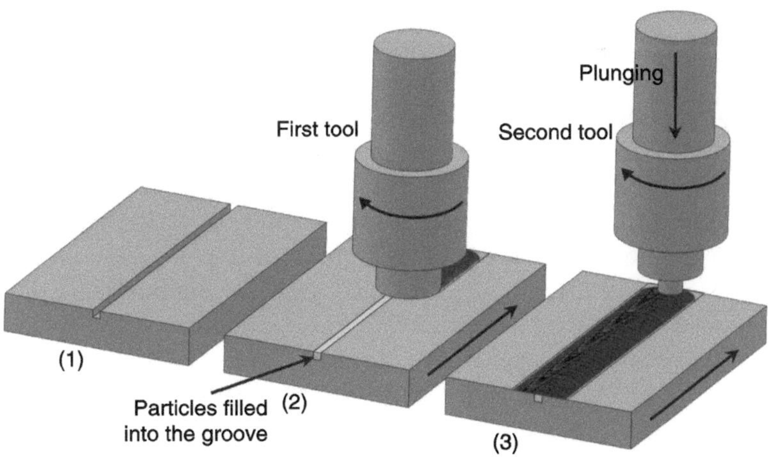

Figure 1.7 Friction stir processing for fabrication of surface composite layer [8]

stirring action of the tool pin can distribute the particles across the processing zone. This can be achieved by the powder being inserted in the processing zone, usually by pouring the powder into a groove or hole created in the sheet. Figure 1.7 shows the steps of creating and filling the groove, and then FSP. As seen in the figure, a pinless tool is required to cover the groove top after filling, to prevent scattering of the powder during FSP.

It is evident that FSW and FSP share the same mechanism (Figure 1.5(b)). That is why in some literature, the two terms are interchangeable. However, they do have different purposes in practical applications. The goal of FSW is to join two plates together, whereas FSP aims at modifying the microstructure of a single component or of multiple workpieces.

1.6 Content outline

The main goal of this book is to provide a relatively complete context of the latest works, advances and applications in industries for FSW, in order to make this technique of welding more familiar and applicable for readers. Not all process details of the different works are presented here, and readers are recommended to follow the references for more detail and information.

Physical properties of FSW/FSP material are significantly affected by complex material movement, plastic deformation and temperature distribution. Chapter 2 provides important physical insights into the mechanism of the process in terms of heat generation, heat transfer and plastic flow of material during the FSW process. These factors are considered from both empirical and mathematical points of view.

Chapter 3 identifies both the microstructural evolution mechanisms during FSW/FSP and the different microstructural zones formed after FSW/FSP of various metals and their alloys. The recrystallization mechanism, resulted from a combination of severe plastic deformation and generated heat during FSW/FSP, is considered carefully and different types of recrystallization mechanisms, such as continuous dynamic recrystallization (CDRX), discontinuous dynamic recrystallization (DDRX), geometrical dynamic recrystallization (GDRX) and static recrystallization (SRX) during FSW/FSP of alloys, are explained in detail.

Chapter 4 addresses the joint efficiency and mechanical properties of different metals and alloys. In this chapter, different mechanical properties of joint involving, microhardness, tensile properties, residual stress and its distribution, fatigue behaviour and fracture will be discussed.

The use of a non-consumable tool is mandatory during the FSW process. The pin, or probe, and the shoulder are its main components. Three main purposes are provided by the tool, namely heating the workpiece, stirring and moving the plasticized material in order to produce the welded joint, and controlling the processed material under the tool shoulder. Thus, the FSW tool is undoubtedly crucial for the quality expected from the FSW process, being essential for improving both the welded joints mechanical performance and the maximum possible welding speed during the process. Therefore, the detailed characteristics of an FSW/FSP tool will be discussed in Chapter 5. Knowing and understanding the tools well is a natural starting point for successful use of FSW/FSP.

The improved mechanical properties of FSW joints make this technique prone to a wide range of industrial applications. The FSW of low melting temperature materials, such as aluminium and its alloys, has been embraced and commercialized; and recent interest is now focused on joining dissimilar metals. However, the full understanding of the inherent properties of dissimilar joints using the FSW process is required to accommodate the huge demand in industries, including the manufacturing and the aerospace industries. Recent developments in joining dissimilar metals of aluminium and its alloys to other metals, such as copper, magnesium, steel and titanium, have been reviewed and are presented in Chapter 6. A detailed study on the material characterization on FSW of aluminium and copper is also presented.

Surface properties, such as corrosion and wear resistances, can determine the service life of components in many industrial applications. Effects of FSW/FSP processes in corrosion and wear resistances of the weld zone are discussed in detail in Chapters 7 and 8. The wear resistance of the FSW/FSP tool is also described in Chapter 8.

FSP as a modification technique can be used for mechanical alloying, along with repairing the welding and casting defects. Whereas defect-free casting of some high-alloyed materials is difficult, it is preferable in some applications to add alloying elements after casting via other techniques such as FSP. In addition, defects appearing in the material as a result of casting or diffusion welding can be modified by FSP. These issues are investigated for different alloys in Chapters 9 and 10.

Different relative motions of the tool and workpiece generate different loads. Forward motion of the tool produces loads parallel to direction of travel, known as the transverse load, while normal load is the load required for the tool shoulder to remain in contact with the workpiece, and finally, the rotational motion of the tool generates the tangential load as well as torque. Detailed consideration of welding forces in FSW is discussed in Chapter 11.

Thermo-mechanical aspects (temperature, thermal history, stress, strain, material flow, etc.) of FSW/FSP are studied widely in Chapter 12, via finite element models that incorporate the mechanical reaction of the tool and thermo-mechanical process of the welded material. Furthermore, in order to provide a quantitative framework for understanding the dynamics of FSW thermo-mechanical processes, the evolution of longitudinal, lateral and through-thickness strains and material flow in the SZ are simulated in numerical models.

Chapter 13 represents the optimization methods used in FSW/FSP, with emphasis on artificial neural networks (ANNs) and gives some examples of optimization for FSW/FSP process parameters.

Beside metals, polymers can also undergo FSW and FSP. The macrostructural and mechanical properties, as well as changes in molecular chains of different polymers after performing FSW/FSP, is considered in Chapter 14.

Finally, the last chapter is dedicated to representing the industrial applications of FSW worldwide. The recent advances in this method of welding have made it feasible and practical in a wide range of industries including automotive, shipbuilding, aerospace and aviation, train manufacturing and so many other different fields, which are all discussed in Chapter 15.

1.7 References

1. Historical Development of Welding (2006), available from: *http://www.keytometals.com/page.aspx?ID=CheckArticle&site=kts&NM=186*
2. Ruifeng, L., Zhuguo, L., Yanyan, Z. and Lei, R. (2011), 'A comparative study of laser beam welding and laser-MIG hybrid welding of To-Al-Zr titanium alloy', *Materials Science and Engineering A*, 528: 1138–42.
3. Yi, L., Jinhe, L. and Hong, Y. (2011), 'Bubble flow and the formation of cavity defect in weld pool of vacuum electron beam welding', *Vacuum*, 86: 11–17.
4. 'Friction welding', available from: *http://en.wikipedia.org/wiki/Friction_welding*
5. 'Friction welding', Gatwick Technologies Ltd. (2013), available from: *http://www.fpe.co.uk/processes/friction-welding*
6. 'Friction welding', available from: *http://www.thompson-friction-welding.co.uk/*
7. Mishra, R.S and Ma, Z.Y. (2005), 'Friction stir welding and processing', *Materials Science and Engineering R*, 50: 1–78.

8. Asadi, P., Besharati-Givi, M.K. and Faraji, G. (2010), 'Producing ultrafine-grained AZ91 from as-cast AZ91 by FSP', *Materials and Manufacturing Processes*, 25(11): 1219–26.
9. Asadi, P., Faraji, G., Masoumi, A. and Besharati Givi, M.K. (2011), 'Experimental investigation of magnesium-base nanocomposite produced by FSP: Effects of particle types and number of FSP passes', *Metallurgical and Materials Transactions A*, 42(9): 2820–32.
10. Tutunchilara, S., Besharati Givi, M.K., Haghpanahia, M. and Asadi, P. (2012), 'Eutectic Al–Si piston alloy surface transformed to modified hypereutectic alloy via FSP', *Materials Science and Engineering A*, 534: 557–67.
11. Asadi, P., Besharati-Givi, M.K., Abrinia, K., Taherishargh, M. and Salekrostam, R. (2011), 'Effects of SiC particle size and process parameters on the microstructure and hardness of AZ91/SiC composite layer fabricated by FSP', *Journal of Materials Engineering and Performance*, 20(9): 1554–62.
12. Asadi, P. et al. (2012), 'On the role of cooling and tool rotational direction on microstructure and mechanical properties of friction stir processed AZ91', *International Journal of Advanced Manufacturing Technology*, 63: 987–97.

Heat transfer and material flow in friction stir welding

S. Bag, D. Yaduwanshi and S. Pal,
Indian Institute of Technology, Guwahati, India

DOI: 10.1533/9780857094551.21

Abstract: The fundamental knowledge of heat transfer and material flow in friction stir welding (FSW) provides a defect free, structurally sound and reliable weld joint. A sophisticated mathematical model of FSW uses high computing performance devices. This chapter introduces the phenomenological modelling approach via experimental investigation, by interpreting governing mathematical equations and boundary interactions of the process. Detailed finite element formulation of heat transfer and material flow equations are demonstrated. An analytical heat transfer model based on friction mechanism is also illustrated. Several practical issues related to heat transfer and material flow are discussed. FSW has been successfully used to join metallic materials, such as copper, titanium, steel, magnesium and composites. Preheating in a hybrid FSW process is an alternate route to enhance material plasticity along with reduction of plunging forces as compared to conventional FSW. The interaction between thermal, mechanical and metallurgical models is complex, necessitating a multi-scale modelling approach to capture microscopic behaviour. The theoretical investigation on various aspects of FSW explores the possibility of further development of the process model.

Key words: heat transfer, material flow, process model, visco-plastic flow, thermo-mechanical analysis, mass-momentum-energy equations, sliding and sticking friction.

2.1 Introduction

Friction stir welding (FSW) has gained significant attention in recent years, as the solid state joining process of primarily aluminium and/or aluminium alloys is now extended to join relatively harder materials and plastics. Understanding the synergistic interaction of the physical phenomena during welding of various similar or dissimilar materials is of practical significance. Models of increasing complexity and sophistication have been used beneficially in exploring and understanding the underlying physics of the process. Hence, this chapter aims to provide important physical insights into the mechanisms in terms of heat generation, heat transfer and plastic flow of material during the FSW process.

The most general configuration of an FSW tool is a solid cylindrical object with a terminating pin. The rotating tool moves along the contact surfaces of two rigidly clamped substrates on a backing plate support. The tool shoulder makes firm contact with the top surface of the workpiece with an applied load. Therefore, heat generated by friction occurs on the surface in contact with the tool. The tool shoulder experiences a larger amount of heat than the pin surface. The process also experiences severe plastic deformation surrounding the tool and flow of the plasticised metal, from around the front of the tool to the trailing edge where it is forged into a joint [1].

The heat transfer aspect during FSW can be described as three phases. The first phase is called the dwelling period, where the material is preheated by a stationary rotating tool in order to achieve a sufficient temperature ahead of the tool to allow for the traverse movement. The second phase is the transient heating when the welding tool begins traversal movement until the pseudo steady-state is reached. Finally, the pseudo steady-state is where the system reaches an effective constant thermal field around the tool. The relative velocity on a symmetrical point of the advancing side (AS) and retreating side (RS) is not the same, due to a combination of similar directional rotational and transverse motions of the tool. This actually leads to asymmetry in heat transfer and material flow on the two sides of the weld joint [2].

FSW involves complex physical interactions between heat generation, plastic deformation and metallurgical aspects [3]. The heat is generated by friction as well as by plastic deformation, altering the microstructure and properties. The kinetics depend mainly on the temperature and strain rate, and microstructure evolution influences the energy transfer within the system. The phenomenological understanding of heat transfer and material flow demands a strong coupling between thermal, mechanical and metallurgical aspects. The centre of the stir zone (SZ) experiences dynamic recrystallization and is followed by recovered microstructure around the SZs at low levels of deformation and temperature. This actually increases the amount of grain surface energy and number of grain edges [4–5]. A part of the plastic deformation energy is also stored within the thermo-mechanically affected zone (TMAZ) in the form of increased dislocation densities within the deformed grains. The heat affected zone (HAZ) is affected only by temperature through diffusion heat transfer and with a microstructure almost similar to the base material.

In FSW, the heat and mass transfer is mainly influenced by material properties as well as welding variables including the rotational and transverse tool speed. However, the joining takes place by extrusion and forging of the metal at high strain rates [1]. Therefore, the mechanical behaviour of material is assumed either as fully plastic or visco-plastic or only viscous fluid. The material flow field and the temperature field are significant, since they influence the recrystallization kinetics during the FSW process. There exist two modelling approaches for material flow [6–9]. The plasticised material can be treated as a high viscous fluid and the flow field is obtained by solving fluid dynamics. On the other hand, the plastic deformation is modelled on the principals of solid mechanics and the displacement field is used to find the velocity or acceleration field.

The understanding of the heat transfer process in the workpiece can be helpful in predicting the thermal history during welding, which subsequently influences the mechanical properties to evaluate weld joint quality. The heat transfer process for both workpiece and tool can be formulated as a standard boundary value problem and then solved by using an inverse approach combining experimental and numerical studies that consider frictional heat at the interface of the tool shoulder and the workpiece [10,11]. It is observed that the maximum temperature ranges from 80 to 90% of the melting temperature of the welding material at the interface between the tool and the workpiece. However, 20% of the total heat is attributed to the pin and the addition of heat due to the pin has little effect on the thermal profile [12–13].

There are two sources of heat generation for FSW. One is due to friction and the other to plastic deformation at the tool–workpiece interface and at the TMAZ [14–15]. The tool–workpiece interface can be further subdivided into shoulder–workpiece and tool pin–workpiece interfaces. The heat generation between tool–workpiece interfaces is greatly influenced by frictional heat, but the effect of heat generated due to plastic deformation is very much less. Moreover, the difference in temperature distribution between the AS and RS due to non-symmetrical heat flux distribution brings complexity to the heat transfer analysis.

This chapter provides a fundamental understanding of heat transfer and material flow mechanisms through mathematical modelling and experimental investigation. The important issues related to the topic provide an overall view of the subject by creating individual sections on their own merits. Section 2.1 provides governing equations and boundary interactions to demonstrate the physical description of the process. Significant effort has been made to explicitly describe the heat generation model and interaction between thermal and mechanical models. A detailed finite element formulation and computational aspects is also included in this section. Finally, the interaction between thermal, mechanical and metallurgical models is described. The future prospects and concluding remarks are outlined in the last section.

2.2 Principles of heat transfer and material flow

The nature of heat transfer and heat generation rate, and flow of plasticised material is primarily affected by tool geometry, traverse movement of the tool, applied torque and thermo-mechanical condition experienced by the tool. More precisely, the shoulder size and surface angle, pin geometry and the nature of tool surfaces influence the heat transfer mechanism and the material flow pattern. The plasticised materials largely establish the material flow field, where both sliding and sticking generate heat and the material flow is caused only from sticking. In FSW, the tool shoulder surface is generally flat, convex and concave in nature, whereas cylindrical, tapered, inverse tapered and triangular pin geometries are generally used [16–19]. The weld joint quality significantly changes if the tool shoulder chosen is concave rather than flat, since the shoulder surface angle affects the axial force and consequently the pressure distribution [18]. The FSW process with a convex scroll shoulder tends to be stable with a nearly constant plunge depth and results in minimum flash and no defects as opposed to a concave shoulder [20].

The shape of the tool pin also influences the flow of plasticised material and finally affects weld joint properties. The tool shoulder facilitates bulk material flow, whereas the pin aids a layer-by-layer material flow [16]. A triangular tool pin increases the material flow compared with a cylindrical pin. The axial force on the workpiece material and the flow of material near the tool are affected by the orientation of threads on the pin surface.

However, the threads and flutes on the pin increase the heat generation rate due to a larger interfacial area, improve material flow and affect the axial and transverse forces. The square tool geometry results in a more homogeneous distribution of particles than other tools, whereas circular tools experience much less wear than flat-faced tools [20]. However, a significant reduction in process forces with a conical shoulderless tool is possible, which is suitable for variable thicknesses plates. An increase in the angle between the conical surface of the pin and its axis leads to a more uniform temperature distribution along the vertical direction and helps in reducing distortion. However, the experimental analysis on heat transfer provides significant information about the actual process.

Most of the experimental investigation in FSW is carried out on a vertical milling machine. The clamping of the test pieces is done so that the movement of the plates is totally restricted under both plunging and translational forces of the FSW tools. For example, a straight cylindrical tool made of SS304 with shoulder and pin diameters of 20 mm and 6 mm, respectively, are used to butted AA1100 plates. The time–temperature history is captured by a thermocouple, the location of which is indicated in Figure 2.1. The average temperature difference between the RS and

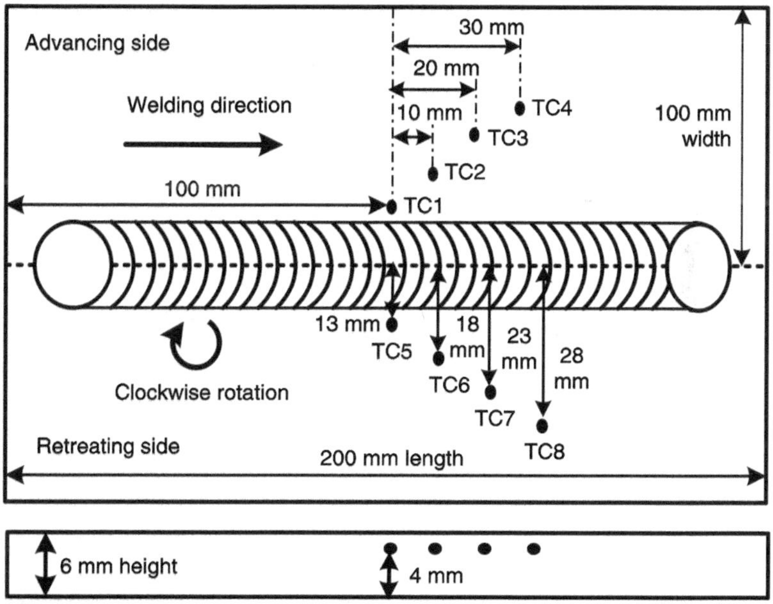

Figure 2.1 **Locations of thermocouples in work piece used for FSW experiments**

AS is found to be 13 K. The transient temperatures recorded from the thermocouples corresponding to FSW at 1.05 mm/s welding speed, 815 RPM and 5.75 kN average plunging force for the AS and RS are depicted in Figure 2.2. Since the thermocouple TC1 is closest to the joint line, its peak temperature is higher than that of the others. So it is assumed from all experimental data that in a defect-free welding, the peak temperature is between 324.4 and 370.1 °C at 3 mm away from shoulder edge on thermocouple point TC1. It is also obvious that the temperature on the AS at all four points is higher than on the RS, although the temperature is measured in similar locations. As it moves away from the centre line, the temperature difference reduces and gains equilibrium at a certain distance. This temperature difference belonging to asymmetrical heat generation and plastic flow of material effectively raises the AS temperature above that of the RS.

Effective and defect-free FSW of dissimilar material is more challenging, as compared to similar material due to different deformation behaviours, the possible formation of detrimental intermetallic compounds, and differences in thermo-physical properties. These factors cause asymmetry in both heat generation and material flow. Moreover, the discontinuity in properties across the butting surfaces has significant influence on the material flow pattern. One of the potential methods of joining aluminium and copper is to introduce an additional local heat source to preheat the copper side up to 200°C. However, the location of the tool must be carefully optimized.

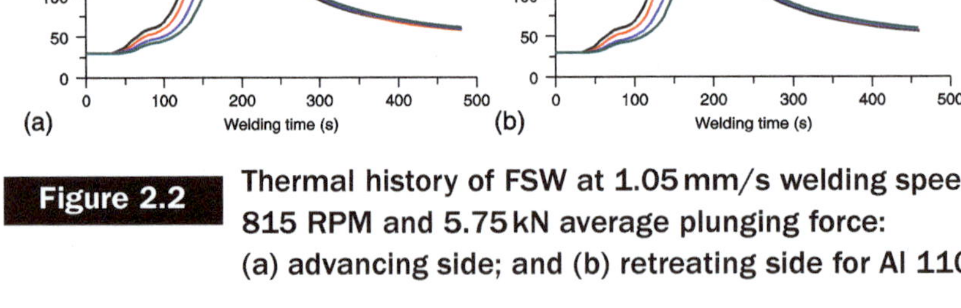

Figure 2.2 Thermal history of FSW at 1.05 mm/s welding speed, 815 RPM and 5.75 kN average plunging force: (a) advancing side; and (b) retreating side for Al 1100

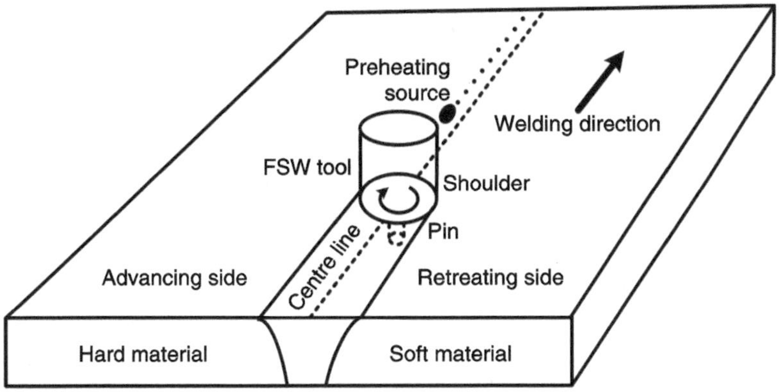

Figure 2.3 **Schematic of dissimilar material joining for butt joint configuration**

Figure 2.3 is a schematic diagram of dissimilar material joining for butt joint configuration. Preheating temperature can make up the lost heat and improve material flow during the FSW process. Among the possible preheating source, plasma arc provides a unique combination of advantages such as high arc stability, concentrated energy density and low equipment cost. The positioning of the workpiece in dissimilar material joining is schematically illustrated in Figure 2.4. The Cu sheet is placed at the AS and aluminium sheet at the RS during the welding process, which attributes to the fact that hard copper acts as a bulk material, which is difficult to transport to the AS [21]. In FSW of dissimilar metals, the pin offset is an important factor. Most of researchers observed that as the pin offset increased towards the Cu-side, there was unsuitable mixing of Al and Cu, which resulted in a tunnelling defect due to an increase in copper percentage and formation of more brittle intermetallic material at the interface with decreased solubility. When the pin offsets are large, Cu pieces get scratched at the metal plate and get mix into the Al matrix [22]. Therefore, the additional heating source to preheat the copper side to 180~250 °C along with optimized offset distance could resolve this problem.

The temperature profile differs in the case of the plasma-assisted FSW (P-FSW) process for joining dissimilar materials. The plasma transfer arc with direct current at a maximum intensity of 50 A is used for butt welding of commercial grade aluminium alloy and pure copper of 6 mm thick plate. The effects of preheating on the thermal history at different points from the weld centre line in AS and RS is shown in Figures 2.5

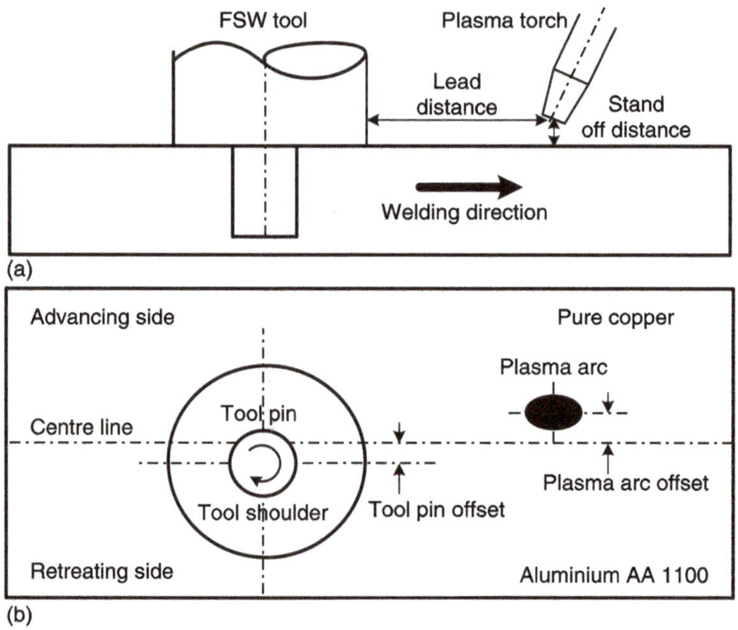

Figure 2.4 Schematic of P-FSW: (a) side view; and (b) top view

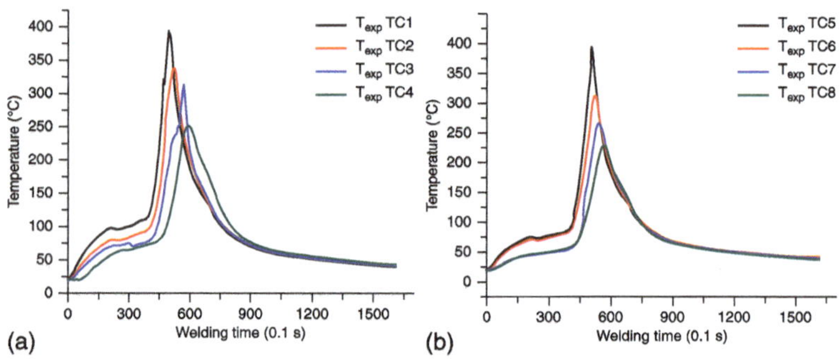

Figure 2.5 Thermal history of P-FSW at 1.05 mm/s welding speed, 815 RPM and 50 amp preheating arc current: (a) advancing side (Pure Cu); and (b) retreating side (AA1100) for Cu-Al

(a–b). In AS, the temperature is higher than RS and confirms the non-symmetrical heat generation. The area under the curve of the thermal history represents total heat input during welding, which is higher in the case of preheating. This additional heat input has an effect on the rate of change in temperature during welding that may improve weld joint properties. This additional heat also increases the temperature of material around the weld zone, so that the material tends to lose some of its mechanical strength and facilities the ease of welding.

A state-of-the-art literature review indicates that the material flow occurring through the RS and the transport of the plasticized material is the combined effect of the relative motion between the tool and the workpiece, the slug of plasticized material flow around the tool, and the rotational motion of the threaded pin that tends to push material downwards close to the pin. The helical motion of a conical pin pushes the material downwards in the front and upwards in the rear, which results in more uniform properties across the workpiece thickness [7]. Hence, tapered tools are preferred when welding thick sheets [23]. For a cylindrical pin tool, material flow is upwards near the pin periphery, whereas the material beneath the shoulder is pushed downwards due to the axial force from the shoulder. Thus, moving away from the pin periphery, the reference line of particles curves upwards and then bends down, resulting in a 'hook' formation. The nature of hook formation is influenced by pin and shoulder geometries [6,24].

There exist several approaches to visualizing material flow patterns in FSW, using a tracer technique as marker or through the welding of dissimilar alloys. Experimentally, the movement of tracer particles and spatial variations of texture have been used to gain an understanding of material flow in FSW. The technique used to visualize the flow patterns uses small steel balls as a tracer and by stopping the forward motion suddenly and simultaneously retracting the tool at a rate that causes the pin to unscrew itself [25]. By sectioning the pin intact keyhole at the end of a weld, the flow of material within the threads of the tool is possible to study. Different marker materials, such as aluminium alloy that etch differently from the base metal [26–28], copper foil [29], small steel shots [25,30], Al–SiC and Al–W composites [31] and tungsten wire [32], have been used to track the material flow during FSW. However, the marker technique provides the data on the initial and final positions of the marker in the welded material and the flow is then reconstructed from these positions.

Steel shots can be indexed in the welding path and X-ray images are used to track secondary position of steel shots. It is observed that the steel shots penetrate from AS to RS, whereas the shots implanted in the RS remain on the same side [33]. In another experiment, steel spheres were inserted into grooves of the threaded pin [30]. The radiography test of the post-weld indicates that the vertical motion of the steel spheres is minimal and tends to move upwards. A study in which a copper foil was inserted at the faying surface of lap joint configuration of dissimilar aluminium alloy of 2xxx and 6xxx series, indicated that the material directly in contact with the pin rotates with the velocity of the tool. However, a longitudinal vortex is generated that creates significant vertical mixing in the weld. Several researchers have investigated the flow patterns in butt welds of similar and dissimilar aluminium alloys [35–38]. These studies show that the soft aluminium alloy placed on the AS of the weld is stirred into a harder alloy and the vortex-like flow patterns are more distinct compared to alloys of relatively identical hardness. However, if the harder alloy is stirred into a much softer alloy, the vortex patterns are minimal [39].

The most simplified way to analyse the material flow in FSW, copper foil of 0.5 mm thickness as a tracer material is embedded at different positions and orientations within the butt joint configuration (Figure 2.6). The trace material is tracked by suddenly stopping the forward motion of the welding tool and sectioning the keyhole at the end of a weld, which can help in studying the flow of material in the region close to the welding tool. The tracer material distribution around the welding tool is preserved at the end of the weld by revealing the path with the help of radiography as it is transitioned from its original position to final position.

The flow of material during FSW is a complex process that is not fully understood, despite numerous investigations and models. Several studies have compared material flow during FSW by treating the process as an extrusion process. As seen from Figure 2.7, the copper of a 0.5 mm thick sheet is used to visualize the material flow in the FSW process. The analysis technique can easily detect copper from the base material and the thinner marker is unlikely to reduce the effectiveness of the analysis. The marker is placed in parts of the weld only and not through the entire weld thickness. If a strip of copper is placed in the top third of the joint-line, a further benefit would be information about the through-thickness flow. [39] reported that the estimated average velocity of material during FSW of aluminium alloy is approximately 0.1 to 0.3 times the shoulder rotational speed, which is comparable with CFD models [9].

Material flow in FSW is largely uncharacterized due to difficulty in measurement and tracing the marker. In this respect, the mathematical model is demanded to analyse the material flow. The fully coupled thermal-mechanical visco-plastic flow models can provide reasonably good results [8,9,40–42]. However, one of the problems with these models is that their rigorous validation with experiments is difficult. The experimental study with copper markers imitates the computational model and indicates that weld pitch and tool pin shape determine the

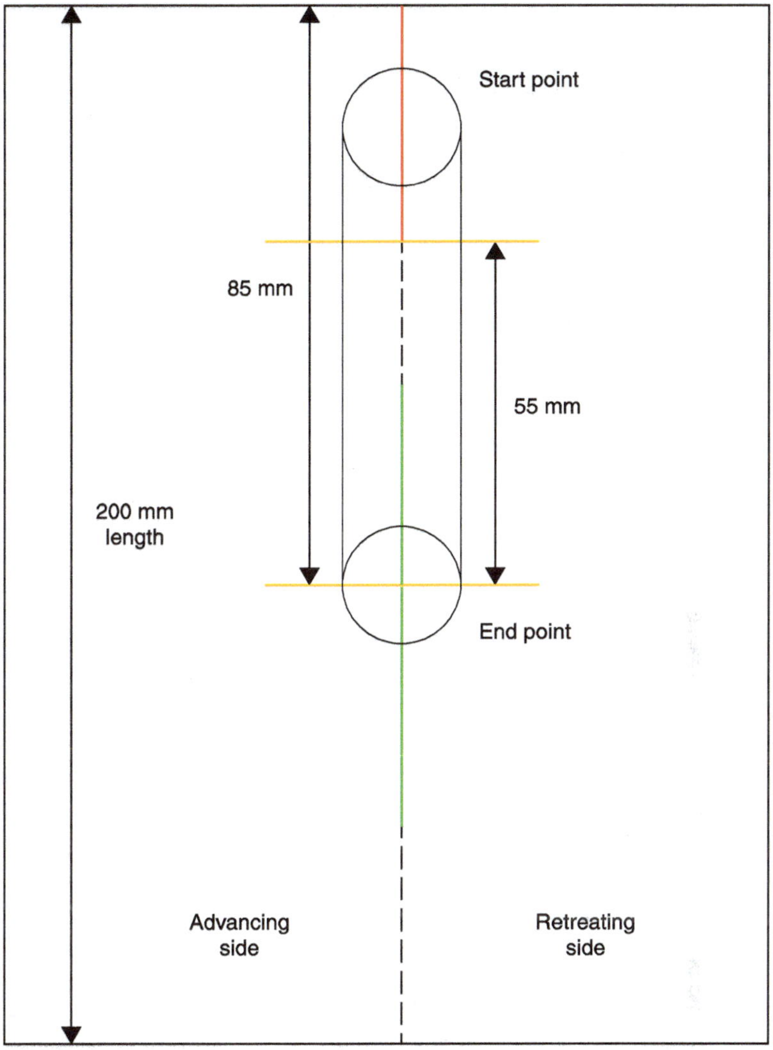

Figure 2.6 **The layout of copper marker of 0.5 mm used in two different orientations. The colour lines represent placement of copper marker material at the weld centre**

complexity of flow [34]. At small pitch values, the marker is continuous rather than discontinuous.

The plastic flow of material produces the velocity field, strain field and strain rate field, which may provide significant information about the

Figure 2.7 Exit-hole damage in a 6 mm thick AA1100 alloy weld. The copper marker can be seen in the joint line on the lower left

process. The stream lines may be indicative of the presence of rotational zones and the recirculating flow of material around the tool pin [8,9]. However, the thickness of the recirculating material flow region is mainly decided by material properties, such as viscosity and rate of heat transfer to the tool. Hence it is important to consider the heat transfer mechanism to the tool material. The typical streamlines of a flow field is also indicative that beyond the region of the recirculating zone, material transfer may occur, mainly on the RS and a flow reversal may occur in the AS close to the pin. This forms a relatively stagnant zone and may indicate the formation of defects, such as a 'wormhole' on the AS [43]. Therefore, in-depth analysis of material flow may predict the defects in FSW due to lack of adequate material flow.

Apart from several phases of heat transfer, i.e. initial and final periods of welding, heat generation occurs almost at a constant rate and the system achieves a quasi-steady state for a uniform rotational speed and at a constant transverse speed of the tool. In this period the weld profile and properties remain almost constant. Analysis is performed in a moving coordinate system, where the time variable is explicitly transformed to the transverse velocity term and actually eliminates the time variable from the governing equations. The temperature and velocity profile explicitly depends on tool transverse velocity and hence the solution at one position of the tool is sufficient to explain the whole process. However, the transient analysis of the whole process can also be performed, which resembles the actual process. In this case, the tool has to move over the time step in the fixed coordinate system and the distance covered at each time step depends on tool transverse velocity. Both approaches are explored in the following sub-sections.

2.2.1 Mathematical foundation

Although transport phenomena-based models in FSW are considered for material flow analysis, the conduction heat transfer based model is often used due to its simplicity, ease of formulation and reduced computational time. However, only the heat conduction model is able to predict the temperature distribution and cooling rate, and is not capable of finding the material flow field. Hence coupled mass, momentum and energy equations are often solved to obtain strain and the strain rate field. The heat transfer model considered here is a three-dimensional (3-D) Cartesian coordinate system that can handle both the transient and quasi-steady state processes. The theoretical formulation to develop heat transfer analysis and material flow, both for transient and quasi-steady state conditions, are presented in the following sections. The governing equations and the boundary conditions are illustrated and in-depth analysis of the mathematical models helps to understand the physical process of FSW.

Heat transfer

Heat generation in FSW is mainly governed by two mechanisms. One is due to friction and other to plastic deformation, where heat is transferred into the workpiece following Fourier's law of heat conduction. In numerical analysis, the transient temperature field T, which is a function of time t and the spatial coordinates (x,y,z), is estimated by the 3-D nonlinear heat transfer equation in which the moving Cartesian coordinate system with a positive y-direction moving tool is written as

$$\frac{\partial}{\partial x}\left(k_x \frac{\partial T}{\partial x}\right)+\frac{\partial}{\partial y}\left(k_y \frac{\partial T}{\partial y}\right)+\frac{\partial}{\partial z}\left(k_z \frac{\partial T}{\partial z}\right)+\dot{Q} = \rho\, C_p\left(\frac{\partial T}{\partial t}-V_T\frac{\partial T}{\partial y}\right) \qquad [2.1]$$

where ρ, C_p and k refer respectively to density, specific heat and directional thermal conductivity of the workpiece material, $\dot{Q}$ is the rate of heat generation and V_T is the transverse speed of the tool. It should be noted that the interface thermal resistance between two plates is neglected here. The boundary condition for the FSW thermal model is specified as the surface interaction of the solution domain. The initial condition for the calculation is expressed as

$$T(x,y,z,0) = T_i \qquad [2.2]$$

where T_i is the ambient temperature at time $t = 0$. The boundary interaction of the solution domain is schematically represented in Figure 2.8. The solution domain is marked by a dotted line (white zone), which has escaped from the tool. In a finite domain of interest, there is convection and radiation heat loss from the top and side surfaces and bottom surface is subjected to convection heat loss. The tool–workpiece interface acts as a source of heat due to friction and plastic deformation. However, the heat transfer to the tool is normally neglected and a zero heat flux normal to the tool surface is considered. Mathematically, the boundary condition is expressed as

$$k\frac{\partial T}{\partial n} = h(T - T_0) + \varepsilon\sigma\,(T_0^4 - T^4) - q_s \qquad [2.3]$$

where n is the normal direction vector of the boundary, h is the convection coefficient, q_s is heat flux between tool and workpiece, ε is the emissivity and σ is the Stefan-Boltzmann constant. The second term on the right-hand side of Eq. 2.3 indicates the radiation heat loss. It should be mentioned that the workpiece is supported by a backing plate and there is direct contact between the workpiece bottom surface and the backing plate. For this conduction heat loss through the backing plate, it is assumed to be equivalent to the surface film conditions on the bottom surface of the workpiece, which is uniform throughout the surface. It actually reduces the complexity of contact surface modelling and

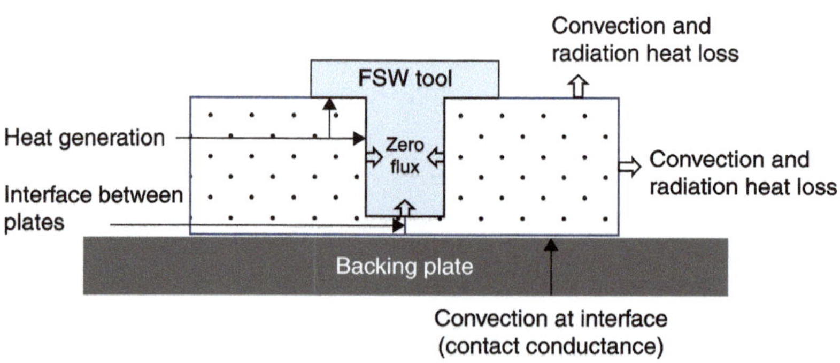

Figure 2.8 **Schematic of solution domain and boundary interactions**

computational time. The heat loss from the bottom surface of the workpiece through the backing plate is ultimately lost due to convection (the term 'contact conductance' is used) from the backing plate surface. Therefore, it is justifiable to consider convective heat transfer from the bottom surface of the plate.

In quasi-steady state heat transfer analysis, the transient heat conduction can be modified in the following way to estimate the constant temperature field independent of the position of rotating tool. Considering the fact that the welding heat source is moving with a constant velocity, V_T, in the y-direction, the moving coordinate system (x, ξ, z) can be chosen in such way that

$$\xi = y - V_T t \qquad [2.4]$$

where ξ is the distance of the heat source from a fixed position, such that at any time t and '$V_T t$' is the distance between the heat source and the origin of the fixed coordinate system (x, y, z) along the y-axis. As y and ξ are explicit functions of time t, it is possible to write:

$$\frac{\partial \xi}{\partial t} = -V_T \qquad [2.5]$$

and

$$\frac{\partial \xi}{\partial t} = \frac{\partial T}{\partial \xi}\frac{\partial \xi}{\partial t} = -V_T \frac{\partial T}{\partial \xi} \qquad [2.6]$$

Therefore, the transient heat conduction equation can be modified in the moving coordinate system as

$$\frac{\partial}{\partial x}\left(k_x \frac{\partial T}{\partial x}\right) + \frac{\partial}{\partial y}\left(k_\xi \frac{\partial T}{\partial \xi}\right) + \frac{\partial}{\partial z}\left(k_z \frac{\partial T}{\partial z}\right) + \dot{Q} = -\rho C_p V_T \frac{\partial T}{\partial \xi} \qquad [2.7]$$

It is obvious that the above equation is not a function of time and represents the quasi-steady state heat conduction equation where the temperature profile changes with transverse motion of the tool. The boundary condition for quasi-steady state analysis is similar to Eq. 2.3.

It is obvious that the boundary condition for heat transfer analysis involves the surface heat flux term, q_s, in Eq. 2.3. Therefore, all the contact surface of the tool with the workpiece material acts as a source of heat. However, the volumetric heat flux, i.e. heat generation due to viscous dissipation, is included through the term $\dot{Q}$ in Eq. 2.1. This is nothing to do with boundary interaction, which consists of three parts, i.e. tool-shoulder interface, tool pin vertical or inclined interface and tool pin bottom surface interface. The typical range of heat flux for stainless steel is that approximately 80% heat is generated at the shoulder, approximately 20% at the tool pin vertical surface and approximately 50% of the mechanical work is converted to heat input [8]. The implementation of heat flux boundary condition is described in the subsequent subsection.

Material flow

The properties of the material, mechanical loading condition and process parameters in the FSW process decide the nature of material flow that can be assumed as elasto-plastic, fully plastic, visco-plastic or fully viscous fluid. Therefore the flow field is obtained by solving fluid dynamics if the material is a viscous fluid or displays visco-plastic behaviour. However, the plastic deformation is modelled from the principal of solid mechanics. In this section, the flow field is explained from the fluid dynamics principle by assuming a non-Newtonian, incompressible and visco-plastic material. The conservation of mass and momentum equations are solved within the solution domain, assuming no density variation. The coupling of the energy equation is also necessary to include the temperature dependent material properties, and the heat conduction equation is modified by the convective material flow. The conservation of momentum and mass in transient state are expressed in vector form as [44]

$$\rho\left(\frac{\partial \vec{V}}{\partial t}+\vec{V}\cdot\nabla\vec{V}\right)=\nabla\cdot\vec{\sigma}+\vec{F} \qquad [2.8]$$

and

$$\nabla\cdot\vec{V}=0 \qquad [2.9]$$

where $\nabla=\left\{\frac{\partial}{\partial x}\frac{\partial}{\partial y}\frac{\partial}{\partial z}\right\}^T$ represents the gradient operator, $\vec{V}=\{u\ \ v\ \ w\}^T$ corresponds to liquid metal velocity, $\vec{\sigma}_s$ is the total stress tensor, ρ is the density of the material, $\vec{F}$ corresponds to the body force vector and t is the time variable. The total stress tensor in Eq. 2.8 can be expressed by using Stoke's law as [56]

$$\vec{\sigma}=\mu\left[(\nabla\vec{V})+(\nabla\vec{V})^T\right]-P\vec{I} \qquad [2.10]$$

where P is the hydrostatic pressure, μ is the viscosity of the material and $\vec{I}$ is the identity matrix. From Eqs 2.8 and 2.9, along with suitable modification due to tool movement, the momentum conservation equations can be modified as

$$\rho\left(\frac{\partial \vec{V}}{\partial t}+\vec{V}\cdot\nabla\vec{V}\right)=-\nabla P+\mu\nabla\cdot\left[(\nabla\vec{V})+(\nabla\vec{V})^{T}\right]+\rho\vec{U}\cdot(\nabla\vec{V})+\vec{F} \qquad [2.11]$$

where $\vec{U}$ is the velocity vector of the tool, which is equivalent to V_T for directional movement of the tool along the *y*-axis. In association with the conservation of mass and the momentum equations, the equation for the conservation of energy is also solved for the heat transfer and fluid flow analysis. The general governing equation for heat conduction in Cartesian coordinates can be stated as

$$\nabla\cdot\left(\vec{k}\,\nabla T\right)+\dot{Q}=\rho C_p\left(\frac{\partial T}{\partial t}+\vec{U}\cdot\nabla T-\vec{V}\cdot\nabla T\right) \qquad [2.12]$$

where $\vec{k}$ is the directional conductivity of an orthotropic solid. The term that corresponds to $\vec{V}$ on the right-hand side of Eq. 2.12 accounts for the convective flow of liquid metal in energy transport. It is noteworthy that for a stationary heat source, Eqs 2.11 and 2.12 can be modified by inserting the null velocity vector and the solution can be obtained over the fixed coordinate system (x,y,z), whereas in quasi-steady state analysis, the transient term should be neglected and the solution exists over a moving coordinate system (x,y,z), whose origin can be assumed to the position of the rotating tool. The thermal boundary condition is same as described in Eq. 2.3.

Figure 2.9 depicts the schematic view of velocity boundary conditions corresponding to a flat tool shoulder and a workpiece interface. Assume a point 'A' on the workpiece having radius r at an angle θ with respect to the direction of tool transverse speed V_T. The material is rotating at an absolute rotational speed of ω. The linear velocity components on the boundary of the top surface of the tool shoulder is expressed as

$$v=\omega r\ sin\theta-V_T;\ u=\omega r\ cos\theta \qquad R_p\le r\le R_s \qquad [2.13]$$

The velocity at the tool pin periphery is defined as

$$v-\omega R_p\ sin\theta-V_T;\ u=\omega R_p\ cos\theta \quad r-R_p;\ 0\le h\le H_p \qquad [2.14]$$

The velocity boundary condition at the bottom of tool pin surface is defined as

$$v=\omega r\ sin\theta-V_T;\ u=\omega r\ cos\theta \quad 0\le r\le R_p \qquad [2.15]$$

where R_s, R_p and H_p are tool shoulder and tool pin radius and tool pin height, respectively for a cylindrical tool. If the transverse tool movement is neglected, i.e. for a stationary tool, the above boundary conditions can be modified by considering $V_T=0$.

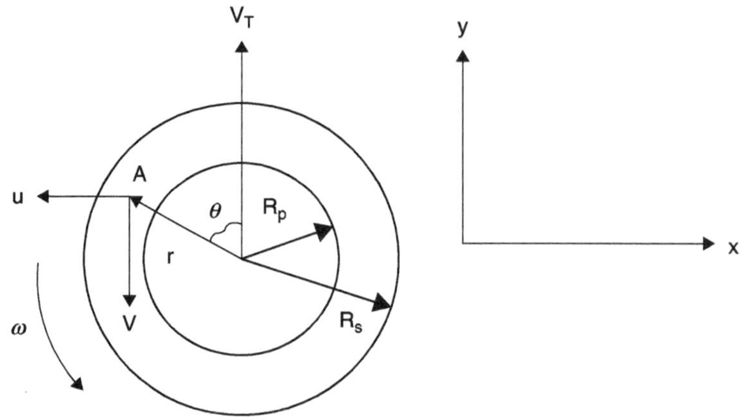

Figure 2.9 **Velocity boundary conditions in FSW (top view)**

From theoretical formulation, it is obvious that the mathematical model involves several parameters. The material behaviour or material model increases the complexity due to temperature or strain rate dependency of material properties. There exist several approaches to predict the material behavior. Since the material flow involves non-Newtonian flow of visco-plastic material, the viscosity can be calculated following Hart's simplified model [45], where flow stress is divided into plastic and viscous parts as

$$\sigma_e = \sigma_p + \sigma_y \qquad [2.16]$$

where the viscous part represents the frictional resistance to dislocation glide and the plastic part represents flow resistance to dislocation entanglement. The empirical correlations for plastic and viscous part are expressed as

$$\sigma_p = K\, exp\left[-\left(\frac{b}{\dot{\varepsilon}}\right)^{\lambda}\right] \qquad [2.17]$$

and

$$\sigma_V = G\left(\frac{\dot{\varepsilon}}{a}\right)^{1/M} \qquad [2.18]$$

where

$$b = b_0\left(\frac{K}{G}\right)^{N} exp\left(-\frac{Q}{RT}\right) \qquad [2.19]$$

and

$$a = a_0 exp\left(-\frac{Q_0}{RT}\right) \qquad [2.20]$$

$\dot{\varepsilon}$ is the deformation rate and G, Q, Q_0, a_0, b_0, λ, N and M are material parameters that can be determined from experiments [46]. However, K is the scalar variable and the saturated value of this variable is used as a function of temperature and strain rate. It is expressed as

$$K = \left(\frac{C}{\varphi}\right)^{m_0} \qquad [2.21]$$

where φ is the Fischer factor [47] and is given by

$$\varphi = Tln\left(\frac{D_0}{\dot{\varepsilon}}\right) \qquad [2.22]$$

where D_0, m_0 and C are experimentally determined parameters. However, the viscosity can be determined from Penrzyna's visco-plasticity model [48] as

$$\mu = \frac{\sigma_e}{3\dot{\varepsilon}} \qquad [2.23]$$

The effective strain rate $\dot{\varepsilon}$ is given by

$$\dot{\varepsilon} = \left(\frac{2}{3}\varepsilon_{ij} \in_{ij}\right)^{1/2} \qquad [2.24]$$

where ε_{ij} is the strain rate tensor and is defined by

$$\varepsilon_{ij} = \frac{1}{2}\left(\frac{\partial u_i}{\partial x_j} + \frac{\partial u_j}{\partial x_i}\right) \qquad [2.25]$$

The above expressions to estimate the viscosity as a function of temperature and strain rate can be used directly in Eq. 2.11. The viscosity generally decreases with increase in both strain rate and temperature, whereas strain rate is the most dominating factor to influence the viscosity.

The above-mentioned boundary conditions do not take into account the contact state variable. The velocity boundary condition in a steady or stationery heat source analysis at the contact interface is modified as

$$u = \delta\omega r\ sin\theta; \qquad v = \delta\omega r\ cos\theta \qquad [2.26]$$

and

$$R_p \leq r \leq R_s \qquad [2.27]$$

where δ is the contact rate variable describing the dimensionless degree of sticking and r is the radial position. The rest of the velocity boundary conditions are extracted by modifying Eqs 2.14 and 2.15 in a similar fashion.

There is no straightforward mechanism to estimate the extent of slip and hence the contact state variable can be calculated by imperial relationships. The extent of slip is estimated by statistical correlation from experimentally measured values at various relative velocities as [8]

$$\delta = 0.2 + 0.6\left[1 - \exp\left(-\delta_0 \frac{\omega r}{\omega_0 R_s}\right)\right] \qquad [2.28]$$

where δ_0 is an adjustable parameter, r is the distance of the point from tool axis, ω is the angular velocity of the tool, R_S is the radius of the tool shoulder, ω_0 is the normalizing rotational velocity, which can be taken as the mid-point of the range of rotational speeds considered in a specific case. In the case of pure friction $\delta = 0$. For joining aluminum 1xxx series, the typical value of δ_0 is considered as 1.014 [8] and other parameters are varied with welding conditions. For a welding speed of 1.63 mm/s, 600 rpm and 6.23 kN average plunging force, and 50 mm shoulder diameter, the typical values of δ at the shoulder interface is approximately 0.49, at the pin's side surface is approximately 0.42 and at the pin's bottom surface is approximately 0.41. Therefore, δ is almost constant for all three contact surfaces and hence the constant value for all surfaces can be taken as approximately 0.44.

The thermal boundary flux for stationary or fully steady state analysis is represented as

$$q_s = \phi[\delta \omega r \tau_y(T) + (1 - \delta)\omega r \tau_f] \qquad [2.29]$$

where $\tau_y(T)$ is the temperature dependent yield shear stress of the material at the contact interface and ϕ the power efficiency factor, i.e. the amount of mechanical energy converted to heat energy. The frictional stress τ_f in Eq. 2.29 is equivalent to the applied plunge pressure on the shoulder, i.e.:

$$\tau_f = \mu_f P_T \qquad [2.30]$$

where P_T is the applied pressure (plunge) on the shoulder and μ_f is the coefficient of friction between the tool and workpiece. During the cross-wedge rolling process, the empirical correlation of the coefficient of friction is expressed as

$$\mu_f = 0.5\ X\ exp(-\delta\omega r) \qquad [2.31]$$

where the linear velocity is in m/s. However, this equation is valid when the velocity varies from 0.1 to 1.6 m/s [49]. If it is assumed that the friction shear stress is at equilibrium with the viscous shear stress at the contact interface in steady–state condition:

$$\tau_f = \tau_v = \sqrt{\tau_x^2 + \tau_y^2} \qquad [2.32]$$

where the viscous stresses τ_x and τ_y are set at the boundary and since the rotational velocity boundary conditions are defined on the x-y plane and the flow in the z-direction is negligible. The non-Newtonian viscous fluid can be represented by the power low viscosity model, which is a function of temperature and strain rate. The shear stress is expressed as

$$\tau = \mu\dot{\gamma}^n = \mu_{eff}\dot{\gamma} \qquad [2.33]$$

where μ_{eff} is the effective viscosity dependent of strain rate and temperature and $\dot{\gamma}$ is the shear strain rate. The effective viscosity is estimated by

$$\tau_r(T) = \eta_{eff,r}(T)\dot{\gamma}_r \qquad [2.34]$$

and

$$\eta_{eff,r}(T) = \frac{\tau_r(T)}{\dot{\gamma}_r} = \frac{\tau_r(T)}{\sqrt{3}\dot{\varepsilon}_r} \qquad [2.35]$$

where $\tau_r(T)$ and $\dot{\varepsilon}_r$ are reference values of shear stress and strain rate and are calculated from experimental data.

2.2.2 Heat generation model

A sophisticated numerical model demands a strong understanding of heat generation mechanism and quantitative estimation, which is complex in nature due to the interaction of thermal and mechanical processes. The heat generated during the welding process is equivalent to the mechanical energy converted to thermal energy by the tool at the tool–workpiece interface and due to plastic deformation in the TMAZ. Local interfacial heat generation due to friction is the product of frictional force and sliding velocity. The interfacial deformation heat is the product of shear stress and the velocity of the workpiece material, which sticks to the tool as it moves. A fraction of the plastic deformation energy is stored within the thermo-mechanically processed region in the form of defects and elongated grains. Part of the mechanical energy is spent through recovery and dynamic recrystallization phenomena that occurs simultaneously. This section aims to describe different analytical models of heat generation, neglecting the effect of kinetics of microscopic phenomena. The heat generated at the interface of tool–workpiece is assumed due to friction as well as plastic deformation. The generated heat can be assumed as a direct product of adhesion and deformation of the material around the tool. Finally, the heat flux calculation is described and it is imposed in any finite element based model as surface heat flux through the boundary interaction.

Stationary heat source

Heat generated from the transverse movement is a significantly smaller amount than from rotation. The material flow and heat generation are characterized by the contact conditions at the interface, and are described as sliding, sticking or partial sliding/sticking conditions [17]. The contact condition under the shoulder can be described by sliding friction, using a friction coefficient μ and interfacial pressure p, or sticking friction, based on the interfacial shear strength at an appropriate temperature and strain rate. The heat generation due to friction on an elemental area dA at the tool–workpiece interface, considering high rotational speed compared to traverse speed of the FSW tool, is given by [17]

$$dQ_f = (1 - \delta)\omega r \tau_c dA \qquad [2.36]$$

where δ (<1) is the extent of sticking and $1 - \delta$ accounts for the amount of frictional work dissipated into the workpiece, ω is angular velocity of the tool and r is the radial distance of elemental area dA from the axis. The contact state variable (δ) is assumed to change linearly with distance from the centre of the pin. Based on these assumptions and geometrical aspects of the deformation zone, the contact state variable can also be expressed as

$$\delta = 1 - \frac{V_{matrix}}{V_{tool}} \qquad [2.37]$$

where V_{matrix} is the peripheral velocity of the workpiece and V_{tool} is the peripheral velocity of the tool. Therefore, when δ is 1, no material sticks to the tool and all the heat is generated by friction. In contrast, when δ is 0, all the heat is generated by plastic deformation. A 100% sticking condition at the pin–material interface is assumed, that is the same as the tool velocity. The heat generated due to shear deformation leading to the workpiece material sticking to the tool is given by

$$dQ_s = \delta\omega r\tau_c dA \qquad [2.38]$$

For sliding friction, the contact shear stress is $\tau_c = \mu p$ and for sticking friction is $\tau_c = \tau_y$ where p is the applied pressure and τ_y is the shear yield stress. The contact shear stress is also defined as

$$\tau_y = \frac{\sigma_y}{\sqrt{3}} \qquad [2.39]$$

where σ_y is the yield stress. Plastic deformation also occurs away from the tool–workpiece interface and its effect on local heat generation rate may be estimated over volume dV as

$$dQ_p = \beta\mu\phi\, dV \qquad [2.40]$$

where β is the fraction of plastic deformation work that is dissipated as heat, μ is the non-Newtonian viscosity of the plasticized material and ϕ is related to velocity gradient. The term ϕ can be expressed from the velocity field as [50]

$$\phi = 2\sum_{i=2}^{2}\left(\frac{\partial u_i}{\partial x_i}\right)^2 + \left(\frac{\partial u_1}{\partial x_2} + \frac{\partial u_2}{\partial x_1}\right)^2 + \left(\frac{\partial u_1}{\partial x_2} + \frac{\partial u_2}{\partial x_1}\right)^2 + \left(\frac{\partial u_2}{\partial x_2} + \frac{\partial u_2}{\partial x_2}\right)^2 \qquad [2.41]$$

The fraction of plastic deformation work that is converted to heat is in the order of approximately 0.6 to 0.8 and becomes higher at high strain rates. However, overall magnitude of the heat generation rate by viscous dissipation is approximately 5% of the total heat generation rate, which is significantly small [18,51]. Neglecting the effect of volumetric plastic deformation work, total elemental heat due to friction and plastic deformation on the tool–workpiece interface is considered as a linear combination of sliding and sticking friction and is defined as

$$dQ_{FSW} = dQ_f + dQ_p = (1-\delta)\omega r\mu p\, dA + \delta\omega r\frac{\sigma_y}{\sqrt{3}}dA \qquad [2.42]$$

Hence it is necessary to define the extent of slip δ to precisely distinguish the heat generation due to friction or plastic deformation (Eq. 2.28). This elemental heat flux calculation is extended to a more general shaped tool. Figure 2.10 describes the typical *FSW* tool having the geometric shape of a truncated cone without any thread along with an angular-shaped shoulder. The geometric parameters of the tool are shown in the figure.

The heat energy generated at the contact interface between a rotating conical (truncated)-shaped *FSW* tool and a stationary workpiece are

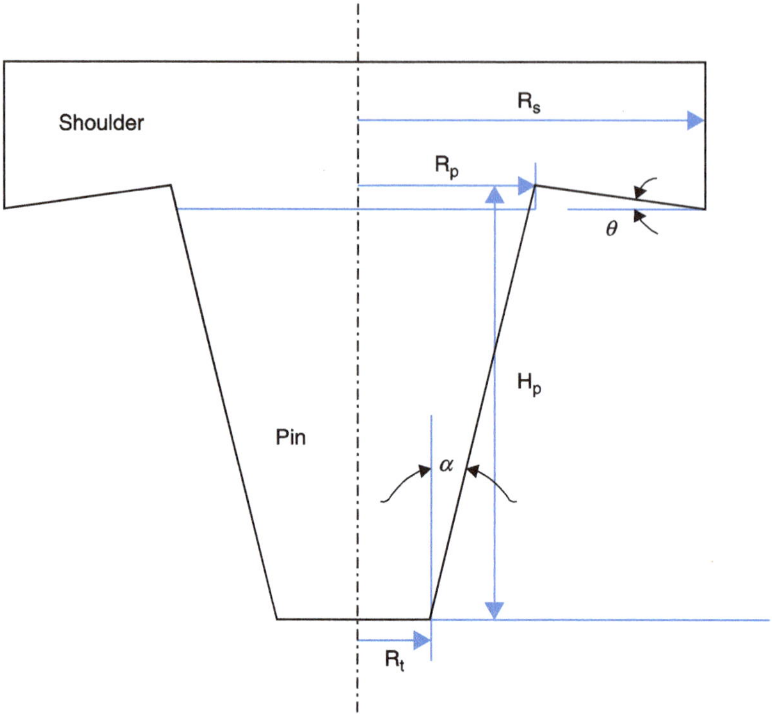

Figure 2.10 Geometry of truncated conical tool

subdivided as Q_S, Q_P and Q_T, i.e. on the tool shoulder's surface, tool pin's side surface and tool pin's tip surface, respectively. The analytical estimation of heat generation from the shoulder is expressed as

$$Q_s = \frac{2}{3}\pi\tau_c\omega\left(R_s^3 - R_p^3\right)(1+\tan\theta) = \frac{2}{3}\pi\omega\left[\delta\tau_y + (1-\delta)\mu p_1\right]\left(R_s^3 - R_p^3\right)(1+\tan\theta) \quad [2.43]$$

where p_1 is the uniform pressure at the tool shoulder–workpiece interface. Heat generation from the probe is expressed as

$$Q_P = \frac{2}{3}\pi\tau_c\omega\left(R_p^3 - R_t^3\right)\left(\frac{1+\tan\alpha}{\tan\alpha}\right) = \frac{2}{3}\pi\omega[\delta\tau_y + (1-\delta)\mu p_2]\left(R_p^3 - R_t^3\right)\left(\frac{1+\tan\alpha}{\tan\alpha}\right) \quad [2.44]$$

where p_2 is the uniform pressure at the pin–workpiece contact interface and $tan\,\alpha = \frac{R_p - R_t}{H_p}$. The heat generation from the probe tip surface is expressed as

$$Q_T = \frac{2}{3}\pi\omega\tau_c R_t^3 = \frac{2}{3}\pi\omega[\delta\tau_y + (1-\delta)\mu p_3]R_t^3 \qquad [2.45]$$

where p_3 is the uniform pressure at the tip–workpiece contact interface Therefore the total heat generation at the contact interface is the summation of all individual components and is expressed as

$$Q_{FSW} = Q_s + Q_p + Q_T \qquad [2.46]$$

The expression of heat generation in Eq. 2.46 is more general and it is possible to customize by considering the effect of only sticking or non-sticking conditions, whether the tool is conical or cylindrical, or whether the shoulder surface is flat or sloping. For a cylindrical tool with a flat shoulder surface, the component of the heat generation term is simplified as

$$Q_s = \frac{2}{3}\pi\tau_c\omega\left(R_s^3 - R_p^3\right) \qquad [2.47]$$

and

$$Q_p = 2\pi\tau_c\omega R_p^2 H_p \qquad [2.48]$$

and

$$Q_T = \frac{2}{3}\pi\omega\tau_c R_p^3 \qquad [2.49]$$

where R_p and H_p are tool shoulder radius and height, respectively. Generally, the tool–shoulder surface generates more heat and hence it is important to consider the sliding to sticking condition to precisely estimate heat generation. In this case, the plunging force applied to the plate surface by the tool creates a uniform pressure over the shoulder surface. For the other two surfaces, the only sliding condition is a reasonably good approximation to estimate heat generation. The weight fraction of heat generated from the three different surfaces with respect to total heat generation depends on the geometry of the tool and the contact condition depicted in Eq. 2.28. The typical weight fraction for the shoulder is approximately 0.85, pin side surface approximately 0.13 and pin bottom surface approximately 0.02 for a cylindrical tool. The typical tool dimensions are shoulder radius 10 mm, pin radius 3 mm and tool shoulder height 5.75 mm. The calculation indicates that heat generation from the probe is more than 15%, but it is typically negligible for a thin plate. Around 85% of the heat is generated from the tool shoulder surface, which dominates over other contact surface heat generation.

The generated heat is distributed over the surfaces of the tool-workpiece as boundary flux. Figure 2.11 depicts the heat intensity over the tool surfaces. It is assumed that the heat flux distributes over the tool shoulder surface and hence is expressed as

$$q(r) - q_m \cdot r \qquad [2.50]$$

where q_m is the heat intensity per unit length and varies linearly along the radial direction. Therefore, the heat flux over the tool shoulder is expressed as

$$q_s = \frac{3Q_s r}{2\pi\left(R_s^3 - R_p^3\right)} \qquad [2.51]$$

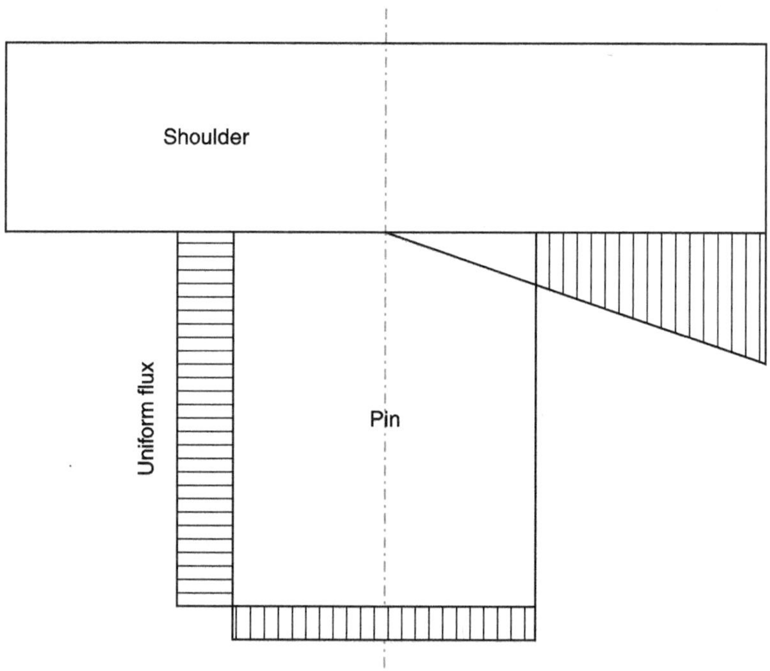

Figure 2.11 **Heat intensity distribution of a cylindrical tool**

where r is the radial distance from the centre and lies in between R_s and R_p. However, if we assume that the heat intensity distributes uniformly, the heat flux on pin–workpiece interface and tool pin bottom surface are expressed as

$$q_P = \frac{Q_p}{2\pi R_p H_p} \quad [2.52]$$

and

$$q_T = \frac{Q_T}{\pi R_p^2} \quad [2.53]$$

It should be noted that volumetric heat generation due to viscous dissipation in the shear layer is included in Eq. 2.1 by the rate of heat generation per unit volume and is defined by

$$\dot{Q} = \frac{dQ_p}{dV} = \beta \mu \dot{\varepsilon}^2 \quad [2.54]$$

where μ is the viscosity, $\dot{\varepsilon}$ is the effective strain rate and β is a constant that indicates the fraction of plastic work that is converted to heat. The spatially variable strain rate $\dot{\varepsilon}$ is computed by Eq. 2.24.

The reliability in calculation of temperature distribution and material flow depends on input model parameters and it is difficult to define these parameters from scientific principles alone. The value of the friction coefficient varies mainly with temperature, relative velocity and surface contact conditions. However, the friction becomes lower at high temperatures due to localized softening of the material and also influences the heat generation rate. The simple approach is to modify the coefficient of friction to adjust the heat generation rate. During plastic deformation, the yield strength decreases with temperature and leads to a reduction in heat generation. The strain rates also diminish rapidly away from the tool and hence most of the heat generation due to plasticity occurs close to the tool–workpiece interface.

Moving heat source

As Eqs 2.1 and 2.2 indicate, the heat generation rate is influenced by the rotational speed but not by the welding speed [40]. In moving heat source problems, it is assumed as a quasi-steady state and the heat generation is modified by relative motion of the tool with respect to the material. Figure 2.7 shows the velocity diagram of the top surface under the tool shoulder. The rate of heat generation per unit area as a function of angle θ is expressed as

$$q_1(\theta) = \eta[\delta\tau_y(\omega r - V\sin\theta) + (1-\delta)\mu p] \quad R_p \le r \le R_s \qquad [2.55]$$

$$q_2(\theta) = \eta[\delta\tau_y(\omega R_p - V\sin\theta) + (1-\delta)\mu p] \quad r = R_p;\ 0 \le h \le H_p \qquad [2.56]$$

$$q_3(\theta) = \eta[\delta\tau_y(\omega r - V\sin\theta) + (1-\delta)\mu p] \quad 0 \le r \le R_p \qquad [2.57]$$

where η is efficiency factor, i.e. the amount of mechanical energy converted to heat energy. The heat flux is directly calculated at each node point at an angle θ and acts as a surface flux over the domain defined by Eqs 2.55 to 2.57. It is obvious the estimated heat flux is non-symmetrical due to the linear transverse movement of the tool and the calculated temperature will also be non-symmetrical for a quasi-steady state analysis.

Of the total heat generated at the shoulder–workpiece interface, some of it is transported into the tool material while the rest enters the workpiece. The total heat generated at the shoulder–workpiece interface can been partitioned between the workpiece (H_w) and the tool (H_t). Assuming a steady-state, one-dimensional (1-D) heat flow from the interface of dissimilar metal joints, the ratio of heat flux is estimated as [52]

$$\eta = \frac{H_w}{H_t} = \frac{(k\rho c_p)_w^{1/2}}{(k\rho c_p)_t^{1/2}} \qquad [2.58]$$

Therefore the heat flux into the workpiece is calculated as $\frac{\eta}{\eta+1}$ of the total heat generation at the interface.

2.2.3 Finite element formulation

To develop the numerical model for heat transfer and material flow, the mathematical tool of finite element method is followed here. This section will provide a working knowledge of the finite element formulation of the problem from governing equations and boundary conditions in FSW processes. The computational issues will also be covered in this section, which is further subdivided into two parts: heat conduction and coupled mass, and momentum and heat transfer analysis. First, the governing equations and the boundary conditions are illustrated, their discretization using the finite element method over the solution domain are being described next. The in-depth analysis of the mathematical models helps understanding of the physical processes of FSW.

Heat transfer model

The solution domain consists of eight-noded isoperimetric brick elements. The temperature variable, T, within the element is expressed in terms of nodal temperatures as

$$T=\sum_{i=1}^{8}[N_i]\{T_i\} \qquad [2.59]$$

The global coordinates can also be interpolated within the element using similar shape functions as

$$x=\sum_{i=1}^{8}[N_i]\{x_i\} \qquad [2.60]$$

and

$$y=\sum_{i=1}^{8}[N_i]\{y_i\} \qquad [2.61]$$

and

$$z=\sum_{i=1}^{8}[N_i]\{z_i\} \qquad [2.62]$$

Applying Galerkin's weighted residue technique to the governing equation yields a general form as

$$\int_{\Omega}\phi\left\{\frac{\partial}{\partial x}\left(k\frac{\partial T}{\partial x}\right)+\frac{\partial}{\partial y}\left(k\frac{\partial T}{\partial y}\right)+\frac{\partial}{\partial z}\left(k\frac{\partial T}{\partial z}\right)+\dot{Q}-\rho C_p\frac{\partial T}{\partial t}\right\}d\Omega=0 \qquad [2.63]$$

where ϕ is a weighting function and Ω refers to the volume of the solution domain. Using thermal boundary conditions and substituting the weighting function as the shape function in Eq. 2.63, it is possible to obtain in matrix form for any specific element 'e' as

$$[H^e]\{T\}+[S^e]\left\{\frac{\partial T}{\partial t}\right\}+[\bar{H}^e]\{T\}-\{f_Q^e\}-\{f_q^e\}-\{f_h^e\}=0 \qquad [2.64]$$

where

$$[H_{ij}^e]=\int_{\Omega^e}k\left(\frac{\partial N_i}{\partial x}\frac{\partial N_j}{\partial x}+\frac{\partial N_i}{\partial y}\frac{\partial N_j}{\partial y}+\frac{\partial N_i}{\partial z}\frac{\partial N_j}{\partial z}\right)d\Omega \qquad [2.65]$$

$$[S_{ij}^e]=\int_{\Omega^e}\rho C_p N_i N_j d\Omega; \qquad [\bar{H}_{ij}^e]=\int_{\Gamma^e}h_{eff}N_i N_j d\Gamma \qquad [2.66, 2.67]$$

and

$$\left\{\left(f_q^e\right)_i\right\}=\int_{\Gamma_1^e} N_i q_s d\Gamma;\ \left\{\left(f_h^e\right)_i\right\}=\int_{\Gamma_2^e} N_i h_{eff} T_0 d\Gamma \quad [2.68, 2.69]$$

$$\left\{\left(f_{\dot{Q}}^e\right)_i\right\}=\int_{\Omega^e} N_i \dot{Q} d\Omega \quad [2.70]$$

Considering the contribution from all the elements, the final algebraic equation is

$$[\bar{\bar{H}}]\{T\}+[S]\left\{\frac{\partial T}{\partial t}\right\}=\{f\} \quad [2.71]$$

where $[\bar{\bar{H}}]=[H]+[\bar{H}]$; $\{f\}=\{f_{\dot{Q}}\}+\{f_q\}+\{f_h\}$

Equation 2.71 is used within each time step to solve the unknown nodal temperature field during the 3-D transient heat transfer analysis. In general, the so-called α-method for time discretization following the one-step method is expressed as [56]

$$[S]\frac{\{T_i^2\}-\{T_i^1\}}{\Delta t}+\alpha[\bar{\bar{H}}]\{T_i^2\}+(1-\alpha)[\bar{\bar{H}}]\{T_i^1\} \quad [2.72]$$
$$=\alpha\{f\}^2+(1-\alpha)\{f\}^1;\ 0\le\alpha\le 1$$

The most common values for α are

$\alpha=0$, *Explicit Eulers scheme or forward difference scheme*
$\alpha=1$, *Implicit Euler scheme or backward difference scheme*
$\alpha=\frac{1}{2}$, *Crank – Nicolson scheme*
$\alpha=\frac{2}{3}$, *Galerkin method*

In time the discretization scheme, the Galerkin method, i.e. $\alpha=2/3$, is unconditionally stable and is generally followed assuming the load vector, {f}, is independent of the state of time, which is basically an explicit scheme. In quasi-steady state analysis, the nodal variable, i.e. temperature, is independent of time and the elemental matrix equation is expressed as

$$[H^e]\{T\}+[\bar{S}^e]\{T\}+[\bar{H}^e]\{T\}=\{f_{\dot{Q}}^e\}+\{f_q^e\}+\{f_h^e\} \quad [2.73]$$

where

$$[\bar{S}_{ij}^e]=-\int_{\Omega^e}\rho c_p V_T N_i \frac{\partial N_i}{\partial y} d\Omega \quad [2.74]$$

The final algebraic equation for the 3-D pseudo-steady state heat transfer analysis is

$$[\bar{\bar{H}}]\{T\}=\{f\} \qquad [2.75]$$

where

$$[\bar{\bar{H}}]=[H]+[\bar{H}]+[\bar{S}] \quad ; \quad \{f\}=\{f_{\dot{Q}}\}+\{f_q\}+\{f_h\} \qquad [2.76, 2.77]$$

Here the algebraic form of Eq. 2.75 is solved in a ramped-up load step for quasi-steady- state analysis [56].

Material flow model

Three-dimensional material flow analysis consists of four degrees of freedom at each node, i.e. three velocity components and one temperature component. The reduced integration penalty method in finite element methodology is followed to solve the conservation equations of mass and momentum equations [56]. The penalty method is opted to devise the continuity equation by penalty parameter (λ) in the following way:

$$\frac{P}{\lambda}+\left(\frac{\partial u}{\partial x}+\frac{\partial v}{\partial y}+\frac{\partial w}{\partial z}\right)=0 \qquad [2.78]$$

Physically, the penalty parameter indicates the dilatational viscosity in fluid flow analysis and the value is set to a large number ($\lambda \rightarrow \infty$), so that it can satisfy the continuity equation. To avoid nonlinearity, the velocities in convective terms are made independent from the velocity variables that linearize the momentum equation. Considering the elemental shape function, the unknown variables (velocity and temperature) within the element are expressed in terms of the corresponding nodal variables as

$$u=\sum_{i=1}^{8}[N_i]\ \{u_i\} \qquad [2.79]$$

and

$$v=\sum_{i=1}^{8}[N_i]\ \{v_i\} \qquad [2.80]$$

and

$$w=\sum_{i=1}^{8}[N_i]\ \{w_i\} \qquad [2.81]$$

and

$$T=\sum_{i=1}^{8}[N_i]\ \{T_i\} \qquad [2.82]$$

Performing the finite element discretization to the momentum conservation, the algebraic equations the x-, y- and z-directions can be written in linearized form:

$$[M^e]\left\{\frac{\partial\{u\}}{\partial t}\right\}+[\bar{C}^e]\{u\}+[\hat{K}_{11}^e]\{u\}+[\hat{K}_{12}^e]\{v\}+[\hat{K}_{13}^e]\{w\}$$
$$+\left\{2[K_{11}^e]+[K_{22}^e]+[K_{33}^e]\right\}\{u\}+[K_{21}^e]\{v\}+[K_{31}^e]\{w\}=\{B^{xe}\} \quad [2.83]$$

and

$$[M^e]\left\{\frac{\partial\{v\}}{\partial t}\right\}+[\bar{C}^e]\{v\}+[\hat{K}_{21}^e]\{u\}+[\hat{K}_{22}^e]\{v\}+[\hat{K}_{23}^e]\{w\}$$
$$+[K_{21}^e]\{u\}+[K_{11}^e]+2[K_{22}^e]+[K_{33}^e]\{v\}+[K_{23}^e]\{w\}=\{B^{ye}\} \quad [2.84]$$

$$[M^e]\left\{\frac{\partial\{w\}}{\partial t}\right\}+[\bar{C}^e]\{w\}+[\hat{K}_{31}^e]\{u\}+[\hat{K}_{32}^e]\{v\}+[\hat{K}_{33}^e]\{w\}+[K_{31}^e]\{u\}$$
$$+[K_{32}^e]\{v\}+\left\{[K_{11}^e]+[K_{22}^e]+2[K_{33}^e]\right\}\{w\}+\{B^{ze}\}$$

[2.85]

Equations 2.83 to 2.85 can be represented in matrix form as

$$\begin{bmatrix}[M^e] & 0 & 0\\ 0 & [M^e] & 0\\ 0 & 0 & [M^e]\end{bmatrix}\begin{Bmatrix}\frac{\partial\{u\}}{\partial t}\\ \frac{\partial\{v\}}{\partial t}\\ \frac{\partial\{w\}}{\partial t}\end{Bmatrix}+\begin{bmatrix}[\bar{C}^e] & 0 & 0\\ 0 & [\bar{C}^e] & 0\\ 0 & 0 & [\bar{C}^e]\end{bmatrix}+\begin{Bmatrix}\{u\}\\ \{v\}\\ \{w\}\end{Bmatrix}$$
$$+\begin{bmatrix}[\hat{K}_{11}^e] & [\hat{K}_{12}^e] & [\hat{K}_{13}^e]\\ [\hat{K}_{21}^e] & [\hat{K}_{22}^e] & [\hat{K}_{23}^e]\\ [\hat{K}_{31}^e] & [\hat{K}_{32}^e] & [\hat{K}_{33}^e]\end{bmatrix}\begin{Bmatrix}\{u\}\\ \{v\}\\ \{w\}\end{Bmatrix}$$
$$+\begin{bmatrix}2[K_{11}^e]+[K_{22}^e]+[K_{33}^e] & [K_{21}^e] & [K_{31}^e]\\ [K_{21}^e] & [K_{11}^e]+2[K_{22}^e]+[K_{33}^e] & [K_{23}^e]\\ [K_{31}^e] & [K_{32}^e] & [K_{11}^e]+[K_{22}^e]+2[K_{33}^e]\end{bmatrix}\begin{Bmatrix}\{u\}\\ \{v\}\\ \{w\}\end{Bmatrix}$$
$$=\begin{Bmatrix}\{B^{xe}\}\\ \{B^{ye}\}\\ \{B^{ze}\}\end{Bmatrix}$$

[2.86]

or

$$[M]\{\dot{U}\}+\{[\bar{C}]+[\hat{K}]+[K]\}\{U\}=\{B\} \quad [2.87]$$

where

$$[\hat{K}^e_{ij}]=\int_{\Omega^e}\lambda\frac{\partial N_i}{\partial x_i}\frac{\partial N_j}{\partial x_j}d\Omega /= \quad [2.88]$$

and

$$[K^e_{ij}]=\int_{\Omega^e}\mu\frac{\partial N_i}{\partial x_i}\frac{\partial N_j}{\partial x_j}d\Omega \quad [2.89]$$

$$[M^e_{ij}]=\int_{\Omega^e}\rho N_i N_j d\Omega \quad [2.90]$$

$$[\bar{C}^e_{ij}]=\int_{\Omega}\rho\left\{u^0 N_i\frac{\partial N_j}{\partial x}+v^0 N_i\frac{\partial N_j}{\partial y}+w^0 N_i\frac{\partial N_j}{\partial z}\right\}d\Omega \quad [2.91]$$

The final matrix form of mass and momentum is simplified as

$$[M]\{\dot{U}\}+[\bar{K}]\{U\}=\{B\} \qquad \text{where} \qquad [\bar{K}]=[\bar{C}]+[\hat{K}]+[K] \quad [2.92]$$

where $[M]$ represents mass, $[\bar{C}]$ represents the velocity dependent convective transport, $[K]$ depicts the viscous diffusion term and $[\hat{K}]$ refers to the penalty term. The velocity field is discretized over the time domain using Galerkin's scheme ($\alpha = 2/3$) as

$$\{U_i^2\}=-\left[\frac{2}{3}[\bar{K}]+\frac{1}{\Delta t}[M]\right]^{-1}\left\{\left[\frac{1}{3}[\bar{K}]-\frac{1}{\Delta t}[M]\right]\{U_i^1\}-\{B\}\right\} \quad [2.93]$$

where $\{U_i^2\}$ is the velocity field after the elapse of time Δt and $\{U_i^1\}$ is the reference velocity at the beginning of time interval Δt. The final form of the matrix after time discretization is

$$[\bar{\bar{K}}]\{U_i^2\}=\{B'\} \quad [2.94]$$

Where

$$[\bar{\bar{K}}]=\left[\frac{2}{3}[\bar{K}]+\frac{1}{\Delta t}[M]\right] \quad [2.95]$$

and

$$\{B'\}=\left\{\{B\}-\left[\frac{1}{3}[\bar{K}]-\frac{1}{\Delta t}[M]\right]\{U_i^1\}\right\} \quad [2.96]$$

It is obvious that the pressure term disappears in the penalty FE approach to compute the velocity field in the solution domain. However, the pressure term is recovered from the velocity field. To avoid computational difficulty due to the penalty term, it is necessary to evaluate the coefficients of the penalty parameter, $[\hat{K}^e_{ij}]$, using the reduced Gauss-Quadrature numerical integration technique [56]. It is performed by under-integrating the penalty term (one point less) with respect to viscous and convective terms. The estimated pressure term is

$$P = -\lambda \sum_{i=1}^{8} \left(\frac{\partial N_i}{\partial x}\{u_i\} + \frac{\partial N_i}{\partial y}\{v_i\} + \frac{\partial N_j}{\partial z}\{w_i\} \right) \quad [2.97]$$

Since the penalty term is evaluated using the reduced Gauss point rule for linear elements, the pressure is computed at interior integration points and the components are extrapolated to the nodes by simple linear extrapolation procedure. It is also necessary to solve the energy equation sequentially using the convective energy transport term. The discretized form of energy equation is

$$[H^e]\{T\} + [C^e]\{T\} + [S^e]\frac{\partial\{T\}}{\partial t} + [\bar{H}^e]\{T\} = \{f^e_Q\} + \{f^e_q\} + \{f^e_h\} \quad [2.98]$$

where

$$[H^e_{ij}] = \int_{\Omega^e} k \left(\frac{\partial N_i}{\partial x}\frac{\partial N_j}{\partial x} + \frac{\partial N_i}{\partial y}\frac{\partial N_j}{\partial y} + \frac{\partial N_i}{\partial z}\frac{\partial N_j}{\partial z} \right) d\Omega \quad [2.99]$$

$$[C^e_{ij}] = \int_{\Omega^e} \rho C_p \left(u^0 N_i \frac{\partial N_j}{\partial x} + v^0 N_i \frac{\partial N_j}{\partial y} + w^0 N_i \frac{\partial N_j}{\partial z} \right) d\Omega \quad [2.100]$$

$$[S^e_{ij}] = \int_{\Omega^e} \rho C_p N_i N_j d\Omega \quad [2.101]$$

Considering the contributions of all elements within the solution domain, the final equation is

$$[S]\{\dot{T}\} + [\bar{\bar{H}}]\{T\} = \{f\} \quad [2.102]$$

where

$$[\bar{\bar{H}}] = [\bar{H}]; + [H] + [C] \quad [2.103]$$

and

$$\{f\} = \{f_Q\} + \{f_q\} + \{f_h\} \quad [2.104]$$

The discretization of Eq. 2.102 in the time domain is followed by a similar procedure to that in the previous section. In quasi-steady state analysis, the matrix form of momentum equation is

$$\{[M]+[\bar{C}]+[\hat{K}]+[K]\}\{U\}=\{B\} \qquad [2.105]$$

where

$$[M_{ij}^{e}]=-\int_{\Omega^e}\rho V_T N_i \frac{\partial N_j}{\partial y}\, d\Omega \qquad [2.106]$$

The energy equation for quasi-steady state is also represented as

$$[\bar{\bar{H}}]\{T\}=\{f\} \qquad [2.107]$$

where

$$[\bar{\bar{H}}]=[\bar{H}]+[H]+[C]+[S] \qquad [2.108]$$

and

$$\{f\}=\{f_{\dot{Q}}\}+\{f_q\}+\{f_h\} \qquad [2.109]$$

$$[S_{ij}^{e}]=-\int_{\Omega^e}\rho C_p V_T N_i \frac{\partial N_j}{\partial y}\, d\Omega \qquad [2.110]$$

All other terms have been defined earlier. It is customary to mention that the detailed finite element formulation of the problem acts as a foundation to develop the computer code in FSW, which helps the research community to write their own programs for heat transfer and material flow involved in FSW or fusion welding processes.

2.2.4 Thermo-mechanical processing

Apart from the fluid dynamics approach, the velocity field can be obtained from thermal and mechanical model interaction. In mechanical analysis, the temperature history can be obtained from the thermal analysis, which is the input for structural analysis. A static mechanical analysis is generally performed where the inertial effect is neglected, since a constant rotational speed and a constant longitudinal speed are used during FSW. Two basic sets of equations relating to the mechanical model are followed; the equilibrium equations and the constitutive equations. The equations of equilibrium are

$$\sigma_{ij,j}+\rho b_i=0 \qquad \sigma_{ij}=\sigma_{ji} \qquad [2.111]$$

where σ_{ij} is the stress tensor and b_i is the body force. The constitutive equations for a thermo-elastoplastic model are expressed in incremental stress-strain mode and a Lagrangian approach is followed. The increment of total strain is the sum of incremental elastic, plastic and thermal strains and is expressed as

$$\{d\varepsilon\} = \{d\varepsilon^{e}\} + \{d\varepsilon^{p}\} + \{d\varepsilon^{th}\} \qquad [2.112]$$

Stress–strain relations can be written as [53]

$$\{d\sigma\} = [D^{ep}]\{d\in\} - [C^{th}]\{dT\} \qquad [2.113]$$

where

$$\begin{aligned} [D^{ep}] &= [D^{e}] - [D^{p}] \\ [C^{th}] &= \{\alpha\}[D^{e}] \end{aligned} \qquad [2.114]$$

where $[D^{ep}]$ is the elasto-plastic matrix, $[D^{e}]$ is the elastic stiffness matrix, $[D^{p}]$ is the plastic stiffness matrix, $[C^{th}]$ is the thermal stiffness matrix, $\{d\sigma\}$ is the stress increment, $\{d\in\}$ is the strain increment and $\{dT\}$ is the temperature increment.

The above formulation is valid if the material behaviour is considered as elastoplastic in nature. However, the elastoplastic model can be improved by considering appropriate hardening mechanisms such as rate dependent or independent, and linear or multi-linear strain hardening. The incremental strain can be modified by incorporating the dislocation hardening, more complex effect of micro-structural changes, creep and transformation induced plasticity. However, the von-Mises yield criteria is followed as

$$\sigma_{av} = \sqrt{\frac{1}{2}[(\sigma_1 - \sigma_2)^2 + (\sigma_2 - \sigma_2)^2 + (\sigma_2 - \sigma_1)^2]} \qquad [2.115]$$

where, $\sigma_1, \sigma_2, \sigma_2$ are principal stresses and σ_{av} is the average 1-D stress. A fully coupled or decoupled thermal and structural analysis can be performed to obtain the displacement field and residual stress and strain generation in FSW. However, this analysis is limited to small displacement fields. The thermo-mechanical modelling approach is more practical when the heat generation due to FSW tool contact conditions is included and the mutual action between the contact conditions and the internal reaction stresses are significant. In this case, a coupled dynamic thermo-mechanical analysis is more important, using the ALE approach for large deformation process associated with FSW. The mechanical response from a force balance is modelled as

$$\rho\bar{u} + c\dot{u} + ku = p \qquad [2.116]$$

where ρ is the density, c is the damping coefficient, k is the stiffness coefficient, p is the body force and u is the displacement vector. The finite element formulation of Eq. 2.116 is

$$[M]\{\bar{u}\} + [C]\{\dot{u}\} + [K]\{u\} = \{P\} \qquad [2.117]$$

where $[M]$ is the mass matrix, $[C]$ is the viscous damping matrix, $[K]$ is the stiffness matrix and $\{P\}$ is the external force vector that includes body force, distributed surface forces and concentrated loads acting on the system. $\{u\}$, $\{\dot{u}\}$ and $\{\bar{u}\}$ are nodal displacement, velocity and acceleration vectors, respectively. To obtain the temperature distribution, the famous Newmark algorithm is used for time domain discretization [54]. The temperature at time $t+\Delta t$ is calculated by

$$\{u^{t+\Delta t}\} = \begin{bmatrix} [K] + \dfrac{1}{\beta(\Delta t)^2}[M] \\ + \dfrac{\delta}{\beta\Delta t}[C] \end{bmatrix}^{-1}$$

$$\left\{ \{P^{t+\Delta t}\} + [M]\left(\frac{1}{\beta(\Delta t)^2}\{u^t\} + \frac{1}{\beta\Delta t}\{\dot{u}^t\} + \left(\frac{1}{2\beta} - 1\right)\{\bar{u}^t\} \right) \right.$$

$$\left. +[c]\left(\frac{\delta}{\beta\Delta t}\{u^t\} + \left(\frac{\delta}{\beta} - 1\right)\{\dot{u}^t\} + \frac{\Delta t}{2}\left(\frac{\delta}{\beta} - 2\right)\{\bar{u}^t\} \right) \right\} \qquad [2.118]$$

where β and δ are parameters that can be determined to obtain accuracy and stability. The matrix relation gives temperature vector $\{u^1\}$ at the end of time step Δt in terms of known temperature given as vector $\{u^2\}$ at the start of the time step and we can proceed for subsequent steps. However, $\{\bar{u}\}$ and $\{\dot{u}\}$ at time $t+\Delta t$ are calculated by

$$\{\bar{u}^{t+\Delta t}\} = \frac{1}{\beta(\Delta t)^2}(\{u^{t+\Delta t}\} - \{u^t\}) - \frac{1}{\beta\Delta t}\{\dot{u}^t\} - \left(\frac{1}{2\beta} - 1\right)\{\bar{u}^t\} \qquad [2.119]$$

and

$$\{\dot{u}^{t+\Delta t}\}\{\dot{u}^t\} + \Delta t(1-\delta)\{\bar{u}^t\} + \delta\Delta t\{\bar{u}^{t+\Delta t}\} \qquad [2.120]$$

Unconditionally, the stable Galerkin method, i.e. $\delta=3/2$, $\beta=8/10$ is used to find the temperature distribution at the end of each time step. However, a temperature and strain rate dependent elastic-plastic Johnson-Cook material law model is the most popular to estimate flow stress value. The flow stress ($\sigma\downarrow y$) is given by

$$\sigma_y = (A + B[\overline{\in}^{pl}]^n)\left(1 + C ln\frac{\dot{\overline{\in}}^{pl}}{\dot{\varepsilon}_0}\right)\left\{1 - \left(\frac{T - T_r}{T_m - T_r}\right)^m\right\} \quad [2.121]$$

where $\overline{\in}^{pl}$ is the effective plastic strain, $\dot{\overline{\in}}^{pl}$ is the effective plastic strain rate, $\dot{\varepsilon}_0$ is the normalizing strain rate and A, B, C, n, T_m, T_r and m are material constants.

2.3 Interaction between thermal, mechanical and metallurgical field

There are several approaches to solve the temperature and material flow field in FSW processes. The coupling between heat transfer and material flow in FSW is computationally intensive and can be solved either by implicit or explicit schemes. It should be noted that the solution domain for the temperature field is different from the strain or deformation fields. A coupled and iterative approach is necessary, due to temperature dependent material properties and also due to dependence of the velocity field for the solution of the energy equation. When discrepancies between all variables in two successive iterations is less than a specified error limit, the solution moves to the next load or time step. The coupling of temperature and velocity fields is performed by following the SIMPLE algorithm based solution procedure, which is generally followed for the finite volume based approach [55].

One of the critical issues of FSW processing is to predict material behaviour by using the concept of internal parameters that cover macroscopic or microscopic features such as work hardening, relaxation and creep, whereas microscopic phenomena include dislocation density, phase transformation and grain properties. The main difficulty is the lack of proper data for a specific material, to include all physical phenomena to model material behaviour. Therefore, a material model has been developed in which an exhaustive number of phenomena are taken into consideration to describe the visco-plastic deformation of metals [45]. Additionally, the internal parameters such as plastic strain tensor and the dislocation density were considered, since these parameters are involved with the measurement of work hardening. The dislocation density values used are averages for each element in a finite element-based modelling approach.

Figure 2.12 illustrates the connection between various physical processes involved in FSW processing. With proper processing conditions

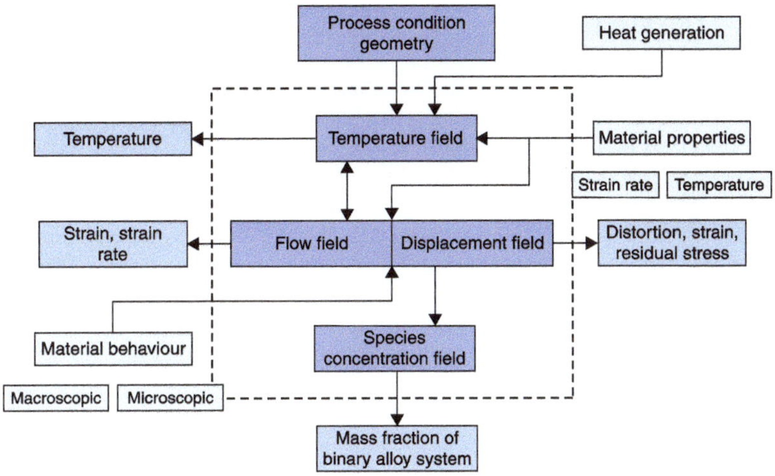

Figure 2.12 Flow chart showing interaction between fields

and geometric parameters, the system of FSW provides the information of temperature distribution, strain and strain rate (or velocity field) distribution only on the plastically deformed zone, the distortion and residual stress field and the mass fraction of individual components of the SZ. It is obvious from Figure 2.12 that there is a complex interaction between temperature field, flow field or displacement field and species concentration field. To estimate the temperature field, a dedicated heat generation model and temperature dependent material properties are required. The heat generation model needs a well-defined friction model and contact conditions.

However, the material properties also depend on strain rate, i.e. material flow and the energy equation relies on the convective flow of material. Hence a coupled approach by solving mass, momentum and energy equations are most suitable to estimate temperature and flow field. On the other hand, static mechanical analysis provides the information of displacement and stress field that rely on the temperature field, i.e. thermal strain and the macroscopic and microscopic behaviour of the materials. The species conservation indicates the transport of the species where material properties depend on temperature. The most complex phenomena in FSW are to predict the microscopic behaviour that needs a multi-scale modelling approach.

2.4 References

1. Thomas, W.M., Nicholas, E.D., Needham, J.C., Murch, M.G., Smith, P. and Dawes, C.J. (1991), 'Friction stir butt welding', International Patent Application No. PCT/GB92/02203.
2. Dawes, C.J. and Thomas, W.M. (1996), 'Friction stir process welds aluminum alloys', *Weld. J.*, 75(3): 41–5.
3. Cho, J.H., Boyce, D.E and Dawson, P.R. (2005), 'Modeling strain hardening and texture evolution in friction stir welding of stainless steel', *Mater. Sci. Eng. A*, 398: 146–63.
4. Su, J.Q., Nelson, T.W., Mishra, R. and Mahoney, M. (2003), 'Microstructural investigation of friction stir welded 7050-T651 aluminium', *Acta. Mater.*, 51(3): 713–29.
5. Frigaard, O., Grong, O. and Midling, O.T. (2001), 'A process model for friction stir welding of age hardening aluminium alloys', *Metall. Mater. Trans. A*, 32: 1189–200.
6. Schmidt, H. and Hatte, J. (2005), 'A local model for the thermo-mechanical conditions in friction stir welding', *Model. Simul. Mater. Sci. Eng.*, 13: 77–93.
7. Buffa, G., Hua, J., Shivpuri, R. and Fratini, L. (2006), 'Design of the friction stir welding tool using the continuum based FEM model', *Mater. Sci. Eng. A*, 419: 381–8.
8. Nandan, R., Roy, G.G., Lienert, T. and DebRoy, T. (2007), 'Three-dimensional heat and material flow during friction stir welding of mild steel', *Acta. Mater.*, 55: 883–95.
9. Nandan, R., Roy, G.G. and DebRoy, T. (2006), 'Numerical simulation of three-dimensional heat transfer and plastic flow during friction stir welding', *Metall. Mater. Trans. A*, 37(4): 1247–59.
10. Chao, Y.J., Qi, X. and Tang, W. (2003), 'Heat transfer in friction stir welding-Experimental and numerical studies', *ASME Trans.*, 125: 138–45.
11. Chao, Y.J. and Qi, X. (1998), 'Thermal and thermo-mechanical modeling of friction stir welding of aluminum alloy 6061-T6', *J. Mater. Process. Manuf. Sci.*, 7: 215–33.
12. Tang, W., Guo, X., McClure, J.C., Murr, L.E. and Nunes, A. (1998), 'Heat input and temperature distribution in friction stir welding', *J. Mater. Process. Manuf. Sci.*, 7: 163–72.
13. Colegrove, P., Pinter, M., Graham, D. and Miller, T. (2000), 'Three-dimensional flow and thermal modeling of the friction stir welding process', *Second International Symposium on Friction stir Welding*, 26–28 June, Gothenburg, Sweden, 26.
14. Askari, A., Silling, S., London, B. and Mahoney, M. (2001), 'Modeling and analysis of friction stir welding processes', *Symposium on Friction stir Welding and Processing*, Warrendale, PA, 43–54.
15. Fonda, R.W. and Lambrakos, S.G. (2002), 'Analysis of friction stir welding using an inverse-problem approach', *Sci. Technol. Weld. Join.*, 7(3): 177–81.
16. Nandan, R., DebRoy, T. and Bhadeshia, H.K.D.H. (2008), 'Recent advances in friction stir welding: process, weldment structure and properties', *Prog. Mater. Sci.*, 53: 980–1023.

17. Schmidt, H., Hattel, J. and Wert, J. (2004), 'An analytical model for the heat generation in friction stir welding', *Model. Simul. Mater. Sci. Eng.*, 12: 143–57.
18. Colegrove, P.A. and Shercliff, H.R. (2004a), 'Development of Trivex friction stir welding tool. Part 1: Two-dimensional flow modelling and experimental validation', *Sci. Technol. Weld. Join.*, 9: 345–51.
19. Colegrove, P.A. and Shercliff, H.R. (2004b), 'Development of Trivex friction stir welding tool. Part 1: Two-dimensional flow modelling and experimental validation', *Sci. Technol. Weld. Join.*, 9: 352–61.
20. Elangovan, K. and Balasubramanian, V. (2007), 'Influences of pin profile and rotational speed of the tool on the formation of friction stir processing zone in AA2219 aluminium alloy', *Mater. Sci. Eng. A*, 459: 7–18.
21. Xue, P., Ni, D.R., Wang, D., Xiao, B.L. and Ma, Z.Y. (2011). 'Effect of friction stir welding parameters on the microstructure and mechanical properties of the dissimilar Al–Cu joints', *Mater. Sci. Eng. A*, 528: 4683–9.
22. Agarwal, S.P., Nageswaran, P., Arivazhagan, N. and Devendranath, K.R. (2012), 'Development of friction stir welded butt joints of a 6063 aluminium alloy and pure copper', *International Conference on Advanced Research in Mechanical Engineering*, 46–50.
23. Zhao, Y.H., Lin, S.B., Qu, F. and Wu, L. (2006), 'Influence of pin geometry on material flow in friction stir welding process', *Mater. Sci. Technol.*, 22: 45–50.
24. Khandkar, M.Z.H. and Khan, J.A. (2001), 'Thermal modeling of overlap friction stir welding for Al-alloys', *J. Mater. Process. Manuf. Sci.*, 10: 91–105.
25. Colligan, K. (1999), 'Material flow behavior during friction stir welding of aluminum', *Weld. J. Research Supplement*, 78(7): 229–37.
26. Reynolds, A.P., Seidel, T.U. and Simonsen, M. (1999), 'Finite element simulation of material flow in friction stir welding', *First International Symposium on Friction stir Welding*, 14–16 June, Thousand Oaks, CA.
27. Reynolds, A.P. (2000), 'Visualization of material flow in autogenously friction stir welds', *Sci. Technol. Weld. Joi.*, 5: 120–4.
28. Seidel, T.U. and Reynolds, A.P. (2001), 'Visualization of the material flow in AA2195 friction stir welds using a marker insert technique', *Metall. Mater. Trans.*, 32A: 2879–84.
29. Guerra, M., McClure, J.C., Murr, L.E. and Nunes. A.C. (2001), 'Material flow during friction stir welding', *Symposium on Friction stir Welding and Processing*, Warrendale, PA, 25–34.
30. Colligan, K. (1999), 'Material flow behaviour during friction stir welding of aluminium', *First International Symposium on Friction stir Welding*, 14–16 June, Thousand Oaks, CA.
31. London, B., Mahoney, M., Bingel, W., Calabrese, M., Bossi, R.H. and Waldron, D. (2003), 'Material flow in friction stir welding monitored with Al-SiC and Al-W composite markers', *Symposium on Friction stir Welding and Processing II*, Warrendale, PA, 3–10.
32. Midling, O.T. (1994), 'Material flow behaviour and microstructural integrity of friction stir butt weldments', *Fourth International Conference on Aluminum Alloys*, Atlanta, GA: Georgia Institute of Technology, 451–8.

33. Pourahmad, P. and Abbasi, M. (2013), 'Materials flow and phase transformation in friction stir welding of Al 6013/Mg', *T. Nonferr. Metal. Soc.*, 23: 1253–61.
34. Guerra, M., Schmidt, C., McClure, J.C., Murr, L.E. and Nunes, A.C. (2003), 'Flow patterns during friction stir welding', *Mater. Charact.*, 49: 95–101.
35. Kazi, S.H. and Murr, L.E. (2001), 'Complex flow phenomena associated with friction stir welding of aluminium alloys', *Symposium on Friction stir Welding and Processing*, Warrendale, PA, 139–50.
36. Murr, L.E., Trillo, E.A., Li, Y., Flores, R.D., Nowak, B.M. and McClure, J.C. (1999), 'Solid-state flow associated with the friction stir welding of dissimilar metals', *Fluid Flow Phenomena in Metals Processing*, Warrendale, PA, 31–40.
37. Li, Y., Murr, L.E. and McClure, J.C. (1999), 'Flow visualization and residual microstructures associated with the friction stir welding of 2024 aluminium to 6061 aluminium', *Mater. Sci. Eng.*, 271A: 213–23.
38. Ouyang, J.H. and Kovacevic, R. (2002), 'Material flow and microstructure in the friction stir butt welds of the same and dissimilar aluminium alloys', *J. Mater. Eng. Perform.*, 11: 51–63.
39. Schmidt, H.B., Dickerson, T.L. and Hattel, J.H. (2006), 'Material flow in butt friction stir welds in AA2024-T3', *Acta. Mater.*, 54: 1199–209.
40. Schmidt, H.B. and Hattel, J.H. (2008), 'Thermal modeling of friction stir welding', *Scripta Mater.*, 58: 332–7.
41. Santiago, D., Urquiza, S., Lombera, G. and de Vedia, L. (2009), '3-D modeling of material flow and temperature in friction stir welding', *Soldag. insp. São Paulo*, 14(3): 248–56.
42. Santiago, D., Urquiza, S., Lombera, G. and de Vedia, L. (2004), 'Numerical modeling of joints welded by friction stir welding process', *Mater. Res.*, 7(4): 569–74.
43. Seidel, T.U. and Reynolds, A.P. (2003), 'Two-dimensional friction stir welding process model based on fluid mechanics', *Sci. Technol. Weld. Join.*, 8: 175–83.
44. Bag, S. and De, A. (2013), *Computational Models for GTA and Laser Welding Processes*, Germany: Lap Lambert Academic Publishing.
45. Hart, E.W. (1976), 'Constitutive relations for the non-elastic deformation of metals', *J. Eng. Mater. Technol.*, 98: 193–202.
46. Cho, J.H., Boyce, D.E. and Dawson, P.R. (2005), 'Modeling strain hardening and texture evolution in friction stir welding of stainless steel', *Mater. Sci. Eng. A*, 398A: 146–63.
47. Fisher, E.S. (1966), 'Temperature dependence of the elastic moduli in alpha uranium single crystals. Part iv: (298° to 923°K)', *J. Nucl. Mater.*, 18(1): 39–54.
48. Zienkiewich, O.C. and Cormeau, I.C. (1974), 'Visco-plasticity – plasticity and creep in elastic solids – a unified numerical solution approach', *Int. J. Numer. Method. Eng.*, 8: 821–45.
49. Kong, H.S. and Ashby, M.F. (1991), 'Friction-heating maps and their applications', *MRS Bull*, 16: 41–8.
50. Simar, A., Lecomte-Beckers, J., Pardoen, T. and de Meester, B. (2006), 'Effect of boundary conditions and heat source distribution on temperature distribution in friction stir welding', *Sci. Technol. Weld. Join.*, 11: 170–7.

51. Bastier, A., Maitournam, M.H., van Dang, K. and Roger, F. (2006), 'Steady state thermo-mechanical modelling of friction stir welding', *Sci. Technol. Weld. Join.*, 11: 278–88.
52. Lienert, T.J., Stellwag, W.L., Grimmett, B.B. and Warke, R.W. (2003), 'Friction stir welding studies on mild steel', *Weld. J. Res. Suppl.*, 82: 1–9.
53. Gupta, O.P. (2002), *Finite and Boundary Element Methods in Engineering*, New Delhi: Oxford and IBH Publications.
54. Reddy, J.H. and Gartling, D.K. (2000), *The Finite Element Method in Heat Transfer and Fluid Dynamics*, 2nd edn, Florida: CRC Press.
55. Patankar, S.V. (1988), *Numerical Heat Transfer and Fluid Flow*, New York: Mc-Graw-Hill.
56. Rebelo, N. and Kobayashi, S. (1980), 'A coupled analysis of visco-plastic deformation and heat transfer, Part I', *Int. J. Mech. Sci.*, 22: 699–705.

3

Microstructural evolution

A. Heidarzadeh, Faculty of Engineering, Azarbaijan Shahid Madani University, Tabriz, Iran; K. Kazemi-Choobi, Sahand University of Technology, Iran; H. Hanifian, Islamic Azad University, Iran and P. Asadi, University of Tehran, Iran

DOI: 10.1533/9780857094551.65

Abstract: Friction stir welding/processing (FSW/FSP) of metals and their alloys has been investigated widely in recent years. The microstructure of alloys affects the mechanical and physical properties of processed materials and joints, therefore studying microstructure evolution during FSW/FSP of alloys is necessary. Recent investigations show that dynamic restoration phenomenon can occur due to the coexistance of strain and heat, during FSW. Also, some investigators believe that static restoration phenomena can occur. The type of restoration phenomena depend on material properties and procedure, whose management can control the microstructure of joints or processed metals, and hence their mechanical properties. In this chapter, the microstructural concepts including different microstructural zones, types of microstructural studies and different restoration phenomena are described. Then, microstructural evolution during FSW/FSP of different metals and alloys is discussed. Moreover, some illustrative examples are added to the end of each section.

Key words: friction stir welding, friction stir processing, micro-structural evolution, recrystallization, recovery.

3.1 Introduction

As discussed in detail in previous chapters, generated heat and deformation cause the base metal (BM) to flow without reaching the melting point and allow traversing of the tool along the weld line. The deformed material is transferred from the advancing side (AS) of the tool pin to the retreating side (RS) and is forged by the tool shoulder, resulting in a solid-state bond between two plates. This severe plastic deformation in the weld zone leads to a general change in microstructural and, subsequently, in mechanical properties.

This chapter identifies briefly both the microstructural evolution mechanisms during friction stir welding/processing (FSW/FSP) and different microstructural zones existing after FSW/FSP of various metals and their alloys.

3.2 Microstructural concepts

3.2.1 Different microstructural zones

Generally, FSW divides the microstructure of joints into distinct zones. A typical macrostructure of a joint welded by FSW is illustrated in Figure 3.1. The distinct zones can be described as the following [1]:

a) *Unaffected material or BM*: This is material far from the weld without deformation; and although it may have experienced a thermal cycle from the weld, it is not affected by the heat in terms of microstructure or mechanical properties.

b) *Heat affected zone*: In this region, which is clearly lying closer to the weld centre, the material has experienced a thermal cycle, which modifies the microstructure or the mechanical properties. However, the

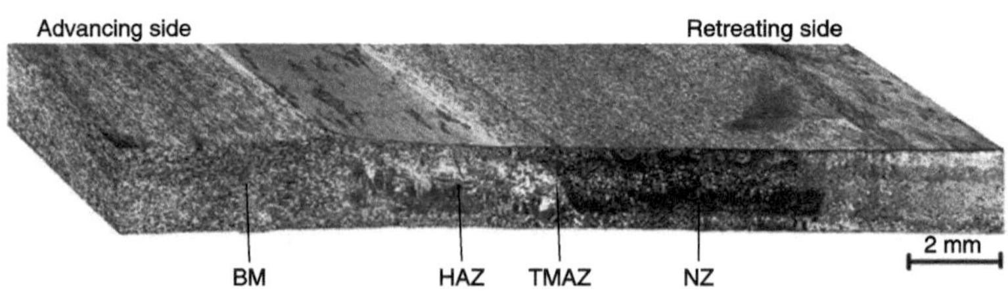

Figure 3.1 **Four typical different distinct zones of joints after FSW [2]**

plastic deformation has not occurred in this area. The most common metallurgical phenomena usually occurring in this area are grain growth, solubilization and coarsening of strengthening precipitates.

c) *Nugget zone (NZ)*: This area is in the weld centre (the location of the pin during welding) and is composed of equiaxed grains. During FSW, coincidence of heat and heavy plastic deformation results in occurrence of dynamic recrystallization at the centre of the weld. The shape of this region depends on the welding parameters such as rotational speed, traverse speed and axial force. The grains within NZ are often smaller than the grains in the BM. A unique feature of NZ is the general occurrence of several concentric rings, which have been likened to an onion ring structure. The exact source of these rings has not been definitely recognized, although differences in particle number density, grain size and texture have all been suggested.

d) *Thermo-mechanically affected zone* (*TMAZ*): In this region, the FSW tool has plastically deformed the material, and the generated heat has had some influence on the material. The common feature of this region is elongated or rotated grains, and recrystallization does not happen in this area because of inadequate heat and plastic deformation. In Figure 3.2, different zones of NZ, TMAZ and HAZ in the FSPed AZ31B magnesium alloy are depicted.

e) According to Figures 3.1 and 3.2, the translation of the rotating tool creates a characteristic asymmetry between the two sides of the processed zone.

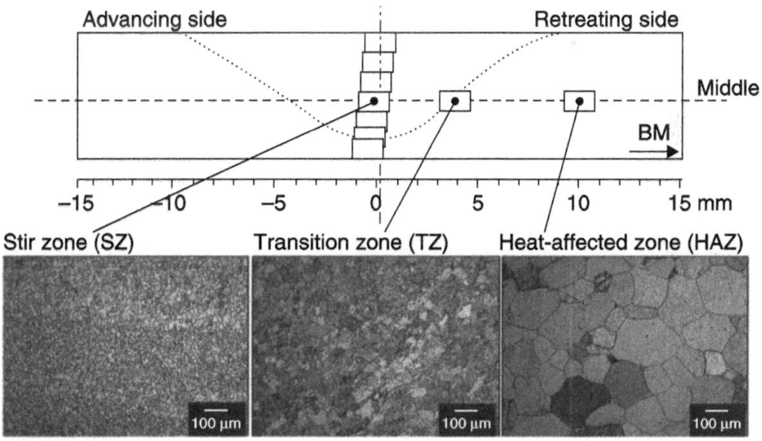

Figure 3.2 **Microstructure of the SZ or NZ, TZ and HAZ of the FSPed AZ31B Mg alloy [3]**

The AS is where the direction of tool rotation is the same as that of tool translation, and the RS is where they are the opposite.

3.2.2 Types of microstructural investigations

Generally, there are two typical methods used to investigate the microstructural evolution during FSW:

a) In order to understand microstructure evolution during welding in the NZ, we can investigate the microstructure and texture from the BM to the NZ. It is believed that the change of strain and temperature from the BM to the NZ is reasonably similar to that of a point located in the nugget during welding.

b) The second method is using the stop action technique. In this method, power is suddenly removed from the welding machine to stop the tool rotating, leaving the tool embedded in the workpiece. This approach uses the resistance of the workpiece material to stop the tool's rotation quickly and thereby protect the complex deformation field that exists around the welding tool during FSW.

3.2.3 Types of microstructural evolution mechanisms

During FSW/FSP, the tool provides both heating and heavy deformation of material. The heat is produced mainly by friction between the rotating shoulder and the workpiece, while the rotating pin deforms the locally heated material. The heated material flows around the rotating pin and then fills the cavity at the rear of the tool. Therefore, the FSW process can be assumed as a hot deformation technique and thus hot deformation microstructural mechanisms can occur during FSW/FSP.

Three types of dynamic recrystallization mechanisms are likely to produce new grains in the weld nuggets:

1. *discontinuous dynamic recrystallization* (*DDRX*): which occurs by nucleation and growth of new grains;
2. *continuous dynamic recrystallization* (*CDRX*): which involves the formation of arrays of low-angle boundaries (LABs) and a gradual increase in boundary misorientations during hot deformation, finally leading to new grain development; and

3. *geometric dynamic recrystallization* (*GDRX*): resulting from the impingement of serrated grain boundaries, which can occur when grains are extremely elongated by severe hot deformation.

All of these mechanisms lead to grain refinement. Some investigators believe that the development of grain structure is a complex process driven mainly by grain subdivision, but also involves the geometrical effect of strain and local grain boundary migration. Generally, the type of occurring mechanism mainly depends on the type of metals or alloys. The following sections will explain the DDRX, CDRX and GDRX mechanisms during hot deformation processes.

Discontinuous dynamic recrystallization

According to Table 3.1, in metals with a low or medium stacking fault energy (copper, nickel and austenitic iron), the recovery processes are slow and DDRX may happen when a critical strain or deformation is reached. DDRX normally begins at the old grain boundaries (Figure 3.3). New grains are then nucleated at the boundaries of the growing grains and in this way, a thickening band of recrystallized grains is shaped.

Table 3.1 **Restoration mechanisms of materials with different stacking fault energies [4]**

T/T_m \ Material	Low to medium SFE	High SFE
0.8–1.0	*Recovery*	*Recovery*
0.6–0.8	*Discontinuous DRX*	
0.4–0.6		*Continuous DRX*
0.2–0.4	*Continuous DRX*	
0–0.2	*Recovery*	*Recovery*

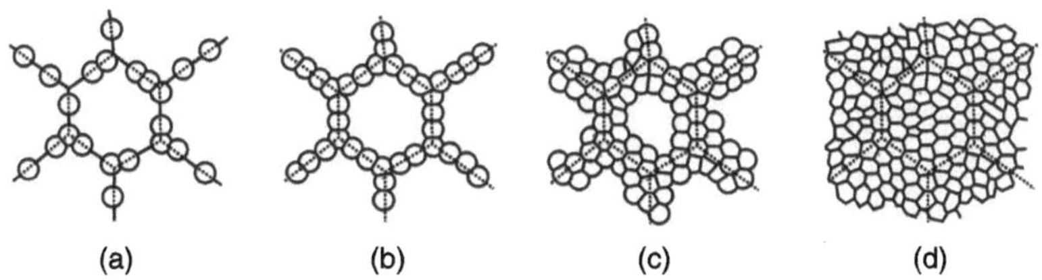

Figure 3.3 **Evaluation of microstructure during DDRX [4]**

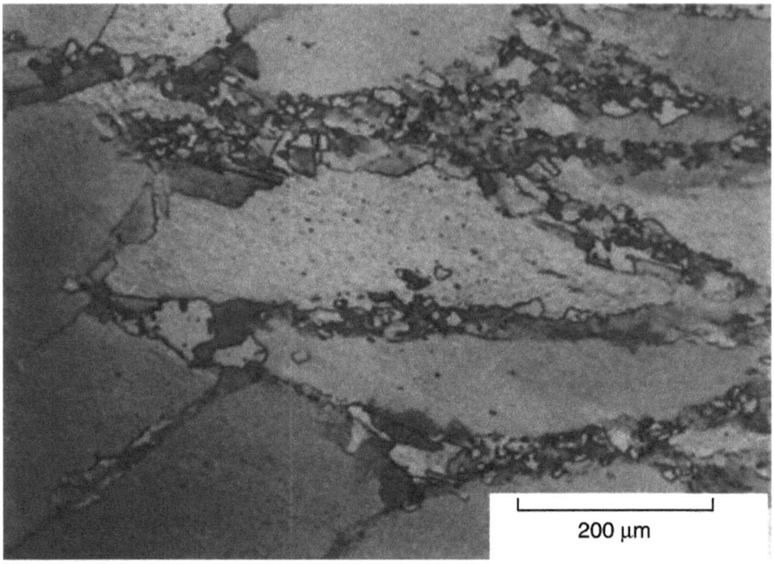

Figure 3.4 Partial DDRX at primary grain boundaries in pure copper [4]

These types of structures have been called necklace structure of grains, and appear when there is a large difference between the initial grain size and the recrystallized grain size. Consequently, all material will become fully recrystallized. Figure 3.4 shows a microstructure of pure copper undergoing partial DDRX. This type of dynamic recrystallization, which has clear nucleation and growth stages, can be classified as a discontinuous process. During DDRX, two factors prevent the growth of recrystallized grain. First, as the material continues to deform, the dislocation density of the new grains increases, and consequently reduces the driving force for more growth. The growth of recrystallized grains is finally stopped. Second, the nucleation of further grains at the migrating boundaries prevents regrowth of the older recrystallized grains [4].

Continuous dynamic recrystallization

According to Table 3.1, in certain materials with high stacking fault energies, new grains with high-angle boundaries (HABs) may form during straining, by progressive rotation of subgrains with little accompanying boundary migration. In this type of dynamic recrystallization (i.e. CDRX), nucleation and then growth happens in a similar way to DDRX.

CDRX involves the progressive rotation of subgrains neighbouring initial grain boundaries during deformation. During this phenomenon, a gradient of misorientation expands from the centre to the edge of initial grains (Figure 3.5). In the centre of an old grain, subgrains may not be well developed or may have very low misorientations. Towards the grain boundary, the misorientations increase, and at larger strains, HABs may develop.

CDRX is normally found in metals and alloys in which dislocation movement is hindered by either a lack of slip systems (e.g. magnesium alloys) or by solute drag (e.g. aluminium-magnesium alloys). In these types of alloys, inhomogeneous plastic deformation and accelerated dynamic recovery occurs in the initial grain boundaries. A schematic diagram of the CDRX mechanism in magnesium is illustrated in Figure 3.6. The suggested mechanism for CDRX in magnesium alloys takes place because of the inadequate independent slip systems required for homogeneous plasticity. In this mechanism, local shearing occurs near initial grain boundaries (Figures 3.6(a)) and then dynamic recovery leads to formation of new subgrains or grains (Figure 3.6(a,b)). It is notable that at higher deformation temperatures, the slip is easier and hence homogeneous plasticity occurs. Therefore, in these cases, the mechanism is similar to DDRX.

In the case of Al-Mg and Al-Zn alloys (aluminium alloys containing solute additions), significant lattice rotations may develop at the grain boundaries during deformation at temperatures in the range 300 to 400 °C (Figure. 3.7).

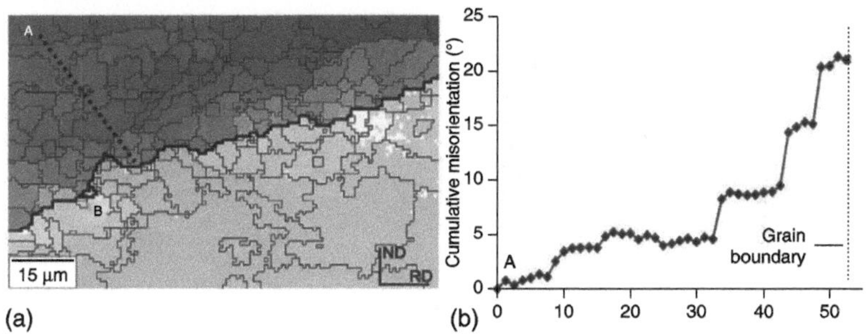

Figure 3.5 **Lattice rotations measured by EBSD along a grain boundary in Al-5%Mg, deformed in plane strain compression at 350 °C: (a) EBSD map showing high-angle boundaries as thick lines and low-angle boundaries as thin lines. Different colours reveal misorientations at the boundary serrations; and (b) the cumulative misorientations along the line from point A to the boundary [4]**

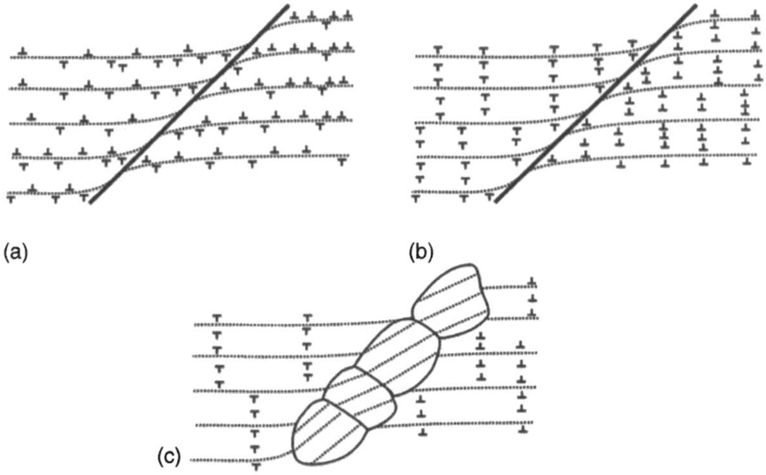

Figure 3.6 Schematic of mechanism proposed to dynamic recrystallization in magnesium [4]

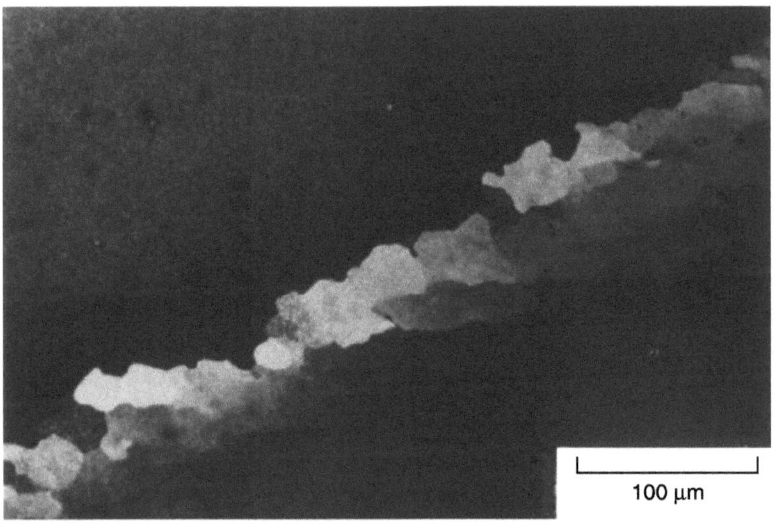

Figure 3.7 Misorientations developed by lattice rotation near the grain boundary during deformation of Al-5%Mg [4]

Geometric dynamic recrystallization

The mechanism of GDRX is illustrated schematically in Figure 3.8. As was shown in Figure 3.7, serrations develop in initial grain boundaries during dynamic recovery. In addition, the wavelength of these serrations is comparable with the subgrain size. When the material is subjected to a large reduction in cross-section, the original grains become flattened (Figure 3.8). Because the subgrain size during high temperature deformation is almost independent of strain, the fraction of boundaries, which have high angles, increases with the strain.

Eventually the size of the boundary serrations will become similar to the grain thickness. The serrated boundaries will touch each other, resulting in a microstructure of fine equiaxed grains with a size similar to that of the subgrain. An example of this procedure in an Al-Mg-Fe alloy is shown in Figure 3.9.

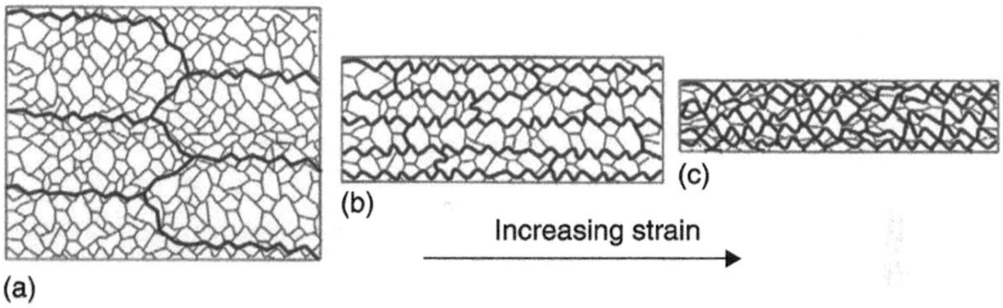

Figure 3.8 GDRX. The serrated HABs (thick lines) become closer as deformation progresses; however, the subgrain size remains approximately constant. Finally, the HABs impinge leads to a microstructure with high angle-boundaries [4]

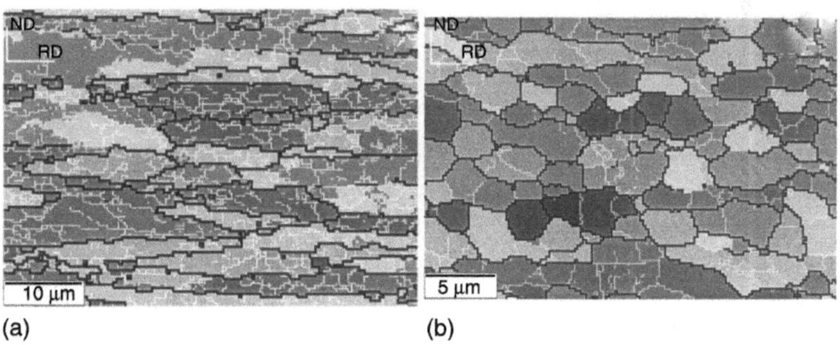

Figure 3.9 Geometric dynamic recrystallization by EBSD in an Al-3Mg-0.2Fe alloy deformed at 350 °C in plane strain compression: (a) $\varepsilon = 0.7$; and (b) $\varepsilon = 3$ [4]

3.3 Microstructural evolution in pure iron and steels

3.3.1 Pure iron

The basic issue of the mechanisms that involve microstructural evolution throughout FSW/FSP of pure iron is now discussed. The presented data is about iron of 99.9 wt.% purity, which was FSWed at a welding speed of 60 mm/min and rotation speed of 200 rpm. To investigate microstructure evolution, the electron back scatter diffraction (EBSD) techniques in conjunction with scanning electron microscopy (SEM) equipped with a field emission (FE) SEM gun were applied [5].

Distinct microstructural zones

EBSD observations indicate that the primary structure of BM involves a single phase, equiaxed α-iron with an average grain size of about 130 μm (Figure 3.10(a)). The majority of the grain boundaries are high-angle in character and only sporadic LABs are collected within some grains. Close similarity is noticeable between misorientation angle distribution for HABs and those predicted for an assembly of randomly orientated grains (random distribution), which is exposed.

By the broken line in Figure 3.10(b), the BM is just weakly textured (no peaks larger than 3.5 random), with the <100> direction aligned roughly parallel to the normal direction (ND) (Figure 3.10(c)). Three distinct zones of the SZ/NZ, BM and TMAZ can be readily recognized in a low magnification overview of the transverse cross-section of the FSPed zone shown in Figure 3.11.

To explore microstructural differences between BM and SZ during FSP, understanding the transition nature of TMAZ microstructure is necessary. The grain structures that develop in TMAZ on AS and RS show that both areas generally have the same characteristics of microstructural evolution. Therefore, studying the microstructural evolution in one of the AS or RS is adequate.

Microstructural evolution

Figure 3.12(a) illustrates a composite EBSD map taken from the RS of the tool, and the chosen regions are shown at higher magnifications in Figures 3.12(b–d). The average grain boundary spacing and HAB fraction

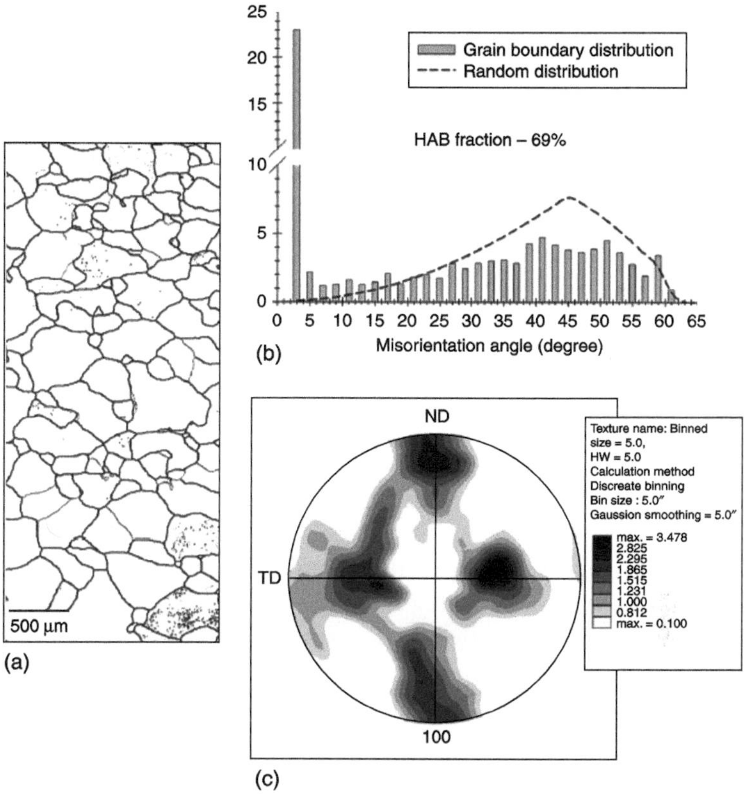

Figure 3.10 **Microstructural characteristics of the BM: (a) EBSD map showing the distributions of high- and low-angle boundaries (black and grey lines, respectively); (b) misorientation angle distribution is revealed by a broken line; and (c) {100} pole figure [5]**

determined in these areas, are summarized in Table 3.2. Figure 3.12(a) demonstrates that TMAZ expands by about 1 mm from the SZ. Close to the outer border of TMAZ (Region 1 in Figure 3.12(a)), because of shear deformation induced by rotating tool, the parent grains tend to be reorientated along a flow pattern around the SZ. By increasing the temperature, initial grain boundaries become wavy (Figure 3.12(b)), and local boundary migration starts. The fraction of LABs increases considerably, by up to 93% (Table 3.2). Formation of fine equiaxed subgrains (Figure 3.12(b)) is due to heterogeneous distribution of LABs within grains

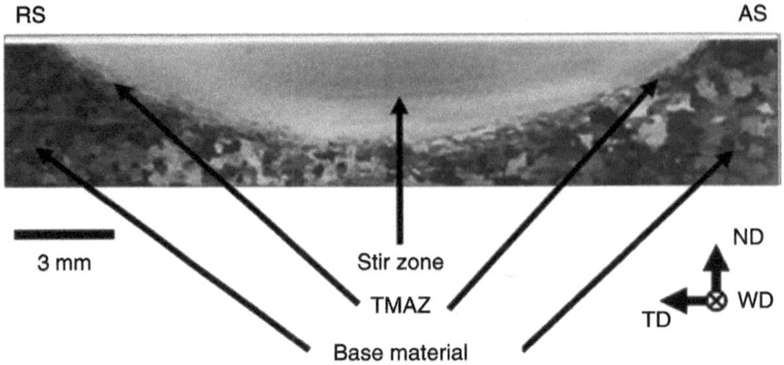

Figure 3.11 Transversal cross-section of friction stir processed zone [5]

and start to come together close to triple joints and along the initial grain boundaries.

However, the extended LABs also form in grain centres, where they are generally organized as regular arrays of nearly parallel bands aligned in a common direction within the grains (Figure 3.12(a)). Such regular arrays are referred to as boundary sets. An interconnecting set of the bands is possibly seen within some grains. Configuration of the boundary sets gives rise to orientation spread within the initial grains, which is seen as variation of orientation contrast in the EBSD map (Figure 3.12(a)).

Due to the strain rising towards the SZ, the original grains become thinner in the area near to the SZ (Region 2 in Figure 3.12(a)). Significant decrease in average LAB spacing is an indication of development of new LABs. The LABs homogeneous distribution in grain centres and arrangement at some interconnecting boundary sets (Figure 3.12(c)) can be seen in grains with high LAB density and the boundary sets become a main characteristic in the microstructure.

Typically, one of the boundary sets is aligned in the longitudinal direction of an elongated grain; this set is typically well-developed and somewhat extended across the grain. This primary boundary set is sheared by interconnecting secondary sets, which creates offsets and ledges along the boundaries (several examples are arrowed in Figure 3.12(c)). Such microstructures, called S-bands, are well described in deformed face centred cubic (f.c.c) and body centred cubic (b.c.c) metals. Some boundary segments have already accumulated misorientation of more than 15 degrees. Decreasing in average

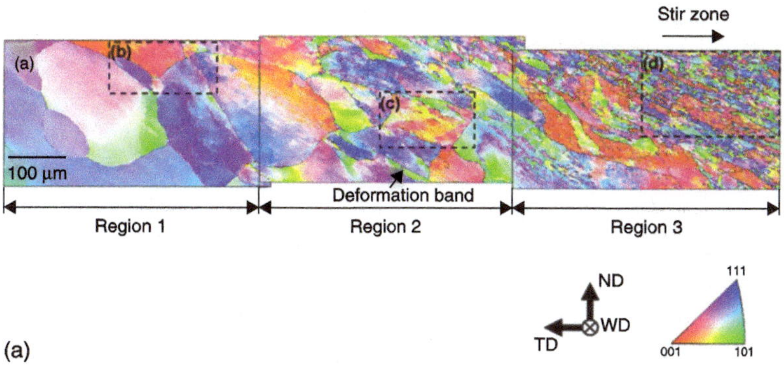

(a)

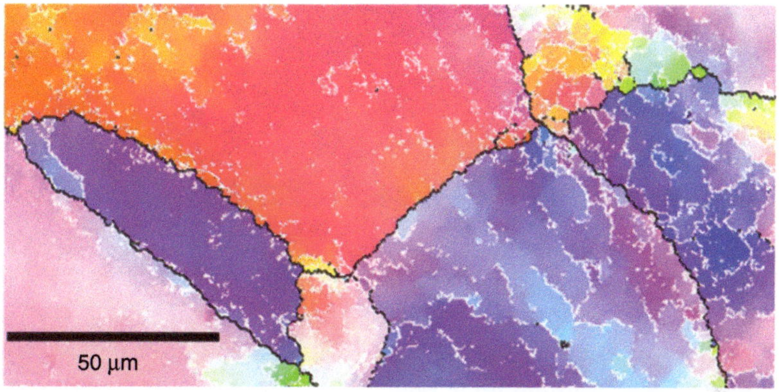

(b)

(c)

Figure 3.12 (a) EBSD map of TMAZ on the RS side of the tool; and (b–d) at higher magnifications. The individual grains are coloured according to their crystallographic directions relative to WD; colour code triangle is shown in the bottom right-hand corner. LABs and HABs are depicted as white and black lines, respectively [5]

(Continued)

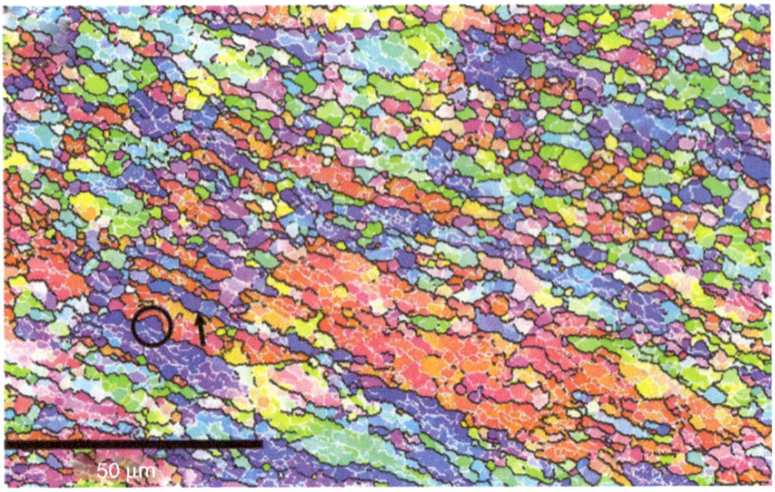

(d)

Figure 3.12 Continued

Table 3.2 Average grain boundary spacing and HAB fraction for different regions shown in Figure 3.12(a) [5]

HAB_s fraction (%)	Average grain boundary spacing in TD (μm)		Location
	HAB_s	LAB_s	
7	18.6	123.0	Region 1
11	4.6	58.4	Region 2
28	2.6	12.3	Region 3

HAB spacing and increasing of HAB fraction (Table 3.2) are led by geometric requirements of strain. The HAB segments specially belong to primary boundary sets. These parts can be found in the grain centres comparatively far from the original grain boundaries and there is no propensity for such segments to group near the original grain boundaries. Some of these deformation-induced HABs are combined, outlining deformation bands (Figure 3.12(a)). Orientation of the material inside such deformation bands has rotated significantly away from the neighbouring matrix.

In the subgrains separated by the LABs of the same boundary set (an example is shown in Figure 3.12(c) as a selected area) an alternative orientation change is regularly obvious. This specifies that the adjacent boundaries within the boundary sets have approached an alternating sense of misorientation, i.e. they have a propensity to compensate for each other on a local scale. However, at a grain scale, the orientation spread may be very large and different areas of the similar parent grains can develop extensively dissimilar orientations (Figure 3.12(c)). In some cases, the parent grains cannot be obviously distinguished, as seen in the top right-hand corner of Region 2 (Figure 3.12(a)). More complicated microstructures can be seen close to the SZ (Region 3 in Figure 3.12(a)). However, the basic microstructural characteristics seen at lower strains can still be recognized. The initial grains are prolonged, becoming thin fibrous grains, and boundaries of the primary boundary sets accumulate significant misorientation. The initial grain structure is broken up, and accordingly a lamella structure is shaped. This structure includes HABs aligned in a general direction cross-linked by LABs (Figure 3.12(d)). Furthermore, it is not homogeneous and includes a small number of larger and more stable grain fragments, which have survived deformation.

On account of localized shearing and bulges associated with local grain boundary migration (an example is encircled in Figure 3.12(d)), longitudinal boundaries of the lamella structure are not straight over a long distance and have offsets. Development of the lamella structure leads to severe reduction of average HAB spacing and a significant increase in HAB fraction (Table 3.2). Nevertheless, average LAB spacing also decreases (Table 3.2), indicating that new LABs persist. The experimental outcomes obviously indicate that microstructural evolution of pure iron in TMAZ shows a pattern roughly similar to the common microstructural behaviour studied for f.c.c and b.c.c metals of medium to high SFE subjected to cold or warm deformation. As a result, grain subdivision and the geometric effects of strain, which decrease the HAB spacing, are motivated forces of grain refinement procedure.

3.3.2 Steels

Due to some significant results, the FSW/FSP of steels has not developed as fast as for aluminium. First, material of the tool has to survive much more strenuous conditions because of the strength of steel. Second, there are also several welding techniques where steel can be acceptably and unfailingly welded. Third, the effects of phase transformations associated with FSW have not been studied in enough depth. Finally, optimizing the weld for a required set of properties needs significant experimentation, because the diversity of steels available is much larger than that of any other alloy system. The microstructural results of research performed at different laboratories using FSW/FSP of steels are reviewed.

Low carbon and Fe-12%Cr steels are FSWed using a tungsten based tool. The weld zone is shown to include a variety of martensite, bainite and ferrite structures, along with tool debris [6]. The possibility of welding 1018 steel has been evaluated. The macrostructure of the cross-section of a weld is illustrated in Figure 3.13. Because of the allotropic transformation on cooling, TMAZ is not readily clear in the microstructure of the weld.

It is found that SZ microstructures consist of ferrite, grain boundary ferrite and fine pearlite. In the SZ, the structure was found to be finer closer to the shoulder and coarser away from the shoulder. The optical micrographs of the SZ and HAZ are shown in Figure 3.14.

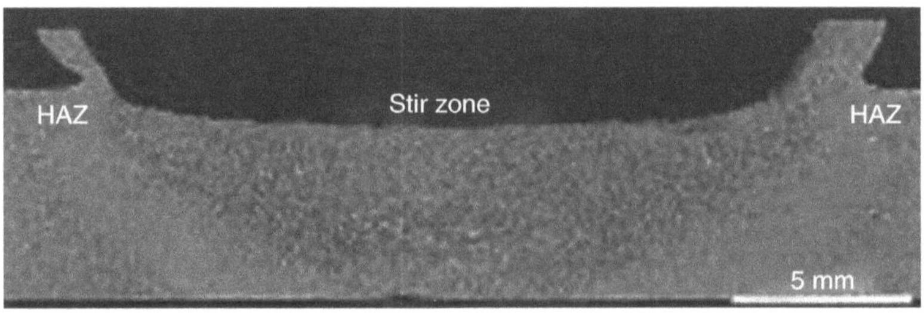

Figure 3.13 Traverse section of a FSWed 1018 steel. The SZ, HAZ and BM areas can be detected [7]

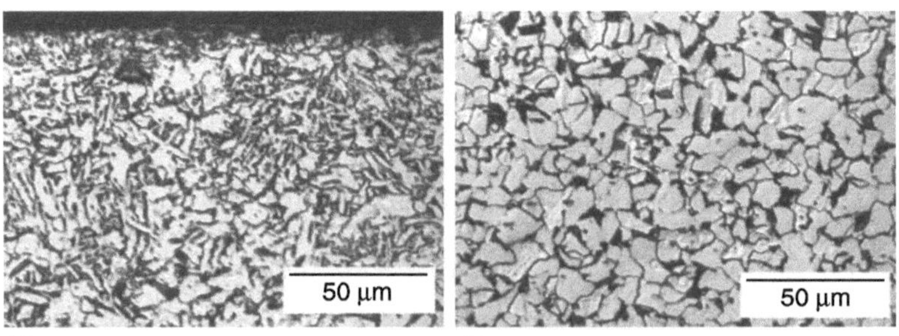

Figure 3.14 Optical micrographs of 1018 steel joint: (a) SZ; and (b) HAZ [7]

Another study reported the FSW of S355 carbon manganese steel sheets, using tungsten rhenium tools. In this study, tempering of the first pass by the heat from the second pass was the important microstructural observation. Equiaxed grains in the stir zone (SZ), with a grain size faintly reduced from the BM, is observable in 304L stainless welding using a tungsten alloy tool. They also observed narrow bands in the SZ, but made no determination as to the origin or complete structure of the bands. Figure 3.15 illustrates characteristic bands observed in this examination. Extremely refined SZ microstructure, with an undisclosed dark banded structure in the SZ, are reported on welding of 304L and AL-6XN stainless steels. It also explains the complexity of achieving sound welds in AL-6XN, because several voids are found in the resulting weld.

Another study welded a 304L sheet with a thickness of 6 mm using polycrystalline cubic boron nitride (PCBN) tools with a rotational speed of 400 rpm and travel speed of 75 mm/min. Extensively varying microstructures were found by reason of different parameters. Figure 3.16 demonstrates a cross-sectional overview of a FSW in 304L stainless steel. SZ and TMAZ reveal dynamically recrystallized microstructure and

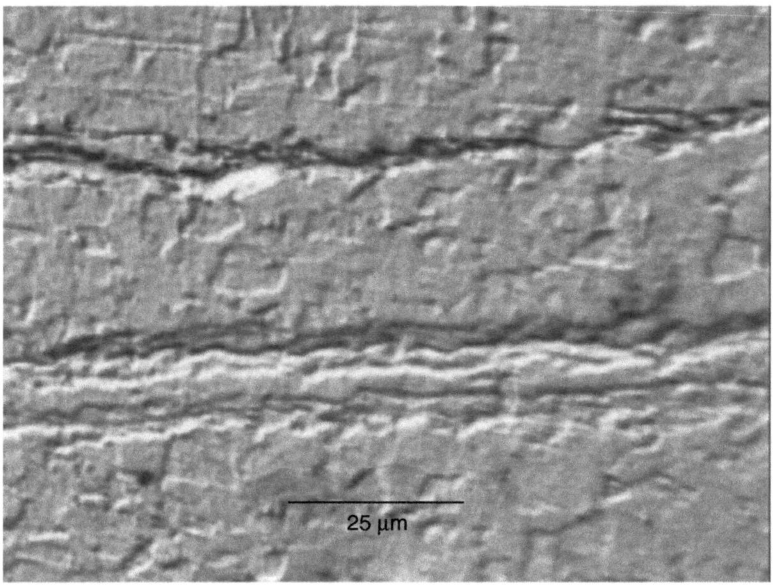

Figure 3.15 Microstructure of a joint in which a typical band is observed [8]

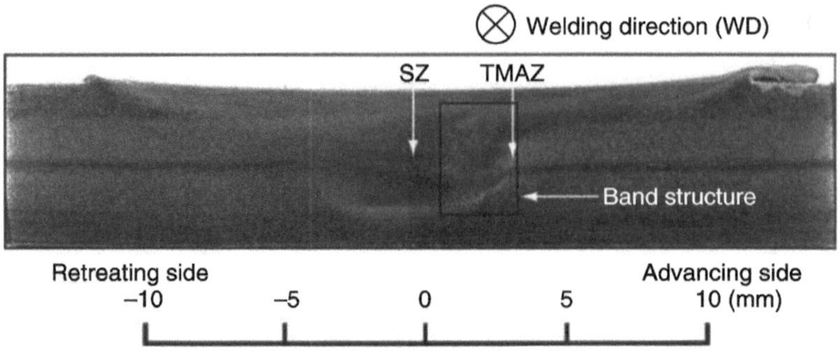

Figure 3.16 304 stainless steel FSW joint. Cross-section is perpendicular to the welding direction [9]

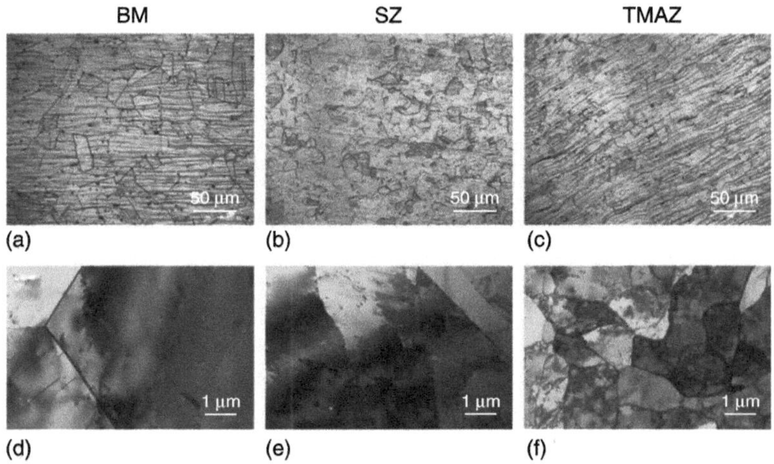

Figure 3.17 Optical micrographs and TEM images of SZ, TMAZ and BM [9]

recovered microstructure, respectively. The microstructural observation shows that the sigma phase, including the several stacking faults, was formed at the AS of the SZ.

From introduction of high strain and dynamic recrystallization during FSW, relating the rapid formation of the sigma phase to the transformation of austenite to delta ferrite in the SZ is a possibility. Figure 3.17 shows the optical microstructures and transmission electron microcopy (TEM) images of different regions in a 304 stainless steel FSWed joint.

Sigma formation in FSW of a variety of stainless alloys with chemical compositions at different distances from the sigma/austenite area of the

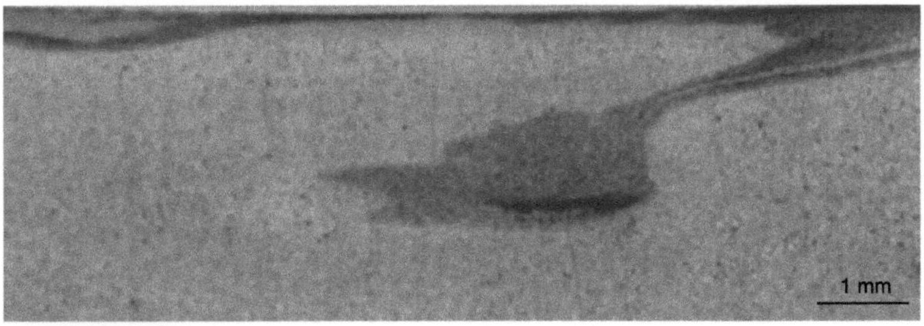

Figure 3.18 **Macrostructure of FSWed 304L [10]**

Fe-Ni-Cr ternary diagram is explored. Researchers predict the tendency for sigma formation and theorize that sigma formation was a pointer for recrystallization in 304L. It is revealed that changes of welding parameters have an effect on the amount and location of sigma. Figure 3.18 shows a FSWed region of 304L, where the dark regions contain a superior amount of the sigma phase.

Quenched and tempered carbon manganese steel with a thickness of 6.4 mm is FSWed by using a PCBN. The results show that tool wear was very low, but not determined quantitatively. Both in the former austenite grains and the transformation product, greatly refined grain structures with the SZ are detected.

The SAF 2507 super duplex stainless steel (DSS) is FSWed using a PCBN tool at a rotational speed of 450 rpm and traverse speed of 60 mm/min, and produces sound joints with an exceptional surface finish. The joint microstructure is equiaxed and small-grained with an average size of 4 μm in the SZ. Ferrite content increased from 40 to 50% across the SZ, in comparison with 45% in the BM.

Also, the effect of the traverse speed on the microstructure and mechanical properties of the SZ in FSW of SAF 2205 DSS is studied at a constant rotational speed of 600 rpm, and by using a tungsten carbide (WC)-based material tool. According to Figure 3.19, X-ray radiography shows that sound joints are productively achieved for the welding speeds within the range of 50 to 200 mm/min, while a groove-like defect is generated at the higher speed of 250 mm/min. Furthermore, increasing the welding speed decreases the grain size of the SZ (Figure 3.20).

The complexity of welding of high carbon steels is due to a greater tendency of martensite formation at the joint, which causes typical welding troubles such as a decrease in toughness.

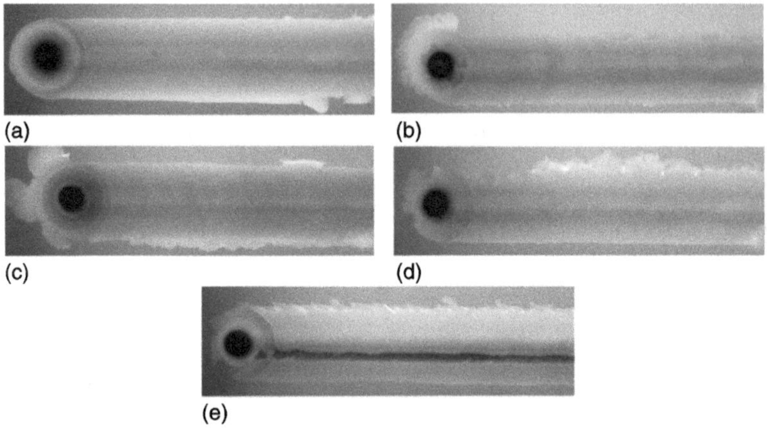

Figure 3.19 X-ray radiographs of different welding speeds: (a) 50 mm/min; (b) 100 mm/min; (c) 150 mm/min; (d) 200 mm/min; and (e) 250 mm/min [11]

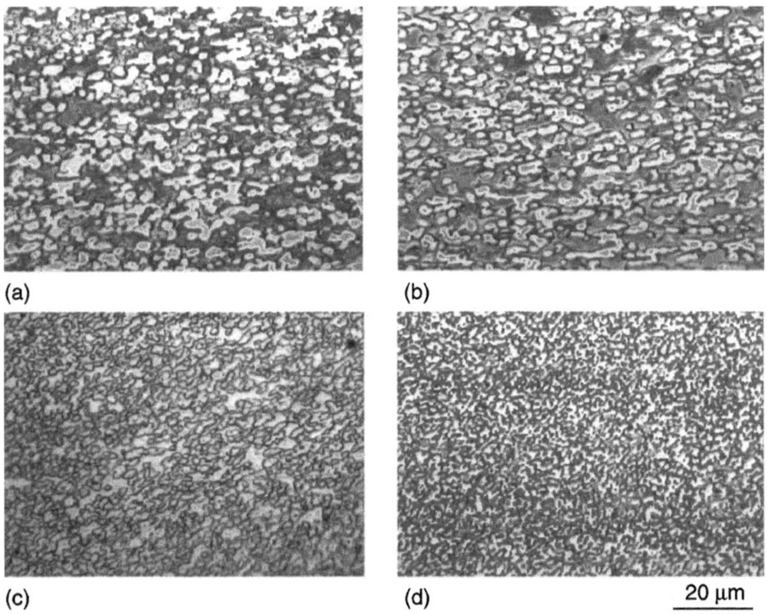

Figure 3.20 Microstructures of the SZ at different traverse speeds: (a) 50 mm/min; (b) 100 mm/min; (c) 150 mm/min; and (d) 200 mm/min [11]

Hyper eutectoid steel (0.85 mass% C) under a maximum temperature below A_1 (726 °C) is FSWed, in order to prohibit martensite transformation throughout the welding process. The results show that welding without the formation of the martensite structure can be successfully reached and the joint shows good toughness [12].

The super-austenitic stainless steel (a typical low SFE material) is FSWed. The textural pattern in the SZ is reported as a superimposition of {111} <uvw> and {hkl} <110> partial fibers with a prominent $(\bar{1}1\bar{1})$ and $[\bar{1}\bar{1}0]$ simple shear texture component. The pole figures of (111) and (110) are depicted in Figure 3.21. The grain structure evolution is a complex procedure (Figure 3.22). At the TMAZ border, misorientation of the original annealing twin boundaries starts to deviate from the ideal $\Sigma3$ twin/matrix orientation relationship, and this development finally transforms the twin boundaries into HABs. Nearer to the SZ, LABs expand widely close to the initial grain boundaries, and local grain boundary migration begins to take place. Creation of the final SZ microstructure was assumed to be primarily ruled by DDRX, but also involves grain subdivision and the geometrical effect of strain [13].

The 2 mm thick plates of SK4 high carbon steel alloy (0.95%C) are welded using a load controlled FSW machine, at a constant welding speed of 100 mm/min and different rotation speeds varied between 100 and 400 rpm. When FSW is done at 100 rpm, a duplex structure of spheroidal cementite and fine ferrite is formed and homogenously distributed in the whole SZ. Conversely, while the FSW is performed at a rotation speed higher than 100 rpm, extremely fine pearlite and martensite structures are formed in the upper part of the SZ and are increased by increasing rotation speed [14,15].

3.3.3 Illustrative example

Duplex stainless steel

Duplex stainless steels (DSS) containing both ferrite (α) and austenite (γ) phases are increasingly used as alternatives to austenitic, ferritic stainless steels and nickel-based alloys. In these alloys, a good combination of properties is reached at a α/γ ratio of 1:1. Melting and rapid solidification in fusion welding procedures affect this desired ratio. Moreover, precipitation of brittle intermetallic phases may happen. Ferrite content, coarse grains and more extensive precipitation of intermetallic phases are observable in the weld metal and HAZ of DSS compared to BM and

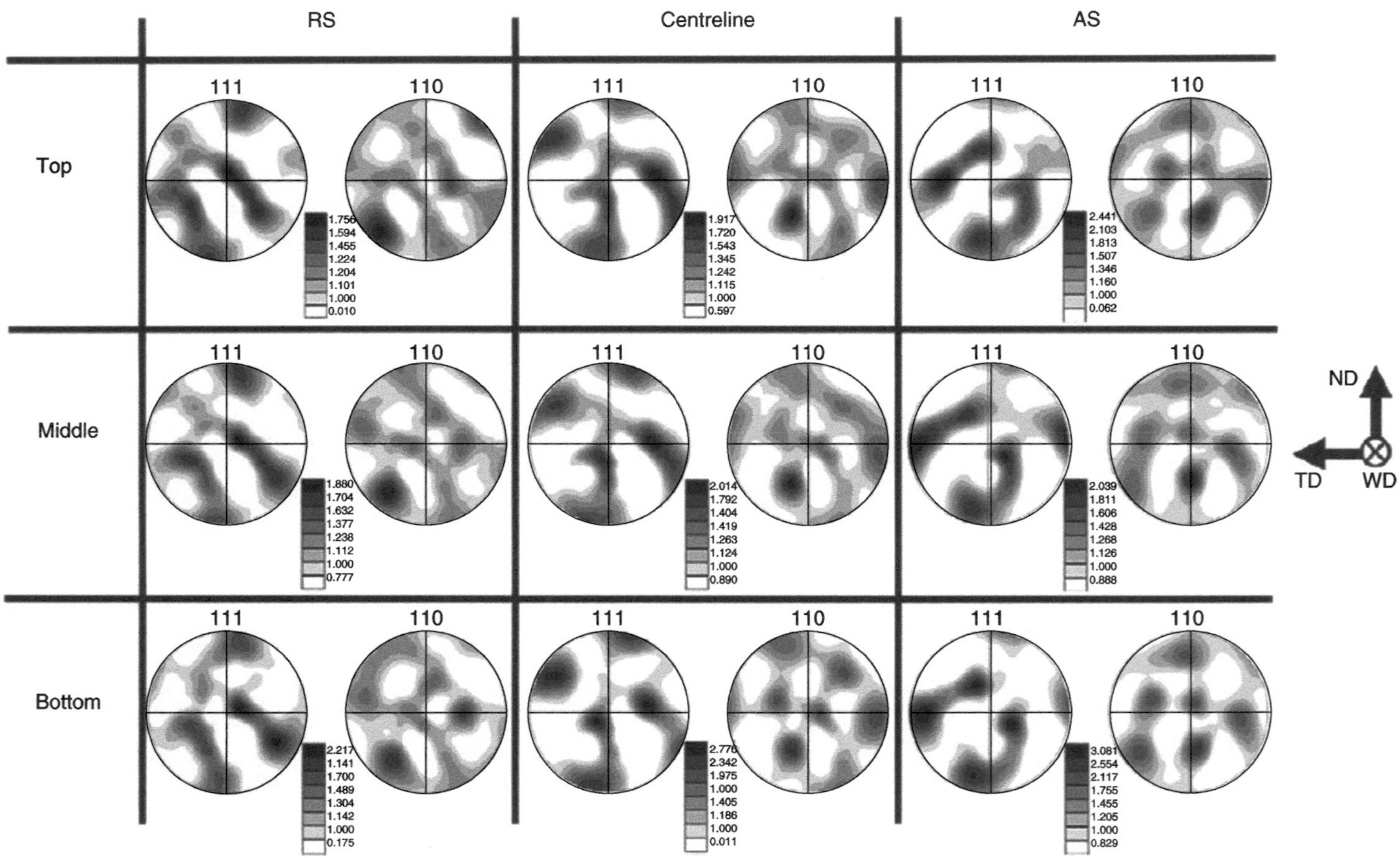

Figure 3.21 {111} and {110} pole figures at various locations within the SZ [13]

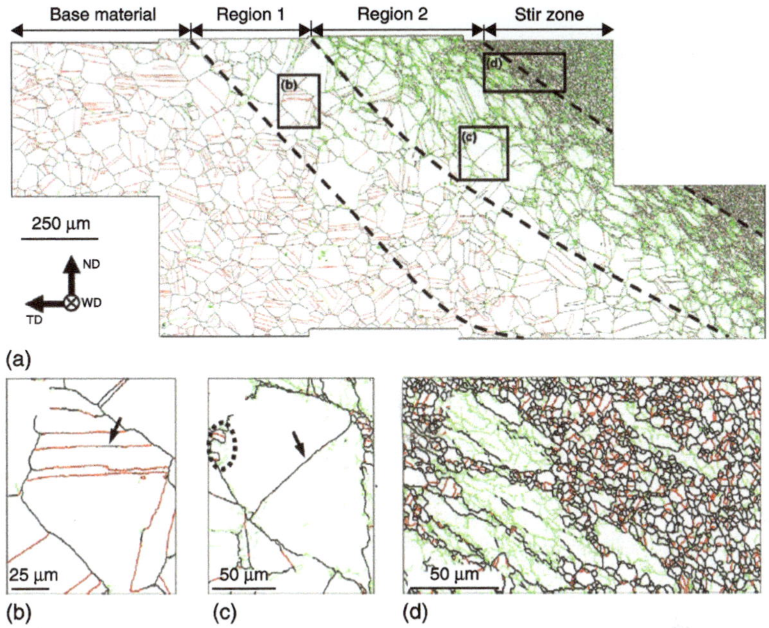

Figure 3.22 **(a) Composite EBSD map of the TMAZ on the RS of the weld; and (b–d) at higher magnifications. LAB, HAB and Σ3 boundaries are depicted as green, black and red lines, respectively [13]**

these features decrease both the corrosion resistance and the mechanical properties of the weld. Solid state joining results in avoidance of problems associated with the fusion welding processes of DSS.

In this section, the data is presented for a 2 mm thick plate of SAF 2205 DSS in a form of rolled and annealed (at 1050 °C for 1 h) plate. The plate was FSWed at a welding speed of 50 mm/min and a rotational speed of 800 rpm, and microstructural experiments were carried out on a transverse cross-section of the weld by optical microscopy (OM) and EBSD. Three different zones of SZ, the BM and a narrow transition area of TMAZ can be readily distinguished in Figure 3.23. Figures 3.24 and 3.25 illustrate the EBSD information corresponding to locations A to C in the cross-section centre line of Figure 3.23, in the form of phase maps and misorientation angle histograms, respectively. Table 3.3 gives microstructural characteristics, such as α and γ grain size and phase fraction.

EBSD analyses shows that the BM includes a 48.5% austenite phase comprising of equiaxed grains with an average size of 5 μm, embedded in ferrite matrix with an average grain size of 7 μm (Figure 3.24(a)). A

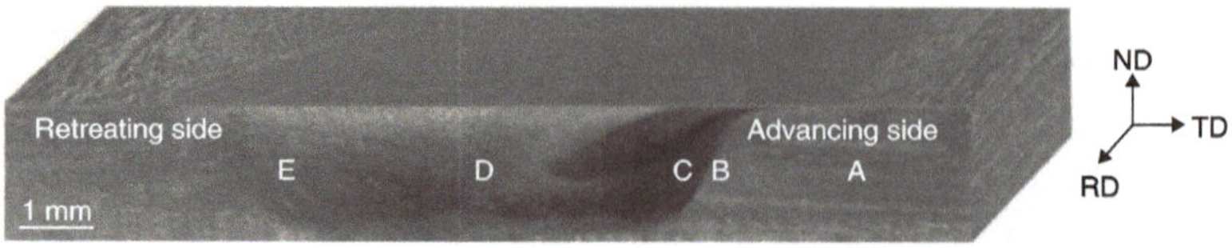

Figure 3.23 Macrograph of SAF 2205 after FSW. Zones A and B stand for the BM and TMAZ; Zones C, D and E lie in the advancing, centre and retreating side of the SZ [16]

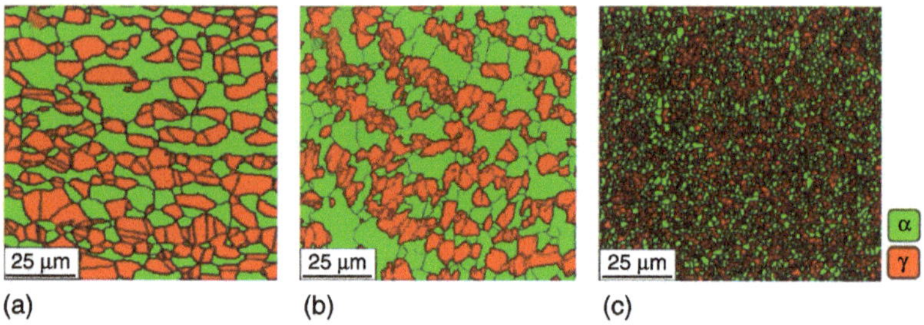

Figure 3.24 Phase maps of various microstructural regions in SAF 2205: (a) BM; (b) TMAZ; and (c) SZ [16]

detailed statistical analysis of the misorientation angle/axis pairs across the γ/γ boundaries (Figure 3.25(a)) demonstrates that about 63% of the HABs exhibit a first-order twin coincidence site lattice (CSL) orientation relationship, characterized by a 60-degree rotation about the <111> axis. Data of twins during recrystallization and grain growth of the starting material [16] can result from the presence of such a high quantity of twin boundaries in the microstructure. Furthermore, the amount of LABs with misorientation angles (θ) between 2 and 15 degrees is very low in austenite, whereas the ferrite phase in Figure 3.25(d) has significant quantities of LABs.

Because of plastic deformation in the direction inclined to the TMAZ/SZ boundary, grains in the TMAZ are noticeably curved (Figure 3.24(b)). Due to such plastic deformation, the primary annealing twins within the austenite were rotated away from the ideal CSL orientation relationship and in the same way, initially straight coherent twin boundaries changed to curved HABs.

Texture analysis of different areas of SZ are shown in Figure 3.26, in the form of {111} and {110} pole figures of γ and α, respectively. The texture generated in the SZ can be described in terms of the shear plane

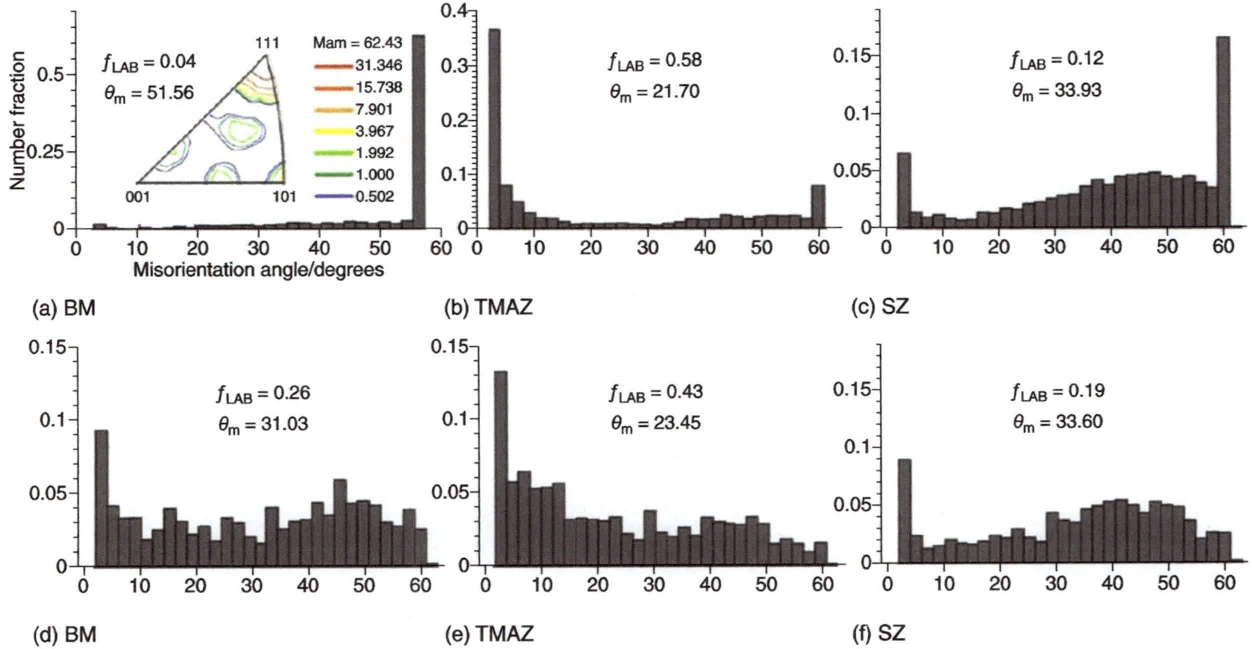

Figure 3.25 Histogram of misorientation angle of γ/γ (a–c) and α/α boundaries (d–f), within the BM, TMAZ and SZ showing the distribution of misorientation axis vector in the crystal lattice coordinates. f_{LAB} and θ_m indicate the fraction of LAB_s and mean misorientation angle, respectively [16]

normal (SPN) and shear direction (SD) aligned about vertical to and peripheral with either cone surface of the pin or the flow lines in the SZ, respectively [17–20].

Corresponding to the texture generated by deformation of bcc and fcc metals throughout torsion testing [9–12], γ and α phases have the typical simple shear texture (Figure 3.26). This obviously reveals that the key effect of the stirring pin is shearing and the effect of the tool shoulder on the plastic flow is insignificant. In addition, deformation textures reveal the occurrence of the cube texture. CDRX in the ferrite phase within the SZ is probable by reason of experimental data. Microstructure evolution during this mechanism consists of two sequential processes [21]:

1. formation of three-dimensional arrays of LABs; and
2. regular transfer of LABs into HABs.

LABs are continuously shaped in the first procedure by dynamic recovery during deformation by rearrangement of accumulating lattice dislocations. The observed enhancement of low-angle boundary fractions

Table 3.3 Grain size and phase ratio of various regions shown in Figure 3.25 [16]

SZ(RS)	SZ(centre)	SZ(AS)	TMAZ(AS)	BM	
1.21	0.84	0.73	4.14	7.04	α grain size (μm)
0.91	0.75	0.66	3.45	5.16	γ grain size (μm)
51.8/48.2	52.3/47.7	51.5/48.5	54/46	51.5/48.5	α/γ ratio

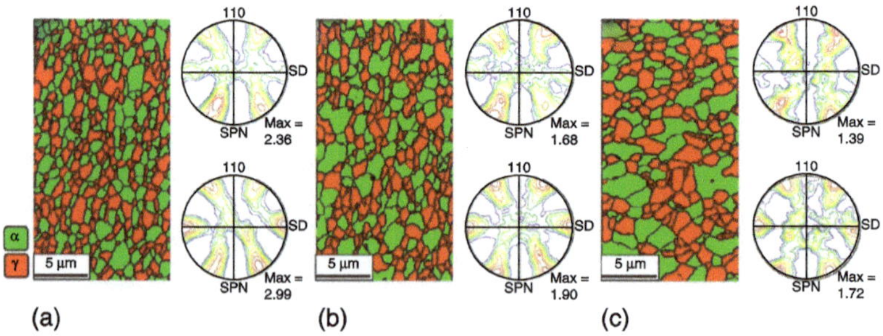

Figure 3.26 Phase maps of areas C (a), D (b) and E (c) in Figure 3.25, in corresponding {111}γ and {110}α pole figures [16]

of both γ and α in the TMAZ can be attributed to this [20]. The progressive increase of LABs misorientation through the absorption of mobile dislocations in the primary boundaries is the main feature of the second process, which causes the conversion of LABs into HABs and brings about the formation of individual segments of an HAB that appears independent of the original HABs. The change of a nucleus into a grain occurs by the conversion of all its LAB facets into HABs.

The above mechanism is commonly associated with the presence of deformation texture [22].

Ferritic stainless steel

Excellent stress corrosion cracking resistance, sufficient high temperature resistance to oxidation, good weldability and formability are some characteristics of ferritic stainless steels (SS) [23]. Due to these advantages, ferritic (SS) have found extensive applications in automotive exhaust systems for manifolds, exhaust pipes, catalytic converters, mufflers, tail pipes and other components. In these steels, grain coarsening occurs in the welded region on account of heat produced throughout the conventional welding process (e.g. resistance welding), so they directly solidify from liquid to ferrite phase without any intermediate phase transformation [24].

Although these steels have positive properties in wrought conditions, a conventional fusion welding technique is recognized to reduce toughness, ductility and corrosion resistance of these steels due to grain coarsening during the process. It is therefore recommended that these steels are welded with a low heat input and at high welding speeds [25]. In this section, data is presented for 2 mm thick 409 SS plates, which were friction stir butt welded using a tool made of PCBN. For microstructural examination of the FSWed joints, the EBSD technique is used, and the specimens are prepared perpendicular to the welding direction (WD). The cross-sectional macrograph of the welded sample at a plunging depth of 1.65 mm is illustrated in Figure 3.27(a). The result of the tensile test indicates that the failure happens in the BM for all samples tested, as revealed in Figures 3.27(b and c), which also demonstrates cross-sectional macrography of the welded sample at a plunging depth of 1.57 mm.

As shown in Figures 3.27(a,c), there is no significant dissimilarity in the macrostructure between two different plunging depths, except for somewhat different vertical deviations from the welding surface. As illustrated in the cross-section, the FSW joint consists of SZ, BM, the narrow TMAZ and the HAZ between the TMAZ and BM, even though it is not easy to distinguish each zone. The SZ has a basin-like profile, which expands noticeably towards the upper surface [26].

As shown in Figures 3.28(a,b), significantly small homogeneous grains develop in the SZ. Dynamic recrystallization induced by severe shear deformation, and the significant amount of heat generated during FSW, result in a fine-grained microstructure in the SZ [26–31]. The average grain size in the SZ (Region 2 in Figure 3.27), is about 1.7 μm, approximately 7 times smaller than that of the BM (Region 1 in Figure 3.27).

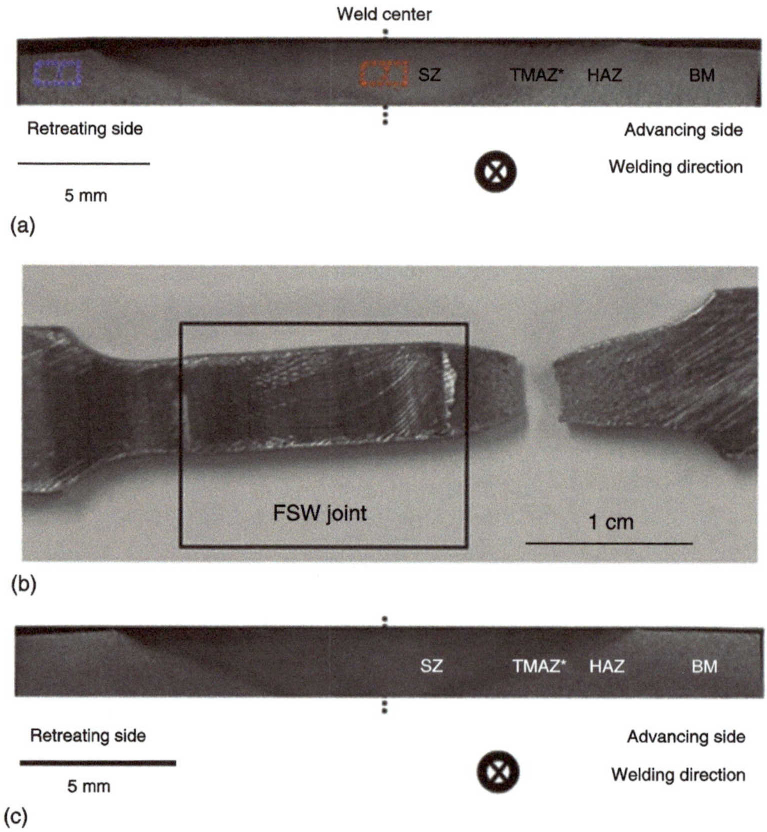

Figure 3.27 (a) Macrograph of the weld with a plunging depth of 1.65 mm (* means approximate region); (b) the FSW specimen with a plunging depth of 1.65 mm after the tensile test; and (c) cross-sectional macrograph of the welded sample with a plunging depth of 1.57 mm [26]

SFE of ferritic SS is larger than that of austenitic SS [4,32], and therefore in similar welding conditions, the degree of grain refinement in the ferritic SS is much higher compared to that in the austenitic SS [26–30]. In low SFE materials, it is difficult for dislocation cross slip to take place, since the partial dislocations are widely separated, while their recombination onto the new slip plane requires a significant amount of energy. The partial dislocations become closer and may be recombined readily to help cross slip by increasing SFE. For ferritic stainless steel with a relatively larger SFE, the CDRX occurs with consequently very fine microstructure.

In the orientation maps in Figures 3.28(a,b), the LABs (2° < LABs < 15°) and the HABs (>15 degrees) are depicted as white and black lines, respectively. The grain boundaries of the BM consist mostly of HABs and only a few LABs. Misorientation angle distribution in the BM is very close to a random distribution, which is shown as the dotted line in Figure 3.28(a). In the SZ, the fraction of LABs is about 54.4%, which is significantly larger than 11.3% in the BM. This shows that due to the significant increase in temperature of the material, in addition to dynamic recrystallization, recovery takes place in the SZ during the FSW process [33,34].

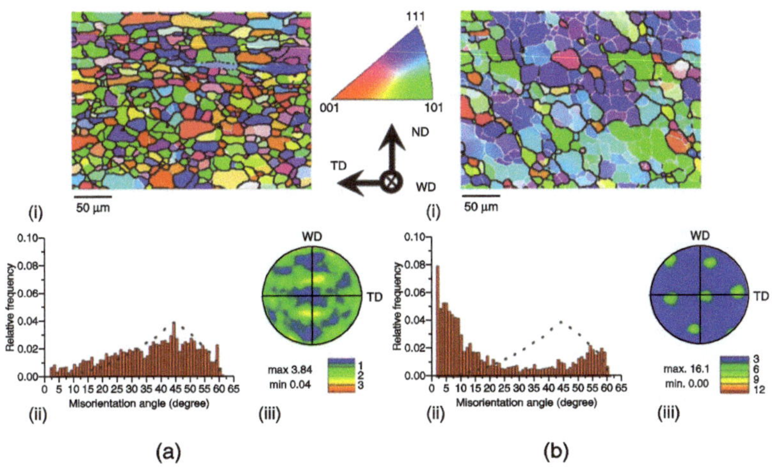

Figure 3.28 (i) Orientation plot; (ii) misorientation angle distribution: (a) dotted line shows the random distribution); and (iii) {110} pole figure of (a) base material (region 1 in Figure 3.29); and (b) FSWed zone (zone 2 in Figure 3.29), respectively. WD, TD and ND correspond to welding, transversal and normal directions, respectively [26]

3.4 Copper and its alloys

Copper and its alloys have excellent properties, such as thermal and electrical conductivity, strength, formability, corrosion resistance, colour, resistance to sparking and non-magnetic behaviour. These characteristics allow copper extensive applications in a wide range of industries. Consequently, there is an increasing demand for the welding of this metal. The main issues affecting weldability of copper and its alloys by fusion welding are thermal conductivity, solidification interval of the alloys and the existence of low melting elements.

Because of some considerable results, copper and its alloys are difficult to join by conventional fusion welding. First, in order to reach localized melting during welding, a high-temperature heat source is required due to the high thermal conductivity of pure copper and its alloys. Second, some alloying elements such as zinc and tin, decrease weldability by increasing susceptibility to cracking. Third, existence of adherent oxides such as aluminium, nickel and beryllium oxides may hinder welding. A variety of volatile elements such as zinc can be present in copper alloys, which can vaporize and change the chemical composition of the alloy during fusion welding. Considering these difficulties, many possible applications of FSW of copper and its alloys have been recognized. In this section, the microstructural features of FSW/FSP of copper and its alloys are summarized.

3.4.1 Pure copper

The thermal and electrical conductivities of copper are at a maximum for the pure metal and decreases drastically with alloying. Pure copper does not experience phase transformations after solidification and remains as f.c.c (α) phase in the solid state. Recently, some investigators have studied the FSW of pure copper and copper alloys and found that the microstructure of joints consists of the four typical zones of BM, HAZ, TMAZ and NZ. Figure 3.29 shows a typical microstructure of a FSW pure copper joint.

The majority of research proves that the nugget has equiaxed recrystallized small grains and the grain size of the NZ decreases after FSW. Similarly, the elongated grains in TMAZ and coarse grains in HAZ are observed [35–38]. For example, when 4 mm thick copper plates with an average grain size of 210 μm are welded at a high rotation speed of 1250 rpm and a traverse speed of 60 mm/min [39–42], the grain size decreases from 210 to 100 μm, and when 2 mm thick copper plates with

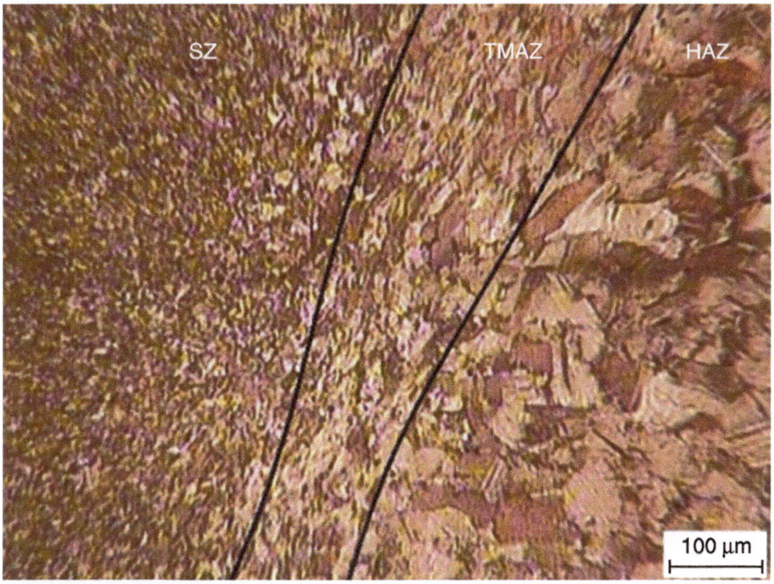

Figure 3.29 Microstructural evolution of a joint welded at a rotation speed of 600 rpm and welding speed of 75 mm/min [35]

average grain size of 30 μm are welded at 1000 rpm and 30 mm/min low traverse speed, nugget grain size reduces to 11 μm. FSW of commercially pure copper plates of 2 and 4 mm thickness is studied. The results show that welding under lower heat input conditions (higher traverse speed, lower rotational speed and lower axial force) result in finer grain sizes of NZ (Figure 3.30) [35–38].

The effect of the tool pin profile at a constant rotation rate of 600 rpm and traverse speed of 75 mm/min on the microstructure of NZ demonstrates that the square pin leads to finer recrystallized grain structure compared to a threaded cylindrical one (Figure 3.31). The results are attributed to higher eccentricity of the square pin profile and its pulsation effect, leading to a higher degree of plastic deformation and higher heat input into the joints [35–38]. Also, the effect of the shoulder geometry on the grain size of NZ in FSW of 1 mm thick copper DHP plates reveals that greater grain refinement is achieved using the scrolled shoulder relative to flat and conical ones [40].

The influence of traverse speed on microstructure of the 3 mm thick copper plate joints is investigated at a low tool rotation rate of 600 rpm and a range of traverse speeds of 25 to 200 mm/min. The results indicate

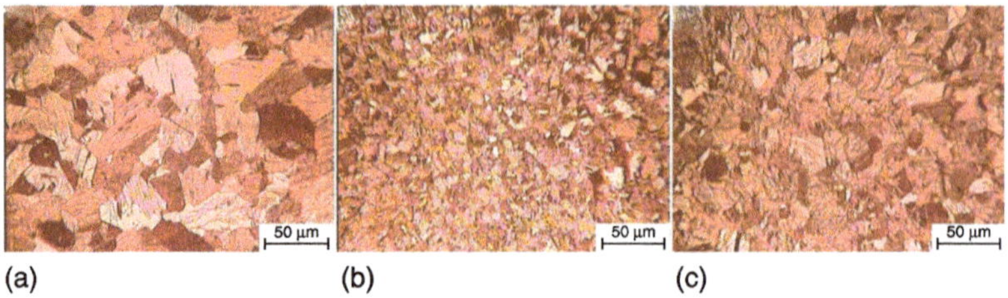

Figure 3.30 Microstructures: (a) BM; (b) specimen welded at lower heat input (700 rpm, 100 mm/min and 2.5 kN); and c) specimen welded at higher heat input (1100 rpm, 100 mm/min and 2.5 kN) [38]

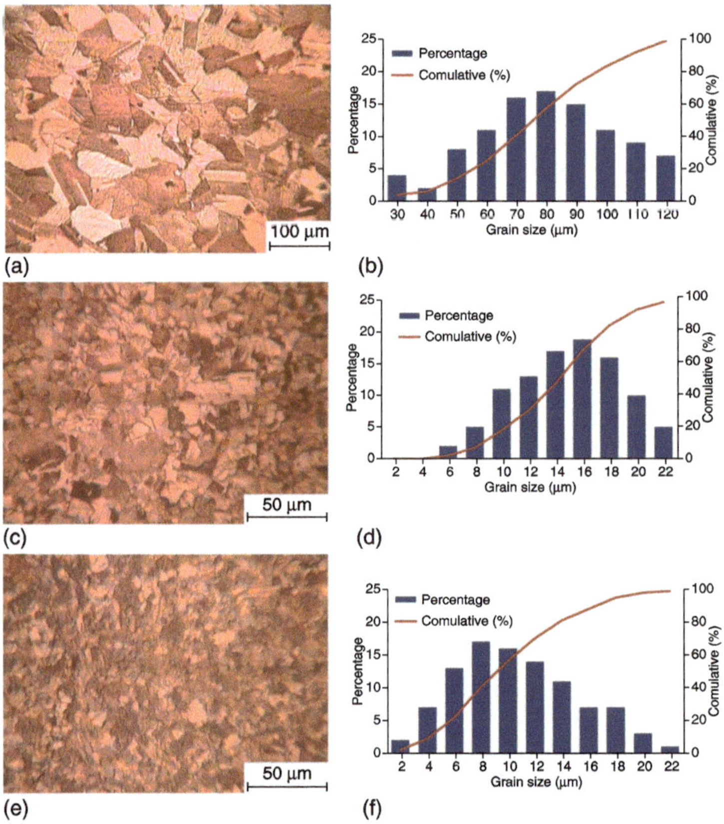

Figure 3.31 Microstructure and grain size distribution graph: (a and b) the BM; (c and d) the SZ of sample welded by threaded cylindrical tool; and (e and f) the SZ of a sample welded by a square tool [36]

that as the welding speed increases, the grain size of NZ first increases and then decreases, TMAZ becomes narrow and the boundary between these two zones becomes more distinct [39].

3.4.2 Brasses

Brass is an alloy of copper and zinc, where the percentage of zinc can alter in the range of 10 to 50%, leading to different microstructural features. Conventional fusion welding techniques of brass is typically hard, since brass possesses high thermal conductivity, approximately 10 to 100 times higher than many steels and nickel alloys. During fusion welding, the temperatures are very high. For example, in TIG welding, the temperature of the electrode or cathode is approximately 4200 °C and almost 3200 °C in the workpiece or anode. These temperatures are higher than the melting and boiling temperatures of both copper and zinc (1083/2590 °C and 419/907 °C, respectively). Therefore, under conventional fusion welding conditions, the materials in brass melt and evaporate and the joints lose their properties due to change of composition. In addition, the evaporation of elements makes the joints porous. In order to prevent the evaporation of the copper and zinc (zinc evaporates more than copper), solid state welding processes are preferable. Researchers also have studied the FSW of brasses. In this section, the microstructural properties of FSW of some brasses are presented [43–45].

The sound FSWs of 60/40 brass are achieved at a wide range of traverse and rotational speeds (Figure 3.32). As shown in Figure 3.33, the microstructures of the welds reveals fine grains of SZ and elongated grains of TMAZ. The material has a double phase structure of α and β phases. The results show that decreasing the heat input causes finer grains in the NZ (Figure 3.34) [43–45].

The microstructure of FSW 70/30 brass shows the four typical distinct zones of the BM, HAZ, TMAZ and the SZ. The SEM data reveals that no evaporation or chemical change occurs during FSW in the different regions (Figure 3.35). Also, the brass colour is spoilt by fusion welding processes [43–45].

The effect of different traverse speeds of 190 to 375 on microstructure and of FSW 70/30Zn brass is studied at a constant rotation speed of 950 rpm. The results show that NZ contains fine grains and TMAZ includes some coarse grains. Also, a unique deformation pattern, i.e. stir band in the NZ area, is recognized. It is reported that the density of the stir band increases by an increase in traverse speed (Figure 3.36) [43–45].

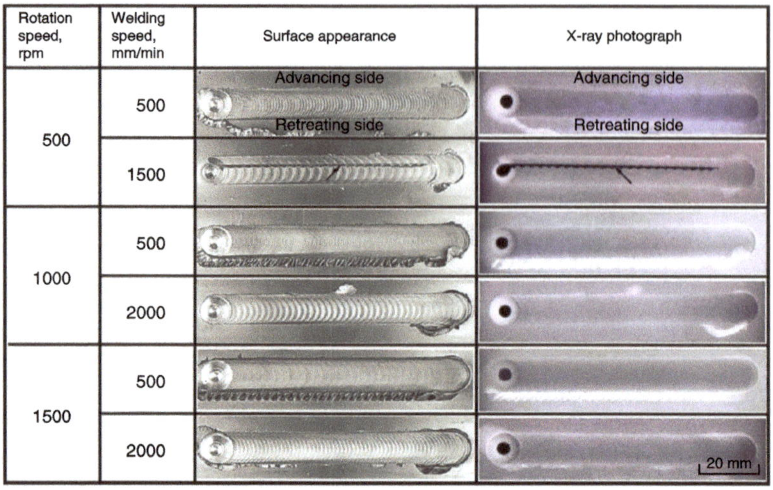

Figure 3.32 Surface and X-ray radiographs of the 60/40 friction stir welds [43]

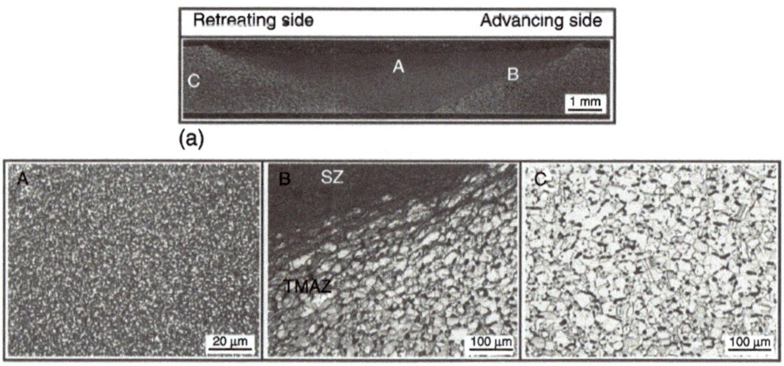

Figure 3.33 Macrostructure and microstructures of the welded samples at rotational speed of 1000 rpm and traverse speed of 500 mm/min [43]

3.4.3 Cast NiAl bronze

Friction stir processing (FSP) has been developed based on FSW principles, which as a rising metal working method can produce modified and refined microstructures near the surface layers of metals and alloys.

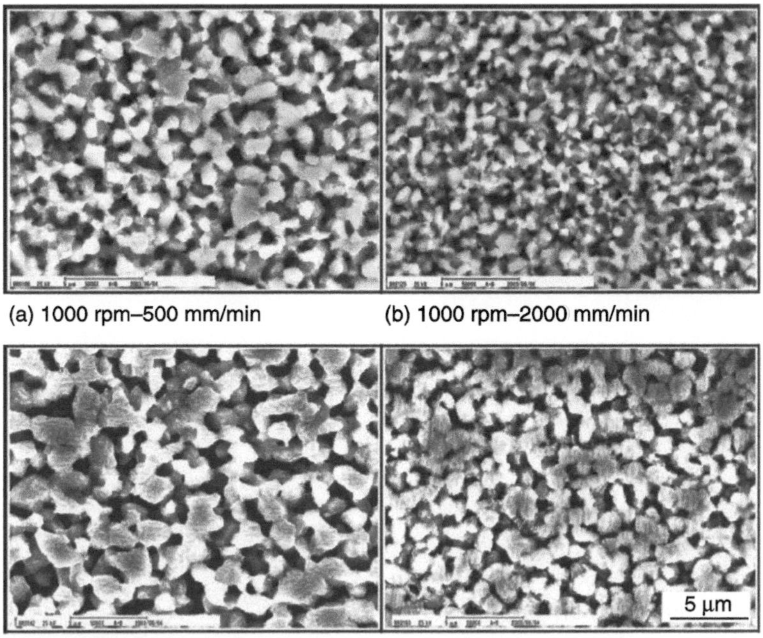

(a) 1000 rpm–500 mm/min (b) 1000 rpm–2000 mm/min

(c) 1500 rpm–500 mm/min (d) 1500 rpm–2000 mm/min

Figure 3.34 SEM analysis of NZ of various welds [43]

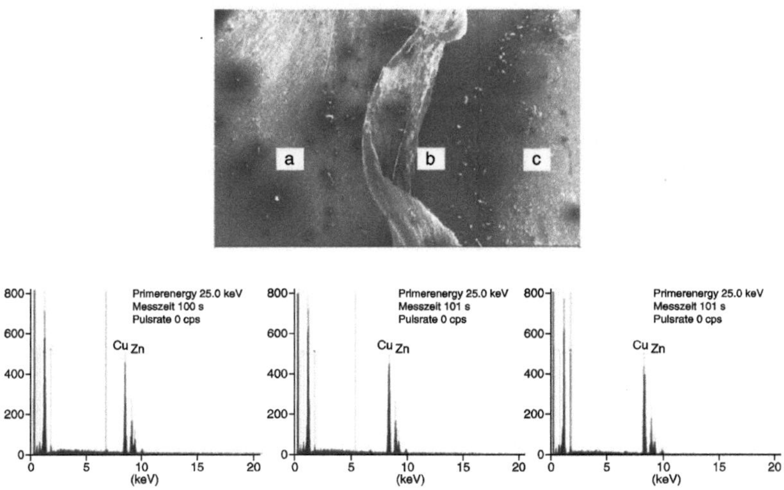

Figure 3.35 SEM and ingredients: (a) the SZ; (b) the HAZ; and (c) the BM (FSW parameters: 2050 rpm and 112 mm/min) [44]

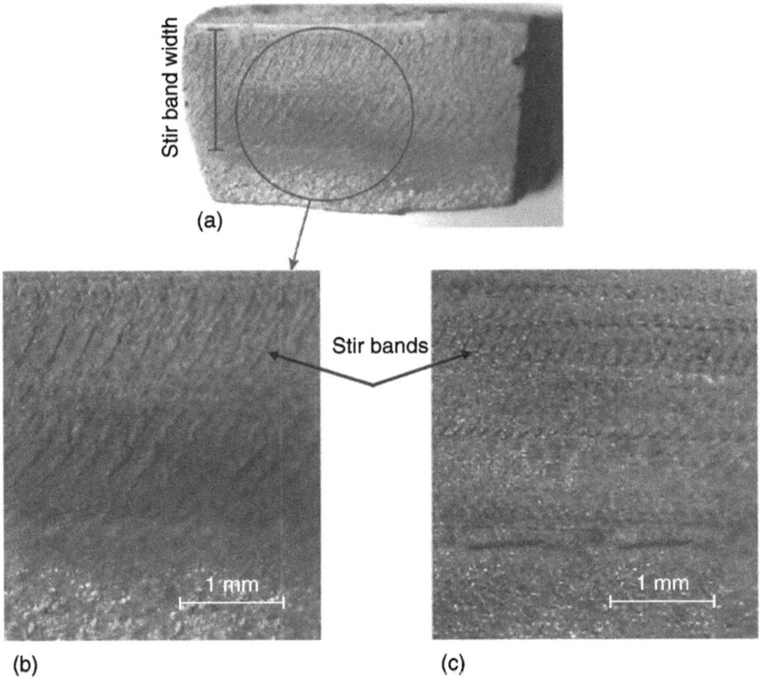

Figure 3.36 Cross-section of weld, stir bands shape and density: (a and b) 235 mm/min; and (c) 375 mm/min [45]

During FSP, the tool traverses on the surface of a workpiece with the purpose of forming modified and refined NZ microstructures, particularly in alloys. In the case of cast metals, the FSP changes the as-cast microstructure to a homogenized and refined wrought microstructure. Also, the elimination of porosities and redistribution of inclusions without any changes of the shape, are other advantages of FSP of the as-cast metals and alloys [46–50].

NiAl bronze (NAB) alloys as quaternary alloys are produced by the addition of nickel and iron to copper-aluminium alloys. These alloys are generally used for naval components because of their good combination of strength, fracture toughness and corrosion resistance. The microstructure of the as-cast NAB is complex and principally includes coarse Widmanstatten α phase, nickel-iron-aluminium κ phases and island martensite β' phase. Furthermore, there are some shrinkage porosity defects in the as-cast NAB structures. FSP is the most used technique for modifying the cast NAB microstructures and improving the mechanical properties. In the past few years, a number of studies have been conducted to understand the microstructure evolution of the cast NAB during FSP. This section summarizes some of these studies [46–50].

The microstructural features of the FSP NAB (Cu-9Al-5Ni-4Fe-1Mn or UNS C95800) are investigated. The results show that NZ of the FSP NAB has inhomogeneous microstructures, including Widmanstatten structures, equiaxed fine-grain structures, and banded or lamellar structures. Also, the welding parameters influence the microstructures and temperature distribution in the FSP NAB. Different microstructures of NZ are illustrated in Figure 3.37, which are related to their dissimilar

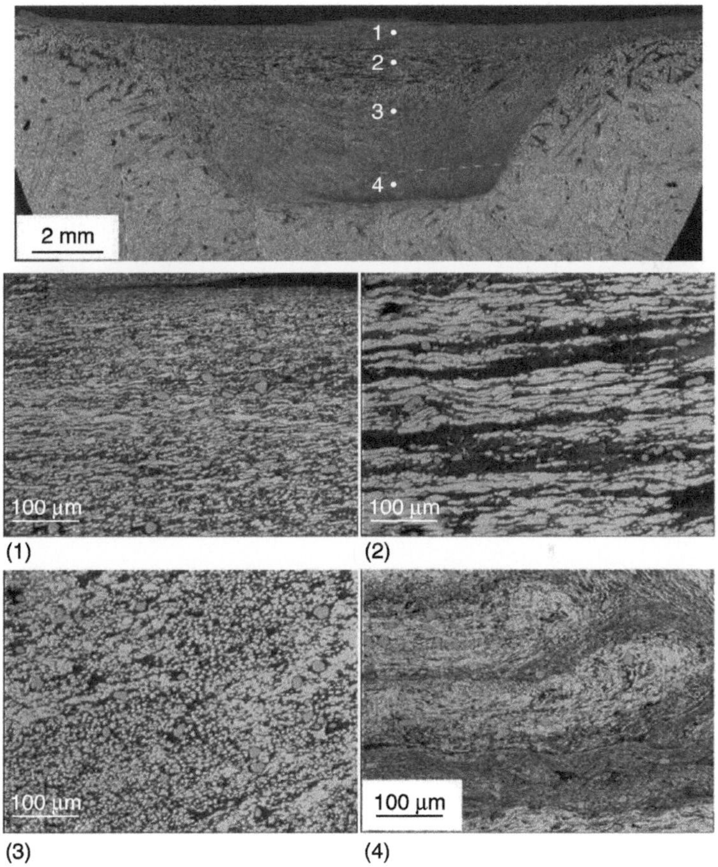

Figure 3.37 Optical microscopy of regions (1) to (4) at higher magnification, as depicted in the macrograph. Light etching areas are the α phase and the dark ones are the β transformation products that involved Widmanstatten α, bainite and martensite [46]

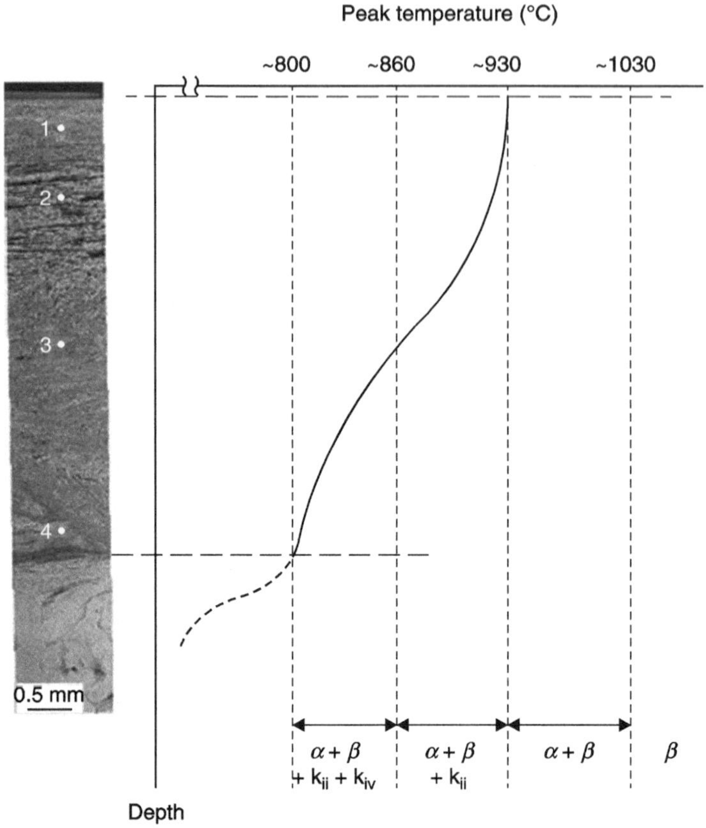

Figure 3.38 **Distribution of peak temperatures vs. depth in NZ [46]**

temperature history during the FSP. The distribution of peak temperatures vs. depth in the NZ are shown in Figure 3.38. The micro-texture reveals that the NZ–TMAZ border is a distinct boundary between recrystallized and deformed regions. The α-phase grain refinement demonstrates DRX and, in areas close to NZ-TMAZ border, particle stimulated nucleation (PSN) at undissolved Fe_3Al particles [46–50].

3.4.4 *Illustrative example*

Development of nanocrystalline structure in copper

The microstructure evolution of copper during development of nanocrystalline structure by FSP is studied by continuous quenching to increase cooling rates. Therefore, different locations around the pin are characterized by OIM. As shown in Figure 3.39, these areas are indicated as R1 to R8, and are at increasingly greater distances from the tool pin–SZ interface. Positions R1 to R7 are under the tool shoulder, and are approximately 0, 0.01, 0.02, 0.06, 0.2, 1 and 2 mm away from the tool pin, respectively. Area R8 is placed beyond the tool shoulder and shows fundamentally the final state of the microstructure [51].

Figure 3.40 illustrates OIM experiments of crystallite structures that develop from the microbands formed around the pin tool. Data is from the areas identified in Figure 3.39. The elongated crystallites appear to change rapidly by coarsening generally in the transverse direction. Equiaxed crystallites form at position R3 at a distance of 0.02 mm from the pin tool–SZ interface. With increasing distance behind the tool, the equiaxed grain structures grow in areas of decreasing temperature (R4–R8 in Figure 3.40). The last grain structures remain within the submicron regime. There is little difference between the misorientation angle distributions achieved in different areas, and the crystallites stay surrounded mainly by HABs [51].

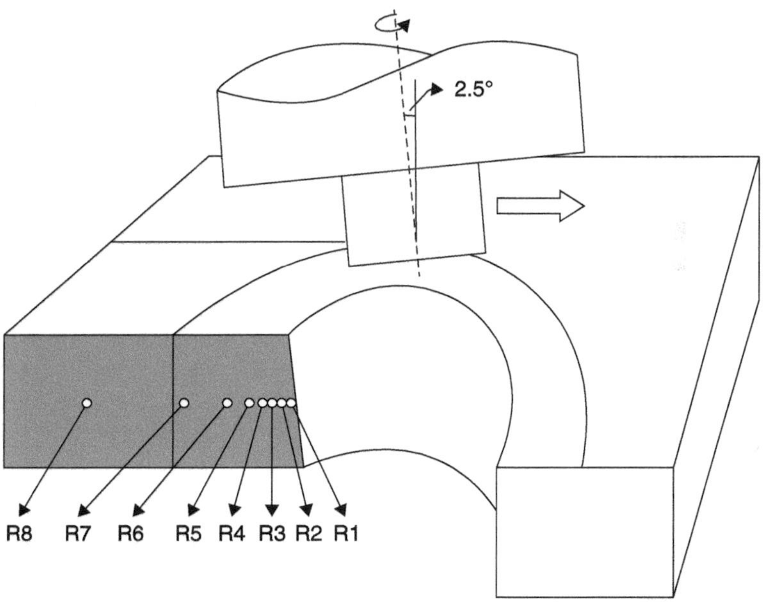

Figure 3.39 FSP and positions scanned by OIM [51]

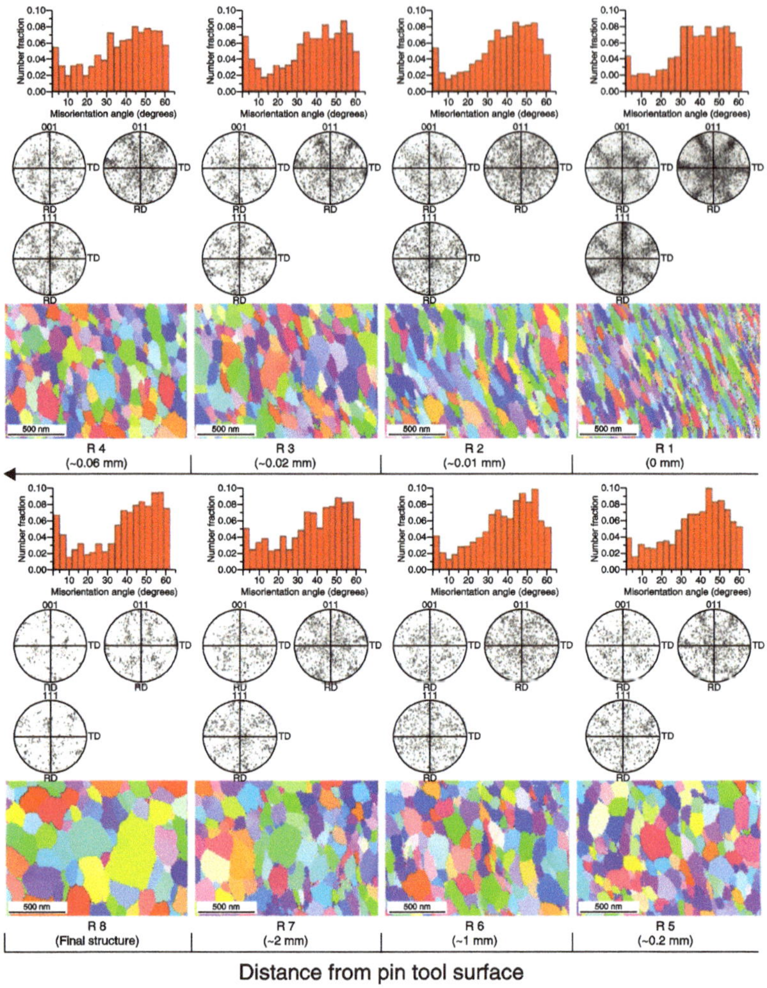

Figure 3.40 **OIM including crystal structures, pole figures and misorientation angle distributions for different areas behind pin tool from R1 to R8, as indicated in Figure 3.39 [51]**

The micro-texture changes during microstructural evolution. All the pole figures are presented with the reference direction (RD) being aligned with the normal to the plate surface and the traverse direction (TD) aligned transverse to the process. As depicted in Figure 3.40 (area R1), the elongated crystal structures near the pin tool display a fibre texture formed by rotation of a shear factor around an axis approximately aligned with the tool axis. With transformation to equiaxed grain structures, the texture change becomes more random. This could be attributed to grain boundary migration or even rotation during microstructure evolution [51].

In summary, during FSP of pure copper, the very small and elongated crystallites substructures form initially around the pin tool. The elongated crystallites convert rapidly to equiaxed grain structures by stable coarsening after passage of the tool. As mentioned in previous sections, the sharp gradients in strain, strain rate and temperature result in strict strain heterogeneity. This phenomenon results in extremely concentrated microplastic deformation in local micro-areas in the material around the pin tool. Under such conditions, a lot of micro-bands are formed in the deformed regions. The micro-band structures are nano-scale elongated crystallites surrounded by HABs and rapidly change to equiaxed and randomly-orientated grain structures behind the tool pin [51].

The microstructure evolutions during FSW/FSP of copper

FSW of pure copper in a wide range of traverse speeds, tool rotation speeds and axial forces has been carried out, and microstructural evolution has been investigated by OM, electron backscattered diffraction (EBSD) and TEM techniques [52]. Figure 3.41 illustrates the boundary maps of NZ of the joints welded at a constant traverse speed of 650 mm/min, and with different rotation speeds and axial forces. The black, grey and pale grey lines designate the high-angle, low-angle and twin boundary, respectively. As shown in Figure 3.41, at a constant axial force, the NZ is more refined by decreasing of the rotation speed. Also the joints welded at a rotation speed larger than 900 rpm reveal larger grains comparable with that of the BM, because of the higher heat input condition. In addition, higher axial forces (1200 and 1500 kg) lead to a much more refined microstructure in comparison with lower axial force (1000 kg). As a result, increasing the axial force has a significant effect on the grain refinement of NZ, in comparison with decreasing tool rotation speed [52].

The changes in the boundary misorientations of different joints NZ are depicted in Figure 3.42. The NZ of the joints, welded with 1000 kg axial force, has a large fraction of HABs. Similarly, the amount of the HABs decreases while the axial load increases to 1200 or 1500 kg. Therefore, NZ of the joints welded at lower axial forces undergoes complete dynamic recrystallization to form new grains (HABs), and higher axial forces

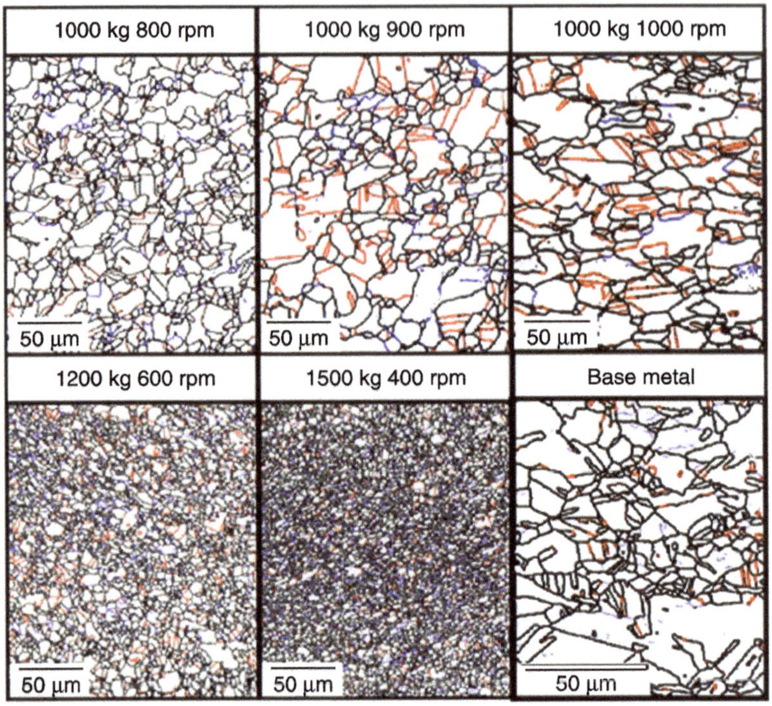

Figure 3.41 Grain boundaries of NZ at different welding conditions, together with that of the base metal. The black, blue and red lines show the high angle, low angle and twin boundary, respectively [52]

result in partial dynamic recrystallization in NZ of the joints. The maximum fraction of the angles is located at 60 degrees, which contributes to the formation of annealing twins. It is moreover remarkable that the amount of twin boundaries (angles located at 60 degrees) decreases with increasing axial force. This is attributed to the formation of the twin boundaries at higher temperatures. Thus, the TEM images of NZ of the joints (Figure 3.43) indicate the existence of the twin boundaries and more DRX at higher heat input conditions [52].

The effect of the welding parameters on microstructure evolution in the welded joints is studied using the EBSD technique. For this purpose, a large area EBSD map is achieved by some small area EBSD measurements in the cross-section of the joints welded under different conditions. For comparison, the microstructure evolutions examined by EBSD technique

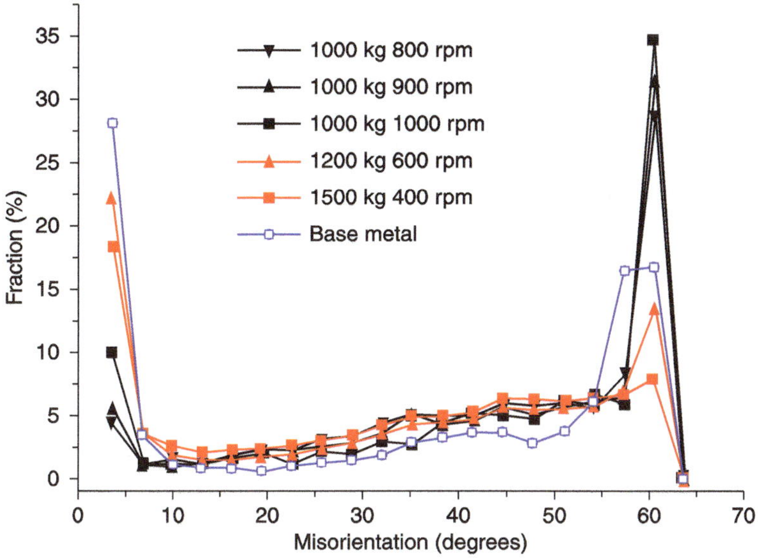

Figure 3.42 Distribution of misorientation angle in NZ of the joints welded under different welding conditions [52]

Figure 3.43 TEM analysis of NZ welded under different conditions: (a) BM; (b) 1000 kg/800 rpm; (c) 1000 kg/900 rpm; and (d) 1500 kg/400 rpm [52]

of the joints welded at 1500 kg 400 rpm and 1000 kg 900 rpm are illustrated in Figures 3.44 and 3.45. The examined areas of the joints are indicated by a dashed rectangle in corresponding macrostructures. Also, the related inverse pole figure (IPF) colour map and corresponding pole figures are illustrated in Figures 3.44 and 3.45.

From Figure 3.44 (sample welded at 1500 kg 400 rpm) it is obvious that there are three distinct areas with different microstructural characteristics, divided by black dotted lines. These distinct regions are namely NZ, HAZ and TMAZ. In the NZ, a clear refinement of the microstructure can be observed. In the HAZ, there are coarse equiaxed grains, while in the TMAZ, there are elongated grains created between NZ and HAZ because of the shear stresses. However, there is no clear border to distinguish different microstructural zones in the large area IPF map shown in Figure 3.45.

As a result, the microstructure in the entire cross-section of the joint reveals similar characteristics, indicated by coarse and equiaxed grains [52]. The illustrated pole figures in Figures 3.44 and 3.45 reveal that the texture strengths in HAZ, TMAZ and NZ are 4.098, 5.013 and 2.860, and 4.131, 3.014 and 2.089 times random for the joints welded at

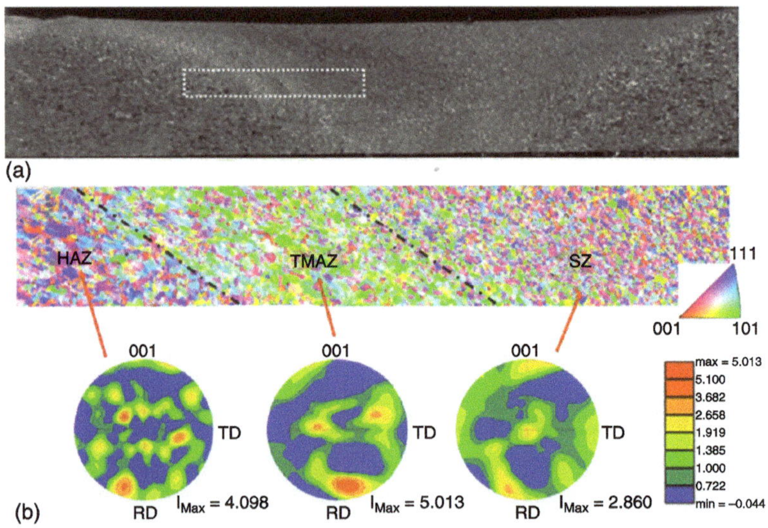

Figure 3.44 Microstructure development of the joint welded at 1500 kg/400 rpm: (a) OM figure showing the location for EBSD measurement; and (b) IPF map and the related pole figures for different specific regions [52]

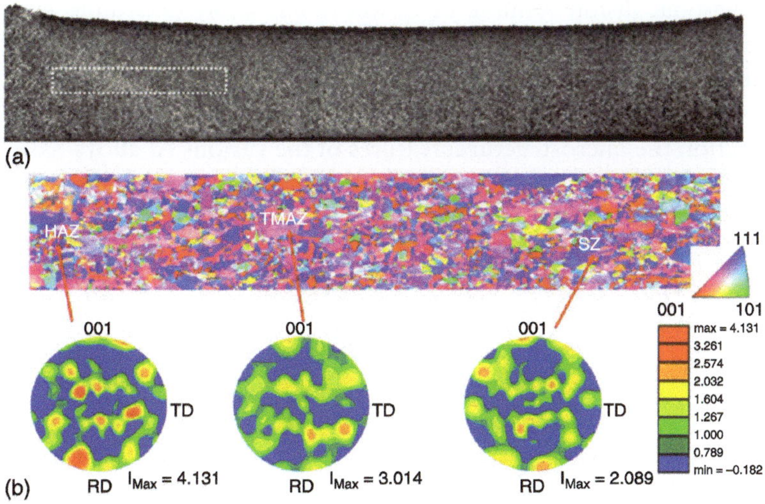

Figure 3.45 **Microstructure development of the joint welded at 1000 kg/900 rpm: (a) OM figure showing the location for EBSD measurement; and (b) IPF map and the related pole figures for different specific regions [52]**

1500 kg 400 rpm and 1000 kg 900 rpm, respectively. The texture strengths of the joint welded at 1500 kg 400 rpm demonstrates that TMAZ has a somewhat stronger texture than other areas. The temperature rising during the welding under these conditions is quite low and thus only partial dynamic recrystallization occurs in TMAZ. From the texture results of the joint welded at 1000 kg 900 rpm, it can be found that dynamic recrystallization occurs during welding. Even though higher strain and strain rate are experienced in the joint welded at 1000 kg 900 rpm, the temperature rise is high. Consequently, all parts of the joint consist of equiaxed grains without any deformed structure and specific distinct regions [52].

3.5 Aluminium and its alloys

The reduced heavy weight and high strength to weight ratio of Al and its alloys make them good alternatives to ferrous materials in particular industrial applications. For example, aluminium alloys are increasingly employed in the automotive industry with the aim of weight saving.

Conventional fusion welding of Al alloys has many difficulties such as oxidation, solidification cracking and porosity. FSW is a solid state welding method invented to join the Al alloys. In recent years, many researchers have investigated the FSW of different types of Al alloys. In this section the microstructural features of the various Al alloys has been summerized briefly.

3.5.1 Non-heat treatable alloys

Non-heat treatable and heat treatable alloys are two main groups of aluminium alloys. Non-heat treatable alloy categories (1xxx, 3xxx and 5xxx) are not strengthened by second-phase particles and usually are called non-precipitation hardening alloys. The strengthening methods for these alloys include cold working (strain hardening) or refining the microstructure (reduction of grain size or formation of substructure). For non-heat treatable alloys, the only appropriate heat treatment is annealing. The annealing process of cold worked metals includes recovery, recrystallization and grain growth, and it is applied to increase the capacity for further deformation. In the case of Al alloys, a completely annealed alloy is assigned usually as 'O temper', while alloys with some level of cold deformation are called 'Hxxx', in which the xxx indicates the amount of cold working [1].

In FSWed joints of non-heat treatable alloys, the only heat treatment possible in different distinct zones is an annealing cycle. The final microstructural features depend on the type of alloys (O or H condition). For example, if the BM is in the O condition, it will be possible that the temperature transient results in grain growth in HAZ. However, if the BM is in the H condition, recrystallization can occur in the HAZ area [1].

3.5.2 Illustrative example

In this section, the data is presented for FSW of 5251 aluminium plate (2.2% Mg, 0.24% Mn, 0.22% Fe and balance Al in wt.%) with O and H initial conditions. Two cold-rolled and annealed plates are characterized from BM to NZ using the EBSD technique to investigate different stages of dynamic recrystallization in NZ. In Figure 3.46, the microstructures of both FSWed plates are illustrated [53].

As illustrated in Figure 3.46, the BM grains are around 60 μm long and 20 μm high in the H14 sample and 100 μm long and 35 μm high in the O

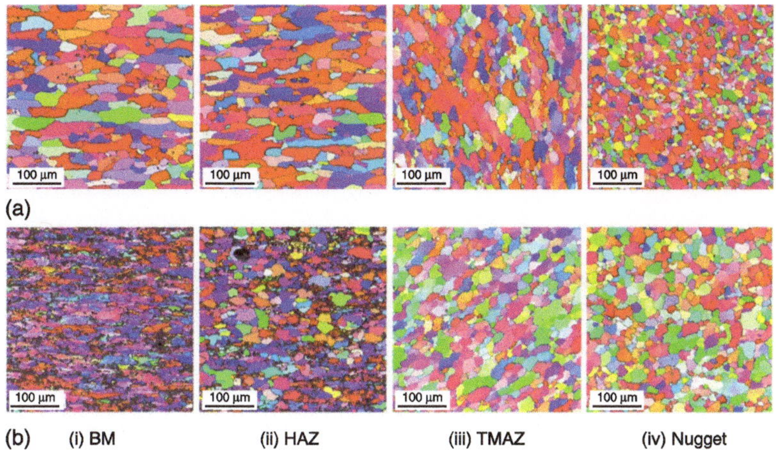

Figure 3.46 EBSD of the BM to NZ: (a) the 5251 O; and (b) the 5251 H14. High-angle boundaries (HABs) are black [53]

sample. In HAZ, in the deformed matrix of the 5251 H14, new equiaxed grains are obvious, but in the 5251 O, the initial grain microstructure of BM is still visible. In TMAZ of the 5251 O, grains are clearly rotated and elongated by plastic deformation. Moreover, in TMAZ of the H14, the recrystallized grains are somewhat elongated and the grain size is close to the NZ grain size. The NZ of both plates is characterized by fine recrystallized and equiaxed grain microstructure, but smaller grains in the O weld (14 μm) is observable when compared to the H14 one (20 μm).

EBSD data in TMAZ of the 5251 O and grain-subgrain distribution in TMAZ and NZ of the 5251 O are illustrated in Figure 3.47. The boundary misorientation angle distribution in TMAZ and NZ of the 5251 O is also displayed in Figure 3,48 [53].

In Figure 3.46(a), from the BM to TMAZ, no noticeable change can be proofed in the microstructure, but grains of TMAZ (misorientated by >15 degrees) become rotated and elongated. Also, according to the Figure 3.47(a), the grains of TMAZ include subgrains (misorientated by <15 degrees) and the HABs are serrated.

As can be seen from Figure 3.47(b), subgrains size in TMAZ are comparable to grain size in NZ, and the number of LABs falls obviously from TMAZ to NZ (Figure 3.48), with the number of HABs becoming the majority. All these observations reveal that the GDRX may occur during FSW, because the grains are adequately flattened by plastic deformation in TMAZ [53].

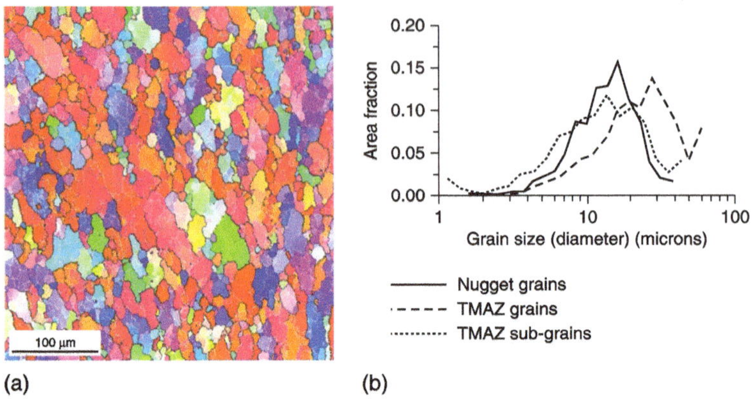

Figure 3.47 (a) EBSD from TMAZ of the 5251 O extended grains with serrated boundaries containing low-angle boundaries (in white); and (b) grain and subgrain size in TMAZ and grain size in NZ [53]

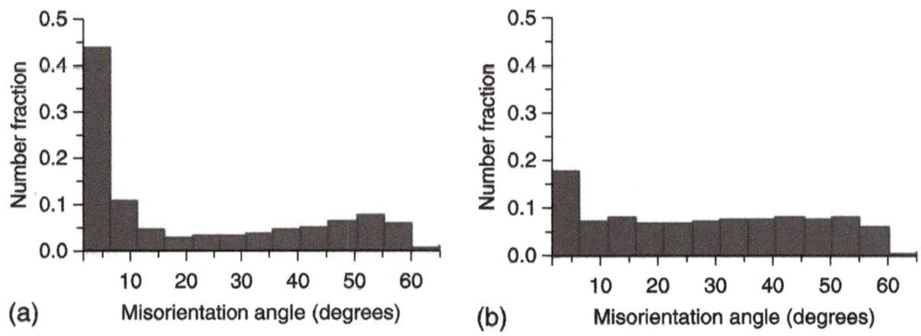

Figure 3.48 Boundary misorientation angle distribution: (a) in TMAZ; and (b) in NZ of the 5251 O [53]

In Figure 3.49, the EBSD map in TMAZ of the 5251 H14 and grain size distribution from HAZ to NZ are depicted. Unlike the 5251 O, no GDRX is observed in TMAZ of the 5251 H14. Indeed, as grains in HAZ are equiaxed after static recrystallization, they do not become sufficiently elongated by plastic deformation in TMAZ to permit the serrations to pinch off. In this region, few grains contain LABs (Figure 3.49(a)), which represent only 28% of boundaries, but it is interesting to note that the mean subgrain size is about 10 µm and that in the nugget an increase of grains around 10 µm large is clearly observed (Figure 3.49(b)).

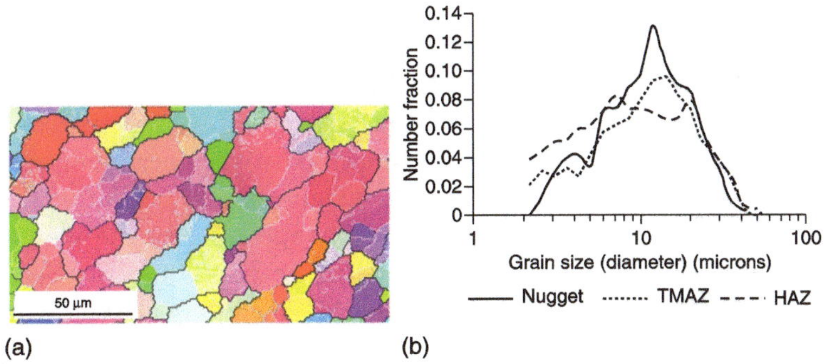

Figure 3.49 (a) EBSD map in TMAZ of 5251 H14 extended grains with serrated boundaries containing low-angle boundaries (in white); and (b) grain size from HAZ to NZ [53]

Subgrains transform into grains from TMAZ to nugget by CDRX. Nevertheless, the grain size or the fraction of LABs (Figure 3.49(a)) do not remarkably change from TMAZ to nugget. In fact, the amount of HABs saturates to a stable value after dynamic recrystallization [53].

3.5.3 Heat treatable alloys

The heat treatable Al alloys (2xxx, 6xxx and 7xxx) are strengthened by second-phase particles and are popular to precipitation hardening of alloys. The primary alloying elements for the three alloy classes are copper (2xxx), magnesium and silicon (6xxx) and magnesium and zinc (7xxx), respectively. Usually the strengthening effect is mainly due to the existence of a single precipitate phase. However, it can be due to several precipitate phases in the microstructure. Therefore, because of the heat input associated with FSW in these alloys, metallurgical phenomena other than annealing can take place in different regions of the joints, such as ageing, over-ageing, precipitation growth or dissolution [1].

In NZ of heat treatable alloys, depending on the alloy type and welding condition, different phenomena may happen. For example, NZ may experience an over-aged condition, a partially solution heat treated condition or a single-phase solid solution. The microstructural features of different zones can be investigated directly by TEM and inferred by its response to a post-weld ageing treatment [1].

Precipitation hardenable aluminium alloy 6063

The Al alloys 6063-T5 and T4 are FSWed at various rotation speeds, and then the microstructure of the joints, are investigated. The thermal history of the welds during FSW and the relationship between peak temperature and rotational speed are illustrated in Figure 3.50, respectively. The thermal results show that the maximum temperature of the joint rises sharply with increase in rotation speed of up to 2000 rpm. Also, the microstructures of NZ (Figure 3.51) reveal that increasing the temperature (higher heat input with increasing rotation speed) leads to larger grain size. The results demonstrate that the relationship between grain size and the maximum temperature of the welds is similar to the static grain growth equation [54].

During FSW, DRX grains nucleate with the shear texture component induced by the plastic deformation. Just after the tool passing, the DRX nuclei (with a very smaller size compared to final NZ grain size) grains grow statically during the cooling of the joints. The shear texture component remains in NZ during the grain growth, and the static grain growth is severely affected by temperature. Therefore, the final grain size of NZ increases with increasing the temperature peak or rotational speed.

The TEM images of the BM of the 6063 T5 alloy and NZ of the joints welded at the 800, 1220 and 2450 rpm are shown in Figure 3.52. The BM contains a high density of needle-shaped precipitates of about 40 nm in length and a low density of rod-shaped β' precipitates of about 200 nm in length. Also it is obvious that there are no precipitates in any of the NZ. Thus, the dissolution of all precipitates happens during the FSW and is related to the thermal cycle [54].

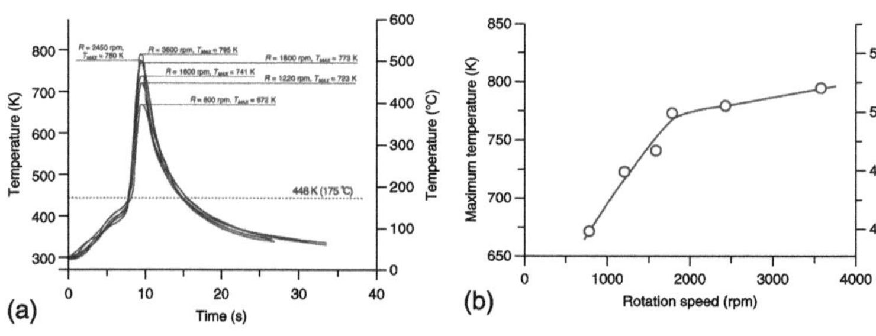

Figure 3.50 **(a) Thermal treatments during FSW at the some rotation speeds of Al alloy 6063; and (b) relationship between the welding rotation speed and the highest temperature of the welding thermal cycle [54]**

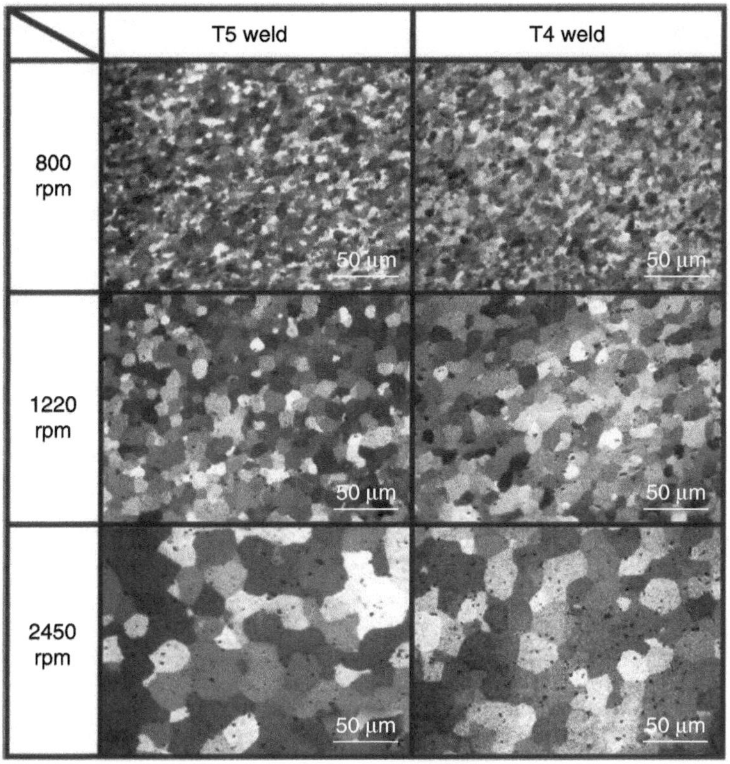

Figure 3.51 NZ microstructures of the distinctive joints welded at special rotation speeds [54]

Figure 3.53 represents NZ TEM images of the Al alloy 6063-T5 joints welded at different rotation speeds after post-weld ageing. As can be seen from Figure 3.53, post-weld ageing considerably increases the density of needle-shaped precipitates [54].

Friction stir welding of 7050-T651 aluminium

The presented data in this section is about FSW of a 7050-T651 Al plate. A typical macrostructure of the welded joint is shown in Figure 3.54. According to the macrostructure, there are four distinct areas in the cross-section of the joints, including the dynamically recrystallized zone (DXZ or NZ), TMAZ and HAZ. The grain structure, dislocation density and second phase particles in different areas are studied and compared to the BM [55].

The TEM images of the grain structures in different weld regions are shown in Figure 3.55. Elongated grains or pancake microstructures,

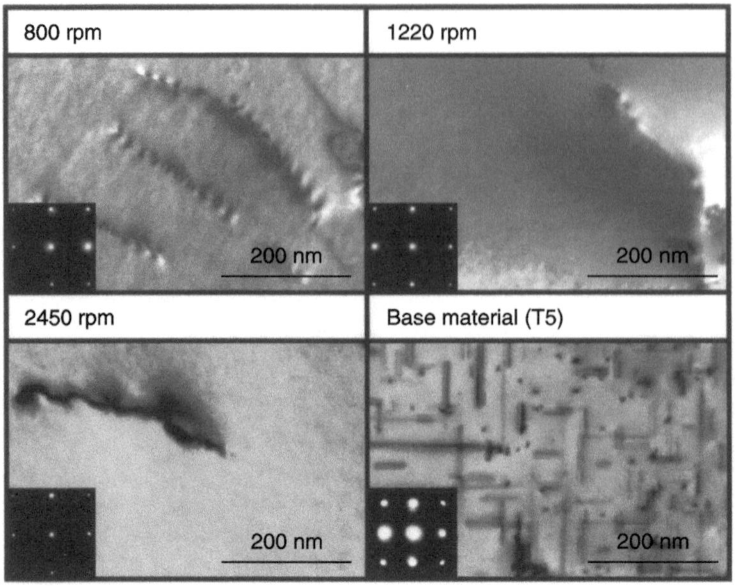

Figure 3.52 TEM of NZ of Al alloy 6063-T5 and the BM of Al alloy 6063-T5 [54]

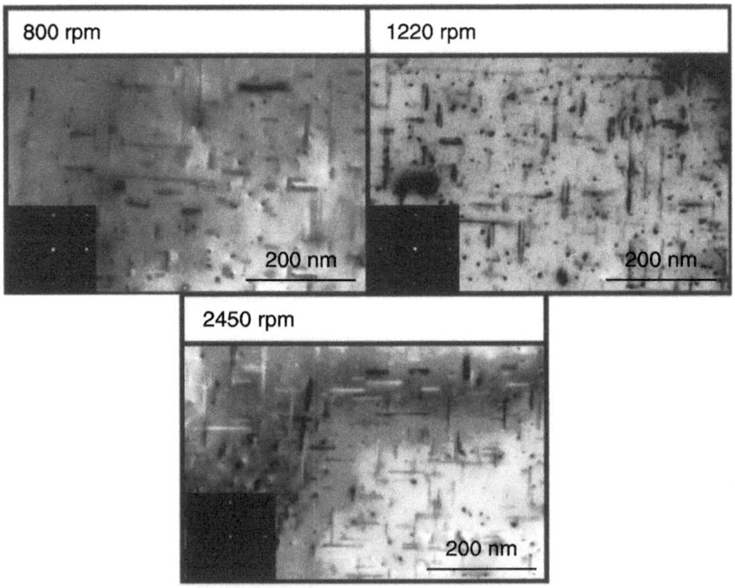

Figure 3.53 TEM of NZ in the post-weld aged welds of Al alloy 6063-T5 [54]

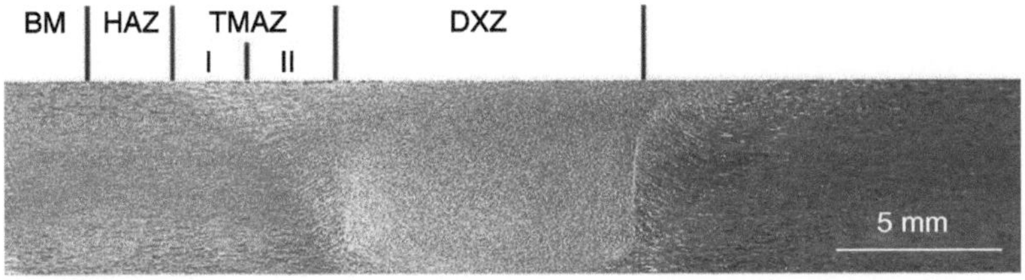

Figure 3.54 Cross-section of FSW in 7050 Al, depicting the special weld regions from base metal to DXZ [55]

Figure 3.55 Grain structure of some weld areas: (a) base metal; (b) dark field image of region (a); (c) HAZ; (d) TMAZ I; (e) TMAZ II; and (f) DXZ [55]

including subgrains of 1 to 5 μm, is the feature of BM (Figures 55(a,b)). From Figure 3.55(c), the influence of FSW on HAZ grain structure is small. Generation of large amount of dislocations and grain growth are apparent in the TMAZ I region (Figure 3.55(d)). According to Figure 3.55(e), in the TMAZ II region, the structure consists of recovered subgrains with a size of 1 to 2 μm. In addition, the NZ (DXZ) includes recrystallized, fine equiaxed grains. NZ grains have high and various dislocation densities with different levels of recovery from grain to grain (Figure 3.55(f)) [55].

In Figure 3.56, the TEM images of the TMAZ I and II regions are illustrated in more detail. The TMAZ I region consists of high dislocation density, dislocation cells and dislocations pinned by second phase particles. It is obvious that the TMAZ II region has a recovered structure. It can be concluded from Figures 3.55 and 3.56 that due to subgrain growth related to absorption of dislocations into the boundaries, DRX in NZ can be believed as CDRX. In addition, absorption

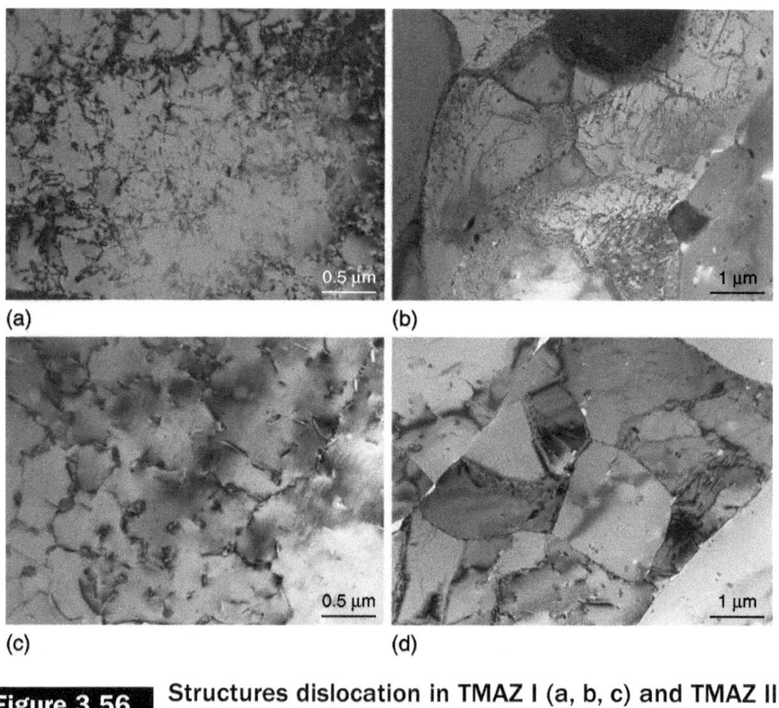

Figure 3.56 Structures dislocation in TMAZ I (a, b, c) and TMAZ II (d) regions: (a) high dislocation density; (b) dislocation cell structure; (c) dislocations pinned by second phase particles; and (d) improved structure [55]

of dislocations into subgrain boundaries increases the misorientation between neighbouring subgrains and changes them into new CDRX grains [55].

In the HAZ region, the precipitates coarsen and the PFZ widens (Figure 3.57(b)). Although the peak temperature in HAZ (250 °C) is higher than the dissolution temperature of η (180–260 °C), a significant amount of η precipitates exist in the HAZ. This result reveals that the

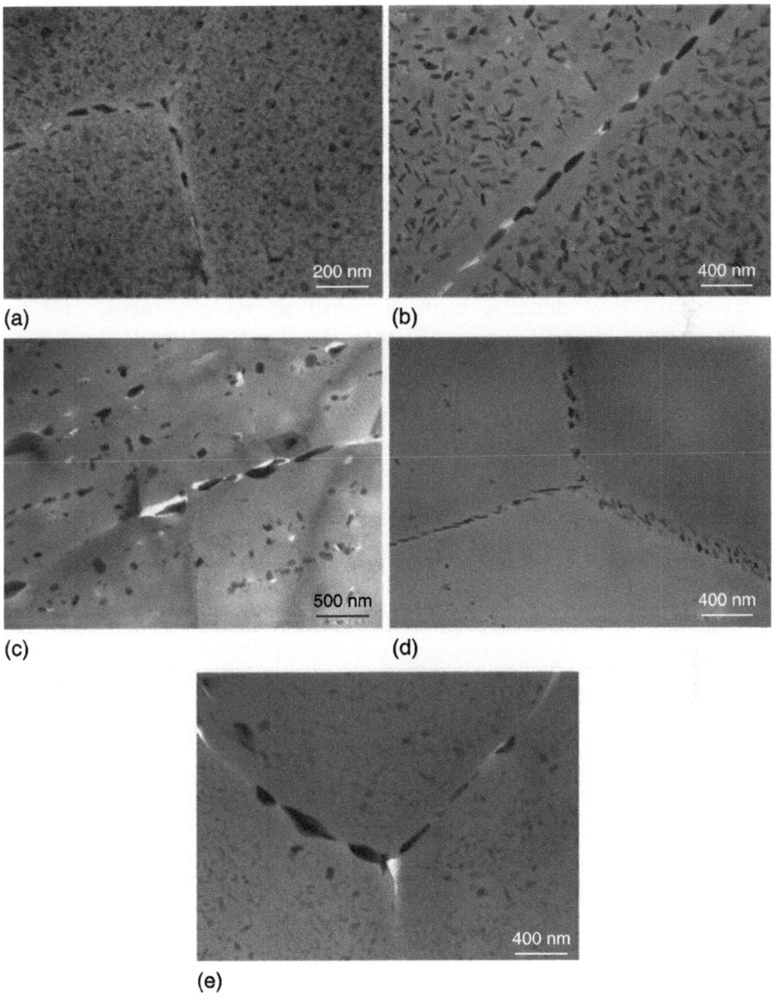

Figure 3.57 Precipitate microstructures inside and along the grain boundaries: (a) parent material; (b) HAZ; (c) TMAZ I; (d) TMAZ II; and (e) DXZ [55]

time of dissolution temperature is inadequate for full dissolution of the coarser η particles. In the TMAZ I region, with peak temperature above 350 °C, η particles are strictly grown and somewhat dissolved while a small amount of very fine η particles are reprecipitated. The temperature in the TMAZ II region (>400 °C) results in the complete dissolution of the strengthening particles and reprecipitates them preferentially along grain boundaries, subgrain boundaries and on dislocations during cooling. In NZ, the η particles dissolve and reprecipitate heterogeneously on dislocations and on matrix Al_3Zr particles. The amount of the particles depends on the density of lattice dislocations [55].

3.6 Magnesium alloys

According to their unique characteristics, magnesium alloys have been appropriate candidates to replace aluminium alloys in many industrial applications such as automobile and aerospace. Some of their unique properties are low density, high strength to weight ratio, and good castability. They are also considered advanced materials in terms of energy conservation and environmental pollution regulations. Thus, welding and joining of magnesium alloys are vital for their applications [56–61].

Joining of the magnesium alloys by conventional fusion welding methods has faced some difficulties, which affects their mechanical properties. For example porosity, oxides and hot crack are some of defects produced by the fusion welding methods. Full removal of the surface oxides and applying proper welding parameters can reduce fusion welding defects. Gas tungsten arc welding (GTAW) and gas metal arc welding (GMAW) are two fusion welding methods used for joining of the parts of magnesium alloy. Also, in the GTAW and GMAW methods, there are some problems such as need for high purity shielding gases, high welding residual stresses and changes in microstructure resulting from melting and solidification [56–61].

FSW is a very suitable method for joining of the magnesium alloys, because no melting occurs during this method, and therefore it can remove defects associated with solidification such as porosity and cracks. As FSW does not require high heat input, the metallurgical defects related to it such as oxides can also be eliminated. In addition, FSW does not result in solute loss by evaporation or segregation during solidification, resulting in homogeneous distribution of solutes in the weld [56–61].

3.6.1 *Illustrative example*

Evolution of grain structure during FSW of AZ31 magnesium alloy

In this section, data is presented for a 2 mm thick sheet of a commercial AZ31 magnesium alloy with a nominal chemical composition of Mg-3.0Al-1.0Zn (wt.%). A bead on plate FSW is carried out at a tool traverse speed of 600 mm/min and rotation speed of 2000 rpm by stop action technique. To study the grain structure evolution during FSW of AZ31 magnesium alloy, the high resolution EBSD technique is applied. The microstructural features of the BM are summarized in Figure 3.58. According to Figure 3.58(a), the BM consists of low aspect ratio grains with a mean grain size of 20 μm. As can be seen from Figure 3.58(b), most of the grain boundaries are HABs (69%) and random LABs are gathered within some grains. Misorientation distribution is characterized by a sharp low-angle peak [60].

Also, some clustering of misorientation axes exists in the central part of the stereographic triangle, as well as near the $\langle 2\bar{1}\bar{1}0 \rangle$ pole. It is considered that the fraction of $\{10\bar{1}2\}$ twin boundaries is less than 1%. In addition, the BM has a moderate texture with the {0002} basal planes

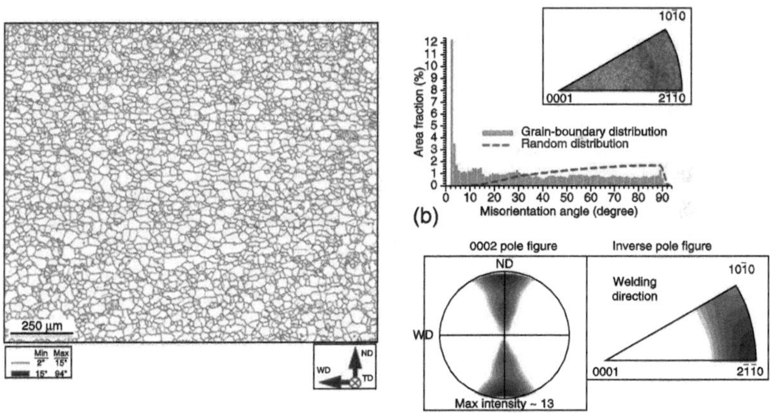

(a) (c)

Figure 3.58 **Microstructure of the BM: (a) EBSD map showing the distributions of HAB_s and LAB_s (black and grey lines, respectively); (b) misorientation distribution; and (c) textural data represented by {0002} pole figure and the converse pole figure for the WD [60]**

orientated nearly parallel to the WD and with the <$2\bar{1}\bar{1}0$> direction tending to align with the WD (Figure 3.58(c)) [60].

Figures 3.59(a) and 3.60(a) indicate the composite EBSD maps taken in front of and behind the tool, respectively. Furthermore, characteristic {0002} pole figures achieved from these different regions (illustrated in Regions 1–10) are represented in Figures 3.59(b) and 3.60(b).

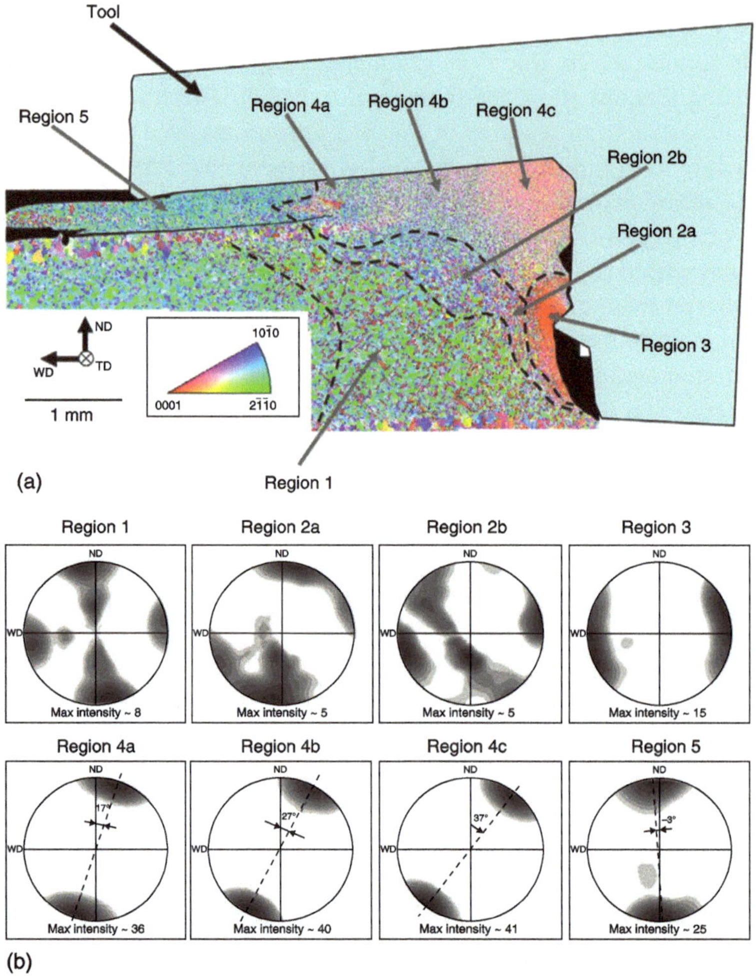

Figure 3.59 (a) EBSD map prepared in front of the tool keyhole (colour code triangle is shown in the bottom left-hand corner) with superimposed contour of the tool section; and (b) {0002} pole figures showing texture in selected zones [60]

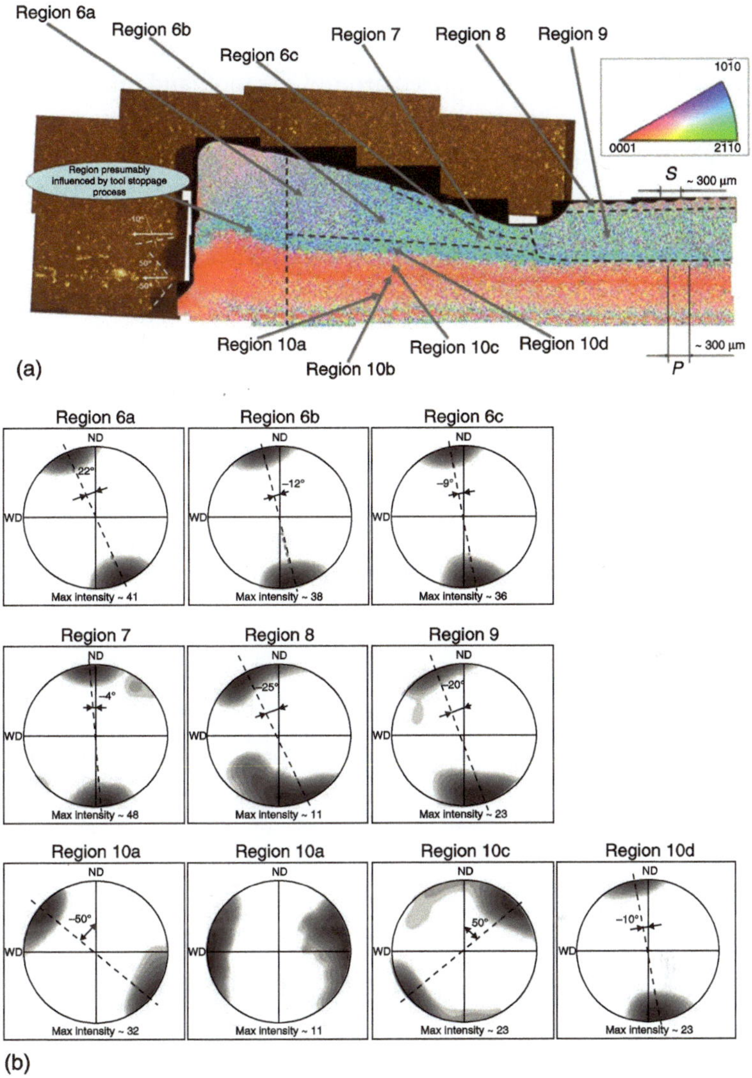

Figure 3.60 (a) EBSD map taken behind the tool keyhole (colour code triangle is shown in the top right-hand corner) with superimposed tool section; and (b) {0002} pole figures showing texture in selected regions [60]

In summary, it is considered that the tool shoulder has a remarkable effect on the texture and material flow in the upper part of the joint. The material first experiences a compound deformation arising from the combined effects of the rotating pin and shoulder. Also, contribution of the shoulder induced component continuously increases with the distance from the pin. Consequently, some surface layer of the material is entrapped by the shoulder, carried around the tool, and finally deposited far behind the pin. During the deposition procedure, the material in the upper part of the joint undergoes extra deformation, which considerably changes the texture. EBSD maps illustrating typical microstructures in Regions 1 to 10 are summarized in Figures 3.61 and 3.62. In the maps, LABs are depicted as grey lines and HABs as black lines [60].

In Region 1 (Figure 3.61(a)), initial grain shape does not tend to transform but the greater part of the grains including twins extending across the grains (an example is arrowed in the high magnification map). The twins normally have an orientation close to [0001] || WD; therefore they are key contributors to the new texture component in Region 1 (Figure 3.59). The average grain size in this region (11 μm) is about half that of the BM. The total HABs fraction decreases to 67%, thus it can be indicated that LABs form. Therefore, the grain refinement in Region 1 is generally ruled by plentiful twinning, which produces additional HABs [60].

In Region 2 (Figure 3.61(b)), because of the geometrical requirements of strain, the initials grains and twins reorient. Also, the initial grain boundaries

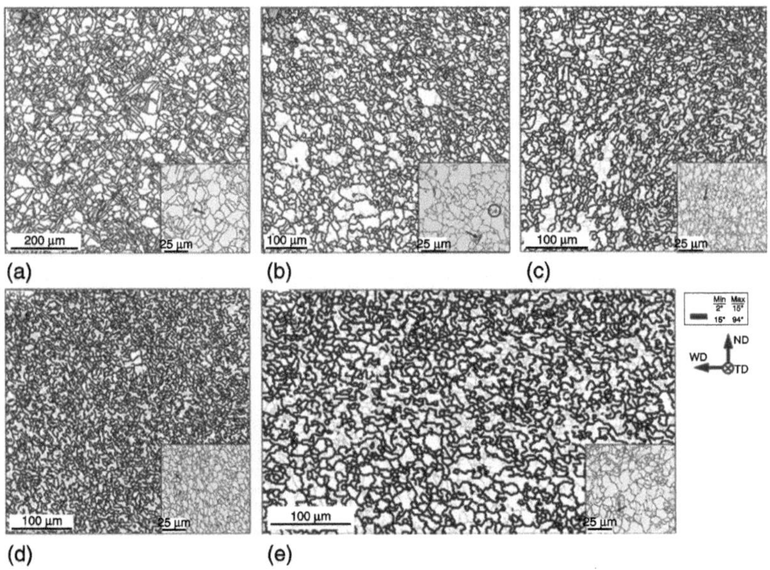

Figure 3.61 (a) EBSD map taken behind the tool keyhole (colour code triangle is shown in the top right-hand corner) with superimposed tool section; and b) {0002} pole figures showing texture in chosen regions [60]

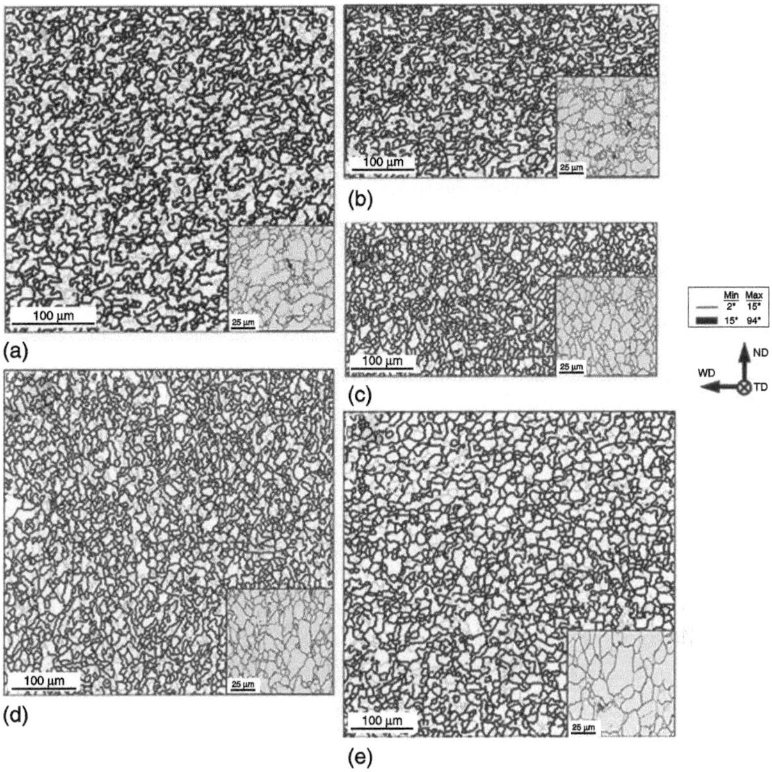

Figure 3.62 **EBSD maps illustrating characteristic microstructure behind the tool: (a) Region 6; (b) Region 7; (c) Region 8; (d) Region 9; and (e) Region 10. High resolution EBSD maps are shown in the bottom right-hand corners of the maps [60]**

and twin boundaries become wavy. Swells appear along the grain boundaries, mostly at triple joints, creating fine equiaxed grains (an example is encircled in the high magnification image). In general, the fine grains are fully surrounded by HABs and possess a random crystallographic orientation. This feature is attributed to the initial stage of DDRX. Moreover, in this region, misorientation across the twin boundaries starts to deviate from the accurate twin/matrix relationship (86 degrees $\langle 11\bar{2}0\rangle$). This process ultimately causes the twin boundaries to transform into random HABs.

However, some fine twins with misorientation of about 86 degrees $\langle 11\bar{2}0\rangle$ (ideal relationship) are found inside the grains. It reveals that a new generation of twins occurs in this region. The average grain size in Region 2 is approximately 9 µm. The reduction of the grain size in Region

2 is generally related to the geometrical thinning of the initial grains and the DDRX nucleation of the new fine grains. The overall HABs fraction in this region decreases down to 59%, which is attributed to the LABs formation. The grain subdivision procedure (subgrain formations) typical for cubic metals is rather difficult in magnesium alloys. This is due to the lack of easily activated slip systems in these materials. Therefore, the microstructural evolution in Region 2 is mainly a result of the geometrical effects of strain and the limited DDRX [60].

The microstructures of Regions 3 to 7 are generally similar to each other. In these regions, according to the area, the grains tend to align either with the pin column surface or with the shoulder plane or have a combined orientation in the regions of the combined pin/shoulder material flow. Also, the fine equiaxed grains normally have almost no substructure and are approximately fully surrounded by HABs; therefore, their structure is comparable to the recrystallization nuclei in Region 2. On the other hand, crystallographic orientation of the fine grains in Regions 3 to 7 is commonly not random. This presumably means that their origin is not associated with the discontinuous recrystallization. Thus, the microstructure evolution in these regions is ruled by the deformation induced by the pin, shoulder or a combination of both, respectively [60].

The developed microstructures in Regions 8 to 10 are a combination of low aspect ratio recrystallized grains and irregularly shaped deformed grains with different levels of subgrain development. The size and volume fraction of recrystallized grains increase. Moreover, some subgrain boundary segments appear to accumulate misorientation surpassing 15 degrees, thus converting the HAB/LAB matrix into irregularly shaped deformed grains. This structure morphology sustains the assumption of the high temperature nature of the deformation in Regions 8 and 9. However, in Region 10, it is difficult to recognize whether the partially recrystallized microstructure originates from dynamic recrystallization or from static annealing, because the material flow directly close to the tool was polluted by the tool stoppage process, and thus the microstructure nearby to the pin is unavailable for analysis [60].

3.7 Magnesium-based composites

The mechanical properties, such as hardness and wear resistance of the magnesium alloys, are not sufficient to widen their applications in high loading conditions, though some processes to fabricate magnesium alloy composites have been employed to improve their mechanical properties. A group of materials with good strength, higher stiffness, higher hardness and better wear and corrosion resistance are particle-reinforced magnesium metal matrix composites, which compensate for the disadvantages of the magnesium alloys.

For most isotropic composites, the micro-sized whiskers or particulates with a reinforcement volume fraction of 15 to 35% are usually added into metallic alloys. Nanosized particle reinforcements can significantly increase the mechanical strength by more effectively enhancing the particle hardening mechanisms than micro-sized particles. A homogeneous distribution of nano-sized particles provides a good balance between reinforcement and inter-particle spacing to maximize the hardness and wear resistance. On the other hand, it is difficult to accomplish uniform distribution of ceramic particles on metallic substrate by usual surface treatments. However, this problem becomes obvious in the case of nanoparticles, since such powders show poor flowability due to the strong interparticle attractive forces associated with small particle sizes. There are some methods to fabricate particulate reinforced Mg-based composites, including stir casting, squeeze casting, molten metal infiltration and powder metallurgy [62–70].

Recently, FSP has been developed based on FSW to incorporate ceramic particles into the metal matrix to form bulk composites. Also, it is possible to produce a surface composite layer by this process [62–70].

3.7.1 *Illustrative example*

AZ91/SiC nanocomposites

In this section, the data is presented for an as-cast AZ91/SiC surface nanocomposite layer produced by FSP. The AZ91 as-cast magnesium alloy with a composition of (in wt%): Al, 9.1; Zn, 0.68; Mn, 0.21; Si, 0.085; Cu, 0.0097; Ni, 0.001; Fe, 0.0029; and Mg, balance and SiC particles with an average diameter of 30 nm and 99.98% purity, were used as reinforcement. The rotational and traverse speeds of the tool were changed from 710 to 1400 rpm and from 12.5 to 63 mm/min, respectively. The tool was rotated with the tilt angle of 3 degrees.

Figure 3.63(a) illustrates the SEM studies of the SZ. The grains in TMAZ are elongated in the interface of SZ and BM (Figures 3.63(b–d)). Figures 3.63(e–f) demonstrate the SZ microstructure of the light and dark regions, respectively. Increasing the volume fraction of SiC particles decreases the grain size. This is why the pinning effect confines the grain

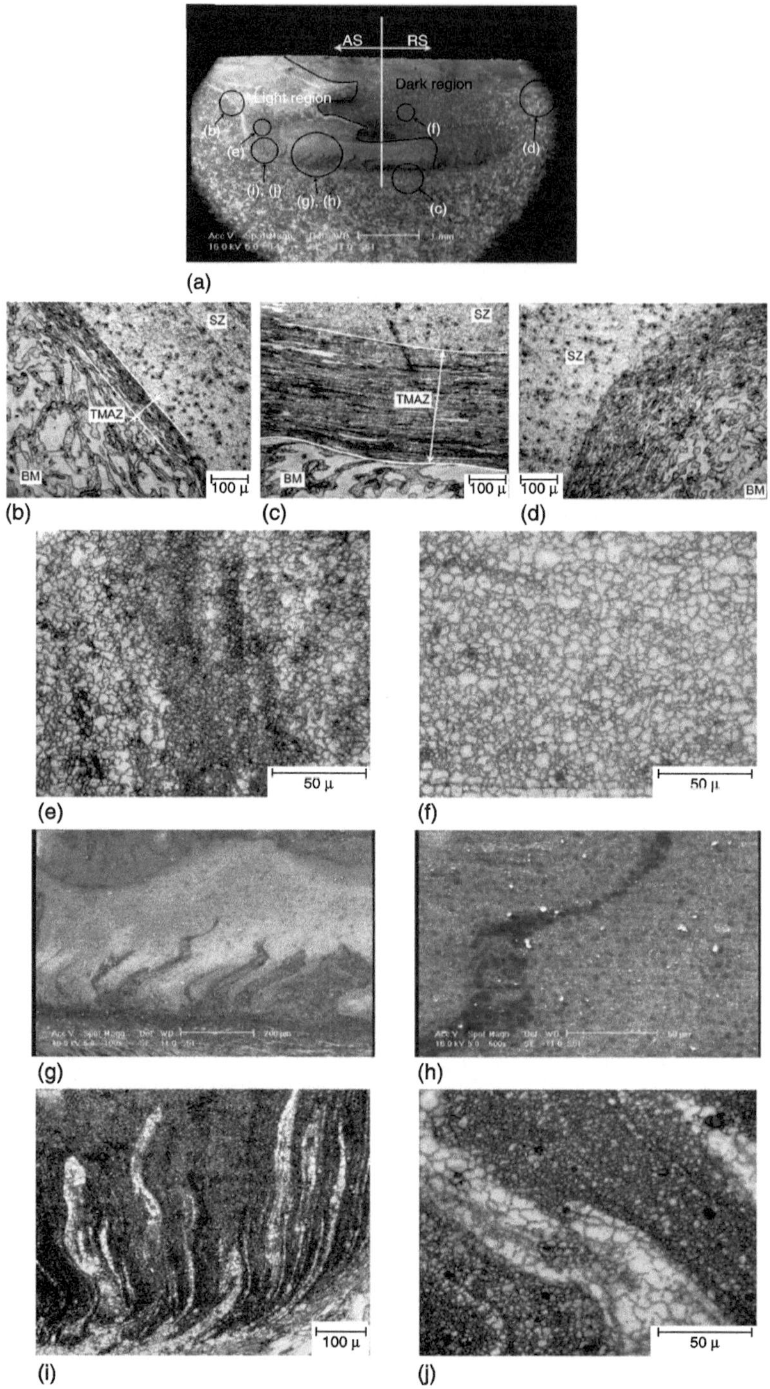

Figure 3.63 (a) SEM macrograph of the SZ of the specimen produced by two FSP passes without changing the tool rotating direction; (b–j) optical and SEM micrographs of the selected regions in (a) [63]

growth during recrystallization. Figures 3.63(g–j) show the optical and SEM micrographs of the inhomogeneous microstructure resulting from non-uniform distribution of the SiC particles [63].

Figure 3.64 shows the effect of rotational speed on the microstructure of light and dark regions. The grain size decreases with decreasing the tool rotational speed. The higher the rotational speed, the more the heat generation, and the more the grain growth during recrystallization, especially when SiC particles are lacking. Figure 3.65 shows microstructure of the dark region in the specimens produced by one FSP pass with different traverse speeds. Increasing the traverse speed decreases the time

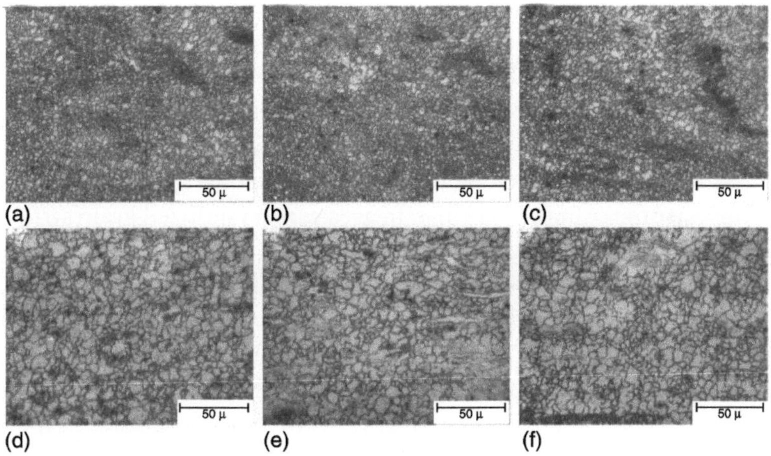

Figure 3.64 Microstructure of the light region (a–c) and dark region (d–f). The rotational speed was (a), (d) 710 rpm, (b), (e) 1120 rpm, and (c), (f) 1400 rpm [63]

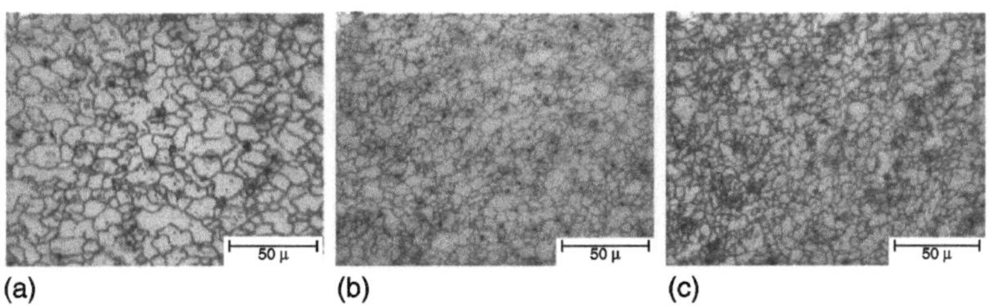

Figure 3.65 Microstructure of the dark region. The traverse speed was: (a) 12.5 mm/min; (b) 40 mm/min; and (c) 63 mm/min. The rotational speed was 900 rpm in all cases [63]

that the material is affected by the process heat. Thus, the grain growth declines during recrystallization [63].

AZ91/Al_2O_3 composites

A different study investigated the AZ91/Al_2O_3 surface composite produced by FSP. The material used was an AZ91 as-cast magnesium alloy with the following composition (wt.%): Al, 8.9; Zn, 0.79; Si, 0.0091; Cu, 0.0014; Ni, 0.0011; Fe, 0.0017; Be, 32 ppm; Mg, balance. Commercially available Al_2O_3 powder with 3 different sizes 3000, 300 and 30 nm, and 99.9% purity was used. A constant tool rotating rate of about 900 rpm was adopted and the constant travel speed was about 63 mm/min. The welding tool was rotated in a clockwise direction. The tilted angle was 3 degrees.

Figure 3.66(a) shows the macrostructure of the joint with the optical microstructures of the BM (Figure 3.66(b)), the SZ (Figure 3.66(c)), TMAZ (Figure 3.66(d)) and particle agglomerated zone (Figure 3.66(e)). The grains in TMAZ are elongated and there is some grain refinement in this region (some equiaxed grains). In area c, recrystallized grain structure and homogeneously distributed particles can be plainly seen. Al_2O_3 particles are distributed within this area, because of the occurrence of vigorous stirring during the FSP.

As shown in Figure 3.67, the grains of the AZ91 matrix are clearly refined because of dynamic recrystallization during the FSP. However, the FSP with the nano-sized Al_2O_3 particles decreases effectively the grain size of the BM in comparison with micron size particles in which some grains are less than 3 μm (Figure 3.67(b–d)). Also, the triangular tool brings about more grain refinement and uniform particle dispersion compared to a square tool [64].

Figure 3.68 shows the effect of pass number and tool shape on microstructure, grain size and cluster size of the samples processed with the rotational and traverse speeds of 900 rpm and 63 mm/min, respectively. The grain and cluster sizes in the samples processed by the triangular tool are smaller than that of the square tool. In addition, increasing the pass number in FSP is an effective method to produce homogenized and ultrafine grain microstructure.

3.8 Titanium and its alloys

Because of high strength to weight ratio and good corrosion resistance in various corrosive environments, titanium and most of its alloys are

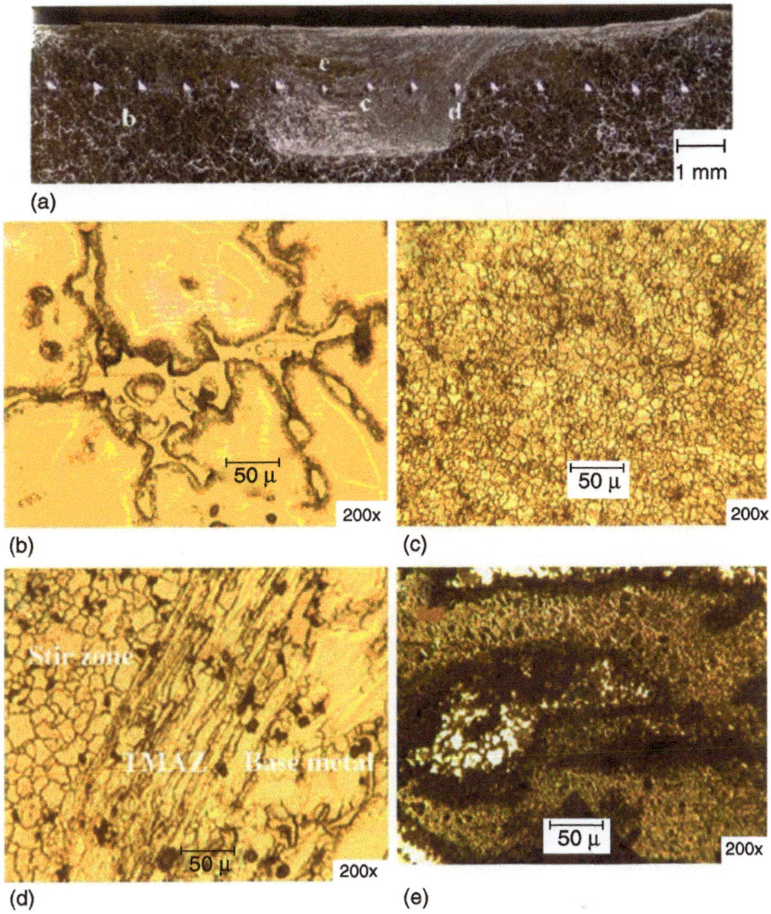

Figure 3.66 **(a) Macro-image of the cross-section of FSPed specimen; (b) BM; (c) SZ; (d) stir zone, TMAZ (equiaxed grains) base metal; and (e) the particle agglomerated zone [64]**

applicable in aerospace structures. Titanium alloys are categorized as commercially pure (generally contain the hexagonal close packed (hcp) α phase), α-β alloys (contain both α and β phases) and metastable β alloys that include the bcc phase.

Only a few accessible titanium alloys have been investigated for FSW, which include pure Ti, Ti-6Al-4V, Ti-15V-3Cr-3Al-3Sn and Beta 21-S.

Pure titanium has a tendency for deformation by mechanical twinning during FSW [65–68]. By using TEM, a large density of dislocations and mechanical twins are observable in the NZ of an FSW joint and reveal an

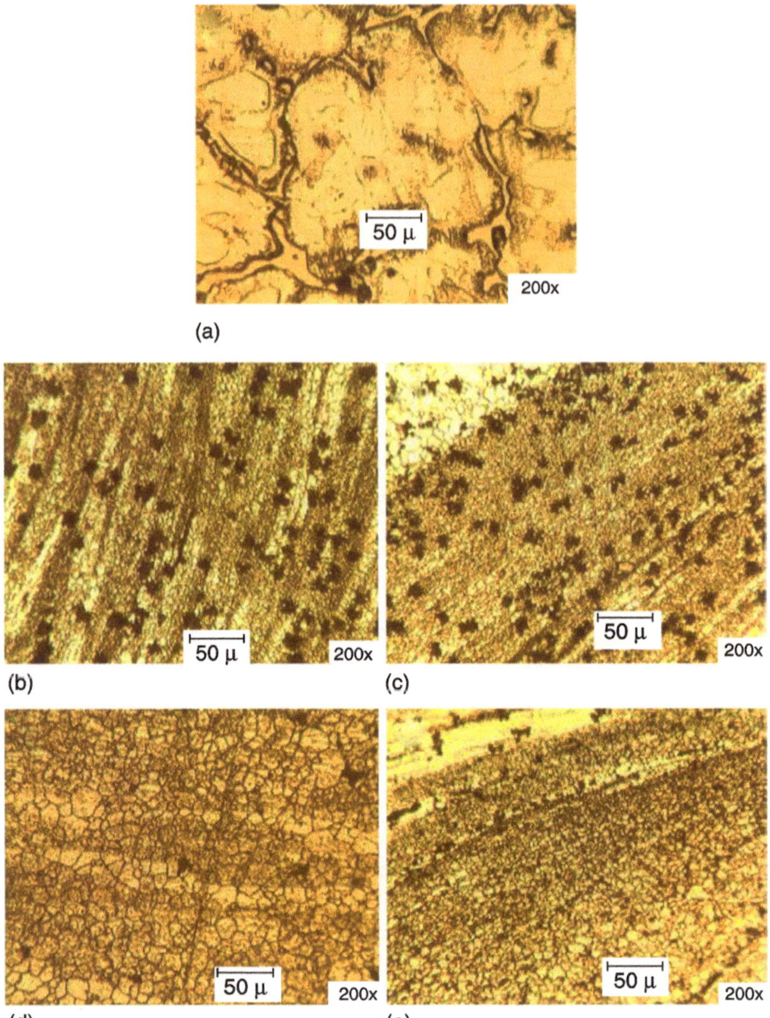

Figure 3.67 Effect of alumina particle size and tool geometry on grain size and cluster size: (a) base metal; (b–d) FSPed specimen with 0.03, 0.3 and 3 μm alumina particle size by square tool, respectively; and (e) FSPed specimen with 0.3 μm alumina particle size by triangular tool [64]

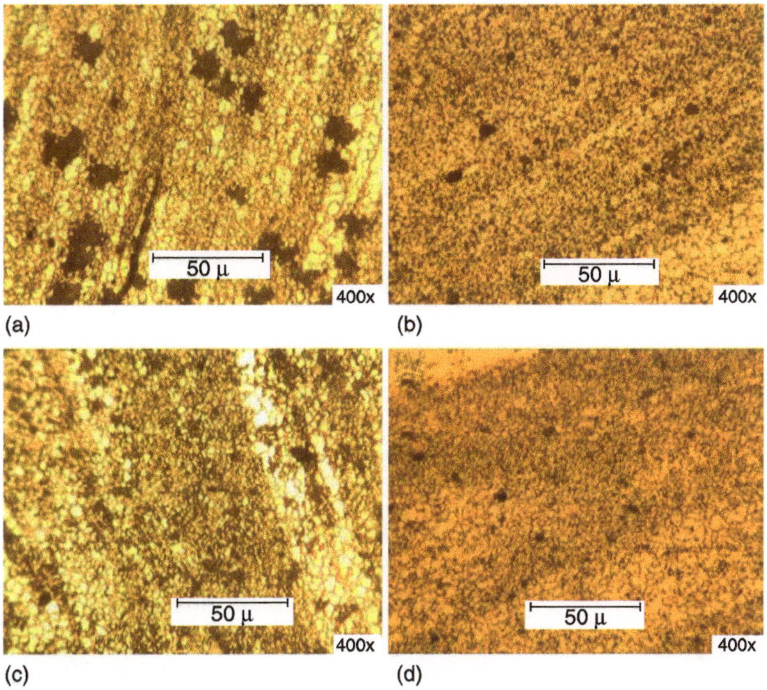

Figure 3.68 Optical micrograph in rotational and traverse speeds of 900 rpm and 63 mm/min, in which the particle size was 0.3 μm: (a) square tool, single passes; (b) square tool, three pass; (c) triangular tool, single pass; and (d) triangular tool, three passes [64]

elongated fine structure. However, the general grain shape seems to remain equiaxed on the scale of OM. Thus, it appears that recrystallization happens during FSW but the recrystallized grains undergo a small amount of plastic deformation. Moreover, TMAZ has not been obviously defined typical in aluminium FSW. The following sections provide illustrative examples for FSW of titanium alloy [71–74].

3.8.1 Illustrative example

Ti-6Al-4V alloys

In this section, the data is presented for FSW of a Ti-6Al-4V plate welded at a rotation speed of 500 rpm and a traverse speed of 60 mm/min. The macrostructure of the welded joint is illustrated in Figure 3.69(a). The SEM micrograph of the NZ is shown in Figure 3.69(b). According to the SEM macrograph, it is obvious that the microstructure consists of bcc β-phase (lighter phase) and hcp α-phase (darker phase). This structure shows that during FSW, the temperature in the NZ goes above the β transformation temperature. Pole figures of α and β phases demonstrate that the <0001> and <11$\bar{2}$0> directions of the α phase are almost parallel to the <110> and <111> directions of the β phase, respectively. As a result, the crystallography of α and β phases is related to the Burgers OR [71].

The Burgers OR between α and β phases indicate that it is possible to reconstruct the parent β-grain structure existing before the β to α phase transformation. The EBSD map of the reconstructed β grains on the RS of the SZ is shown in Figure 3.70(a), in which the selected regions are illustrated at higher magnifications in Figures 3.7(b–d). The SZ border on the RS (the broken line in Figure 3.70(a)) reveals that the reconstructed β grains tend to be aligned roughly parallel to this dotted line. According to the Figure 3.70(b), some of the swells create fine grains. Also, the EBSD maps in Figures 3.70(c and d) reveal that some β grains are extremely elongated, which include subgrain boundaries. These sub-

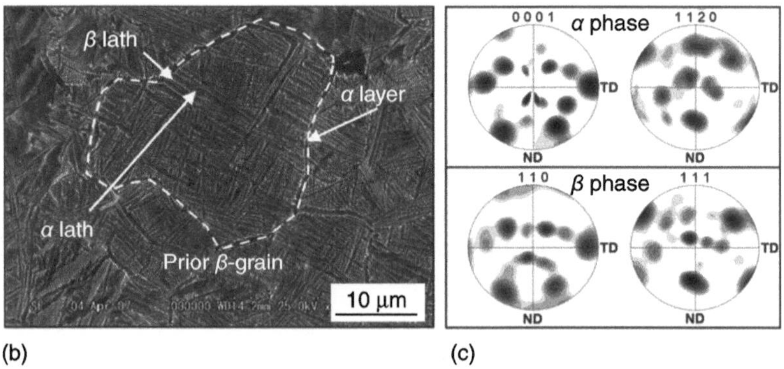

Figure 3.69 (a) Low magnification outline of the SZ; (b) SEM figure showing characteristic structure morphology in the SZ; and (c) pole figures obtained from the α and β phases [71]

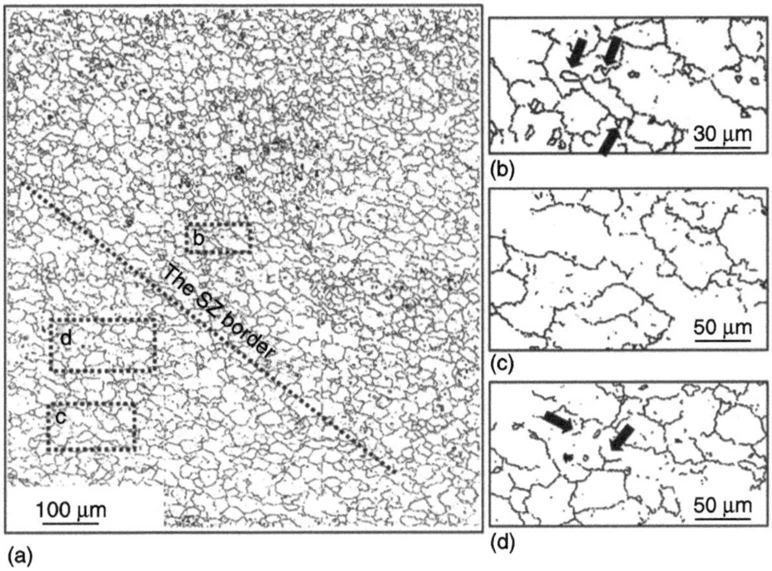

Figure 3.70 Grain boundary graphs of the reconstructed β grain structure on the RS of the SZ; and (b–d) magnified maps of the selected areas shown by the dotted rectangles in (a) [71]

boundaries divide the elongated grains into several almost equiaxed grains [71].

3.9 References

1. Mishra, R.S. and Ma, Z.Y. (2005), 'Friction stir welding and processing', *Materials Science and Engineering R*, 50(1–2): 1–78.
2. Farrokhi, H., Heidarzadeh, A. and Saeid, T. (2013), 'Friction stir welding of copper under different welding parameters and media', *Science and Technology of Welding and Joining*, 18(8): 697–702.
3. Wooa, W., Chooa, H., Prime, M.B., Feng, Z. and Clausen, B. (2008), 'Microstructure, texture and residual stress in a friction stir processed AZ31B magnesium alloy', *Acta Materialia*, 56(8): 1701–11.
4. Humphreys, F.J. and Hatherly, M. (2004), *Recrystallization and Related Annealing Phenomena*, 2nd edition, UK: Pergamon.
5. Mironov, S., Sato, Y.S. and Kokawa, H. (2008), 'Microstructural evolution during friction stir processing of pure iron', *Acta Materialia*, 56(11): 2602–14.

6. Thomas, W.M., Threadgill, P.L. and Nicholas, E.D. (1999), 'Feasibility of friction stir welding steel', *Science and Technology of Welding and Joining*, 4(6): 365–72.
7. Leinert, T.J., Stellwag, W.L., Grimmett, B.B. and Warke, R.W. (2003), 'Friction stir welding studies on mild steel', *Welding Journal*, 43: 1–9, available from: *http://www.aws.org/wj/supplement/01-2003-LIENERT-s.pdf*
8. Reynolds, A.P., Tang, W., Gnaupel, T. and Prask, H. (2003), 'Structure, properties, and residual stress of 304L stainless steel friction stir welds', *Scripta Materialia*, 48(9): 1289–94.
9. Park, S.H.C., Sato, Y.S., Kokawa, H., Okamoto, K., Hirano, S. and Inagaki, M. (2003), 'Rapid formation of the sigma phase in 304 stainless steel during friction stir welding', *Scripta Materialia*, 49(12): 1175–80.
10. Sterling, C.J., Nelson, T.W., Sorensen, C.D., Steel, R.J. and Packer, S.M. (2003), 'Friction stir welding of quenched and tempered C-Mn steel', *The Friction stir Welding and Processing II, TMS Annual Meeting*, San Diego, CA.
11. Saeid, T., Abdollah-Zadeh, A., Assadi, H. and Malek Ghaini, F. (2008), 'Effect of friction stir welding speed on the microstructure and mechanical properties of a duplex stainless steel', *Materials Science and Engineering A*, 496(2): 262–8.
12. Chung, Y.D., Fujii, H., Ueji, R. and Tsuji, N. (2010), 'Friction stir welding of high carbon steel with excellent toughness and ductility', *Scripta Materialia*, 63(2): 223–6.
13. Mironov, S., Sato, Y.S., Kokawa, H., Inoue, H. and Tsuge, S. (2011), 'Structural response of super-austenitic stainless steel to friction stir welding', *Acta Materialia*, 59(14): 5472–81.
14. Jafarzadegan, M., Feng, A.H., Abdollah-Zadeh, A., Saeid, T., Shen, J. and Assadi, H. (2012), 'Microstructural characterization in dissimilar friction stir welding between 304 stainless steel and st37 steel', *Material Characterization*, 74: 28–41.
15. Khodir, S.A., Morisada, Y., Ueji, R. and Fujii, H. (2012), 'Microstructures and mechanical properties evolution during friction stir welding of SK4 high carbon steel alloy', *Materials Science & Engineering A*, 558: 572–8.
16. Saeid, T., Abdollah-Zadeh, A., Shibayanagi, T., Ikeuchi, K. and Assadi, H. (2010), 'On the formation of grain structure during friction stir welding of duplex stainless steel', *Materials Science and Engineering A*, 527(24–25): 6484–8.
17. Miura, H., Sakai, T., Hamji, H. and Jonas, J.J. (2004), 'Preferential nucleation of dynamic recrystallization at triple junctions', *Scripta Materialia*, 50(1): 65–9.
18. Field, D.P., Nelson, T.W. and Hovanski, Y. (2001), 'Heterogeneity of crystallographic texture in friction stir welds of aluminium', *Metallurgical and Materials Transactions A*, 32(11): 2869–77.
19. Mironov, S., Sato, Y.S. and Kokawa, H. (2008), 'Microstructural evolution during friction stir-processing of pure iron', *Acta Materialia*, 56(11): 2602–14.
20. Fonda, R.W., Bingert, J.F. and Colligan, K.J. (2004), 'Development of grain structure during friction stir welding', *Scripta Materialia*, 51(3): 243–8.

21. Gourdet, S. and Montheillet, F. (2003), 'A model of continuous dynamic recrystallization', *Acta Materialia*, 51(9): 2685–99.
22. Barnett, M.R. and Montheillet, F. (2002), 'The generation of new high-angle boundaries in aluminium during hot torsion', *Acta Materialia*, 50(9): 2285–96.
23. Mohandas, T., Reddy, G.M. and Naveed, M. (1999), 'A comparative evaluation of gas tungsten and shielded metal arc welds of a "ferritic" stainless steel', *Journal of Materials Processing Technology*, 94(2–3): 133–40.
24. Folkhard, E. (1998), *Welding Metallurgy of Stainless Steels*, New York: Springer-Verlag Wien.
25. Villafuerte, J.C. and Kerr, H.W. (1990), 'Grain structures', *Metallurgical and Materials Transactions A*, 21(3): 979–86.
26. Cho, H.H., Han, H.N., Hong, S.T., J.H., Kwon, Y.J., Kim, S.H. and Steel, R.J. (2011), 'Microstructural analysis of friction stir welded ferritic stainless steel', *Materials Science and Engineering A*, 528(5): 2889–94.
27. Reynolds, A.P., Tang, W., Gnaupel-Herold, T. and Prask, H. (2003), 'Structure, properties and residual stress of 304L stainless steel friction stir welds', *Scripta Materialia*, 48(9): 1289–94.
28. Sato, Y.S., Nelson, T.W. and Sterling, C.J. (2005), 'Recrystallization in type 304L stainless steel during friction stirring', *Acta Materialia*, 53(3): 637–45.
29. Sato, Y.S., Nelson, T.W., Sterling, C.J, Steel, R.J. and Pettersson, C.-O. (2005), 'Microstructure and mechanical properties of friction stir welded SAF 2507 super duplex stainless steel', *Materials Science and Engineering A*, 397(1–2): 376–84.
30. Fujii, H., Cui, L., Tsuji, N., Maeda, M., Nakata, K. and Nogi, K. (2006), 'Friction stir welding of carbon steels', *Materials Science and Engineering A*, 429(1–2): 50–7.
31. Sato, Y.S., Yamanoi, H., Kokawa, H. and Furuhara, T. (2007), 'Microstructural evolution of ultra-high carbon steel during friction stir welding', *Scripta Materialia*, 57(6): 557–60.
32. Schramm, R.E. and Reed, R.P. (1975), 'Stacking fault energies of seven commercial austenitic stainless steels', *Metallurgical and Materials Transactions A*, 6(7): 1345–51.
33. Kang, S.-H., Bang, W.H., Cho, J.-H., Han, H.N., Oh, K.H. et al. (2005), 'Microtexture analysis of friction stir welded Al 6061-T651 plates', *Materials Science Forum*, 495: 901–6.
34. Kang, S.-H., Han, H.N., Oh, K.H., Cho, J.-H., Lee, C.G. and Kim, S.-J. (2009), 'Investigation of the material flow and texture evolution in friction stir welded aluminum alloy', *Metals and Materials International*, 15(6): 1027–31.
35. Khodaverdizadeh, H., Mahmoudi, A., Heidarzadeh, A. and Nazari, E. (2012), 'Effect of friction stir welding (FSW) parameters on strain hardening behaviour of pure copper joints', *Materials and Design*, 35: 330–4.
36. Khodaverdizadeh, H., Heidarzadeh, A. and Saeid, T. (2013), 'Effect of tool pin profile on microstructure and mechanical properties of friction stir welded pure copper joints', *Materials and Design*, 45: 265–70.

37. Heidarzadeh, A., Saeid, T., Khodaverdizadeh, H., Mahmoudi, A. and Nazari, E. (2013), 'Establishing a mathematical model to predict the tensile strength of friction stir welded pure copper joints', *Metallurgical and Materials Transaction B*, 44(1): 175–83.
38. Heidarzadeh, A. and Saeid, T. (2013), 'Prediction of mechanical properties in friction stir welds of pure copper', *Materials and Design*, 52: 1077–87.
39. Shen, J.J., Liu, H.J. and Cui, F. (2010), 'Effect of welding speed on microstructure and mechanical properties of friction stir welded copper', *Materials and Design*, 31(8): 3937–42.
40. Xie, G.M., Ma, Z.Y. and Geng, L. (2007), 'Development of a fine-grained microstructure and the properties of a nugget zone in friction stir welded pure copper', *Scripta Materialia*, 57(2): 73–6.
41. Surekha, K. and Els-Botes, A. (2011), 'Development of high strength, high conductivity copper by friction stir processing', *Materials and Design*, 32(2): 911–16.
42. Sakthivel, T. and Mukhopadhyay, J. (2007), 'Microstructure and mechanical properties of friction stir welded copper', *Journal of Material Science*, 42(19): 8126–9.
43. Park, H.S., Kimura, T., Murakami, T., Nagano, Y., Nakata, K. and Ushio, M. (2004), 'Microstructures and mechanical properties of friction stir welds of 60% Cu–40% Zn copper alloy', *Materials Science and Engineering A*, 371(1–2): 160–9.
44. Meran, C. (2006), 'The joint properties of brass plates by friction stir welding', *Materials and Design*, 27(9): 719–26.
45. Sarvghad Moghaddam, M., Parvizi, R., Haddad-Sabzevar, M. and Davoodi, A. (2011), 'Microstructural and mechanical properties of friction stir welded Cu-30Zn brass alloy at various feed speeds: Influence of stir bands', *Materials and Design*, 32(5): 2749–55.
46. Oh-ishi, K. and McNelley, T.R. (2004), 'Microstructural modification of as-cast NiAl bronze by friction stir processing', *Metallurgical and Materials Transactions A*, 35(9): 2951–61.
47. Oh-ishi, K., Cuevas, A.M., Swisher, D.L and McNelley, T.R. (2003), 'The influence of friction stir processing on microstructure and properties of a cast nickel aluminum bronze material', *Materials Science Forum*, 426–32: 2885–90.
48. Oh-ishi, K. and McNelley, T.R. (2005), 'The influence of friction stir processing parameters on microstructure of as-cast NiAl bronze', *Metallurgical and Materials Transactions A*, 36(6): 1575–85.
49. Oh-ishi, K., Zhilyaev, A.P. and McNelley, T.R. (2006), 'A microtexture investigation of recrystallization during friction stir processing of as-cast NiAl bronze', *Metallurgical and Materials Transactions A*, 37(7): 2239–51.
50. McNelley, T.R., Oh-ishi, K. and Zhilyaev, A.P. (2007), 'Microstructure evolution and microstructure-property relationships in friction stir processing of NiAl bronze', *Materials Science Forum*, 539–43: 3745–50.
51. Su, J.Q., Nelson, T.W., McNelley, T.R. and Mishra, R.S. (2011), 'Development of nanocrystalline structure in Cu during friction stir processing (FSP)', *Materials Science and Engineering A*, 528(16–17): 5458–64.

52. Sun, Y.F. and Fujii, H. (2010), 'Investigation of the welding parameter dependent microstructure and mechanical properties of friction stir welded pure copper', *Materials Science and Engineering A*, 527(26): 6879–86.
53. Etter, A.L., Baudin, T., Fredj, N. and Penelle, R. (2007), 'Recrystallization mechanisms in 5251 H14 and 5251 O aluminum friction stir welds', *Materials Science and Engineering A*, 445–6: 94–9.
54. Sato, Y.S., Urata, M. and Wa, H. (2002), 'Parameters controlling microstructure and hardness during friction stir welding of precipitation-hardenable aluminum alloy 6063', *Metallurgical and Materials Transaction A*, 33A(3): 625–35.
55. Su, J.Q., Nelson, T.W., Mishra, R. and Mahoney, M. (2003), 'Microstructural investigation of friction stir welded 7050-T651 aluminium', *Acta Materialia*, 51(3): 713–29.
56. Park, S.H.C., Sato, Y.S. and Kokawa, H. (2003), 'Basal plane texture and flow pattern in friction stir weld of a magnesium alloy', *Metallurgical and Materials Transaction A*, 34(4): 987–94.
57. Mansoor, B. and Ghosh, A.K. (2012), 'Microstructure and tensile behaviour of a friction stir processed magnesium alloy', *Acta Materialia*, 60(13–14): 5079–88.
58. Esparza, J.A., Davis, W.C., Trillo, E.A. and Murr, L.E. (2002), 'Friction stir welding of magnesium alloy AZ31B', *Journal of Materials Science Letters*, 21(12): 917–20.
59. Padmanaban, G. and Balasubramanian, V. (2010), 'An experimental investigation on friction stir welding of AZ31B magnesium alloy', *International Journal of Advanced Manufacturing Technology*, 49(1–4): 111–21.
60. Suhuddin, U.F.H.R., Mironov, S., Sato, Y.S., Kokawa, H. and Lee, C.W. (2009), 'Grain structure evolution during friction stir welding of AZ31 magnesium alloy', *Acta Materialia*, 57(18): 5406–18.
61. Motalleb-nejad, P., Saeid, T., Heidarzadeh, A., Darzi, Kh. and Ashjari, M. (2014), 'Effect of tool pin profile on microstructure and mechanical properties of friction stir welded AZ31B magnesium alloy', *Materials and Design*, 59: 221–6.
62. Asadi, P., Faraji, G. and Besharati, M.K. (2010a), 'Producing of AZ91/SiC composite by friction stir processing (FSP)', *The International Journal of Advanced Manufacturing Technology*, 51: 247–60.
63. Asadi, P., Besharati-Givi, M.K. and Faraji, G. (2010b), 'Producing ultrafine-grained AZ91 from as-cast AZ91 by FSP', *Materials and Manufacturing Processes*, 25(11): 1219–26.
64. Faraji, G., Dastani, O. and Akbari Mousavi, S.A.A. (2010), 'Effect of process parameters on microstructure and micro-hardness of AZ91/Al_2O_3 surface composite produced by FSP', *Journal of Materials Engineering and Performance*, 20(9): 1583–90.
65. Asadi, P., Faraji, G., Masoudi, A. and Besharati Givi, M.K. (2011), *Metallurgical and Materials Transactions A*, 42(9): 2820–32.
66. Tutunchilara, S., Besharati Givib, M.K., Haghpanahia, M. and Asadi, P. (2012), 'Eutectic Al–Si piston alloy surface transformed to modified hypereutectic alloy via FSP', *Materials Science and Engineering A*, 534: 557–67.

67. Lee, C.J., Huang, J.C. and Hsieh, P.J. (2006), 'Mg-based nano-composites fabricated by friction stir processing', *Scripta Materialia*, 54(7): 1415–20.
68. Faraji, G. and Asadi, P. (2011), 'Characterization of AZ91/alumina nanocomposite produced by FSP', *Materials Science and Engineering A*, 528(6): 2431–40.
69. Asadi, P., Besharati-Givi, M.K., Abrinia, K., Taherishargh, M. and Salekrostam, R. (2011), 'Effects of SiC particle size and process parameters on the microstructure and hardness of AZ91/SiC composite layer fabricated by FSP', *Journal of Materials Engineering and Performance*, 20(9): 1554–62.
70. Asadi, P. et al. (2012), 'On the role of cooling and tool rotational direction on microstructure and mechanical properties of friction stir processed AZ91', *International Journal of Advanced Manufacturing Technology*, 63: 987–97.
71. Mironov, S., Zhang, Y., Sato, Y.S. and Kokawa, H. (2008), 'Development of grain structure in β-phase field during friction stir welding of Ti-6Al-4V alloy', *Scripta Materialia*, 59(1): 27–30.
72. Mironov, S., Sato, Y.S. and Kokawa, H. (2009), 'Development of grain structure during friction stir welding of pure titanium', *Acta Materialia*, 57(15): 4519–28.
73. Liu, H. and Zhou, L. (2010), 'Microstructural zones and tensile characteristics of friction stir welded joint of TC4 titanium alloy', *Transactions of Nonferrous Metals Society of China*, 20(10): 1873–8.
74. Sanders, D.G., Ramulu, M., Klock-McCook, E.J., Edwards, P.D., Reynolds, A.P and Trapp, T. (2008), 'Characterization of superplastically formed friction stir weld in titanium 6AL-4V: Preliminary results', *Journal of Materials Engineering and Performance*, 17(2): 187–92.

4

Mechanical properties

D.F.O. Braga, A.C.F. da Silva and P.M.G.P. Moreira, University of Porto, Portugal

DOI: 10.1533/9780857094551.141

Abstract: Friction stir welding (FSW) and friction stir processing (FSP) are two solid state processes that involve heating by friction and plastic deformation. As a result of these inputs, mechanical properties differ from those of base materials. For application of these techniques, it is necessary to know the hardness, static, fracture and fatigue properties associated with materials welded or treated. This chapter summarizes studies on each technique for several alloys. Process parameters have a high influence on mechanical properties achieved and residual stress and distortion are also of high importance. Due to their nature, residual stresses are unavoidable and play a fundamental role on properties such as fatigue behaviour. Residual stress and its associated distortion in FSW and FSP are discussed. These techniques are applicable to a high number of different alloys and applications of multi-material joining applications. Combined with the effect of process parameters, it is virtually impossible to study the mechanical properties for each case, therefore property prediction is highly important. A summary of studies regarding this issue is also presented.

Key words: hardness, ultimate tensile strength, elongation, property prediction, residual stress, fatigue, S-N curves.

4.1 Introduction

Friction stir welded (FSW) joints improved mechanical behaviour makes this technique prone to a wide range of industrial applications, where weight and cost savings are a prime design priority. As an example, the aeronautic and aerospace industries have pushed the development of this technology, targeting its use in fuselage, propulsion, wings, nose and other critical structures and components.

In order to achieve a technological level of confidence, which allows the widespread industrial application of FSW, mechanical characterization studies are required. For safety critical structures, the need to fulfil strict certification requirements, such as JAA-FAA (Joint Aviation Authorities – Federal aviation administration) [1] for aeronautic structures and ESA-ES TEC (European Space Agency – European Space Research and Technology Centre) [2] for space systems, there is a need for extensive characterization of fracture and fatigue properties along with quasi-static mechanical properties.

Due to the complexity of the mechanisms involved in these processes, not only different materials will have different properties when friction stir welded (FSW) or friction stir processed (FSP), but different parameters will also induce changes to the mechanical properties of joints. Therefore, this chapter will attempt to summarize mechanical properties of FSW joints and materials modified by FSP. The corrosion theme will be discussed in further chapters.

As this process was first implemented in aluminium alloys, there has already been extensive work on the characterization of joints manufactured with these materials. With the development of this technique, characterization efforts were extended to other materials, such as copper, magnesium, steel and titanium.

4.2 Hardness

FSW/FSP involves complex material flow under intensive plastic deformation. This is dependent on several process parameters, which result in highly modified microstructures. Several studies correlate the resultant microstructures with hardness profiles and use the hardness profile data as an initial evaluation of variation in mechanical properties. Hardness profiles for different materials, including dissimilar materials and joint types, will be presented in this section, as well as comparison with results obtained using other welding techniques.

The material experiences different combinations of strain and temperature along the stir zone (SZ), due to the non-symmetrical material flow around the tool. So non-uniform plastic deformation or strain gradient in the SZ during the process is presented, and as a result a variation in the microstructure along the weld is found [3]. Besides the base material (BM) zone, FSW joints generally present three distinguishable zones. Close to the weld the material is subjected to temperature gradients that alter properties without inducing plastic deformation; this area is designated by the heat affected zone (HAZ). This zone presents changes in static strength, ductility, corrosion susceptibility, hardness and toughness. The thermo-mechanically affected zone (TMAZ) appears between the HAZ and the nugget. In this region sufficient temperature and mechanical load results in plastic deformation. The nugget is the central zone of the weld, representing material recrystallized due to strong plastic deformation [4]. Mechanical properties are usually related with the microstructure, in particular the hardness profiles across the joints. Hardness profiles have been taken as an indicator of changes in microstructure and respective mechanical aspects [4].

4.2.1 Aluminium welds

Aluminium alloys are classified into heat-treatable alloys (precipitate-strengthened) and non-heat-treatable (solid-solution-hardened), which result in different hardness profiles when FSW [5].

In aluminium alloys, the heating in HAZ is generally high enough for the recovery of cold work and coarsening of precipitates, which leads to changes in mechanical properties in this region. The TMAZ may present a significant size characterized by lower hardness and increased corrosion susceptibility. The nugget is generally composed of fine grain size and is considered to have severe plastic deformation, and in some cases can assume probe geometry [4].

Heat-treatable aluminium alloys

In precipitation-strengthened aluminium alloys, the hardness profile depends mostly on the dissolution and/or coarsening of strengthening precipitates, than on the grain size. During welding, the temperature that leads to dissolution and growth of the precipitates is achieved, which leads to a decrease of hardness in the weld zone [4,6].

A hardness reduction in the weld zone is common in FSWs of the artificially aged aluminium 6xxx series. In the HAZ, the temperature achieved during the welding process has a great impact in over-ageing and in decreasing dislocation density, consequently lowering the local hardness. Typical hardness profiles show that minimum hardness is found in the TMAZ. This is due to the loss of elongated formed grains occurring, as well as ageing and annealing processes. There is usually considerable evidence of partial recrystallization close to the interphase TMAZ nugget, which is very marked in the advancing side (AS), while the retreating side (RS) shows a smoother change in microstructure. The minimum hardness is commonly found in the RS [4,5,7,8].

An Al6082-T6 FSW hardness profile along the joint cross-section is presented in Figure 4.1. It can be observed that the weld average hardness is lower than the BM hardness, with 33% softening when compared to the parent material. The minimum hardness is identified in the TMAZ RS [8,9]. Through thickness hardness profiles of 3 mm thick specimens of Al6082-T6, Al6061-T6 and dissimilar Al6082-T6 + Al6061-T6 FS welds were evaluated in [9]. The dissimilar Al6082-T6 + Al6061-T6 joints show a pronounced variation of hardness at the specimen middle plane, due to the presence of both materials in this area. Results of this study are presented in Figures 4.2.

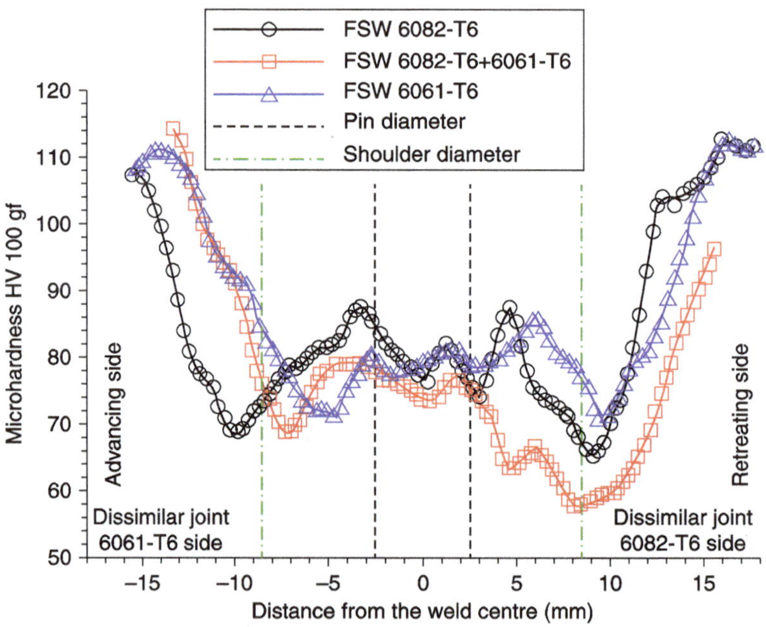

Figure 4.1 **Microhardness profile of FSW Al6061-T6 and dissimilar joints [9]**

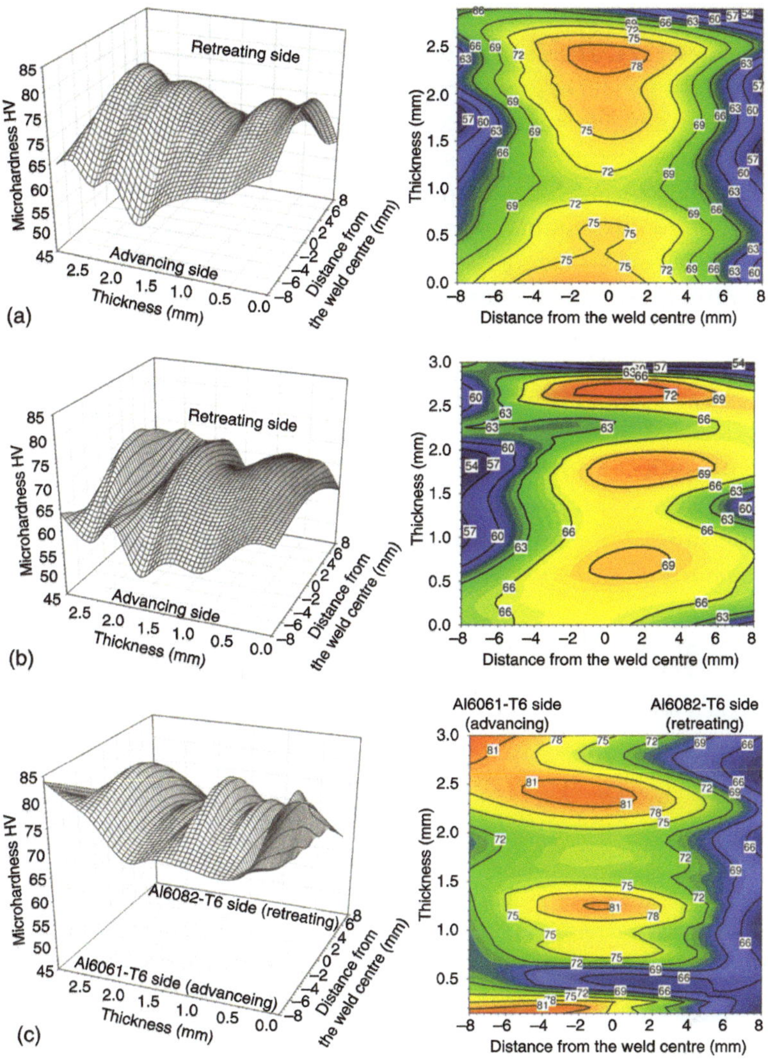

Figure 4.2 **Hardness profile through the FZW joint thickness: (a) Al6082-T6; (b) Al6061-T6; and (c) similar joint of Al6082-T6 with Al6061-T6 [9]**

In FSW 2xxx aluminium alloys, a softened region originates, located in the HAZ, where failure usually occurs. This type of alloy naturally ages at room temperature, which leads to a hardness increase and corresponding improvement in mechanical properties. The highest effect occurs in the first ageing week, stabilizing to an equilibrium condition [10].

The properties of Al2019 aluminium alloy along the thickness were investigated [11]. The zone close to the welded surface presented a larger equiaxed recrystallized grain structure and finer second-phase precipitates, and therefore higher hardness compared to the middle and bottom of the weld. This behaviour is due to the effect of the shoulder in the zone near to the welded surface. A softening of approximately 40% concerning BM was reported. In aluminium alloys, the shoulder influence is limited, typically less than 2 mm of depth, depending on the shoulder design and their capacity to generate heat [10,11].

Al2024 aluminium alloy welds obtained using gas tungsten arc welding (GTAW) and FSW were compared. GTAW resulted in lower hardness values than those obtained with FSW [10].

The 7xxx aluminium alloys present the commonly observed 'W' shape welding profile. An increase in hardness is verified with an increase of duration of natural ageing (Figure 4.3). Formation of a high volume fraction of smaller GP(II) zones produced the majority of the strength improvement [12].

Non-heat treatable aluminium alloys

A softened weld zone is not verified in solid-solution hardened (non-heat-treatable) aluminium alloys. For FSW Al1080-O, slightly higher hardness

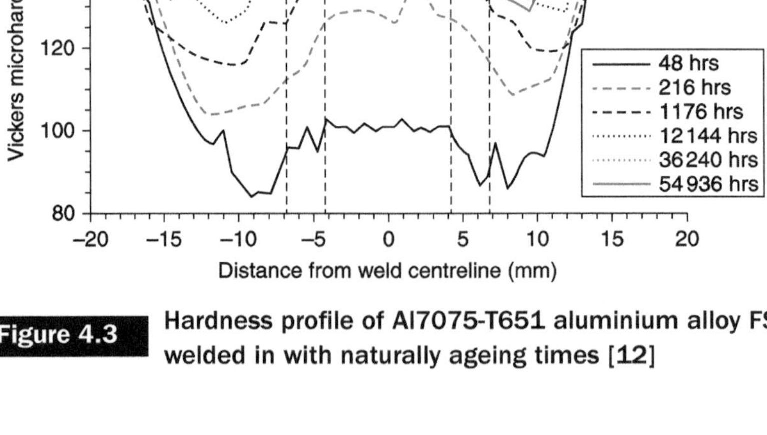

Figure 4.3 Hardness profile of Al7075-T651 aluminium alloy FS welded in with naturally ageing times [12]

values are found in the nugget zone (NZ) when compared with its parent material. This fact can be explained by the presence of recrystallized grains with a low density of dislocations [5]. TMAZ presents recovered grains with a sub-grain structure and thereby the maximum hardness was achieved in this zone [5]. The hardness profile in Al5083-O FSW presents no variation in hardness along the weld. The hardness in this type of alloy is dominantly governed by the dispersion strengthening, due to distribution of small particles presenting an inter-particle spacing much lower than the grain size [5,6,10].

Friction stir processing of aluminium alloys

Commercially pure aluminium was subjected to FSP, and the effects of secondary phase or precipitates did not emerge [3]. Grain size refinement, from 84 to 3 μm, was achieved in a single pass using 640 rpm, 150 mm/min and an axial force of 5 kN. The BM was found to have 29 HV and when subjected to FSP an increase of 34% was observed, due to grain refinement and dislocations interaction. The hardness peak was found on the AS, coinciding with the presence of a higher fraction of high-angle grain boundaries (HABs). This leads to the conclusion that a significant role in hardness may be played by the dislocation hindrance mechanism.

FSP enables the incorporation of SiC particles into aluminium alloys surface layers originating from composites aluminium alloy with a significant increase of hardness. An Al5083 aluminium alloy FS processed with 27% SiC particles in the surface layer has an increased hardness of 85 HV (substrate plate) to 173 HV [5].

A surface of as-cast eutectic LM13 Al–Si alloy was successfully transformed to a modified hypereutectic by multi-pass FSP and addition of 10% Si powder [13]. An increase in tensile strength, elongation and hardness was achieved due to the elimination of porosities and dendritic branches, and also break-up of needle-like eutectic silicon. The addition of 10% Si powder with only one pass of FSP increases the hardness when compared to the as-cast alloy. Nonetheless, the following FSP passes reduce the hardness, due to the dissolution of precipitates such as Mg_2Si.

4.2.2 Other alloys

FSW has been demonstrated as able to weld different alloys besides aluminium alloys, such as copper, magnesium, steel and titanium [4].

Copper alloy friction stir welds

FSW copper alloys show greater dissipation of heat through the workpiece, due to their higher thermal diffusivity, requiring a higher heat input during welding [6]. For this alloys, appropriate temperatures for a successful FSW joint were defined to be between 460 and 530 °C [14]. The NZ presents fine recrystallized grains, the TMAZ has deformed large grains, and the HAZ presents equiaxed grains larger than those of the base metal (BM) [5,6]. In the hardness profiles, a softening in the weld zone was noted due to lower density of dislocation compared to BM (Figure 4.4) [15].

In a FSW 60% Cu to 40 % Zn copper alloy joint, the hardness values were higher than those of the BM (1.4–1.7 times), due to refined grains in the SZ. An increase in the welding speed tends to increase the hardness [17]. Typical hardness profiles for these alloys are shown in Figure 4.5.

FSP was used to prepare copper base composites reinforced with 30 nm and 5 μm SiC particles on the surface of pure copper [18]. A sample of pure copper was FSP without reinforced particles, demonstrating a hardness decrease in comparison to BM. However, the addition of reinforced

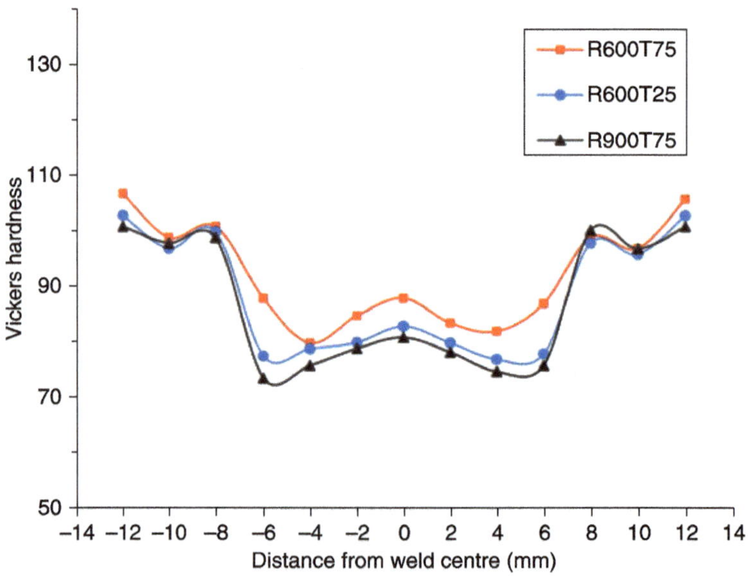

Figure 4.4 **Microhardness profiles of pure copper FSW butt-joint with different welding parameters [16]**

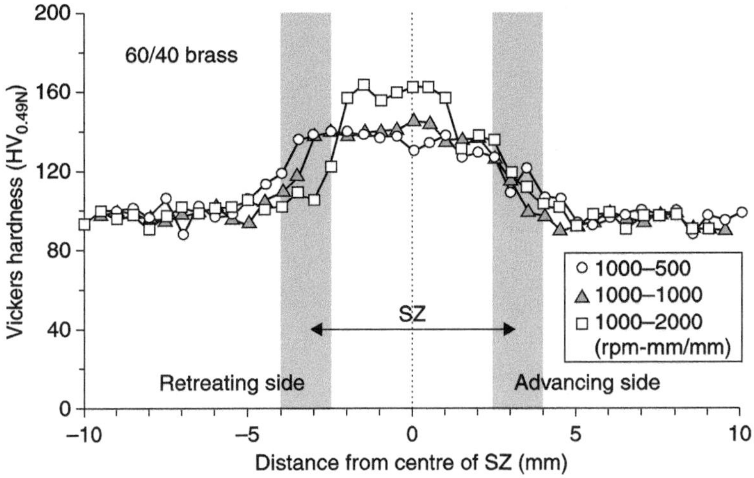

Figure 4.5 **Hardness profiles in cross-section of the 60% Cu–40% Zn alloy FSW welds as a function of welding parameters [17]**

particles increases significantly the hardness values of all composites. Composites hardness was showed to increase by increasing the volume fraction and decreasing the size of reinforcing particles. The hardness profiles of the FSP pure copper without reinforced particles and with different percentage and sizes of reinforced particles obtained in this study is presented in Figure 4.6. The increase in number of passes has a direct effect on the improvement of hardness, due to better stirring of the material and thereby reduction in SIC particles and copper matrix grain size [19].

An improved surface strength of 3 mm thick pure copper plate is FSP without affecting the electrical properties [20]. The nugget grain size achieved (3 μm) was significant smaller than the BM (19 μm). A decrease in grain size with increasing traverse speed at constant rotation speed (lower heat input – 300 rpm and 250 mm/min) was also observed. A hardness improvement, when compared to BM (85 HV), was verified following a decrease in grain size; also 114 HV was achieved in the nugget of the sample with 3 μm grain size.

Magnesium alloy FS welds

In FSW magnesium alloys joints, peak temperatures of 370 °C up to 500 °C may be achieved, which will likely result in the occurrence of

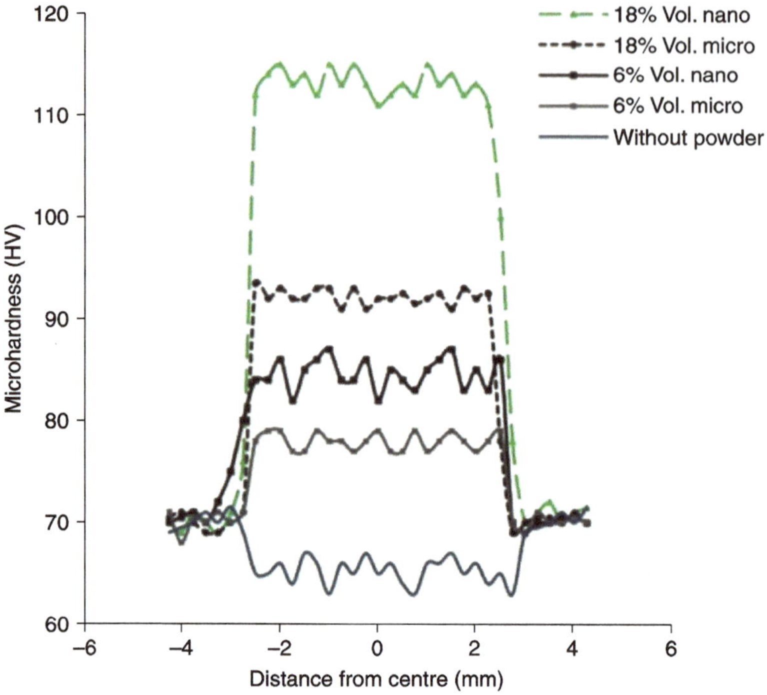

Figure 4.6 **Hardness profiles of specimens FS processed without powder, with micro- and nano-sized SiC particles. Volume fractions in the specimens with powder were 6 and 18% [18]**

liquid phases, and thereby generate a complex microstructure in the weld [5,21]. The hardening effect of the precipitates may not be modified during FSW, as the hardness evolution is related mainly to the grain size evolution observed [21]. The variations in hardness follow the Hall–Petch relationship (Figure 4.7) and tend to be less pronounced for a low heat input. Therefore, in general, FSW magnesium alloy joints present higher hardness than that of the BM due to a refined grain structure [5,6].

The welding parameters have a great influence on the heat generated. The increase of shoulder diameter, tool rotation speed, applied pressure or the decrease of welding speed produced an increase of heat input during welding and thereby promoted grain growth. The larger grain refinement, due to the lower nugget temperature achieved during welding, showed the best mechanical properties [21]. However, an increase in shoulder diameter from 18 to 24 mm does not exhibit any differences in hardness profiles of FSW of AZ31 magnesium alloy [22].

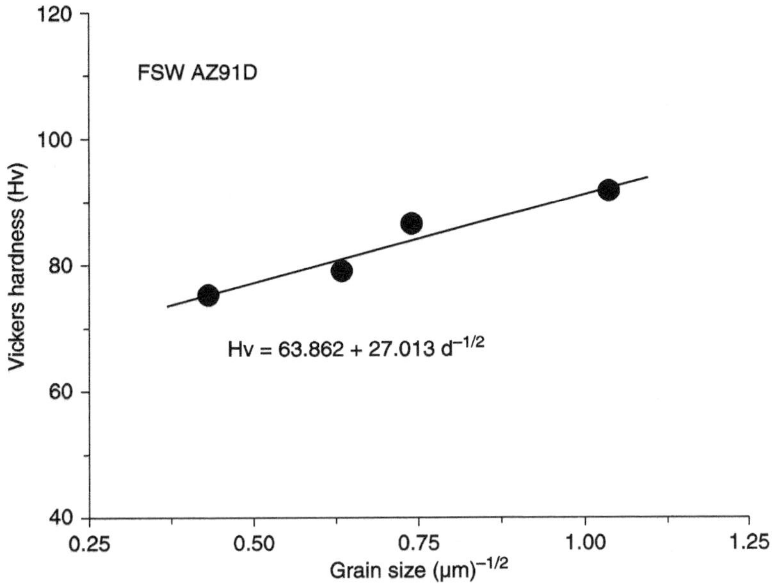

Figure 4.7 **Hardness variation with grain size in FS welded magnesium AZ91D [5]**

Regarding FSW AZ31B-H24 magnesium alloy, it was reported that hardness increases with the use of higher welding speeds [23,24]. Axial force has a significant effect on grain size and thereby on hardness. When welding 6 mm thick butt joints of AZ61 magnesium alloy, a hardness value of 84 HV was observed [25]. It was also found that using an axial force of 5 kN (with 1200 rpm and 90 mm/min) provides improved mechanical properties, due to the finer grains (12 μm) in the nugget.

Several studies report a decrease in hardness in AZ31 magnesium alloys joints made with the FSW technique [23,24,26]. In AZ31 alloy (Mg–3Al–1Zn wt.% wrought), the hardness variation in the joint was in the range of 45 to 65 HV [6].

Hardness measured in rolled Mg–Zn–Y–Zr alloy 6 mm thick FSW joints showed that the hardness values of the nugget were higher than that of the BM [27]. Also, the decrease of rotational speed provided fine recrystallized grain size and thereby improved hardness.

FSP was used to produce AZ91 Mg alloy/alumina particle reinforced nano-composite [28]. The use of a square tool, 900 rpm rotational speed and 40 mm/min traverse speed were found to be the optimum conditions for producing sound materials. The results show an improvement in hardness due to grain refinement and pinning effect of nano-alumina particles when compared to the BM. High rotational speed/welding speed (ω/v) ratio and FSP number of passes also increases the hardness values.

Steel friction stir welds

FSW steel joints are subjected to high temperature during welding (>1000 °C), supporting transformation, recrystallization and grain growth, and therefore presenting a more complex microstructural evolution than aluminium alloys [5,29]. Generally, the TMAZ is not evident; however, its existence was identified in FSW 304 and 316L stainless steel. Generally, the hardness profiles demonstrate that the hardness at the central zone is much higher than that of the BM [5].

A study on FSW AISI 409M ferritic stainless steel joints is presented in [30]. During FSW, coarse ferrite grains of parental material change to a very fine duplex structure of ferrite and martensite. This transformation provided 57% higher tensile strength and an increase in hardness from 170 HV (BM) to 320 to 382 HV in the NZ. An AISI 409M ferritic stainless steel FSW hardness profile and the corresponded macro-section are presented in Figure 4.8. The increase of hardness regarding the BM was verified.

The hardness profiles obtained for carbon steels FSW at different speeds are presented in Figure 4.9 [29]. For the ultra-low-carbon interstitial-free

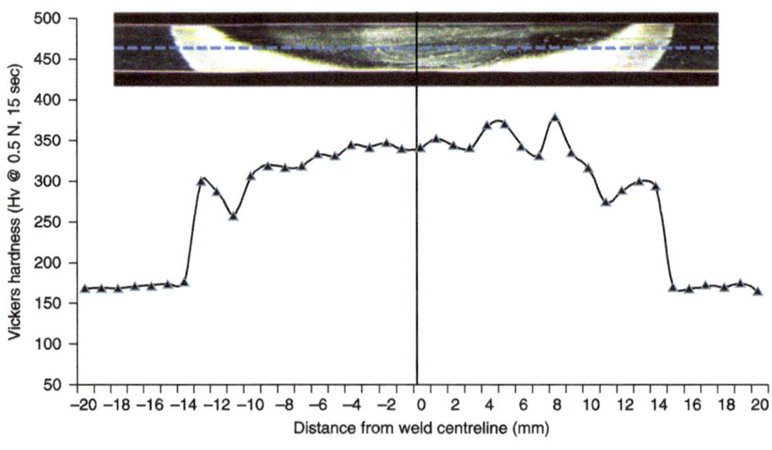

Figure 4.8 **Hardness profile of a FS welded AISI 409M ferritic stainless steel joint [30]**

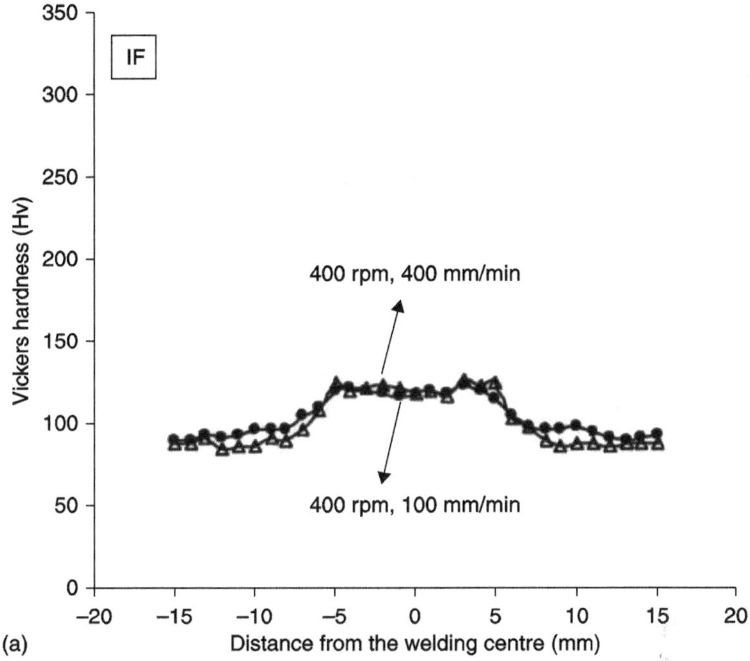

(a)

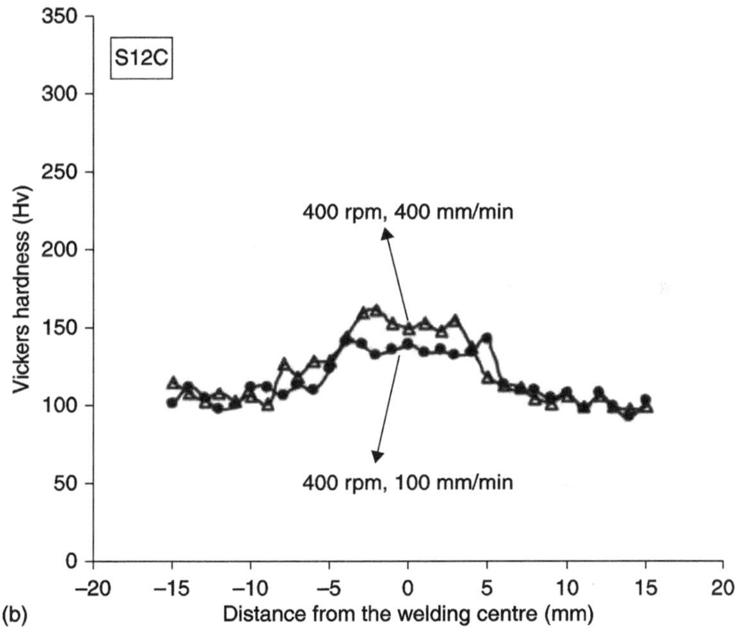

(b)

Figure 4.9 **Microhardness profile of different carbon steels friction stir welded: IF, S12C and S35C [29]**

(Continued)

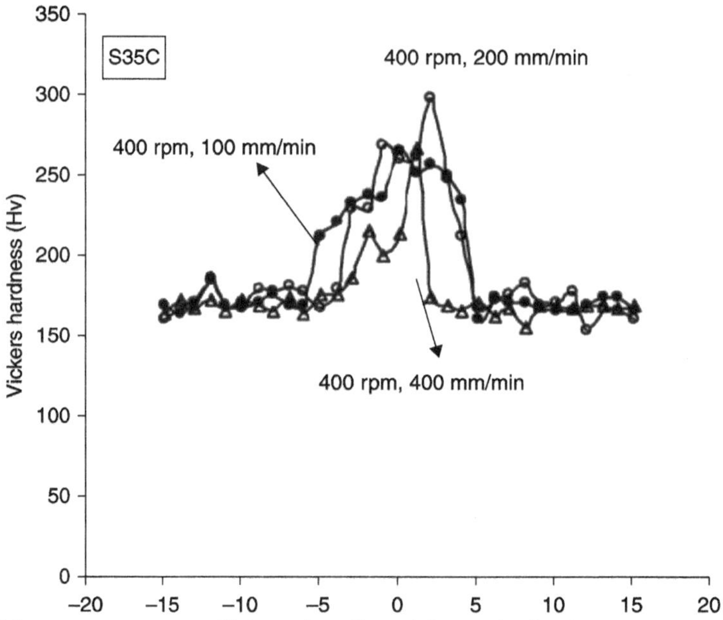

(c)

Figure 4.9 Continued

(IF) steel, the welding speed seems not to significantly affect the hardness profile, while the S12C steel (with 12wt.%C) FSW hardness profiles are affected by welding speed, i.e. they increase with an increasing welding speed. It was concluded that the optimum welding speed, for FSW S35C steel (with 0.34wt.%C), is approximately 200 mm/min.

FSW of martensitic mild steel causes a large reduction in hardness within the SZ. This is presumably due to the cooling rate not being sufficient to reproduce a fully martensitic microstructure. As such, final grain size in the SZ is always coarser and more isotropic than that in the ultrafine starting microstructure [6]. Results of this study are presented in Figure 4.10.

Titanium alloy friction stir welds

In titanium alloys, allotropic phase transformation together with deformation and continuous cooling, produces a complex weld microstructure compared to those observed in FSW aluminium alloys. Some authors describe a very narrow TMAZ of approximately 30 μm

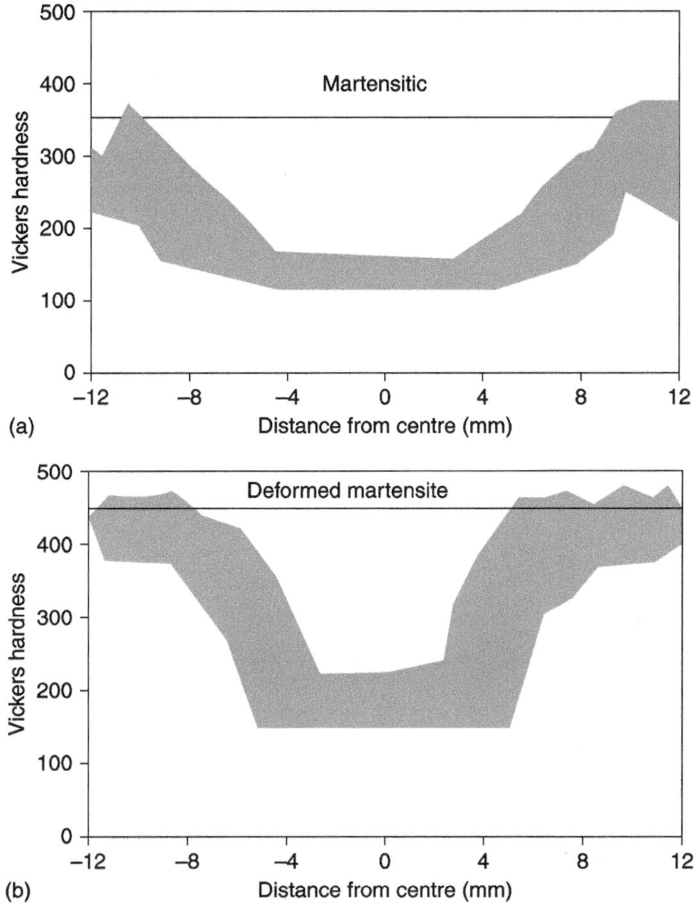

Figure 4.10 **Hardness profiles of FS welded martensitic mild steel: (a) welded in the as-quenched martensitic state; and (b) welded after rolling the martensite to a 35% cold reduction; the horizontal line indicates the starting hardness [6]**

and others their absence [5,6,31]. Ti-6Al-4V FSW joints were analysed [31] and identified as structures without TMAZ and no nuggets, having only SZs and HAZs. The microstructure, like the hardness data, is not clearly defined in the literature [5]. The hardness profiles across the Ti–6Al–4V FSW joints were different for various studies. The softening usually observed in the HAZ of titanium alloys fusion welds, due to growth of β grains, is not observed in Ti–6Al–4V FSW joints [5]. The hardness profile results for FSW Ti-6Al-4V welded [31] are presented in Figure 4.11. It can be observed that the FSW-affected zone achieved higher hardness than the BM and the highest value decreases with increasing tool rotational speed. This fact was also proved and reported elsewhere [32,33]. However, hardness profile measurements, shown in [34], revealed a scattered profile (Figure 4.12).

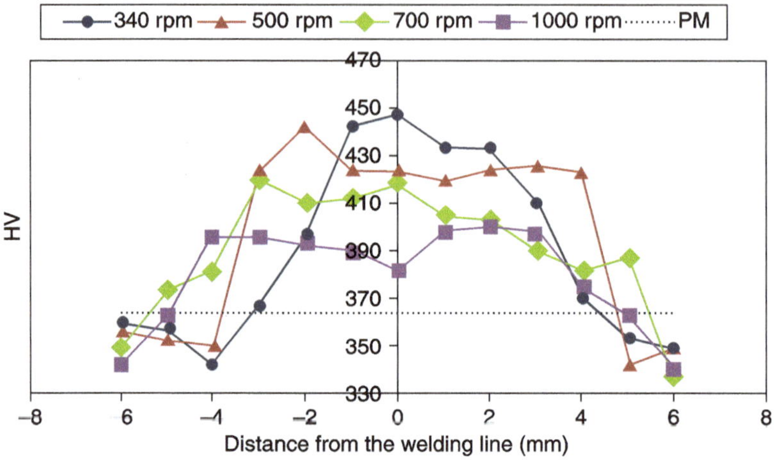

Figure 4.11 Ti-6Al-4V FSW microhardness profiles using a welding speed of 50 mm/min [31]

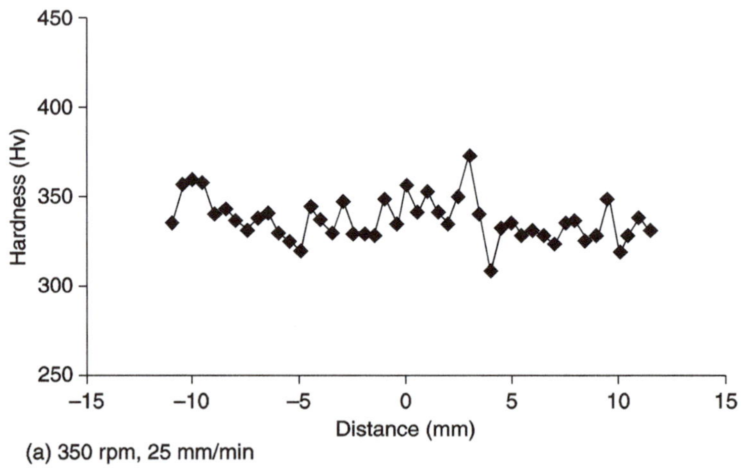

Figure 4.12 Ti-6Al-4V FSW microhardness profile (350 rpm, 25 mm/min) [34]

4.3 Tensile properties

The microstructural modifications induced by FSW/FSP, beyond modifying the hardness profile, are also reflected in different mechanical properties, such as strength and ductility. In this section, a brief review of material properties, which have endured FSW/FSP, will be made. A benchmark analysis of tensile properties of joints obtained with FSW and competing fusion welding techniques will also be discussed. Along with properties for butt-joints, a review including properties for other types of joint configuration, such as lap- and T-joints, will be made. Also, materials beyond aluminium alloys will be discussed in terms of common mechanical properties.

4.3.1 Aluminium alloys

Friction stir welding

Aluminium alloys FSW joints show a reduction in both strength and ductility compared to unwelded parent metal. It is also noted that the different zones have different resistances to deformation due to differences in grain size, precipitate size and distribution.

Heat-treatable aluminium alloys FSW efficiency ranges varies from 65 to 96%. Nonetheless, this efficiency is considerably higher than conventional fusion welds [5]. During tensile tests, the failure of particular age-hardened aluminium alloys tends to occur in the HAZ on the RS, where a lower hardness zone was observed [5,6].

Tensile properties of Al2024-T4 FSW improved with increasing rotational and advancing speeds, keeping a constant ratio of tool traverse speed/rotation rate (of 10) [5,10]. Also, higher strength and joining efficiency were observed in thinner plates than in thicker plates that were FSW [5].

Tensile properties for different zones along the thickness were obtained, by slicing specimens of the joint [11]. The joint zone close to the welded surface presented slightly higher strength and ductility than in the bottom or middle (Figure 4.13). Nevertheless, entire joint tensile properties are higher than in the different individual parts tested.

Comparison of the mechanical properties of Al6082-T6 and Al6061-T6 butt welded through FSW and MIG is shown in Figure 4.14 [8]. The FSW joints show an efficiency of 51 and 70% for yield strength (YS) and ultimate tensile strength (UTS), respectively. The tensile tests failure

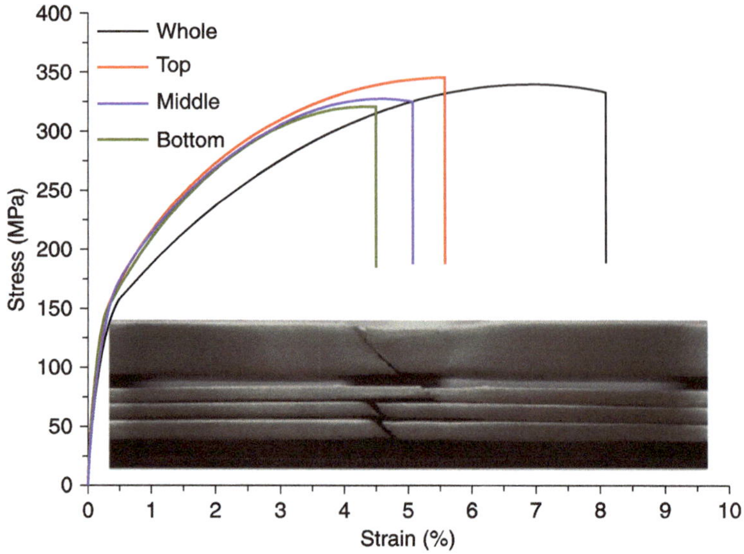

Figure 4.13 Al2019 aluminium alloy FSW weld slice specimen tensile curves [11]

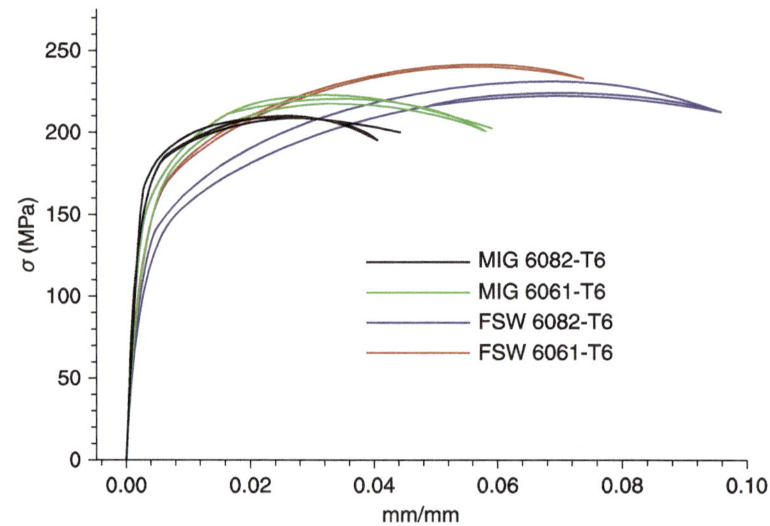

Figure 4.14 6082-T6 and 6061-T1 MIG, FSW and LBW welds tensile deformation curve [8]

occurred in the softened region, which happened to be the weakest point of the specimen [9].

Slightly higher tensile strengths and ductility than the BM were found for FSW Al6063-T5, when subjected to solution heat treatment and ageing (SHTA) 530 °C/1 h + 175 °C/12 h. The solution heat treatment produces a supersaturated solid solution and the subsequent ageing leads to the homogenous reprecipitation of precipitates. A strain homogeneous distribution through the weld is achieved, increasing the strength and leading to failure in the BM region [5]. It was observed that BM heat treatment has a significant effect on the joint efficiency of FSW joints. This is observed when comparing mechanical test results of FSW Al6013 under both conditions of T4 and T6, as it is possible to observe the difference in UTSs [5]. Some studies aimed at predicting tensile properties of FSW joints, which were attempted using analysis of variances (ANOVA) in 6xxx aluminium alloys [35,36].

In Al7075-T651 welded samples, the failures occurred as shear fractures in the HAZ. This is because this zone presents the lowest strength, which is due to significantly coarsened precipitates and the development of the precipitate-free zones (PFZs) [5]. An ageing treatment (121 °C/24 h) on longitudinal tensile specimens of the same alloy, containing only the NZ, results in a recovery of YS which is at the expense of UTS and ductility, the latter one in particular. That is attributed to the increase in the volume fraction of fine hardening precipitates and the development of PFZs at grain boundaries [5].

Natural ageing has a significant impact on the microhardness and tensile properties of 7XXX FSWs. Tensile strengths improvements of 24 and 29% measured for the 7050-T7651 and 7075-T651 FSW with the natural ageing time before welding were reported [12].

To better understand the intrinsic tensile properties and localized microstructure, some studies using Al7075 alloy FSW mini-tensile specimens were performed. Constant strengths in the nugget and nugget/TMAZ transition region were reported, with UTS values close to the BM values (~80% YS). Ductility was also significantly improved, due to the fine-grained microstructure. TMAZ retains the deformed structure, presenting a ductility decrease. In TMAZ/HAZ transition, UTSs start to drop until the lowest strength (60% of BM) is achieved in the HAZ [5]. The variation of properties throughout the FSW Al7075 alloy is presented in Figure 4.15.

Metal matrix composites of aluminium reinforced with 10 to 30 vol.% ceramic particles are effectively joined with a high quality through the FSW technique. The joints present a uniform ceramic particle distribution

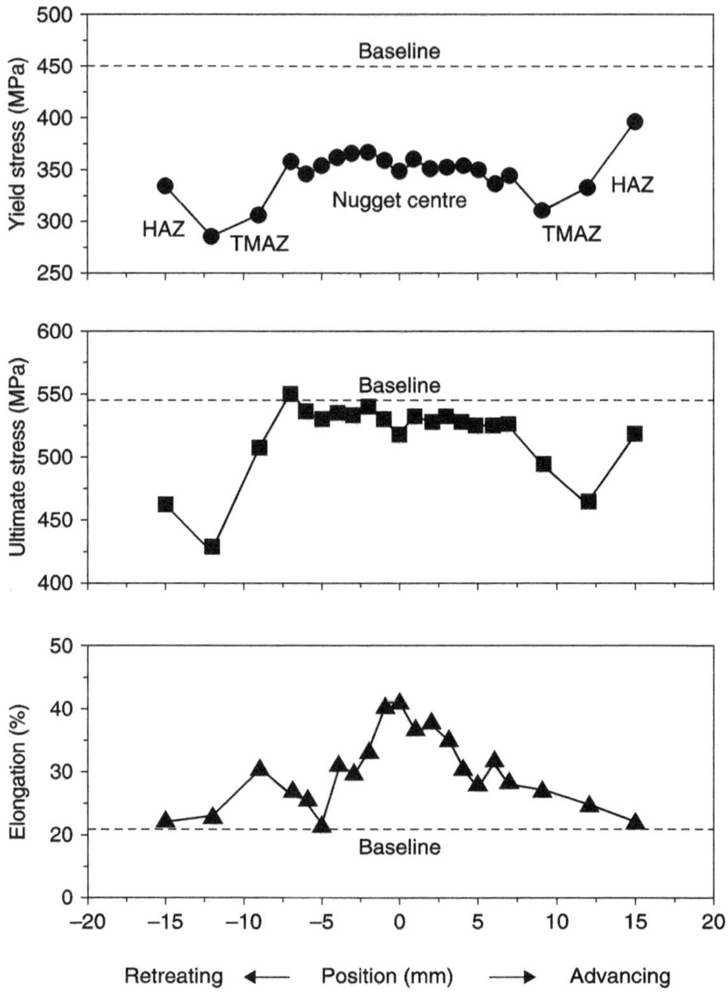

Figure 4.15 FSW Al7075 alloy tensile properties across the weld [5]

and do not show visible defects or evidence of any chemical reaction between reinforcements and matrix alloy. Comparing the mechanical properties of FSW with the tungsten inert gas (TIG) technique, FSW exhibited an improvement of UTS and YS, with similar or slightly improved ductility [5].

Although butt-joint configurations are more common in FSW, other joint configurations may be implemented as well. Studies on FSW lap-joints produced from Al2024-T3 and Al7075-T6, with several welding and rotational speeds, different tool designs, and both single- and double-pass welds may also be found in the literature [37]. Joint efficiencies of up to 86% were achieved, which may be comparable to FSW butt-joints. Double-pass welding shows consistently higher strength than single-pass welds, as the interface shape along with the reduction of effective sheet thickness are detrimental to the joint's mechanical properties. Double-pass welding allows selection of the loaded side of the weld, and as such the most critical interface shape may be unloaded. FSW lap-welding of Al2024-T3 with varying process parameters has also been performed [38]. It was concluded that tensile strength is improved when using multiple weld lines, if correct tuning of the process, static performance could be very close to the BM, and also that multiple passes over the same weld line do not improve the performance of the weld.

Currently, some published works may be found regarding T-joints welded through FSW. However, these studies mainly focus on aluminium alloys. Mechanical properties of Al6082-T6 and Al2024-T4 T-joints welded through FSW have been studied [39–42]. For Al6082-T6 FSW T-joint studies, it has been reported that the tilt angle plays a relevant role in static performance. A strong decrease of FSW T-joint performance was verified for high tilt angles, revealing the best performance at 3 degrees of tilt angle. In these works, several defects were verified, revealing the necessity of an appropriated choice of tools and set of process parameters [40,42]. Material flow in two-part Al2024-T4 FSW T-joints was also investigated [41]. A study on Al2024-T4 and Al6082-T6 FSW T-joints plasto-mechanics was performed by [39], applying the same process parameters, and varying only the rotational and welding speed.

It was noted that the two aluminium alloys present different behaviours in the plastic field. For the Al6082-T6 it was shown to acquire effective joints, whereas the Al2024-T4 presented poor results. Also, it was stated that the highest revolutionary pitch values (0.30 mm/r in this case) results in high feed, but with low heat generated by friction providing high strain values and possibly the development of defects, such as tunnels. However, the lowest revolutionary pitch values (0.07 mm/r in this case) result in high heating and low feed, resulting in high strain values and development of defects. This same work reported that the geometric tool characteristic, such as the shoulder diameter and the tool probe geometry, significantly affects the FSW process. This may be due to the fact that the shoulder surface provides most of the thermal flux through friction and the probe geometry affects the local material flux during the FSW process. The use of large shoulder diameters was claimed to provide the best welds [43,44], since these easily achieve the high heat input necessary to plasticize the material, thus allowing fulfilment of the joint fillets [43]. Al6013-T4 FSW T-joints were studied [45] and it was found that a rounded fillet radius may increase the propensity to crack formation when the joint is subjected to tensile stress.

It is important to take into account that the parameter effects referred to may not be applied to other applications, due to the effect of others parameters and their interactions.

Friction stir processing

A significant improvement in strength was verified due to refinement grain size, being in accordance with the Hall–Petch relationship, when studying a commercially pure aluminium subjected to FSP [3]. The major contribution comes from the finer grain size, but high dislocation density and sub-grain boundaries also contribute to improvement in strength as they act as obstacles for dislocation motion. However, a small loss in ductility was verified. Tensile properties are presented in Tables 4.1 and 4.2.

Cast alloys, such as Al–7 wt.% Si–Mg alloys, have mechanical properties, in particular ductility, toughness and fatigue resistance limited by porosity, coarse acicular Si particles, and coarse primary aluminium dendrites. To improve these, it is necessary to refine the microstructure and eliminate the porosity, although effective elimination is difficult when using heat-treatment techniques. In Al–Si–Mg castings, FSP allowed for a redistribution of the Si particles uniformly into the aluminium matrix. FSP induces intense material deformation, effectively modifying the microstructure of Al–Si–Mg castings. FSP in A356 results in a significant break-up of coarse acicular Si particles and primary aluminium dendrites, creating a homogeneous distribution of Si particles in the aluminium matrix, and almost eliminating casting porosity. These microstructural modifications improve mechanical properties, particularly ductility and fatigue threshold stress (by >80%). An increase from 169 MPa tensile strength, 132 MPa YS and 3% elongation to 251 MPa, 171 MPa and 31%, respectively, can be achieved [5].

4.3.2 Other alloys

Copper alloys

Pure copper FSW joint tensile strength is slightly lower than that of BM; however, it has been shown to have higher tensile strength when compared to electron beam welded (EBW) joints [5]. Failure occurs near the HAZ, which also shows the least hardness [15]. An example of tensile curves with different welding parameters is presented in Figure 4.16. Strength of FSW copper can improve with decreasing tool shoulder diameter and tool rotation rate or increasing tool traverse speed [5].

By comparing FSW 60/40 brass tensile strength to the BM, it showed improved strength. With increasing welding speed, the tensile strength increased, although with degradation of the elongation [5].

The tensile mechanical properties of a surface composite layer of pure copper with SiC particles obtained through FSP were studied [18]; tensile curves were mapped (Figure 4.17). In [20], it was reported that YS and UTS increased with decreasing grain sizes in the NZ in a 3 mm thick copper plate FSP.

Compared with BM, an increase in elongation percentage and decrease in tensile strength were verified for FSP pure copper without reinforcing particles. Independent from the size or percentage of the reinforcing particles, the tensile strength and elongation of the composites were lower than those of the FSP specimen without powder or the as-received BM. The addition of SiC in a pure copper by FSP does not improve the tensile properties but improves the hardness and the wear resistance of the composite layer.

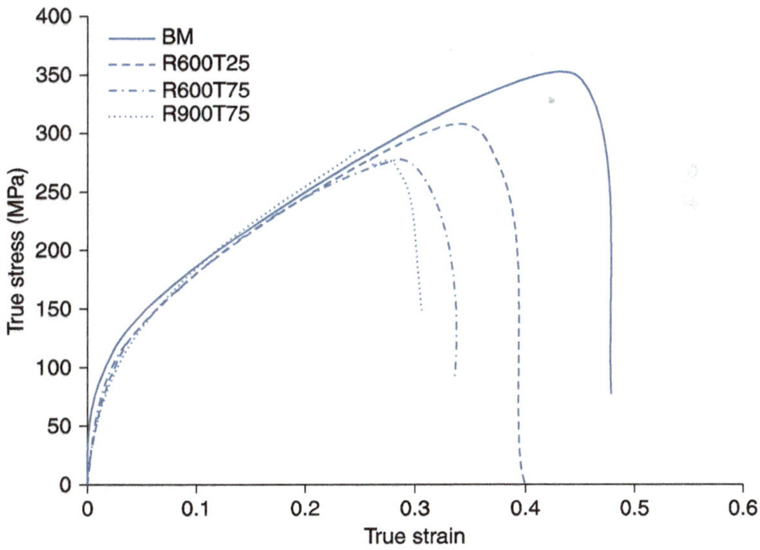

Figure 4.16 Tensile curve of a pure copper FSW joint [16]

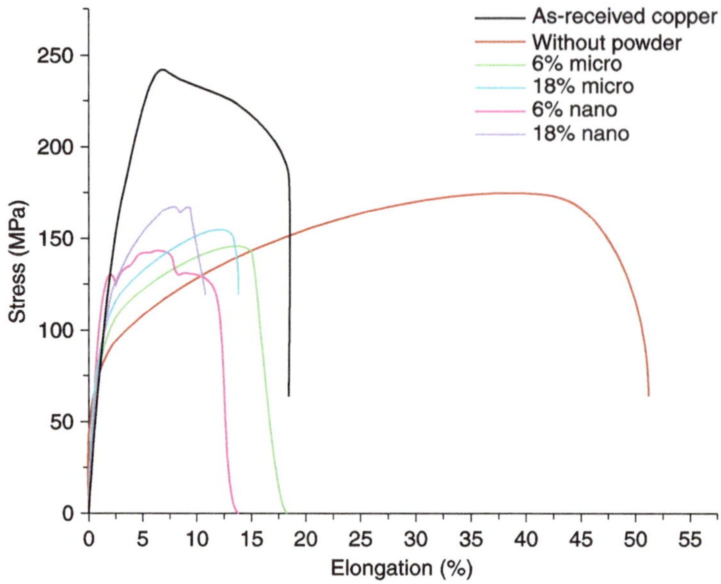

Figure 4.17 **Stress-strain curves of as-received pure copper and FS processes pure copper with and without SiC particles [18]**

Cu/SiC composites were also successfully fabricated by FSP [19]. Mechanicals properties were found to improve with an increase in the number of passes. These improvements may be justified by the reduced size of the SiC particles and improvement in their dispersion, as well as the reduction in grain size of the copper matrix.

Magnesium alloys

FSW cast magnesium alloys joints (e.g. AZ91) tensile properties, when compared to the BM, show improvement although, in wrought magnesium alloys such as AZ31B-H24 and AZ61, a decrease in these properties is reported [5,24]. Failure is reported to occur in the BM [5]. However, it is also reported that the occurrence of the failure in the TMAZ/nugget transition on the AS, except for the butt-joint weld with 1400 rpm and 700 mm/min, failed on the RS in the weld [21].

An increase in tensile and YSs with the welding speed in FSW AZ31B-H24 magnesium alloy may be found [23]. Nevertheless, the tensile strength remains constant when using 15 to 30 mm/s. The Hall–

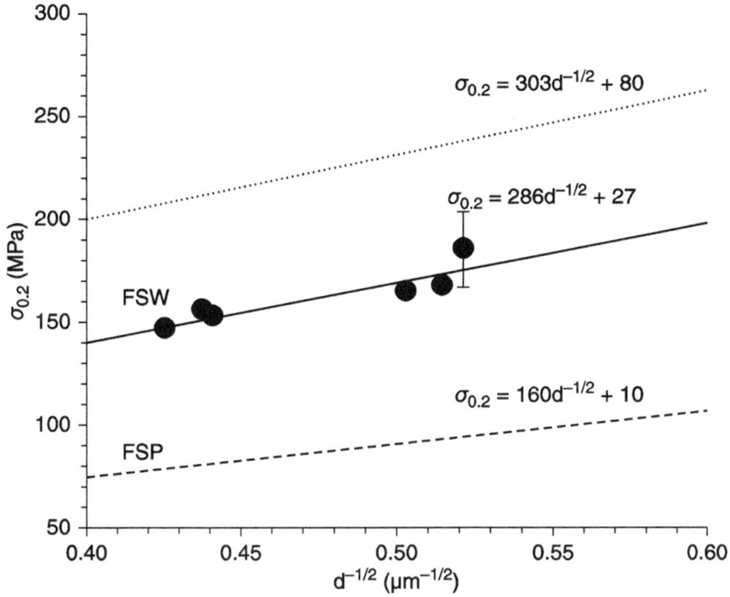

Figure 4.18 Yield strength and grain size relation in FSW AZ31B-H34 magnesium alloy [23]

Petch relationship between YS and the inverse root of SZ grain size is reported (Figure 4.18).

The grain size has a direct influence on the final properties, being mostly influenced by heat input [21,25]. However, in the case of an AZ61 magnesium FSW joint welded with 5 kN of axial force, it was shown to exhibit superior tensile properties [25], which leads to the conclusion that improvements in the joint properties are not only due to the formation of finer grains but also due to the absence of macro-defects. Low axial forces (3 or 4 kN) may lead to coarse grain and formation of defects such as voids, wormholes or lack of surface fill. Nevertheless, the use of high axial force may lead to turbulent material flow and subsequently defects in the NZ besides the grain coarsening.

A minimum welding speed of 200 mm/min is required for AZ31 magnesium 2 mm FSW butt welds, in order to achieve sound welds [21]. Lower welding speeds induce intense stirring that may be responsible for weld defects. Above this value, combining the use of lower rotation speed, the frictional heat generated may not be sufficient to promote the necessary material flow, resulting in voids or tunnel formation. Excessive rotation speeds may result in cracks appearing due to expelling of the material. An improvement of AZ31 magnesium 6.3 mm joints FSW tensile strength and elongation with the increase of shoulder diameter from 18 to 24 mm has also been studied [22]. The failure location changed from the TMAZ to the nugget with the increase of shoulder diameter. In the same material, a slight decrease of tensile and YS of the different zones of the weld was observed with the increase of rotational/welding speeds ratio [26]. This increase in the rotational/welding speeds ratio also increased the weld nugget size and decreased the incomplete root penetration defect. This is due to increased heat input and softening of material providing improved material flow.

Steels

FSW steel joints present, in general, higher yield and UTS when compared to BM failure in the parent metal. The ductility is similar to that of the BM [5,29].

The weld nugget UTS of FSW carbon steels with different percentages of carbon is presented in Figure 4.19 [29]. An increase of welding speed influence on UTS is verified as the carbon percentage increases (from IF

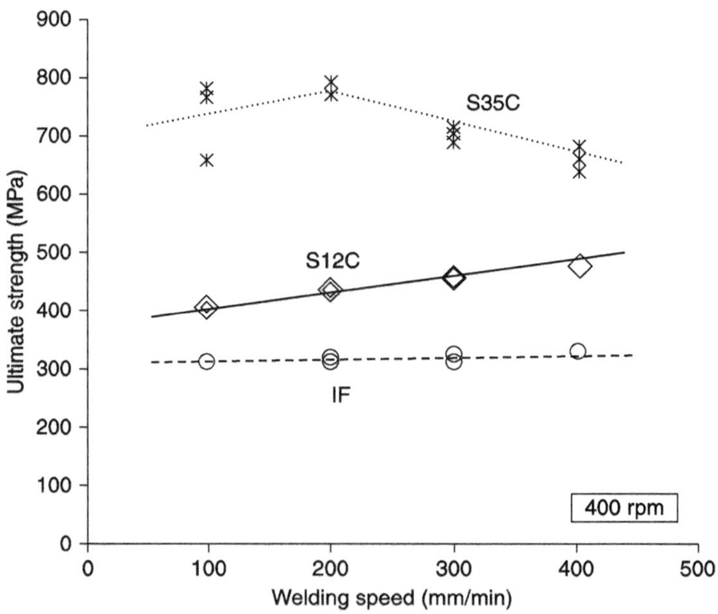

Figure 4.19 Tensile strength of FS welds carbon steels with different carbon percentage [29]

to SC35C). In SC35C (the alloy with the most carbon percentage), the strength of the joints showed a peak close to 200 mm/min, which also occurs with the hardness, as previously mentioned.

When comparing steel HSLA-65 FS welds with different thicknesses, 6.4 and 12.7 mm, similar tensile strength was identified. A slight improvement in YS, with a large increase in elongation (18.5–30%), was verified [5].

A quenched and tempered C–Mn steel, joined through FSW, resulted in a decrease in tensile properties (Table 4.1), with failure occurring in the HAZ. However, strengths of FSW C–Mn steel were superior to those observed in gas metal arc welding (GMAW) joints UTS of 136 MPa, YS of 126 MPa and 5.5% elongation [5].

Titanium alloys

Concerning titanium alloy welds, the yield and UTS exhibit almost 100% joint efficiency and the ductility is also favourably enhanced [5]. In FSW pure titanium, the fracture has been observed to occur only in the HAZ RS [33]. For FSW Ti-6Al-4V alloy, transverse tensile tests showed that failure occurs in the HAZ [32]. It is also reported that FSW Ti-6Al-4V alloy samples exhibited lower strength and elongation than the BM. This statement may be confirmed in Figure 4.20, together with the effect of rotational speed in tensile properties. As seen in this figure, a decrease in tensile properties of the joint with the use of higher rotational speeds may be verified. In the same study, the SZ tensile tests results revealed that the SZ showed higher mechanical properties than the BM, and that the HAZ presented the lower values.

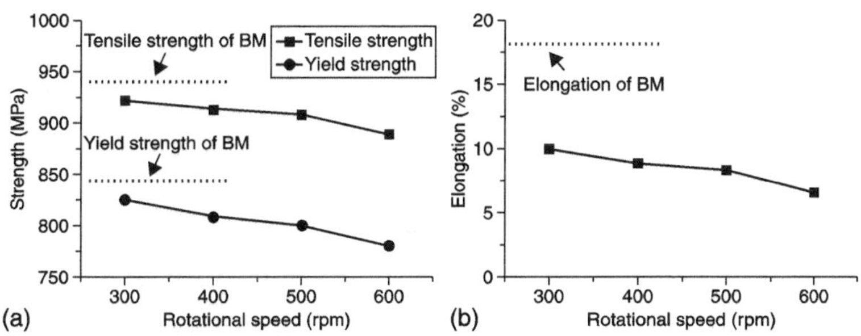

Figure 4.20 **Rotational speed effect in tensile properties of FSW Ti-6AL-4V: (a) yield and tensile strengths; and (b) elongations [32]**

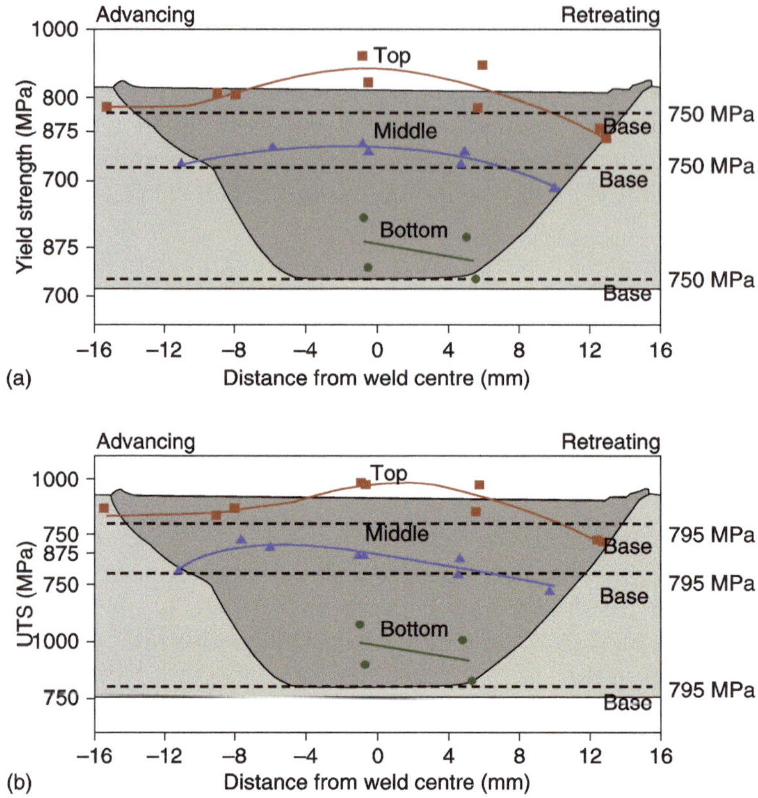

Figure 4.21 Ultimate and yield strengths variation through the FSW Ti-5111 joint on schematic picture [46]

The local mechanical properties of a 12.7 mm thick plate of Ti-5111 were examined using micro-samples [46]. An increase in strength and ductility in the SZ was shown. Some results obtained in this study showed an increase in YS and UTS in the SZ, as this zone experiences the greatest amount of deformation (Figure 4.21).

Tensile properties summary

The tensile properties of several materials, when FSW, are presented in Table 4.1.

Along with the summary of tensile properties of FSW joints presented in Table 4.1, Table 4.2 provides a summary of tensile properties of FSP alloys found in the literature.

Table 4.1 Tensile properties to different material FS welded

Material and conditions		UTS (MPa)	Efficiency (%)	YS (MPa)	Elongation (%)	References
		Aluminium alloys				
Al1100	BM	110	84	110	32	[47]
	FSW	92		–	–	
2017-T351	BM	428	82	–	–	[48]
	FSW	354		–	–	
2024-T351	BM	483–493	83–90	–	–	[5,49]
	FSW	410–434			12	
2219-T87	BM	470–475.8	62–66	390	15	[5,47]
	FSW	297–310.3		–	–	
2024-T3	BM	478	89–90	424	–	[5,10]
	FSW	425–441		280–305	–	
2019-T851	BM	499	80	450	15	[50]
	FSW	395		266	11	
5083-0 (6–15 mm)	BM	285–298	95–119	–	–	[5]
	FSW	271–344		–	–	
6061-T6	BM	342	64	306	17	[9]
	FSW	231		148	6	

(continued)

Table 4.1 Tensile properties of different material FS welded (*continued*)

Material and conditions		UTS (MPa)	Efficiency (%)	YS (MPa)	Elongation (%)	References
			Aluminium alloys			
6082-T6	BM	323	68	276	17.5	[9]
	FSW	224		134	6.5	
7039-T6	BM	414	86	328	15	[51]
	FSW	354		217	18.7	
			Copper alloys			
Material and conditions		**UTS (MPa)**	**Efficiency (%)**	**YS (MPa)**	**Elongation (%)**	**References**
Pure copper	BM	234	94	178	47	[16]
	FSW	216–221		118–127	36–43	
			Magnesium alloys			
Material and conditions		**UTS (MPa)**	**Efficiency (%)**	**YS (MPa)**	**Elongation (%)**	**References**
MB3	BM	245	97	145	12	[52]
	FSW	238		–	–	
AZ31 B	BM	309	65	–	–	[24]
	FSW	201		–	–	
AZ61A (6 mm)	BM	271	83	217	8.40	[25]
	FSW	224		177	7.2	

Steels						
Material and conditions		**UTS (MPa)**	**Efficiency (%)**	**YS (MPa)**	**Elongation (%)**	**References**
409L stainless steel	BM	536	107	364	31	[30]
12%Cr	FSW	574		382	33	
304L	BM	483	129	172		[5]
	FSW	621		340		
Titanium alloys						
Material and conditions		**UTS (MPa)**	**Efficiency (%)**	**YS (MPa)**	**Elongation (%)**	**References**
Pure titanium	BM	440	97.5	–	–	[33]
	FSW	429		–	–	
Ti-6Al-4V	BM	957.7	106	897.0	12.7	[5]
	FSW	1013.6		912.9	12.7	

Table 4.2 Tensile properties of different materials FS processed

Material and conditions		UTS (MPa)	YS (MPa)	Elongation (%)	References
LM13 eutectic Al–Si as-cast aluminium alloy	BM	173	–	4	[13]
	FSP	242	–	26.6	
LM13 eutectic Al–Si as-cast aluminium alloy with addition of Silicon power	BM	173	–	4	[13]
	FSP	327	–	19.2	
Pure aluminium (99.2%)	BM	72	35	39	[3]
	FSP	90	82	35	
Aluminium alloy 6061-T6/ (SiC+Gr) surface hybrid composite	BM	303	279	12.5	[53]
	FSP	221	187	10.2	
Pure copper (99.86 %)	BM	270	209	22	[20]
	FSP	328	261	23	
Copper base composite reinforced with 30 nm nano-sized SiC particles	BM	174	112	48	[18]
	FSP	142–165	104–126	9–12	
Magnesium alloy AZ61	BM	190	140	13	[54]
	FSP	242	147	11	
Magnesium alloy AZ61 composite reinforced with SiO_2 particles	BM	190	140	13	[54]
	FSP	251	225	4	

4.4 Residual stresses and distortion

Residual stresses are definitively a relevant and open issue in welding processes, due to their influence on a component's life and overall performance. Although FSW is a solid-state joining process, which produces low-distortion welds of high quality, significant levels of residual stresses may arise. While the heat input in FSW is lower than in other welding techniques, the clamping is usually more rigid, restraining the contractions of the weld nugget and HAZ. These residual stresses can influence in-service performance of the welded components with respect to fatigue and corrosion behaviour.

Several authors have studied the residual stress field due to welding and its relative fatigue crack growth rate. It has been observed that in FSW joints, the crack growth behaviour is predominantly influenced by residual stresses and microstructure, whereas hardness changes only have a minor influence.

Residual stress as a result of FSW depends on various process parameters, with some of them showing a larger influence than others. When analysing the residual stress field in FSW Al5083 [55], it was concluded that the peak longitudinal stresses increase as the traverse speed increases, which was probably due to steeper thermal gradients and reduced time for stress-relaxation. However, the transverse stresses did not show a direct relationship with rotational speed.

A major concern associated with residual stresses is distortion, which exists in six primary forms: transverse shrinkage, longitudinal shrinkage, rotational distortion, angular distortion, bending distortion and buckling. In welding, the movement of material towards the weld zone causes transverse and longitudinal shrinkage and rotation distortion. As during heating there is significant plastic flow, and during cooling the thermal strain is mostly accommodated through elasticity, there are larger displacements in the cooling process than during the heating, causing the displacement of the material towards the weld zone. This results in transverse shrinkage. Along with transverse shrinkage and caused by the same phenomenon the weld zone will become shorter, resulting in longitudinal shrinkage. Rotational distortion may be explained by the difference in material being displaced from one end of the weld to the other, although this type of distortion is usually small in FSW. Angular and bending are two very common types of distortion in welding, as they are due to longitudinal residual stresses through the thickness of the workpiece. Buckling distortion happens when induced compressive residual stresses surpass the critical buckling load, which is dependent on the stiffness and geometry of the workpiece [4]. An example of distortion in FSW is show in Figure 4.22.

Several techniques have been developed and studied in order to mitigate the effects of the residual stress in FSW. These may be techniques applied during welding, such as weld sequencing, or they may be post-weld treatments. Post-weld treatments work by introducing more stresses, which will counteract with the residual stresses derived from the welding process.

Global mechanical tensioning has been studied as an approach to control residual stress and distortion in FS during welding. FSW Al7749-W51 subjected to this treatment were measured for residual stresses with synchrotron X-ray diffraction (XRD), showing significant reduction of peak stress [56]. Also, when applying this technique to FSW Al2024, by stretching between 35 and 70% of the yield stress of the material during welding, buckling distortion was eliminated [57].

Another approach at controlling residual stress in FSW is local mechanical tensioning. This is carried out by means of roller tensioning directly in the weld line and can be performed during the welding process, although it shows better results when applied upon welding [58]. The application of this technique to the FSW process has been investigated [59]. The distortion of a large integrally stiffened panel was virtually eliminated through post-weld direct rolling. The effect of roller tensioning during FSW and as a post-weld treatment, has been compared [60].

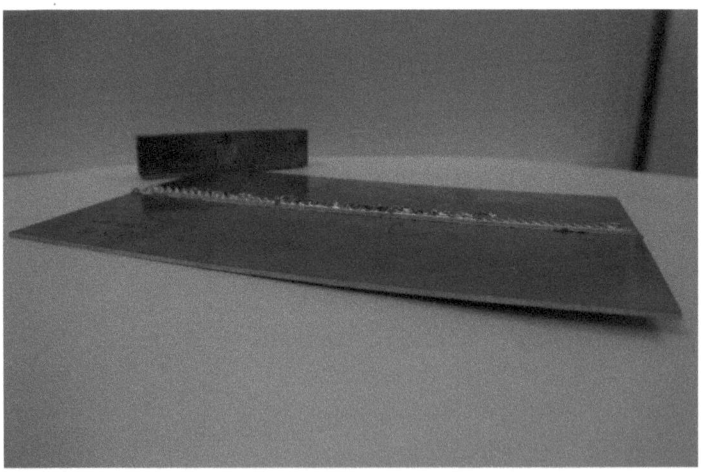

Figure 4.22 Example of distortion in a plate

Again, post-weld rolling showed a significant effect, while roller tensioning during welding did not present a significant improvement.

Shot peening, which is a well-established process for forming aluminium components in the aerospace industry, has also been studied as a post-weld treatment for FSW [4]. This process has been modelled and it was demonstrated that the technique significantly improves the structural performance of welded joints [61]. Particularly in FSW, several studies regarding the effect of shot peening and laser peening may be found. For example, FSW Al2195 subjected to shot peening was studied by measuring residual stresses along with fatigue crack growth testing [62], or laser shot peened FSW Al7075-T7351 sheets were investigated for their fatigue performance related to superficial treatment [63].

4.4.1 Aluminium alloys

Residual stress field measurements studies in FSW may be found in the literature. The through thickness residual stress field was measured in several aluminium alloys, using the hole drilling method [64]. It showed that unlike traditional welding processes, the residual stresses are negative on the surface, and increase through the thickness up to values of about 100 to 150 MPa at a depth of 0.5 to 1.0 mm.

Various measurements of residual stresses contemplating various materials, and even dissimilar material joints, may be found in the literature [5]. XRD measurement of FSW Al-Li alloy Al7050-T7451 show that the overall levels of residual stress is lower than those achieved by conventional fusion welding. Also, it was observed that residual stresses were higher in the transition between fully and partially recrystallized regions when compared to other regions of the weld. Regarding the variations within the weld, parallel to welding direction (longitudinal) tensile residual stresses were found, while in the normal to welding direction (transverse), the residual stresses were compressive.

FSW joints of Al2024-T3 to Al6013-T6 were measured through several measurement techniques [65]. These techniques were XRD, neutron diffraction and high-energy synchrotron radiation, and good agreement was obtained, allowing for higher levels of confidence and as such more sound findings. It was found that, independently of the pin diameter, tool rotational and traverse speeds parameters, the longitudinal residual stresses were consistently higher than the transverse residual stresses. Also, as observed in studies [66,67], an 'M'-like shape of the residual stresses was distributed across the weld, with high points in the transitions from fully to partially recrystallized regions. Also, the diameter of the tool shoulder was found to have an influence on the width of the 'M'-like shape, with larger diameter shoulders resulting in a wider shape. Also, the speed parameters (welding and rotational speeds) were found to have an influence on the peek value of residual stresses, with lower speeds resulting in lower magnitudes of residual stresses.

Beyond diffraction measurement techniques, destructive or semi-destructive techniques were used to study FSW. The contour method was, for example, used to assess the residual stresses in a cross-section of a 3 mm thick friction stir butt-welded aluminium alloy Al6082-T6 plate, comparing the results of the contour with ones obtained using incremental hole drilling [67]. With the contour method it was possible to obtain a full contour map of the longitudinal residual stresses [68]. The profile obtained by the contour method was in line with previously shown profiles for aluminium FSW with an 'M'-like shape distribution. In a similar specimen, the incremental hole drilling technique (iHDT) [69] showed similar results, with regard to maximum values, low compressive stresses outside the weld nugget, and the overall profile shape. The obtained profiles are shown in Figure 4.23.

The contour method may also be used beyond single alloy FSW, as the measurements in a dissimilar joint of thick (25.4 mm) plates of aluminium alloys Al7050-T451 and Al2024-T351 show [66]. The joint was produced at 50.8 mm/min, and upon welding the panel was aged at 121 °C for 24 h to stabilize the weld nugget. The residual stress profile followed traditional patterns seen in other measurements of FSW, despite the increased thickness of the specimen when regarding typical measurements. Neutron measurements were also performed and results were compared. Figure 4.24 shows both contour method and neutron diffraction

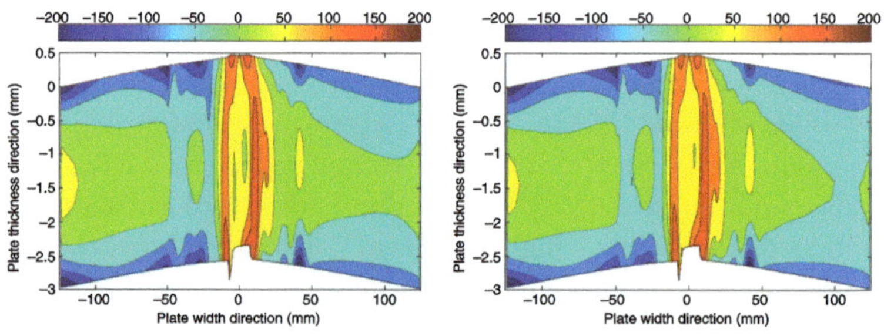

Figure 4.23 Residual stresses perpendicular to the cross-section of the cutting plane measured by the iHDT compared to the obtained contour method results [67]

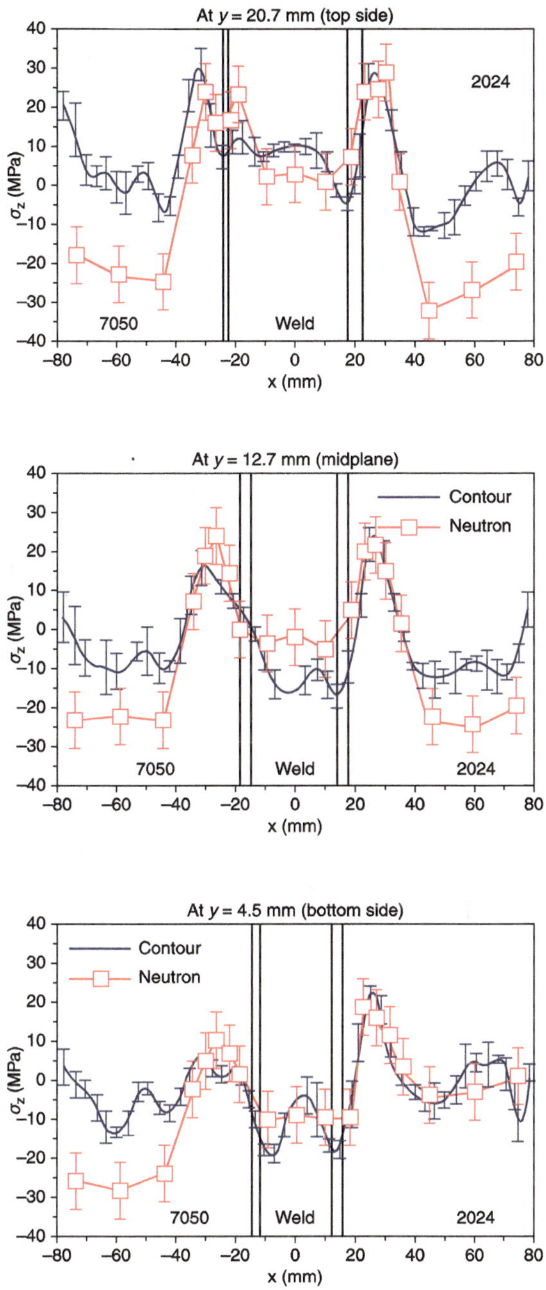

Figure 4.24 Contour method and neutron diffraction measurements at three different locations along the thickness [66]

measurements of residual stresses at three positions along the thickness (top side, mid-plane and bottom side), where it is possible to observe that the peek tensile stresses are observed at the top surface, where the contact between the workpiece and the shoulder occurred. Both results show some asymmetry, with higher stresses on the AS. Also, the asymmetry may arise from the use of different aluminium alloys.

Even though the joint analysed was a thick dissimilar FSW, the peek residual stresses remained low (20% of material YS).

A major concern that goes along with residual stresses is distortion. For this purpose, distortion was measured by the stereo vision based system, while residual stresses were measured with the contour method, and an instrumented clamping device was developed [50]. This clamping system is based on four instrumented beams measuring the vertical load and four resistive load cells for the horizontal load measurement. A scheme of the developed system may be seen in Figure 4.25.

It was possible to observe that higher clamping forces in both directions lead to a lower overall distortion (11.8 and 14.5 mm), although with increasing gap distances, the effect is mitigated. When compared with the welds with an initial gap, distortions were lower for the lower clamping forces, and higher for the higher clamping forces than the welds with no initial gap. The link between the clamping forces and the residual stresses was demonstrated. With higher clamping forces, the residual stresses through the thickness were slightly higher, but presented a more uniform distribution. In conclusion, the author suggests that higher clamping forces (in the order of 2.5 kN) lead to better joint properties, with moderately high clamping forces resulting in lower distortions and more uniformity through the thickness residual stresses. Excessively rigid clamping will lead to higher residual stresses, and so equilibrium must be achieved.

4.4.2 Steels

Beyond different aluminium alloys, residual stress field measurements were made on other materials. FSW 304L stainless steel was studied through neutron diffraction [70]. Residual stresses, in both longitudinal and transverse directions, were measured for welds obtained at two different rotational speeds, 300 and 500 rpm, showing that the region subjected to higher tensile residual stress (>200 MPa) was approximately 5 mm wider when using the higher speed. This difference was explained by the differences in temperature fields, which were also measured. It was

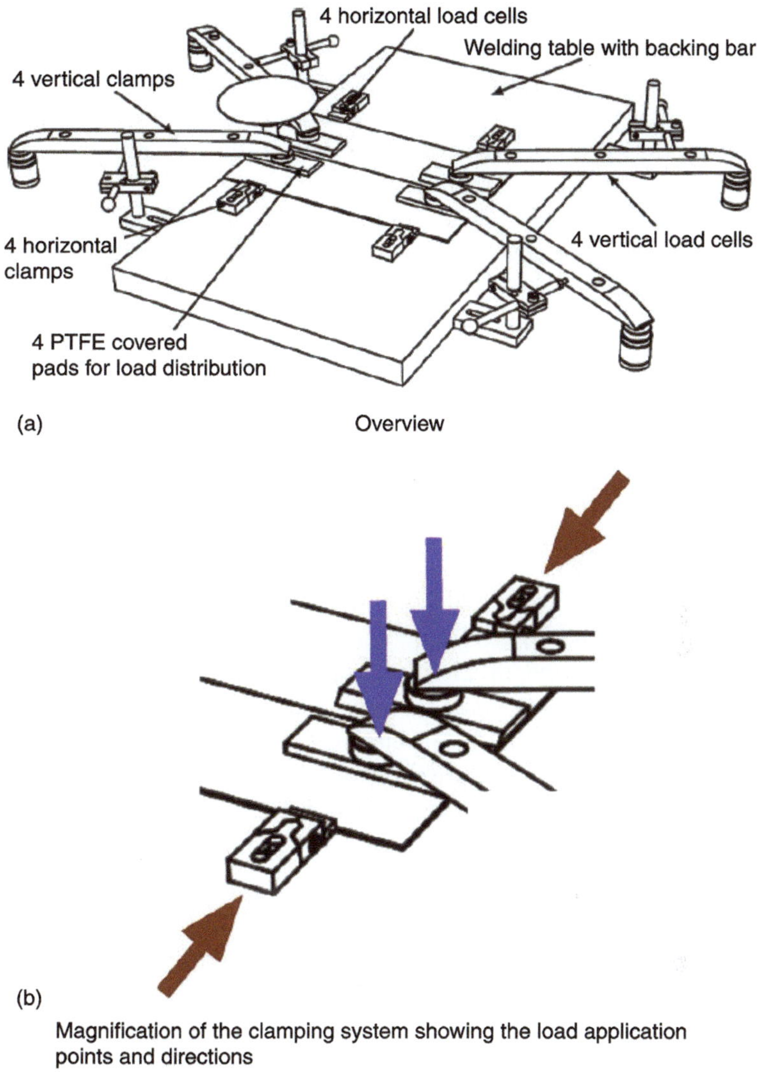

Figure 4.25 **Schematic of instrumented clamping system: (a) overview; and (b) system detail with representations of load application points and direction [50]**

observed that when comparing longitudinal stresses with transverse ones, the average values of the latter are much smaller. Another observation made was that transverse residual stresses are contrary to longitudinal stresses, varying significantly through the thickness. This difference from the zone close to the weld surface to the root surface, causes a bending moment on the plate about the longitudinal axis of the weld. This bending will then be responsible for distortions, upon the release of the piece from the clamping system. The author explains this difference in residual stresses along the thickness with the faster cooling in the root, due to intimate contact with the backing plate.

When compared with typical residual stress field studies performed on aluminium, stainless steel FSW was shown to have higher values of residual stresses relative to the material YS. The author explains this by relating it to the austenitic steel properties, such as coefficient of thermal expansion, thermal conductivity and elevated flow stress, or with the simpler hardness/strength distribution observed in stainless steels that are welded in an annealed condition when compared to FSW high-strength aluminium alloys. This last explanation is based on the observation that high-strength aluminium alloy FSW may be significantly under-matched rather than over-matched with regard to the BM and that the minimum strength region occurs in the HAZ, and as such the residual stress levels are limited by the YS of the HAZ, which is relatively lower than that of the base aluminium alloy.

Along with experimental characterization of residual stresses in FSW, there have been some efforts in numerical simulation. Three-dimensional nonlinear thermal and thermo-mechanical numerical simulations for the FSW of 304L stainless steel, are presented in [71]. Two rotational speeds were simulated, 300 and 500 rpm, and transient temperatures and residual stresses were studied. For the thermal numeric model, an inverse analysis of experimental data of transient temperature history at specific locations during FSW was performed. The inverse analysis resulted in a good match between temperature fields of the experimental data and the numeric simulation results. The maximum temperature achieved with the simulation (900–1000 °C) was significantly lower than the melting temperature of 304L stainless steel at 1450 °C. The obtained residual stresses were compared with experimental data, also showing good agreement and small differences between the two rotational speeds. Fixture release was also modelled, showing a significant reduction in residual stress when compared to the already existing ones. Regarding energy, it was concluded that 50% of total energy in FSW of stainless steel is related to the temperature increase of the workpiece. This value is smaller than that obtained for FSW aluminium alloy 2195 (75–80%) [71].

FSW has been presented as the solution for obtaining PM2000 steel joints with optimized oxides size distribution [72]. FSW welds of this steel were measured for residual stresses through neutron diffraction, and the obtained profile showed the 'M'-like shape and asymmetry typical of FSW. The peek values observed were 320 MPa in the RS and 250 MPa in the AS.

4.5 Fatigue behaviour

For many applications, fatigue is a critical concern that requires much characterization effort. In order to characterize FSW fatigue behaviour, the stages of crack initiation, crack growth and final rupture should be taken into consideration. Testing performed at constant load levels, as described in standard [73], is used to obtain the fatigue crack growth for different load ratios. Typically stress amplitude/number of cycles (S-N) is presented to compare fatigue performance

4.5.1 Aluminium alloys

As FSW was initially developed for aluminium alloys, the welds produced with these alloys have been extensively studied. A series of reports on the S-N fatigue behaviour of several FSW aluminium alloys (Al6006-T5, Al2024-T351, Al2024-T3, Al2024-T3, Al6013-T6, Al7475-T76, Al2219-T8751 and Al2519-T87) have been presented [5]. From these studies, it was observed that FSW presented lower fatigue strength at 10^7 cycles when compared to BM. Also, it was verified that FSW transverse specimens presented lower fatigue strength than longitudinal FSW specimens. The surface quality of FSW was shown to have a significant effect on the fatigue strength of the welds, while on the other hand the speed/rotation rate ratio seems not to have a significant effect.

Comparisons of different properties, such as fatigue behaviour between FSW and other technologies, are an obvious form of presenting the technique as a viable option in design, and so a few studies may be found in the literature. Two aluminium alloys (Al6061-T6 and Al6082-T6) were welded by both metal inert gas (MIG) and FSW [8], and then subjected to S-N tests, at stress ratio $R = 0.1$. The obtained fatigue lives of the different specimens studied are presented in Figure 4.26.

It was found that FSW Al6082-T6 had an infinite fatigue life (superior to 10^7 cycles) for a remote load, which corresponds to 65 and 60% of yield stress. FSW Al6061-T6 presented a lower fatigue life than FSW

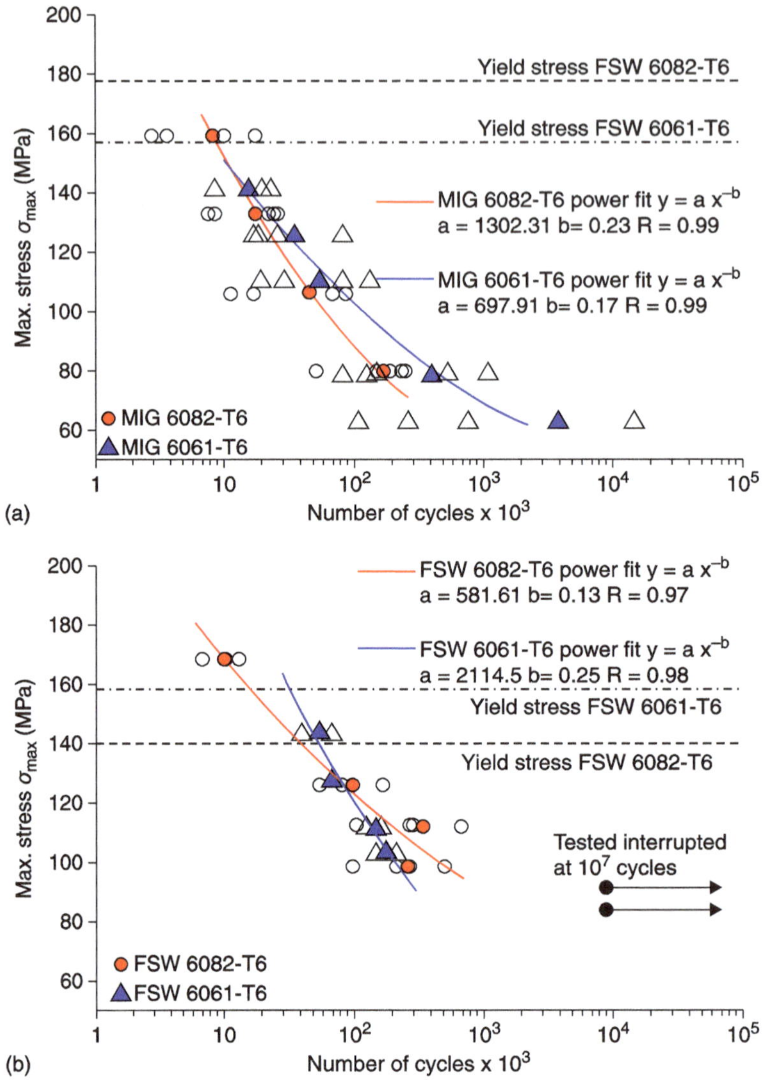

Figure 4.26 S-N fatigue date of welded specimens: (a) MIG; and (b) FSW [26]

Al6082-T6 at stresses lower than 130 MPa. For both tested aluminium alloys, FSW specimens showed higher fatigue lives when compared with MIG specimens.

Fatigue crack behaviour between FSW, MIG welding and TIG has been compared [74]. Al6082 plates were welded with these three techniques. S-N curves for the three techniques in Al6082-T6 are shown in Figure 4.27.

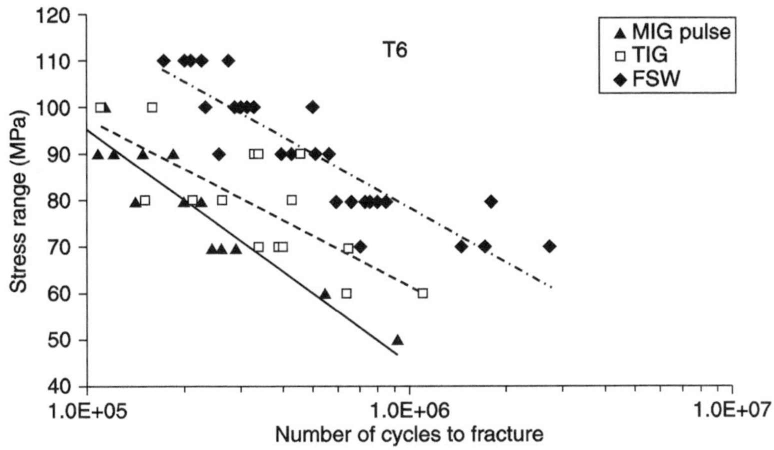

Figure 4.27 S-N fatigue data for MIG-pulsed, TIG and FSW [74]

It was verified that FSW presented higher fatigue strength than MIG and TIG welding for Al6082 aluminium alloy. An interesting conclusion was that although an extra low welding speed gave improved properties, fatigue crack propagation was relatively independent of welding speed. This finding has obvious consequences, as faster welding is more cost-effective.

As previously mentioned, several post-weld treatments may be used to mitigate the effect of residual stresses in fatigue crack growth. Fatigue crack growth of aluminium Al2195, when laser peened and shot peened, has been investigated [62], making the immediate connection between residual stresses and fatigue crack growth. Crack propagation was measured in specimens subjected to different peening processes, as well as as-welded specimen, at two stress ratios R, 0.1 and 0.7, with all the cracks at the weld centre and perpendicular to the weld direction. Laser peening resulted in slower crack growth rates, but shot peening did not show any significant difference when compared to the untreated specimens. The resultant crack growth rate was similar to that shown for the unwelded material, which leads to the conclusion that laser peening restores the crack growth rate to the unwelded level.

A similar study may be found for FSW Al7075-T7351 butt joints [63]. In this study, specimens with defects parallel to the weld zone, as well as specimens with defects at the weld centre and perpendicular to the welding direction, were submitted to cyclic loading at R = 0.1. Again shot peening did not show a significant improvement, while laser peening showed a substantial decrease in the crack growth rate. The dramatic effect of laser peening on fatigue life, where for the same crack length in the shot peening with three layers the number of cycles is significantly higher, is presented in Figure 4.28.

As FSW joints are shown to have lower fatigue strength than the BM, they are expected to be the site of fatigue cracking in structures that implement this technology. As such, fatigue testing of FSW components is important in the implementation of the technology in structural design. An example of FSW implementation in structural design and consequent fatigue behaviour study of the component focused on an aluminium Al6005C-T5 bridge deck segment subjected to static and fatigue testing [75]. The fatigue strength for the given structural detail was determined and compared to MIG welding. From the analysis of the component, it was noted that although FSW had a negligible effect on static material

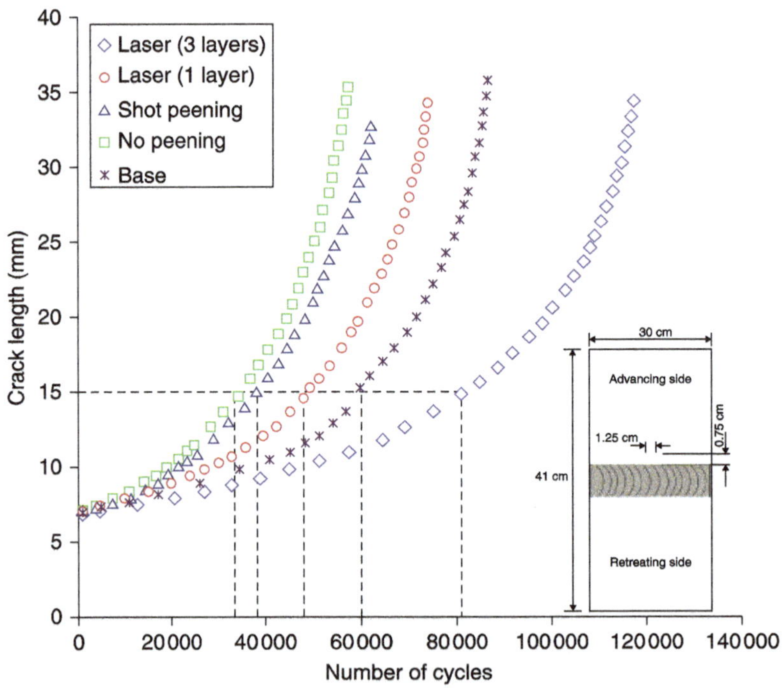

Figure 4.28 Number of cycles vs. crack length in FSW aluminium Al7075-T7351, under R = 0.1 loading [63]

behaviour, the fatigue strength and endurance limit was lower in comparison to the parent material. Also, as crack initiation was observed in the FSW regions, and especially at geometrical surface defects, it was concluded that eliminating the surface defects would result in better performance.

4.5.2 Other alloys and dissimilar joints

As FSW usage expanded to non-aluminium alloys, the interest in studying fatigue properties of these welds has increased. An example of investigations on metal (aluminium) matrix composite FSW fatigue properties is the assessment of fatigue resistance of FSW joints on an as-cast particulate reinforced aluminium-based composite (Al6061/22 vol.%/Al_2O_{3p}) [76].

FS dissimilar welds between Al6013-T4 and stainless steel X5CrNi18-10 have also been investigated for fatigue, among other properties [77]. The S-N curve was compared with Al6013-T6 BM, and the obtained fatigue lifetime was about 30% lower for the welded material. The crack initiation site was at the weld root, where lack of stirred material regions was found, which led to the conclusion that the welding parameters needed to be optimized.

4.5.3 Friction stir processing

Al–Si–Mg alloys, although widely used due to their good casting characteristics and high strength, their porosity, coarse acicular Si particles and coarse primary aluminium dendrites cause limited fatigue resistance. Although many heat-treatment techniques have been developed for these cast alloys, none has the ability to eliminate porosity. FSP was then attempted, as it induces microstructural modifications [5].

The effect on fatigue life of FSP in cast A356 alloy was studied [78]. The S-N curve comparing the A356 subjected to FSP along with the as-casted condition is shown in Figure 4.29.

When subjected to FSP, the fatigue stress threshold showed an improvement higher than 80% in the SZ. FSP of A356 casted alloy is also addressed in [79], with the resultant smaller Si particles of the processed alloy leading to lower fatigue crack growth rates outside of the Paris region.

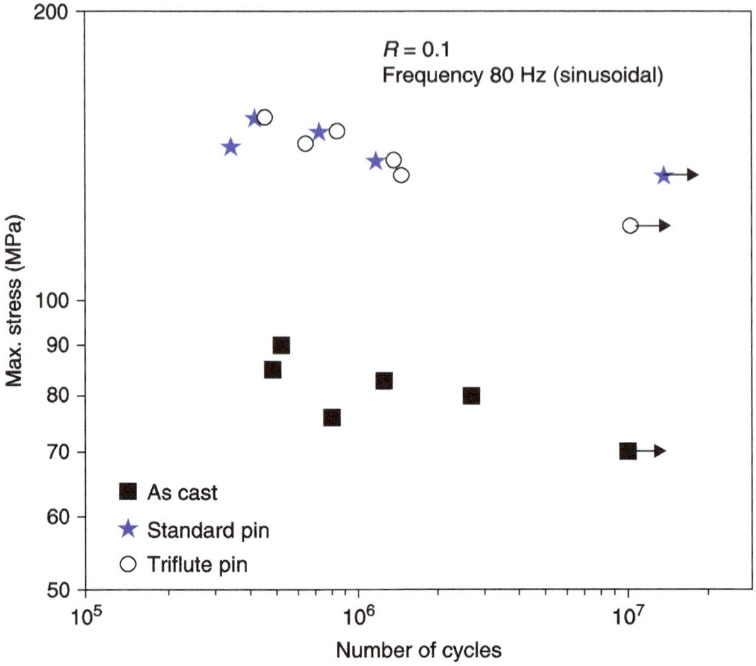

Figure 4.29 S-N fatigue life of A356 cast alloy as-casted and subjected to FSP [78]

4.6 Fracture mechanics

It is commonly accepted that all structures present flaws, which may have been introduced in the manufacturing procedures. In order to define if these flaws are acceptable or not, inspection is required. Regularly, these inspections are based on parameters such as stress intensity factor at the crack tip, J integral or crack tip open displacement (CTOD) when elastic-plastic loads are present. Efforts have been made to characterize the fracture mechanics of FSW/FSP materials, in order to raise the level confidence in the application of this technique in critical structural applications.

A summary of different aluminium alloys studied for fatigue crack propagation, including the testing method as crack propagation may be tested under several different forms (e.g. compact tension, middle crack tension, etc.), may be found in [5]. From these studies some observations were made. One observation was that at lower loads and *R-ratio* of 0.1, the fatigue crack propagation properties of Al2024-T3 and Al6013-T6 are superior to the BM. However, for higher loads and higher *R*-ratios (0.7–0.8), fatigue crack propagation properties of BM and welded material show similar d*a*/d*N*-ΔK curves.

These findings lead to the conclusion that the residual stress profile in the crack tip region in FSW plays a crucial role in crack propagation. As residual stresses are compressive in this region, the effective stress intensity (ΔK_{eff}) at the crack front is reduced, which explains the good performance of the welds at low *R*-ratios. At high *R*-ratios, the effect of residual stresses plays a less important role and therefore the d*a*/d*N*-ΔK curves for both base and welded material tend to be similar. These improvements in fatigue crack propagation properties in low loads and *R*-ratios for FSW were also verified for Al2519-T87 and Al2024-T351. Another finding was that the quality of the FSW had only a limited effect on the d*a*/d*N*-ΔK curve. However, strong dependency of the data obtained and specimen geometry was found, as the residual stresses distribution varies with specimen geometry. By considering this, investigations on the residual profiles of different types of specimens used in fracture strength characterization may be found in [80].

The effect of residual stress in crack propagation was further investigated in Al2024-T351 by studying as welded specimens as well as stress relieved specimens (with application of 2% plastic strain) [49]. This study made use of compact tension specimens and surface crack tension (SCT) specimens, with variations in crack positioning and orientation regarding the weld. A comparison of fatigue crack growth on the SCT specimens between welded material and BM is shown in Figure 4.30.

Fatigue crack growth for FSW Al2024-T351 was shown to be highly influenced by location and orientation of defects regarding the weld line. In cracks perpendicular to the weld, the fastest crack growth was obtained closer to the weld line (11 mm), while in the parallel direction, crack growth rates were higher at larger distances (28 mm). Crack growth rates for defects in both directions were lower with the application of mechanical stress relief, resulting in crack growth behaviour similar to the BM, regardless of orientation and location of defects [5].

Most studies of fracture mechanics regarding FSW focus on mode I, but investigations on mixed mode I/II fracture of FSW specimens may also be found. Mixed mode I/II experiments were performed on Al2524-T351 alloy FSW and in BM, and the CTOD was calculated for each situation [81]. From these experiments it was verified that particle volume fraction and FSW process parameters have significant influence on fracture resistance for flaws in the under-matched FSW zone. Fracture strength of FSW is smaller than the BM for all levels of process parameters (Figure 4.31)

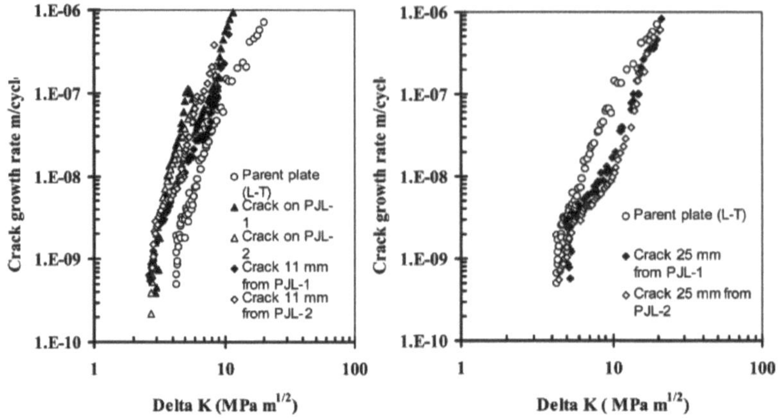

Figure 4.30 Crack-growth rates of FSW Al2024-T351 butt-welded, with cracks propagating orthogonal to the weld [49]

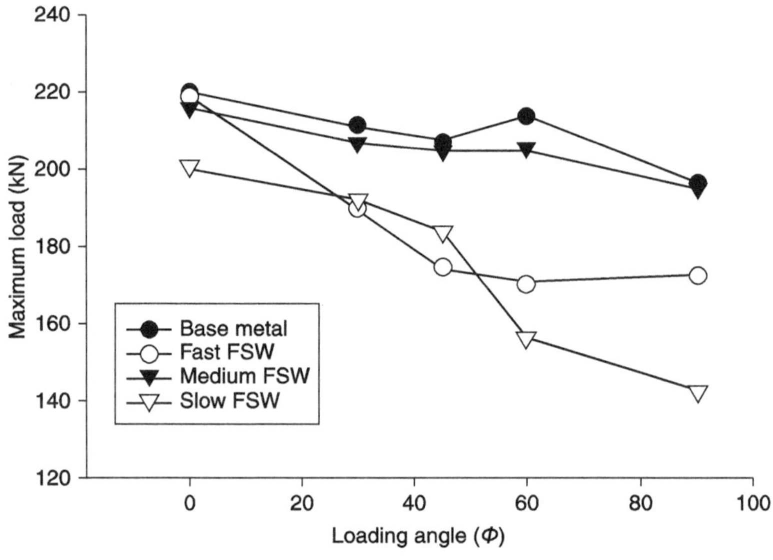

Figure 4.31 Maximum load for different loading angles in Al2524 [81]

An approximately 10% reduction in strength was obtained for medium FSW process parameters and a stable tearing behaviour that is nearly identical to the BM was observed. With slow FSW process parameters, the reduction in strength was about 30% and load-bearing capacity showed a reduction of approximately 20% during stable tearing. Fracture behaviour of FSW Al2524 was also compared to FSW Al2024, which are metallurgically similar, but showed differences in local fracture mode and fracture path. It was verified that Al2524 had faster transitions from local mode I to mode II in the FSW zone. It was concluded that in the terminal steady-state, ductile crack growth process, the macroscale differences (e.g. yield stress, residual stress and grain size) will not have a significant effect, as crack paths for BM specimens and welded ones tend to coincide. Observations transferable to larger structures, as indicated by the author, are:

a) the potential for strong interactions between the initial crack path and microstructural features;
b) the potential for large-scale features (e.g. varying yield stress zones) to have a major impact on the stable tearing path; and
c) the likelihood for either (c-1) unstable crack growth or (c-2) complex fracture paths and fracture surfaces (slanting at varying angles) in FSW joints under far-field mixed mode loading conditions, due to the interactions noted previously.

Fatigue crack growth under 0.33 *R*-ratio of FSW Al7050-T7451 as welded, as well as subjected to post-weld heat treatment (PWHT) (as-FSW+T6), was investigated [82]. In the as-FSW+T6 condition, the lowest near-threshold resistance was obtained in the NZ, and the highest near-threshold resistance was observed in the HAZ. For higher values of *R*-ratio (0.7), the differences between welded and BM were almost non-existent. The author suggested that this behaviour was due to the intergranular failure mechanism within the NZ, with residual stresses playing a more dominant role within the HAZ. Other studies confirmed this behaviour for FSW Al7050-T7451 in as-FSW under heat-treated conditions [5].

Along with crack propagation, fracture toughness has also been emphasised by many studies on FSW properties. When comparing crack open displacement of FSW Al5005-H14, Al6061-T6, Al7020-T6, Al2014-T651, Al7075-RRA, Al7108 and Al5083-O within the NZ, with the BM, the welded material showed higher fracture toughness. FSW A6082 showed similar fracture toughness to the BM [5].

In order to study the fracture mechanics of FSW aluminium alloy 2195, upon S-N fatigue testing, specimens were loaded up to fracture [83]. During loading, a remotely applied load was measured with a load cell, and displacement was measured through a clip gage, producing a load/displacement curve. Crack propagation for different stress ratios and different types of specimens is presented in (Figure 4.32).

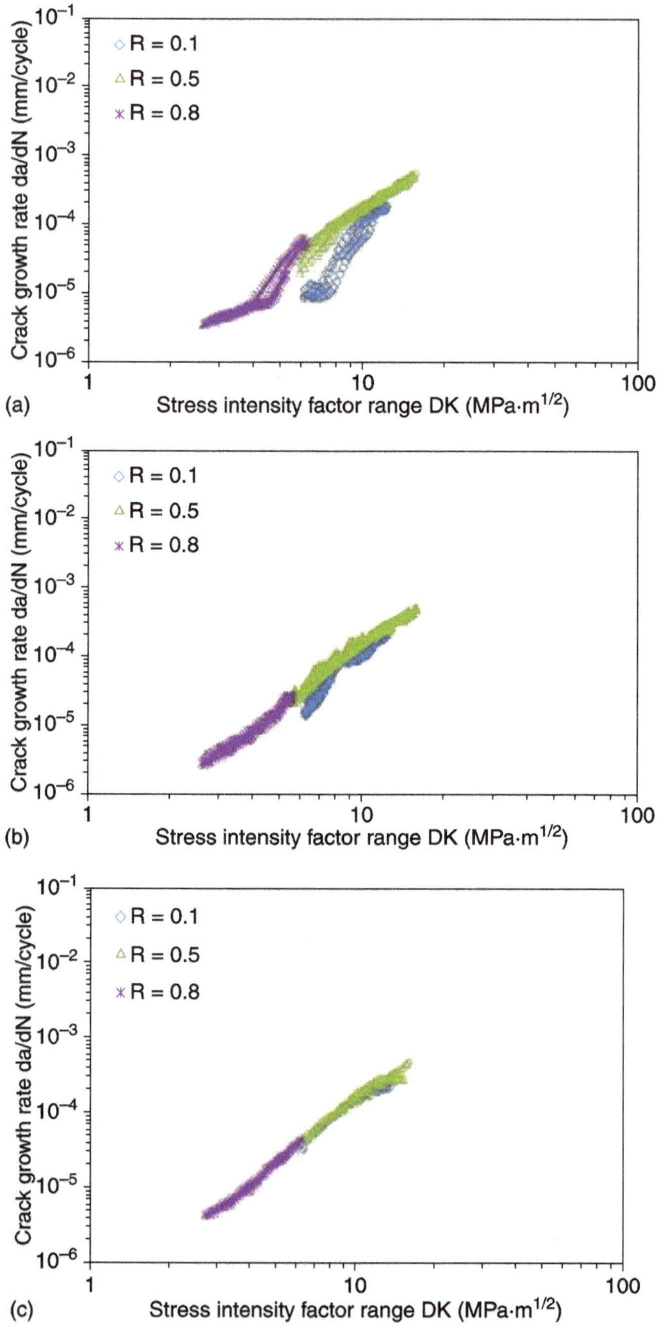

Figure 4.32 Crack growth for different stress ratios for: (a) base material; (b) HAZ; and (c) weld material [83]

Again welded material showed higher fracture toughness than the BM. The welded material crack propagation did not show significant stress ratio dependency (Figure 4.32). The high fracture strength may be attributed to the microstructure modifications, as when only very small particles (nanometric size range) are found in the microstructure, the deformation behaviour becomes highly influenced by shearing of these particles, and so leads to strain localization [6]. Along with particles size, grain boundaries are also presented as a relevant factor in fracture toughness. Higher fracture toughness may be obtained with lower energy grain boundaries, as these will force the failure mode to transition from intergranular to transgranular fracture.

As already mentioned, studies on FSW aluminium alloys are more common as the welding technique was initially developed for these materials, but other materials may also be welded by this method and, as such, studies on the effect of FSW in the properties of these materials are of great interest. Thick plates of ISO 3183 X80M (API 5L X80) steel were FSW with two passes on both sides, and consequently tested for fracture toughness through CTOD [81]. The results showed toughness variations with welding parameters. Rotational speeds of 300 and 350 rpm showed higher fracture toughness than the BM for SZ and HAZ specimens, while 500 rpm showed much lower toughness values for the SZ specimens. The obtained fracture toughness with rotational speed is shown in Figure 4.33.

Fracture toughness of A356 cast aluminium alloy was shown to be improved with FSP [79], due to the more ductile aluminium matrix, which exhibits more uniform plastic deformation and the absence of localized shear bands in the matrix. The homogenization of the material, a consequence of the FSP, is the reason for the absence of localized shear bands in the matrix. The FSP also contributes to reduction of micro-voids, where nucleation of cracks can occur. The crack closure effect at high ΔK values is believed to be increased with the reduction of Si particles size, which happens in the SZ. By comparing fractography of the FSP samples and as-casted specimens, the former showed consistent crack paths at both high and low values of ΔK in the matrix material, while with as-casted material crack paths at low ΔK values were in the crack paths in the matrix material and at high ΔK values cracks propagated partly through the silicon particles and at the interface of the silicon particles and matrix. Cracks propagating in FSP A356 were deflected upon encountering silicon particles, showing higher resistance to fatigue crack growth.

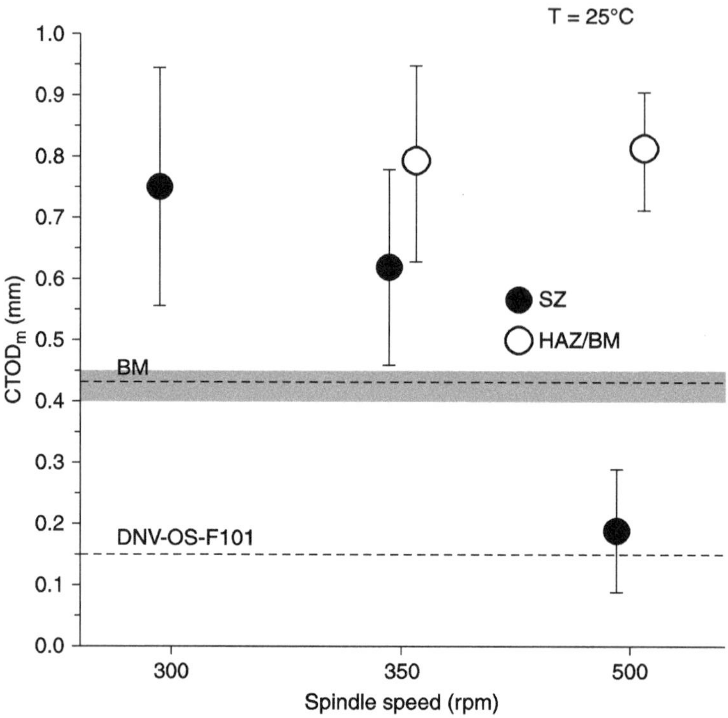

Figure 4.33 Fracture toughness vs. rotational speed [84]

4.7 References

1. JAA-FAA (1998), 'Damage tolerance and fatigue evaluation of structure', *JAR*, 25: 571.
2. ESA-ESTEC (1999), 'Fracture control', *ECSS-E-30-01A*, ECSS Secretariat – European Cooperation for Space Standardization.
3. Yadav, D. and Bauri, R. (2012), 'Effect of friction stir processing on microstructure and mechanical properties of aluminium', *Materials Science and Engineering A*, 539(0): 85–92.
4. Lohwasser, D. and Chen, Z. (2010), *Friction stir Welding: From Basics to Applications*, Cambridge: Woodhead Publishing.
5. Mishra, R.S. and Ma, Z.Y. (2005), 'Friction stir welding and processing', *Materials Science and Engineering R: Reports*, 50(1–2): 11–78.
6. Nandan, R., DebRoy, T. and Bhadeshia, H.K.D.H. (2008), 'Recent advances in friction stir welding – Process, weldment structure and properties', *Progress in Materials Science*, 53(6): 980–1023.

7. Moreira, P.M.G.P., de Oliveira, F.M.F. and de Castro, P.M.S.T. (2008), 'Fatigue behaviour of notched specimens of friction stir welded aluminium alloy 6063-T6', *Journal of Materials Processing Technology*, 207(1–3): 283–92.
8. Moreira, P.M.G.P., de Figueiredo, M.A.V. and de Castro, P.M.S.T. (2007), 'Fatigue behaviour of FSW and MIG weldments for two aluminium alloys', *Theoretical and Applied Fracture Mechanics*, 48(2): 169–77.
9. Moreira, P.M.G.P. et al. (2009), Mechanical and metallurgical characterization of friction stir welding joints of AA6061-T6 with AA6082-T6', *Materials & Design*, 30(1): 180–7.
10. Mishra, R.S. and Mahoney, M.W. (2005), *Friction stir Welding and Processing*, ASM International.
11. Xu, W. et al. (2013), 'Influence of welding parameters and tool pin profile on microstructure and mechanical properties along the thickness in a friction stir welded aluminum alloy', *Materials & Design*, 47(0): 599–606.
12. Fuller, C.B. et al. (2010), 'Evolution of microstructure and mechanical properties in naturally aged 7050 and 7075 Al friction stir welds', *Materials Science and Engineering A*, 527(9): 2233–40.
13. Tutunchilar, S. et al. (2012), 'Eutectic Al–Si piston alloy surface transformed to modified hypereutectic alloy via FSP', *Materials Science and Engineering A*, 534(0): 557–67.
14. Hwang, Y.M., Fan, P.L. and Lin, C.H. (2010), 'Experimental study on Friction stir welding of copper metals', *Journal of Materials Processing Technology*, 210(12): 1667–72.
15. Lee, W.-B. and Jung, S.-B. (2004), 'The joint properties of copper by friction stir welding', *Materials Letters*, 58(6): 1041–6.
16. Khodaverdizadeh, H. et al. (2012), 'Effect of friction stir welding (FSW) parameters on strain hardening behavior of pure copper joints', *Materials & Design*, 35(0): 330–4.
17. Park, H.S. et al. (2004), 'Microstructures and mechanical properties of friction stir welds of 60% Cu–40% Zn copper alloy', *Materials Science and Engineering: A*, 371(1–2): 160–9.
18. Barmouz, M. et al. (2011), 'Investigation of mechanical properties of Cu/SiC composite fabricated by FSP: Effect of SiC particles' size and volume fraction', *Materials Science and Engineering A*, 528(3): 1740–9.
19. Barmouz, M. and Givi, M.K.B. (2011), 'Fabrication of *in situ* Cu/SiC composites using multi-pass friction stir processing: Evaluation of microstructural, porosity, mechanical and electrical behavior', *Composites Part A: Applied Science and Manufacturing*, 42(10): 1445–53.
20. Surekha, K. and Els-Botes, A. (2011), 'Development of high strength, high conductivity copper by friction stir processing', *Materials & Design*, 32(2): 911–16.
21. Commin, L. et al. (2009), 'Friction stir welding of AZ31 magnesium alloy rolled sheets: Influence of processing parameters', *Acta Materialia*, 57(2): 326–34.
22. Yang, J. et al. (2010), 'Effects of heat input on tensile properties and fracture behavior of friction stir welded Mg–3Al–1Zn alloy'. *Materials Science and Engineering A*, 527(3): 708–14.

23. Cao, X. and Jahazi, M. (2009), 'Effect of welding speed on the quality of friction stir welded butt joints of a magnesium alloy', *Materials & Design*, 30(6): 2033–42.
24. Afrin, N. et al. (2008), 'Microstructure and tensile properties of friction stir welded AZ31B magnesium alloy', *Materials Science and Engineering A*, 472(1–2): 179–86.
25. Razal Rose, A., Manisekar, K. and Balasubramanian, V. (2011), 'Effect of axial force on microstructure and tensile properties of friction stir welded AZ61A magnesium alloy', *Transactions of Nonferrous Metals Society of China*, 21(5): 974–84.
26. Abbasi Gharacheh, M. et al. (2006), 'The influence of the ratio of "rotational speed/traverse speed" (ω/v) on mechanical properties of AZ31 friction stir welds', *International Journal of Machine Tools and Manufacture*, 46(15): 1983–7.
27. Xie, G.M. et al. (2011), 'Effect of rotation rate on microstructures and mechanical properties of FSW Mg-Zn-Y-Zr alloy joints', *Journal of Materials Science & Technology*, 27(12): 1157–64.
28. Faraji, G. and Asadi, P. (2011), 'Characterization of AZ91/alumina nanocomposite produced by FSP', *Materials Science and Engineering A*, 528(6): 2431–40.
29. Fujii, H. et al. (2006), 'Friction stir welding of carbon steels', *Materials Science and Engineering A*, 429(1–2): 50–7.
30. Lakshminarayanan, A.K. and Balasubramanian, V. (2010), 'An assessment of microstructure, hardness, tensile and impact strength of friction stir welded ferritic stainless steel joints', *Materials & Design*, 31(10): 4592–600.
31. Fratini, L. et al. (2010), 'A new fixture for FSW processes of titanium alloys', *CIRP Annals – Manufacturing Technology*, 59(1): 271–4.
32. Zhang, Y. et al. (2008), 'Microstructural characteristics and mechanical properties of Ti–6Al–4V friction stir welds', *Materials Science and Engineering A*, 485(1–2): 448–55.
33. Lee, W.-B. et al. (2005), 'Microstructural investigation of friction stir welded pure titanium', *Materials Letters*, 59(26): 3315–18.
34. Kitamura, K. et al. (2013), 'Flexible control of the microstructure and mechanical properties of friction stir welded Ti–6Al–4V joints', *Materials & Design*, 46(0): 348–54.
35. Heidarzadeh, A. et al. (2012), 'Tensile behavior of friction stir welded AA 6061-T4 aluminum alloy joints', *Materials & Design*, 37(0): 166–73.
36. Elangovan, K., Balasubramanian, V. and Babu, S. (2009), 'Predicting tensile strength of friction stir welded AA6061 aluminium alloy joints by a mathematical model', *Materials & Design*, 30(1): 188–93.
37. Cederqvist, L. and Reynolds, A. (2001), 'Factors affecting the properties of friction stir welded aluminum lap joints', *Welding Journal*, 80(12): 281.
38. Papadopoulos, M. et al. (2013), 'Mechanical behaviour of AA 2024 friction stir overlap welds', *International Journal of Structural Integrity*, 4(1): 8.
39. Fratini, L., Buffa, G. and Shivpuri, R. (2009), 'Influence of material characteristics on plastomechanics of the FSW process for T-joints', *Materials & Design*, 30(7): 2435–45.

40. Fratini, L. et al. (2007), 'Experimental characterization of FSW T-joints of light alloys', *Key Engineering Materials*, 344: 751–8.
41. Buffa, G. et al. (2008), 'Material low in FSW of T-joints: Experimental and numerical analysis', *International Journal of Material Forming*, 1(1): 1283–6.
42. Fratini, L. et al. (2006), 'Friction stir welding of AA6082-T6 T-joints: Process engineering and performance measurement', *Proceedings of the Institution of Mechanical Engineers part B Journal of Engineering Manufacture*, 220: 669–76.
43. Acerra, F. et al. (2010), 'On the FSW of AA2024-T4 and AA7075-T6 T-joints: An industrial case study', *The International Journal of Advanced Manufacturing Technology*, 48(9–12): 1149–57.
44. Cui, L. et al. (2012), 'Characteristics of defects and tensile behaviors on friction stir welded AA6061-T4 T-joints', *Materials Science and Engineering A*, 543(0): 58–68.
45. Erbslöh, K. Donne, C.D. and Lohwasser, D. (2003), 'Friction stir welding of T-joints', *Materials Science Forum*, 426–32: 2965–70.
46. Nimer, S., Wolk, J. and Zupan, M. (2013), 'Local property characterization of friction stir welded Ti-5111: Transverse orientation measurements', *Acta Materialia*, 61(8): 3050–9.
47. Rajakumar, S. and Balasubramanian, V. (2012), 'Establishing relationships between mechanical properties of aluminium alloys and optimised friction stir welding process parameters', *Materials & Design*, 40(0): 17–35.
48. Liu, H.J. et al. (2003), 'Tensile properties and fracture locations of friction stir-welded joints of 2017-T351 aluminum alloy', *Journal of Materials Processing Technology*, 142(3): 692–6.
49. Bussu, G. and Irving, P.E. (2003), 'The role of residual stress and heat affected zone properties on fatigue crack propagation in friction stir welded 2024-T351 aluminium joints', *International Journal of Fatigue*, 25(1): 77–88.
50. Richter-Trummer, V. et al. (2012), 'Influence of the FSW clamping force on the final distortion and residual stress field', *Materials Science and Engineering A*, 538(0): 81–8.
51. Sharma, C., Dwivedi, D.K. and Kumar, P. (2012), 'Effect of welding parameters on microstructure and mechanical properties of friction-stir welded joints of AA7039 aluminum alloy', *Materials & Design*, 36(0): 379–90.
52. Wang, K.-S. et al. (2006), 'Evaluation of microstructure and mechanical property of FSW welded MB3 magnesium alloy', *Journal of Iron and Steel Research, International*, 13(4): 75–8.
53. Devaraju, A., Kumar, A. and Kotiveerachari, B. (2013), 'Influence of rotational speed and reinforcements on wear and mechanical properties of aluminum hybrid composites via friction stir processing', *Materials & Design*, 45(0): 576–85.
54. Ma, Z.Y. (2008), 'Friction stir processing technology: A review', *Metallurgical and Materials Transactions A*, 39(3): 642–58.
55. Peel, M. et al. (2003), 'Microstructure, mechanical properties and residual stresses as a function of welding speed in aluminium AA5083 friction stir welds', *Acta Materialia*, 51(16): 4791–801.

56. Altenkirch, J. et al. (2008), 'The effect of tensioning and sectioning on residual stresses in aluminium AA7749 friction stir welds', *Materials Science and Engineering A*, 488(1–2): 16–24.
57. Price, D. et al. (2007), 'Distortion control in welding by mechanical tensioning', *Science and Technology of Welding & Joining*, 12(7): 620–33.
58. Coules, H. et al. (2012), 'Effect of high pressure rolling on weld-induced residual stresses', *Science and Technology of Welding & Joining*, 17(5): 394–401.
59. Wen, S. et al. (2010), 'Rolling to control residual stress and distortion in friction stir welds', *Science and Technology of Welding & Joining*, 15(6): 440–7.
60. Altenkirch, J. et al. (2009), 'Residual stress engineering in friction stir welds by roller tensioning', *Science and Technology of Welding & Joining*, 14(2): 185–92.
61. Meo, M. and Vignjevic, R. (2003), 'Finite element analysis of residual stress induced by shot peening process', *Advances in Engineering Software*, 34(9): 569–75.
62. Hatamleh, O. (2009), 'A comprehensive investigation on the effects of laser and shot peening on fatigue crack growth in friction stir welded AA 2195 joints', *International Journal of Fatigue*, 31(5): 974–88.
63. Hatamleh, O., Lyons, J. and Forman, R. (2007), 'Laser and shot peening effects on fatigue crack growth in friction stir welded 7075-T7351 aluminum alloy joints', *International Journal of Fatigue*, 29(3): 421–34.
64. Fratini, L. and Zuccarello, B. (2006), 'An analysis of through-thickness residual stresses in aluminium FSW butt joints', *International Journal of Machine Tools and Manufacture*, 46(6): 611–19.
65. Donne, C.D., Wegener, J., Pyzalla, A. and Buslaps, T. (2001), 'Investigations on residual stresses in friction stir welds', *Third International Symposium on Friction stir Welding*, Kobe, Japan.
66. Prime, M.B. et al. (2006), 'Residual stress measurements in a thick, dissimilar aluminum alloy friction stir weld', *Acta Materialia*, 54(15): 4013–21.
67. Richter-Trummer, V. et al. (2011), 'The contour method for residual stress determination applied to an A6082-T6 friction stir butt weld', *Materials Science Forum*, 681: 177–81.
68. Prime, M.B. (2001), 'Cross-sectional mapping of residual stresses by measuring the surface contour after a cut', *Journal of Engineering Materials and Technology*, 123(2): 162–8.
69. ASTM (1999), 'Standard test method for determining residual stresses by the hole-drilling strain-gage method, in ASTM Standard E837-991999', *Amercian Society for Testing and Materials*: Philadelphia, PA.
70. Reynolds, A.P. et al. (2003), 'Structure, properties, and residual stress of 304L stainless steel friction stir welds', *Scripta Materialia*, 48(9): 1289–94.
71. Chao, Y.J., Qi, X. and Tang, W. (2003), 'Heat transfer in friction stir welding—experimental and numerical studies', *Journal of Manufacturing Science and Engineering*, 125(1): 138–45.
72. Mathon, M.H. et al. (2009), 'Study of PM2000 microstructure evolution following FSW process', *Journal of Nuclear Materials*, 386–8(0): 475–8.
73. ASTM (2008), 'Standard test method for measurement of fatigue crack growth rate', in E6472008, ASTM International.

74. Ericsson, M. and Sandström, R. (2003), 'Influence of welding speed on the fatigue of friction stir welds, and comparison with MIG and TIG', *International Journal of Fatigue*, 25(12): 1379–87.
75. Vigh, L.G. and Okura, I. (2013), 'Fatigue behaviour of friction stir welded aluminium bridge deck segment', *Materials & Design*, 44(0): 119–27.
76. Minak, G. et al. (2010), 'Fatigue properties of friction stir welded particulate reinforced aluminium matrix composites', *International Journal of Fatigue*, 32(1): 218–26.
77. Uzun, H. et al. (2005), 'Friction stir welding of dissimilar Al 6013-T4 To X5CrNi18-10 stainless steel', *Materials & Design*, 26(1): 41–6.
78. Sharma, S.R., Ma, Z.Y. and Mishra, R.S. (2004), 'Effect of friction stir processing on fatigue behavior of A356 alloy', *Scripta Materialia*, 51(3): 237–41.
79. Sharma, S.R. and Mishra, R.S. (2008), 'Fatigue crack growth behavior of friction stir processed aluminum alloy', *Scripta Materialia*, 59(4): 395–8.
80. Richter-Trummer, V. and de Castro, P. (2011), 'The through-the-thickness measurement of residual stress in a thick welded steel compact tension specimen by the contour method', *The Journal of Strain Analysis for Engineering Design*, 46(4): 315–22.
81. Sutton, M.A. et al. (2006), 'Microstructure and mixed mode I/II fracture of AA2524-T351 base material and friction stir welds', *Engineering Fracture Mechanics*, 73(4): 391–407.
82. Jata, K.V., Sankaran, K.K. and Ruschau, J.J. (2000), 'Friction stir welding effects on microstructure and fatigue of aluminum alloy 7050-T7451', *Metallurgical and Materials Transactions A*, 31(9): 2181–92.
83. Moreira, P.M.G.P. et al. (2012), 'Fatigue and fracture behaviour of friction stir welded aluminium–lithium 2195', *Theoretical and Applied Fracture Mechanics*, 60(1): 1–9.
84. Santos, T.F.A. et al. (2010), 'Fracture toughness of ISO 3183 X80M (API 5L X80) steel friction stir welds', *Engineering Fracture Mechanics*, 77(15): 2937–45.

5

Tool and welding design

V. Infante and C. Vidal, University of Lisbon, Portugal

DOI: 10.1533/9780857094551.199

Abstract: Friction stir welding (FSW) is a solid-state joining technology developed by The Welding Institute (TWI) in 1991. This environmentally-friendly and energy-efficient technique has become an attractive joining process for a wide variety of industries, offering a number of attractive advantages when compared to other welding processes. The basic concept of FSW is a rotating tool with a specially designed probe and shoulder inserted into the material to be joined, and traversed along the joint line. In the beginning, the tool design was simple, limiting both the material stirring and the welding speed, but has become progressively more complex, and its design features have evolved to move material more efficiently. Currently, the welding speeds practised are those required by commercially applications production. Owing to the tool features, the material flow around the tool probe can be extremely complex and significantly different from one tool to the other. The present chapter provides an overview of the most commonly-used friction stir tool designs and materials and welded joint configurations.

Key words: friction stir welding, friction stir tools, tool materials, weld properties, welding design.

5.1 Introduction

Over the last few years, many efforts have been dedicated to the development of the friction stir welding (FSW) process. In order to achieve this, attention must focus on the mechanical and microstructural characteristics of the welded joints, by conducting research at different levels in terms of FSW parameters. This chapter discusses FSW tools, their geometry and materials (steels, nickel and cobalt-based alloys, refractory metals, metal composites and cubic boron nitride), as well as their influence on material flow and weld properties, and the welding design considering the welded joint configurations.

This chapter is organized into three main sections in order to fulfil the purpose of its content, namely: 'Friction stir tools', 'Effect of tool geometry on material flow and weld properties' and 'Welding design'.

Problems such as the geometry of probes and shoulders, their evolution and consequences on the welded joints performance, are addressed. The tool geometry effects on the material flow during the welding process and afterwards on the weld properties are presented and discussed, supported by several scientific published studies. Concerning the subject of welding design, the joint configurations are presented and some welding procedures are pin-pointed. Sometimes the conventional FSW procedure is not suitable for complex joint configurations, so some modifications are required.

The different sections of the chapter are a summary of those found in the literature.

5.2 Friction stir tools

The use of a non-consumable tool is mandatory during the FSW process. The pin, or probe, and the shoulder are its main components. Three main purposes are provided by the tool, namely heating the workpiece, stirring and moving the plasticized material in order to produce the welded joint and controlling the processed material under the tool shoulder. Therefore, the FSW tool is undoubtedly crucial to the achievement of the FSW process, being essential for improving both the welded joint's mechanical performance and the maximum possible welding speed during the technological process.

The thermo-mechanical deformation and the workpiece frictional heating, essential for friction stirring, are influenced by the FSW tool design [1]. The shoulder is responsible for generating most of the heat during the process, applying a downward pressure on the workpiece surface, constraining the plasticized material around the probe and preventing the escape of the plasticized material from the workpiece joint line. However, the rotating probe mainly drags along, plasticizes and mixes the adjacent material presents in the stir zone (SZ), creating a welded joint without the need of fusion [2].

In recent years, new geometrical features have been developed concerning FSW tool design [3]. Owing to the various tools features, the material movement around the probe can be extremely complex and considerably different from one tool to the next [2].

The shoulder diameter and profile and the probe length, diameter and profile are important parameters in determining the welding speed and the quality of the weld. Another important parameter in the determination of the suitability of a tool for a specific application is the tool material itself [4].

Even though most of the common problems of fusion welding technique are avoided during the FSW process, since there is no bulk melting of the workpiece, which is the reason why this welding process is extensively and successfully applied for welding aluminium and other soft alloys, the FSW tool is subjected to severe stress and high temperatures mainly during the welding of hard alloys such as steels and titanium alloys. Indeed, the commercial application of the FSW process to these alloys is now limited by the high cost and short life of FSW tools [5–7]. Recently, many significant efforts have been made to develop cost-effective and suitable tool designs but further work is needed to ensure the application of the FSW process to hard alloys [8].

An efficient tool material selection, design and cost are required in the pursuit of commercially useful tools for the welding of hard materials. The Re-stir tool applies periodic reversal of tool rotation, in order to prevent the inherent asymmetry of conventional FSW. With the exception of FSW with the Re-stir tool, material flow is essentially asymmetrical about the joint interface where the material flows primarily on the retreating side (RS) [9–13]. Understanding this asymmetry in material flow is important to optimize tool design [3].

5.2.1 Tool geometry

Regarding tool geometry, there are three types of FSW tools: fixed, adjustable and self-reacting. The fixed probe tool corresponds to a single component comprising the shoulder and the probe. Due to its fixed probe length, this kind of tool can only be used to weld workpieces/components with a specific and constant thickness [2]. In adjustable tools, the shoulder and the probe are independent components, allowing adjustments in the probe length and making up different configurations of shoulder/probe facilitating the welding of a large number of varied components [14,15]. In adjustable tools, the shoulder and the probe can be manufactured using different materials and the probe can be easily replaced or its length modified. Self-reacting tools are composed of three different components, the top and bottom shoulders and the probe [16,17]. These tools can only operate perpendicularly to the workpiece surface, in contrast with the fixed and adjustable tools that can be tilted longitudinally and laterally to the workpiece [2].

Tool shoulders

Tool shoulders are designed to promote frictional heating of the surface and subsurface regions of the workpiece that is to be welded. The tool shoulder is responsible for the majority of the deformational and frictional heating in the welding of thin metal sheets [1]. Figure 5.1 summarizes the most typical shoulder outer surfaces [18].

Concave shoulder

The first and most common shoulder design in friction stirring was the concave shoulder [13,19–21]. The concave shoulder has become popular for restricting material extrusion from the sides of the shoulder [3]. This simple shape is easy to produce and capable of performing sound and good welds. The shoulder concavity is defined by a small angle between the edge of the shoulder and the axis of the probe, typically from 6 to 10 degrees. During the tool plunge time, the material displaced by the tool probe is fed into the cavity within the tool shoulder. The concave surface profile of the tool shoulder works as an escape volume for its forging action.

Owing to a downward pressure on the tool, the displaced material carried in the concave shoulder profile renders a forging action on the

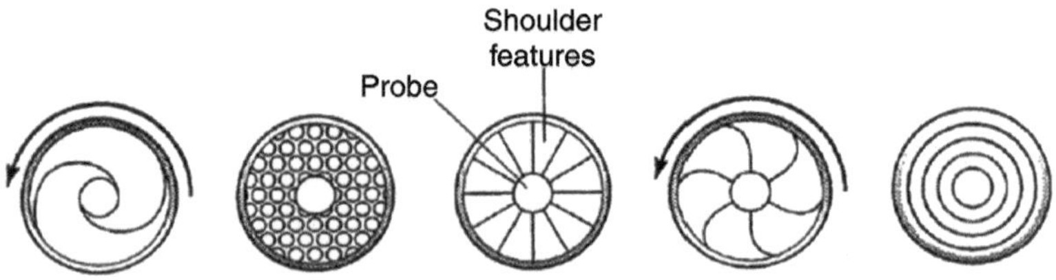

Figure 5.1 Typical shoulder outer surfaces [18]

material behind the tool. In addition, the forward movement of the tool forces the new material into the cavity under the shoulder and pushes the existing material behind the probe. The correct use of this shoulder requires tilting of the tool as much as is needed to retain the material and to enable the rear edge of the tool shoulder to produce a compression force on the forging welding. Values from 2 to 4 degrees are the most reported in the literature as suitable tilt angles [22].

Shoulder features

The shoulder can also contain some geometric features that increase the material friction. These features can be scrolls, ridges, knurling, grooves or concentric circles and can be embedded into any tool shoulder profile such as concave, flat or convex ones. Scrolls are the most commonly-used shoulder features [23,24]. Scrolls and concentric circles are another two types of shoulder feature that have been presented in the literature [3].

Scroll shoulder

A typical scrolled shoulder consists of a flat end surface with a spiral channel cut from the edge towards the centre. The channels help the material to flow from the edge of the shoulder to the probe, thus eliminating the need to tilt the tool [3]. The concave smooth shoulder end tends to be pushed away from the workpiece top surface during the FSW process at high welding speeds, due to the stirred material that is continuously trapped in the store/cavity under the shoulder [2,3]. Removing the tool tilt simplified the friction stirring machine design and made it possible to produce complex nonlinear weld paths. Concave shoulder tools also have a tendency to lift away from the workpiece surface when the tool welding speed increases. Replacing the concave shoulder with a scrolled shoulder reduces the tool lift to increase the welding speed. The use of a scrolled shoulder tool eliminates the undercut produced by a concave tool and reduces the flash production during the welding process.

Because these tools work normal to the workpiece, the normal forces are lower than for concave shoulder tools, which must apply load in both the normal and transverse directions to keep the shoulder in contact with the workpiece. In addition, the material within the channels is continually sheared from the plate surface, thereby increasing the deformation and frictional heating at the surface [3]. Thus, the use of a scrolled shoulder requires more positional care than the use of a concave shoulder. Scrolled shoulder tool limitations can be identified as the inability to weld two components with different thicknesses, the inability to accommodate workpiece thickness variation in the length of the weld line, and to weld complex curvatures (especially tight curvatures). Scrolled shoulder tools can weld two plates with different thicknesses, but some amount of material from the thicker plate is expelled in the form of flash [3].

However, a concave shoulder combined with a scrolled feature can reduce the tool lift during high welding speed processes [25]. Combining the scroll end surface with a convex shoulder design prevents material displacement away from probe and takes advantage of the greater flexibility in the contact area between the shoulder and the workpiece, which then improves the mismatch tolerance of the joint, increases the ease of joining different thicknesses and promotes the ability to weld complex curvatures [2].

Convex shoulders

Another possible shoulder shape is a convex profile. Early attempts for the convex end surface were unsuccessful because the convex profile tended to push the material away from the probe [25]. The main advantage of the convex shoulder profile is that the outer edge of the tool need not be engaged with the workpiece, so the shoulder can be engaged with the workpiece at any location along the convex surface [26]. Thus, a sound weld is produced when any part of the scroll is engaged with the workpiece, moving material towards the probe. This shoulder design allows a greater flexibility in the contact area between the shoulder and the workpiece (amount of shoulder engagement can be changed without any loss in weld quality), improves the joint mismatch tolerance, increases the ease of joining of different-thickness workpieces, and improves the ability to weld complex curvatures as reported before. The profile of the convex shoulder can be either tapered or curved [2,3].

Tool probes

The tool probe has the ability to produce deformational and frictional heating. This component is designed in such a way as to disrupt the contacting surfaces of the components to be welded, to shear the material in its front and to move the material behind the tool. The depth of deformation and tool welding speed are mainly controlled by the tool probe [27], as well as its geometry influencing the flow of plasticized material and affecting the performance of the welded joints [3,8,28,29].

An example of a tool probe design is the Whorl design that uses a tapered probe with re-entrant features or a variable pitch thread in order to improve the downward flow of the plasticized material. Triflute and Trivex probes are other typical probe geometries. The Triflute design is complex, comprising a system of three tapering, threaded re-entrant flutes that promote and amplify the material flow around the tool. The Trivex tool design is simpler than the Triflute one, using a non-cylindrical probe that has been found to reduce the forces acting on the tool during the welding process. However, a Triflute (or triangular) probe increases the material flow around the tool when compared with a cylindrical probe [30]. The thread orientation of the probe surface influences the axial downward force acting on the workpiece material and the material flow around the tool [31]. For thicker plates, more complex features on the probe have been added in order to favour material flow and mixing, reducing the load values during the welding process [2].

The most common probe designs are presented and described below:

Cylindrical probes

A cylindrical threaded probe with a round bottom (Figure 5.2) is the probe geometry presented in the very first FSW patent [32]. This probe design is commonly referred to as the 5651 tool in the friction stir community [33]. The threads embedded on the probe surface are used to transport the material from the surface (near the shoulder) down to the probe bottom. Therefore, a clockwise tool rotation requires left-handed threads. The conventional cylindrical threaded probe has been successfully used for butt welding of aluminium alloys up to 12 mm in thickness [3]. A round or domed end to the tool probe is used to reduce its wear, eliminating local stress concentration, mainly during the tool plunge time, and improves the quality of the welded joint root under the bottom of the probe. A dome radius of about 75% of the probe diameter was reported as the best dome radius. It was claimed that its reduction (up to a flat bottom tool) increases the probability of obtaining a poor-quality weld, especially directly underneath the probe [33]. Although a dome radius offers some advantages, the flat bottom probe design is the most commonly used probe design [3].

Changing from a round bottom to a flat bottom probe is attributed to a geometrical argument. This can be reasoned on the basis of the surface velocity of a rotating cylindrical probe that increases from zero at the centre to a maximum value at the edge. The local surface velocity coupled with the friction coefficient between the probe and the material determines the deformation during the welding process. The higher surface velocity

Figure 5.2 Cylindrical threaded probe with a round bottom

at the probe edge can increase its stirring power and hence promote the material flow under the probe end [2]. The lowest point of a round bottom probe has the lowest velocity and the least stirring action. A larger round bottom probe radius will reduce the differential velocity, while a smaller probe radius will increase the differential velocity. The increased surface velocity at the bottom of the probe will increase the throwing power of the probe, or the ability of the probe to affect the material flow below the end of the probe. In addition, the flat bottom probe is easier to produce and its weaker points can be smoothed by a higher downward forging force [3]; although this high force is critical during plunging [2]. The main advantage of the cylindrical probe design is that its length and diameter can be modified in order to pursue the user's needs.

Truncated cone probes

A simple modification of a cylindrical probe turns it into a truncated cone probe (Figure 5.3). The use of a truncated cone probe leads to the application of lower transverse loads, when compared to a cylindrical

Figure 5.3 Truncated cone probe for FSW

probe. The largest moment load on a truncated cone probe is at the base of the cone. One variation of this probe is the stepped spiral probe, which was the design developed for high-temperature materials [3,10]. Cylindrical probes were found to be suitable to weld aluminium alloys plates of up to 12 mm thick, but not the better geometry to FSW thicker plates at faster welding speeds.

Whorl and MX-Triflute probes

The Whorl probe design was developed by TWI [34,35]. With a Whorl probe tool (Figure 5.4), the volume of material displaced compared with that displaced by a cylindrical probe of the same diameter is reduced by 60%. Reducing the displaced volume makes it possible to decrease the traverse loads, which enables higher welding speeds. The main difference between truncated cone probes and Whorl probes is the design of the helical ridge on the probe surface. In the case of the Whorl probe, the helical ridge is more than an external thread, acting as an auger and producing a clear downward movement. Circular, oval,

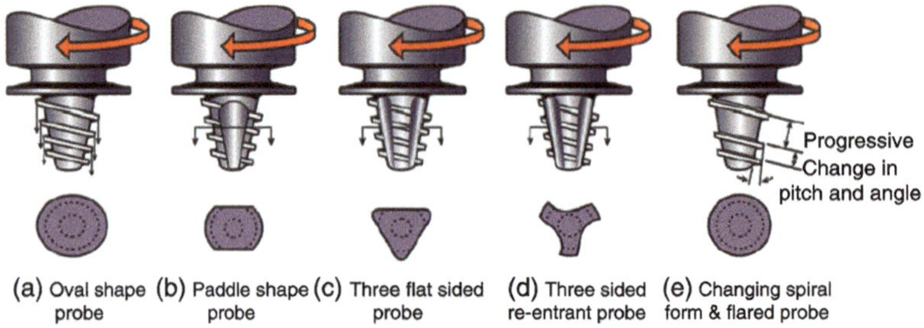

Figure 5.4 Whorl type probes [42]

flattened and re-entrant probe cross-sections are some of the variations of the Whorl probe design [35]. The significant advantage of the Whorl probe is the higher ratio between the dynamic volume (the volume of material swept by the tool probe during one rotation) and the static volume (the volume of the probe itself) [3]. A tool design with a higher swept rate is reported to reduce the voids and allow the surface oxides to be more effectively disrupted and dispersed within the nugget zone (NZ), due to the stronger stirring and mixing action for the material flow. In conventional FSW, the dynamic/static volume ratio can be increased via the use of the re-entrant features, threads with flutes and/or flats machined into the probe [36].

The MX-Triflute probe (Figure 5.5) is an evolution of the Whorl probe [34,35]. The MX-Triflute probes contain, in addition to the helical ridge, three flutes cut into the helical ridge. These three flutes reduce 70% of the volume of material displaced by a cylindrical probe and supply additional deformation at the weld joint line. Thus, it is possible that Whorl and MX-Triflute probes have smaller probe volumes than tools with cylindrical probes [3]. In Whorl probes, the tapered threads facilitate the plastic flow, since they induce a vertical component of velocity. Also, the flute in the MX-Triflute probe increases the interfacial area between the tool and the workpiece, leading to an increase in the heat generation, softening and material flow during the welding process. As a consequence of intense stirring, both the traversing force and the torque are reduced [37]. Although cylindrical, Whorl and Triflute probes are suitable for butt welding; they are not useful for lap welding, where excessive thinning of the upper plate can occur together with trapping of adherent oxides between the overlapping surfaces. For solving this problem, Flared-Triflute and A-skew™ tools were developed to ensure fragmentation of the interfacial oxide layer and a wider weld than is

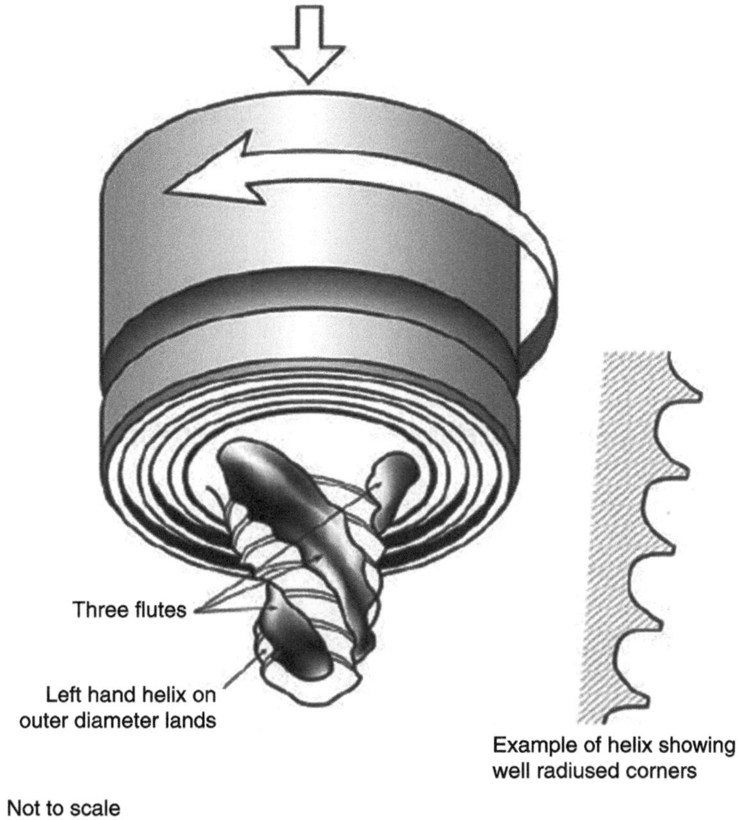

Figure 5.5 MX-Triflute probe [42]

usual for butt welding [38]. The Flared-Triflute tool is similar to the MX-Triflute tool, having an expanded flute, and the A-skew™ tool is a threaded tapered tool with its axis sloped to that of the spindle of the welding equipment.

Both of these tools increase the swept volume relative to that of the probe, thus expanding the stir region and resulting in a wider weld and successful lap joints [3]. Studies concerning FSW of 5 mm thick plates of AA5251 [39] and up to 50 mm thick plates of copper [40] were reported using Triflute probes. By changing to an MX-Triflute probe, it is possible to increase the welding speed [40]. In addition to being suitable for welding thick copper plates, the MX-Triflute probe has also shown promise in welding thick aluminium alloy plates. It was demonstrated that an MX-Triflute probe is more effective in breaking up silicon particles and healing casting porosity, during friction stir processing (FSP) [41], than either cylindrical or truncated cone probes [3]. A Whorl tool was used to successfully weld 75 mm thick 6082-T6 aluminium alloy in two steps: one pass for the upper surface and the other pass for the lower surface, each side with a penetration depth higher than 35 mm [42,43].

Trivex probes

Two-dimensional (2-D) computational fluid dynamics simulations were performed in order to study the material flow around the tool probe. A series of probe designs was used as well as a novel slip model, which was implemented on the 2-D probe profiles to establish the profile that produced the minimum traverse force [44,45]. With the optimal 2-D probe profile, generated simulations of two versions of new probe designs were created by The Welding Institute (TWI): the featureless Trivex probe (Figure 5.6(a)) and threaded MX-Trivex probe (Figure 5.6(b)). Investigations have demonstrated that the Trivex and MX-Trivex probes produce an 18 to 25% reduction in traversing forces and a 12% reduction in forging (perpendicular) forces in comparison to an MX-Triflute probe of similar dimensions, in welding tests of 6.35 mm thick plates of 7075-T7351 aluminium alloy [44,45]. However, both the Trivex and the Triflute probes produce FSW with comparable mechanical performance in terms of tensile strength [3].

(a) (b)

Figure 5.6 (a) Trivex; and (b) MX-Trivex tools [44]

Threadless probes

Threadless probes are useful in specific FSW applications, namely when thread features would not survive without fracture or severe wear. Threadless probes are chosen for high strength or highly abrasive alloys, as the threaded features can be easily worn away [2]. As such, tools operating under aggressive environments cannot retain threaded tool features without excessive probe wear. Therefore, probes for these extreme conditions typically consist of simple designs with robust features. In the FSW of thin sheets, the tools commonly have probes with a smaller surface area for features, because the embodiment of any thread would severely weaken the tool probe, causing its premature failure. Thus, thin sheets are commonly FSW using threadless probe tools. Threadless probes have also been employed to purposely produce defective welds [23], as well as to study the material flow during the welding process [46].

5.2.2. Tool materials

The production of a good FSW joint requires the proper tool material to be selected for the desired application. New challenges have emerged with the requirement of welding materials, such as metal matrix composites, titanium or steel [1]. It is important to have a FSW tool that keeps its dimensional stability, and does not lose its features or fractures during the welding process [3]. It is desirable that the tool material has:

- *Good mechanical strength*: at room temperature and, mainly, at elevated temperatures. The welding process is carried out around 70 to 90% of the material melting point, so it is important that the tool material has sufficient strength at this temperature, otherwise it will failure [4];
- *Food wear resistance*: despite some potentially undesirable effects on the microstructure of the welded joint, significant tool wear increases the cost of the FSW process [8];
- *No reactivity with the workpiece material*: the formation of undesirable phases during the welding process, if the tool and workpiece material reaction can lead to tool material degradation [8];
- *Good fracture toughness*: it is mandatory that the tool resists the damage carried out during the plunge and dwell periods [2]. As well as difficulties in tool production leading to increase in cost, it is

often desirable to sacrifice some durability for an acceptable tool price [4];

- *Low coefficient of thermal expansion*: during the welding process the workpiece temperature field depends on the tool material's thermal conductivity. The appropriate value of thermal conductivity depends on the process variables, the workpiece material and some other tool material properties [8]. The coefficient of thermal expansion may affect the thermal stresses in the tool [8]. A low coefficient of thermal expansion between the probe and the shoulder materials reduces the thermal stresses [2]. However, a high thermal conductivity may be undesirable if excessive removal of heat from the tool–workpiece interface requires a high tool rotation speed to soften the workpiece material and to reduce tool stress;
- *Good machinability*: A good machinability is important in order to facilitate the manufacture of complex features on the shoulder and probe surfaces [3].

The properties of some tool materials used for FSW are summarized in Tables 5.1 to 5.5 [8].[1]

In the beginning of the development of the FSW process, the first focus was on the welding of aluminium alloys followed by the combination of other resistance metals such as copper and magnesium. The welding of these materials is suitable and sufficient if the tool is manufactured from steel, which is afterwards hardened in order to provide a good resistance to wear. However, for strength alloys such as titanium, nickel and steel, the effective welding cost and the tool lifetime remain as research topics, being the tool material that is one of the most important issues to be solved [2].

Next, the entire published tool materials listed for FSW process are presented and described.

1 SD: shoulder diameter; PD: pin diameter; PL: pin length; PS: pin shape; SS: shoulder shape; PD_t: pin diameter at the top (larger diameter) for tapered pin; PD_b: pin diameter at the bottom (smaller diameter) for tapered pin; SC: straight circular; TC: tapered circular; SCT: straight circular threaded; TCT: tapered circular threaded; SSq: square; TSq: tapered square; SHex: hexagonal; THex: tapered hexagonal; TOct: tapered octagonal; LHT (RHT): left (right) handed thread; 3F: three flats; TC3F: tapered circular with three flats; UTS: ultimate tensile strength; FSSW: friction stir spot welding; RSW: resistance spot welding. Joint efficiency is the ratio of the tensile strength of the joint to that of the base metal (BM) [8].

Table 5.1 Tool materials, geometries and welding operating parameters used for FSW of magnesium alloys [8]

Workpiece material	Tool material	Tool shape and size	Operating parameters	Remarks
AZ31 Mg, 1·5 mm thick	H13 steel	SD: 10 mm; PD: 4 mm; PL: 1·8 mm; PS: SCT, 3F with M4 threads	1000–3000 rev min^{-1}; dwell time: 1, 4 s; plunge rate: 0–10 mm s^{-1}; FSSW	Welds with 3F/threaded superior to those with SCT
AZ31 Mg, 1·5 mm	H13 steel, 46–48 HRC	SD: 10 mm; PD: 4 mm; PL: 1·8 mm; PS: SCT, and threaded and unthreaded 3F	1000–3000 rev min^{-1}; dwell time: 1 s; plunge rate: 2·5 mm s^{-1}; FSSW	Joint efficiencies: 74–83%
AZ31B-H24 Mg alloy, 2 mm	Mild steel, stainless steel, armour steel, high carbon steel, high speed steel	PD: 3·175 mm; PL: 1·65 mm; PS: SC, LHT, RHT	1000–2000 rev min^{-1}; 300–1800 mm min^{-1}	Joint efficiencies: 48·8–96·7%
AZ31B Mg alloy, 6 mm	H13 steel	SD: 15, 18, 21 mm; PS: SC, TC, SCT, triangular and square; PL: 5·7 mm; PD: 6 mm	1600 rev min^{-1}; 40 mm min^{-1}; 0° tilt	Joint efficiencies: up to 62%
AZ31B-H24 Mg alloy, 2 mm		SD: 19 mm; PL: 2–3·5 mm; PD: 6·35 mm	1200 mm min^{-1}; 500–2000 rev min^{-1}	

Table 5.2 Tool materials, geometries and welding operating parameters used for FSW of aluminium alloys [8]

Workpiece material	Tool material	Tool shape and size	Operating parameters	Remarks
6111-T4 Al alloy, 0·9 mm thick	H13 steel	SS: flat with scroll; SD: 10 mm; PL: 0–1·6 mm	2000 rev min^{-1}; dwell time: 2·5 s; plunge rate: 2·5 mm s^{-1}; FSSW	Better quality with pinless tool
7075-T7351, 6·35 mm	1. MP159; 2. Dievar tool steel; 3. MP159 pin, H13 shoulder	PS: Triflute, Trivex	394 and 457 rev min^{-1}; 300–540 mm min^{-1}	Weld UTS: 470–488 MPa
7075-T7351; 6·35 mm, 16 mm	Steel	PS: threaded	190–457 rev min^{-1}; 0·3–1·4 mm rev^{-1}	Surface scaling and voiding problems
Al alloys, 5 mm	H13 steel	SS: concave; SD: 15 mm; PS: SC, SCT, triangular; PL: 4·7 mm, 6 mm	600–1500 rev min^{-1}; 25–1000 mm min^{-1}; 3° tilt	Peak joint efficiencies: 70–100%
7020-T6 Al alloy, 4 mm	H13 steel	SD: 10–20 mm, flat; PD: 3–8 mm; PL: 4·2 mm; PS: frustum and SC	1400 rev min^{-1}; 80 mm min^{-1}	Peak joint efficiency: 92%
6082-T6 Al, 1·5 mm	Tool steel	SS: scroll, cavity, fillet; PD: 1·7 mm; PS: SC; PL: 1·2 mm	1810 rev min^{-1}; 460 mm min^{-1}; 2° tilt	Joint efficiencies: ~76%
6061-T6 Al, 9·5 mm and 12·7 mm	High carbon steel	SD: 25?4 mm; PD: 5·2–7·6 mm; PL: 1·8–7·1 mm	650 rev min^{-1}; 150 or 200 mm min^{-1}; 3° tilt	No property degradation in weld metal
6061-T6 Al, 6·3 mm		SS: concave; SD: 26 mm; PD: 5·6 mm; PL: 5·9 mm; PS: SCT	286–1150 rev min^{-1}; 30–210 mm min^{-1}	
5754 Al, 1·32 mm		SS: concave, convex, flat; SD: 12 mm; PD: 5 mm; PL: 1·6 mm	1500 rev min^{-1}; dwell time: 2 s; plunge rate: 20 mm min^{-1}; FSSW	
A319 and A413 Al alloy, 6 mm		PD: 6 mm	1000 rev min^{-1}; 120 mm min^{-1}	
7020-T6 Al, 4 mm		SS: concave; SD: 13 mm; PS: SC, TC3F; PL: 3·19 mm: PD: 5 mm	300–1620 rev min^{-1}; 100–900 mm min^{-1}; 2·5° tilt	

Table 5.3 Tool materials, geometries and welding operating parameters used for FSW of metal matrix composites [8]

Workpiece material	Tool material	Tool shape and size	Operating parameters	Remarks
6061-T6 Al+ 20%Al_2O_3, 5 and 6 mm thick	AISI oil hardened Tool steel (62 HRC)	SD: 19 mm; PS: SCT; PD: 6·3 mm	500–2000 rev min^{-1}; 60–540 mm min^{-1}; 1° tilt	No wear after some distance (150–300 mm) depending on process parameters
Al 359+ 20%SiC, 4 mm	AISI oil hardened tool steel (62 HRC)	SD: 19 mm; PS: SCT; PD: 6·3 mm; PL: 3·6 mm	500–1000 rev min^{-1}; 360 and 660 mm min^{-1}	Joint efficiencies: 78·9–99·5%
Al 359+ 20 vol.-%SiC, 4 mm Al–10 wt-%TiB_2, 6 mm	AISI oil hardened steel	SD: 19 mm diameter; PD: 6·3 mm	1000 rev min^{-1}; 60–540 mm min^{-1}	Joint efficiencies: 80–98%
Al–15wt-% Mg_2Si, 6 mm	High C high Cr Steel (60–62 HRC) H13 steel	SD: 16 mm; PS: SSq, TSq, SOct, TOct, SHex, THex,	2000 rev min^{-1}; 30 mm min^{-1}	Joint efficiencies: 72–114%
AA 6061–(3–7)%TiC, 6 mm	High C, high Cr steel	SD: 18 mm; PS: TCT; PL: 5·7 mm PS: SSq, TSq, SHex, THex, TOct	710–1400 rev min^{-1}; 125 mm min^{-1} 30–135 mm min^{-1}	

Table 5.4 Tool materials, geometries and welding operating parameters used for FSW of titanium and its alloys [8]

Workpiece material	Tool material	Tool shape and size	Operating parameters	Remarks
cp-Ti, 3 mm	pcBN	SS: concave; SD: 15 mm;	200 rev min^{-1}; 50 mm min^{-1}; Ar shield	Severe tool wear
Ti, 3 mm	1. HSS; 2. WC pin, HSS shoulder; 3. WC pin, W shoulder	PS: tapered at 45° and truncated; PL: 1·7 mm; PD_t: 5·1 mm	(1250 rev min^{-1}; 32 mm s^{-1}), (1500 rev min^{-1}; 60 mm min^{-1}); tilt angle: 1, 3°	Up to 100% joint efficiency obtained with W–WC tool with low wear; low strength and high wear with other tools
Ti–6Al–4V, 3–12 mm	W–La alloy	SD: 18 mm; PS: SC; PD: 5 mm; PL: 2·85 mm	150–750 rev min^{-1}; 50–200 mm min^{-1}	Joint efficiency: 100%
Ti, 2 mm	WC–Co	SD: 19–32 mm; PS: tapered; PL: 2·8–13·3 mm	200–350 rev min^{-1}; 50–150 mm min^{-1}	Joints that failed in BM for some cases
Timetal 21 S, 1·59 mm	W alloy	SD: 15 mm; PL: 2 mm; PD: 6 mm	200 rev min^{-1}; 51–305 mm min^{-1}; Ar shield	No volumetric defects found
Ti, 5·6 mm	Sintered TiC	Proprietary	1000 rev min^{-1}; 500 mm min^{-1}	Joint efficiency: 97%
Ti–6Al–4V, 2 mm	W–3 wt-%Re	SD: 11 mm; PL: 1?8 mm; PD_t: 6 mm; PD_b: 4 mm	400 rev min^{-1}; 50 mm min^{-1}; 2·5u tilt; Ar shield	No volumetric defects found
Ti-5111 plate, 12·7 mm	W alloy	PL: 12·7 mm; PD_t: 25·4 mm; PD_b: 9·5 mm	140 rev min^{-1}; 51 mm min^{-1}	
Ti–15V–3Cr–3Al–3Sn, 3 mm	Mo based alloy	SS: convex; SD: 15 mm; PD_t: 5?1 mm; PD_b: 3 mm	400 rev min^{-1}; 60 mm min^{-1}; Ar shield	

Table 5.5 Tool materials, geometries and welding operating parameters used for FSW of ferrous alloys [8]

Workpiece material	Tool material	Tool shape and size	Operating parameters	Remarks
Fe–1·02C–0?24Si–0·37Mn–1·42Cr, 2·3 mm thick	pcBN	SD: 14 mm; PL: 2 mm; PD_t: 5·8 mm; PD_b: 4 mm	400–800 rev min^{-1}; 76 mm min^{-1}; Ar	Defect free welds produced at all rates
NSSC 270 superaustenitic SS, 6 mm	pcBN	Convex scrolled shoulder step spiral (CS4) pin tool	400 and 800 rev min^{-1}; 30–60 mm min^{-1}	Strength and ductility comparable with that of the base metal at 400 rev min^{-1}; more intermetallic phases at 800 rev min^{-1} caused poor joints
SAF 2507 super duplex SS, 4 mm	pcBN	SD: 25 mm; PL: 3·8 mm	450 rev min^{-1}; 60 mm min^{-1}; 3?5u tilt	Joint strength similar to base metal
DP 780 carbon steel, 1·5 mm	pcBN	SS: concave; PS: tapered, various step geometries; PL: 2 mm	800–1600 rev min^{-1}; dwell time: 1–10 s; FSSW	Lap shear strengths greater than RSW achieved for dwell time 8 s or greater
430 ferritic, 329J4L duplex, 304, 316L and 310 steels, 6 mm	pcBN	PL: 4·29 mm	550 rev min^{-1}; 80 mm min^{-1}; 3·5° tilt angle; Ar	Significant tool wear
Hot stamped boron steel, 1·4 mm	pcBN	SS: concave; SD: 10·2 mm; PL: 2·3 mm;	35 mm overlap welds; 800–2000 rev min^{-1}; 1?9–10·5 s welding time	'Hundreds' of welds made without significant wear
304L SS, 3·2 mm	W alloy	TC3F	300 and 500 rev min^{-1}; 102 mm s^{-1}; Ar	UTS of weld lager than UTS of base metal
15-5PH, 2·6 mm	W–25%Re	SD: 19 mm	300–450 rev min^{-1}; 60–350 mm min^{-1}; tilt angle: 3°; Ar	Joint efficiencies: 80–98%; tool wear at pin tip and shoulder edge
DP 600, 1·22 mm	W–25%Re	SS: concave; SD: 16 mm; PD: 6 mm; PL: 2·1 mm	3000 rev min^{-1}; plunge rate: 30–60 mm min^{-1} (FSSW)	Properties similar to RSW
Low carbon steel, 0·6 mm	1. WC–13% Co; 2. WC–13%Coz6%Ni, 1·5%Cr_3C_2	SD: 10 mm; PS: TC;		
Carbon steel, 1·6 mm	WC based			
SK5 steel, 1·6 mm	WC based			
AISI 1018 mild steel, 6·3 mm	Mo and W based tools			
DP 590 steel, 1·2 mm	Si_3N_4, with and without TiC, TiN coating			

(Continued)

Table 5.5 Continued

Workpiece material	Tool material	Tool shape and size	Operating parameters	Remarks
		PL: 1·7 mm; PD: 4–5·1 mm PS: SC	1600 rev min^{-1}; plunge rate: 15 mm min^{-1} (FSSW)	Acceptable strengths for all 500 welds; self-optimized tool after high initial wear
		SD: 12 mm; PD: 4 mm; PS: SC; PL: 1·4–1·5 mm	100–800 rev min^{-1}; 25–400 mm min^{-1}; 3° tilt	Joints stronger and more ductile than base metal
		SD: 12 mm; PD: 4 mm; PL: 1·5 mm	100–400 rev min^{-1}; 100–200 mm min^{-1}; 3° tilt; Ar shield 25·4–102 mm min^{-1}	Joints strengths similar to or higher than base metal
		SS: concave; SD: 10 mm; PL: 1·3 mm; PD: 4 mm	3000 rev min^{-1}; Ar (FSSW lap joint)	Defect-free welds were obtained and failure occurred in base metal; greatest tool wear occurred during plunge stage
				Contaminations with Si and N from tool caused reduction in strength

Tool steels

Tool steel is the most widely used tool material for welding aluminium alloys. Within the tool steels [3,47–55], AISI H13, a chromium–molybdenum hot worked air hardening steel, has been the most commonly-used material for this purpose [56,57]. This steel has good elevated temperature strength, thermal fatigue and wear resistance [3], easy availability and machinability, low cost and established material characteristics. In addition to FSW aluminium alloys, H13 tools have been used to FSW both oxygen-free copper (Cu-OF) and phosphorus-deoxidized copper with high residual phosphorus (Cu-DHP) [58]. However, the lower welding speed required for the welding of Cu-DHP limits the use of H13 in industrial applications.

Other tool steels used for FSW tools include oil hardened 0-1 [3], D2 [59], SKD61 [60], Orvar Supreme [61] and Divar [62]. The maximum use temperature of tool steels depends on the type of tool steel: oil- and water-hardened tool steels can be used up to 500 °C (930 °F) and secondary-hardened tool steels can be used up to 600 °C (1110 °F) [3]. Steel tools have also been used for joining dissimilar materials in both lap and butt joints configurations [8]. Tool wear during FSW of metal matrix composites is greater when compared with FSW of soft alloys, due to the presence of hard, abrasive phases in the composites. For FSW of aluminium matrix composites, some studies [48,50,62] have shown that the tool suffers an initial wear after which it obtains a self-optimized shape, leading to much less pronounced wear afterwards. This self-optimized final shape, which depends on the process parameters and is generally smooth with no threads, can reduce wear when used as the initial tool shape. Total wear was found to increase with rotational speed and decrease at lower welding speed, which suggests that process parameters can be adjusted to increase tool lifetime [48,50].

Nickel and cobalt-based alloys

Nickel- and cobalt-based super alloys were initially created for aircraft engine components applications for offering high strength, ductility, good creep and corrosion resistance. Their strength comes from the precipitates where the working temperature must be kept below the precipitation temperature (typically 600 to 800 °C, or 1110 to 1470 °F) to prevent precipitates suffering from over-ageing and dissolution. However, super alloys are difficult to machine, which makes it not possible to manufacture complex features such as flutes and/or flats on the tool profile [2]. Nickel-based alloys, such as IN738LC, IN939, MAR-M-002, Stellite 12, IN-100, PM 3030, Nimonic 90, Inconel 718, Waspalloy and Nimonic 105 [40], have been used to FSW copper alloys. The cobalt nickel-based alloy MP 159, readily machinable [3], has been used as tool material in the FSW of aluminium alloys, particularly in the 7xxx series, to reduce tool wear during the welding process [63,64]. Although harder tool materials exhibit improved wear performance, their use encourages conservative parameter selection in such a way that concerns about tool fracture can potentially narrow the operational window for the FSW process [65].

Refractory metals

Tungsten, molybdenum, niobium and tantalum are used as tool materials due to their high temperature capabilities and high densities [2]. Because the majority of these alloys are produced as a single phase, their strength is maintained to nearly the melting point (1000 to 1500 °C, or 1830 and 2730 °F). However, tantalum and niobium have high solubility of oxygen at elevated temperatures, which quickly degrades the ductility. The refractory metals have limited availability, long lead times, high costs and are difficult to machine (typically grinding processes are required).

Tungsten-based alloys have been used in the friction stirring of copper alloys, nickel aluminium bronze, titanium alloys and steels [3]. Pure tungsten is strong at elevated temperatures but has poor toughness at room temperature, and has a fast wear when used as a tool material for FSW of steels and titanium alloys. However, the durability of tool materials such as tungsten, tungsten carbide (WC) or molybdenum can be improved by alloying to the base tool material more ductile materials such as cobalt or rhenium [66]. Four tungsten-based materials have been specifically reported for friction stirring tools, namely W, W-25%Re, Densimet [11] and W-1%LaO_2 [3]. Tungsten-rhenium alloy has a high operational temperature and machining features require the application of a grinding process; beyond that, tungsten-rhenium alloy has a high production cost.

Densimet consists of small spheres of tungsten bound in a matrix containing either nickel-iron or nickel-copper combinations [11]. However, in contrast to other tungsten-based alloys, Densimet is readily machinable by conventional methods and has a lower raw material cost. The high thermal conductivity of Densimet has been cited as a reason to use this material for the shoulder of FSW tools [39] used to weld thick copper [3].

Carbides and metal-matrix composites

Carbide materials that are commonly used in machining tools offer superior wear resistance and reasonable fracture toughness as a probe/shoulder material for FSW at room temperature. Ceramic particle reinforced metal-matrix composites (MMCs) have also been used as tool materials, but the brittle nature of the composite can result in fracture during the tool plunging phase [2]. Carbines have a better performance at elevated temperatures, because they are more suitable for producing machining tools. Friction stirring tools made from tungsten carbide are reported to have smooth and uniform thread surfaces [67]. However, severe wear is observed when the tools contain threads.

The high-temperature strength of WC and WC-Co tools was used to weld interstitial-free steel and carbon S45C steel to 6064 Al [68,69]. MMCs using TiC as the reinforcing phase have also been used as tool materials for copper alloys. Both sintered TiC:Ni:W and hipped TiC:Ni:Mo alloys were used to friction stir copper alloys [19]. However, both TiC-containing alloys produced brittle tools that fractured during the tool plunge [3]. An option for wear reduction in FSW of MMCs is the use of functionally gradient materials (FGM), a type of composite having a high percentage of reinforcement everywhere, except near the joint line [70].

Cubic boron nitride

Polycrystalline cubic boron nitride (PCBN) was originally developed for the turning and machining of tool steels, cast irons and super alloys [3,71]. Recently, PCBN has gained acceptance as a friction stir tool material, especially for high-temperature alloys [3]. The PCBN was chosen as a friction stir tool based on its prior success in extreme machining applications [3]. Furthermore, it has a low coefficient of friction that results in a smooth weld surface [72]. However, due to high temperatures and pressures required in the manufacturing process of PCBN, the tool production costs are very high. Owing to its low fracture toughness, it also has a tendency to fail during the initial plunge stage [8]. PCBN is only used in the shoulder and/or in the probe; the tool body is made from other materials such as tungsten carbide. The PCBN tools require a low eccentricity spindle to minimize tool fracture. Successful PCBN FSW have been made with ferritic steels [71], dual phase steels, austenitic stainless steels, type 430 stainless steel, 2507 super duplex stainless steel, nickel-based alloys, Narloy Z, nickel aluminium bronze,

Invar, class 40 grey cast iron [73], copper [40], sonoston [74], ultrafine-grained steels and nitinol [13,60]. PCBN is the preferred tool material for FSW of hard alloys such as steels and titanium alloys [8]. Tool wear in PCBN is insignificant compared with titanium-rhenium alloy.

5.3 Effect of tool geometry on material flow and weld properties

Tool geometry is an essential parameter in the FSW process, where the knowledge of its influence on heat generation, material flow and weld properties needs more research [75]. Indeed, most of the studies consider the effect of the tool shoulder and probe separately, since the tools geometry is complex and difficult to characterize.

The effect of tool probe design was studied in terms of the weldability and mechanical properties of FSW joints of 2014 aluminium alloy [76]. When tapered tools with threads were used, defect-free welds were obtained, in contrast with the cylindrical and tapered tool probes that did not promote effective mixing in the vertical direction leading to wormholes at the base of the TMAZ. Other studies have also confirmed that tools with screw threads generate more heat and improve flow of the softer material by employing a downward force.

On the advancing side (AS), a wormhole is generated due to the insufficient plasticity and material flow, since the material streams mainly on the RS [59,77]. At low temperatures this effect becomes more prominent, due to the slower flow of materials. Another important parameter of the FSW process is the angle between the conical surface of the probe and its axis. A more uniform temperature distribution along the vertical direction is observed by increasing the probe angle, which helps in reducing distortion [78]. An angle of around 40 degrees is assumed to be optimum for 7xxx aluminium alloys [78].

The influences of tool probe profiles were related to the formation of the FSP zone in an aluminium alloy [79]. Mechanically sound and metallurgically defect-free welds were produced using a square probe profiled tool in comparison with welds fabricated by other tool probe profiles. It was found that the joints manufactured by square tool profiles with straight cylindrical, tapered cylindrical, threaded cylindrical, square and triangular probes exhibited more superior tensile properties than those fabricated by others tool probe profiles. Therefore, the correct selection of the probe profile is fundamental in the material flow and in turn regulates the welding speed of the FSW process. The triangular and square probe profiles produce pulsating stirring action in the flowing material, due to flat faces in contrast with the case of cylindrical, tapered and threaded probe profiles, where no such pulsating action was observed.

The influence of probe geometry on the macrostructure, microstructure and mechanical properties of FSW, reinforced with SiC nano-particles, was studied using five geometrically different probe tools [80]. The most uniform particle distribution was found in the case of using a threaded tapered probe tool. However, by using a four-flute cylindrical probe tool, it was found that reinforcements had harshly accumulated in the SZ of the FSW specimen. Furthermore, threaded tapered specimens showed the highest microhardness, and the lowest microhardness was found in the four-flute cylindrical specimens. In spite of four-flute cylindrical specimens, other specimens failed in base metal (BM) during tensile testing. The highest ultimate tensile strength (UTS) was recorded for the specimen FSW with a triangular probe tool. Fracture surface of tensile specimens were also observed using the scanning electron microscope (SEM) technique.

The most uniform particle distribution within the SZ was found in joints produced with a threaded tapered probe. However, in the triangular probe specimens were observed clusters in the range of 250 nm. However, inappropriate material flow in a four-flute cylindrical specimen resulted in severe accumulation of reinforcements, as well as the formation of large clusters in the order of 480 nm.

Although threaded tapered, triangular, square and four-flute square specimens had almost identical strength, the UTS of the four-flute cylindrical specimen was drastically reduced by voids and the highest UTS value was obtained with the triangular probe tool.

A study concerning the influence of the tool shoulder geometry on properties of FSW in thin copper sheets pointed out that the flat shoulder tool is inadequate for performing welds, because several defects were observed for all welding conditions [81]. On the other hand, the scrolled shoulder tool is more effective than the conical one in the production of defect free welds. However, to avoid internal defects, both geometries required a minimum rotational speed. In the weld nugget, a greater grain refinement, higher hardness and improved strength were also obtained, using the scrolled tool and the same welding parameters.

The effect of tool geometry on 5 mm thick AA5086-O and AA6061-T6 joints revealed that a concave shoulder and a conical probe with three grooves provided more homogeneous SZs compared to other tool profiles due to higher heat input [82].

The material flow in FSW is complex and not clearly understood. Research in material flow have been performed in different materials using distinct techniques such as the introduction of marker materials, stop action, microstructural observations, welding dissimilar materials, computational modelling or a combination of two or more of these techniques. The insertion of the marker material in the weld line can change the nature of material flow in the weld, due to different material flow characteristics of the base and marker materials and introduction of additional interfaces [83]. The material flow in FSW of similar materials cannot be compared with the material flow in dissimilar materials of different flow properties.

In FSW joint formation, there are two different modes of material flow regimes involved [84], the probe driven and the shoulder driven flows. It is possible to obtain defect-free welded joints by merging these material flow regimes.

The probe design should be such that the maximum quantity of transferred material is retained in the weld cavity. The shoulder driven material flow can be defined as the effectiveness of the shoulder to retain the material in the weld cavity. Consequently, the tool shoulder should be designed to maximize the amount of the ejected material from the weld cavity by the probe, reflecting back into the weld cavity [84].

The main differences between the two flow mechanisms are related to the deep differences in the shoulder driven material flows around the tool. The shoulder influence area is discontinuous and has different characteristics at the leading and trailing sides of the tool with a conical cavity. For a tool with a scrolled shoulder, it was verified that the shoulder influence area spreads through the entire thickness, all around the tool, with a straight interface with the surrounding base materials (BM) [85].

The flow of material in the thickness direction, mainly in thick plates, is related to the tool probe. The increased flow of material in the thickness direction in complex geometries is associated with the increased interface area between the tool and the workpiece. The presence/absence of the tool probe affects the material flow. The use of a scrolled shoulder without tool tilting allows the welding of aluminium alloys, improving the surface finish and the mechanical properties of the welds and allows higher welding speeds [86]. The cavity in a concave shoulder is required namely:

- to form a compressed annular ring of workpiece material around the tool probe;
- to prevent the escape of plasticized material; and
- to direct the material flow towards the probe [87].

The effect of tool shoulder geometry on the material flow during heterogeneous FSW of thin plates of two automotive aluminium alloys was developed using a tool shoulder with a conical cavity and a scrolled tool shoulder [85]. The shoulder with a conical cavity produces welds with exceptional appearance, but with some thickness reduction. For these welds, the characteristics of the weld nugget depends of the probe driven flow, yielding an onion ring structure promoted by the flux of plasticized material around the probe and through the plate thickness. However, the influence of the shoulder on weld formation is also important, since it drags material into the shear layer that creates the weld nugget, through two different mechanisms acting simultaneously at the leading side of the weld and at the weld crown. Welds performed with a scrolled shoulder tool were less smooth and did not display any reduction in thickness. In these welds, a substantial interaction between probe driven and shoulder driven material flow occurs, the latter being determinant to the weld structure.

The material flow is described as a chaotic-dynamic mix in FSW of dissimilar aluminium alloys [88]. The material flow pattern of a FSW aluminium alloy was studied via a stop action technique using a FSW probe tool with left-hand thread (LHT) in a clockwise rotation. Superior welded joints were obtained and it was stated that the material flow was dependent on the exact tool geometry and welding parameters used [89].

The most relevant part of the heat flux generated during FSW can be attributed to the frictional force at the tool shoulder–workpiece interface [90]. By increasing the shoulder diameter, a strong beneficial effect on both ductility and strength of the joint was obtained [91].

5.4 Welding design

The square butt joint and the lap joint are the most used FSW joint configurations. These two and other joint geometries commonly used in the FSW process are shown in Figure 5.7. Other typical joint types, not shown in the figure, are pipe welds, hemispherical welds, multiple lap welds, double T-joint welds, corner welds and fillet welds.

FSW makes the elimination of rivets and fasteners possible, as well as the demanding of an overlap joint configuration. The butt joint configuration also aids joint evaluation and quality control, because crack formation is minimized due to joint homogeneity and its full penetration. Butt joints can be carried out in the closed square style, even in thick pieces larger than 6 mm, and FSW is normally tolerant to light oxide layers [66,93]. To weld overlap and butt joints is possible in all positions; mixed welds can also be performed in different thicknesses or

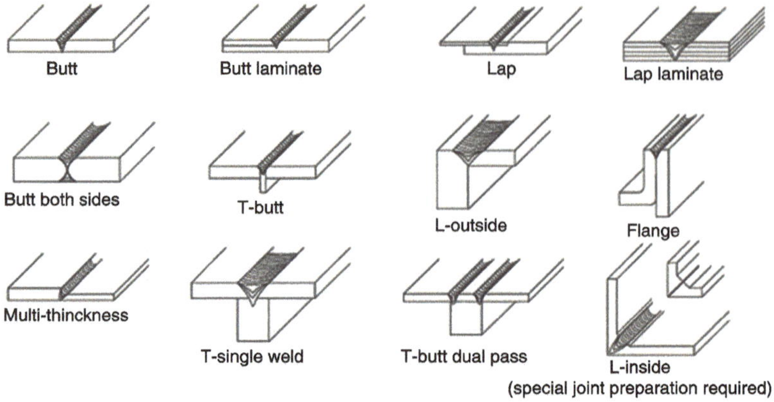

Figure 5.7 Friction stir joint geometries [92]

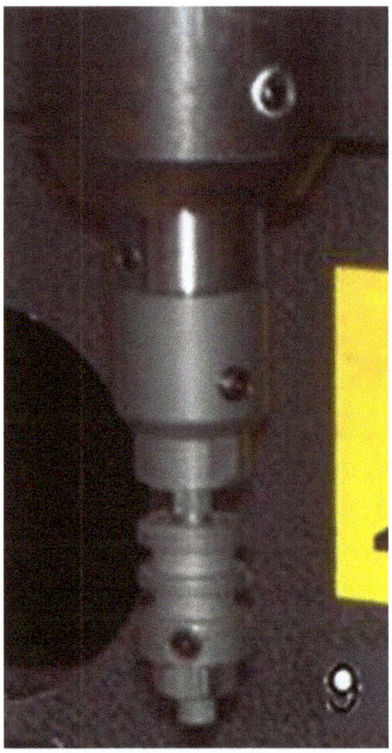

Figure 5.8 Double-sided bobbin tool [95]

different materials [94]. To FSW a lap joint, a double-sided bobbin tool (Figure 5.8) can be used instead of a conventional tool [95].

One of the main advantages of FSW is that the welded products are ready to use. With adequate design, the elements are ready to use immediately after welding. However, it is important to remember that designs suitable for MIG or TIG welding are not necessarily appropriate for FSW. A fillet joint geometry, commonly used with MIG, may not be suitable for FSW, for which T-joint geometry is much more appropriate (Figure 5.9).

(a)

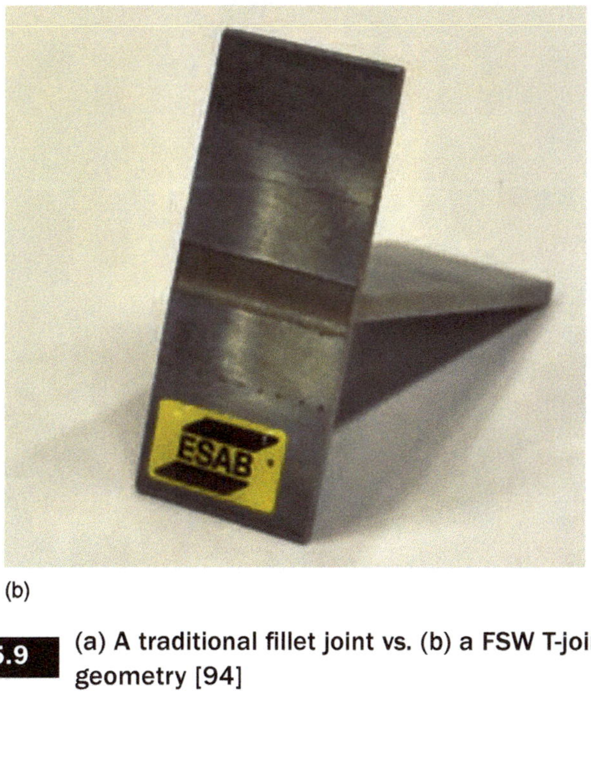

(b)

Figure 5.9 (a) A traditional fillet joint vs. (b) a FSW T-joint geometry [94]

To assemble a butt joint configuration, two plates or sheets are positioned on a backing plate and properly clamped to avoid the abutting joint faces from being forced apart. The role of the backing plate is to resist the normal forces associated with FSW and the workpiece. In comparison, for a lap joint configuration, two lapped plates or sheets are clamped, and a backing plate may or may not be required, depending on the lower plate thickness. However, the tool designed for a butt joint, with the faying surfaces parallel to the tool rotation axis, is not ideal for a lap joint, with the faying surfaces normal to the tool rotation axis [1].

A method to solve the fixture problem is to use an FSW-friendly design, eliminating the need for a backing bar. One example where this approach is well used is in the manufacture of extruded profiles (Figure 5.10).

As referred to above, with the FSW process, plates of different thicknesses can be joined in a proper way, obtaining high-quality sets (Figure 5.11). Overlap joints produced by FSW provide a new alternative method to resistance-spot-welded or seam-welded pieces. An optimal option to spot-welding to achieve a watertight seal may be seen in Figure 5.11. A friction spot joint (Figure 5.12(a)) can be obtained using a tool such that shown in Figure 5.12(b) [96].

Figure 5.10 Designs which make it possible to FS weld hollow profiles [94]

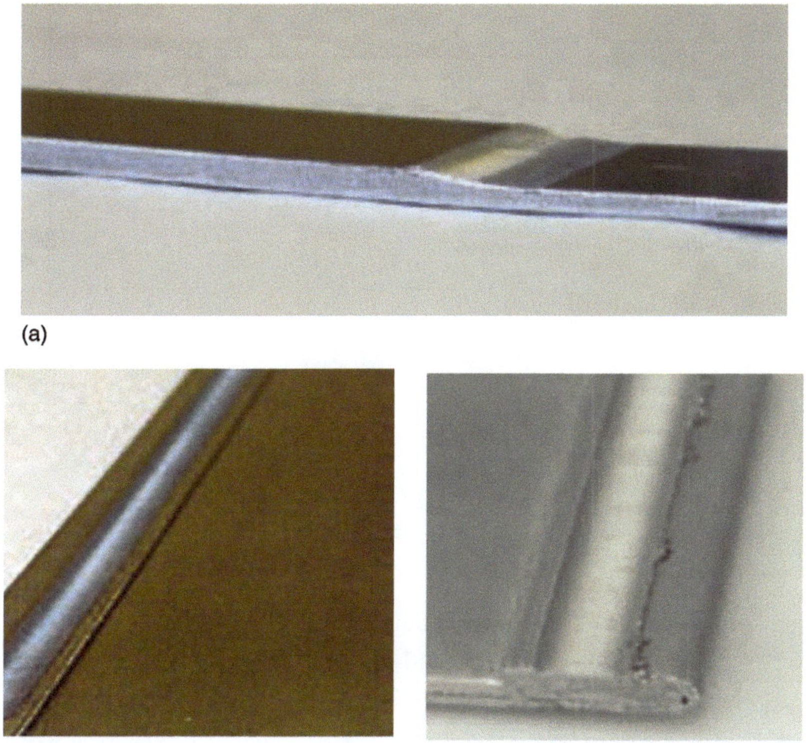

(a) (b) (c)

Figure 5.11 **Potential automotive applications produced by FSW: (a) mixed joint between two thicknesses (1 + 2 mm); (b) overlap joint on 1 mm thickness; and (c) folded seal weld (©ESAB) [94]**

Simple butt-joint configurations are the most used in commercial FSW applications. Other alternative designs, such as T-sections and corner welds are rarely considered. An alternative method of the Stationary Shoulder Friction stir Welding (SSFSW) was established to produce high-quality internal corner welds [97].

The SSFSW method was developed with the objective of welding low heat conductivity materials such as titanium alloys, where a more uniform heat input into the weld is beneficial. The SSFSW technique allowed joining of plates, which are placed in distinct angular planes such as T-joints by using a shaped stationary shoulder. This method may overcome some of the problems of welding T-joints using conventional FSW tools [98] of eliminating joint interfaces and the formation of small cracks in the corners [99], and also eliminating undercutting of the weld

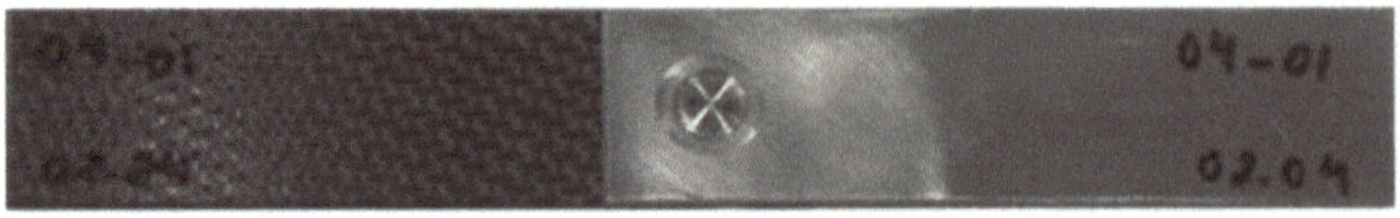

(a)

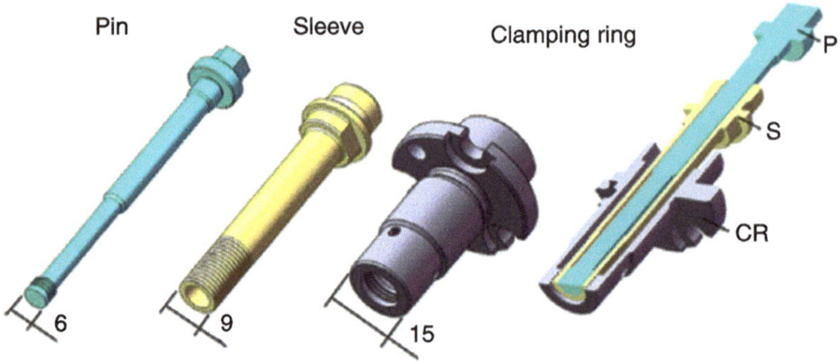

(b)

Figure 5.12 **(a) Top view of a sound friction spot joint of AA2024-T3 (bare)/CF-PPS; and (b) the tool comprised of a pin, sleeve and clamping ring (dimensions are in mm) [96]**

section. Small bulk void defects were observed in unreinforced T-butt joints manufactured by the SSFSW technique [100].

Initial trials manufacturing T-sections samples in 8 mm thick wrought aluminium alloy 6082-T6 plates, and also in a dissimilar alloy combination of AA7075-T6 and AA2014-T6, produced welds of excellent quality and components with low distortion (Figure 5.13).

Friction stir Lap Linear Welding (FSLLW) has been widely used in automotive and aeronautic industries to replace self-pierce rivet joining, as well as to manufacture firm panels and beam assemblies where attachment of stringers and frames to the main body of a structure is required [101]. However, material selection has an effect on the FSLLW process. When it is applied to two intrinsically different materials that have great differences in their melting points, and/or form brittle intermetallic compounds (e.g. aluminium and magnesium, aluminium/magnesium and steel, aluminium and copper), the welding tool probe often penetrates only into the upper plate rather than through the interface of the overlapped metal plates [102,103].

T-lap joints are widely used in back-supported seats for cars, bridge structures and support frames for pressure vessels [104]. Three different joint geometries of T-lap/T-butt-lap/T-butt were successfully produced by

(a)

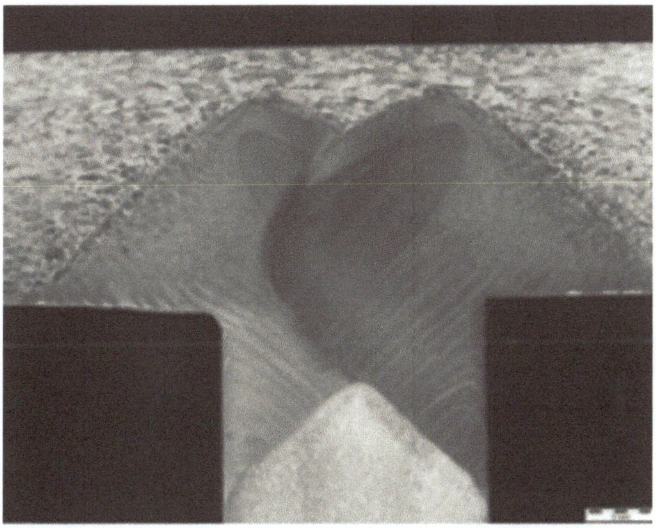

(b)

Figure 5.13 **(a) General appearance; and (b) macrograph of a T-section fabricated by Corner SSFSW in 8 mm thick AA7075-T6 and AA2014-T6 [96]**

FSW using the AA6061-T4 aluminium alloy and the tool presented in Figure 5.14 [105]. The T-lap joints had the higher tensile properties along the skin direction, while the T-butt joints showed the best mechanical performance along the stringer direction.

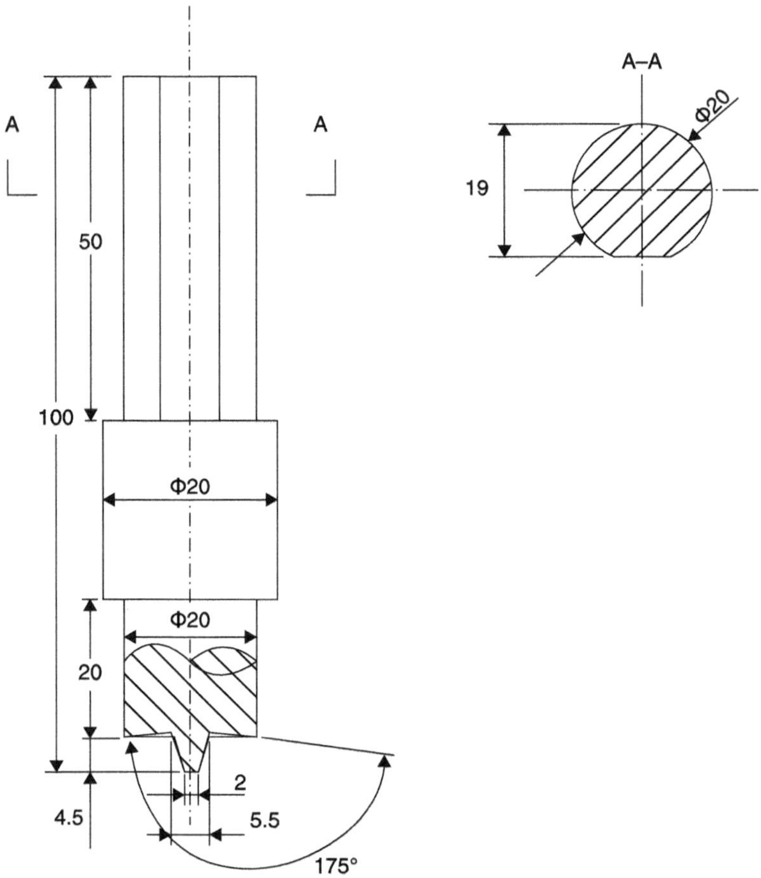

Figure 5.14 Schematic profile of the FSW tool used to produce T-lap, T-butt-lap and T-butt joints [105]

5.5 Summary

For the past 20 years, significant developments have been achieved concerning the FSW process. Numerous welding tools, shoulders and probes have already been designed and proposed throughout the entire process evolution. To date, low cost and long-life FSW tools have been well-developed for low-strength materials such as aluminium and magnesium alloys. However, long-life tools with reasonable costs are still unavailable for abrasive materials such as particle reinforced metal matrix composites, and high strength materials such as titanium, nickel and steels. To this end, further efforts should be made to develop new tool materials and design new effective special tools. Tool material properties such as strength, fracture toughness, hardness, thermal conductivity and thermal expansion coefficients affect the weld quality, tool wear and performance.

The tool design influences heat generation, plastic flow, the power required, and the uniformity of the welded joint. The material flow during the FSW process is still not fully understood due to its complexity. FSW production technology is being considered for future designs. A number of different applications in the commercial and military aircraft industry are under evaluation.

5.6 Acknowledgements

The authors would like to acknowledge the Portuguese Foundation for the Science and Technology (FCT) for its financial support through the PhD scholarship FCT SFRH/BD/62963/2009.

5.7 References

1. Mishra, R.S. and Mahoney, M.W. (2007), Chapter 1: 'Introduction, friction stir welding and processing', *ASM International*.
2. Zhang, Y.N., Cao, X., Larose, S. and Wanjara, P. (2012), 'Review of tools for friction stir welding and processing', *Science and Technology of Welding and Joining*, 51(3): 250–60.
3. Fuller, C.B. (2007), Chapter 2: 'Friction stir tooling: Tool materials and designs, friction stir welding and processing', *ASM International*.
4. Rowe, C.E.D. and Thomas, W.M. (2006), 'Advances in tooling materials for friction stir welding (Cedar Metals Ltd, TWI Cambridge)', *Materials Congress – Disruptive Technologies for Light Metals*, London, April.
5. Thomas, W.M., Threadgill, P.L. and Nicholas, E.D. (1999), 'Feasibility of friction stir welding steel', *Science and Technology of Welding and Joining*, 4(6): 365–72.
6. Bhadeshia, H. and DebRoy, T. (2009), 'Critical assessment: friction stir welding of steels', *Science and Technology of Welding and Joining*, 14(3): 193–6.
7. Çam, G. (2011), 'Friction stir welded structural materials: beyond Al-alloys', *International Materials Reviews*, 56(1): 1–48.
8. Rai, R., De, A., Bhadeshia, H.K.D.H. and DebRoy, T. (2011), 'Review: Friction stir welding tools', *Science and Technology of Welding and Joining*, 16(4): 325–42.

9. Konkol, P. (2003), 'Characterization of friction stir weldments in 500 Brinell hardness quenched and tempered steel', *Proceedings of the Fourth International Conference on Friction stir Welding*, Park City, UT, TWI, September.
10. Mahoney, M.W., Bingel, W.H., Sharma, S.R. and Mishra, R.S. (2003), 'Microstructural modification and resultant properties of friction stir processed cast NiAl bronze', *Materials Science* Forum, 426–32: 2843–8.
11. Densimet Data Sheet, Trade Literature, Plansee GmbH, Germany.
12. Attallah, M. and Salem, H.G. (2004), 'Effect of friction stir welding process parameters on the mechanical properties of the as-welded and post-weld heat treated AA2095', *Proceedings of the Fifth International Conference on Friction stir Welding*, Metz, France, TWI, September.
13. Charit, I. and Mishra, R.S. (2005), 'Low temperature superplasticity in a friction stir processed ultrafine-grained Al-Zn-Mg- Sc alloy', *Acta Materialia*, 53: 4211–23.
14. Ding, R.J. and Oelgoetz, P.A. (1999), 'Auto-adjustable probe tool for friction stir welding', US Patent No. 5893507.
15. Ding, R.J. (2000), 'Force characterization on the welding pin of a friction stir welding retractable pin-tool using aluminium-lithium 2195', *Proceedings of the Second International Conference on Friction stir Welding*, Gothenburg, Sweden, TWI, June.
16. Thomas, W.M., Nicholas, E.D. and Smith, S.D. (2001), 'Friction stir welding-tool developments', *Proceedings Aluminium Automotive and Joining Sessions*, Warrendale, PA, TMS, 213–24.
17. Skinner, M. and Edwards, R.L. (2003), 'Improvements to the FSW process using the self-reacting technology', *Materials Science Forum*, 426: 2849–54.
18. Thomas, W.M. (1996), 'Friction stir welding', UK Patent Application 2, 306–66.
19. Sutton, M.A., Reynolds, A.P., Yan, J., Yang, B. and Yuan, N. (2006), 'Microstructure and mixed mode I/II fracture of AA2524-T351 base material and friction stir welds', *Engineering Fracture Mechanics*, 73(4): 391–407.
20. Sutton, M.A., Reynolds, A.P., Yang, B. and Taylor, R. (2003), 'Mode I fracture and microstructure for 2024-T3 friction stir welds', *Materials Science Engineering A*, 354, 6–16.
21. Yan, J., Sutton, M.A. and Reynolds, A.P. (2005), 'Process structure property relationships for nugget and heat affected zone regions of AA2524-T351 friction stir welds', *Science and Technology of Welding and Joining*, 10(6): 725–36.
22. Baskoro, A.S. et al. (2013), 'Effects of welding parameters in micro friction stir lap welding of aluminium A1100', *Advanced Materials Research*, 789: 356–9.
23. Brinckmann, S., Strombeck, A., Schilling, C., dos Santos, J.F., Lohwasser, D. and Koçak, M. (2000), 'Mechanical and toughness properties of robotic-FSW repair welds in 6061-T6 aluminium alloys', *Proceedings of the Second International Conference on Friction stir Welding*, Gothenburg, Sweden, TWI, June.
24. Colligan, K.J., Xu, J. and Pickens, J.R. (2003), 'Welding tool and process parameter effects in friction stir welding of aluminium alloys', *Proceedings of*

the Conference: Friction stir Welding and Processing II, 181–90, Warrendale, PA, TMS.
25. Colligan, K.J. and Pickens, J.R. (2005), 'Friction stir welding of aluminium using a tapered shoulder tool', *Proceedings of the Conference: Friction stir Welding and Processing III*, 161–70, San Francisco, CA.
26. Colligan, K. (2003), 'Tapered friction stir welding tool', US Patent No. 6669075.
27. Thomas, W.M. et al. (1991), 'Friction stir butt welding', GB Patent No. 9125978-8.
28. Lorrain, O., Favier, V., Zahrouni, H. and Lawrjaniec, D. (2010), 'Understanding the material flow path of friction stir welding process using unthreaded tools', *Journal of Materials Processing Technology*, 210(4): 603–9.
29. Choi, D.H., Ahn, B.W., Lee, C.Y., Yeon, Y.M., Song, K.U. and Jung, S.B. (2010), 'Effect of pin shapes on joint characteristics of friction stir spot welded AA5J32 sheet', *Materials Transactions*, 51(5): 1028–32.
30. Hirasawa, S., Badarinarayan, H., Okamoto, K., Tomimura, T. and Kawanami, T. (2010), 'Analysis of effect of tool geometry on plastic flow during friction stir spot welding using particle method', *Journal of Materials Processing Technology*, 210(11): 1455–63.
31. Chowdhury, S.M., Chen, D.L., Bhole, S.D. and Cao, X. (2010), 'Tensile properties of a friction stir welded magnesium alloy: Effect of pin tool thread orientation and weld pitch', *Materials Science Engineering A*, A527(21–22): 6064–75.
32. Thomas, W.M. et al. (1991), 'Friction stir butt welding', International Patent Application PCT/GB92/02203 and GB Patent Application 9125978-8.
33. Dawes, C.J., Threadgill, P.L., Spurgin, E.J.R. and Staines, D.G. (1995), 'Development of the new friction stir technique for welding aluminium: Phase II', TWI member report 5651/35/95.
34. Thomas, W.M., Nicholas, E.D. and Smith, S.D. (2001), 'Friction stir welding-tool developments', *Proceedings of the TMS Aluminium Automotive and Joining Sessions*, 213.
35. Thomas, W.M., Staines, D.G., Norris, I.M. and de Frias, R. (2002), 'Friction stir welding-tools and developments', *FSW Seminar*, Porto, Portugal: Instituto de Soldadura e Qualidade.
36. Aissani, M., Gachi, S., Boubenider, F. and Benkedda, Y. (2010), 'Design and optimization of friction stir welding tool', *Materials and Manufacturing Processes*, 25(11): 1199–205.
37. Mitsuo, H. (2000), 'Friction agitation joining method and friction agitation joining device', Japan Patent Application 2000-094156, 4 April.
38. Atsuo, K., Yoshinori, O. and Mutsumi, Y. (2003), 'Friction stir welding method', Japan Patent Application 2003-112271 and 2003-112272, 15 April.
39. Strangwood, M., Davis, C. and Attallah, M. (2004), 'Microstructural development and thermal fields modelling in friction stir welds of strain-hardenable Al-*alloys*', *Proceedings of the Fifth International Conference on Friction stir Welding*, Metz, France.
40. Cederqvist, L. (2003), 'A weld that lasts for 100 000 years: FSW of copper canisters', *Proceedings of the Fourth International Conference on Friction stir Welding*, Park City, UT.

41. Ma, Z.Y., Sharma, S.R., Mishra, R.S. and Mahoney, M.W. (2003), 'Microstructural modification of cast aluminium alloys via friction stir processing', *Materials Science Forum*, 426–32: 2891–6.
42. Thomas, W.M., Lockyer, S.A., Kalee, S.W. and Staines, D.G. (2003), 'Friction stir welding: an update on recent developments', *Proceedings of the Conference on Stressed Components in Aluminium Alloys*, Birmingham, UK.
43. Thomas, W.M., Johnson, K.I. and Wiesner, C.S. (2003), 'Friction stir welding recent developments in tool and process technologies', *Advanced Engineering Materials*, 5(7): 485–90.
44. Colegrove, P.A., Shercliff, H.R. and Threadgill, P.L. (2003), 'Modeling and development of the Trivex friction stir welding tool', *Proceedings of the Fourth International Conference on Friction stir Welding*, Park City, UT.
45. Colegrove, P.A. and Shercliff, H.R. (2004), 'Development of Trivex friction stir welding tool, Part I: Two-dimensional flow modeling', *Science and Technology of Welding and Joining*, 9(4): 345–51.
46. Zettler, R., Lomolino, S., dos Santos, J.F., Donath, T., Beckmann, F. et al. (2004), 'A study of material flow in FSW of AA2024-T351 and AA 6056- T4 alloys', *Proceedings of the Fifth International Conference on Friction stir Welding*, Metz, France.
47. Prado, R.A., Murr, L.E., Soto, K.F. and McClure, J.C. (2003), 'Self-optimization in tool wear for friction stir welding of Al 6061z20% Al_2O_3 MMC', *Materials Science Engineering A*, A349(1–2): 156–65.
48. Prado, R.A., Murr, L.E., Shindo, D.J. and Soto, K.F. (2001), 'Tool wear in the friction stir welding of aluminium alloy 6061z20% Al_2O_3: A preliminary study', *Scripta Materialia*, 45(1): 75–80.
49. Fernandez, G.J. and Murr, L.E. (2004), 'Characterization of tool wear and weld optimization in the friction stir welding of cast aluminium 359z20% SiC metal-matrix composite', *Materials Characterization*, 52(1): 65–75.
50. Lee, W.B., Yeon, Y.M. and Jung, S.B. (2003), 'Joint properties of friction stir welded AZ31B-H24 magnesium alloy', *Materials Science and Technology*, 19(6): 785–90.
51. Rodriguez, N.A., Almanza, E., Alvarez, C.J. and Murr, L.E. (2005), 'Study of friction stir welded A319 and A413 aluminium casting alloys', *Journal of Materials Science*, 40(16): 4307–12.
52. Nami, H., Adgi, H., Sharifitabar, M. and Shamabadi, H. (2011), 'Microstructure and mechanical properties of friction stir welded Al/Mg2Si metal matrix cast composite', *Materials and Design*, 32(2): 976–83.
53. Singh, R.K.R., Sharma, C., Dwivedi, D.K., Mehta, N.K. and Kumar, P. (2011), 'The microstructure and mechanical properties of friction stir welded Al–Zn–Mg alloy in as welded and heat treated conditions', *Materials and Design*, 32(2): 682–7.
54. Rajakumar, S., Muralidharan, C. and Balasubramanian, V. (2011), 'Influence of friction stir welding process and tool parameters on strength properties of AA7075-T6 aluminium alloy joints', *Materials and Design*, 32(2): 535–49.
55. Thomas, W.M., Nicholas, E.D., Watts, E.R. and Staines, D.G. (2002), 'Friction based welding technology for aluminium', *Materials Science Forum*, 396–402: 1543–8.

56. Feng, A.H. and Ma, Z.Y. (2007), 'Formation of Cu_2FeAl7 phase in friction stir welded SiCp/Al-Cu-Mg composite', *Scripta Materialia*, 57: 1113–06.
57. Najafi, M., Nasiri, A.M. and Kokabi, A.H. (2008), 'Microstructure and hardness of friction stir processed AZ31 with SiCp', *International Journal of Modern Physics B*, 22B(18–19): 2879–85.
58. Savolanen, K., Mononen, J., Saukkonen, T., Hänninen, H. and Koivula, J. (2004), 'Friction stir weldability of copper alloys', *Proceedings of the Fifth International Conference on Friction stir Welding*, Metz, France.
59. Guerra, M., Schmidt, C., McClure, J.C., Murr, L.E. and Nunes, A.C. (2002), 'Flow patterns during friction stir welding', *Materials Characterization*, 49: 95–101.
60. Fujii, H. et al. (2006), 'Friction stir welding of ultrafine grained interstitial free steels', *Materials Transactions*, 47(1): 239–42.
61. Midling, O.T. and Rorvik, G. (1999), 'Effect of tool shoulder material on heat input during friction stir welding', *Proceedings of the First International Conference on Friction stir Welding*, Thousand Oaks, CA.
62. Colegrove, P.A. and Shercliff, H.R. (2003), 'Experimental and numerical analysis of aluminium alloy 7075-T7351 friction stir welds', *Science and Technology of Welding and Joining*, 8(5): 360–8.
63. Shindo, D.J., Rivera, A.R. and Murr, L.E. (2002), 'Shape optimization for tool wear in the friction stir welding of cast Al359-20% SiC MMC', *Journal of Materials Science*, 37(23): 4999–5005.
64. Andrews, D., Russell, M. and Martin, J. (2012), 'Recent developments in FSW at TWI', *Proceedings of the 9th International Symposium on Friction stir Welding*, Cambridge, UK.
65. Reynolds, A. and Tang, W. (2012), 'Thermal management for production of very high strength alloy 7050 friction stir welds', *Proceedings of the 9th International Symposium on Friction stir Welding*, Cambridge, UK.
66. Gibson et al. (2013), 'Friction stir welding: Process, automation, and control', *Journal of Manufacturing Processes*, available from: *http://dx.doi.org/10.1016/j.jmapro.2013.04.002*
67. Colligan, K. (2000), 'Dynamic material deformation during friction stir welding of aluminium', *Proceedings of the Second International Conference on Friction stir Welding*, Gothenburg, Sweden.
68. Yasui, T., Ishii, T., Shimoda, Y., Tsubaki, M., Fukumoto, M. and Shinoda, T. (2004), 'Friction stir welding between aluminium and steel with high welding speed', *Proceedings of the Fifth International Conference on Friction stir Welding*, Metz, France.
69. Fukumoto, M., Yasui, T., Shimoda, Y., Tsubaki, M. and Shinoda, T. (2004), 'Butt welding between dissimilar metals by friction stirring', *Proceedings of the Fifth International Conference on Friction stir Welding*, Metz, France.
70. Lee, J.A., Carter, R.W. and Ding, J. (1998), 'Friction stir welding of aluminium metal matrix composites', NASA technical report.
71. Sorenson, C.D., Nelson, T.W., Packer, S.M. and Steel, R.J. (2004), 'Innovative technology applications in FSW of high softening temperature materials', *Proceedings of the Fifth International Conference on Friction stir Welding*, Metz, France.

72. Sato, Y.S., Harayama, N., Kokawa, H., Inoue, H., Tadokoro, Y. and Tsuge, S. (2009), 'Evaluation of microstructure and properties in friction stir welded super austenitic stainless steel', *Science and Technology of Welding and Joining*, 14(3): 202–9.
73. Ramasubramanian, U., Arbegast, W.J., Stone, G.A. and Grant, G. (2005), 'Friction stir processing of class 40 grey cast iron', *Friction stir Welding and Processing III*, 115–22.
74. Lynch, S., Edwards, D.P., Majumdar, A., Moutsos, S. and Mahoney, M.W. (2003), 'Friction stir processing of a high-damping mn-cu alloy used for marine propellers', *Materials Science Forum*, 426–32: 2903–8.
75. Nandan, R., DebRoy, T. and Bhadeshia, H.K.D.H. (2008), 'Recent advances in friction stir welding process, weldment, structure and properties', *Progress in Materials Science*, 53: 980–1023.
76. Zhao, Y.H., Lin, S.B., Wu, L. and Qu, F.X. (2005), 'The influence of pin geometry on bonding and mechanical properties in friction stir weld 2014 Al alloy', *Materials Letters*, 59: 2948–52.
77. Schmidt, H. and Hattel, J. (2005), 'A local model for the thermo-mechanical conditions in friction stir welding', *Modelling and Simulation in Materials Science and Engineering*, 13: 77–96.
78. Buffa, G., Hua, J., Shivpuri, R. and Fratini, L. (2006), 'Design of the friction stir welding tool using the continuum based FEM model', *Materials Science and Engineering A*, 419: 381–8.
79. Elangovana, K. and Balasubramanianb, V. (2008), 'Influences of tool pin profile and welding speed on the formation of friction stir processing zone in AA2219 aluminium alloy', *Journal of Materials Processing Technology*, 200: 163–75.
80. Bahrami, M., Givi, M.K.B., Dehghani, K. and Parvin, N. (2014), 'On the role of pin geometry in microstructure and mechanical properties of AA7075/SiC nano-composite fabricated by friction stir welding technique', *Materials and Design*, 53: 519–27.
81. Galvão, I., Leal, R.M., Rodrigues, D.M. and Loureiro, A. (2013), 'Influence of tool shoulder geometry on properties of friction stir welds in thin copper sheets', *Journal of Materials Processing Technology*, 213: 129–35.
82. Aval, H.J., Serajzadeh, S., Kokabi, A.H. and Loureiro, A. (2011), 'Effect of tool geometry on mechanical and microstructural behaviours in dissimilar friction stir welding of AA5086–AA6061', *Science and Technology of Welding & Joining*, 16: 597–604.
83. Zhang, H., Lin, S.B., Wu, L., Feng, J.C. and Ma, S.L. (2006), 'Defects formation procedure and mathematic model for defect free friction stir welding of magnesium alloy', *Materials and Design*, 27: 805–9.
84. Kumar, K. and Kailas, S.V. (2008), 'The role of friction stir welding tool on material flow and weld formation', *Materials Science and Engineering A*, 485: 367–74.
85. Leal, R.M., Leitão, C., Loureiro, A., Rodrigues, D.M. and Vilaça, P. (2008), 'Material flow in heterogeneous friction stir welding of thin aluminium sheets: Effect of shoulder geometry', *Materials Science and Engineering A*, 498: 384–91.

86. Dawes, C.J. and Thomas, W.M. (1999), 'Development of improved tool designs for friction stir welding of aluminium', *Proceedings of 1st International Symposium on Friction stir Welding*, Thousand Oaks, CA.
87. Scialpi, A., De Filippis, L.A.C. and Cavaliere, P. (2007), 'Influence of shoulder geometry on microstructure and mechanical properties of friction stir welded 6082 aluminium alloy', *Materials and Design*, 28: 1124–9.
88. Li, Y., Murr, L.E. and McClure, J.C. (1999), 'Solid-state flow visualization in the friction stir welding of 2024 Al to 6061 Al', *Scripta Materialia*, 40: 1041–6.
89. Colligan, K. (1999), 'Material flow behaviour during friction stir welding of aluminium', *Welding Research*, 78: 229–37.
90. Schmidt, H., Hattel, J. and Wert, J. (2004), 'An analytical model for the heat generation in friction stir welding', *Modelling and Simulation in Materials Science and Engineering*, 12: 143–57.
91. Forcellese, A., Gabrielli, F. and Simoncini, M. (2012), 'Mechanical properties and microstructure of joints in AZ31 thin sheets obtained by friction stir welding using "pin" and "pinless" tool configurations', *Materials and Design*, 34: 219–29.
92. Website: available from: *http://www.mtiwelding.com/equipment/2/Friction Stir-Joint-Geometries*
93. Shultz, E.F., Cole, E.G., Smith, C.B., Zinn, M.R., Ferrier, N.J. and Pfefferkorn, F.E. (2010), 'Effect of compliance and travel angle on friction stir welding with gaps', *ASME Journal of Manufacturing Science and Engineering*, 132: 1–9.
94. ESAB, *Friction stir Welding*, Technical Handbook,
95. Andrade, Eládio (2009), 'Desenvolvimento de Ferramenta Bobbin-Tool para o Processo de Soldadura por Fricção Linear, Caracterização e Análise da Liga de alumínio AA6061-T4', MSc Thesis, Instituto Superior Técnico, Technical University of Lisbon.
96. Goushegir, S.M., dos Santos, J.F. and Amancio-Filho, S.T. (2014), 'Friction spot joining of aluminium AA2024/carbon-fiber reinforced poly (phenylene sulfide) composite single lap joints: Microstructure and mechanical performance', *Materials and Design*, 54: 196–206.
97. Martin, J.P., Stanhope, C and Gascoyne, S. (2011), 'Novel techniques for corner joints using friction stir welding', *TMS Annual Meeting & Exhibition*, San Diego, CA.
98. Steel, R.J., Nelson, T.W., Sorensen, C.D. and Packer, S.M. (2005), 'Friction stir welding of steel T-joint configurations', *Proceedings of the International Conference on Offshore and Polar Engineering*, Seoul, Korea.
99. Ocenásek, V., Slámová, M., dos Santos, J.F. and Vilaça, P. (2005), 'Microstructure and properties of friction stir welding aluminium alloys', *Proceedings of the 14th International Metallurgical & Materials Conference*, Czech Republic.
100. Penalva, M., Arregi, B., Rivero, A., Buffa, G. and Fratini, L. (2010), 'Development of corner fillet joints', *Proceedings of the 8th International FSW Symposium*, Timmerdorfer Strand, Germany.
101. Dubourg, L., Merati, A. and Jahazi, M. (2010), 'Process optimization and mechanical properties of friction stir lap welds of 7075-T6 stringers on 2024-T3 Skin', *Materials and Design*, 31(7): 3324–30.

102. Abdollah-Zadeh, A., Saeid, T. and Sazgari, B. (2008), 'Microstructural and mechanical properties of friction stir welded aluminium/copper lap joints', *Journal of Alloys and Compounds*, 460: 535–8.
103. Chen, Y.C. and Nakata, K. (2008), 'Friction stir lap joining aluminium and magnesium alloys', *Scripta Materialia*, 58(6), 433–6.
104. Teng, T.L., Fung, C.P., Chang, P.H. and Yang, W.C. (2001), 'Analysis of residual stresses and distortions in T-joint fillet welds', *International Journal of Pressure Vessels Piping*, 78(8): 523–8.
105. Hou, X., Yang, X., Cui, L. and Zhou, G. (2014), 'Influences of joint geometry on defects and mechanical properties of friction stir welded AA6061-T4 T-joints', *Materials and Design*, 53: 106–17.

Friction stir welding of dissimilar metals

E.T. Akinlabi, University of Johannesburg, South Africa and S.A. Akinlabi, University of the Witwatersrand, South Africa

DOI: 10.1533/9780857094551.241

Abstract: This chapter focuses on advances in friction stir welding (FSW) of dissimilar materials. Since the invention of the FSW technique, the process has been continually improved and its scope of application expanded. Most of the work done to bring FSW to production applications has not only been practical in nature, but has also been driven primarily by the pressing need of industry. The FSW of low melting temperature materials, such as aluminium and its alloys, has been embraced and commercialized and recent interest is now focused on joining dissimilar metals. However, a full understanding of the inherent properties of dissimilar joints using the FSW process is required to accommodate the huge demand in industry, including the manufacturing and aerospace industries. Recent developments in joining dissimilar metals of aluminium and its alloys to other metals such as copper, magnesium, steel and titanium have been reviewed and presented. A detailed study on the material characterization on FSW of aluminium and copper is also presented.

Key words: friction stir welding, dissimilar joining, dissimilar metals, tool displacement, microhardness, microstructure, intermetallics.

6.1 Introduction

Dissimilar metal joining in general has great potential in practical applications to replace riveted joints leading to huge cost and weight saving, while friction stir welding (FSW), in particular being a solid state welding process, is of great advantage with respect to producing better joint integrity. Over the past few years, the manufacturing industry has developed an interest in joining dissimilar materials, but the differences in the mechanical and thermal properties of such materials make this difficult. Furthermore, the successful welding of dissimilar metals is a challenge due to the differences in the physical, chemical and metallurgical properties of the base materials (BM) used, and finding the optimum process parameter window continues to be the basic prerequisite [1–13].

Generally, the joining of dissimilar metals and alloys using the FSW process is usually achieved by severe plastic deformation (SPD) of both metals being joined together. SPD can lead to grain dynamic recrystallization, which permits the flow of plasticized material occurring in solid state. Thus, the resulting recrystallized, equiaxed and usually submicron grains form in the weld zone after being frozen [4,5]. When a nonferrous material such as aluminium (Al) is involved, brittle intermetallic phases often form at the weld interface of dissimilar joints [6]. This is a common phenomenon in dissimilar welds with Al as one of the BMs, not only in fusion welding and irrespective of the energy input either global as in arc welding or local as in resistance spot welding, but also in solid state welding [7,8], for example, in friction and FSWs [9–11]. Inherently, weld interfaces in dissimilar joints are associated with sharp changeovers in the resulting properties, due to the heterogeneous nature of the welds.

6.2 Experimental practice

FSW of dissimilar metals presents enormous challenges when compared to the welding of similar materials; this is due to the difference in the chemical, mechanical, physical and thermal properties of the metals being joined. Against this background, apart from the optimization of the operating process parameters, research studies [12–14] have proved that FSW of dissimilar materials with good joint integrities are better achieved when the tool pin is offset and when the material with the high melting temperature is placed on the advancing side (AS) during the

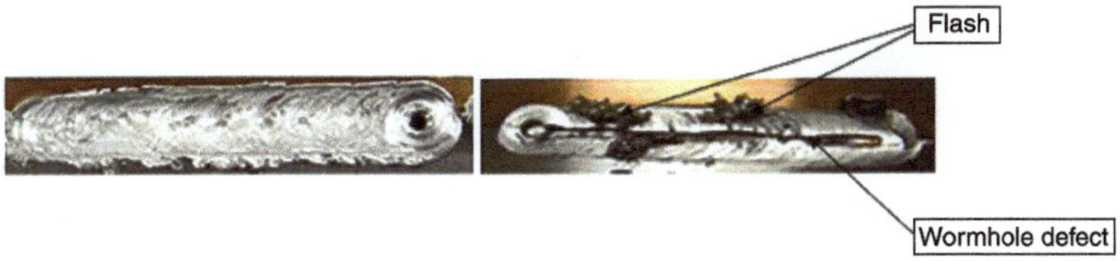

Figure 6.1 **(a) Copper on the advancing side; and (b) copper on the retreating side [14]**

welding procedure. A typical study to investigate this fact was conducted on FSW of Al and copper (Cu) conducted at a constant rotational speed of 600 rpm and a feed rate of 150 mm/min. The Cu sheet was placed in the AS (Figure 6.1(a)) and also placed in the retreating side (RS) (Figure 6.1(b)).

Physical observation of the resulting welds revealed that a weld with a good surface appearance was achieved with the Cu placed on the AS, while the weld conducted with the Cu in the RS was characterized with heavy flash and wormhole defects. This can be attributed to the fact that when the Cu sheet is placed on the RS, material flow is more difficult to achieve due to the higher melting temperature of Cu compared to Al, thus resulting in weld defects and poor surface quality.

The appropriate linear tool displacement position was also investigated by conducting dissimilar FSW of Al and Cu sheets. The welds were produced at a constant rotational speed of 1200 rpm and 150 mm/min with the tool pin plunged in:

- copper (Cu);
- at the weld centre line; and
- in the Al alloy (AA) (Figures 6.2(a–c)).

It was found that the welds produced with the tool pin plunged into Cu, and at the weld centre line, resulted in wormhole formation, while the weld produced with the tool pin offset into the Al sheet has a good surface appearance. Further examination of the metallographic cross-sections of the welds produced at the same input process parameters showed that metallurgical bonding and the mixing of both materials were achieved more effectively at the joint interface with the tool pin plunged into the AA (Figures 6.3(a–f)).

It was observed that in the welds produced with the tool pin plunged into Cu, (Figure 6.3(a)), the Cu deforms at a slower rate compared to the

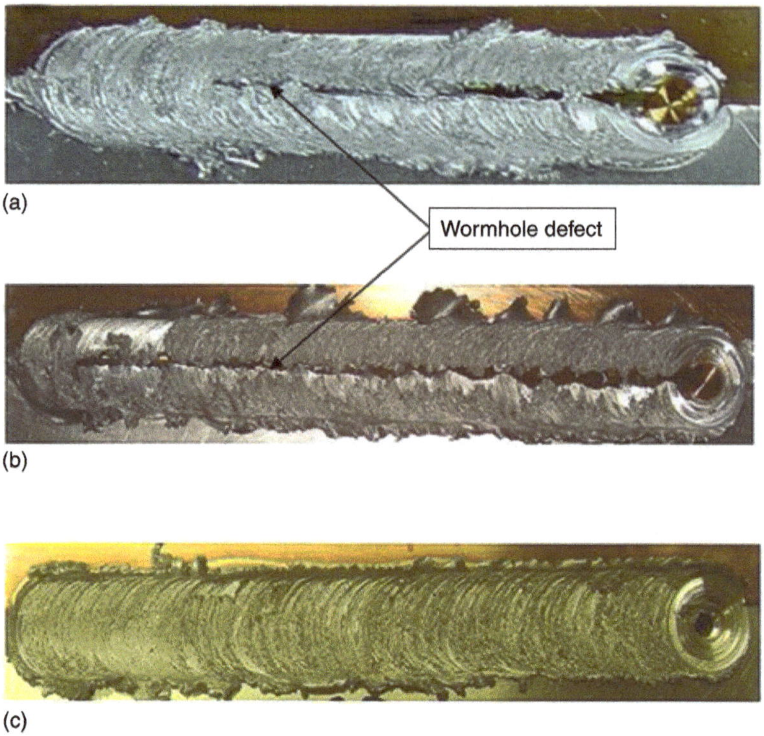

Figure 6.2 Welds produced with the tool pin plunged in (a) Cu; (b) at centre line; and (c) in AA [15]

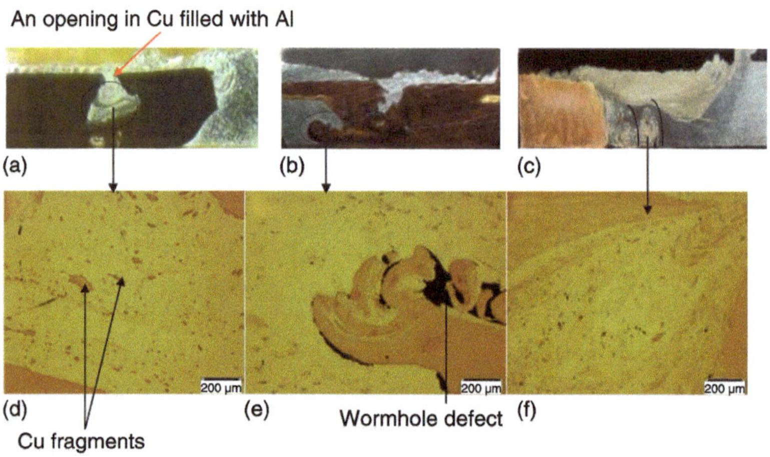

Figure 6.3 Macro- and micrographs of welds produced at 1200 rpm and 150 mm/min with the tool pin plunged in (a) and (d) Cu; (b) and (e) at centre line; (c) and (f) in AA [14]

AA due to its higher melting point, and it then moves towards the AA at the joint interface. The macrograph shows that during the welding process, there was an opening in the Cu (indicated by the grey arrow) filled with the AA, characterized with some Cu fragments at the centre of the weld, as shown in the micrograph in Figure 6.3(d) [14].

When the tool pin was plunged at the weld centre line (Figure 6.3(b)), both materials deformed at different rates, due to the large difference in their melting temperatures. Therefore, the stirring at the centre was not sufficient to achieve the required mixing, subsequent to the bonding of the materials being joined; this then resulted into a wormhole defect, as shown in the micrograph in Figure 6.3(e) [14].

The resulting good top surface and better mixing achieved in the welds produced with the tool pin plunged into the AA (Figure 6.3(c)), can be attributed to the fact that the AA with its lower melting point deforms plastically faster compared to the Cu. Hence, it deforms and mixes with the Cu, and then fills the gap behind the stirring pin. Also, it was observed that no other critical weld defects were found by examining the weld cross-section [14].

Thus it can be concluded that the established facts with respect to the experimental process of FSW of dissimilar materials proves that the material with the higher melting point should be placed at the AS and the tool offset should be made into the material with the lower melting temperature, in order to achieve welds with desirable properties.

6.3 Friction stir welding of dissimilar aluminium alloys

Al overcomes two of the disadvantages of steel. It is lighter in weight and is stable in environments that corrode steel. It has a wide range of properties that are used in engineering structures. Its high strength-to-weight ratio makes it very suitable in the aircraft and automotive industries. The AAs, AA6XXX and AA5XXX are extensively used in the fabrication of aircraft structures and other structural applications [16]. Dissimilar joining of these two alloys is encountered in industrial structures. The choice of FSW, being a solid state joining process, is preferably suitable for the welding of Al and its alloys; this eliminates the problem of weld solidification cracking. In addition, the FSW process has overcome the problems associated with fusion welding of Al and its alloys, such as porosity, segregation and heat affected zone (HAZ) liquidation cracking. Hence, FSW is suitably employed for dissimilar welding of Al and its alloys [17]. The uniqueness of FSW of dissimilar materials has attracted extensive research interest because of the potential engineering importance and problems associated with the conventional welding process. In view of this, different dissimilar AA combinations have been successfully FSW with excellent joint efficiencies [18–24].

6.3.1 Microstructural characterization of dissimilar aluminium alloys

Microstructural evaluation of the joint interfaces of FSW of dissimilar AAs revealed that significant mechanical mixing of the two alloys joined are noticed in the stir zone (SZ) or weld nugget with complex vortex, whorl and swirl features characteristic of chaotic-dynamic mixing.

It is important to note that the processing parameters play a vital role in the effective joining of dissimilar welds. In a research study on SEW of AA2219 and AA5083 [25], the rotational speed, transverse speed, tool geometry and the ratio between the tool shoulder diameter and the pin diameter were the parameters taken into consideration. The optimum process parameters were determined with reference to the resulting ultimate tensile strength (UTS) of the joints. Microstructural studies of dissimilar AAs revealed that the material placed on the AS dominates the nugget region [26]. Compared to alloy 2219, alloy 5083 contained fewer and finer second-phase particles. Three distinct microstructural zones, the stir zone (SZ), thermo-mechanically affected zone (TMAZ) and heat-affected zone (HAZ), were present in the weld (Figure 6.4).

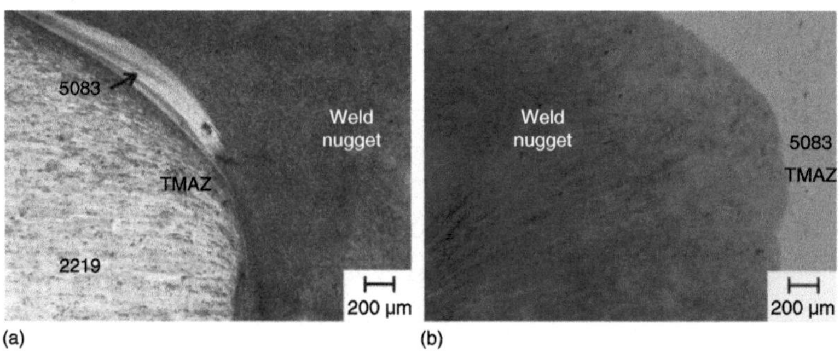

Figure 6.4 Thermo mechanical affected zone (TMAZ) microstructure: (a) advancing side; and (b) retreating side [25]

Figure 6.4 shows that the microstructures of the weld; the TMAZ on the AS showed highly deformed grains, with a clearly discernible SZ/TMAZ and TMAZ/HAZ boundaries (Figure 6.4(a)). However, on the RS, these interfaces were rather diffused, especially the latter (Figure 6.4(b)). In the HAZ, on either side of the weld nugget, there were no noticeable changes in the grain structure compared to the respective unaffected BMs.

During the FSW process, the material undergoes intense plastic deformation at elevated temperatures, resulting in the generation of fine and equiaxed recrystallized grains [26–31]. The fine microstructure in FSWs produces good mechanical properties. These studies show that dissimilar defect-free, high efficiency welded joints can be produced using a wide range of process parameter combinations and recommends parameter windows for producing the best joint with good tensile properties in dissimilar joining of Al and its alloys. Analysis of variance (ANOVA) of the tool geometries employed showed that the ratio between the tool shoulder diameter and pin diameter is the most dominant factor in deciding the joint soundness, while the pin geometry and the welding speed also play significant roles in the evolving weld quality.

In a study conducted on the effects of tool rotational speed and pin profile on microstructure and tensile strength of dissimilar FSW AA5083-H111 and AA6351-T6 AAs [32], it was found that different regions evolved in the resulting microstructures. The microstructure of the weld zone obtained using different tool pin profiles at tool rotational speed of 950 rpm and the straight square (SS) pin profile is presented in Figure 6.5.

It is evident from this figure that the microstructure of the dissimilar joint is dependent to a large extent on the position of the tool pin within the weld zone during the welding process. Different microstructures were

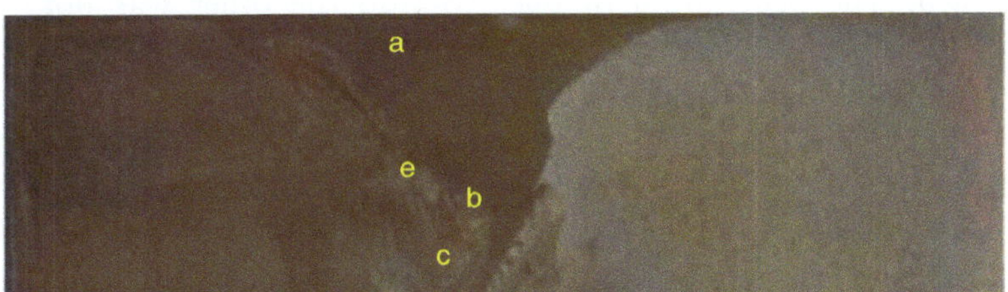

Figure 6.5 Microstructure of various regions in the weld zone, as marked in the macrostructure using SS tool pin profile at 950 rpm [32]

observed while moving across the breadth and depth of the weld zone. All the observed microstructures were categorized into three different regions, namely the unmixed region, mechanically mixed region and the mixed flow region [32]. The mixed flow region is absent in the joints fabricated using the tapered pin profiles. The same region was also not present in the joints fabricated using the straight pin profiles at tool rotational speed of 650 rpm. The formation and characterization of the interfacial regions of the welds consists of the unmixed region, mixed flow region and the mechanically mixed region. Detailed microstructural characterization of the different zones observed in the weld is further discussed in detail. The region designated (a) in Figure 6.5 is presented in Figure 6.6(a).

The unmixed region (Figure 6.6(a)) contains the microstructure of either the AA5083 or AA6351. This region is observed at the top of the joints closer to the shoulder region. The fine-grained structure indicates that this unmixed region experienced dynamic recrystallization during the welding process. Second, the mixed flow regions are presented in Figures 6.6(b–d).

It was found that these regions contain structures of alternative lamellae of both the AAs joined. A complex vortex-like flow pattern is visible. The mixed flow region presents extreme super plastic flow of plasticized material, which creates chaotic intercalation patterns. The stirring action of the tool causes intense plastic deformation and *in situ* extrusion of the AAs, AA6351 and AA5083. The plasticized material is transported layer by layer, which forms such a lamellae structure. The penetration and mixing of both the AAs in this region is intense.

Finally, the mechanically mixed region (Figure 6.6(e)) contains the microstructure of both the AAs.

The etchant reveals the AA5083 as darker and the AA6351 as lighter. The plasticized dissimilar alloys are mechanically coupled to each other as shown. The penetration of one AA into the other was not fully accomplished but the dynamic recrystallization of the grains is evident. It was reported that the two processing parameters, viz the rotational speed and the feed rate, affected the joint strength due to variations in the behaviour of the material flow, loss of cold work in the HAZ of the AA5083 side, dissolution and over-ageing of precipitates of the AA6351 side and the formation of macroscopic defects in the weld zone [32].

Similarly, dissimilar welding of AAs (AA2024-T351 and AA6056-T4) was achieved by FSW [33]. The butt joints were achieved by varying the process parameters, such as the rotational speed (500–1200 rpm) and the welding speed (150–400 mm/min), while the axial force and the tool geometry were kept constant. Parameter optimization, which has been

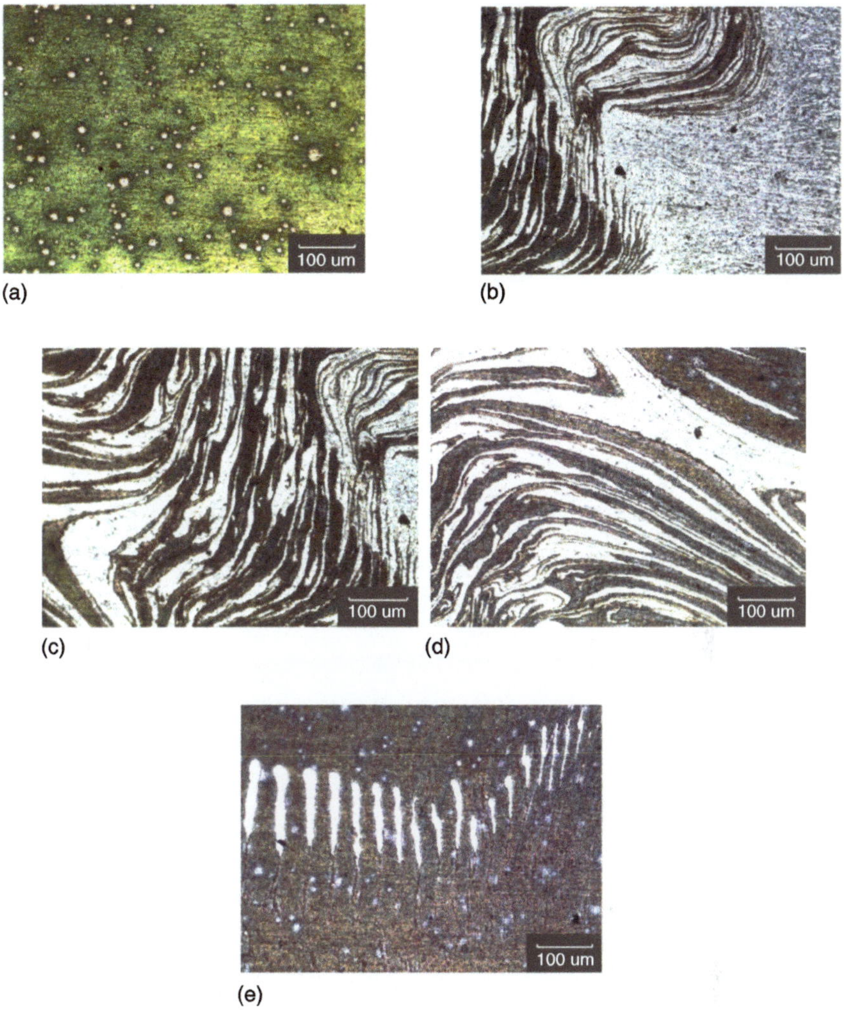

Figure 6.6 **(a) Unmixed region; (b–d) mixed flow regions; and (e) mechanically mixed region [32]**

based on the results of the macroscopic analysis and microhardness testing, indicated that sound joints can be obtained in the parameter range of rotational speed of 800 rpm and welding speed of 150 mm/min.

A cross-section of the weld (Figure 6.7(a)) reveals a well-formed SZ with a lamellar pattern formed by dark and bright lamellae, as was also reported by the earlier investigation [33]. In addition, no internal defects were observed in the samples examined within different points of the weld seam from the investigated joints.

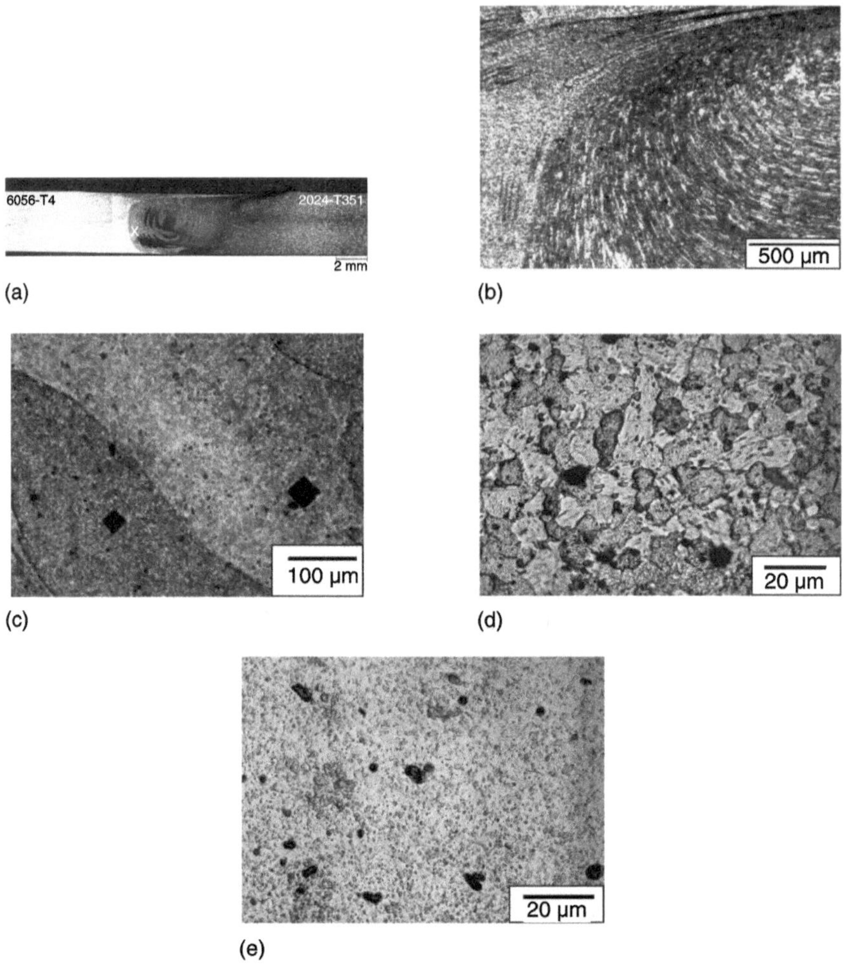

Figure 6.7 (a) Cross-section of dissimilar AA2024-T351 and AA6056-T4 weld; (b) transition area between TAFZ and stir zone; (c) higher magnification of the weld cross-section shown in (a) in the position marked 'X' in the nugget area; (d) dark lamella formed at the interface; and (e) bright lamella formed at the interface [33]

The microstructural evolution developed from the cross-section was observed through an optical microscopy (Figure 6.7(b)). It was observed that a change in the grain orientation begins in the transition between the TMAZ and the SZ. As shown in Figure 6.7(b), it is observed that the grains are aligned according to the direction of the tool rotation and approaching about 90 degrees tilt, thereby achieving a good degree of recrystallization but with large decrease in the grain sizes. However, it was observed that recrystallization was fully developed within the lamellae pattern in the SZ; this is shown by the very fine grain sizes shown in Figure 6.7(c). It was also found that different lamellae (dark and bright) were formed in the interfacial region (Figures 6.7(d and e)). The different lamellae formed are due to an essentially non-chemical mixture of the parent metals joined. The scanning electron micrographs (SEM) of the SZ is presented in Figures 6.8(a and b).

It is important to note that the lamellae colour in the BSE image is reversed (i.e. dark lamella is bright and vice versa), due to the atomic number contrast mechanism.

The SEM-EDS micrographs in Figures 6.8(a and b) show the presence of the peaks of Cu (bright regions of the picture) and peaks of Si (dark regions of the picture). This leads to the conclusion that the dark lamella originates from the Al–Cu alloy-AA2024-T351, while the bright lamella originates from the Al–Si–Mg alloy-AA6056-T4.

The FSW of dissimilar 2024 and 7075 AAs was also conducted [34]. The FSW tool comprises of a threaded pin of 4 mm diameter and 2.85 mm long with a flat shoulder diameter of 12 mm. The 7075-T6 AA was placed in the AS; the tool feed rate was kept constant at 254 mm/min, while the force and rotation speed have been varied according to three rotational speeds (400, 1000 and 2000 rpm) and forces (13.4, 12 and 9.8 kN), respectively. An MTS ISTIR PDS FSW machine in force control with a tilt angle of 3 degrees was used.

The microstructural characterization of the welds produced at 1000 rpm is presented in Figure 6.9. It was observed that the FSW joint has no defects or any other imperfections, such as lack of penetration or porosity.

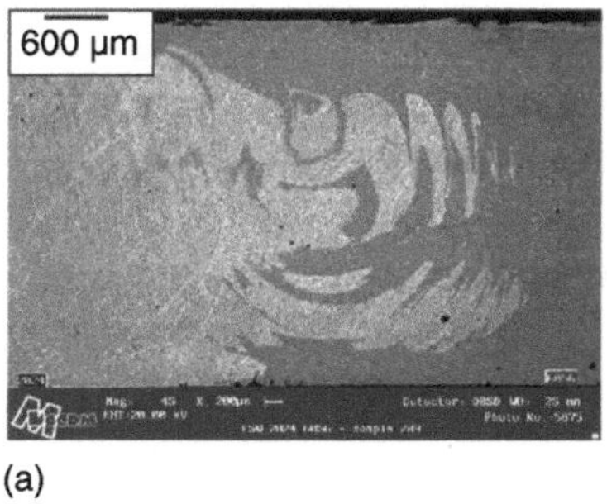

(a)

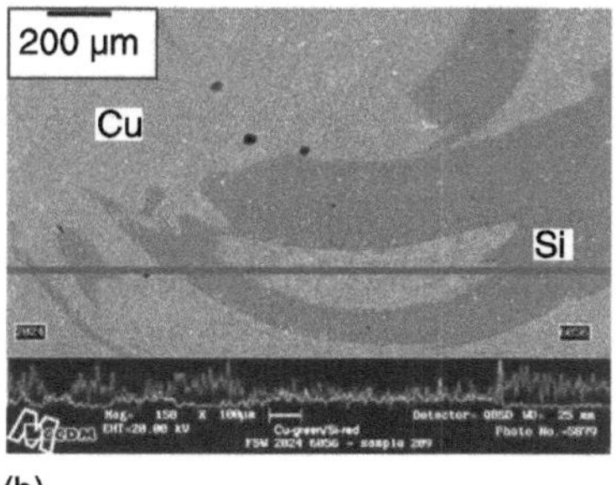

(b)

Figure 6.8 (a) SEM-BSE image of the stir zone (45×); and (b) SEM-BSE stir zone (150×) [33]

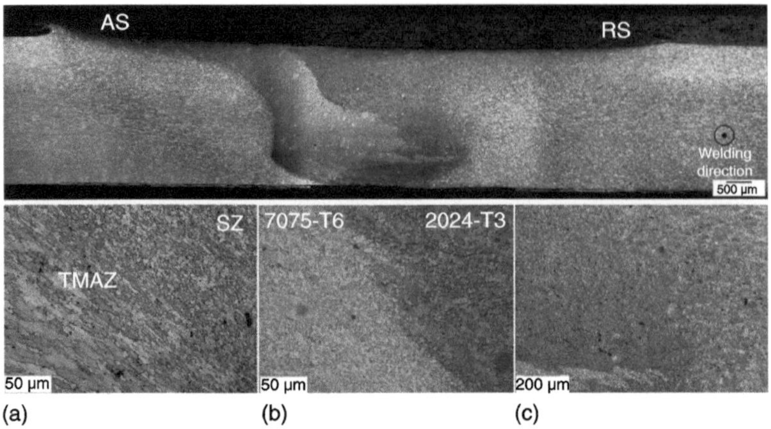

Figure 6.9 **Optical macrograph of the cross-section of condition 7075/2024: (a) TMAZ-SZ boundary, advancing side; (b) SZ-SZ interface between 7075-T6 and 2024-T3; and (c) TMAZ-SZ boundary, retreating side [34]**

Typical microstructural features of FSW, such as the SZ, TMAZ and HAZ were identified at the joint interfaces. The HAZ microstructure on both materials is very similar to that of the BM. The TMAZ region experienced both heating and plastic deformation during FSW, being characterized by a deformed structure induced by the mechanical stirring of the tool. The SZ was formed by two regions consisting of recrystallized fine grains of both 2024 and 7075 AAs. The measured grain size was approximately 5 to 7 μm in the 2024 region and 3 to 5 μm in the 7075 region, as shown in Figures 6.10(a and b), respectively. The boundary separating the two regions is very narrow, suggesting that there would not be any significant diffusion phenomena between both alloys.

No onion ring formation has been observed in this weld, suggesting a lack of intermixing between both alloys joined. However, some evidence of intermixing initiation can be seen in the bottom of the SZ, as some small regions of the alloy located in the AS are pushed into the RS. That is not the case for the top part of the SZ, where a narrow and continuous boundary separates both alloys. It was found that the resulting microstructure has evidence of the fine grain structure and precipitates distribution differences originated by the process. In addition, the welded joint properties were investigated through the tensile and fatigue tests.

The presence of the mixing layers in FSW reduces the crack initiation points in the welded samples, thereby enhancing the fatigue behaviour in

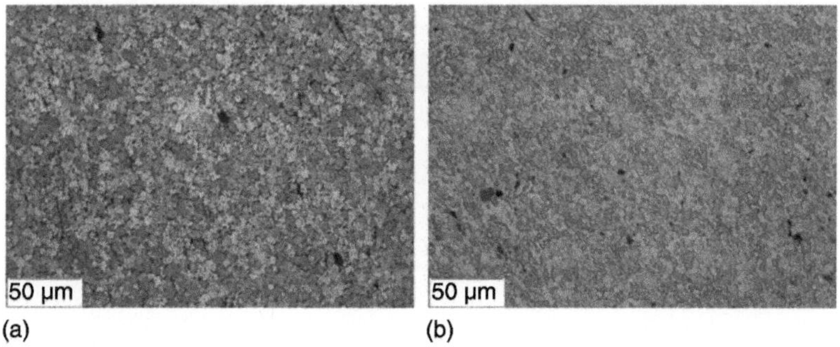

Figure 6.10 Optical micrographs of cross-sections of both stir zones: (a) AA7075-T6; and (b) AA2024-T3 [34]

comparison to the parent materials. The fatigue behaviour of the welded joints is acceptable and allows consideration of the FSW as an alternative joining technology for the Al sheet alloys considered in this research work. The specimens fracture surfaces after testing were analysed by using a FEG/SEM microscope, revealing the defects typology and location after the FSW process, and it was observed that the microscopic mechanisms occurred during high stress deformations and finally failure in the welded samples [34].

The FSW of ultra-thin materials were also investigated. Ultra-thin dissimilar FSW of 0.8 mm 2024 and 6082 AAs were studied and reported [35]. The dissimilar weld was produced with 2024 AA (higher strength) positioned on the RS and 6082 AA on the AS. The FSW tool was made from a 56NiCrMoV7-KU material with a cylindrical non-threaded probe with a 1.7 mm diameter, 0.6 mm height and a shoulder of 6 mm diameter.

The resulting microstructures were investigated by optical microscopy. The obtained joints showed no porosity or other defects in both the top and the root of the weld surface in the two welding conditions. Unlike FSW of similar alloys that exhibit four distinct regions, namely parent material, HAZ, TMAZ and SZ, the dissimilar welds exhibit eight distinct regions (Figure 6.11). The SZ is the region that experiences the highest strain and undergoes recrystallization (Figures 6.11(d and e)). Its evolving microstructure is due to the mechanical action of the tool probe that generates a continuous dynamic recrystallization process.

The higher temperature and the SDP during the welding in the SZ result in a new equiaxed fine-grain structure with an estimated grain dimension of less than 3 μm both for the 2024 and 6082 AAs. By moving towards the base metals (BMs), adjacent to the SZ, there is the TMAZ

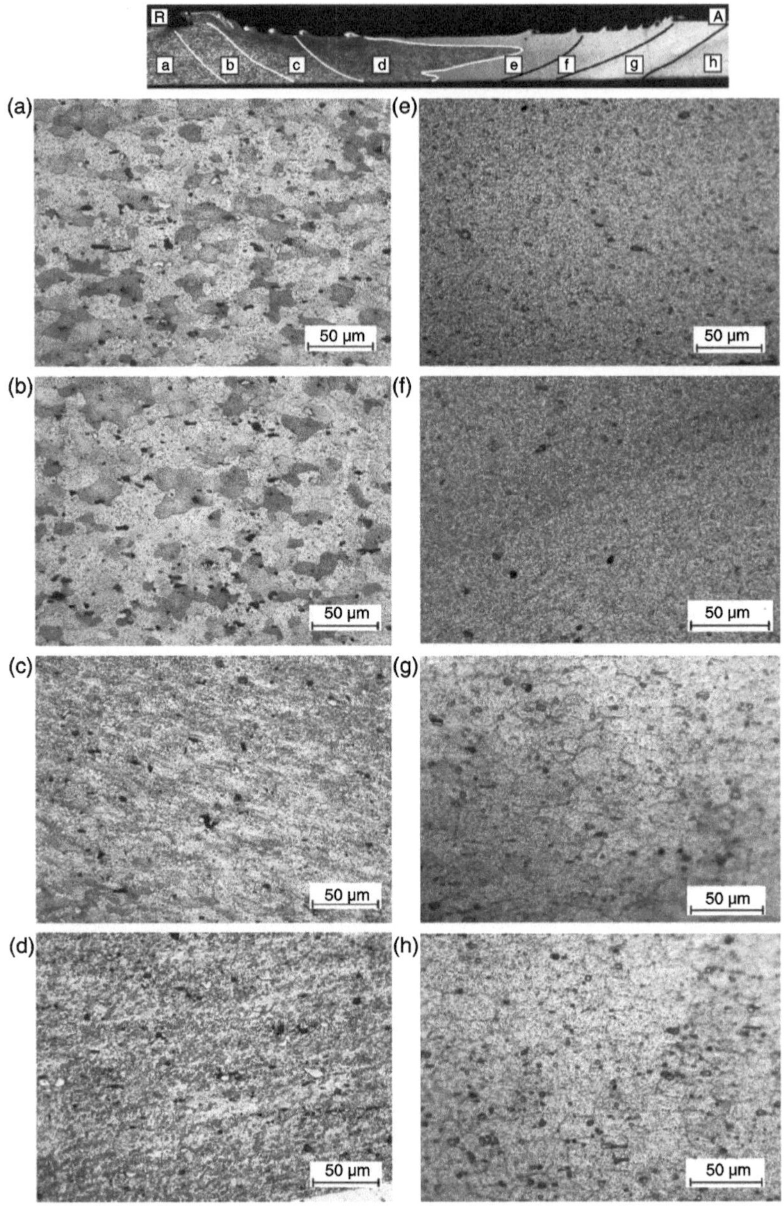

Figure 6.11 Macrograph with the main distinct regions of a dissimilar IFS weld and optical microstructures of the eight regions: (a) Parent material 2024; (b) HAZ 2024; (c) TMAZ 2024; (d) Stir zone 2024; (e) Stir zone 6082; (f) TMAZ 6082; (g) HAZ 6082; and (h) Parent material 6082 [35].

(Figures 6.11(c and f)) where no recrystallization was observed. It seems that the temperature, derived from the process, was not high enough and deformation was not so adequate as to cause recrystallization. The region adjacent to the TMAZ is the HAZ (Figures 6.11(b and g)), where the grain size is similar compared to the BM. The HAZ experienced a peak temperature, which results in a decrease in the hardness. Adjacent to the HAZ is the BM (Figures 6.11(a and h)).

Studies on lap welds of dissimilar Al and it alloys have also been conducted. A typical study includes a research study of friction stir lap welding (FSLW) of a 4 mm thick AA6063 and AA5052. The plates were FSW vertical to their rolling directions. The so-called lap-butt joint of dissimilar AA6063/AA5052 AAs in this study consists of three plates. In this joint design, one butt-side is a single AA6063 plate of 4 mm thick, and the other butt-side is two overlap AA5052 plates of 2 mm thick [36]. The result of the study showed that the lap-butt joint configuration was achieved using a designed tool of $W_9Mo_3Cr_4V$ with some geometric modifications. It was confirmed that the parameter of the pin offset played an important role and exerted stronger influence upon the joint quality. Three typical micro-flow patterns of plasticized materials were found in the weld zones: circumfluence, laminar-flow and turbulent-flow. Some weld-defects and the morphologies of 'onion rings' were also detected. The studies reviewed showed similar evolving microstructures at the joint interfaces and it was found that different optimum process parameter windows exist for different materials joined.

6.3.2 Mechanical properties of dissimilar aluminium alloys

This subsection presents the mechanical properties of dissimilar AA welds; the common material characterization conducted in this respect is the microhardness profiling and the tensile behaviour of the welds.

Microhardness profiling of dissimilar aluminium alloys welds

The micrograph of Vickers microhardness profiles, conducted on the weld produced at 800 rpm and 150 mm/min [33], is presented in Figures 6.8(a and b). The Vickers microhardness measurements conducted proved the origin of the lamellae pattern formation observed in the weld, where dark lamellae showed higher values of microhardness compared to

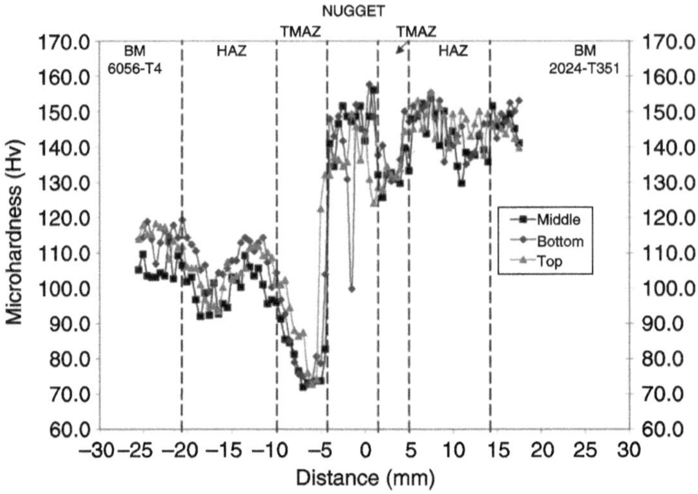

Figure 6.12 Microhardness profiles of the weld cross-section [33]

the bright lamellae; this is shown in the microhardness profiles presented in Figure 6.12.

The microhardness profiles were divided into four typical microstructural regions based on previous study: BM, HAZ, TMAZ and SZ.

A close observation of the HAZs (–20 to –10 mm and 5 to 15 mm, respectively) revealed a decrease in microhardness due to accelerated ageing and recovering processes caused by the weld thermal cycle [37,38]. In these regions, the microhardness peaks were observed to be unstable, disrupting the declining trend of the microhardness profiles. On the TMAZs (–10 to –5 mm and 2 to 5 mm, respectively) the decrease in the microhardness for both alloys can be explained by the local dissolution of strengthening particles and a reduction in the dislocation density. Nevertheless, closer to the transition between the TMAZ and the SZ, an increase in microhardness was observed and this is thought to be due to a degree of work hardening that took place close to the tool centre [37,38]. On the SZ, the increase in microhardness is associated with the re-precipitation of fine particles.

Vickers microhardness profiles were also measured across the welds [25], to ascertain possible microstructure/property variations among the various regions of the weldment. The results are pictorially shown in Figure 6.13.

The hardness studies revealed that the lowest hardness in the weldment occurred in the HAZ on alloy 5083, where tensile failures were observed to have taken place [25].

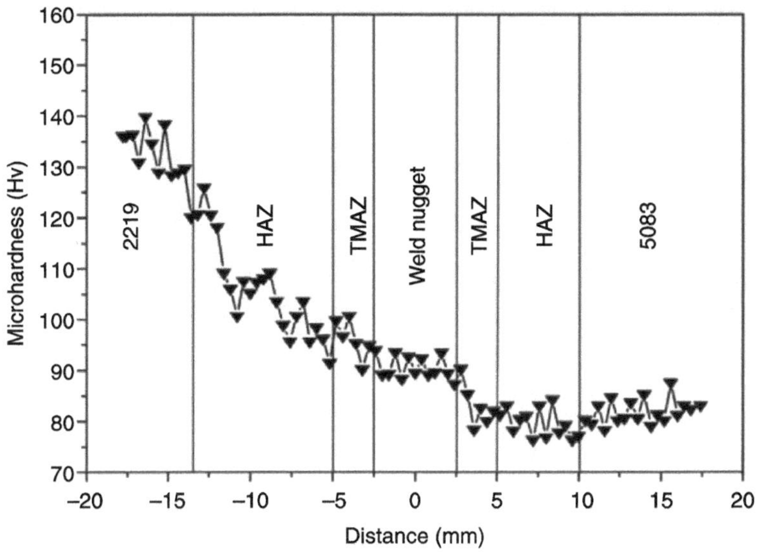

Figure 6.13 **Microhardness profile across the weld [25]**

FSW, being an appropriate solid state welding process, was also employed to successfully join 6 mm thick dissimilar AAs, AA5083-H111 and AA6351-T6, with the aim of studying the effect of tool rotational speed and pin profile on the microstructure and the tensile strength of the joints [32]. This result shows that the dissimilar joints were successfully made using three different tool rotational speeds of 600, 950 and 1300 rpm and five different tool pin profiles of straight square (SS), straight hexagon (SH), straight octagon (SO), tapered square (TS) and tapered octagon (TO). Three different regions, namely the unmixed region, mechanically mixed region and the mixed flow region, were observed in the welded zone.

The microhardness profile of the weld conducted at 1000 rpm and a feed rate of 254 mm/min performed at mid-thickness line as well as at 1 mm above and below the mid-thickness line of 4 mm dissimilar 2024 and 7075 AAs, is presented in Figure 6.14.

The microhardness of the BMs, 2024-T3 and 7075-T6, is approximately 137 and 172 HV0.2 respectively. Minimum hardness values have been found at the HAZ region of the RS with values close to 123 HV0.2, corresponding to approximately 90% of the hardness value found in the BM. We can clearly observe that the microhardness distribution is heterogeneous, particularly at the SZ. It should be noted that in the

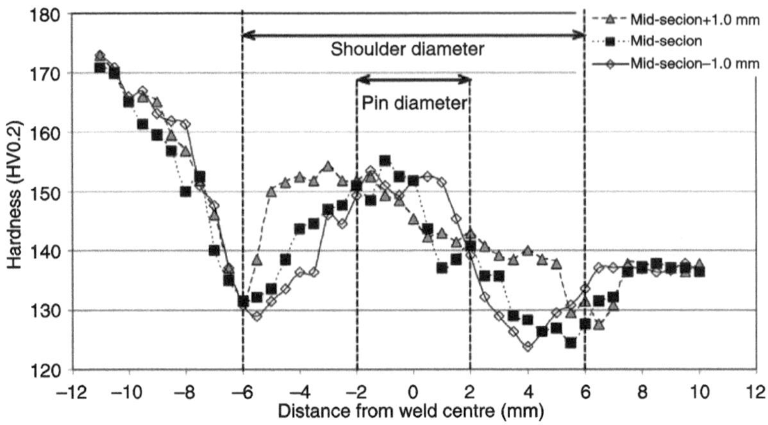

Figure 6.14 **Mircohardness distributions of FS welded cross-section joints performed at mid-thickness as well as at 1 mm below and above the mid-thickness line [34]**

case of mid-thickness and 1 mm below the mid-thickness lines, the SZ has approximately the size of the pin diameter; while at 1 mm above the mid-thickness line, it is relatively larger. This hardness distribution has been attributed to the absence of material mixing between both AAs, in which the interface between both SZs is clearly defined. Therefore, consecutive measurements, which have been taken inside the SZ but in different BMs close to the interface, can differ substantially [34].

Tensile properties of dissimilar aluminium alloys

Tensile strength of FSW samples is usually assessed to compare the strength and the ductility of the welded samples to the BMs. In quantifying the integrity of joints in dissimilar welding, a rule-of-thumb for the quality of the welds is that failure should occur in the weaker of the two BMs joined, at least some distance away from the weld interface. Tensile properties of dissimilar AAs are presented in this subsection.

In a research study conducted by [32] using different tool geometries, the welds were produced by varying the rotational speed while the feed rate was kept constant at 60 mm/s. It was reported that the tool rotational speed and the pin profile considerably influenced the tensile strengths of the joints produced. The tensile behaviour of the welds produced is presented in Figure 6.15.

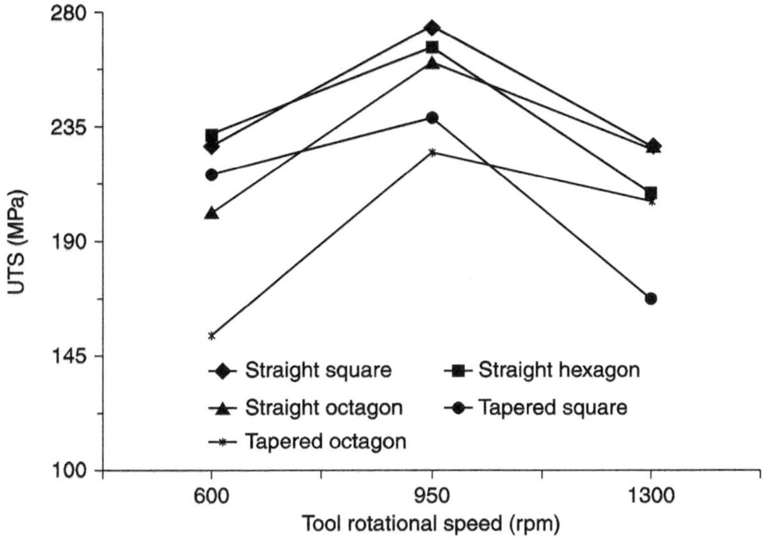

Figure 6.15 **Tensile behaviour of the welds [32]**

It was found that the joints failed at different zones of the welds without an apparent trend. The joint which was fabricated using the tool rotational speed of 950 rpm and SS pin profile, yielded the highest tensile strength of 273 MPa and was noted to be the optimum [32].

Furthermore, the tensile behaviour of the welds conducted by [33] was characterized. It was reported that the tensile testing results show that strength up to 90% of the weakest joining partner 6056-T4 can be achieved. The fracture of the tensile samples took place in the TMAZ of the alloy 6056-T4, while the annealed structure led to a decrease in the microhardness (Figure 6.12). This behaviour was confirmed by micro-flat tensile testing, which indicated a drop in the tensile strength and an associated increase of the strain in the regions of microhardness drop. This study has shown that in a dissimilar FSW, the weaker component dictates the performance of the joint, where failure happens in the region of the greatest strength reduction related to annealing phenomena. Microscopic investigation as well as the evaluation of local mechanical properties has suggested that mechanical mixing is the major material flow mechanism in formation of the SZ [33].

Tensile mechanical properties of the joints produced, as reported in [35] were evaluated by means of standard tensile tests in the transversal direction; the tests were conducted at room temperature with deformation velocities below 10^{-4} s^{-1}. In particular, the 2024 welded specimens showed very high UTS values (~91% of the BM strength, i.e. 470 MPa) and elevated yield strength. The 6082–6082 and the 2024–6082 joints showed similar strength with a UTS of about 69% of the 6082-T6 BM (330 MPa). All the studied welds showed very limited data spreads for yield and ultimate stresses and elongations [35].

Research and applications of different joint configurations of dissimilar Al and its alloys has been reviewed and reported. However, a lot of research work is still ongoing, in order to achieve the desirable properties for commercialization purposes. The next section will present recent research studies on FSW of dissimilar Al and magnesium (Mg).

6.4 Friction stir welding of aluminium and magnesium alloys

AAs are widely used in automotive, aerospace and ship building industries, due to the combination of mass reduction and high strength, while Mg alloys are also attractive in these fields, also due to their lower density and high specific strength. Therefore, it can be expected that the joining of AAs and Mg alloys needs to be studied for potential industrial applications. However, it is difficult to obtain sound dissimilar welds of these two kinds of alloys through conventional fusion welding techniques, since large intermetallic compounds, which have a strong negative effect on the mechanical properties of the weld, are easy to form in the welds [39].

6.4.1 Microstructural characterization of dissimilar aluminium and magnesium

Several studies were investigated with the aim of consolidating the application of FSW of dissimilar joints for Al and Mg [40–42]. One of such studies was the investigation of the evolving microstructure of AZ31 Mg alloy and 1060 AA [40]. The results confirmed that the microstructure was characterized by an intercalation lamellae band of both materials joined. A similar study was also conducted into dissimilar joining of AZ31 Mg alloy and 1050 AA [43]. The results confirmed the presence of intermetallic compounds at the joint interfaces of the welds.

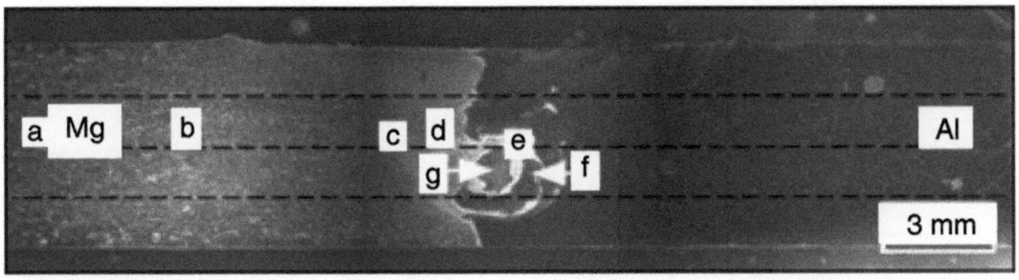

Figure 6.16 Optical macrograph of the cross-section of dissimilar 5052 Al alloy and AZ31 Mg alloy weld [44]

Another research study on FSW of 5052 AA and AZ31 Mg alloy was conducted [44]; the typical microstructural zones resulting from the welds is shown in Figure 6.16; this includes the BM, HAZ, TMAZ and the SZ.

The SEM microstructures of these zones are further shown in Figure 6.17. The BM shown in Figure 6.17(a) consists of large equiaxed grains (50 μm) and fine grains (10 μm).

It was noted that dynamic recrystallization occurred in the SZ, which was attributed to the plastic deformation and thermal cycle effect caused by the stirring tool during the welding process. It is also important to note that the fine equiaxed recrystallized grains presented different structures in the different locations of the SZ, such as regions 'd', 'e' and 'f' in Figure 6.17. The grains with an average size of about 5.4 and 6.9 μm, which were much smaller comparer to the BM, were obtained in regions 'd' and 'e', located on the Mg side and Al side of the SZ, respectively. Intercalated structures were formed and the average grain size of AZ31 Mg alloy in these structures was about 2.8 μm (Figure 6.17(f)).

Furthermore, it is believed that joining Al to Mg might be advantageous in order to achieve optimum performance in practical applications. However, the joining of Al to Mg by welding is often hampered by the unavoidable formation of intermetallic compounds at the Al–Mg interface, which is typical of dissimilar joining. This formation of Al–Mg intermetallics results in a high hardness and low ductility region in weld joints [45–48]. The Al–Mg dissimilar FSW exhibit an S-shaped intermixing at the interface at the lower rotational speeds. At higher rotational speeds (>900 rev min-1), the welded joints developed an interpenetrating feature along the curved interfaces [48].

Another investigation involving the welding of 12-mm thick plates of 5083 Al and AZ31 Mg was conducted at a feed rate ranging from 60 to

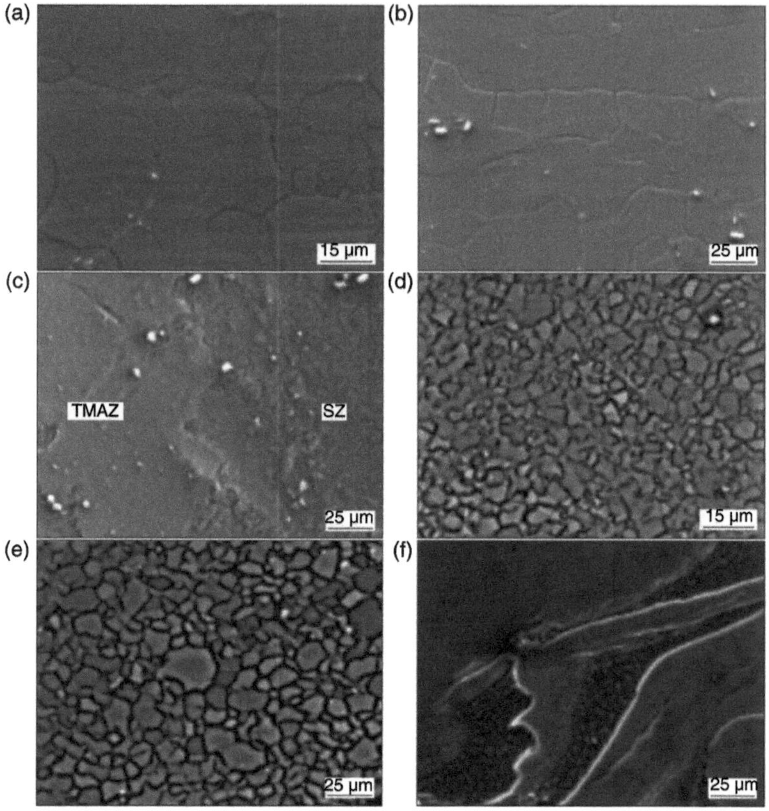

Figure 6.17 SEM images of AZ31 in different regions, marked with letters: (a) BM; (b) HAZ; (c) interface of TMAZ/SZ; (d) SZ in Mg side: (e) SZ in Al side; and (f) intercalated microstructure [44]

100 mm/min and the rotational speed from 300 to 400 rpm [49]. The results also confirmed the formation of intermetallics (Figure 6.18) at the joint interfacial regions of the butt joints.

The report claimed that intermetallic formation occurs by the liquid–solid and not by solid state transformation. Hence, the key difficulty is the ability to control the liquation (i.e. liquid formation) and the growth of Mg–Al intermetallic compounds during the welding process. This is then necessitated by lowering the heat input during the welding process. Therefore, submerged friction stir welding (SFSW) is recommended, being a new variation to FSW. The SFSW is carried out under water, either in a bath or in a situation where water is constantly flowing across

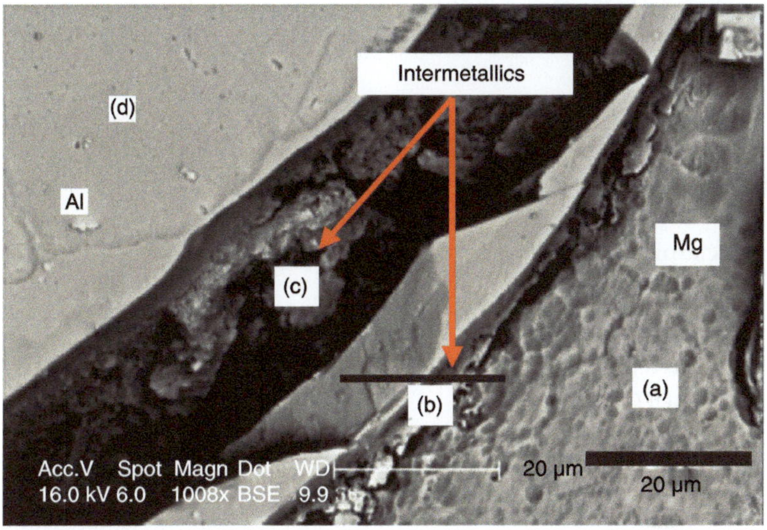

Figure 6.18 SEN micrograph of the interface of Al/Mg alloy for the aid welded specimen [49]

the surface of the sample. This technique has been recommended for alloys that are sensitive to overheating during the welding process [49].

SEM observations indicated that the interface zone of the Al/Mg joint included the transition region on Mg substrate (Figure 6.17, region b), and the transition region on Al substrate (Figure 6.17, region c). The Al and Mg distributions in these regions analysed by the energy dispersive spectroscopy (EDS) are shown in Table 6.1.

The EDS results indicated that regions 'b' and 'c' apparently have different chemical compositions, i.e. region 'b' has lower Al and higher Mg contents compared to region 'c'. The interface of the Al/Mg FSW

Table 6.1 Elemental analysis results of the Al/Mg interface layer of air welded specimen [49]

Region of interest	Al	Mg	Si	Inference composition
a	5.82	93.52	0.66	Mg alloy
b	44.66	55.13	0.21	$Al_{12}Mg_{17}$
c	59.14	39.71	1.15	Al_3Mg_2
d	92.97	6.69	0.34	Al alloy

joint produced the obvious diffusion between the Mg substrate and the Al substrate in the condition of stirring.

6.4.2 Mechanical properties of dissimilar aluminium and magnesium

The mechanical properties of dissimilar joints of Al and Mg are presented in this section.

Microhardness evaluation of dissimilar aluminium and magnesium welds

In a research study on FSW of 5052 AA and AZ31 Mg alloy [44], it was reported that high peaks were measured in the microhardness profile. It was found that some of the values of the Vickers microhardness found in the SZ were twice as high as that of the BMs [44] (Figure 6.19).

The microhardness profiling was conducted at 1.5, 3 and 4.5 mm at the weld cross-section. The hardness measured along the 1.5 mm line was

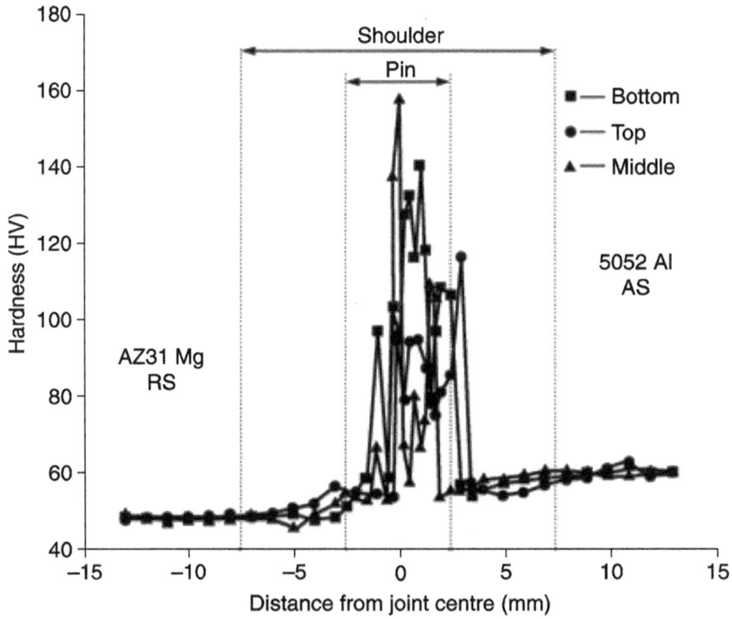

Figure 6.19 Microhardness profiles of microstructures from magnesium to aluminium with different locations [44]

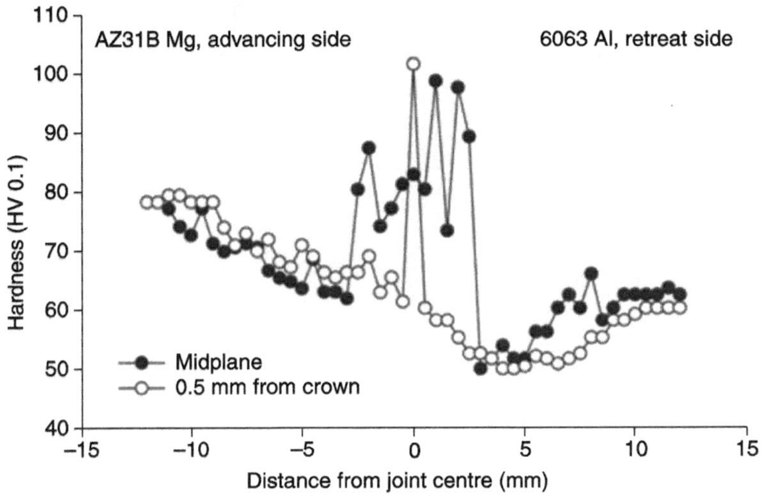

Figure 6.20 Microhardness profile of 6063 Al and AZ31B Mg [48]

generally lower than that of 3 and 4.5 mm, due to the fact that almost no intermixing occurred between the 5052 AA and AZ31 Mg alloy in the top region at 1.5 mm. The microhardness profile was correlated to the evolving microstructure and it was found that the onion ring structure and the intercalated microstructure (Figure 6.17) at the SZ were responsible for the sharp variations of hardness in the welded zone. Vickers microhardness profiling (Figure 6.20) was also conducted in characterizing the factors affecting the properties of dissimilar Al and Mg alloy welds [48].

It was found that the presence of a narrow intermetallic layer at the joint interface was responsible for the sharp hardness values measured. Generally, the interface regions of welds between dissimilar Al and Mg are usually characterized with intermetallic compounds, which is typical of dissimilar joints.

Tensile properties of dissimilar aluminium and magnesium welds

Tensile tests were conducted on dissimilar FSWs between 5052 AA and AZ31 Mg alloy [44]. The tensile fracture locations were correlated to the microhardness profile and it was reported that the tensile samples failed at the location corresponding to sharp high hardness value due to brittleness [44]. The tensile behaviour of dissimilar joints between 6063

AA and AZ31B Mg alloy were also characterized; it was found that the interpenetrating feature observed at the joint interface promotes microvoid coalescence on the tensile fracture surfaces due to mechanical interlocking. It was also found that transverse tensile strength of the weld joint is strongly influenced by the interfacial features: the strength increases with increasing interface length due to improved bonding, interpenetrating feature thickness and the area fraction of microvoid coalescence on the tensile fracture surfaces and decreases with increasing maximum intermetallic layer thickness [48].

Since the formation of intermetallic compounds in the interface of dissimilar welds affects the mechanical properties of the joint significantly, efforts are made to produce dissimilar joints with low heat input to control the formation of the intermetallic compounds, thereby improving the evolving properties.

6.5 Friction stir welding of aluminium alloys and steel

Employing environmental-friendly techniques and processes in engineering and industrial applications has become a must, because of environmental regulations and fast growing fuel prices. Among other things, the automotive industries are now inclined towards low consumption automobiles. The reduction in energy consumption may be obtained by lightening of vehicles by using Al or Mg in various parts of the vehicle instead of steel. However, the use of dissimilar joints of Al and steel in the industries is desirable, but may result in some problems due to the difference in the properties of the materials. Joining of Al to steel is generally difficult because of the differences between their physical and chemical properties. Both alloys have incomparable melting temperatures, thermal conductivity, coefficient of linear expansion and heat capacities [50]. Research investigations have confirmed the phenomena around the formation of brittle intermetallic compounds; Fe_2Al_5 and $FeAl_3$ in the weld zone during the joining of Al to steel. This has been attributed to the low solubility of iron in the Al [51]. This challenge consequently has made the formation of a strong Al-steel joint very difficult when the conventional fusion welding technique is employed. Several attempts were made towards joining Al to steel, some of which are now recommended. This includes diffusion bonding, laser welding, ultrasonic and FSW techniques [53–55].

Past research showed that the use of a combination of steel and AA has been increasing in fabrication of vehicles, due to the inherent advantage of reducing the weight of the assembled vehicles. However, investigation reports from many trials found it difficult to make sound welds, because of the formation of hard and brittle intermetallic compounds [56].

6.5.1 Microstructural evaluation of dissimilar alloys and steel

It was found that joining 2 mm thick SS400 mild steel sheets to AA5083 in a butt-joint configuration was achieved with the FSW process. A small volume of intermetallic compounds was formed at the upper part of the steel–Al interface, while no intermetallic compounds were observed in the middle and bottom parts of the interface. The regions where the intermetallic compounds formed seemed to be fracture paths in the joint. Many fragments of the steel were scattered in the AA matrix and the oxide film removed from the faying surface of the steel by the rubbing motion of a rotating pin, which was observed at the interface between the steel fragments and the AA matrix [56].

The study of the microstructure, hardness and fatigue properties of friction stir butt welded 4 mm thick Al 6013-T4 to X5CrNi18-10 stainless steel was successfully conducted. A complete and crack-free bonding between both materials was achieved. The complex and non-smooth interface consists of very fine a-Fe (ferrite) grains and thin strips of Fe_2Al_5 intermetallic compounds. The grain refinement and defect formation in the Fe-TMAZ results in significant strain hardening. In addition, the joint efficiency showed an UTS equal to 80% of the AA BM. No evidence of plastic deformation was observed on the high strength steel side in either joint. All deformation during the tensile test took place solely in the AA, this being the weaker material [57].

Furthermore, several characterizations were investigated with the aim of improving the process of joining dissimilar Al to steel. Some of the studies reported include FSW of Al to high strength steels [58], analysing the effect of rotational speed on temperature rise and joint strength of 7075-T65 AA and mild steel under the condition of constant welding speed, the effect of heat input on the formation of intermetallic compounds and resultant tensile strength [59] and the feasibility of joining Al 6061 to AISI 1018 steel with the aim of establishing the effect of pin position on the temperature distribution and microstructure of weld zone at constant tool rotation and traverse speed [60]. The evolving

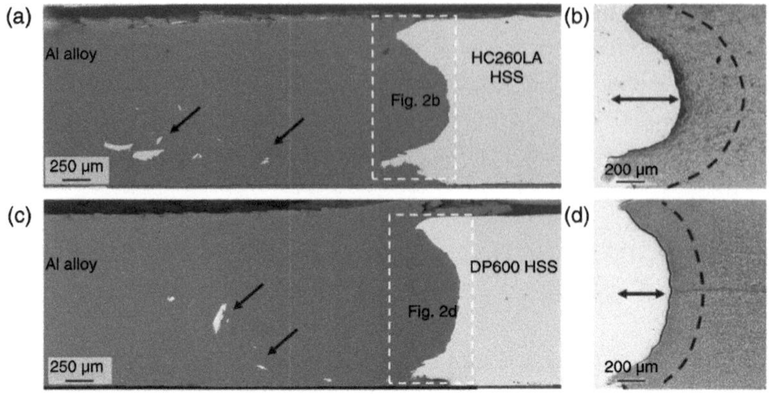

Figure 6.21 **Cross-section overview of the investigated joints: (a) Al alloy to HC260LA HSS and (b) thermo-mechanically deformed interface; (c) Al alloy to DP600 HSS; and (d) thermo-mechanically deformed interface [58]**

microstructures of dissimilar joints between Al and steel are typical of dissimilar FSWs [58].

It was found that the cross-section of the joint is characterized with vortex-like features at the Al side and fragments of the steel material [58]. In addition, because dissimilar joints are usually characterized with intermetallics, the examination of the type of the intermetallic compound formed during the FSW of 6056-T4 AA and 304 austenitic steel was also conducted and it was concluded from the study that the formation of intermetallic compounds decreases noticeably because of the solid-state welding (FSW) process employed [61].

Another interesting study on the formation of a steel-Al hybrid butt weld was also reported. Part of the findings was that the FSW process needed some modification for effective formation of the hybrid butt joint. This became necessary because of the application of steel and Al in the automotive industry. The choices of these two materials are currently employed in the automotive industry, because of the unique property of Al for significant overall weight reduction of the structure and steel, which contributes rigidity and stiffness to the structure. This is a critical component because of the safety of passengers. The hybrid welded joint for thin sheets was achieved through friction stir knead welding (FSKW) [62]; Figure 6.22 is a schematic of the modified FSW process.

First, the softer Al sheet is positioned in a lap joint on the steel blank. The materials investigated were DC04 steel, as well as Al AA5182 and

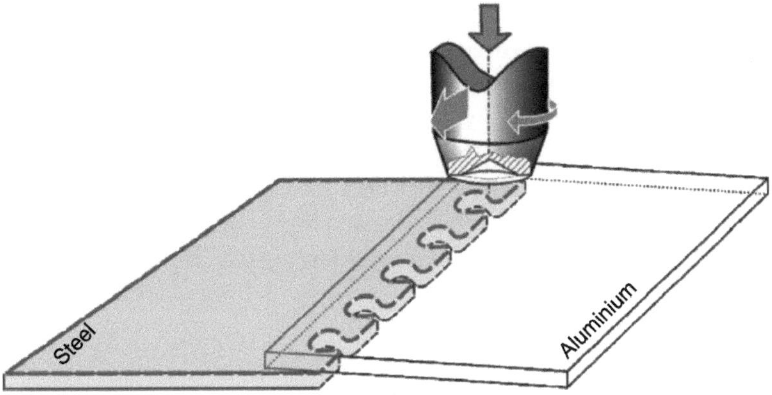

Figure 6.22 **Schematic of the modified friction stir knead welding [62]**

AA6016, all of 1 mm thickness. These materials are typically used for the body-in-white, as well as for the outer panels of automotive vehicles. The steel blank dimension is $130 \times 100\,mm^2$ and the Al blank is $180 \times 120\,mm^2$. These dimensions are necessary to allow the rotating tool to cover the material. The steel sheet exhibits a special contoured geometry at the cutting edge, while the Al sheet is a conventional cut by a linear cutting edge. An axially symmetrical tool rotates in the normal position or at a slight tilt to the plane of the sheets and is slowly pushed into the upper sheet. The second characteristic is that a tool without a pin was employed in this process; this is contrary to the FSW process. Hence, the shoulder generates frictional heat as the tool rotates and a vertical force is applied. This consequently plasticizes the Al at the upper blank and kneads it into the cavity of the countered steel blank. Metallographic investigations showed a characteristic deformation of the Al in between the steel tooth due to the stirring effect during FSKW, as well as a layer of chemical reaction between steel and Al. Furthermore, microhardness tests present a local hardening of AA5182 and DC04, whereas the Martens Hardness of AA6016 softens up to 10% [62].

A research study into interfacial reactions during FSW of steel and Al was also reported [63]. From the joint between a FSW 4 mm sheet of 6056-T4 AA and 304 austenitic steel, it was reported that a complex interface structure was observed, which is made up of a swirl or vortex-like intercalation structure and a thin intermetallic compound layer. The thickness of the intermetallic compound layer was approximately 250 nm and was identified as the Al_4Fe phase with a hexagonal close-packed structure. Adjacent to the intermetallic compound layer, are the ultra-fine grain structure and elongated grains. The elongated grains appeared to consist of the ferrite phase, which can be identified on the basis of the strong TEM astigmatism by ferromagnetism, the body centred cubic (b.c.c.) structure and the inclusion of a ferrite stabilizer element observed in the region. The effect of the stirring action and high strain rate on the stainless steel part after the FSW partially contributed to the ferrite transformation from the austenite phase, even at a relatively low temperature [63].

6.5.2 Mechanical properties of dissimilar aluminium and steel

The mechanical properties of dissimilar welds between Al and steel are presented in this subsection.

Microhardness evaluation of dissimilar aluminium and steel

Typically, every dissimilar weld is usually characterized with intermetallic compounds resulting in high hardness values at the joint interfaces; FSW of Al and steel with huge differences in their properties also result in intermetallics when joined, hence high peaks in the microhardness profiles. The microhardness profiles of Al and high strength steels is presented in Figure 6.23.

The microhardness profiles were measured at three different depths in cross-sections at the top, middle and bottom (Figure 6.23) of the two

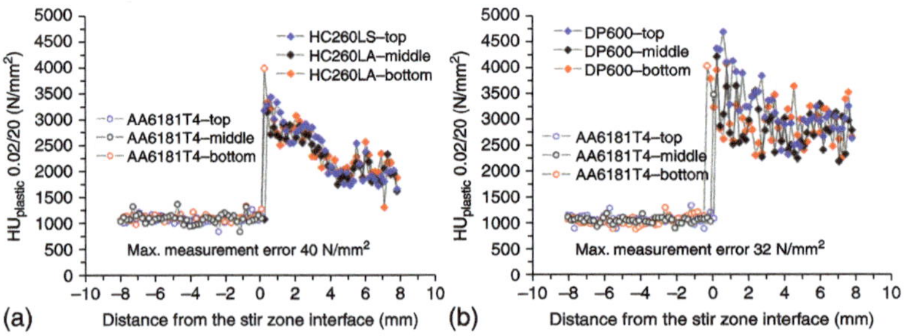

Figure 6.23 Microhardness profiles: (a) Al alloy to HC260LA HSS; and (b) Al alloy to DP600 HSS [58]

high strength steels employed in the study. Highest values were measured at the joint interfaces in both cases, which have resulted from the presence of intermetallics [58].

Tensile properties of dissimilar aluminium and steel

In a study on the effects of welding parameters and tool geometry on the properties of 3003-H18 AA to mild steel conducted using FSW [64], the tensile behaviour of the welded samples were correlated to the heat input and the process parameters. Welds were conducted at constant welding speed, while the rotational speed was kept constant. It was found that the UTS decreased as the rotational speed increased. Welds conducted at high rotational speed were characterized with large defects due to lack of consolidation at the joint, hence the low UTS measured [64]. In correlating the tensile behaviour of the joints to the heat input, it was reported that the combined effect of heat flow and the deformation of the material during the welding procedure eliminates all aspects of cold working through dynamic recrystallization. Therefore, the strength of the welds decreased [64]. These behaviours are typical of dissimilar joints between Al and steel and the joints are usually weaker [58,64]. Advances in FSW of dissimilar Al and steel have been reviewed and reported, with the main focus on the evolving microstructure and the mechanical properties. The next section will focus on the FSW of Al and titanium (Ti).

6.6 Friction stir welding of aluminium and titanium

Titanium and its alloys are attractive materials to use in a variety of applications. Their high strength, low density and corrosion resistance make it the best material to use for various applications. Ti is used extensively in aircraft and biomedical applications, because of its low density combined with high strength. It is also used for turbine and turbojet blades, because of its good creep resistance. A combination of these exciting properties and that of Al are desirable in some practical applications. Against this background, the dissimilar FSW of Al and Ti finds application in the aeronautics and automotive industries, where current efforts to ensure a high strength material with low weight properties are achieved at a reasonable cost reduction. For example, in the field of aircraft manufacturing, the welding of skin-stringer joints is progressively replacing riveted fuselage structures. Though the dissimilar jointing of Al to Ti alloys is recognized as a challenge, owing to the significant differences in the physical properties of the Al and Ti alloys involved very few researches has been conducted in this particular field of study. Some of the common problems associated with the dissimilar Al to Ti welds using the conventional fusion techniques are poor mechanical properties, which are due to the lopsided complex weld pool profiles and formation of brittle intermetallic compounds [65].

6.6.1 Microstructural characterization of dissimilar aluminium and titanium

One of the most recent research studies of dissimilar welding of Al and Ti involves a 4 mm wrought sheet of Al-6Mg alloy and Ti-6Al-4V alloy. The grade 5-Ti6Al4V alloy is the most widely-used Ti alloy and is referred to as the workhorse of the Ti industry. A cuboid-shaped groove of 2.2 mm depth was removed from the AA sheets, while the Ti alloy sheets were placed into the grooves and butted with the AA sheets (Figure 6.24) during the welding procedure [66].

A study into the dissimilar FSW of AA and Ti alloy grade 5 using the modified butt joint configuration showed that a good formability between

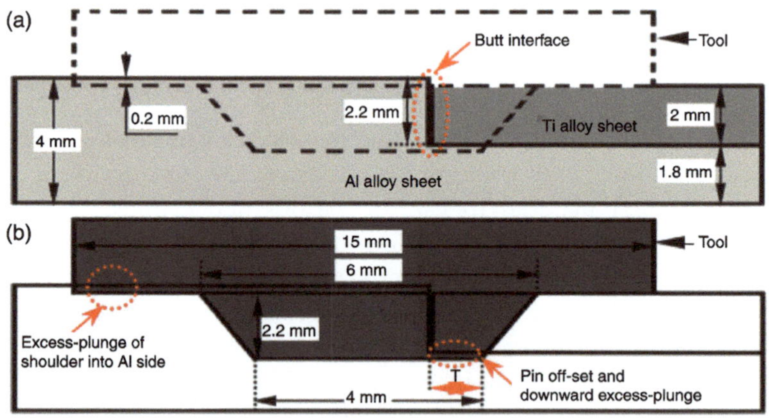

Figure 6.24 Schematic of the aluminium and titanium modified butt joint: (a) design of parent material; and (b) set-up of the FSW tool [66]

Al and Ti joints can be achieved without weld defects. The modified joint configuration and pin plunge set-up helps to reduce the tool shoulder wear and tear and also eliminates butt flaw. Also, the Al and Ti diffusion interlayer of about 2 micron was produced at the butt welded interface, using z lower tool rotating speed and lower pin-offset towards the Ti alloy butt side. However, the Al–Ti butt joint was also characterized by a bulky intermetallic compounds interlayer of about 10 microns. This layer increases as the friction heat input increases and a higher plasticized deformation is also observed [66]. A typical cross-section macrostructure of the modified Ti–Al joint is shown in Figure 6.25, with a process parameter of 1200 rpm, 60 mm/s and pin offset of 0.5 mm.

On the cross-section view, it was observed that a composite-like structure of Al-matrix material mixed with the Ti particles was formed in the stir nugget zone (SNZ) at the AA butt side. Also found in the SNZ is the plasticized Al-matrix states of 'the vortex flow' and 'the turbulent flow' at the AA butt side. The flow characteristics of the solid state Al-matrix material are a function of the tool rotating traversing action during the welding process. This consequently results in plasticized deformation, dynamic recrystallization, dissolution and precipitation of the AA. Conversely, the detached Ti particles were dispersed according to the corresponding flow pattern of the SNZ. This implies that the distribution laws of the broken and dispersed Ti-6Al-4V particles also indicated or implied the vortex/turbulent movement features in the SNZ of the Al-matrix material [66].

Another unique way of joining Al to Ti alloy by FSW using a cutting pin was also reported [67]. The study involves 3 mm thick Al 1060 sheets

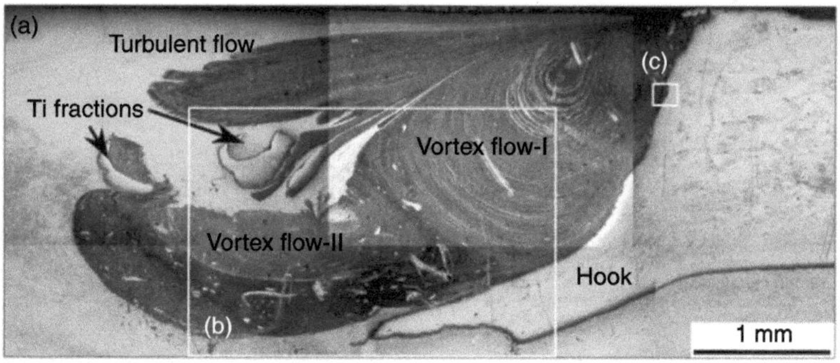

Figure 6.25 Joint cross-section macrostructure by optical microscope [66]

and 3 mm thick Ti alloy Ti-6Al-4V sheets. The tool used consists of a concave shoulder with a 25 mm diameter and a cutting stir pin made of tungsten carbide with a 6 mm diameter and 3.2 mm long. The designed friction stir tool with a cutting pin allows the stir pin to plunge into the Ti to promote deformation. The schematic diagram of the lap joint is shown in Figure 6.26. During the FSW process, the cutting motion of the pin partly replaced stir motion, and then the frictional heat was reduced. Therefore, the welding temperature on the interface decreases when compared to that of the pure friction without cutting edges, and consequently may prevent the formation of intermetallic compounds.

In addition, the cutting pin is further regarded as very advantageous, because the strong Ti alloy could easily be deformed using the cutter and the rate of wear of the cutting tool is greatly reduced. The result showed that Al and Ti alloy sheets were successfully lap-joined by FSW. During the welding process, the surface layer of the Ti plate was cut off by the pin, and intensively mixed with the Al situated on the Ti plate [67].

6.6.2 Mechanical properties of dissimilar aluminium and titanium

The resulting hardness and the tensile behaviour of joints between Al and Ti are reported in this subsection.

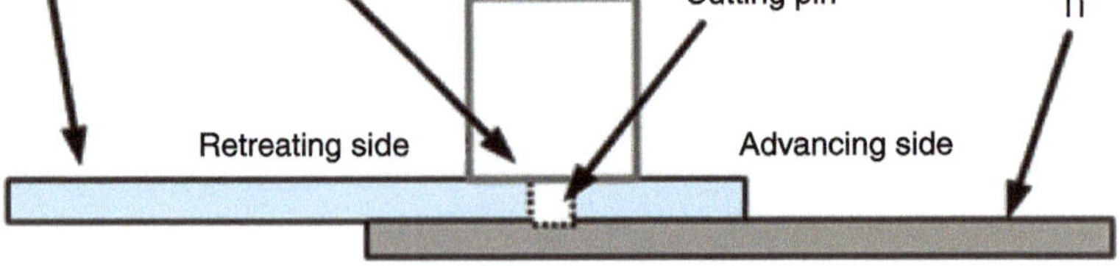

Figure 6.26 Schematic diagram of the lap joint set-up and configuration [67]

Microhardness evaluation of dissimilar aluminium and titanium

A typical study conducted on interface characteristics of FSW lap joints of Ti/Al dissimilar alloys [68], LF6 AA and TC1, were used as the BMs, while the rotational speed and the feed rate were varied. The resulting hardness behaviour of the weld interface was characterized (Figure 6.27).

It was reported that the high peaks at the SZ were due to Ti-Al intermetallic compounds at the joint interface [68]. This distribution is typical of dissimilar welds.

Tensile properties of dissimilar aluminium and titanium

In a typical study on FSW of Al and Ti, it was reported that the resulting Al–Ti interfacial structure has an advantage over the Al–Ti bulky layer of intermetallic compound formed with respect to the effect on the joint tensile properties. When the diffusion bonding interlayer is achieved, the joint mechanical tensile strength reached more than 92% of the parent material – Al–6 Mg alloy strength, with a ductile fracture mode and a tensile fracture path within the AA butt side [66]. In another study on

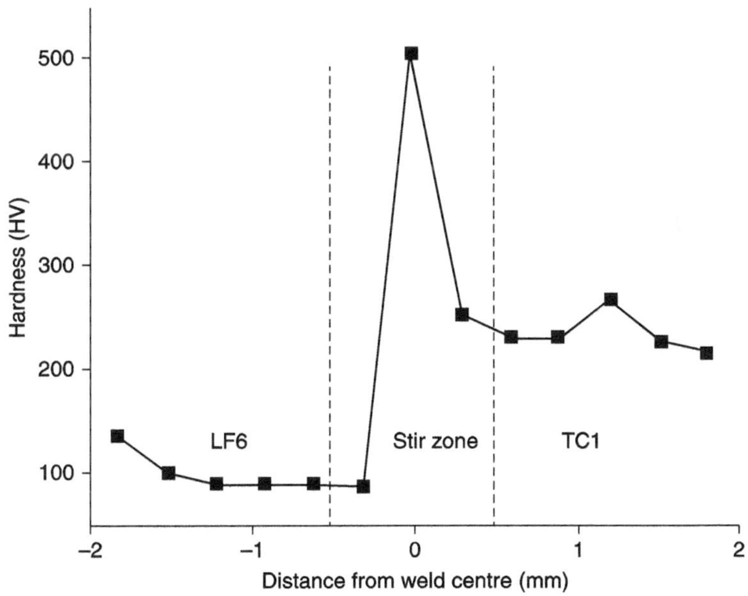

Figure 6.27 Microhardness distribution of Ti/Al lap joint made by FSW [68]

2024-T3 AA and Ti alloy grade 5 Ti6Al4V, the UTS of the joints reached 73% of the AA2024-T3 BM strength [69].

The evolving microstructure and the mechanical properties of dissimilar joints between Al and Ti have been reported, and the next section will focus on FSW of Al and Cu.

6.7 Friction stir welding of aluminium and copper

Al and Cu alloys are the most widely-used metals in the electrical and electronic industries. The need to make electrically stable bolted joints between these two materials as well as welding them together has been of great interest in the past decade. However, the welding of these two dissimilar materials is difficult to perform due to the differences in chemical, physical and mechanical properties. The joining of Al and Cu dissimilar materials is of great interest and in demand for industrial applications. The need to join these materials arises from the thermal and mechanical properties they possess, such as a high corrosion resistance and a high electrical conductivity. The use of FSW to join these two materials has resulted in improved contact surface, improved current flow and less resistance compared to the conventional use of mechanical fasteners. The improvements will ultimately lead to energy savings; this will also lead to a global energy consumption decrease if the method is implemented on a global scale in the electrical industry.

Similar to other dissimilar FSWs, the joining of these two materials tends to form brittle intermetallic compounds. This can be avoided or minimized through the use of solid-state joining technologies such as FSW. The Al to Cu joints are generally difficult to achieve and in past experiences during mechanical tests, the joints failed at the NZ or along the joint line of the two materials resulting from poor joint integrity [70,71]. To produce defect free welds of good quality between Al and Cu, the copper specimen has to be placed on the AS, while the Al should be placed on the RS during the welding procedure. Detailed characterization of friction stir butt welds of Al and Cu are discussed in the following sub-sections.

6.7.1 Experimental set-up

FSWs between 5754 AA and C11000 Cu, 3.175 mm thick were produced using an Intelligent Stir Welding for Industry and Research Process

Figure 6.28 Experimental set-up during the welding process

Development System (I-STIR PDS) platform available at the Nelson Mandela Metropolitan University, Port Elizabeth South Africa. The set-up during the welding process is shown in Figure 6.28.

In order to optimize the process, the welds were produced using three different shoulder diameter tools, i.e. 15, 18 and 25 mm with a constant tool pin diameter of 5 mm. The features on the tools were threaded pins and concave shoulder (8 degrees). The tools were hardened to 52 HRC. The Cu sheet was placed at the AS while the tool pin was plunged in the AA and made to touch Cu during the welding process; this is an optimized tool displacement setting [15]. The rotational speeds of 600, 950 and 1200 rpm were chosen to produce the welds; these represent a low, medium and high setting, respectively, while 50, 150 and 300 mm/min were the feed rates employed, also representing a low, medium and high setting, respectively.

Other parameters, such as the tool tilt angle was kept constant at 2 degrees, while a dwell time of 2 seconds was used for all the welds. Each process parameter combination was used to produce a 160 mm weld length. Temperature measurement was conducted by using the USB TC-08 Thermocouple Data Logger. The TC-08 software is a data acquisition tool that collects, records, displays and analyses the temperature data. The data were collected through the K-type thermocouples fitted into the machined grooves made on the backing plate to contact the welded plates. Two thermocouples were located at the abutting joint of the two plates being welded. The welding process was achieved using the position control system on the FSW machine, the torque and force feedbacks – F_x, F_y and F_z were captured on the system. For consistency and accuracy, the welding tools were cleaned by soaking them in a solution of 100 g of sodium hydroxide and 100 ml of water after every welding procedure to remove the plasticized Al stuck in the tool geometry.

Some of the tools were also re-machined to achieve the geometry. The welds were sectioned at specific places for characterization. The electrical resistances of the joints were measured using a Signatone Four-Point probe meter with 1.6 mm probe spacing and the sample cross-sectional area was 127 mm^2. An Olympus BX51M upright reflected-light microscope was used to observe the microstructures of the welded samples according to ASTM E3-11 [72]. The AA side was etched with flicks reagent, and the Cu was etched with a solution of 25 ml distilled water, 25 ml ammonia water and 15 ml hydrogen peroxide 3%. The tensile samples were produced and tested in accordance with the ASTM E-8 standard [73]. A servo-hydraulic Instron 8801 tensile testing machine was used to conduct the tests. An extension rate of 5 mm/min and a gauge length of 50 mm were used. The MH-3 Vickers microhardness testing instrument developed by Metkon was employed to measure the microhardness according to ASTM E384 standard [74]. The Vickers microhardness profiles were measured 1.5 mm below the weld surface along the cross-sections of the welds, with a load of 200 gf and a dwell time of 15 seconds [14]. The indentations were taken at an interval of 0.7 mm. The X-ray diffraction was conducted using an X-ray diffractometer (Bruker D8 Advance) equipped with standard Bragg-Brentano geometry with Cu (Ka) radiation and a Nickel filter at the detector. The results and discussion of the results obtained will be presented in the following sub-sections.

6.7.2 Temperature measurements

Two thermocouples, T_1 and T_2, were aligned along the abutting joint of the two plates that correspond to the weld joint. The specific data reported here was captured for one of the hottest welds produced at 600 rpm and 50 mm/min. The temperatures recorded by T_1 and T_2 were

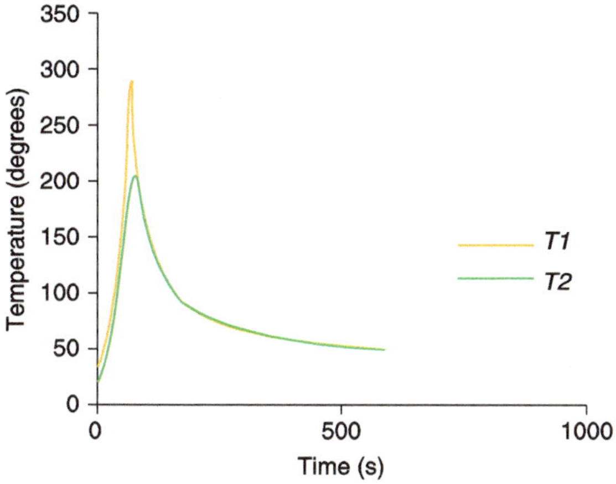

Figure 6.29 Time-temperature plot along the weld line

284 and 204 °C, respectively. The time–temperature plot is shown in Figure 6.29.

T_1 was placed at 80 mm and T_2 at 20 mm of the weld length. A general trend of increasing temperature was observed in the measurement along the weld line. This trend shows that as the tool traverses through the material during the welding process, it gains more energy that consequently translates to an increase in the welding temperature. The melting temperatures of Al and Cu are 660 and 1085 °C, respectively, and the recorded maximum temperature of 284 °C further confirmed that both the Al and Cu were not melted during the process, hence the formation of the solid phase joints.

6.7.3 Characterizing the Input/output processing parameters

The weld numbers, the input process parameters, feedback forces (Fx, Fy and Fz), torques and the calculated heat input are presented in Tables 6.2(a to c) for the 15, 18 and 25 mm shoulder diameter tools employed in this research work, respectively [14].

The torque values reported are steady state response values and the heat input was calculated using [15]:

$$Q = \eta \frac{2\pi\omega T}{f} \tag{6.1}$$

Table 6.2(a) Input/output FSW data for the 15 mm shoulder diameter tool

Weld No.	Rotational speed (rev/min)	Feed rate (mm/min)	Fx (kN)	Fy (kN)	Fz (kN)	Torque (kNm)	Q_{Input} (J/mm)
S15_01	600	50	2.11	-1.17	10.14	16.5	1119.92
S15_02	600	150	2.6	-0.04	13.87	19.97	451.69
S15_03	600	300	3.02	-0.9	17.94	23.63	289.86
S15_04	950	50	1.63	-1.04	6.97	8.59	923.18
S15_05	950	150	2.02	-0.27	10.67	12.09	432.83
S15_06	950	300	2.63	-0.67	12.25	13.12	219.31
S15_07	1200	50	1.52	-0.76	7.45	8.01	1087.67
S15_08	1200	150	3.14	-0.65	11.59	12.67	524.3
S15_09	1200	300	3.54	-0.77	13.07	15.24	299.59

Table 6.2(b) Input/output FSW data for the 18 mm shoulder diameter tool

Weld No.	Rotational speed	Feed rate	Fx (kN)	Fy (kN)	Fz (kN)	Torque (kNm)	Q_{Input} (J/mm)
S18_01	600	50	2.87	-0.93	10.47	17.47	1185.71
S18_02	600	150	3.42	-0.6	14.27	20.96	474.02
S18_03	600	300	4.06	-0.09	18.53	26.2	296.37
S18_04	950	50	2.85	-0.72	11.56	12.8	1374.86
S18_05	950	150	3.12	-0.75	14.47	15.15	542.63
S18_06	950	300	3.47	-0.12	16.69	17.37	293.06
S18_07	1200	50	2.26	-0.18	10.6	12.95	1765
S18_08	1200	150	2.68	-0.23	12.51	15.12	683
S18_09	1200	300	3.16	-0.01	14.91	17.44	405

where Q = heat input (J/mm), η = efficiency factor = 0.9 for Al and Cu, ω = rotational speed (rev/min), T = torque (Nm) and f = feed rate (mm/min).

It was observed from the data obtained for all the weld settings that the advancing force (Fx) and the torque (T) increase as the vertical downward force (Fz) increases. The heat input to the welds, (Q), decreases as the

Table 6.2(c) **Input/output FSW data for the 25 mm shoulder diameter tool**

Weld No	Rotational speed	Feed rate	Fx (kN)	Fy (kN)	Fz (kN)	Torque (kNm)	Q_{input} (J/mm)
S25_01	600	50	3.26	-0.08	20.94	25.51	1731.02
S25_02	600	150	3.9	-0.64	22.49	26.12	590.92
S25_03	600	300	3.99	-0.75	24.28	28.91	326.91
S25_04	950	50	2.33	-0.09	12.14	14.74	1583.49
S25_05	950	150	4.43	0.12	26.08	29.26	934.11
S25_06	950	300	5.2	0.26	32.24	36.74	577.29
S25_07	1200	50	3.34	0.17	15.23	20.51	2067.58
S25_08	1200	150	4.54	0.48	21	33.02	950.15
S25_09	1200	300	5.91	-0.07	24.64	35.81	557.24

feed rate increases. This is typical of FSWs, as at low feed rates, the tool traverses at a slow pace leading to a high interaction between the tool and the materials being joined, hence an increase in the heat input resulting from the stirring of the tool. This is vice versa for high feed rates.

6.7.4 Microstructural characterization of aluminium and copper

Microstructural characterization was conducted on the parent materials and the welded samples. The microstructures of the AA and the Cu are presented in Figures 6.30 (a and b), respectively [14].

Fairly equiaxed grains were observed in both the Al and Cu microstructure. Typical micrographs of welds produced at 600 rpm and 50, 150 and 300 mm/min are presented in Figures 6.31(a–c).

It was observed that the stirring of the dissimilar metals was achieved in all the welds. Fragments of Cu and Al extrusions were observed in the weld produced at 50 mm/min, while heavy flow-lines of both metals were observed in the SZ of the weld produced at 150 mm/min. Concentrated flow line and evidence of the onion ring structure was found in the SZ of the weld produced at the 150 mm/min. Similar observations have been reported by many authors [75–78].

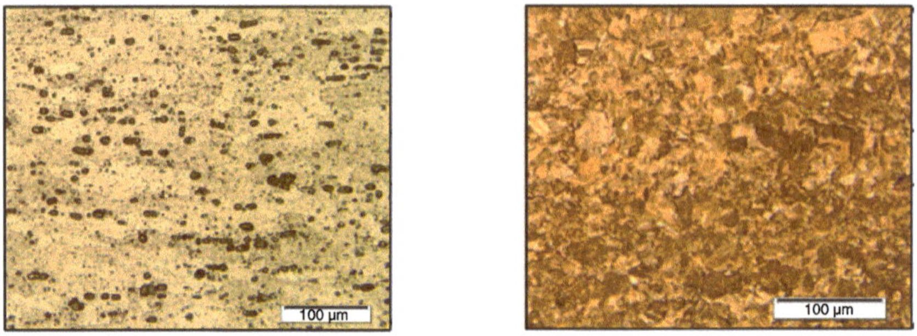

(a) (b)

Figure 6.30 (a) Micrograph of aluminium; and (b) micrograph of copper

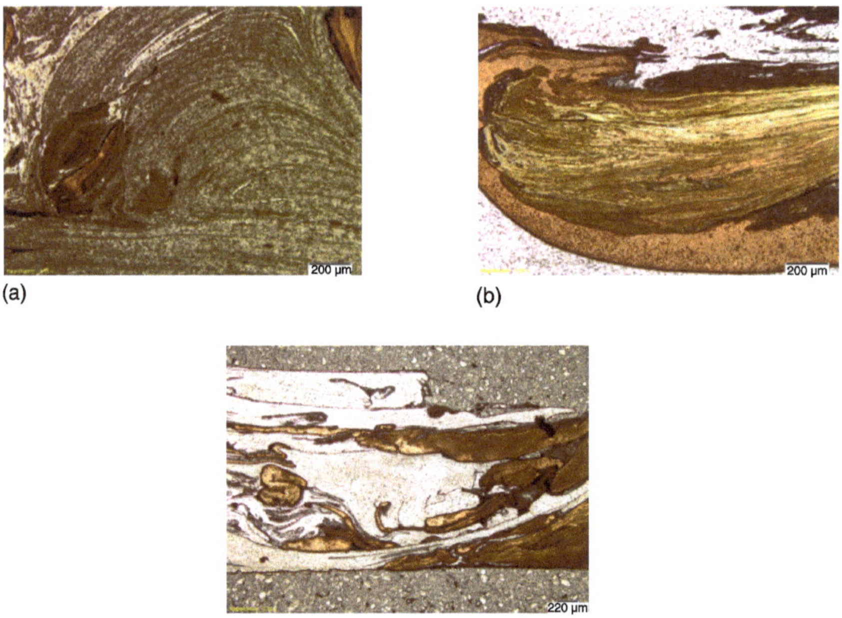

(a) (b) (c)

Figure 6.31 (a) Micrograph of weld produced at 600 rpm and 50 mm/min; (b) micrograph of weld produced at 600 rpm and 150 mm/min; and (c) micrograph of weld produced at 600 rpm and 300 mm/min

6.7.5 Tensile tests

Typical force/extension graphs of the welds produced at 600 rpm and varied feed rates of 50, 150 and 300 mm/min using the 18 mm shoulder diameter tool are presented in Figure 6.32.

The tensile behaviour was correlated to the process parameters and the resulting microstructure. It was observed that the weld produced at the 50 mm/min with the highest heat input has exhibited the most ductile behaviour compared to the other two welds produced at higher feed rates of 150 and 300 mm/min. This same weld has the highest degree of mixed layers of both materials joined. This observation can be related to the heat input, which is lower at a higher traverse speed and thus causes less vertical transport and mixing of materials during the welding process.

6.7.6 Microhardness characterization profiling

The microhardness profiling of all the welds produced was carried out at a depth of 1.5 mm below the weld top surface. The average Vickers microhardness values of the parent materials, AA and Cu, are HV 60 and HV 95, respectively. Typical microhardness profiles of welds produced at a constant of 600 rpm and varied feed rates are presented in Figure 6.33 [14].

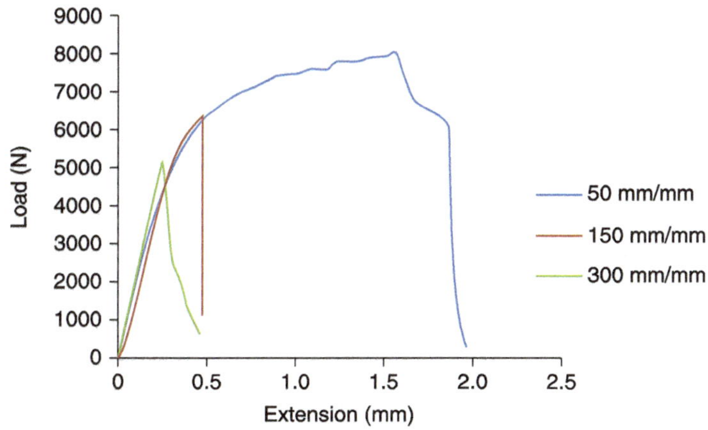

Figure 6.32 **Load-extension of welds produced at 600 rpm and varied feed rates (50, 150 and 300 mm/min)**

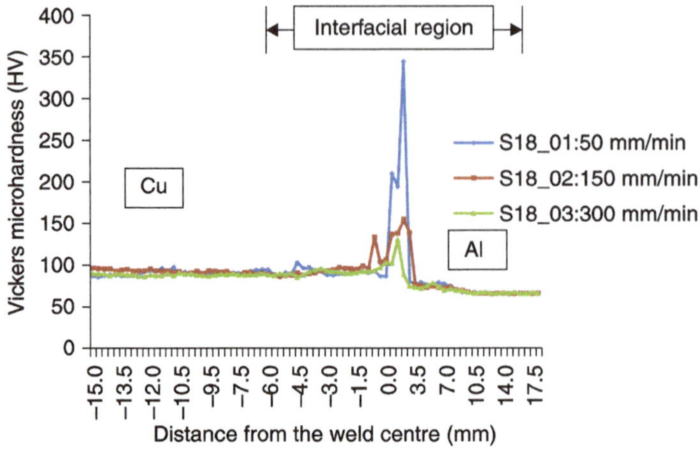

Figure 6.33 **Microhardness profiles of welds produced at a constant rotational speed of 600 rpm with varied feed rates [14]**

It was observed that the joint interface of the welds are characterized by high peaks resulting from the presence of intermetallics compared to the softening trend observed in typical FS welds of Al. The microhardness variation was low in the HAZ regions of both metals. This is in the range between 90 HV and 110 HV for Cu, and between 60 HV and 80 HV (HV_{min} and HV_{max}) for the Al. It was observed in all the welds, the higher Vickers microhardness values in a range of between 130 HV and 350 HV were measured in the TMAZ and the SZ of the weld. These are regions previously occupied by the tool pin and the shoulder during the welding process. This region is referred to in the figure as the *interfacial region*. The increase in the microhardness values at the interfacial regions can be attributed to plastic deformation that has occurred in the regions during the welding process, and also due to the presence of intermetallic compounds.

6.7.7 X-ray diffraction analyses

The high peaks Vickers microhardness values corresponding to the presence of intermetallics were further investigated by X-ray diffraction analysis. The diffraction patterns of the interfacial regions of the weld produced with 600 rpm and varied feed rates are shown in Figures 6.34(a–c).

The additional diffraction peaks shown in the figures are due to the presence of intermetallic compounds such as Al_2Cu or Al_4Cu_9. These compounds were identified as far as possible by matching the PDF data files [79]. Although very low peaks, the common identified intermetallic compounds formed at the joint interfaces of these welds were Al_2Cu and Al_4Cu_9. Similar results have been reported [80,81]. The low peaks indicate low volume fraction of the intermetallics in the welds, hence the weld process parameter combinations employed in this research work can be recommended.

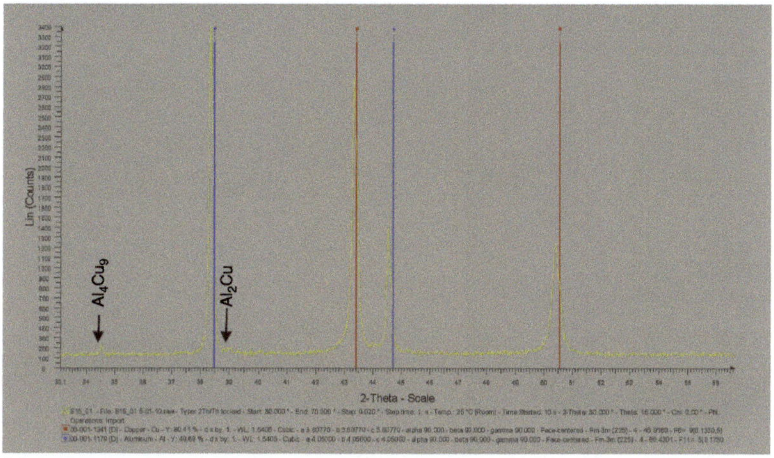

(a)

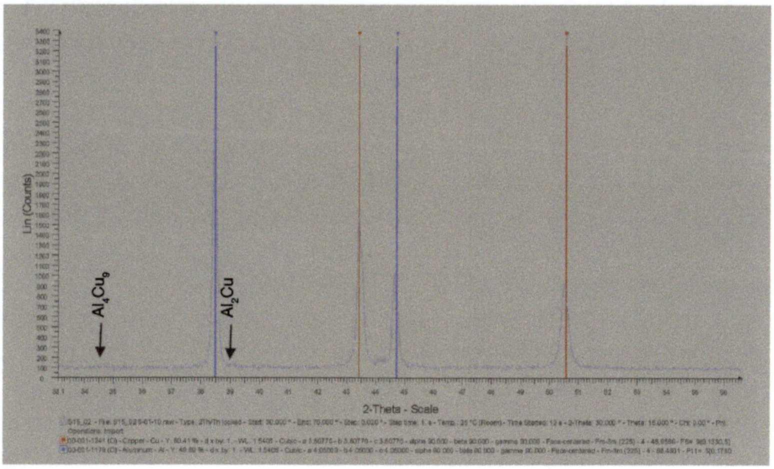

(b)

Figure 6.34 **(a) Diffractogram of weld produced at 600 rpm and 50 mm/min; (b) diffractogram of weld produced at 600 rpm and 150 mm/min; and (c) diffractogram of weld produced at 600 rpm and 300 mm/min [14]**

(Continued)

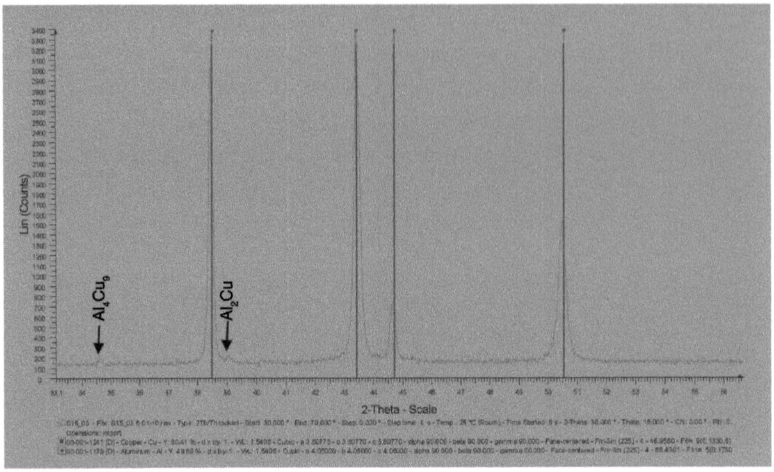

(c)

Figure 6.34 Continued

6.7.8 Electrical resistivity measurements

Electrical resistivity, an inverse of electrical conductivity, is a measure of how strongly a material opposes the flow of electric current. A low resistivity indicates a material/joint that readily allows the movement of electrical charge. Despite the advances in dissimilar joining, particularly joints between Al and Cu that is relevant to the electrical industry, limited information exists on electrical resistivity measurements of weld cross-sections produced [63,75,82,83]. Intermetallic compounds are usually formed as a result of increased temperature during the welding procedure. Against this background, the electrical resistivity of the welds conducted using various parameter combinations was measured and correlated to the heat input into the welds. The electrical resistivities were measured using a four-point probe meter, while the heat input was calculated using Eq. 6.1.

The torque values were steady state values obtained during the welding procedure. The results are presented in Table 6.3 [14].

The electrical resistivities of the welds ranged between 0.088 and 0.1 μΩ. The percentage increase or decrease in the resistivity of each weld compared to the average joint resistance are as indicated using the arrows. The arrows pointing upwards indicate an increase, while those pointing downwards indicate a decrease in the electrical resistivities compared to

Table 6.3 **Electrical resistivities and corresponding heat input [14]**

Weld No.	Resistivity ρ (μΩ)	Percentage change in resistivities compared to average resistivity of parent material	Heat input (J/mm)
PM AL	0.88		
PM CU	0.96		
Average ρ	0.92		
S15 01	0.101	9.8 ↑	1119
S15 02	0.097	5.4 ↑	451
S15 03	0.090	2.1 ↓	289
S15 04	0.101	9.8 ↑	923
S15 05	0.098	6.5 ↑	432
S15 06	0.095	3.3 ↑	219
S15 07	0.101	9.8 ↑	1087
S15 08	0.098	6.5 ↑	524
S15 09	0.088	4.3 ↓	299

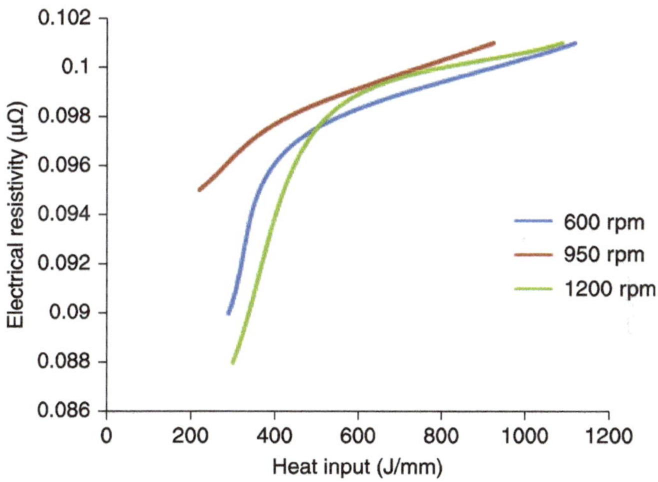

Figure 6.35 **Electrical resistivity vs. heat inputs into the welds [70]**

the parent materials. It was observed that the welds with the highest percentage increase of 9.8% were measured in those welds produced with high heat input. A graphical representation of the trend observed is presented in Figure 6.35 [70].

It was observed that for all the weld settings, the joint electrical resistivities increase as the heat input increases at constant spindle speeds. This trend can be attributed to the presence of intermetallics in the hot welds. Very little variation compared to the average joint resistance of the parent materials was observed in the electrical resistivity readings from one weld to the other. The weld settings employed in this research work can therefore be recommended to produce welds with low or no electrical resistivity. Many research investigations into FSW of dissimilar welds have confirmed that joints of various combinations have been successfully produced but characterized with the formation of intermetallics.

6.8 References

1. Lee, C.Y., Lee, W.B., Kim, J.W., Choi, D.H., Yeon, Y.M. and Jung, S.B. (2008), 'Lap joint properties of FSW dissimilar formed 5052Al and 6061Al alloys with different thickness', *Journal of Materials Science*, 43: 3296–304.
2. Naffakh, H., Shamanian, M. and Ashrafizadeh, F. (2008), 'Weldability of dissimilar welds between type 310 austenitic stainless steel and alloy 657', *Journal of Materials Science*, 43: 5300–4.
3. Chen, T. (2009), 'Process parameter study on FSW joint of dissimilar metals for aluminium-steel', *Journal of Materials Science*, 44: 2573–80.
4. Kazi, S.H. and Murr, L.E. (2001), 'Complex flow phenomena associated with friction stir welding of aluminium alloys', in: Mishra, R.S. et al. (eds), *Friction stir Welding and Processing*, Warrendale, PA: TMS, 139–51.
5. McNelley, T.R., Swaminathan, S. and Su, J.Q. (2008), 'Recrystallization mechanisms during friction stir welding/processing of aluminium alloys', *Scripta Materialia*, 58: 349–54.
6. Bailey, N. (1986), 'Welding dissimilar metals', in: Bailey N (ed.), Cambridge, UK: The Welding Institute, 1–6.
7. Song, J.L., Lin, S.B., Yang, C.L., Ma, G.C. and Liu, H. (2009), 'Spreading behaviour and microstructure characteristics of dissimilar metals TIG welding-brazing of aluminium alloy to stainless steel', *Materials Science and Engineering A*, 509: 31–40.
8. Qiu, R., Shi, H., Zhang, K., Tu, Y., Iwamoto, C. and Satonaka, S. (2010), 'Interfacial characterization of joint between mild steel and aluminium alloy welded by resistance spot welding', *Materials Characterisation*, 61: 684–8.
9. Sahin, M. (2009), 'Joining of stainless steel and aluminium by friction welding', *International Journal of Advanced Manufacturing Technology*, 41(4–5): 487–97.
10. Lee, W.B., Schmuecker, M., Mercardo, U.A., Biallas, G., and Jung, S.B. (2006), 'The joint properties of dissimilar formed Al alloys by friction stir welding according to the fixed location of materials', *Scripta Materialia*, 55: 355–8.

11. Akinlabi, E.T. and Akinlabi, S.A. (2011), 'Effects of process parameters on mechanical and electrical properties of dissimilar aluminium-copper friction stir welds', *ASME McMat.*, 30 May–2 June, Chicago, IL
12. Okamura, H., Aota, K., Satou, K. and Sakamotom, M. (2003), 'Friction stir welding method and component part welded by the method,' US patent 20030102354A1.
13. Agarwal, S.P., Prashanna, N.N., Arivazhagan, K. and Devendranath, R. (2012), 'Development of friction stir welded butt joints of AA 6063 aluminium alloy and pure copper', *International Conference on Advanced Research in Mechanical Engineering* (*ICARM*), 46–50.
14. Akinlabi, E.T. (2011), *Characterisation of Dissimilar Friction stir Welds between Aluminium and Copper*, Germany: Lambert Academic Publishing.
15. Akinlabi, E.T., Els-Botes, A. and Lombard, H. (2010), 'Effect of tool displacement on defect formation in friction stir welding of aluminium and copper', *Proceedings of the 8th International Conference on Friction stir Welding in Hamburg*, 18–20 May, Germany.
16. Heinz, A., Haszler, A., Keidel, C., Moldenhauer, S., Benedictus, R. and Miller, W.S. (2000), 'Recent development in aluminium alloys for aerospace applications', *Materials Science Engineering A*, 280: 102–7.
17. DebRoy, T. and Bhadeshia, H.K.D.H. (2010), 'Friction stir welding of dissimilar alloys – A perspective', *Science and Technology of Welding and Joining*, 15(4): 266–70.
18. Luijendijk, T. (2000), 'Welding of dissimilar aluminium alloys', *Journal of Materials Processing Technology*, 103: 29–35.
19. Bala Srinivasan, P., Dietzel, W., Zettler, R., dos Santos, J.F. and Sivan, V. (2005), 'Stress corrosion cracking susceptibility of friction stir welded AA7075–AA6056 dissimilar joint', *Materials Science Engineering A*, 392: 292–300.
20. Sivashanmugam, M., Ravikumar, S., Kumar, T., Seshagiri Rao, V. and Muruganandam, D. (2010), 'A review on friction stir welding for aluminium alloys', *IEEE*, 216–21.
21. Khodir, S.A. and Shibayanagi, T. (2008), 'Friction stir welding of dissimilar AA2024 and AA7075 aluminium alloys', *Materials Science Engineering B*, 148: 82–7.
22. Cavaliere, P., De Santis, A., Panella, F. and Squillace, A. (2009), 'Effect of welding parameters on mechanical and microstructural properties of dissimilar AA6082–AA2024 joints produced by friction stir welding', *Materials and Design*, 30: 609–16.
23. Priya, R., Subramanya Sarma, V. and Prasad Rao, K. (2009), 'Effect of post-weld heat treatment on the microstructure and tensile properties of dissimilar friction stir welded AA 2219 and AA 6061 alloys', *Transactions Indian Institute Metals*, 62: 11–19.
24. Mishra, R.S. and Ma, Z.Y. (2005), 'Friction stir welding and processing', *Materials Science and Engineering*, 50: 1–78.
25. Koilraj, M., Sundareswaran, V., Vijayan, S. and Koteswara Rao, S.R. (2012), 'Friction stir welding of dissimilar aluminium alloys AA2219 to AA5083 – Optimization of process parameters using Taguchi technique', *Materials and Design*, 42: 1–7.

26. Akinlabi, E.T. and Akinlabi, S.A. (2012), 'Microstructural characterization of dissimilar friction stir welds', *WCE, International Conference on Mechanical Engineering* (*ICME*), London, 4–6 July.
27. Shigematsu, I., Kwon, Y.J, Suzuki, K., Imai, T. and Satio, N. (2003), 'Joining of 5083 and 6061 aluminium alloys by friction stir welding', *Journal of Materials Science Letters*, 22: 353–6.
28. Li, Y., Murr, L.E. and McClure, J.C. (1999), 'Flow visualization and residual microstructures with the friction stir welding of 2024 aluminium to 6061 aluminium', *Materials Science Engineering A*, 271: 213–323.
29. Ouyang, J.H. and Kovacevic, R. (2002), 'Material flow during friction stir welding (FSW) of the same and dissimilar aluminium alloys', *Journal of Materials Engineering Performance*, 11(1): 51–63.
30. Cavalierem, P. and Panella, F. (2008), 'Effect of tool position on the fatigue properties of dissimilar 2024–7075 sheets joined by friction stir welding', *Journal of Materials Processing Technology*, 206: 249–55.
31. Murr, L.E, Li, Y. and McClure, J.C. (1991), 'A comparative study of friction stir welding of aluminium alloys', *Aluminium Transactions*, 1: 141–54.
32. Palanivel, R., Koshy Mathews, P., Murugan, N. and Dinaharan, I. (2012), 'Effect of tool rotational speed and pin profile on microstructure and tensile strength of dissimilar friction stir welded AA5083-H111 and AA6351-T6 aluminium alloys', *Materials and Design*, 40: 7–16.
33. Amancio-Filho, S.T., Sheikhia, S., dos Santos, J.F. and Bolfarini, C. (2008), 'Preliminary study on the microstructure and mechanical properties of dissimilar friction stir welds in aircraft aluminium alloys 2024-T351 and 6056-T4', *Journal of Materials Processing Technology*, 206: 132–42.
34. Da Silva, A.A.M., Arruti, E., Janeiro, G., Aldanondo, E., Alvarez, P. and Echeverria, A. (2011), 'Material flow and mechanical behaviour of dissimilar AA2024-T3 and AA7075-T6 aluminium alloys friction stir welds', *Journal of Materials and Design*, 32: 2021–7.
35. Scialpi, A., De Giorgi, M., De Filippis, L.A.C., Nobile, R. and Panella, F.W. (2008), 'Mechanical analysis of ultra-thin friction stir welding joined sheets with dissimilar and similar materials', *Materials and Design*, 29(5): 928–36.
36. Li, B. and Shen, Y. (2012), 'A feasibility research on friction stir welding of a new-typed lap – butt joint of dissimilar Al alloys', *Materials and Design*, 34: 725–31.
37. Threadgill, P. (1997), 'Friction stir welds in aluminium alloys: Preliminary microstructural assessment', *Industrial Report 513/2/97 of The Welding Institute*, TWI.
38. Yuh, J.C. and Xinhai, Q. (1999), 'Heat transfer and thermo-mechanical analysis of friction stir joining of AA 6061-T6 plates', *Proceedings of the First Symposium on Friction stir Welding*. Thousand Oaks, CA.
39. Pourahmad, P. and Abbasi, M. (2013), 'Materials flow and phase transformation in friction stir welding of Al 6013/Mg', *Transactions of Nonferrous Metals Society of China*, 23: 1253–61.
40. Kostka, A., Coelho, R.S, Dos Santos, J. and Pyzalla, A.R. (2009), 'Microstructure of friction stir welding of aluminium alloy to magnesium alloy', *Scripta Materialia*, 60: 953–6.

41. Yan, J-C., Xu, Z-W., Li, Z-Y., Li, L. and Yang, S-Q. (2005), 'Microstructure characteristics and performance of dissimilar welds between magnesium alloy and aluminium formed by friction stirring', *Scripta Materialia*, 53: 585–9.
42. Kwon, Y.J., Shigematsu, I. and Saito, N. (2008), 'Dissimilar friction stir welding between magnesium and aluminium alloys', *Materials Letters*, 62: 3827–9.
43. Sato, Y.S., Park, S.H.C. Michiuchi, M. and Kokawa, H. (2004), 'Constitutional liquation during dissimilar friction stir welding of Al and Mg alloys', *Scripta Materialia*, 50: 1233–6.
44. Yan, Y., Zhang, D., Qiu, C. and Zhang, W. (2010), 'Dissimilar friction stir welding between 5052 aluminium alloy and AZ31 magnesium alloy', *Transactions of Nonferrous Metals Society China*, 20: 619–23.
45. McLean, A.A, Powell, G.L.F., Brown, I.H. and Linton, V.M. (2003), 'Friction stir welding of magnesium alloy AZ31B to aluminium alloy 5083', *Science and Technology of Welding and Joining*, 8: 462–4.
46. Yan, J., Xu, Z., Li, Z., Li, L. and Yang, S. (2005), 'Microstructure characteristics and performance of dissimilar welds between magnesium alloy and aluminium formed by friction stirring', *Scripta Materialia*, 53: 585–9.
47. Chen, Y.C. and Nakata, K. (2008), 'Friction stir lap joining aluminium and magnesium alloys', *Scripta Materialia*, 58: 433–6.
48. Venkateswaran, P. and Reynolds, A.P. (22012), 'Factors affecting the properties of friction stir welds between aluminium and magnesium alloys', *Materials Science and Engineering A*, 545: 26–37.
49. Mofid, M.A., Abdollah-Zadeh, A. and Malek Ghaini, F. (2012), 'The effect of water cooling during dissimilar friction stir welding of Al alloy to Mg alloy', *Materials and Design*, 36: 161–7.
50. Cardarelli, F. (2008), *Materials Handbook: A Concise Desktop Reference*, 2nd edn, London: Springer-Verlag London Ltd., 32.
51. Baker, H. (1992), *ASM Handbook*, vol. 3: *Alloy Phase Diagrams*. 10th edn, Ohio: ASM International, 44.
52. Elliott, S. and Wallach, E.R. (1981), 'Joining aluminium to steel. Part I: Diffusion bonding', *Metal Construction*, 3: 167–71.
53. Paventhan, R., Lakshminarayanan, P.R. and Balasubramanian, V. (2011), 'Prediction and optimization of friction welding parameters for joining aluminium alloy and stainless steel', *Transactions of Nonferrous Metals Society of China*, 21(7): 1480–5.
54. Watanabe, T., Sakuyama, H. and Yanagisawa, A. (2009), 'Ultrasonic welding between mild steel sheet and Al-Mg alloy sheet', *Journal of Materials Processing Technology*, 209(15–16): 5475–80.
55. Sierra, G., Peyre, P., Deschaux-Beaume, F., Stuart, D. and Fras, G. (2007), 'Steel to aluminium key-hole laser welding', *Materials Science and Engineering A*, 447(1–2): 197–208.
56. Watanabe, T., Takayama, H. and Yanagisawa, A. (2006), 'Joining of aluminium alloy to steel by friction stir welding', *Journal of Materials Processing Technology*, 178: 342–9.
57. Uzun, H., Dalle Donne, C., Argagnotto, A., Ghidni, T. and Gambaro, C. (2005), 'Friction stir welding of dissimilar Al 6013-T4 to X5CrNi18-10 stainless steel', *Materials and Design*, 26(1): 41–6.

58. Coelho, R.S., Kostka, A., dos Santos, J.F. and Kaysser-Pyzalla, A. (2012), 'Friction stir dissimilar welding of aluminium alloy to high strength steels: Mechanical properties and their relation to microstructure', *Journal of Materials Science and Engineering A*, 556: 175–83.
59. Tanaka, T., Morishige, T. and Hirata, T. (2009), 'Comprehensive analysis of joint strength for dissimilar friction stir welds of mild steel to aluminium alloys', *Scripta Materialia*, 61(7): 756–9.
60. Chen, C.M. and Kovacevic, R. (2004), 'Joining of Al 6061 alloy to AISI 1018 steel by combined effects of fusion and solid state welding', *International Journal of Machine Tools and Manufacture*, 44(11): 1205–14.
61. Lee, W.B., Schmuecker, M., Alfaro Mercardo, U., Biallas, G. and Jung, S.B. (2006), 'Interfacial reaction in steel–aluminium joints made by friction stir welding', *Scripta Materialia*, 55(4): 355–8.
62. Geiger, M., Micari, F., Fratini, L., Contorno, D., Giera, A. and Contorno, D. (2008), 'Friction stir knead welding of steel aluminium butt joints', *International Journal of Machine Tools and Manufacture*, 48: 515–21.
63. Lee, W-B., Schmuecker, M., Mercardo, U.A., Biallas, G. and Jung, S-B. (2006), 'Interfacial reaction in steel–aluminium joints made by friction stir welding', *Scripta Materialia*, 55: 355–8.
64. Dehghani, M., Akbari Mousavi, S.A.A and Amadeh, A. (2013), 'Effects of welding parameters and tool geometry on properties of 3003-H18 aluminium alloy to mild steel friction stir weld', *Journal of Transactions of Nonferrous Metals Society of China*, 23: 1957–65.
65. Cam, G. and Kocak, M. (1998), 'Progress in joining of advanced materials', *International Material Review*, 43: 1–44.
66. Li, B., Zhang, Z., Shen, Y., Hu, W. and Luo, L. (2014), 'Dissimilar friction stir welding of Ti–6Al–4V alloy and aluminium alloy employing a modified butt joint configuration: Influences of process variables on the weld interfaces and tensile properties', *Materials and Design*, 53: 838–48.
67. Wei, Y., Li, J., Xiong, J., Huang, F., Zhang, F. and Raza, S.H. (2012), 'Joining aluminium to titanium alloy by friction stir lap welding with cutting pin', *Materials Characterization*, 71: 1–5.
68. Chen, Y., Ni, Q. and Ke, L. (2012), 'Interface characteristic of friction stir welding lap joints of Ti/Al dissimilar alloy', *Journal of Transactions of Nonferrous Metals Society of China*, 22: 299–304.
69. Dressler, U., Biallas, G. and Mercado, U.A. (2009), 'Friction stir welding of titanium alloy TiAl64V to aluminium alloy AA2024-T3', *Journal of Material Science and Engineering A*, 526(2): 113–17.
70. Akinlabi, E.T. and Akinlabi, S.A. (2012), 'Effect of heat input on the properties of dissimilar friction stir welds of aluminium and copper', *American Journal of Material Science*, 2(5): 147–52.
71. Akinlabi, E.T. (2012), 'Effect of shoulder size on weld properties of dissimilar metal friction stir welds', *Journal of Materials Engineering Performance*, 21: 1514–19.
72. ASTM E3-11 (2011), *Standard Guide for Preparation of Metallographic Specimens, ASTM international Book of Standards*, 3: 1.
73. ASTM E8-11 (2011), *Standard Test Method for Tensile Testing of Materials, ASTM International Book of Standards*, 3: 1.

74. ASTM E384-11e1 (2011), *Standard Test Method for Knoop and Vickers Hardness of Materials, ASTM International Book of Standards*, 3: 1.
75. Ratnesh, K. and Pravin, K. (2010), 'Investigation of joint properties of friction stir welding of aluminium 6061 alloy to copper', *International Journal of Engineering Research and Technology*, 3: 613–20.
76. Akinlabi, E.T. and Akinlabi, S.A. (2011), 'Effect of heat input on the electrical resistivity of dissimilar friction stir welded joints of aluminium and copper', *Proceedings of the IEEE African International Conference, Livingstone*, Zambia, 13–15 September.
77. Galvao, I., Leal, R.M., Loureiro, A. and Rodrigues, D.M. (2010), 'Material flow in heterogeneous friction stir welding of aluminium and copper thin sheets', *Science and Technology of Welding and Joining*, 15(8): 654–60.
78. Akinlabi, E.T. and Akinlabi, S.A. (2012), 'Effects of process parameters on the evolving properties of dissimilar AL-CU friction stir welds', *Trends in Welding Research, Proceedings of the 9th International Conference*, Chicago, IL, 4–6 June.
79. Powder Diffraction Files (PDF-2) (2002), *International Centre for Diffraction Data*, Pennsylvania, Release.
80. Liu, P., Shi, Q., Wang, W., Wang, X. and Zhang, Z. (2008), 'Microstructure and XRD analysis of FSW joints for copper T2/aluminium 5A06 dissimilar materials', *Materials Letters*, 62: 4106–8.
81. Xue, P., Xiao, B.L., Ni, D.R. and Ma, Z.Y. (2010), 'Enhanced mechanical properties of friction stir welded dissimilar Al–Cu joint by intermetallic compounds', *Materials Science and Engineering A*, 527: 5723–7.
82. Savolainen, K., Mononen, J., Saukkonen, T. and Hänninen, H. (2006), 'A preliminary study on friction stir welding of dissimilar metal joints of copper and aluminium', *Proceedings of the 6th International FSW symposium*, Montreal, Canada. 10–13 October, TWI (UK). Retrieved: CD-ROM.
83. Esmaeili, A., Givi, M.K. and Rajani, H.R. (2011), 'A metallurgical and mechanical study on dissimilar Friction stir welding of aluminium 1050 to brass (CuZn30)', *Materials Science and Engineering A*, 528: 7093–102.

7

Corrosion behaviour in friction stir processed and welded materials

H.S. Arora and S. Mukherjee, University of Texas, USA; H.S. Grewal, H. Singh, and B.K. Dhindaw, Indian Institute of Technology, Ropar, India

DOI: 10.1533/9780857094551.295

Abstract: This chapter presents a comprehensive study on the influence of friction stir processing/welding (FSW/FSP) on corrosion behaviour. It briefly discusses the different aspects of corrosion including corrosion types, measurement techniques and data analysis. The corrosion behaviour of a wide range of friction stir processed materials, including light weight metals such as magnesium and aluminum alloys, as well as high strength metals such as steel, has been discussed in detail. The influence of FSP parameters on the microstructural evolution, comprising grain-size and precipitate refinement along with its correlation with the corrosion properties, has been described for different materials.

Key words: friction stir processing/welding, corrosion, grain refinement, precipitate fragmentation.

7.1 Introduction

Corrosion is the destruction or deterioration of a material following a reaction with its environment, stimulated by chemical, physical and/or electrochemical factors. Metal corrosion occurs predominantly via electrochemical means at the interface between a metal surface and an electrolyte. The equilibrium between the opposing anodic and cathodic reactions determines the rate of corrosion. Metal is oxidized during an anodic reaction, rejecting electrons into an electrolyte, whereas the solution species are reduced during a cathodic reaction. Metal corrosion can occur in different ways. A brief overview of some of the major types of metal corrosion is given below:

1. *Uniform corrosion*: when the whole surface of the material is subjected to nearly the same dissolution rate, it is uniform corrosion. It is considered to be the least dangerous form of corrosion, as it is easily predictable. Rust formed on iron is a typical example of uniform corrosion.
2. *Galvanic corrosion*: galvanic corrosion occurs when two metal surfaces with different electrochemical potential come into contact with each other in the presence of an electrolyte. A metal with higher electrode potential acts as an anode, while the other metal surface acts as a cathode. An electrochemical reaction between the two metals can result in dissolution of the anode. Even a single material with second-phase precipitates can form a mini-galvanic cell and be subject to galvanic corrosion.
3. *Pitting corrosion*: pitting corrosion occurs when only some localized regions are subjected to dissolution, while the remaining areas remain unaffected. The major reasons leading to pitting are a broken oxide layer and the presence of second-phase precipitates. An unstable oxide layer can break off at certain locations leading to localized corrosion. Similarly, the presence of second-phase precipitates can also lead to localized dissolution of the matrix by forming mini galvanic cells.
4. *Crevice corrosion*: crevice corrosion is another type of localized corrosion that is caused by a stagnant liquid present in a crevice over a long period of time. Typical examples are riveted plates, welded structures and threaded areas.
5. *Stress-corrosion cracking*: stress corrosion cracking (SCC) refers to cracking of the material under stress in the presence of a corrosive environment. It may be the result of an externally applied load or some residual stresses present in the material from prior deformation. SCC can initiate in the form of fine cracks that propagate along the grain boundaries, leading to catastrophic failure of the material.
6. *Corrosion fatigue*: corrosion fatigue refers to the synergistic effect of fatigue loading in a corrosive environment. Similar to SCC, the presence of such a corrosive environment can significantly reduce the endurance limit of the material. The failure of the material could be catastrophic.

7.2 Major factors affecting corrosion

1. *Composition*: the alloy composition significantly affects the corrosion behaviour of the material. Addition of certain elements retards corrosion susceptibility, whereas others can enhance it. For instance, Al addition in Mg-based alloys is known to increase the overall corrosion resistance of these alloys. Similarly, addition of rare-earth elements also has beneficial effects on the corrosion behaviour of Mg-based alloys. However, the presence of a small amount of silver can deteriorate their corrosion behaviour in 3% NaCl solution [1].
2. *Secondary processing*: heat treatment or thermo-mechanical processing such as equal channel angular extrusion (ECAE) and friction stir processing (FSP) can produce structural modifications in materials. It can alter the grain structure as well as the type and distribution of secondary phases, both of which can significantly influence the corrosion performance of a material.
3. *Electrolyte pH*: electrolyte pH influences the corrosion behaviour by affecting the stability of the oxide layer formed on the BM. For instance, the oxide layer ($Mg(OH)_2$) formed on the magnesium surface during corrosion is stable at a pH value above 11. Therefore, Mg-based alloys do not show aggravated corrosion in an alkaline solution with pH 11 or above. However, they are subjected to severe corrosion in neutral and acidic electrolyte.
4. *Environmental influence*: environmental conditions including temperature and humidity also affect corrosion behaviour. In general, the corrosion is accelerated in the presence of higher humidity and higher temperature. For instance, at a 9.5% relative humidity, neither pure magnesium nor any of its alloys exhibit surface corrosion for a long time. At 30% humidity, only minor corrosion may occur, whereas at 80% humidity, the surface shows considerable corrosion [2].

7.3 Corrosion test methods

Corrosion testing is performed taking into consideration the type of service environment to which the material will be subjected. The testing could be done in the presence of a liquid or gaseous medium, under varying degrees of temperature and pressure, under both static and dynamic conditions. Although the details related to test techniques could be discussed at length, considering the focus of this chapter, only a brief introduction is given here. For detailed information regarding corrosion test methods, readers can consider specific references, such as [3]. For example, specific test standards have also been developed by the American Society for Testing and Materials (ASTM) and National Association of Corrosion Engineers (NACE). The various types of tests that can be used for evaluating the corrosion performance of the materials are as follows.

7.3.1 Immersion test

Immersion is perhaps the most common and widely used type of test. In this test, the test coupon is immersed in the electrolytic medium. The selection of electrolyte is based on the environment to which the material will be subjected during its service. The sample is cleaned thoroughly in order to remove any dirt and other possible contaminants. It is placed inside a known quantity of electrolyte for a predetermined time interval. The corrosion rate is determined from the amount of mass loss measured during this time interval. It is advisable to clean and completely dry the sample before measuring the difference in mass. The sample can be washed in a solution that will help to remove the weak oxide layer. Specific details about this test method can be found in ASTM standard G31 and NACE standard TM-01-69.

Great care and control should be exercised upon various test parameters (temperature, air content, pH of the electrolyte) while conducting corrosion tests. Volume of the electrolyte is selected so that its corrosiveness is not affected significantly during the testing time. Generally, the minimum recommended solution volume to specimen surface is 20 ml/cm^2. Other than mass loss measurement, corrosion rates could also be interpreted by collecting and monitoring the amount of hydrogen evolved during the reaction.

7.3.2 Electrochemical tests

Electrochemical tests are highly productive in nature and can provide an easy route for evaluating the corrosion performance of materials. In this method, similar to the immersion test, the specimen is placed inside the electrolytic medium under controlled test parameters. In addition to the specimen, which acts as a working electrode, two more electrodes are used. One of the electrodes acts as a reference, while the other acts as a counter. After attaching these electrodes to the potentiostat and completing the circuit, the open circuit potential is measured. Thereafter, various different variants of electrochemical tests can be conducted. For conducting these tests, either current or voltage is varied through a specified range simultaneously measuring the other. The plot of these quantities could be used for interpreting the corrosion rates by fitting them to a desired model, such as Tafel's model. This method could also be used to determine the pitting resistance, polarization resistance, electrochemical impedance spectroscopy (EIS), galvanic corrosion and for conducting cyclic potentiodynamic studies.

7.3.3 Atmospheric test method

Often a laboratory test technique is able to completely replicate actual working conditions, such as the effect of pollutants. Various other factors, such as large variation in the ambient temperature during the service, humidity and effect of rain, etc., are mostly ignored. These factors also contribute significantly in the corrosion performance of the materials. The atmospheric test method could help in testing in actual working environments. The test sample can be placed in a static location for a fixed time interval or might be exposed to dynamic conditions such as on vehicles. Quantitative measurements of the atmospheric parameters, such as temperature, pressure, humidity and level of contaminant, etc., could provide means for scientific evaluation of the test condition. Electrical signals from the test specimen could also be recorded for continuous evaluation of the level of the damage. Various ASTM standards could be found, which would provide guidance for conducting tests and selection of test specimen size.

7.3.4 Application based-tests

There are various other types of corrosion tests that can be compiled under this category, depending upon the type of working environment. These include the hot corrosion test, specifically used to determine the corrosion properties of different materials and coatings under high temperature conditions, such as in boiler and gas turbines. Sea water corrosion is another such test, which could prove to be useful for studying the effect of different parameters, particularly related to sea water exposure. These include the effect of variation in temperature, oxygen content and most importantly the effect of biological actives. Biological fouling could significantly affect the corrosion performance of the material. ASTM standard G52 provides the necessary guidelines for conducting this test, which could involve long test duration, even up to 20 years.

Stress corrosion and erosion-corrosion are other such specific test techniques that are highly application-based methods. In the former, a specimen under static stress conditions is confronted with the corrosive media. This type of corrosion is encountered significantly in the welded section and in components that have retained some residual stresses during its manufacture. Specific details on conducting stress corrosion studies can be found in ASTM standards G30, 38 and 39.

The synergy of corrosion phenomena with erosion results in a completely different form of degradation. Erosion is a form of surface degradation wherein loss of material takes place due to mechanical action of the abrasive particles and/or the minute jets of liquids. The combination of corrosion with erosion possesses a great challenge. The oxide layer formed during the corrosion process is easily removed by the mechanical action exposing a fresh material surface for further corrosion, eventually leading to complete failure of the component.

7.3.5 Salt fog test

The salt fog test is conducted mainly for automotive parts. Test coupons to be tested are placed inside a closed cabinet. A mist of solution of salt and liquid medium is formed after atomization. The condensation of the mist takes place on the test specimen initiating the chemical activity. The temperature of the cabinet should be maintained at 35°C as specified in ASTM standard B117; however, the temperature could be varied using water jackets. Another variant of this test method includes a variation in the pH of the solution using acetic acid (ASTM G85), cyclic acid test, and conducting tests in synthetic sea water conditions involving high humidity levels (> 98%).

7.4 Corrosion behaviour of friction stir processes and friction stir weldings

7.4.1 Mg-based alloys

In a large number of investigations [4–7], the beneficial effects of friction stir processing (FSP) technology on the grain refinement and enhancement of surface properties of the FSP specimens have been demonstrated. It is known that in addition to high strength requirements, good corrosion resistance is a prerequisite for many applications. It is also important to understand the factors and parameters that influence the corrosion resistance of a material.

During analysis of the corrosion behaviour of AXJ550 magnesium alloys, it was observed that higher aluminium concentration as well as presence of $(MgAl)_2Ca$ particles enhances the corrosion resistance of the investigated magnesium alloy [8]. In another study, the corrosion behaviour of friction stir welded (FSW) AM50 was investigated [9]. It was found that corrosion morphology of the AM50 alloy was predominantly controlled by the β-$Mg_{17}Al_{12}$ phase distribution. Inter-granular corrosion was observed in the BM, whereas pitting corrosion occurred in the welds. The polarization curves for the as-cast AM50 and for the weld are shown in Figure 7.1(a).

The corrosion rates for welded joint and the base metal (BM) were found to be 22.2 l and 95.1 $\mu A/cm^2$, whereas the corrosion potential was reported as −1520 mV and −1553 mV, respectively. It can be seen that the corrosion resistance of the welded joint is higher than that of the BM, which is attributed to the finer grain structure. The influence of the microstructure on the atmospheric corrosion behaviour of magnesium alloys AZ91D and AM50 was studied [10]. It was observed that the atmospheric corrosion process was strongly related to the aluminium content in the different phases of the alloys. Addition of 1.5% Nd was found to enhance the corrosion resistance of AZ91 alloy [11]. In another investigation [12], the corrosion properties of high-pressure die cast (HPDC) magnesium-rare earth (RE)-based alloys have been studied. It was reported that increasing RE alloying additions systematically increased corrosion rates.

In an investigation on FSW of magnesium alloy AZ31, it was found that FSW resulted in the reduction of corrosion resistance of as-received alloy, whereas it increased the corrosion resistance of heat-treated AZ31 [13]. The difference in the corrosion behaviour was found to be related to the grain size of AZ31 alloy. A finer grain structure was found to enhance the corrosion resistance by forming denser and uniform oxide layer. Corrosion behaviour of FSW magnesium alloy AM50 also showed higher corrosion resistance compared to the as-cast alloy [14]. Finer grain structure and large volume fraction of β-phase mainly contributed towards enhanced corrosion resistance. Higher corrosion resistance was explained on the basis of large volume fraction of β-phase acting as a cathodic barrier to inhibit corrosion. Inter-granular corrosion was observed in the BM, whereas pitting corrosion occurred in the welds.

Stress corrosion cracking (SCC) susceptibility of FSW Mg alloy AZ31 has also been evaluated [15]. Slow strain rate tensile test (SSRT) showed lower ductility and strength for the FSW samples compared to the as-received AZ31 alloy. In EIS and salt spray tests, the stir zone (SZ) of the welded specimen showed higher corrosion resistance compared to the base AZ31 alloy. However, in an SSRT test performed in a corrosive environment, the welded samples showed higher SCC susceptibility. This could be explained on the basis of hydrogen embrittlement of the SZ, due to presence of a large density of dislocation network. In investigation of the corrosion behaviour of FSP Mg-Y-RE alloy, the corrosion properties were studied under four different conditions:

1. A: as-receive;
2. B: as-received samples subjected to peak age hardening treatment of 210°C for 48 h;
3. C: FSP of B-type material; and
4. D: FSP of A-type material subjected to subsequent peak age hardening treatment of 210°C for 48 h.

The potentiodynamic polarization plots for all four conditions of Mg-Y-RE alloy in a 3.5 wt.% NaCl solution are shown in Figure 7.1(b). The results are compiled in Table 7.1. It can be seen that among all the microstructural conditions, type-D showed the highest positive value of E_{corr} and the lowest value of i_{corr}. It was proposed that with coarsening of the precipitates, the number density of precipitates in the matrix would reduce, resulting into lesser availability of effective sites for initiation of pits. This might be the cause of the positive shift of pitting potential observed in type D.

The influence of pH value, chloride ion concentration and immersion time on the corrosion rate of FSW AZ61A magnesium alloy weldments has been reported [17], for which extruded Mg alloy plates of 6 mm thick of AZ61A grade were butt welded using FSW process. The corrosion rate was evaluated from weight loss measurements. An empirical relationship to predict the effect of pH value, chloride ion concentration and immersion time was developed. The FSW AZ61A magnesium alloy welds showed an increase in the corrosion resistance with the increase in pH. The corrosion rate was found to be higher in the acidic media compared to the alkaline and neutral media for the same concentrations and time. An increase in the Cl– concentrations resulted in an enhanced corrosion rate for the FSW AZ61A magnesium alloy. With the increased immersion period, the corrosion resistance was found to increase. Formation of the

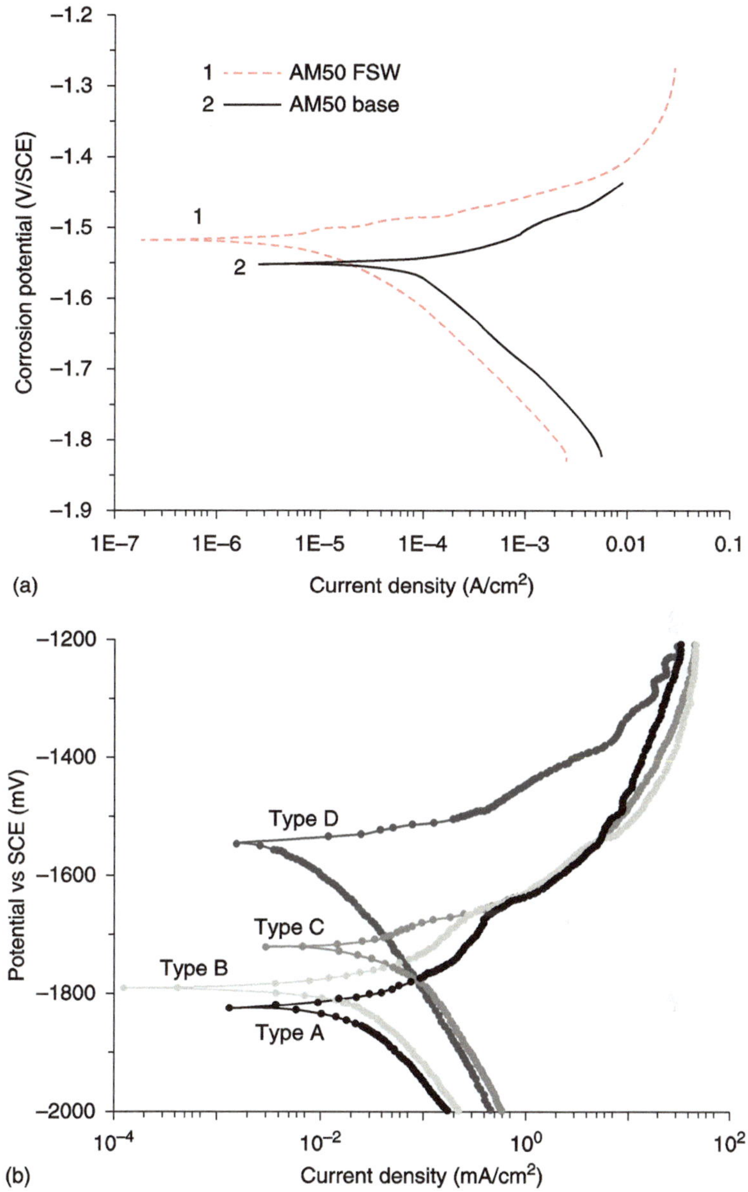

Figure 7.1 Polarization curves: (a) as-cast AM50 and friction stir welded AM50 [9]; and (b) Mg-Y-RE alloy under different conditions; A: as-received; B: as-received samples subjected to peak age hardening treatment of 210 °C for 48 h; C: FSP of B-type material, and D: FSP of A-type material subjected to subsequent peak age hardening treatment of 210 °C for 48 h [15]

Table 7.1 Results of potentiodynamic polarization test for Mg-Y-RE alloy processed under different conditions [16]

Sample conditions	Corrosion potential E_{corr} (mV vs. SCE)	Current density i_{corr} (A/cm^2)
Type A	-1823 ± 22	$5.9 \pm 0.4 \times 10^{-5}$
Type B	-1788 ± 15	$2.7 \pm 0.2 \times 10^{-5}$
Type C	-1719 ± 25	$8.7 \pm 0.4 \times 10^{-5}$
Type D	-1544 ± 28	$1.3 \pm 0.1 \times 10^{-5}$

hydroxide layer increased the corrosion resistivity and was found to be the dominant factor for better corrosion resistance. Studying of *in-vitro* degradation behaviour of a FSP AZ31 magnesium alloy showed that FSP marginally enhanced the degradation resistance of the alloy, which could be attributed to the dissolution of secondary phase [18]. Dissolution of the Mg matrix was found to be reduced by homogenization of the microstructure.

It was proposed that the beneficial effect would be more pronounced for magnesium alloys that contain a high volume fraction of secondary phase particles. The Nyquist plot for the processed as well as unprocessed Mg alloys is shown in Figure 7.2(a), along with the equivalent circuit elements. Higher diameter loop for the processed specimen indicates its higher corrosion impedance. The polarization curves for both the specimens are shown in Figure 7.2(b). The corrosion current (i_{corr}) for the FSP alloy (190 $\mu A/cm^2$) was slightly lower than for the untreated alloy (210 $\mu A/cm^2$).

It has been reported that the behaviour of a refined grain structure with respect to corrosion resistance depends on test conditions such as electrolyte/corrosive medium composition [19]. It has been found that a fine-grained material can show higher corrosion resistance in a dilute corrosive media (i.e. 0.1 M NaCl solution), whereas in a relatively aggressive media (i.e. 3.5% NaCl solution) the fine-grained material became incapable of showing good corrosion performance. Another investigation on the grain size effect on corrosion [20] revealed that the material's behaviour to corrosion is broadly classified into those showing active behaviour and those exhibiting passive behaviour in a corrosive environment. The borderline that differentiates these two types of materials is the magnitude of the corrosion current, I_{corr}. Those materials that show passivating behaviour against corrosion phenomena generally have $I_{corr} < 10\ \mu A/cm^2$. Thus, it is considered that a fine-grained material

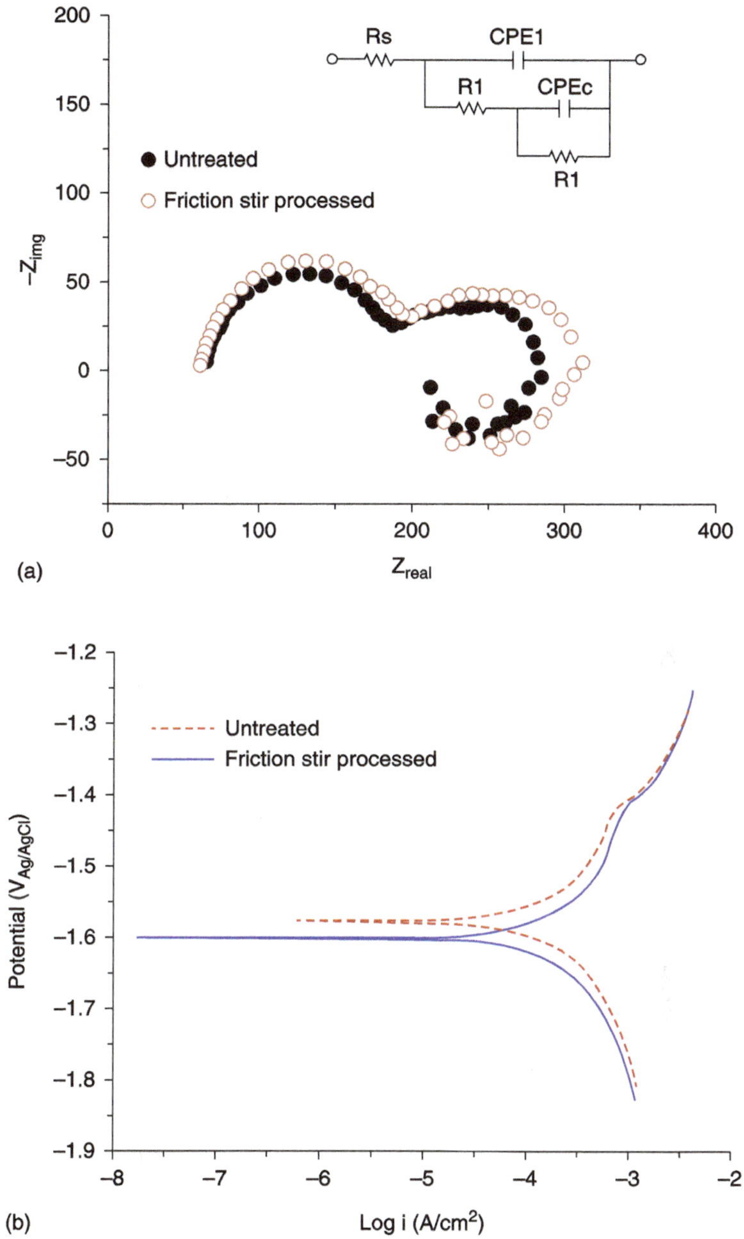

Figure 7.2 Electrochemical results of untreated and friction stir processes magnesium alloy in simulated body fluid: (a) Nyquist plots and equivalent circuit; and (b) polarization curs [18]

with I_{corr} <10 μA/cm² can enhance the corrosion resistance of the material and vice-versa. A plot between corrosion rate and grain size$^{-0.5}$ can be helpful in predicting the correlation between these two factors [20].

Corrosion immersion tests have been reported on the corrosion behaviour of Mg-based AE42 alloy [21], which were conducted in two phases. In the first phase, the corrosion test was performed in accordance with Taguchi's design of experiment (DOE) technique to evaluate those FSP parameters that can maximize corrosion resistance of AE42 alloy. In the second phase, corrosion kinetics for a FSP AE42 specimen was evaluated, which had shown the highest corrosion resistance in the first phase. A solution of 100 ml of 3.5% NaCl with a pH of 7.58 was used for the corrosion testing of samples in the first phase. During the second phase of the immersion test, specimens were subjected to 300 ml of 3.5% NaCl solution to determine their corrosion behaviour under more severe conditions. In both phases of corrosion testing, samples were initially cleaned by dipping in a mixture of 200 g/L CrO_3 + 10 g/L $AgNO_3$ solution for 7 minutes followed by acetone washing before measuring the weight loss. Electrochemical corrosion studies viz potentiodynamic polarization and EIS were conducted on the as-cast as well as FSP AE42 specimens. The tests were performed on a Gamry's electrochemical workstation using a three electrode cell consisting of a saturated calomel electrode (SCE) as a reference electrode, graphite as a counter electrode and an as-cast/FSP AE42 specimen as a working electrode in a 3.5% NaCl solution. FSP tool rpm, cooling temperature and number of FSP passes were found to be the most influential parameters affecting the corrosion behaviour of Mg alloys.

The variation of corrosion rate with respect to the influential FSP parameters is shown in Figure 7.3. It is perceptible from this figure that the corrosion rate is considerably smaller at −10°C and increases as temperature increases from −10°C to 30°C through to 10°C. Furthermore, it can be observed that with the decrease in tool rpm, the corrosion rate appears to decrease. Although no definite trend can be predicted for the number of FSP passes, the corrosion rate appears to reduce with the increase in the number of FSP passes.

Some of the selective scanning electron microscope (SEM) images of the FSP AE42 specimens processed under different conditions in accordance with Table 7.2 and subjected to 100 ml of 3.5% NaCl solution for period of 24 h are shown in Figure 7.4 [21]. From these figures, it can be observed that corroded surfaces of FSP specimens consist of a combination of fine circular pits and fissures. Furthermore, it was noticed that the fine-grained specimens mostly have circular-shaped pits, whereas

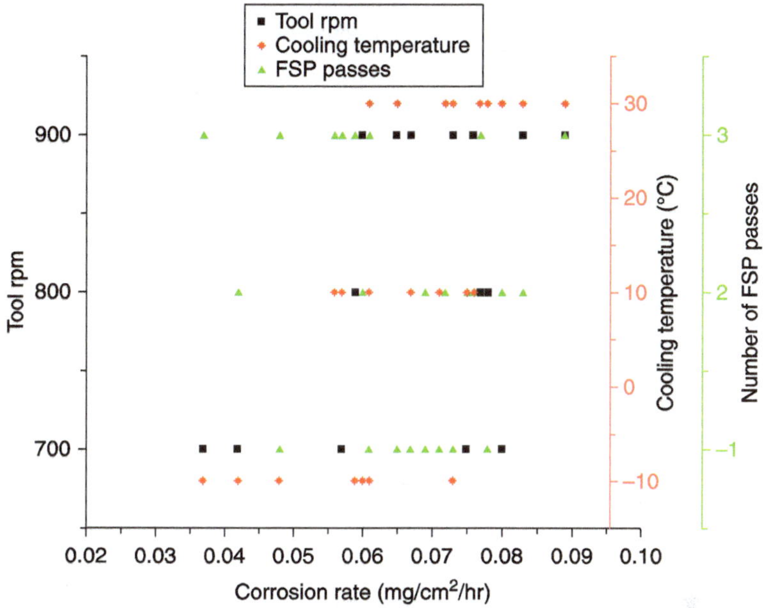

Figure 7.3 **Plot showing the influence of parameter A (tool rpm), parameter B (cooling temperature) and parameter E (FSP passes) on the corrosion rate of friction stir processed Mg-based AE42 alloy [21]**

the relatively coarse-grained specimens are characterized by the presence of corrosion fissures. The schematic showing the possible mechanism responsible for this observation is shown in the Figure 7.5. It appears that in a coarse-grained specimen, the fine corrosion nucleation sites grow in the direction of atom orientations within the grains and result in the sharp corrosion fissures (Figure 7.5(a)), whereas in the case of fine-grained specimens, the extremely fine corrosion nucleation sites grow in a number of different directions, owing to fine grain size, which later coalesce together to produce nearly circular pits (Figure 7.5(b)). In order to obtain quantitative ideas about the pitting, pit density and intensity were measured for some of the corroded specimens in terms of the area fraction and average pit size (Table 7.3). It can be observed that the area fraction of the pits as well as the average pit size is smaller for the FSP specimens processed at the lower cooling temperature of −10°C [21].

Furthermore, it was observed that the corrosion rate of the FSP AE42 specimen, processed at parameters corresponding to the third row of

Table 7.2 Actual values of parameters, corrosion rate and grain size in different experimental runs of Taguchi's optimization experiment for the corrosion rate of Mg-based AE42 alloy subjected to 100 ml of 3.5% NaCl solution for 24 h [21]

Sr. No.	Tool rpm (A)	Cooling temperature (B)	Linear speed (C)	Plunge depth (D)	Number of passes (E)	Corrosion rate* (mg/ cm^2/hr)
1	700	−10	40	0.25	1	0.048 ± 0.0026
2	700	−10	50	0.3	2	0.042 ± 0.0015
3	700	−10	60	0.35	3	0.037 ± 0.0019
4	700	10	40	0.3	2	0.075 ± 0.0018
5	700	10	50	0.35	3	0.057 ± 0.0006
6	700	10	60	0.25	1	0.069 ± 0.0023
7	700	30	40	0.35	3	0.061 ± 0.0012
8	700	30	50	0.25	1	0.073 ± 0.001
9	700	30	60	0.3	2	0.080 ± 0.0023
10	800	−10	40	0.3	3	0.059 ± 0.0038
11	800	−10	50	0.35	1	0.061 ± 0.0011
12	800	−10	60	0.25	2	0.072 ± 0.0046
13	800	10	40	0.35	1	0.071 ± 0.0032
14	800	10	50	0.25	2	0.069 ± 0.0014
15	800	10	60	0.3	3	0.061 ± 0.0015
16	800	30	40	0.25	2	0.072 ± 0.0008

17	800	30	50	0.3	3	0.077 ± 0.0025
18	800	30	60	0.35	1	0.078 ± 0.0014
19	900	–10	40	0.35	2	0.060 ± 0.0015
20	900	–10	50	0.25	3	0.048 ± 0.002
21	900	–10	60	0.3	1	0.073 ± 0.0021
22	900	10	40	0.25	3	0.056 ± 0.002
23	900	10	50	0.3	1	0.067 ± 0.0015
24	900	10	60	0.35	2	0.076 ± 0.0009
25	900	30	40	0.3	1	0.065 ± 0.0025
26	900	30	50	0.35	2	0.083 ± 0.0024
27	900	30	60	0.25	3	0.089 ± 0.0015
28	As-cast AE42					0.109 ± 0.005

Table 7.2, has shown the lowest corrosion rate amongst all the investigated cases [21]. Thus, the corrosion behaviour of the as-cast AE42 and the above-mentioned FSP AE42 specimen was further investigated with respect to time, so as to establish their corrosion kinetics. The plot showing the variation of corrosion rate with respect to time for the as-cast as well as the FSP AE42 alloy is shown in Figure 7.6. It can be observed from this figure that during the initial hours of exposure, the corrosion rate is higher in both cases, the extent of which is comparatively higher in the case of as-cast Mg alloy. After the tenth hour, the corrosion rate of FSP alloy has shown a tendency to become nearly uniform with respect to time, which is an indication of a better corrosion resistance for any alloy; however, the as-cast AE42 alloy has shown some random fluctuations in corrosion rate.

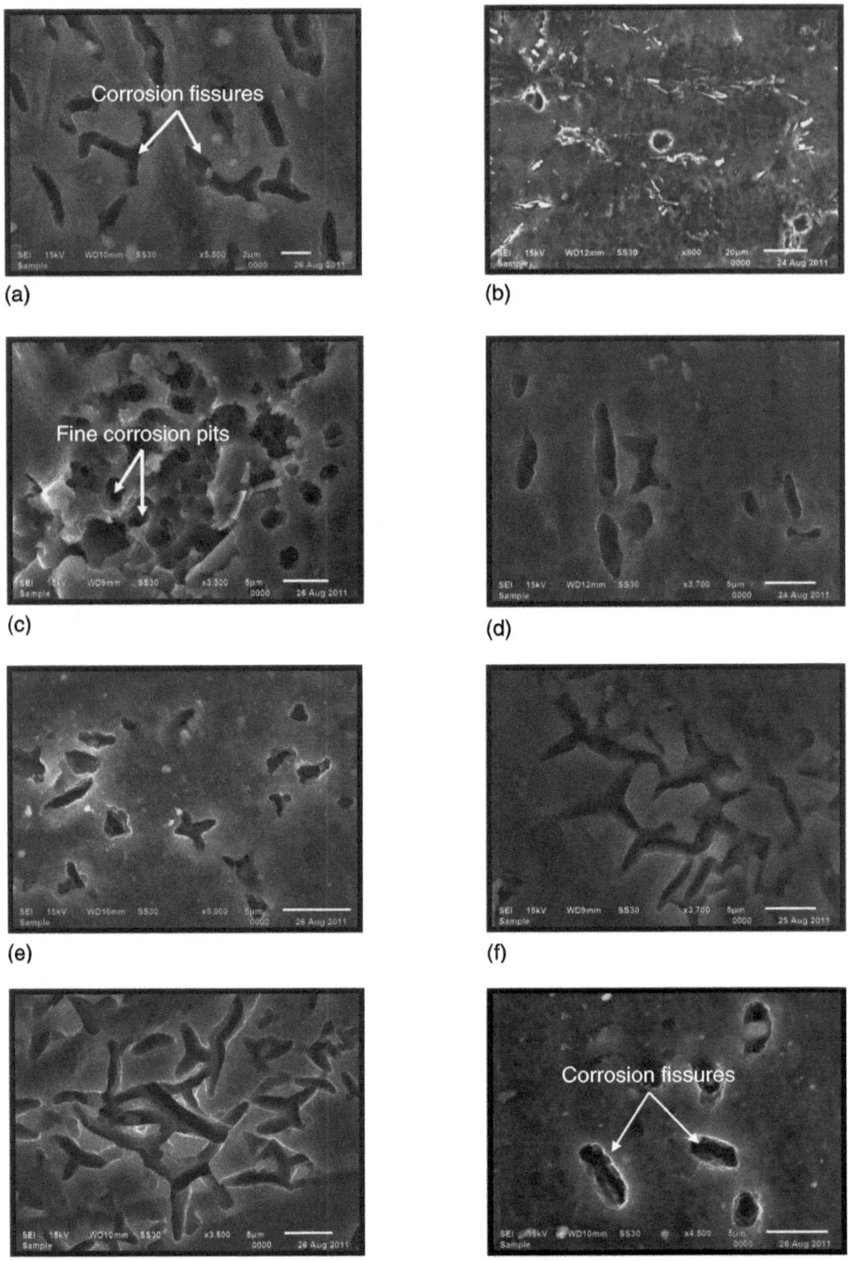

Figure 7.4 SEM images of the corroded Mg-based AE42 alloy specimen friction stir processed at different FSP parameters corresponding to: (a) 1st; (b) 2nd; (c) 6th; (d) 8th; (e) 11th; (f) 13th; (g) 17th; (h) 18th; (i) 20th; (j) 22nd; (k) 26th; and (l) 27th row of Table 7.2, subjected to 100 ml of 3.5% NaCl solution for 24 h [21]

(*Continued*)

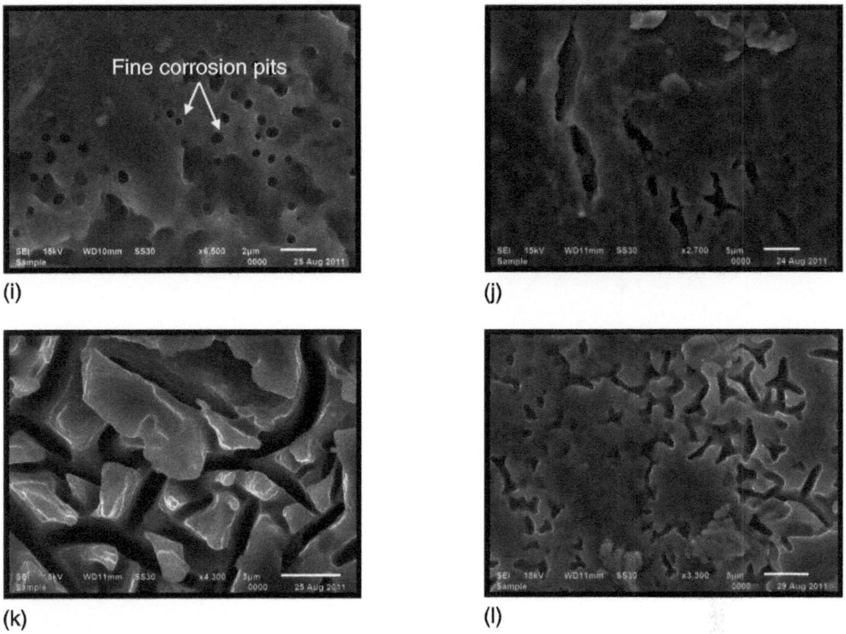

Figure 7.4 Continued

The SEM images for both the specimens at different time intervals (Figure 7.7) revealed that while localized corrosion began and rapidly propagated to greater depths in the as-cast alloy, the FSP specimen was relatively more corrosion resistant during the initial periods [21]. The SEM images taken after 2 and 5 hours (Figure 7.7(a–d)), depict that the locations surrounding larger size precipitates in the as-cast alloy have begun to corrode vigorously, while the FSP AE42 surface is only marginally corroded. The corrosion damage of the as-cast alloy became very severe towards the later part of the corrosion test, as evident from the SEM images taken after 37 and 48 hours, whereas the corrosion loss of the FSP specimen seems to be relatively less. It is evident from Figure 7.6 that the corrosion rate of the FSP AE42 specimen was reduced by nearly 50% as compared to the as-cast AE42 alloy, which is evidence of the usefulness of the FSP process in the context of corrosion phenomenon [21].

The polarization curves for the as-cast as well as the FSP AE42 alloy are shown in Figure 7.8(a) [21]. It is evident from the figure that the anodic polarization curve for FSP AE42 shifts towards more noble

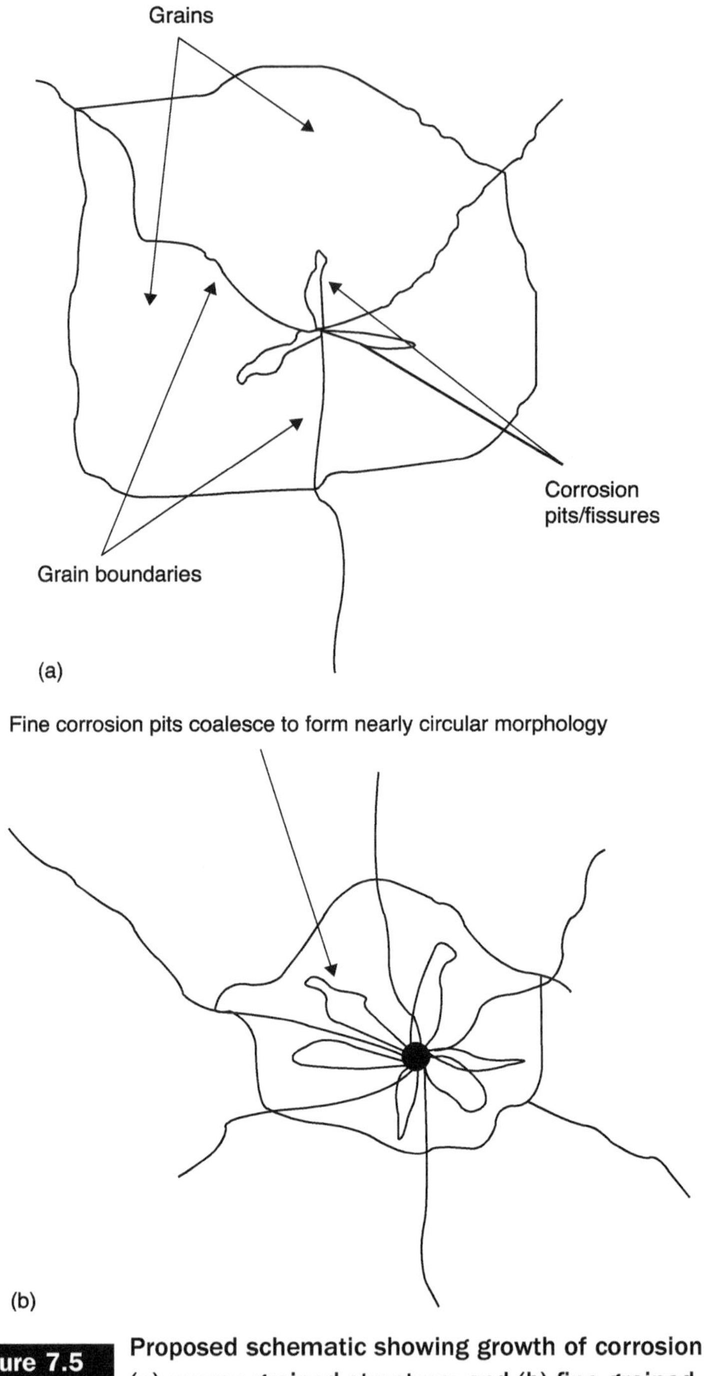

Figure 7.5 Proposed schematic showing growth of corrosion pits: (a) coarse-grained structure; and (b) fine-grained structure in Mg-based AE42 alloy subjected to 3.5% NaCl solution [21]

Table 7.3 Analysis of the pit density and average pit size for some of the selected friction stir processed specimens subjected to 3.5% NaCl solution for 24 h

Sr. No.	Specimen (row no. corresponding to Table 7.3)	Pit Density (area fraction %)	Average Pit Size (μm^2)
1	3	7.5±1.0	1.4±0.20
2	6	10.6±2.6	3.6±0.40
3	11	8.33±1.6	1.8±0.33
4	16	16.3±0.6	4.4±0.66
5	20	11.2±1.4	2.5±0.66
6	26	15.5±0.6	2.6±0.46

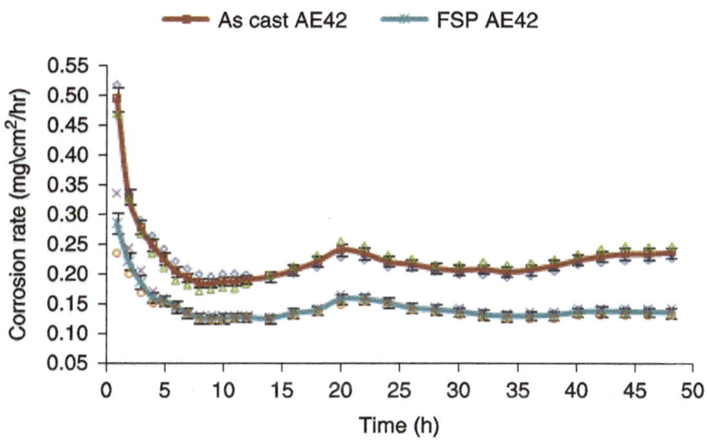

Figure 7.6 Plot showing corrosion rate of the as-cast as well as friction stir processed Mg-based AE42 alloy at different time intervals subjected to 300 ml of 3.5% NaCl solution. The specimen was processed at the optimized set of FSP parameters comprising 700 rpm, −10 °C cooling temperature and 3 FSP passes [21]

potential, implying the inhibition of anodic dissolution, which may again be attributed to the presence of finer grain structure in the FSP alloy. The cathodic reaction rate can be expressed as [22]

$$I_{cathodic} = i_{cathodic} * V_r \qquad [7.1]$$

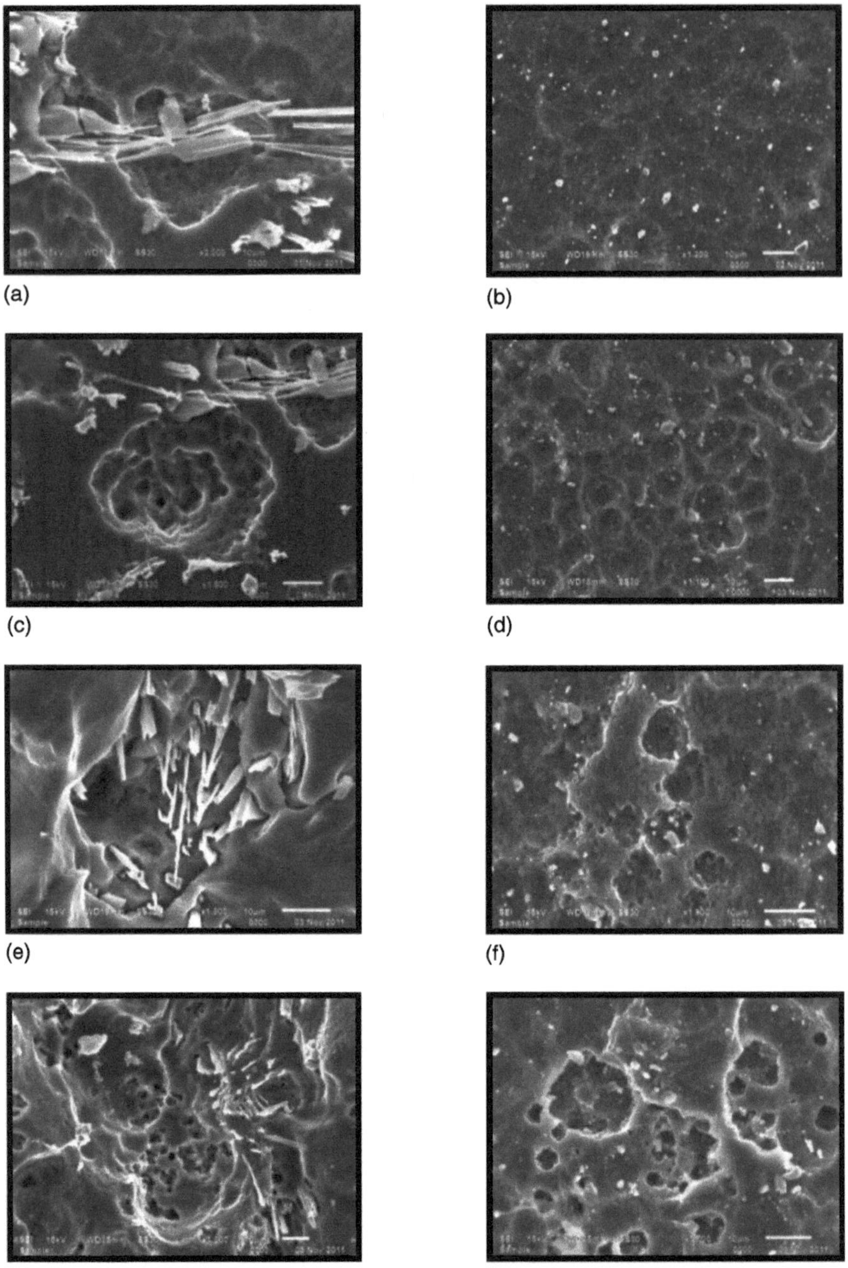

Figure 7.7 SEM images showing morphological changes taking place in as-cast and friction stir processed Mg-based AE42 alloy specimens subjected to corrosion test in 300 ml of 3.5% NaCl solution for 48: (a), (c), (e), (g), (i), and (k) for as-cast AE42; and (b), (d), (f), (h), (j) and (l) for FSP AE42 after 2, 5, 18, 20, 37 and 48 hrs, respectively [21]

(*Continued*)

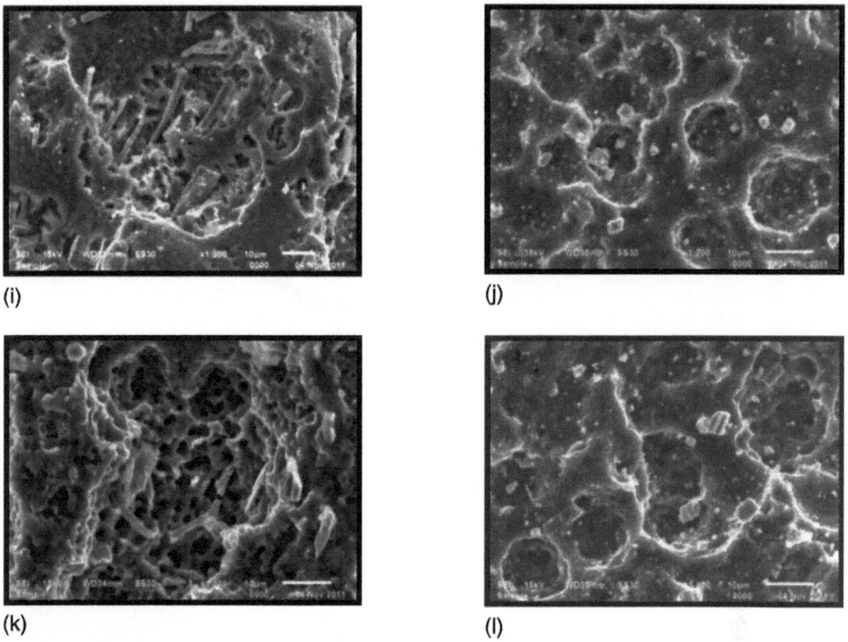

(i) (j) (k) (l)

Figure 7.7 Continued

where $I_{cathodic}$ is the cathodic reduction rate, $i_{cathodic}$ is the cathodic reduction current density on the second-phase elements and V_r is the volume fraction of second-phase elements.

It can be observed from the polarization curve that the cathodic Tafel slope for the FSP AE42 alloy is higher than that of the as-cast alloy, which implies that the cathodic activity of the fine *in-situ* precipitates produced during FSP is smaller than those of the elongated β-phase precipitates present in the as-cast alloy. Smaller cathodic activity of FSP AE42 implies less density of hydrogen reduction current, $i_{cathodic}$. However, during FSP, the precipitates present in the as-cast AE42 alloy were fragmented into finer ones and some new *in-situ* precipitates also developed, thus the volume fraction of second-phase precipitates may have increased. Due to increase in the volume fraction of second-phase precipitates, the cathodic activity of the FSP alloy tends to increase. The decreased anodic activity due to finer grain size, as well as lower cathodic activity attributed to low cathodic reduction current density for the FSP alloy, resulted in lower corrosion rates, as evident from the lower i_{corr} value of 0.32 mA/cm^2 for the FSP alloy, in comparison to 0.54 mA/cm^2 for

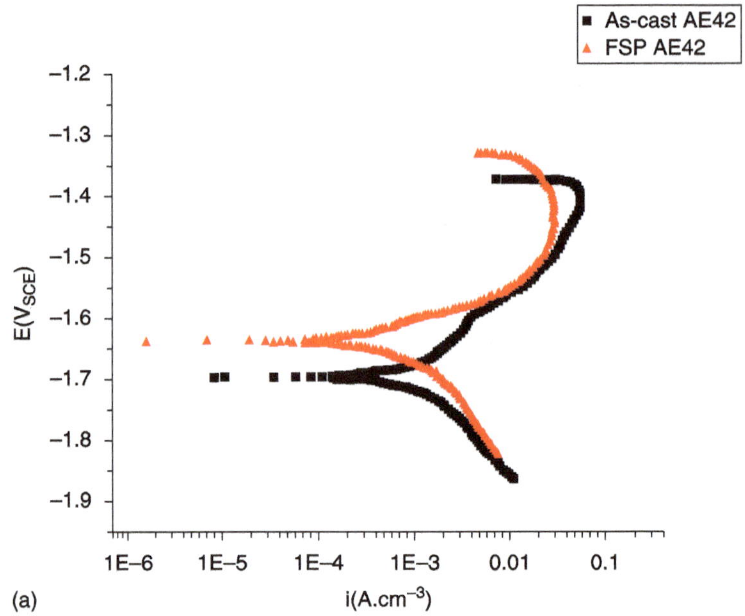

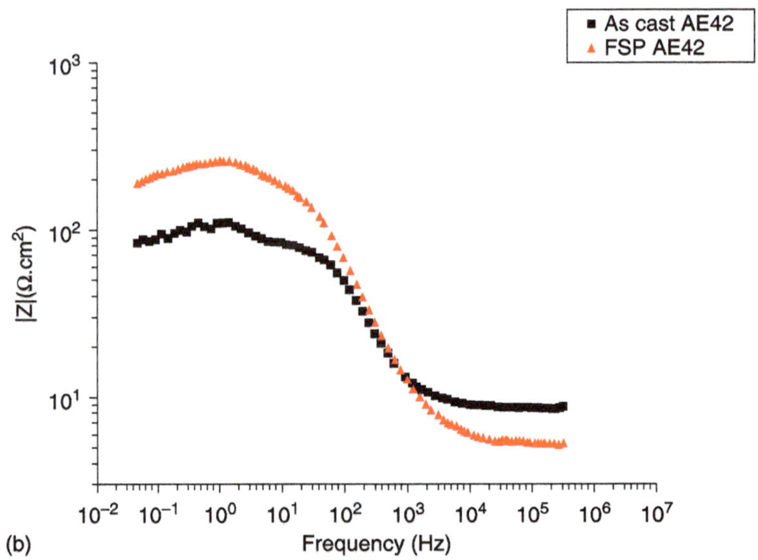

Figure 7.8 (a) Potentiodynamic polarization curves for the as-cast as well as friction stir processed Mg-based AE42 alloy in 3.5% NaCl solution; (b) Nyquist plot;

(*Continued*)

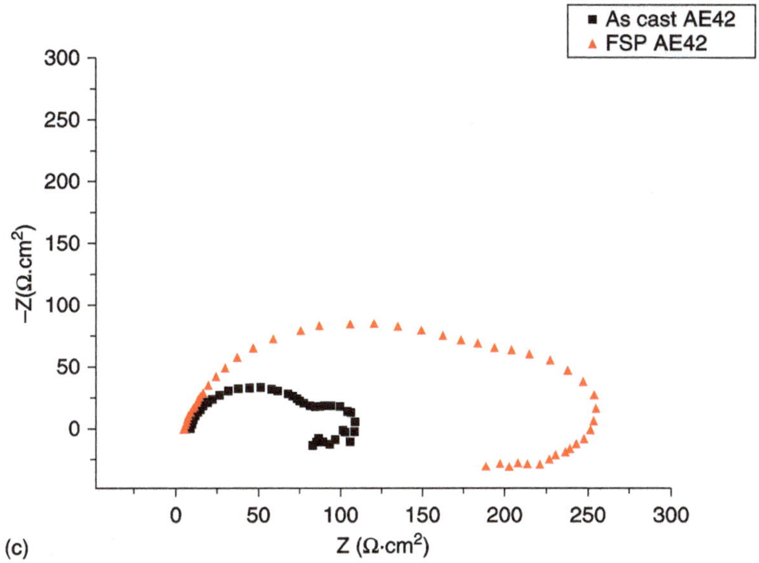

Figure 7.8 **Continued (c) Bode plot for the as-cast and friction stir processed Mg-based AE42 alloy subjected to 3.5% NaCl solution during electrochemical impedance spectroscopy (EIS) testing. The specimen was processed at the optimized set of FSP parameters comprising 700 rpm, −10 °C cooling temperature and 3 FSP passes [21]**

the as-cast alloy [21]. The results of the EIS test are shown as the Nyquist and Bode plot in Figures 7.8(b and c). These also endorse higher corrosion resistance of processed AE42 alloy.

Different investigations have shown opposing results on the influence of fine grain size on the corrosion behaviour of Mg-based alloys. Some investigations have reported that corrosion rates decrease with grain size in neutral and alkaline NaCl solution [23], whereas others have observed that if large residual stresses remain in the material after grain refinement, cathodic kinetics will increase even if overall grain size has decreased [24]. However, some investigators have observed that grain size is relatively more important factor than residual stresses, at least in 3.5% NaCl solution [25]. Grain boundaries have higher energy than the bulk metal and are therefore more chemically active [19]. Thus high density of grain boundary area in a fine-grained material increases the surface reactivity and results in more rapid formation of protective passive film. However, the Pilling-Bedworth ratio for Mg is less than 1 [19], so it appears that the oxide layer formed may not be able to provide efficient corrosion protection to the underlying material, because of the compressive stresses that are generated as a result of oxide/metal mismatch [19]. An increased grain boundary density in a fine-grained Mg alloy helps compensate for the oxide/base metal mismatch by decreasing the compressive stresses [24].

A fine-grained material can show higher corrosion resistance in a dilute corrosive media (i.e. 0.1M NaCl solution), whereas in a relatively aggressive media (i.e. 3.5% NaCl solution), the fine-grained material became incapable of showing good corrosion performance [19]. However, fine-grained FSP AE42 alloy has been shown to have a higher corrosion resistance in 3.5% NaCl solution [21]. Higher corrosion resistance of fine-grained AE42 alloy may be attributed to the higher stability of the oxide layer as a result of reduced geometrical mismatch. The presence of uniformly distributed fine second-phase precipitates can also increase the adherence of the oxide layer to the underlying Mg substrate.

7.4.2 Al-based alloys

While investigating the effect of FSP parameters on the corrosion behaviour of AA2219, it was observed that tool rotational speed significantly affects the corrosion properties of the investigated alloy, whereas linear speed does not have an appreciable influence [26]. The prominent reason for improved corrosion resistance of the FSP specimen was the dissolution of second-phase precipitates, leading to a reduction in the number of available sites for galvanic coupling. Further increase in the number of FSP passes also considerably influenced the corrosion rates [27]. Corrosion studies were conducted using potentiodynamic polarization as well as EIS in 3.5% NaCl solution buffered with KOH to maintain a pH value of 10. It was observed that corrosion rates of the alloy decreased drastically after FSP and the multipass FSP alloy showed even higher corrosion resistance. EIS analysis also showed similar results, that is, higher corrosion resistance for the FSP specimen. The major reason for the improved corrosion resistance of the investigated alloy was found to be dissolution of the $CuAl_2$ precipitate phases during FSP, which resulted in decreased potential sites for the galvanic coupling between the precipitate phases and Al matrix [27].

The sensitization of the microstructure during welding is responsible for the corrosion susceptibility of the nugget's heat-affected zone (HAZ) region of the weldments [28]. The microchemistry of the parent alloy as well as temperature assisted transformations during welding control localization, distribution, size of precipitates and precipitate-free zones. Short-term (PWHT) may be an efficient method to rehomogenize the sensitized microstructure and increase the corrosion resistance of the welds. A comparative study on the corrosion behaviour of FSP and laser melted 2219 aluminium alloy revealed that both FSP and laser processing led to an improvement in the corrosion behaviour of the investigated alloy [29]. FSP was done with single pass (FS1), as well as double pass (FS2) options. Laser treatment of the investigated alloy was done at two different scan speeds, 10 mm/min (L10) and 20 mm/min (L20). Corrosion behaviour was evaluated using electrochemical corrosion testing. The results of potentiodynamic polarization for the FSP and laser-processed samples are summarized in Table 7.4. It can be seen that the E_{pit} values increased from −585 mV in the BM to −485 and −463 mV for FS1 and FS2 samples respectively. Similarly, I_{corr} values decreased from 868.4 μA to 11.0 and 2.9 μA in FS1 and FS2 samples, respectively. The increase in corrosion resistance for the FSP specimens may be attributed to the redistribution of the second-phase precipitates.

Figure 7.9(a) shows the effect of reduction/redistribution of second-phase precipitates on the pitting potential. It can be observed that the corrosion resistance and the amount of dissolved second phase precipitates demonstrated almost a linear relationship. Bodes plots for FSP and laser melted (LM) samples are shown in Figure 7.9(b). It can be observed that the corrosion impedance increases with the decrease in frequency. Furthermore, it can be observed that the impedance values for the BM are on the order of few ohms, whereas for the FSP specimens, the impedance increased to a few kilo-ohms.

During a study [30] of the effect of laser and shot peening on SCC susceptibility of FSW 7075 aluminium alloy joints, shot peened samples were subjected to slow strain rate testing in 3.5% NaCl solution. The specimens did not reveal any signs of corrosion pitting or SCC during

Table 7.4 Results of the electrochemical corrosion test for the base metal (AA2219) and friction stir processed specimens [29]

Specimen	Epit Vs SCE (mV)	Icorr (μA)
Base metal	−585	868.4
FS1	−485	11
FS2	−463	2.9

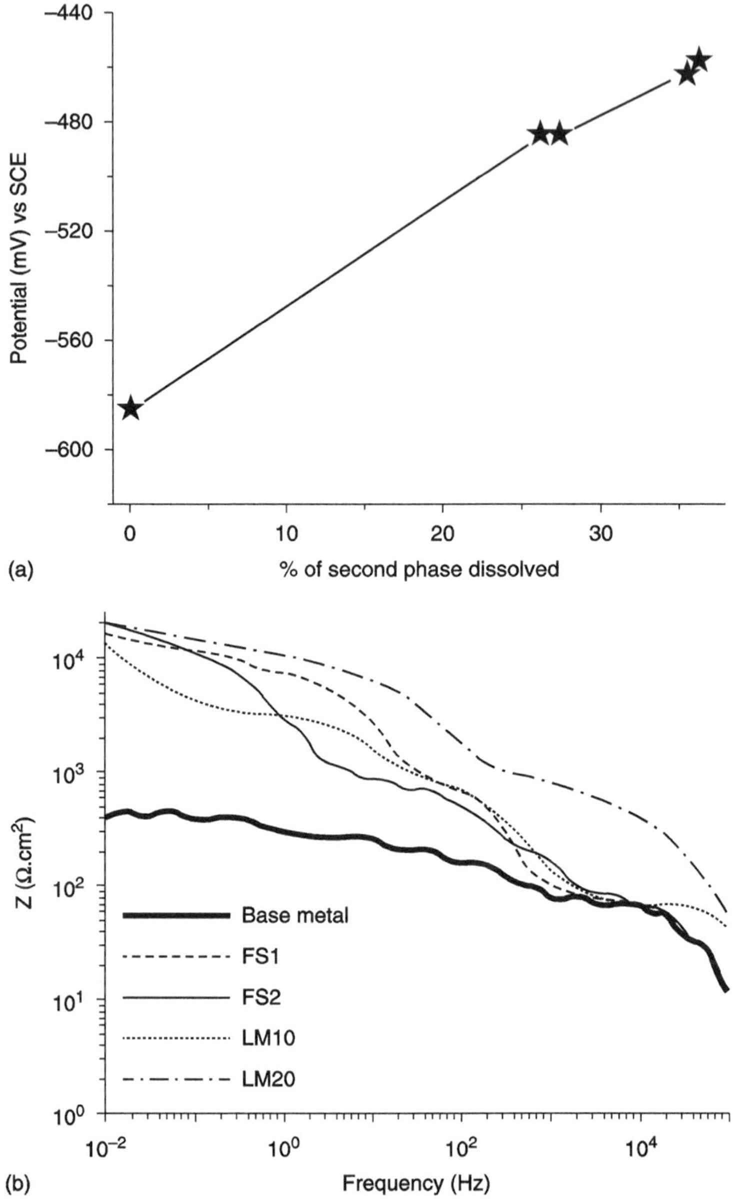

Figure 7.9 (a) Influence of amount of second phase redistributed on E_{pit} values; and (b) EIS curves for surface treated samples along with the base metal [29]

slow strain rate testing. The FSW plates were exposed to 3.5% NaCl solution for 60 days. Pitting corrosion was seen on the samples, but even after 60 day exposure no SCC was detected on any of the peened or unpeened samples. Results showed that the corrosion rates were very small for all surface treated samples. Pits size and number of pits were larger on the unpeened surfaces, whereas they were much smaller on the shot peened surface. The laser treated surface also had fewer pits than the unpeened surface. During the investigation on the effect of FSP on corrosional behaviour of AA6063-T6 Al alloy, it was reported that FSP increased the corrosion rate of AA6063-T6 aluminium alloy [31]. It was found that corrosion resistance was further reduced with the increase in the rotational speed. Figure 7.10 shows the corrosion rate for AA6063-T6 Al alloy in 1M HCl solution, FSP at different tool rotational speeds. It can be observed that there is a sharp decrease in corrosion rate with increase in exposure duration up to 48 hours. Furthermore, the friction processed material showed a decrease in corrosion resistance compared with the BM at all the investigated tool rotational speeds. It was concluded that increasing the tool rotational speed reduced the corrosion resistance of the AA6063-T6 Al alloy. The investigators attributed this to the increase in size of the precipitates with increasing tool rotational speed.

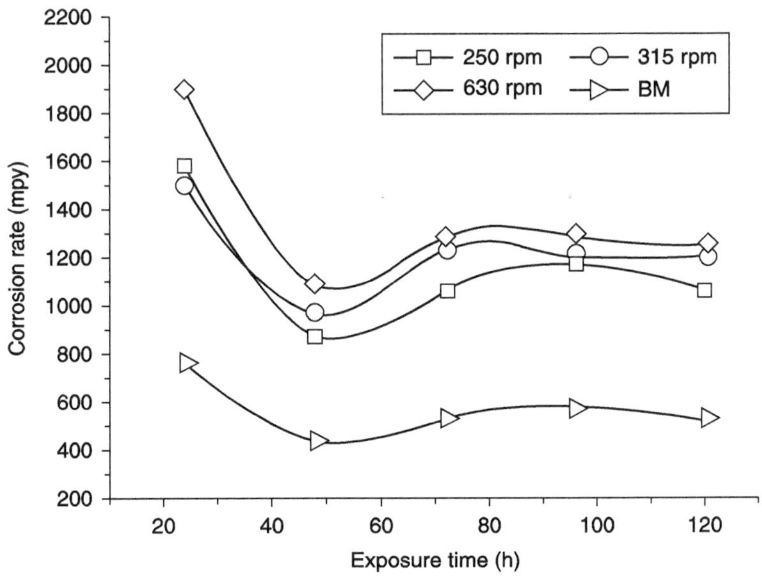

Figure 7.10 Graphs of corrosion rates of friction stir processed AA6063-T6 Al alloy against the exposure time in 1M HCl solution for various tool rotational speeds [31]

The corrosion behaviour of a FSW AA7108 T79 aluminium alloy was investigated by using accelerated testing (ASTM G34 EXCO) and electrochemical measurements [32]. Corrosion testing revealed that the edge regions of the thermo-mechanically affected zone (TMAZ) were most susceptible to corrosion. It was reported that localized corrosion occurred intergranularly due to the non-uniform distribution of $MgZn_2$ precipitates within the TMAZ, whereas the parent alloy had a relatively uniform distribution of the precipitates. Figure 7.11 shows the variation of the open circuit potential profile at various locations across the weld, measured over the cross-section. It can be observed that the potential increased from the edges of the TMAZ to the parent regions; the highest potential of −945 mV within the parent region and the lowest potential, approximately −980 mV at the edges of the TMAZ.

The appearance of lowest open circuit potential at the edges of TMAZ was attributed to the increased anodic reaction rate within the TMAZ. Furthermore, the investigation also revealed that the edge regions of the TMAZ are most susceptible to corrosion. It is perceptible that such regions in the weld are sensitized by the thermal transient caused by heat generated during the welding process. It was proposed that the $MgZn_2$ phase precipitated at grain boundaries resulted in localized corrosion at the grain boundaries compared with the grain matrix. Microstructural studies of the corroded laser-treated FSW aluminium alloy 2024 revealed that the melted zone comprised epitaxially grown columnar grains [33]. The corrosion properties of the FSW and laser-treated welds were studied

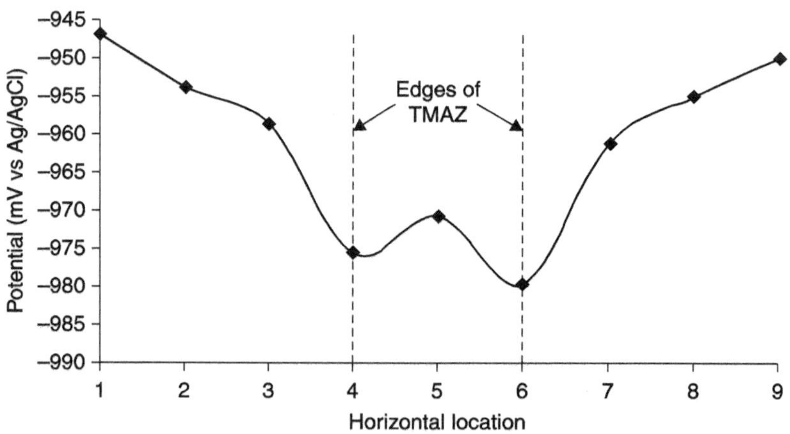

Figure 7.11 Potential profile across the cross-section of the weld [32]

in 0.5 M sodium chloride solution using open circuit potential and cyclic potentiodynamic polarization.

The results of the immersion test showed the presence of large pits in the untreated welds, while no such pits were observed for the laser-treated regions. A noticeable increase in the pit nucleation resistance (390 mV) after the laser surface treatment was reported. Another investigation on the corrosion behaviour of FSW of 2024 aluminium alloy and 2195 (lithium-aluminium) alloy [34] reported similar corrosion behaviour for the BMs as well as for the FSW zone. The results were explained on the basis of the stability of the oxide layer. In the presence of aggressive Cl^- ions, a stable oxide layer could not be formed on the surface of the aluminium, resulting in higher dissolution rates. Corrosion behaviour of the different layers of the FSW 2219 aluminium alloy at different depths along the specimen cross-section was also evaluated [35]. The top surface of the nugget zone (NZ) showed highest resistance and it decreased with depth. Higher resistance at the top surface was attributed to the presence of fine second-phase precipitates, which became coarser deeper in the specimen. Larger-sized second-phase precipitates led to dissolution of the matrix by forming strong galvanic cells with the matrix. FSW region was found to have higher resistance against pitting compared to the BM. Furthermore, corrosion resistance reduced with the increase in transverse speed for constant tool rpm.

Higher tool rpm lead to coarsening of Al_2Cu precipitates, increasing the pitting corrosion in the NZ. Further effects of precipitates on the corrosion performance were observed during FSW of 2024 Al alloy. Microstructural analysis of 2024 Al alloy revealed the presence of two different types of precipitate phases in the base alloy, Al_2CuMg and Al-Cu-Fe-Mn phase. The distribution of Mg in Al_2CuMg phase was found to be heterogeneous, with higher concentration at the center and decreasing towards the edge of the particle. Corrosion studies were conducted by immersion in EXCO solution (4.0 M NaCl + 0.5 M KNO_3 + 0.1 M HNO_3) for different time intervals. Fragmented and redistributed of these second-phase precipitates took placed during FSW. It was found that pitting corrosion density was higher in the shoulder region [36]. During corrosion testing, dissolution of Al_2CuMg phase occurred due to lower self-corrosion potential compared to the aluminium matrix. With increased dissolution, the amount of Al and Mg in the particle decreased, raising their self-corrosion potential and dissolution of the surrounding matrix. Optimization of the FSW parameters was suggested to improve the corrosion resistance.

7.4.3 Miscellaneous alloy systems

Corrosion behaviour of FSP Ti-6Al-4V alloy in HCl solution showed strong dependence on the processing parameters [37]. During this study, processing parameters were varied, so as to produce the peak temperature above as well as below the β-transus. Corrosion testing was done using electrochemical and corrosion immersion tests. Active–passive behaviour for all the samples at the open circuit was reported, which indicates incomplete passivity of the specimens. Samples processed at temperature below the β-transus demonstrated highest corrosion rates, whereas the samples processed at temperature above the β-transus showed superior corrosion resistance. This difference in the corrosion behaviour was attributed to the presence of different microstructures. At temperatures lower than β-transus, the microstructure comprised of high volume fraction (>90%) of equiaxed α grains. The remainder of the microstructure consisted of a combination of retained β phase and fine, acicular $\alpha + \beta$ transformation product.

However, when the peak temperature during processing exceeded the β transus, the microstructure consisted entirely of fine β grains containing fine α colonies. It was found that the α phase, which was relatively rich in V, was attacked preferentially. In another comparative study, corrosion behaviour of gas-tungsten arc welded (GTAW) Ti-6Al-4V alloy was evaluated against FSP Ti-6Al-4V in 5-pct HCl solution. FSP specimen demonstrated higher corrosion resistance, whereas arc welded specimens showed worst corrosion resistance. It was concluded that presence of acicular α/β microstructure as well as element partitioning during welding resulted in higher corrosion rate for the welded specimens [38]. In the case of NiAl bronze, FSP specimens showed considerably lower corrosion rates compared to the as-cast material in immersion corrosion tests. The improved corrosion performance of NiAl bronze was attributed to microstructural refinement, including finer grain size and elimination of casting porosities [39]. However, the polarization curves obtained from electrochemical testing revealed nearly similar corrosion behaviour for the two specimens.

Beneficial effects of FSP can be utilized for improving the properties of steels. During FSP of AISI 316L and 304 steel, it was observed that the precipitation of the sigma phase is deleterious and could severely affect the corrosion performance [40–42]. The sigma phase is rich in Cr, resulting in depletion of the same in the surrounding region. The content of Cr in the surrounding region makes it more vulnerable to corrosion attacks. In the case of AISI 304 steel, the presence of intergranular corrosion was observed in the heat affected zone (HAZ); however, the degree of damage was not severe in comparison to the BM. The advancing side (AS) of the processed zone suffered a greater degree of corrosion due to precipitation of the sigma phase. It was observed that lowering the heat input could significantly reduce the volume fraction of this phase. This means that processing should be carried out at lower rotation speeds, normal loads and higher transverse speeds.

The weld zone of ferritic steel HSLA-65 showed significant improvement in tensile properties. Salt spray testing conducted on this steel showed no preferential corrosion site [43]. Sensitization and SCC studies conducted on austenitic 304 stainless steel also did not show any such preferred corrosion sites. SCC conducted in a boiling solution of 25% NaCl showed almost similar results for both the weld zone and the BM. The high processing temperatures during the welding process did not deteriorate the properties of the weld zone [44]. However, slightly high corrosion susceptibility of the HAZ of the weld was observed for the FSW 301L steel during salt spray testing [45].

However, the FSW section of the super duplex stainless steel (SAF 2507) showed higher pitting corrosion temperature (65°C) in comparison to 40 to 55°C of the arc weld. Significant improvement in the yield and ultimate strength of the weld section was observed in comparison to the BM [46]. FSP of Cu-Mn alloy (52Mn-4Al-3Fe-1.5Ni-39Cu) did not show any improvement in its corrosion performance when tested in sea water at -200mV [47]. Severe cracking and delamination was observed, which was attributed to the SCC. A significant amount of residual stresses are introduced into the material during FSP. These retained stresses could lead to SCC as observed in the case of Cu-Mn alloy. Attempts at stress relieving using heat treatment at different temperatures and durations were made. A heat treatment at a temperature of 450°C for 24 h was sufficient enough to improve corrosion resistance. The fine-grained microstructure of the FSP Cu-Mn alloy showed significant improvement in corrosion; however, no such effect was observed for as-cast alloy after heat treatment [43].

7.5 References

1. Hanawalt, J.D., Nelson, C.E. and Peloubet, J.A. (1942), 'Corrosion studies of magnesium and its alloys', *Trans. AIME*, 147: 273–99.
2. Froats, A., Aune, T.K., Hawke, D., Unsworth, W. and Hillis, J. (1987), *Metals Handbook*, 13th edn, Ohio: ASM.
3. Babolan, R. (2005), *Corrosion Test and Standards: Application and Interpretation*, 2nd edn, ASTM International.

4. Freeney, T.A. and Mishra, R.S. (2010), 'Effect of friction stir processing on microstructure and mechanical properties of a cast-magnesium–rare earth alloy', *Metall. Mater. Trans A*, 41: 73–84
5. Ni, D.R., Wang, D., Feng, A.H., Yao, G. and Ma, Z.Y. (2009), 'Enhancing the high-cycle fatigue strength of Mg–9Al–1Zn casting by friction stir processing', *Scr. Mater.*, 61: 568–71.
6. Cavaliere, P. (2006), 'Effect of friction stir processing on the fatigue properties of a Zr-modified 2014 aluminium alloy', *Mater. Character.*, 57: 100–4.
7. Yazdipour, A., Shafiei, M.A. and Dehghani, K. (2009), 'Modeling the microstructural evolution and effect of cooling rate on the nanograins formed during the friction stir processing of Al5083', *Mater. Sci. Eng. A*, 527: 192–7.
8. Amira, S., Dubé, D., Tremblay, R. and Ghali, E. (2008), 'Influence of the microstructure on the corrosion behaviour of AXJ530 magnesium alloy in 3.5% NaCl solution', *Mater. Character.*, 59: 1508–17.
9. Zeng, R.C., Chen, J., Dietzel, W., Zettler, R., Santos, J.F. et al. (2009), 'Corrosion of friction stir welded magnesium alloy AM50', *Corros. Sci.*, 51: 1738–46.
10. Jönsson, M. and Persson, D. (2010), 'The influence of the microstructure on the atmospheric corrosion behaviour of magnesium alloys AZ91D and AM50', *Corros. Sci.*, 52: 1077–85.
11. Zhang, T., Meng, G., Shao, Y., Cui, Z. and Wang, F. (2011), 'Corrosion of hot extrusion AZ91 magnesium alloy. Part II: Effect of rare earth element neodymium (Nd) on the corrosion behaviour of extruded alloy', *Corros. Sci.*, 53: 2934–42.
12. Birbilis, N., Easton, M.A., Sudholz, A.D., Zhu, S.M. and Gibson, M.A. (2009), 'On the corrosion of binary magnesium-rare earth alloys', *Corros. Sci.*, 51: 683–9.
13. Jang, Y.H., Kim, S.S., Yim, C.D., Lee, C.G. and Kim, S.J. (2007), 'Corrosion behaviour of friction stir welded AZ31B Mg in 3·5%NaCl solution', *Corros. Eng. Sci. Technol.*, 42: 119–22.
14. Zeng, R.C., Chen, J., Dietzel, W., Zettler, R., Santos, J.F. et al. (2009), 'Corrosion of friction stir welded magnesium alloy AM50', *Corros. Sci.*, 51: 1738–46.
15. Kannan, M.B., Dietzel, W., Zeng, R., Zettler, R. and Santos, J.F. (2007), 'A study on the SCC susceptibility of friction stir welded AZ31 Mg sheet', *Mater. Sci. Eng. A*, 460–1: 243–50.
16. Argade, G.R., Kandasamy, K., Panigrahi, S.K. and Mishra, R.S. (2012), 'Corrosion behaviour of a friction stir processed rare-earth added magnesium alloy', *Corros. Sci.*, 58: 321–6.
17. Dhanapal, A., Boopath, S.R. and Balasubramanian, V. (2012), 'Influence of pH value, chloride ion concentration and immersion time on corrosion rate of friction stir welded AZ61A magnesium alloy weldments', *J. Alloys Compd.*, 523: 49–60.
18. Kannan, M.B., Dietzel, W. and Zettler, R. (2011), '*In vitro* degradation behaviour of a friction stir processed magnesium alloy', *J. Mater. Sci. Mater. Med.*, 22: 2397–401.
19. Ralston, K.D. and Birbilis, N. (2010), 'Effect of grain size on corrosion: A Review', *Corros. Sci.*, 66: 501–13.

20. Ralston, K.D., Birbilis, N. and Davies, C.H.J. (2010), 'Revealing the relationship between grain size and corrosion rate of metals', *Scr. Mater.*, 63: 1201–4.
21. Arora, H.S., Singh, H. and Dhindaw, B.K. (2012), 'Corrosion behaviour of Mg based AE42 alloy subjected to friction stir processing', *Corrosion NACE*, 69(2): 122–35.
22. Ambat, R., Aung, N.N. and Zhou, W. (2000), 'Evaluation of microstructural effects on corrosion behaviour of AZ91D magnesium alloy', *Corros. Sci.*, 42: 1433–55.
23. Hamu, G.B., Eliezer, D. and Wagner, L. (2009), 'The relation between severe plastic deformation microstructure and corrosion behaviour of AZ31 magnesium alloy', *J. Alloy Comp.*, 468: 222–9.
24. Hoog, C.O., Birbilis, N. and Estrin, J. (2008), 'Corrosion of pure Mg as a function of grain size and processing route', *Adv. Eng. Mater.*, 10, 579–82.
25. Jang, Y.H., Kim, S.S., Yim, C.D., Lee, C.G. and Kim, S.J. (2007), 'Corrosion behaviour of friction stir welded AZ31B Mg in 3·5%NaCl solution', *Corros. Eng. Sci. Technol.*, 42: 119–22.
26. Surekha, K., Murty, B.S. and Prasad Rao, K. (2009), 'Effect of processing parameters on the corrosion behaviour of friction stir processed AA 2219 aluminium alloy', *Solid State Sciences*, 11: 907–17.
27. Surekha, K., Murty, B.S. and Prasad Rao, K. (2008), 'Microstructural characterization and corrosion behaviour of multipass friction stir processed AA2219 aluminium alloy', *Surf. Coat. Technol.*, 202: 4057–68.
28. Paglia, C.S. and Buchheit, R.G. (2008), 'A look in the corrosion of aluminium alloy friction stir welds', *Scr. Mater.*, 58: 383–7.
29. Surekha, K., Murty, B.S. and Prasad, R.K. (2011), 'Comparison of corrosion behaviour of friction stir processed and laser melted AA 2219 aluminium alloy', *Materials and Design*, 32: 4502–8.
30. Hatamleh, O., Singh, P.M. and Garmestani, H. (2009), 'Corrosion susceptibility of peened friction stir welded 7075 aluminium alloy joints,' *Corrosion Sci.*, 51: 135–43.
31. Mahmoud, T.S. (2008), 'Effect of friction stir processing on electrical conductivity and corrosion resistance of AA6063-T6 Al alloy', *J. Mechanical Engineering Science*, 222: 1117–23
32. Wadeson, D.A., Zhou, X., Thompson, G.E., Skeldon, P., Oosterkamp, L. et al. (2006), 'Corrosion behaviour of friction stir welded AA7108 T79 aluminium alloy', *Corrosion Sci.*, 48: 887–97.
33. Kalita, S.J. (2011), 'Microstructure and corrosion properties of diode laser melted friction stir weld of aluminium alloy 2024 T351', *Applied Surface Science*, 257: 3985–97
34. Corral, J., Trillo, E.A., Li, Y. and Murr, L.E. (2000), 'Corrosion of friction stir welded aluminium alloys 2024 and 2195', *J. Mater. Sci. Lett.*, 19: 2117–22.
35. Xu, W. and Liu, J. (2009), 'Microstructure and pitting corrosion of friction stir welded joints in 2219-O aluminium alloy thick plate', *Corros. Sci.*, 51: 2743–51.
36. Kang, J., Fu, R-D., Luan, G.H., Dong, C.L. and He, M. (2010) '*In-situ* investigation on the pitting corrosion behaviour of friction stir welded joint of AA2024-T3 aluminium alloy', *Corros. Sci.*, 52: 620–6.

37. Atapour, M., Pilchak, A.L., Frankel, G.S. and Williams, J.C. (2010), 'Corrosion behaviour of investment cast and friction stir processed Ti-6Al-4V', *Corr. Sci.*, 52: 3062–9.
38. Atapour, M., Pilchak, A.L., Frankel, G.S. and Williams, J.C. (2010), 'Corrosion behaviour of friction stir-processed and gas tungsten arc-welded Ti-6Al-4V', *Metall. Mater. Trans. A*, 41: 2318–27.
39. Ni, D.R., Xiao, B.L., Ma, Z.Y., Qiao, Y.X. and Zheng, Y.G. (2010), 'Corrosion properties of friction stir processed cast NiAl bronze', *Corr. Sci.*, 52: 1610–17.
40. Chen, Y.C., Fujii, H., Tsumura, T., Kitagawa, Y., Nakata, K. et al. (2012), 'Banded structure and its distribution in friction stir processing of 316L austenitic stainless steel', *Journal of Nuclear Materials*, 420: 497–500.
41. Park, S.H.C., Sato, Y.S., Kokawa, H., Okamoto, K., Hirano, S. and Inagaki, M. (2005), 'Corrosion resistance of friction stir welded 304 stainless steel', *Scr. Mater.*, 51(2): 101–5.
42. Park, S.H.C., Sato, Y.S., Kokawa, H., Okamoto, K., Hirano, S. and Inagaki, M. (2005), 'Corrosion properties in friction stir welded 304 austenitic stainless steel', *Welding in the World*, 4(3–4): 63–8.
43. Konkol, P.J., Mathers, J.A., Johnson, R. and Pickens, J.R. (2003), 'Friction stir welding of HSLA-65 steel for shipbuilding', *J. Ship Prod.*, 19(3): 159–64.
44. Clark, T.D. (2005), 'An analysis of microstructure and corrosion resistance in underwater friction stir welded 304l stainless steel', MSc thesis, Brigham Young University, Provo, UT.
45. Okamoto, K., Hirano, S., Inagaki, M., Park, S.H.C., Sato, Y.S. et al. (2003), 'Metallurgical and mechanical properties of friction stir welded stainless steels', *Proceedings of the Fourth International Symposium on Friction stir Welding*, 14–16 May, Park City, UT, TWI, paper on CD.
46. Steel, R.J., Pettersson, C.O., Sorensen, C.D., Sato, Y., Sterling, C.J. and Packer, S.M. (2003), 'Friction stir welding of SAF 2507 (UNS S32750) Super Duplex Stainless Steel', Paper PO346, *Proceedings of Stainless Steel World*, KCI Publishing.
47. Mahoney, M.W. and Lynch, S.P. (2006), 'Friction stir processing', *OMB*, Report No. 0704-0188.

8

Tribological aspects in friction stir welding and processing

V.N. Malyshev, Gubkin Russian State University of Oil and Gas, Russia

DOI: 10.1533/9780857094551.329

Abstract: Tribology is the science of friction that tries to answer the most basic questions on the methods of friction stir welding (FSW) and friction stir processing (FSP) technology. This chapter attempts to examine the main aspects related to friction, tool wear, friction coefficient, wear mechanisms and other issues in FSW and FSP, of various materials such as aluminum alloys, magnesium alloys, titanium alloys, copper alloys, and various structural steels. The tribological characteristics and behavior of materials under various loads typical of FSW are considered. One advantage of the method is the possibility of FSW to weld various dissimilar materials such as steel-aluminum, titanium-copper, etc. The questions of friction of such materials are also discussed. The FSP method is now becoming an effective method of strengthening of work surfaces by the addition of various reinforcing particles, such as Al_2O_3, ZrB_2, SiC, etc. The tribological characteristics of similar processes are carefully examined. Of course, a lot of attention is paid to the issues of tool wear in FSW and FSP.

Key words: friction, wear, friction coefficient, wear resistance, tribological characteristics, wear mechanism.

8.1 Introduction

Surface properties such as wear resistance can determine the service life of components in many industrial applications. Components produced by aluminium and its alloys exhibit poor tribological properties, even leading to seizure under detrimental conditions. Hence, there is a strong drive to develop a new modification method for Al-based materials and others which provide greater resistance to wear and better tribological properties [1,2,5,6]. Recently, much attention has been paid to friction stir processing (FSP), which is known as a surface modification technique [3,4].

The study of friction stir welding (FSW) and FSP, play an important role associated with tribology, tool wear, coefficient of friction (COF), etc., since friction is an essential feature of these processes. Friction is one of the major sources of parasitic energy losses in machines and industrial processes. Frictional losses will always occur in components and systems that involve moving parts in which surfaces are in contact and relative motion. Examples include gears, bearings, seals and various moving parts in engines and motors. In addition to energy loss, as a result of friction, the resulting shear stresses at the contact interface will lead to wear in its many forms. Wear of machine components results in poor reliability and durability, which may require frequent replacement of parts or systems. But friction plays another role in the process of FSW. Here, friction generates the necessary heat which together with mixing process of workpiece materials creates a strengthened weld.

Friction heat is generated between the wear resistant welding tool shoulder and pin, and the material of the workpieces. This heat, along with the heat generated by the mechanical mixing process and the heat within the material, causes the stirred materials to soften without reaching their melting point. Thus the friction process is the main generator of heat in FSW.

8.2 Main definitions in field of friction and wear

Tribology is the field of science and technology concerned with 'interacting surfaces in relative motion'. Wear and friction are not material properties, but the properties of a tribological system.

Wear is the progressive loss of material due to interacting surfaces in relative motion. It is quantitatively measured as the specific wear rate W_s, defined as volume loss per sliding distance and load [mm^3/Nm]) of a material. It is 'The removal of material from the impairment of a solid surface resulting from friction or impact' [7]. When surfaces slide over one another, they not only experience friction, but also wear and material is lost from both surfaces, even if one is much harder than the other. This is true even for very hard materials such as diamond; a diamond will scratch all other materials but still loses some of its mass as it wears the other material. Otherwise we could not polish diamonds, as polishing is simply controlled wear to produce an apparently smooth and flat surface.

Wear is the direct result of the same processes that causes friction, the movement of the asperities on the surfaces over one another. *Friction* is a measure of the resistance to motion (loss of energy) of two interacting surfaces and is quantitatively described by the *coefficient of friction* μ (dynamic/static). It is therefore a function of the real contact area between the two surfaces and the character and strength of the interaction, which can be described as:

- adhesive
- ploughing
- deforming.

The COF, often symbolized by the Greek letter μ, is a dimensionless scalar value that describes the ratio of the force of friction between two bodies and the force pressing them together. The COF depends on the materials used, for example, ice on steel has a low COF, while rubber on pavement has a high COF. COFs range from near zero to greater than one.

For surfaces at rest relative to each other, $\mu = \mu_s$, where μ_s is the *coefficient of static friction*. This is usually larger than its kinetic counterpart. For surfaces in relative motion, $\mu = \mu_k$, where μ_k is the *coefficient of kinetic friction*.

The Coulomb friction is equal to F_f, and the frictional force on each surface is exerted in the direction opposite to its motion relative to the other surface.

The COF between surfaces normally increases with increasing temperature and decreasing load. The energy lost in the friction phenomena can lead to an increase in temperature, to the emission of noise, and/or to deformation of the contact area. In almost all cases, a lower COF will lead to a lower wear rate. The COF is defined as friction force F divided by the applied load N ($\mu = F/N$) at a given sliding speed.

Investigations [10,12,13] have indicated that a fine-grained structure affects the tribological properties of a surface layer. It is reported that grain refinement leads to the improvement of wear resistance and load bearing capacity of Al-7Si alloy [14,15]. Other authors [16] also report similar results for different Al-based alloys.

8.3 Mechanisms of wear

Wear has been defined as the progressive loss of material from rubbing surfaces. The rate and effect of this removal depend upon the characteristics of the materials involved and environmental conditions. While wear may appear to be simple, there are a number of different mechanisms that can cause it. A broad classification of the main wear mechanisms is shown below and each of these mechanisms results in material removal and wear of different types. Thus, numerous distinct and independent mechanisms are involved in the wear of materials (Figure 8.1). These include:

- *Abrasive wear*: caused by the cutting or ploughing of one surface by another, harder surface. The harder material may simply be a harder alloy or hard irregularities on the counter surface; it may also be an oxide, carbide, work-hardened particle or other foreign material.
- *Fatigue wear*: failure of the material due to repeated stressing from hard irregularities on the counter surface. It is commonly encountered in rolling or sliding contact, where high localized forces are cyclically imposed on a small contact area. This type of wear occurs most commonly in gears and bearings.
- *Adhesive wear*: loss of material by transfer and adhesion to the counter surface. It is the most common and difficult to control. It results from the welding or adhesion and subsequent rupturing of contacting asperities on a surface. It is particularly severe between sliding pairs of metals that are chemically similar and/or mutually soluble. Adhesive wear is closely related to surface friction in such a way that an increase in frictional forces generally causes a large increase in resulting wear.
- *Erosion wear*: the wear of material under a flow of any environment (gas, liquid or mixtures thereof in the presence or absence of mechanical particles), which are the result of changing the geometry, of the weight loss, etc.
- *Corrosive wear*: the wear of material that occurs due to some type of chemical or electrochemical reaction with the environment. It may be due to general oxidation, chemical fumes, humidity or combustion products, and may also be inadvertently caused by ingredients in lubricants. It usually exists simultaneous in combination with one of the types of mechanical wear, therefore it is often called corrosion-mechanical wear.

While the mechanisms and forms of wear are important in understanding the phenomena (and there are many other factors including material properties, states of stress and environmental conditions), it is the final

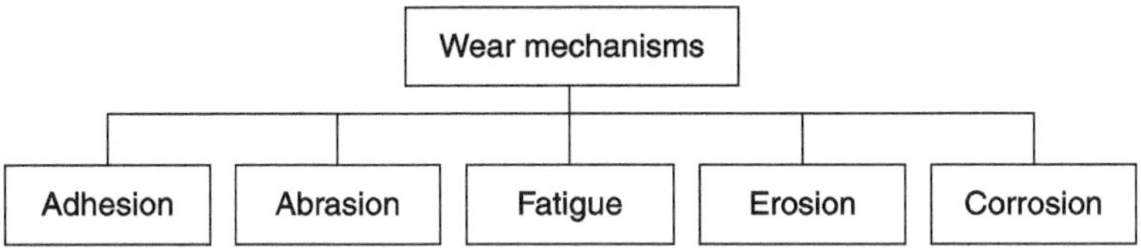

Figure 8.1 Various types of wear mechanisms [7]

result of wear that is of immediate concern. The principal result of wear is loss of mechanical functionality due to:

- dimensional changes; and
- fatigue failure.

Excessive tool wear changes the tool shape (normally by removing tool features), thus changing the weld quality and increasing the probability of defects. In friction stirring, tool wear can occur by adhesive, abrasive or chemical wear (which is addressed subsequently as reactivity) mechanisms. The exact wear mechanism depends on the interaction between the workpiece and tool materials and the selected tool parameters. For example, in the case of polycrystalline cubic boron nitride (PCBN) tools, wear at low tool rotation speed is caused by adhesive wear (also known as scoring, galling or seizing), while wear at high tool rotation speed is caused by abrasive wear [17]. For example, some wear mechanisms such as adhesion, delamination and abrasion are illustrated on Figures 8.2 to 8.4 [18].

The signs of adhesion mechanism can be seen in different parts of the surface of an as-received metal sample. The lower hardness of the

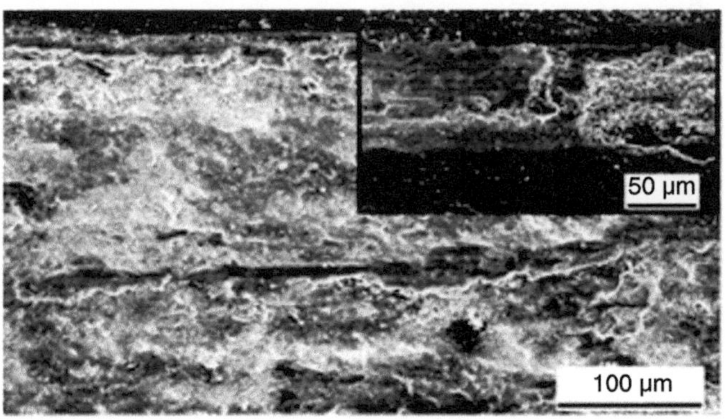

Figure 8.2 SEM micrograph of surface of base metal sample worn under 5 KN applied force [18]

as-received metal has led to the adhesion mechanism. The high COF of the as-received sample may be due to an increase in the friction force needed to overcome the highly adhesive contact between tested surface and counterpart.

The high wear weight loss of the as-received sample can be the result of the adhesion mechanism operating on the surface during dry sliding. The results indicated that the as-received sample has a higher level of surface damage among test samples. These may be attributed to lower hardness and higher mean grain size of this sample.

Figure 8.3 shows the worn surfaces of FSP samples. As can be seen in the figure, delamination and abrasion are active wear mechanisms for the samples produced by 1 to 4 FSP passes and worn under 5 KN applied force. When the number of FSP passes increases, the hardness of samples increases and the amount of grooves and cracks, which are associated with delamination mechanism, also decrease and hence the 4-pass sample shows the lowest level of surface damage (Figure 8.3(d)). The decrease in the signs of delamination mechanism with increasing FSP passes may be attributed to the improvement of metal resistance to crack nucleation and growth. This may be due to the increase in toughness of material resulting from grain refinement [12,13].

As mentioned above, COF of samples decreases by increasing the number of passes. This can lead to a decrease in the level of shear stresses

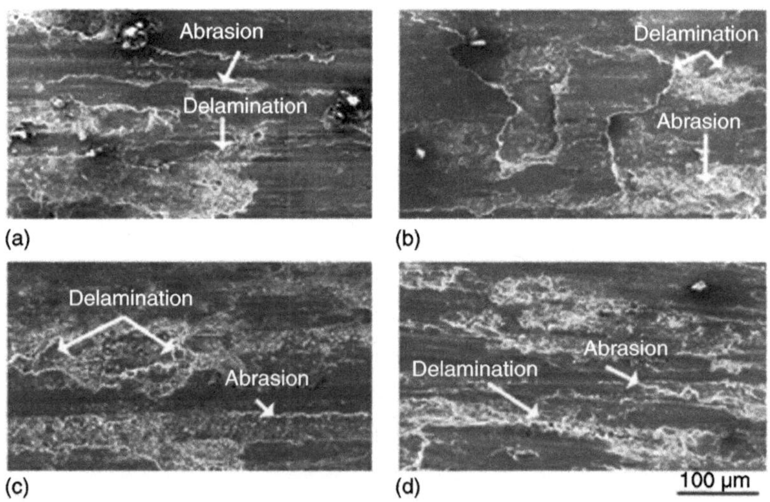

Figure 8.3 SEM micrographs of surfaces worn under 5 KN applied force for the specimens made from: (a) 1-pass; (b) 2-pass; (c) 3-pass; and (d) 4-pass samples [18]

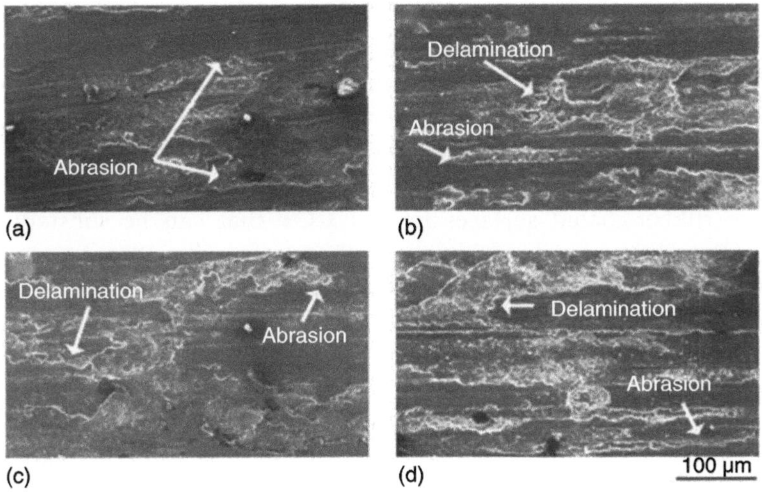

Figure 8.4 **SEM micrographs of the surface of 4-pass samples worn under different applied forces: (a) 1KN; (b) 2KN; (c) 3.5KN; and (d) 5KN samples [18]**

applied to subsurface layers of material during sliding. This phenomenon results in decreasing the probability of nucleation and growth of subsurface cracks during sliding and diminishes the signs of the delamination mechanism. Therefore, the wear weight loss decreases with increasing FSP passes, due to the higher resistance of material to crack propagation.

The surfaces of 4-pass samples worn under different applied forces are illustrated in Figure 8.4. Decreasing the applied force during sliding has led to a decrease in the existence of subsurface cracks and width and depth of worn tracks. However, for the samples worn under 1 KN applied force, the signs of the delamination mechanism disappeared and abrasion was the only active wear mechanism.

8.4 Friction coefficient

The coefficient of friction (COF) is an empirical measurement – it has to be measured experimentally, and cannot be found through calculations. Rougher surfaces tend to have higher effective values. Both static and kinetic COFs depend on the pair of surfaces in contact.

Most dry materials in combination have COF values between 0.3 and 0.6. Values outside this range are rare, but Teflon, for example, can have a coefficient lower than 0.04. A value of zero would mean no friction at all, an exclusive property – even levitation vehicles have drag. Rubber in contact with other surfaces can yield COFs from 1 to 2. Occasionally, it is maintained that μ is always less than 1, but this is not true. While in most relevant applications, $\mu < 1$, a value above 1 merely implies that the force required to slide an object along the surface is greater than the normal force of the surface on the object. For example, silicone rubber or acrylic rubber-coated surfaces have a COF that can be substantially larger than 1.

While it is often stated that the COF is a 'material property', it is better categorized as a 'system property'. Unlike true material properties (i.e. conductivity, dielectric constant, yield strength), the COF for any two materials depends on system variables such as temperature, velocity, atmosphere and also what are now popularly described as ageing and de-ageing times; as well as on geometric properties of the interface between the materials. For example, a copper pin sliding against a thick copper plate can have a COF that varies from 0.6 at low speeds (metal sliding against metal) to below 0.2 at high speeds when the copper surface begins to melt due to frictional heating. The latter speed, of course, does not determine the COF uniquely; if the pin diameter is increased so that the frictional heating is removed rapidly, the temperature drops, the pin remains solid and the COF rises to that of a 'low speed' test.

It is usual to link the vertical force (F_z) to the total mechanical power (P) through a COF μ, assuming that all the heat is generated by friction. The surface heat generated by the friction of the tool on the material can be simply estimated by Eq. 8.1, assuming constant pressure equal to $p_{shoulder}$ [Pa] [19]. Note that this expression neglects the pin lateral surface contribution:

$$Q_s = \frac{2}{3}\pi\mu p_{shoulder}\omega r_S^2 = \frac{2}{3}\mu F_z \omega r_s \qquad [8.1]$$

where r_s is the radius of the tool shoulder. If we assume that the power is only generated by friction (hence $P = Q_s$), we can estimate the COF with

$$\mu = \frac{3P}{2F_{z\omega r_s}} \qquad [8.2]$$

Figure 8.5 shows the evolution of the COF calculated by Eq. 8.2 for different welding parameters.

The COF is clearly more influenced by the rotational speed than by the advancing speed, since the vertical force F_z and the power P depend more on the advancing speed than on the rotational speed, and they both increase

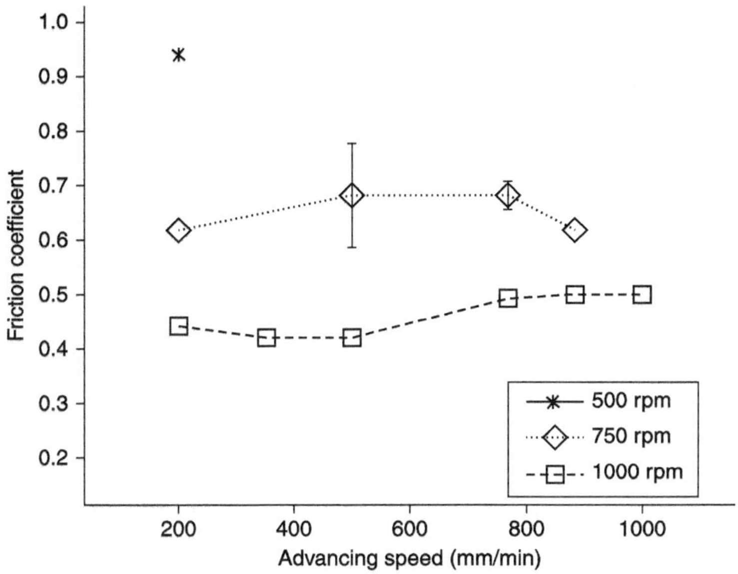

Figure 8.5 **Evolution of the friction coefficient with the welding parameters [19]**

when the advancing speed increases. In the thermal modelling literature, we find all kinds of values for the COF. Some are realistic, like the one proposed by [20] (μ=0.4 for welding parameters of 300 mm/min and 1500 rpm and a shoulder diameter of 20 mm). However, the value proposed in [21] sounds very low (μ=0.15 for welding parameters of 139.7 mm/min and 500 rpm and a shoulder diameter of 25.4 mm). Nevertheless, the hypotheses on which these calculation is based are certainly too simplistic. The influence of material deformation is clearly not negligible.

8.5 Tribotechnical properties of materials processed by friction stir welding and processing

FSP, as a variant of FSW, is used to modify the microstructure of the surface layer of metallic materials [9–11]. Although FSP is still technically in its infancy, it also has the potential to be an effective surface hardening technique for the improvement of friction and wear performance. FSP has been applied to a variety of metallic materials and alloys, and has been demonstrated to produce beneficial effects on improvements in ductility, resistance to fatigue, corrosion and surface defect elimination [3,7,14,21]. Most of the investigations towards application of FSP have concentrated on low melting point alloys, such as aluminium, magnesium and bronze. However, there have been many more limited efforts towards application on steel [22]. Furthermore, its impact on friction and wear behaviour is unknown in applications where parts are in relative motion.

The stirring action and frictional heat generated by the FSP tool can be used to distribute ceramic particles as reinforcement on the surface of light metals such as aluminium and magnesium [23–29].

Most of the main friction and wear tests were conducted with a ball-on-flat contact configuration in reciprocating sliding and with a pin-on-disk configuration in unidirectional sliding [30]. Figure 8.6 shows the test rig and typical specimen contact. Ball specimens are made of polished 12.7 mm (0.5″) diameter hardened 52100 bearing steel with a surface roughness of 0.035 μm Ra. Rectangular flat specimens with 50 × 30 × 0.25 mm nominal dimensions were made from 4140 steel under different conditions: the annealed baseline, heat treated (HT) water quenched, HT salt quenched and FSP. The surfaces of all the flat specimens were ground to the same surface finish of 0.14 μm Ra.

Tests were conducted at different loads from 1 N up to 50 and 75 N, which impose Hertzian contact pressure from 0.04 GPa up to 1.0, and 1.15 GPa, respectively, for a ball-on-flat contact. A reciprocating frequency of 1 Hz was used, with a stroke length of 10 mm, translating to an average sliding velocity of 1 cm/s; test duration was 1 hour, under ambient room temperature. All the tests were conducted with dry sliding and with different lubricants (technical water, different oils, etc.). Other friction and wear tests were also conducted [30] in unidirectional sliding contact test using the pin-on-disk tribometer (Figure 8.7).

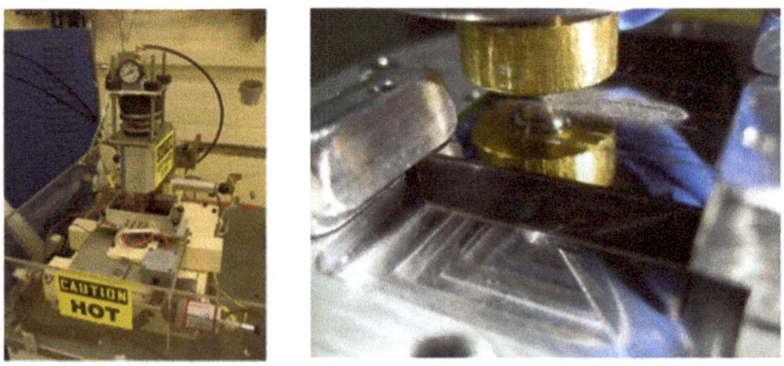

(a) (b)

Figure 8.6 **Reciprocating ball on flat friction and wear test rig [30]**

Figure 8.7 **Pin-on-disc tribometer used for unidirectional sliding friction and wear testing [30]**

Tests were conducted at a constant sliding velocity of 1 cm/s for the duration of 1 hour, at ambient room temperature. All the tests were conducted without lubricant (dry sliding tests) and by using different lubricants. The COF was continuously measured during each test. At the conclusion of the test, the wear was measured in the ball-and-flat test specimens by optical profilometry technique. Worn surfaces were also characterized by both optical and SEM microscopy to assess wears mechanisms.

8.5.1 Aluminium alloys

FSP has been proved to be an effective way to refine the microstructure of aluminium alloys, and thereby improve their mechanical properties [8,31]. Some investigations [12,13] have indicated that a fine-grained structure affects the tribological properties of the surface layer.

In [18] there is a description of tests conducted on friction and wear, in which the base metal (BM) was Al5083 rolled plates of 3 mm thickness with a nominal composition of 4.3 Mg–0.68 Mn–0.15 Si–bal. Al (in wt. pct.). FSP was applied on the surface of the alloy by a tool made of steel H-13, with a shoulder of 20 mm diameter and a pin of 6 mm diagonal length and 2.8 mm height. Samples were subjected to 1 to 4 FSP passes along the same direction, with rotation rate of 1250 rpm and travel speed of 50 mm/min at room temperature.

Wear behaviour of the specimens was evaluated by using a pin-on-disk tester at room temperature. Pin specimens with 5 mm diameter were cut from the processed zone of each sample and ground on emery paper up to grade 320. Counterparts were disks made of AISI D3 steel with a hardness of 58 HRC. The tests were carried out at normal loads of 1 to 5 KN and a rotation speed of 60 rpm. The friction force was recorded automatically against sliding distance by the tester software.

Figure 8.8 shows that increasing the number of FSP passes has resulted in decreasing grain size of the processed zone and increase in hardness. As can

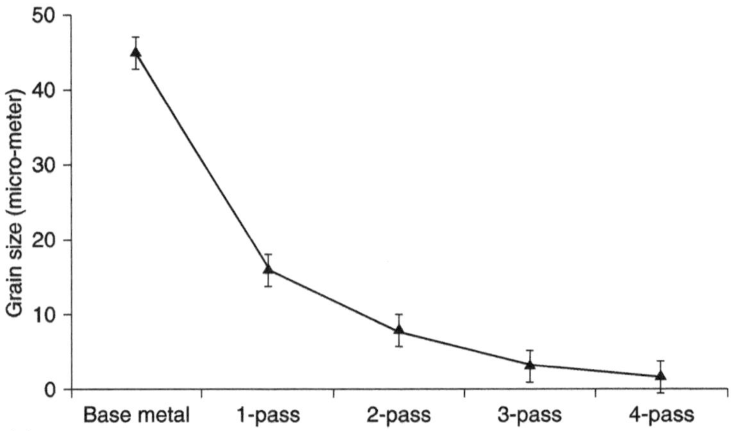

(a)

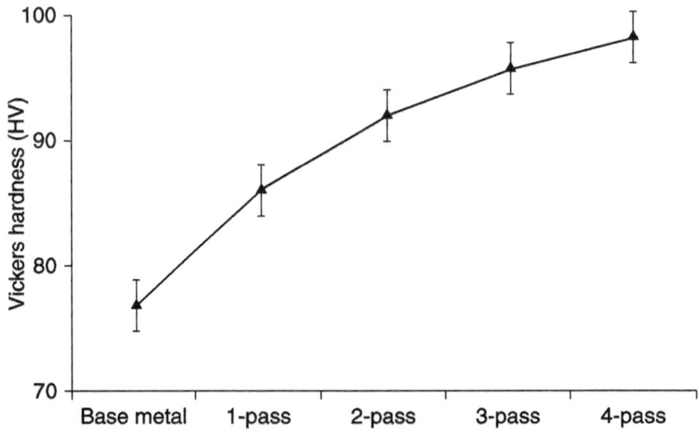

(b)

Figure 8.8 Variations of grain size (a) and mean Vickers hardness (HV) (b) of processed zone of base metal and samples processed under 1 to 4 FSP passes [18]

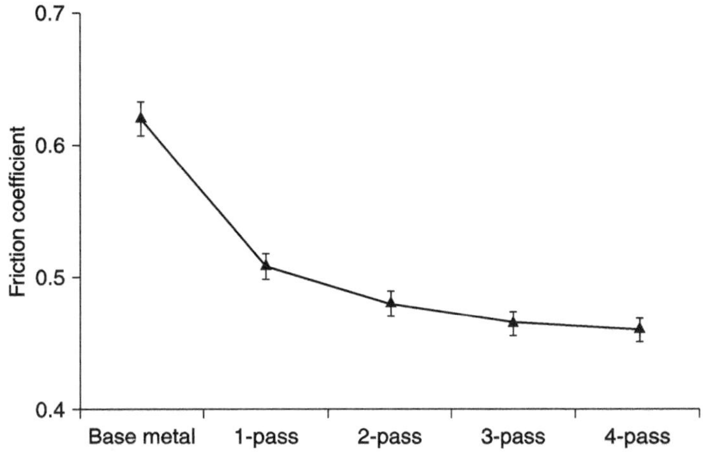

Figure 8.9 Variations of mean friction coefficient of processed zone of base metal and samples produced by different number of FSP passes [18]

be seen in Figure 8.9, the wear weight loss of samples decreases with the increase in the number of FSP passes. This observation is in good agreement with the trend observed for the hardness of the samples (Figure 8.8(b)).

Increasing the hardness of metal can lead to the occurrence of less plastic deformation during sliding [14,15]. The results indicate that FSP can lead to the improvement of wear resistance as well as hardness, especially for a higher number of FSP passes due to the improvement of microstructural characteristics. FSP has led to the decrease in COF (Figure 8.9) of samples in comparison to BM. Moreover, increasing the number of FSP passes has led to a decrease in the amount of mean COF.

These results can be attributed to the differences in the extent of localized plastic deformation at real contact areas of samples and counter-face. Harder surfaces of FSP samples have led to less plastic deformation and lower COF. It seems that a higher number of passes results in decreasing plastic deformation on the surface, which may be attributed to the microstructural refinement due to FSP [14].

The variations of weight loss of 4-pass samples (Figure 8.10 (a)), worn under different applied forces, shows that the wear weight loss decreases with decreasing applied force. Also, as can be seen from Figure 8.10(b), with a decrease in the applied force, the COF decreases. These observations are in good agreement with the wear laws and may be related to the decrease in contact area between counter surfaces, which results in a decrease in the weight loss and COF [7,32].

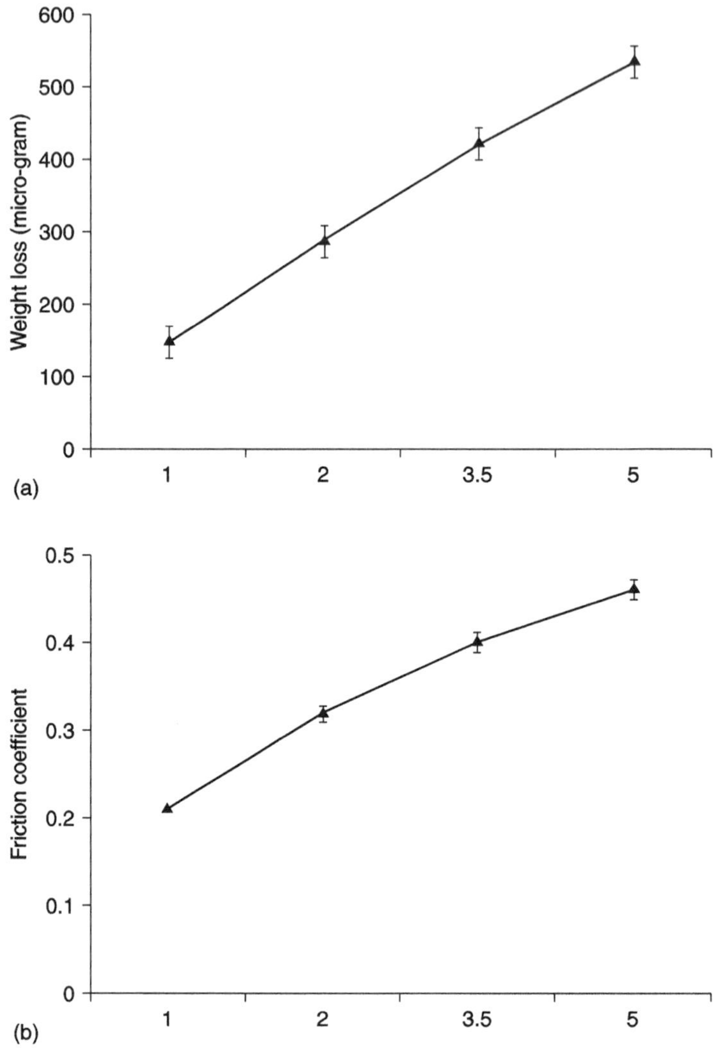

Figure 8.10 Variations of (a) weight loss and (b) friction coefficients of 4-pass samples worn under different applied forces (KgF) [18]

8.5.2 Magnesium alloys

Magnesium alloys offer lightweight alternatives to conventional metallic alloys, and as a result are finding structural applications in the automotive and light truck industry, for manufacturing Mg alloys have some important advantages over Al alloys. Their latent heat is only about two-thirds that of Al alloys. They have lower solubility for iron when molten, and higher fluidity. For die-casting processes, this means that Mg alloys have shorter cycle times and are less reactive with steels, so casting dies last longer and they can be cast in thinner sections [8]. Mg alloys have not been conventionally used for high performance applications, due to their low mechanical properties at room and elevated temperatures [34]. AZ91, a magnesium-based alloy, is commonly used to produce cast components, due to its superb castability and reasonable mechanical properties [35].

Mg alloys also have excellent damping properties, which makes them attractive for improving noise-vibration-harshness (NVH) characteristics [36]. As magnesium alloys generally have inferior formability, sheet material of magnesium alloys is made commercially by casting or die-casting processes, except for some wrought alloys such as AZ31. It is usually difficult to weld these cast magnesium alloys, due to the porosity formation in the weld [8,37].

During FSP, partial melting of the material takes place due to frictional heating between the workpiece and the FSP tool shoulder. Solidification of the partially melted material results in the characteristic nugget zone (NZ) of the FSP region. This method has been successfully used to modify the microstructures of as-cast Al [38,39] and Mg alloys [40]. FSP of magnesium alloy AE42, conducted under different FSP and cooling conditions with rapid external cooling, has resulted in enhancement of the microhardness value nearly by 60%. Thus, FSP can be considered as a significant technique for material strengthening through microstructural refinement [41].

One of the recent applications of FSP is fabrication of hybrid composites. Hybrid composites are engineering materials reinforced by a combination of two or more different types and/or forms of substances, in order to achieve the combined advantages of both. This gives a high degree of freedom in material design [42]. It is observed that the wear rate of FSP AE42 alloy had been reduced under all conditions for the range of normal loads and velocities. Microstructural refinement of FSP AE42 alloy results in greater hardness, greater work hardening capability and improved ductility, thus resulting in reduced wear rates [43].

The aim of investigations [33] is to fabricate AZ91-based hybrid composites using FSP with TiC as reinforcements, in an attempt to enhance its work hardening capability and mechanical properties. FSP of AZ91 alloy was also performed without any particle addition.

The material used in the investigation was Mg alloy AZ91 and commercially available TiC powder. The average particle size of the available TiC powder was about 40 μm. FSP was carried out on a

Computer Numerical Control (CNC) vertical milling machine (5 HP) with the help of an FSP tool and fixture. Rectangular specimens having dimensions 5 mm × 5 mm and 10 mm in length were prepared from the AZ91 ingot. FSP was performed at ambient cooling conditions. FSP was then carried out using an FSP tool, which is a commonly used cylindrical element without threads with 15 mm shoulder diameter, 4 mm pin diameter and 0.5 mm pin length. The material used for the FSP tool was mild steel. The parameters used for FSP consists of a tool rotational speed of 900 rpm, linear speed of 40 mm/min and plunge depth of 0.3 mm, which were kept constant in this investigation. For microstructural characterization, specimens were sectioned, mounted and polished using standard metallurgical procedures.

The tribological properties of the as-cast, as well as FSP AZ91 alloy, processed at the optimized FSP parameters, were evaluated using pin-on-disk wear testing. Rectangular pin specimens with cross-section dimensions of 5 mm × 5 mm and 10 mm in length were machined from the as-cast as well as FSP AZ91 alloy for the wear tests. Dry sliding wear tests were carried out on a universal tribometer (CETR UMT3), using a pin-on-disk test configuration.

The counter-face material used was a stainless steel disk with a hardness value of nearly 220 HV. Before the wear tests, the cross-section of all the specimens were ground using different grades of emery papers down to 1000 grit. All the wear tests were conducted at sliding speeds of 1 m/s under normal loads of 5 and 10 N for a total sliding distance of 2000 m. The test was conducted for 11 cycles. Sliding distance for the first 5 cycles was taken as 100 m each and 250 m for 6 cycles. Weight loss after each test cycle was measured to a precision of 0.01 mg and the wear rate was determined as a function of sliding distance. Before and after each test, both the disk and specimens were cleaned with acetone and dried in air to avoid contamination.

It is evident from the Figure 8.11 that the unprocessed zone contains a coarse grain structure, which becomes finer as we move to the processed zone, i.e. the NZ. The line chart showing the weight loss per unit cross-sectional area for the as-cast AZ91 alloy and FSP AZ91 with TiC powder during the pin-on-disk test on universal tribometer for variable loads, is shown in Figures 8.12(a and b).

The weight loss per unit cross-sectional area was more in the case of the as-cast material compared to the FSP metal. It is evident from the chart that the wear of the FSP material is gradual, as compared to the as-cast material when tested for different loads at constant velocity of disk.

Figure 8.11 Micrograph of Mg-based AZ91Alloy, subjected to friction stir processing, showing base metal, transition zone and the nugget zone [33]

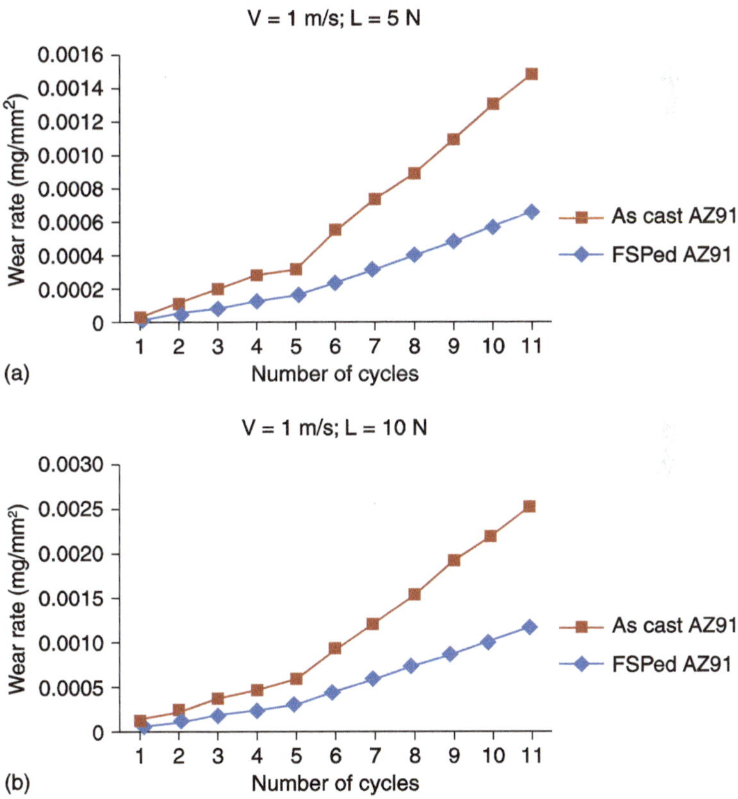

Figure 8.12 Curve representing the wear rate: (a) V = 1 m/s and L = 5 N; and (b) L = 10 N [33]

Thus, the AZ91-based surface nanocomposite was fabricated with TiC reinforcement using FSP successfully. The fabricated nanocomposite exhibited improved wear resistance, which may be attributed to the evolution of finer-grained structure and presence of fine reinforcement particles. Under the action of intense shearing and high axial pressure during FSP, TiC particles were able to disperse beyond the NZ. The similarity between semisolid processing and the current technique is limited to formation of fine-grain size and better dispersion of reinforcement. Since there is no pool of liquid melt, the grains do not show characteristic sphericity.

8.5.3 Titanium alloys

Titanium alloys present interesting possibilities as tribomaterials, but they have not been widely investigated as bearing materials. They are harder and stiffer than Mg and Al alloys, and have a better resistance to exposure to heat and aqueous corrosion. Like Al and Mg, their high affinity for oxygen results in the formation of an adherent surface oxide, but sub-stoichiometric TiO_2 can act as a solid lubricant. While preliminary investigations indicate that FSW is potentially an effective welding technique for the joining of titanium alloys, obviously more research is needed to understand the microstructural evolution of titanium alloys during FSW and operative mechanisms.

Experiments were carried out on two titanium alloys, Ti–6Al–4V and Ti–6Al–2Sn–4Zr–2Mo [44]. Friction and wear tests were conducted using a pin-on-disk apparatus. The diameter of the fixed ball sliders was 9.53 mm. Metallic, ceramic and polymeric materials were represented by the selection of 440C stainless steel, silicon nitride, alumina and polytetrafluoroethylene (PTFE). The Ti alloy disks were 63.5 mm in diameter and 12.7 mm thick. The disk surfaces were polished by 600 grit wet SiC abrasive paper and the pre-test arithmetic average surface roughness (*Ra*), measured with a Taylor-Hobson Talysurf™10 stylus profilometer with a 2.5 μm tip radius, was 0.11 ± 0.02 μm.

Concentric wear tracks ranging from 18 to 52 mm in diameter were used, and the disk rotation rate was adjusted accordingly to provide 0.3 or 1.0 m/s sliding speeds. A 10 N normal load was applied and the test was run for 500 m sliding distance. The friction force was monitored by a load cell-based force measurement system. The wear volumes of the sliders and disks were determined by weight change measurements with an accuracy of 0.1 mg. The wear factor is defined as the wear volume

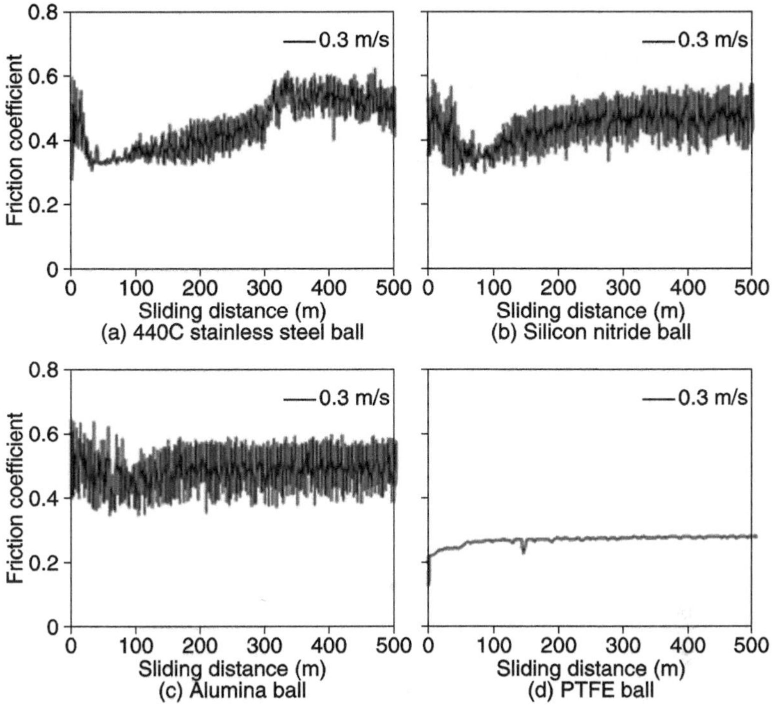

Figure 8.13 **Frictional traces of different sliders against Ti64 disks [44]**

normalized by the applied load and the sliding distance of the pin. All the tests were conducted under ambient air conditions with temperature and humidity in the range of 18 to 22 °C and 52 to 62%, respectively. At least two duplicates were run at each test condition. Good repeatability was obtained in both friction and wear results.

The average COF and its fluctuation was at steady-state for each test condition. Selected friction traces of the four different sliders against Ti64 disks are shown in Figure 8.13. The PTFE slider generated a fairly smooth friction trace (Figure 8.13(d)), due to its self-lubricating nature. The metal and ceramic sliders produced COF in the range of 0.34 to 0.50 with relatively large fluctuations (Figure 8.13(a–c)). The large fluctuation of the COF was thought to be caused by formation of periodic, localized fracture of a transfer layer. Titanium alloy commonly transfers to the counter-face when rubbing against other metals or ceramics [45–47]. The study of surface morphology and surface analysis confirmed this tendency.

A transfer layer was easily identified on the wear scar of the 440C stainless steel ball (Figure 8.14(a)). The energy dispersive spectroscopy

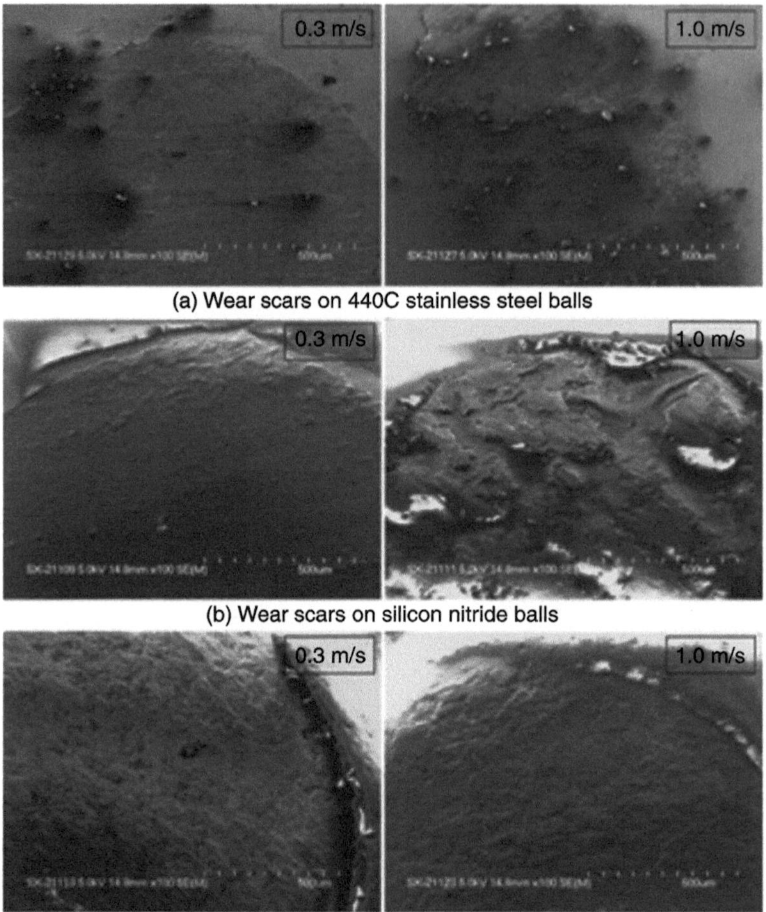

(a) Wear scars on 440C stainless steel balls

(b) Wear scars on silicon nitride balls

(c) Wear scars on alumina balls

Figure 8.14 **SEM images of wear scars on the balls sliding against Ti64 disks [44]**

(EDS) analysis detected Ti and/or Al on the worn surfaces of the metal and ceramic balls (Figure 8.15). The Ti disks suffered high wear rates, in the order of 10^{-4} mm^3/N m, against the metal and ceramic balls, and harder sliders generated relatively more (or at least comparable) wear on the Ti disks. Although harder balls were expected to have higher wear resistance, the results in Tables 8.1(a and b) show a reverse order: the alumina ball wore more than the silicon nitride ball, which in turn wore more than the stainless steel ball.

Remarkably, the wear factors of the ceramic balls were at least five times higher than those of the steel balls. Higher wear rate of an alumina

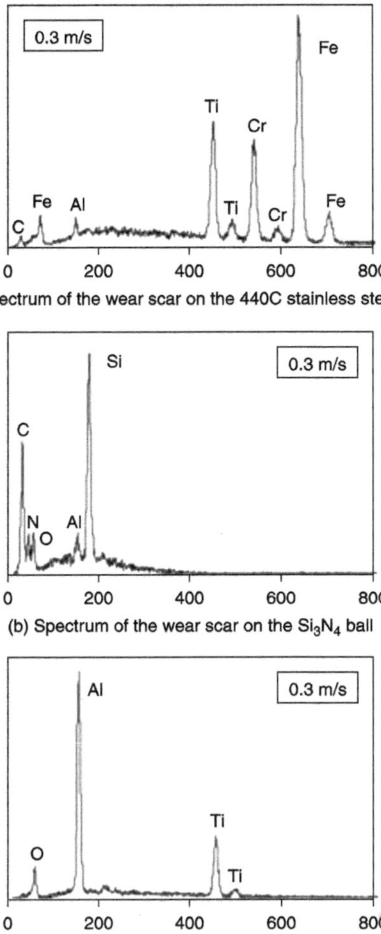

(a) Spectrum of the wear scar on the 440C stainless steel ball

(b) Spectrum of the wear scar on the Si_3N_4 ball

(c) Spectrum of the wear scar on the Al_2O_3 ball

Figure 8.15 **EDS analysis of the wear scars on the sliders [44]**

ball than that of a steel ball was observed [47] when sliding against a Ti64 disk. Figure 8.14 shows the wear scars on the metal and ceramic balls with features of abrasive wear, adhesive wear and plastic deformation. Abrasive wear seemed to dominate the wear process at 0.3 m/s. Those wear scars were larger and flatter, corresponding to their higher wear factors.

The wear scars generated at 1.0 m/s were smaller but much rougher with larger patches of transferred material, implying more severe adhesive wear, possibly due to higher temperature at the contact area. EDS analysis (Figure 8.15) also showed higher Ti and/or Al concentration on the wear

Table 8.1 Friction and wear results [44]

Slider material	Sliding speed					
	0.3 m/s			1.0 m/s		
	Friction coefficient	Wear factor (mm^3/N m)		Friction coefficient	Wear factor (mm^3/N m)	
		Ball	Disk		Ball	Disk
(*a*) Ti–6Al–4V disks						
440C stainless steel	0.50±0.05	6.9×10^{-6}	1.7×10^{-4}	0.35±0.05	1.6×10^{-6}	1.5×10^{-4}
Silicon Nitride	0.47±0.07	3.8×10^{-5}	3.5×10^{-4}	0.36±0.07	6.2×10^{-6}	1.3×10^{-4}
Alumina	0.49±0.07	5.7×10^{-5}	5.7×10^{-4}	0.44±0.07	1.6×10^{-5}	2.0×10^{-4}
PTFE	0.28±0.001	8.4×10^{-4}	N/M[a]	0.29±0.001	6.1×10^{-4}	N/M[a]
(*b*) Ti–6Al–2Sn–4Zr–2Mo disks						
440C stainless steel	0.48±0.05	5.1×10^{-6}	1.3×10^{-4}	0.34±0.04	1.22×10^{-6}	1.1×10^{-4}
Silicon nitride	0.47±0.08	4.4×10^{-5}	3.5×10^{-4}	0.37±0.02	9.40×10^{-6}	1.1×10^{-4}
Alumina	0.49±0.08	1.2×10^{-4}	3.4×10^{-4}	0.42±0.04	2.33×10^{-5}	2.2×10^{-4}
PTFE	0.27±0.001	9.9×10^{-4}	N/M[a]	0.29±0.001	7.92×10^{-4}	N/M[a]

[a] N/M, not measurable

scars at 1.0 m/s. The PTFE slider had the highest wear factor (10^{-3} mm^3/N m). Its counter-face (Ti disk) was protected by the polymeric layer transferred from the PTFE ball and had almost no surface damage except a few shallow circular groves ground by third body particles, probably some metal debris, embedded in the PTFE ball.

The wear resistance of 440C stainless steel, silicon carbide and alumina pins is in the reverse order as their relative Vickers hardness numbers, but in the same relative order as their fracture toughness. Recognizing that the current work was on sliding wear, the abrasive wear of ceramics has been proposed as a function of both hardness and fracture toughness [48]. Further studies on ceramic wear mechanisms showed that fracture toughness may play a dominating role in wear resistance.

The pin-on-disk apparatus is mainly intended to evaluate sliding wear, but in practice, the load history can consist of sliding and impact, since vibrations may occur if the disk surface is not perfectly normal to the axis of rotation. The tendency for impact to occur for small errors in alignment depends also on the sliding speed and how far the contact is from the centre of rotation. Unlike sliding that usually causes plastic shearing in materials, impact may introduce catastrophic failures, such as cracking and crushing of the contact surfaces, leading to faster material removal and the production of sharp ceramic debris fragments that can in turn cause three-body abrasion. Brittle materials, like ceramics, are more sensitive to such repeated impact effects than are tougher metals. In another experiment [49], the wear rate and mechanism of compound impact (impact and sliding) on metals and superalloys were studied. The material with lower fracture toughness had a higher wear rate and gave strong evidence for subsurface damage.

The tribological behaviour and responsible wear mechanisms for titanium alloys Ti–6Al–4V and Ti–6Al–2Sn–4Zr–2Mo, sliding against 440C stainless steel, silicon nitride, alumina and PTFE, showed that higher COF with larger fluctuation and higher wear rate were observed at the lower sliding speed.

8.5.4 Copper alloys

Copper and copper-based alloys are widely used in many engineering applications, because of high thermal and electrical conductivity, very good corrosion and wear resistance and self-lubrication properties [50,51]. However, copper, which has much higher thermal diffusivity than steel, cannot easily be welded by conventional fusion welding techniques. Heat input required for copper is much higher than conventional FSW, because of the greater dissipation of heat through the workpiece. Recently, FSW has been successfully used to weld 50 mm thick copper canisters for containment of nuclear waste [52].

FSW in copper alloys have all the typical zones found in other materials: the nugget, TMAZ, HAZ and base structure. The nugget has equiaxed recrystallized small grains and its hardness may be higher or lower than the BM, depending on the grain size of the BM. When 4 mm thick copper plates with an average grain size of 210 μm were welded at high rpm (1250) and low welding speed (1.01 mm/s) [53], the nugget had a lower hardness (60–90 HV), compared to the BM (105–110 HV). Even though grain size decreased from 210 to 100 μm, hardness decreased slightly due to reduction in dislocation density relative to the BM [53].

The copper alloys have several good properties such as high load capacity, low COF, high corrosion resistance, high wear resistance and high heat conductivity. All of these properties significantly affect the fatigue and wear life [54]. In the high load regime, COF decreases with load for many metallic pairs. It is believed that due to a large amount of wear debris and increased surface roughening, friction force decreases [55,56]. At loads from micro- to nano-Newton range, COF may be very low when the contacting surfaces are very smooth [57]. For different material combinations, friction may increase or decrease when the sliding velocity is increased. During the friction process, because of increased adhesion of the counter-face pin material on the disk, friction increases with the increase in sliding velocity [58]. Nowadays, different nonferrous material combinations are widely used for sliding/rolling applications where low friction is required.

The main tribological properties of copper alloys have been received by way of their trials in friction contact with other materials, such as mild steel [59]. Experiments were carried out using a pin-on-disk set-up. A cylindrical pin of 6 mm in diameter (both ends flat) from mild steel was slid over a horizontal surface (disk), which rotated using a motor. The test disk (from copper) is fixed on a horizontal plate, which rotated and this rotation (rpm) can be varied by an electronic speed control unit. The COF of copper varies with duration of rubbing and these variations at different normal loads are shown in Figure 8.16.

Experiments were carried out at sliding velocity 1.0 m/s and relative humidity 70%. In the experiments, a copper disk was slid against a mild steel counter-face. Curve 1 for normal load 10 N shows that at the early stage of rubbing, COF is 0.27 and after that it increases steadily up to 0.35. Over a duration of 22 minutes, COF becomes steady and for the

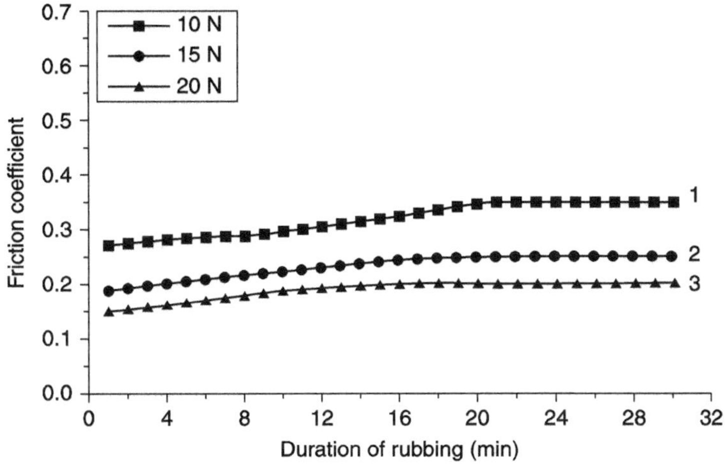

Figure 8.16 **Friction coefficient as a function of duration of rubbing at different normal loads (sliding velocity: 1 m/s; relative humidity: 70%; test sample: copper; pin: mild steel) [59]**

rest of the experiment time it remains constant. Friction is low at the early stage of rubbing, because of a layer of foreign material on the copper disk. This layer generally comprises of lubricating material, metallic oxide and some moisture. At an early stage of rubbing, the oxide layer separates the contacting surfaces and after initial rubbing, the deposited layer breaks up to make true metallic contact.

On the other hand, because of the ploughing effect, trapped wear particles between the contacting surfaces and surface roughening of the disk, friction force increases with rubbing time. After the running-in process for a certain duration, the surface roughness and other parameters reached a steady state value and for this reason there is no change in friction with time. For normal load 15 and 20 N, results are also shown by curves 2 and 3. It is apparent that at different normal loads, the copper disk takes different times to stabilize, which are 22, 19 and 15 minutes when the applied normal load is 10, 15 and 20 N, respectively. During the friction process, roughness and other parameters may reach a certain steady level earlier when the applied load is increased.

The COF varies with rubbing time and this variation at different sliding velocities is shown in Figure 8.17. In the experiment, a copper disk is mated with a mild steel pin at a normal load 15 N. Results are shown by curves 1, 2 and 3 for 1.0, 1.5 and 2.0 m/s, respectively. Curve 1 for 1.0 m/s shows that at the start of rubbing, COF is 0.19 and after that it increases

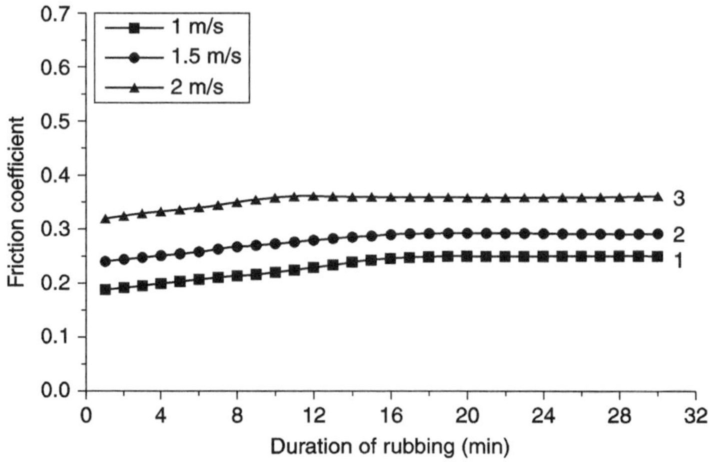

Figure 8.17 **Friction coefficient as a function of duration of rubbing at different sliding velocities (normal load: 15 N; relative humidity: 70%; test sample: copper; pin: mild steel) [59]**

steadily up to 0.25. In the experiments, it was found that after 19 minutes of running-in operation, friction became steady. Due to the ploughing effect and surface roughening, friction increases. After a certain duration of the running-in process, roughness and other parameters reached a steady state value, and therefore there was no change in frictional thrust until the experimental time.

As for the tool materials, the sintered tungsten-based alloy tool with improved geometry was much more effective for FSW of 10 to 25 mm thick copper plates. The tool with a new pin design from the high temperature tool material was strong enough to weld copper plates of more than 30 mm thickness [60,61]. However, in [62,63], it is reported that steel (QRO90) and Inconel were suitable tool materials for FSW of copper, but sintered carbide (K40UF) was too brittle for FSW of copper.

8.5.5 Steels (FSW and FSP of industrial steels)

Surface hardening in conjunction with lubrication is one of the main approaches used to reduce friction and wear in tribological components. There are a variety of methods currently used for surface hardening. In steel, which is perhaps the most common tribological material, surface hardening is done by a variety of methods. These include various forms of thermal heat treatments (e.g. carburizing, nitriding), induction hardening, laser surface hardening, etc. [4,8,22,25]. Numerous kinds of thin-film hard coatings and thermal spray coatings are also commonly used for surface hardening to enhance friction and wear performance. All such surface hardening techniques are energy or cost intensive.

While most of FSW efforts to date have focused on aluminium alloys, there is a considerable interest in it for steels. The lower heat inputs associated with FSW (relative to fusion welding processes) are expected to produce less metallurgical changes in the HAZ and to minimize distortion and residual stresses in steels, which is extremely important in welding of thick-section components. Furthermore, problems with hydrogen cracking in steels would be eliminated due to the solid-state nature of the FSW process. In addition, the solid-state FSW process eliminates the welding fumes, especially those containing hexavalent Cr, to allow compliance with OSHA standards [64]. These advantages are likely to make FSW attractive for joining steel in many applications.

The temperature of the tool shoulder is usually over 1000 °C [65] and the ensuing weld track behind the trailing edge of the rotation tool is 900 to 1000 °C. Similarly, peak temperatures of more than 1000 °C were observed in [66], just above the tool shoulder by both thermocouples and infrared camera. Based on extrapolation of measured temperature and microstructural evidence, it was suggested [66] that the peak temperature of the stir zone (SZ) exceeded 1100 °C and likely surpassed 1200 °C.

Generally, argon was used as the shielding gas to protect both the tool and the weld area from oxidation in FSW of steels. However, some investigations did not reveal if shielding gas was used [66,69]. Furthermore, no study was reported on the effect of shielding gas on the quality of FSW steel joints.

While most 3.2 to 6.4 mm thick steel plates can be successfully welded in a single pass, welds in 6.4 mm thick 304 L steel plate and steel plates of more than 6.4 mm were usually made with two passes from two sides, because the range of influence of the tool is relatively small in the steels compared to aluminium alloys [66–69]. Preheating of both the tool and steel plates to approximately 300 °C facilitated joining of 6.4 mm thick HSLA-65 steel plates in a single pass [70].

Some tribological trials [30] were carried out on surfaces of 4140 steel hardened by FSP. The processed area with the martensitic phase transformation can be clearly seen. Optical microscopy showed that the original microstructure consists of pearlite (left-hand side of Figure 8.18) as expected for annealed 4140 steel. The microstructure of the processed area consists of martensite (right-hand side of Figure 8.18).

(a)

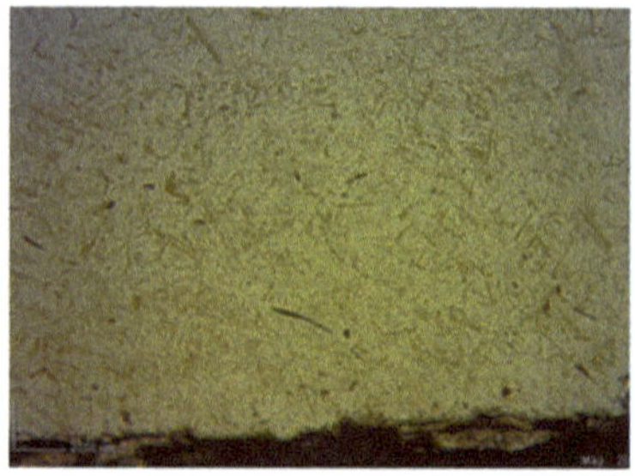

(b)

Figure 8.18 **(a) Pre-FSP microstructure (pearlite) and (b) post-FSP microstructure (martensite) [30]**

Microhardness measurement with a Vickers indenter at a load of 1 kg showed that the baseline material (pearlite) has a hardness of about 200 VHN, while the hardness of the FSP processed area is about 594 VHN (58.6 HRC).

This represents a significant hardening, comparable to the case carburization for the same material. A piece of the baseline 4140 steel was also subjected to conventional heat treatment. The sample was austenitized at 850 °C for 1 hour, and then water quenched. The microstructure of heat hardened material consists of martensite and the measured microhardness was about 450 VHN (45.7 HRC). These results showed that FSP produced the same microstructure in terms of phase and significantly higher surface microhardness compared to conventional heat hardening, perhaps reflective of grain refinement by FSP.

Figure 8.19 shows the variation of COF for the duration of the test with the various flat specimens. Friction behaviour is essentially the same for all the materials at the three different loads tested. The magnitude of the COF is between 0.1 and 0.12, which is typical for boundary lubrication regime. The friction of the baseline material is also noisier compared to the hardened surface, perhaps indicative of more local plastic flow at the asperity contact.

Wear in each flat specimen at the conclusion of the test is shown in Figure 8.20. Although the friction behaviour was similar for all the flats, wear behaviour was very different. Significant wear reduction was observed in all the hardened surfaces compared to the baseline material. FSP reduced the wear the most for all the three hardened techniques.

In the FSP specimens, tests were run with the reciprocating sliding direction perpendicular (FSP$^{\perp}$) and parallel (FSP||) to the tool translation direction. The least amount of wear was observed when sliding in the

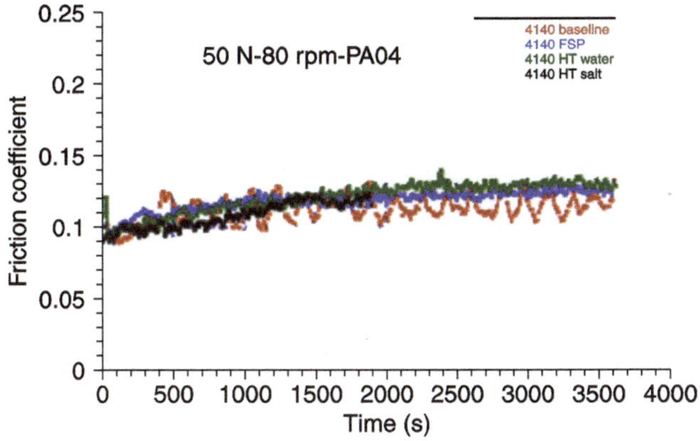

Figure 8.19 Variation of friction coefficient with time as normal contact load of 50 N [30]

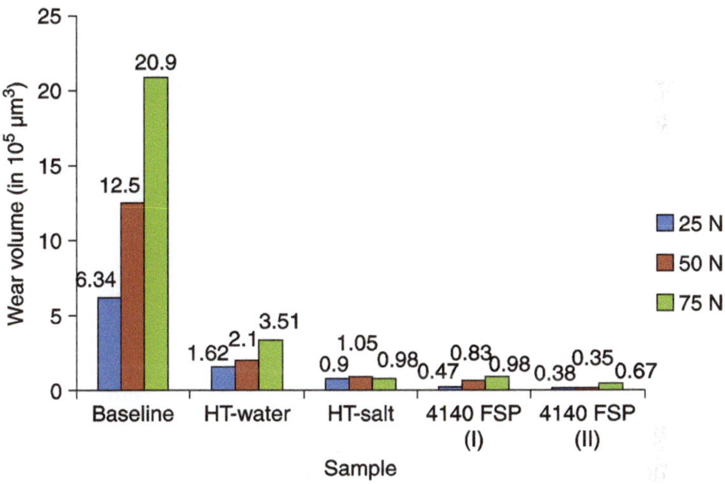

Figure 8.20 Wear in the various 4140 steel flats at the three test loads [30]

parallel direction (Figure 8.20). It should also be noted that the amount of wear increases with increasing load, but much less so in the FSP material, especially when sliding in the parallel direction.

Examination of the worn surface showed that wear in the baseline material involves a significant amount of plastic deformation, as indicated by material pile-up at the edge of the track in Figure 8.21(a). SEM analysis indicates that the wear mechanisms in the baseline material consist of abrasion, as indicated by scratches in the direction of sliding and fatigue

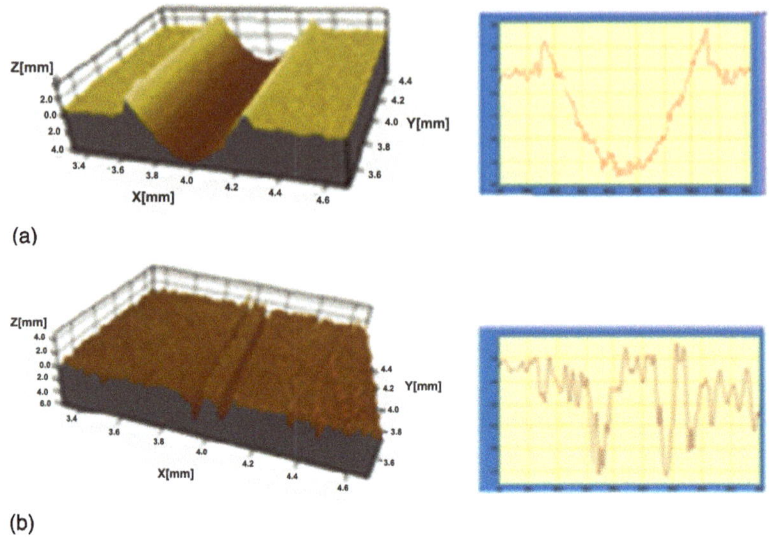

Figure 8.21 Optical 3-D and 2-D profilometry of wear track: (a) in baseline 4140 steel; and (b) FSP 4140 steel flat [30]

indicated by crack formation and material loss. In heat hardened flats, less plastic deformation occurred. There is no evidence of material pile-up at the edge and material removal occurred primarily by abrasion. The increase in hardness is expected to reduce the amount of plastic deformation, which also reduces the extent of fatigue damage. Nonetheless, a considerable amount of material removal did occur.

In the FSP flat, minimal surface damage and material removal occurred. There were only a few scratches and minimal plastic deformation (Figure 8.21(b)). Indeed, SEM analysis shows the occurrence of only superficial damage and material removal. An original grinding mark running across the wear track can still be clearly seen in Figure 8.21(b).

Results of the study of the initial samples showed that FSP has an excellent potential to enhance the wear resistance of 4140 steel. Compared to the baseline material, FSP reduced wear by more than a factor 20 and a factor of 2 compared to traditional heat hardening treatment. The wear mechanism was also changed from abrasion and fatigue to superficial surface damage by FSP. This improvement in wear behaviour can be attributed to microstructural changes in the near-surface material as a result of FSP, consisting of phase transformation and grain refinement.

More friction and wear performance of FSP 4140 steel flats using different variables of rotation and travel speeds and tool designs in lubricated sliding contact with 52100 steel balls were then evaluated. The variations of the COF with test time for various FSP plates are similar to each other. The average COF at different normal loads of 25, 75 and 150 N is shown in Figure 8.22. The average COFs for the different variants of FSP treatment are similar to each other and the conventional heat treatment. There are four variants of FSP with slightly higher COFs, namely FSP 1-14, 1-16, 2-7 and 2-9.

It should be noted that even the higher friction in these four groups is still within the range of a typical COF under a boundary lubrication regime. The wear in various flat specimens at the conclusion of the test is shown in Figure 8.23. The amount of wear in the FSP treated flat is substantially reduced compared to the untreated baseline material. When compared to the HT samples, the amount of wear in the FSP flats is either similar or lower, except for FSP 2-7 and 2-9, in which the wear is slightly higher. This indicates that there may be an effect of FSP parameters on resulting friction and wear properties. This also implies that there may be opportunities to optimize resulting friction properties via modification of FSP parameters and tool designs.

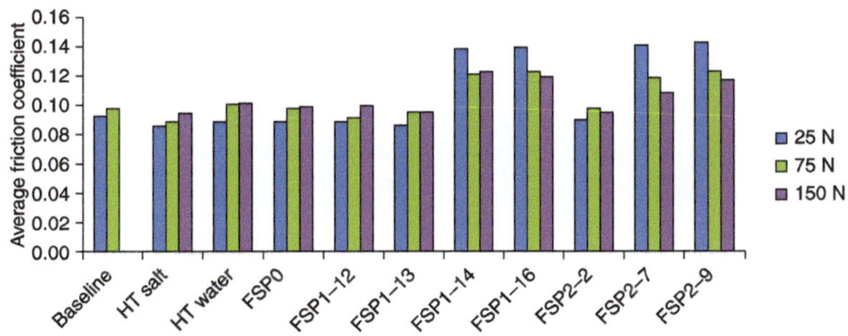

Figure 8.22 Average friction coefficient for lubricated reciprocating 52100 balls and FSP 4140 steel [30]

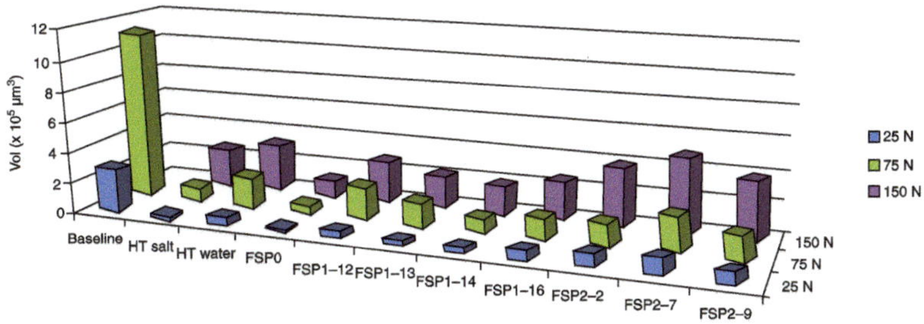

Figure 8.23 Wear volume after lubricated reciprocating sliding wear test [30]

Figure 8.24 shows the variation of the COF for the duration of the test with the various flat specimens. For the baseline material, the COF increased gradually from an initial value of 0.10 to a near steady value of about 0.16 (Figure 8.24(a)). Samples that were HT and water quench showed similar trends in behaviour, but smaller magnitude with test starting with 0.08 and increased gradually to about 0.13. For the initial FSP trial treatment, the COF ranged from 0.11 to about 0.12 for the duration of the test and was much less noisy compared to baseline and heat hardened specimens.

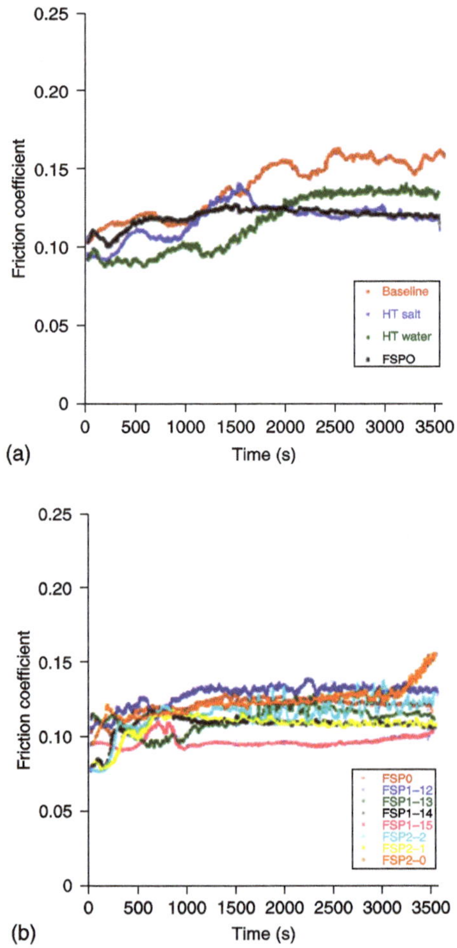

Figure 8.24 Variation of friction with time during unidirectional sliding: (a) comparison of FSP with other treatment; and (b) comparison between FSP variants [30]

The friction variation for all the different variants of the FSP treatments evaluated thus far is shown in Figure 8.24(b). The friction trends are similar with COF values ranging from 0.09 for FSP1-16 to 0.13 for FSP 1-12, values which are typical for boundary lubrication regime. These results suggest optimization of FSP processing for a desired frictional behaviour or values may be possible.

Wear in each flat specimen at the conclusion of the test is shown in Figure 8.25. All the surface hardening processes substantially reduced the wear in the 4140 steel flat. The wear in the various variants of FSP treatment is comparable to the wear in the traditionally surface hardening treatment. Friction and wear performance of FSP 4140 steel flats with different heat treatment, including FSP in dry sliding contact with 52100 steel balls, were evaluated. Tests were conducted with baseline, HT (water and salt quenched), and FSP1-14 and 2-2 FSP trials.

Tests were conducted with normal load of about 5 N, reciprocating speed of 1 Hz and a stroke length of 1 cm (linear speed 2 cm/s) for a duration of 30 minutes. The variations of COF with test time for dry sliding tests for various flats are shown in Figure 8.26. The friction of hardened flat by either heat treatment of FSP is higher than the unhardened baseline material.

The two variants of FSP tested thus far showed similar friction behaviour with a COF higher than HT surfaces. Friction and wear tests were also conducted in unidirectional sliding contact under dry condition using the pin-on-disk tribometer. A ball-on-flat contact geometry was used. Ball specimens are made of polished 12.7 mm (0.5') diameter hardened 52100 bearing steel with a surface roughness of 0.035 μm Ra. Rectangular flat specimens with 50 × 30 × 0.25 mm nominal dimensions were made from 4140 steel under four different conditions, the annealed baseline, HT water quenched, HT salt quenched and different variants of FSP. The surfaces of all the flat specimens were ground to the same surface finish of 0.14 μm Ra.

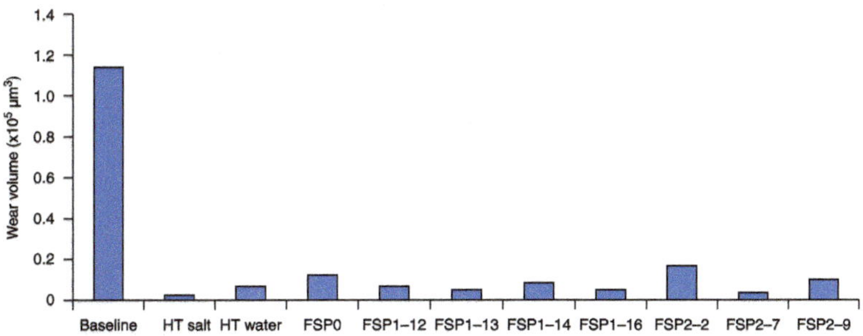

Figure 8.25 Wear in 4140 steel flats with various treatments [30]

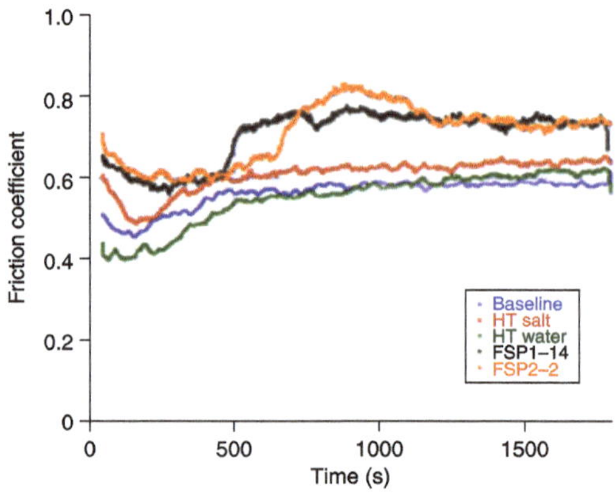

Figure 8.26 Variation of friction with time during dry reciprocating sliding [30]

After the initial run-in period of about 600 seconds, the COF for all the tested pairs became nearly constant for the remaining duration of the test. The magnitude of COF ranged from about 0.45 to about 0.85. A summary of the average COF after the run-in for the various flat specimens is shown in Figure 8.27. A variant of the FSP (1-12) has the lowest average friction of all the tested material, while another variant (1-14) has the highest friction.

Although details of the reasons and mechanisms for the vastly different behaviour in FSP treatment materials is not yet fully understood, the observations suggest that FSP can be optimized for different frictional applications, especially under dry sliding contacts.

The wear in various flats at the conclusion of the test is shown in Figure 8.28. All the surface hardening processes (heat treat and FSP)

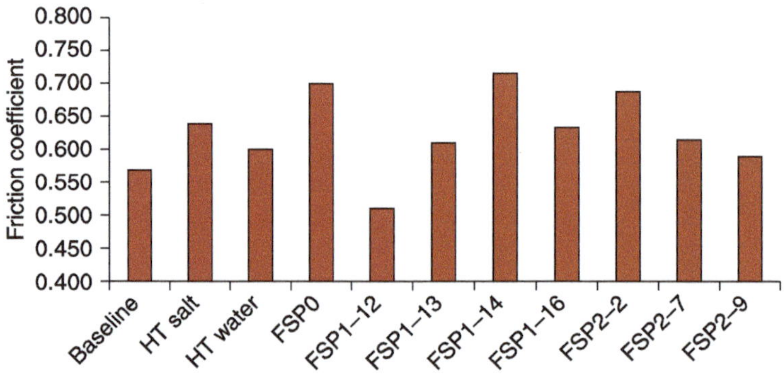

Figure 8.27 Average friction coefficient during dry unidirectional sliding of different flats [30]

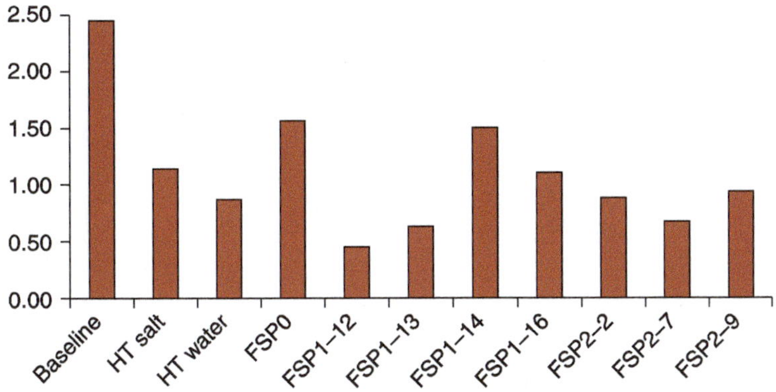

Figure 8.28 Wear volume in flat specimens after dry unidirectional sliding test [30]

reduced wear compared to the baseline. However, significant differences can be seen in the wear of different variants of FSP.

This observation again suggests that just as with friction, there is an opportunity to optimize FSP processing for wear control. Indeed, more work is needed to develop adequate understanding and the role of FSP processing parameter on friction and wear behaviour, especially under dry sliding contact.

8.5.6 Friction stir welding and processing of other materials, dissimilar alloys and metals

FSW is generally identified as a new welding technology that can be used to weld dissimilar alloys and metals. A few studies have been undertaken on FSW dissimilar aluminium alloys, copper alloys or aluminium alloys to other metals [71–76]. For example, FSW of 5083Al to 6082Al and 6061Al to Cu showed that the low-strength material should be placed on the advancing side (AS) to produce better welds, whereas in [77] it was demonstrated that superior welds of 2024Al/D357 were obtained when high-strength 2024Al was placed on the AS of the weld. Although sound FSWs of 2024Al/D357 were produced, very little interpenetration between the AS 2024Al and retreating side (RS) D357, which is of a 'cold weld' [77], was observed for the tool design and FSW parameters used in this study. In this case, FSW of 2024Al/D357 exhibited reduced strength and poor ductility [77].

Two recent studies on FSW of A356/6061Al and 2024Al/7075Al showed promise for joining dissimilar aluminium alloys via FSW [78,79].

FSW of dissimilar A356 and 6061Al alloys were conducted at a tool rotation rate of 1600 rpm, traverse speeds of 87 to 267 mm/min, and a 38 tool tilt angle. Defect-free FSWs of A356/6061Al were demonstrated. Microstructural examinations and property evaluations revealed following important observations.

The microstructure of weld nugget was mainly governed by RS materials. When A356 was at the RS, the Si particles were dispersed over the weld centre, whereas in the case of 6061Al at the RS, the microstructure of 6061Al in the weld centre showed fine and equiaxed recrystallized grains.

Hardness of the NZ was lower than that of 6061Al BM due to the dissolution or coarsening of precipitates and it was higher than that of A356 BM because of dispersion of Si particles.

Although some investigations [81,84] showed that FSW of dissimilar metals, such as aluminium to copper, did not result in generation of sound welds, some recent attempts have demonstrated a success in joining dissimilar metals using FSW, such as aluminium to steel and aluminium to magnesium [80]. For example, in an investigation [81,82] the feasibility of joining 6061Al to mild steel via FSW were studied and similarly in [81] it was shown that an offset of 68% of the pin diameter into the aluminium side led to a better weld quality. In such FSW aluminium/steel joints, steel fragments and intermetallics between iron and aluminium were identified (Figure 8.29).

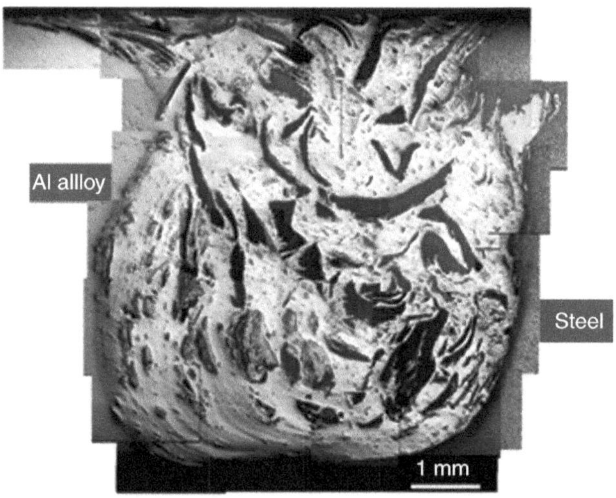

Figure 8.29 A magnified view of the nugget zone [82]

It can be seen that many relatively coarse, dark pieces in the form of a short ribbon are spread through the nugget. EDAX demonstrates that these sheared-off pieces are mild steel. The steel pieces are well distributed within the nugget, except in the lower portion near the base Al alloy. Besides the steel pieces, a number of grey blocky particles are visible, the size of which varied.

Figure 8.30 shows a typical reaction at the surface of the steel pieces and precipitates in the matrix of the Al alloy. It can be seen that the precipitates assumed various morphologies, i.e. sphere-like, needle, flower petal and blocky shapes; their size also varied. Those near the Al alloy side are generally sphere-like and smaller, while those close to the steel side are in the form of a needle and a flower petal, and are larger.

At the boundary zone between the nugget and base steel, the intermetallic compound layer thickness is less than 5 mm. Therefore, it is believed that it would not worsen the mechanical properties of the weld.

This microstructural feature is attributed to strong mechanical stirring and rapid cooling, as well as a relatively low weld temperature. Evidently, it is substantially different from the fusion welding of Al alloys and steels.

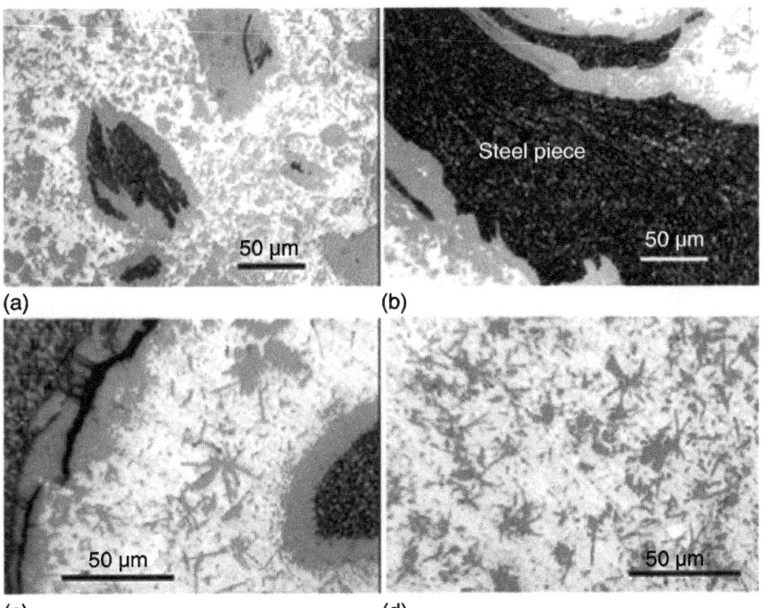

Figure 8.30 Micrographs showing a strong reaction between steel pieces and an Al alloy matrix in the nugget [82]

It is expected that the presence of discrete, well-distributed Al–Fe intermetallic compounds in the nugget, may not be detrimental to the mechanical properties of the weld. The nugget can be taken as an aluminium matrix composite reinforced by steel and Al–Fe intermetallic compounds.

8.5.7 Metal matrix composites

Metal matrix composites offer increased stiffness, strength and wear resistance over monolithic matrix materials. However, the weld ability of these composites is significantly reduced, due to the addition of ceramic reinforcements. An enormous diversity of materials is used for surface reinforcements, the majority being hard ceramic particles as SiC, Al_2O_3 and AlN, to improve surface properties as hardness, superplasticity, formability, corrosion and wear resistances [81–87].

The wear of the tool during FSW occurs at high temperature. Therefore, tools made from alloys with high-temperature wear resistance would reduce damage to the tool. Furthermore, the design of tool geometry is also important to reduce tool wear. Active heating of a composite workpiece before welding may also contribute to reducing tool wear, due to improved flow properties of composites at high temperature.

There are FSP used to incorporate nano-sized Al_2O_3 into AA6082 aluminium alloy to form particulate composite surface layer (Figure 8.31(a)) [83]. The Al_2O_3 particles have an average size of about 50 nm. Perfect bonding between the surface composite and the aluminium alloy substrate was achieved, as shown by the defect-free interface in Figure 8.31(b). Mechanical properties, including microhardness and wear resistance, were tested. The results show that the increase in number of FSP passes causes more uniform distribution of nano-sized alumina particles. The microhardness of the surface improves by three times as compared to that of the as-received Al alloy. A significant improvement in wear resistance in the nanocomposite surfaced Al was observed as compared to the as-received Al alloy. The wear rate is reduced to one-third of that of the as-received Al alloy when the FSP composite layer was incorporated, as shown by the curves in Figure 8.31(c).

Microscopically, the SEM images of the centres of the nugget cross-section (Figure 8.32) suggest that the particle distribution in the matrix was almost uniform under all processing conditions. These areas represent the most important area in the NZ; this is because surfacing is done by applying many adjacent parallel passes to cover the area of interest. The adjacent pass will overlap the advancing or RS of the previous one, which

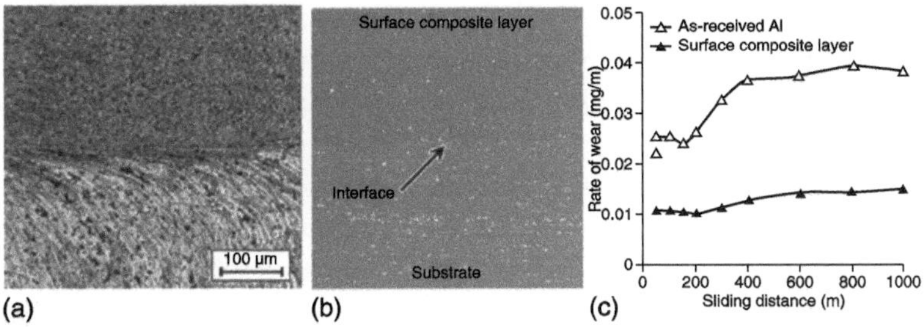

Figure 8.31 Microstructure and wear property of friction stir processing alumina particle reinforced A6082 aluminium alloy [83]: (a) optical micrograph showing the composite top layer and the bottom layer of matrix alloy; (b) SEM image showing the well bonded interface between the two layers; and (c) comparison of wear rate of the as-received alloy and the material with a surface nanocomposite layer produced by four FSP passes

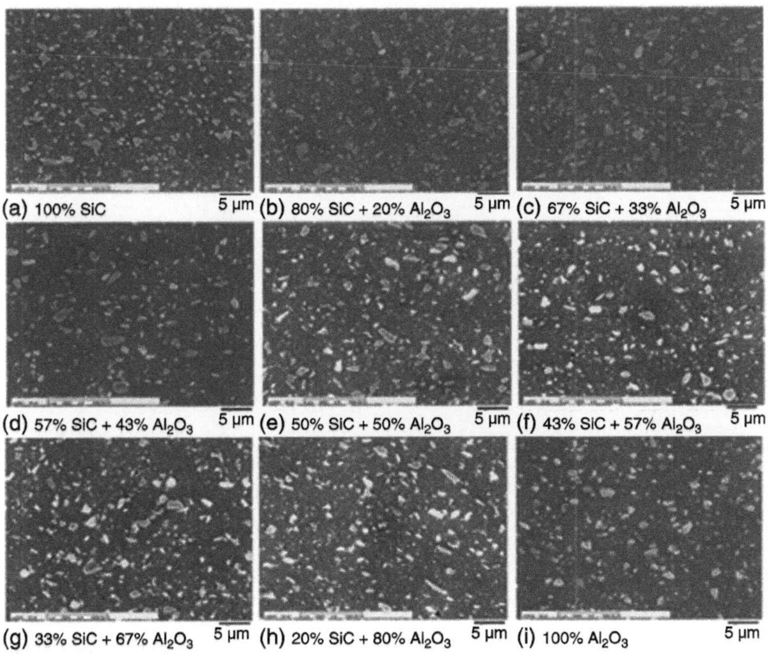

Figure 8.32 SEM micrographs of the nugget centre with different hybrid composites [84]

share in more particle distribution [84]. The central portion of the nugget tends to remain as it is without changes. However, the presence of small voids around some particles or at the particle–matrix interface was observed in the NZs of higher Al_2O_3 contents.

By EDS analyses, it was found that these microvoids were mainly located around the Al_2O_3 particles, while it was less noticeable around SiC particles (Figure 8.33). This may be related to the better wetability of the SiC with aluminium matrix than that of Al_2O_3 [85,88].

As shown in Figure 8.34, the average COF values exhibited a general tendency to decrease with increasing of the relative content of Al_2O_3 at normal loads of 2 and 5 N. It was reported that the Al_2O_3 powder had a lower COF 0.4 [87] than that of SiC 0.7 [88], when they were tested under the same conditions. Therefore, it may be expected that the composites that contained higher ratios of Al_2O_3 particles will have lower COF. Furthermore, it is clear that the average COF decreased with increase of the normal applied load from 2 to 5 N at all composite ratios. The decrease in the average COF with normal applied load was in agreement with many works [89–91].

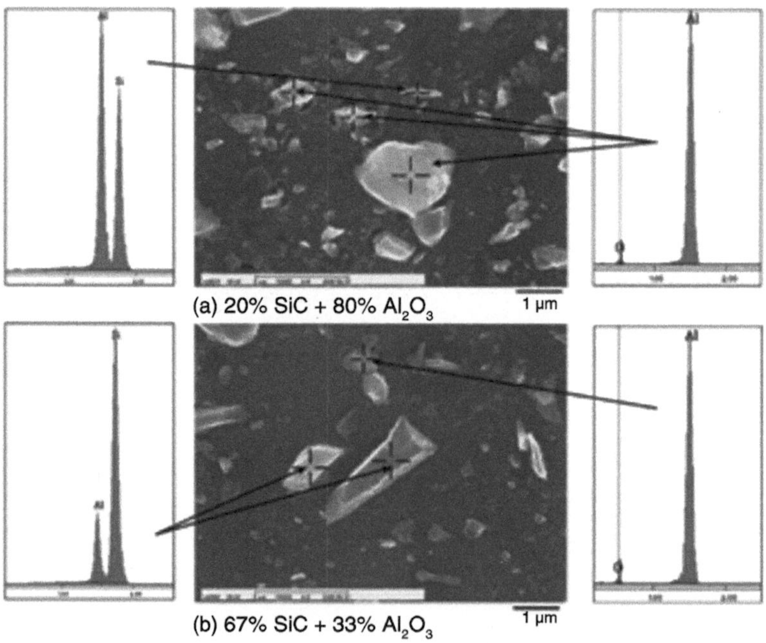

Figure 8.33 SEM images of particles of SiC and Al_2O_3 incorporated in the nugget zone and EDS spectra for some of them: (a) 80% Al_2O_3; and (b) 33% Al_2O_3 [84]

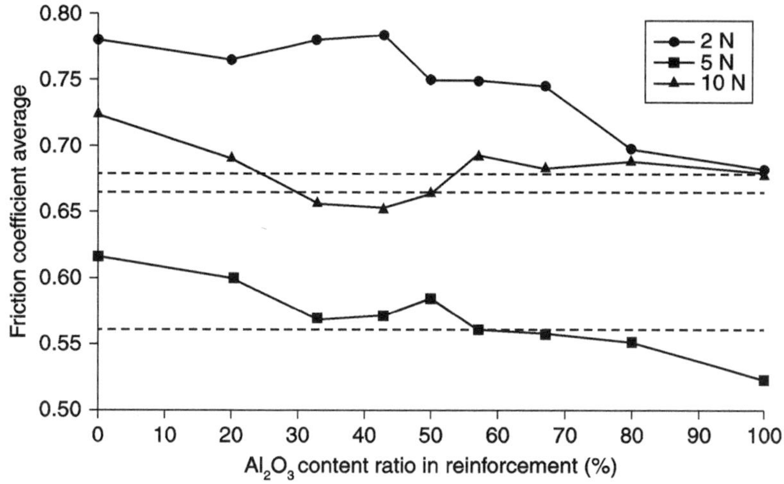

Figure 8.34 **Effect of hybrid ratio of reinforcements on the average friction coefficient at normal applied loads of 2, 5 and 10 N [84]**

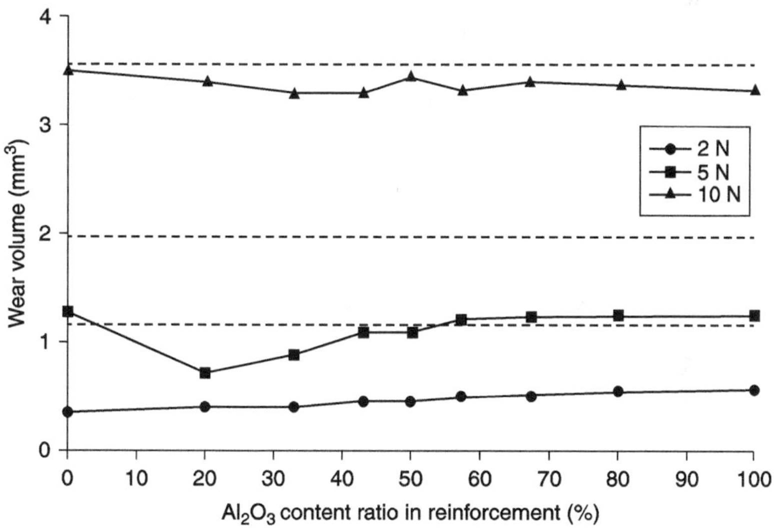

Figure 8.35 **Effect of hybrid ratio of reinforcements on wear volume losses at normal applied loads of 2, 5, 10 N [84]**

Figure 8.35 shows that the addition of reinforcement powder (SiC, Al_2O_3 or their mixture) to an aluminium matrix was beneficial in reducing the wear volume losses at normal loads of 2 and 5 N. The wear volume values, at normal loads of 2 N, increased with the Al_2O_3 particles relative ratio in the hybrid composites. However, the hybrid composites containing Al_2O_3 particles relative ratio of 20% were superior in wear resistance to other composites at a normal load of 5 N, suggesting that the hybrid composites exhibited better wear resistance than those reinforced with only one type of reinforcement.

The attempts of development empirical relationships to predict the sliding wear behaviour of FSW AA6061/0-10 wt.% ZrB_2 butt joints and analyze the influence of process parameters on wear rate and wear resistance of the welded joints were carried out [92]. The alloy AA6061 has been used as a matrix. Experiments were conducted according to central composite rotatable design. Specimens of size 6 mm × 6 mm × 50 mm were extracted from each welded plate. The dry sliding wear behaviour was measured using a pin-on-disk wear apparatus (DUCOM TR20-LE) at room temperature according to ASTM G99-04 standard. The polished surface of the pin was slid on a hardened chromium steel disk. The test was carried out at a sliding velocity of 1.5 m/s, normal force of 25 N and sliding distance of 2500 m. The wear parameters were selected to yield an appreciable steady-state wear based on trial experiments. A computer system was used to monitor the loss of height. The volumetric loss was computed by multiplying the cross-section of the test pin with its loss of height.

The wear rate (W) and wear resistance (R) were calculated [92] as:

- W (mm^3/m) = Volumetric loss/Sliding distance;
- R (m/mm^3) = 1/Wear rate.

Figure 8.36 shows the sliding wear behaviour of FSW AA6061-ZrB_2 *in-situ* composites as a function of weight percentage of ZrB_2 particles. The wear rate decreases with increase in ZrB_2 content, while the wear resistance follows an inverse trend as estimated. The joints fabricated at N = 1150 rpm, S = 50 mm/min and F = 6 kN show higher wear resistance. Good consolidation of material with optimum stirring and adequate heat generation may be taking place at this combination of parameters.

It is evident from Figure 8.36 that addition of *in-situ* formed ZrB_2 particles improved the wear resistance of AA6061. This can be attributed to the hardness imparted by the *in-situ* formed ZrB_2 particles due to the creation of strain fields around the particles during solidification. The increase in wear resistance of welded joints with increased ZrB_2 particles is due to increased presence of ZrB_2 particles in the weld zone. This leads to the conclusion that FSW resulted in homogeneous distribution of ZrB_2 particles, irrespective of their weight percentage.

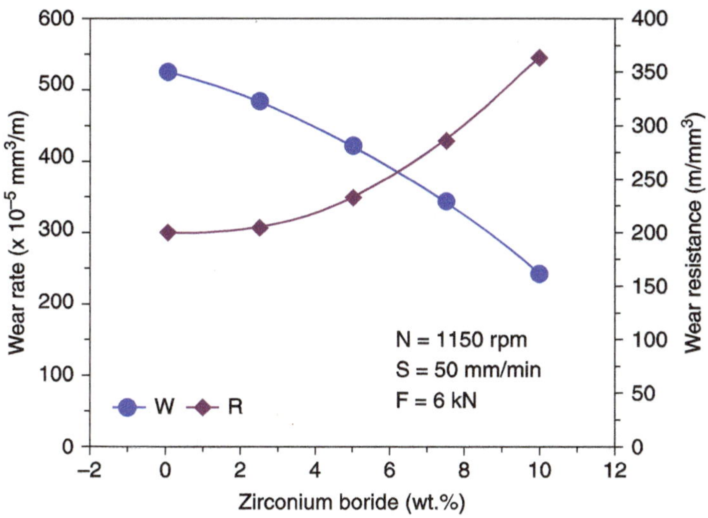

Figure 8.36 **Effect of weight percentage of ZrB_2 on sliding wear behaviour [92]**

The effect of FSW on the developed AMCs is indicated by the relative wear rate. The wear resistance of the composites improved subsequent to FSW, which is described as follows. FSW closes the presence of microporosities in the cast composite. The grain size of aluminium in the weld zone is reduced by dynamic recrystallization.

Figure 8.37 reveals the SEM micrographs of the wear surfaces of selected specimens with a different weight percentage of ZrB_2 particles. A change in wear mode is observed with an increase in weight percentage of ZrB_2 particles, which can be attributed to the homogeneous distribution of ZrB_2 particles in the weld zone as a result of FSW. The wear mode changes from adhesion (Figure 8.37(a–b)) to abrasive wear (Figure 8.37(c–e)). Welded matrix alloy and composite containing 2.5 wt.% of ZrB_2 particles exhibit adhesion wear mode. The frictional heat increases the wear surface temperature, which causes plastic deformation and dislocation in the inner surface of the composites.

The congestion of dislocation results in stress concentration and initiation of cracks. Welded composites containing weight percentage of ZrB_2 particles above 2.5 exhibits abrasive wear mode. The abrasive wear is the result of ZrB_2 particles on the wear surface and the abrasive dusts between two surfaces. ZrB_2 particles bear the load initially. As sliding wear proceeds, the frictional heat softens the surface layer. The difference in the thermal expansion coefficient between the matrix and the ZrB_2

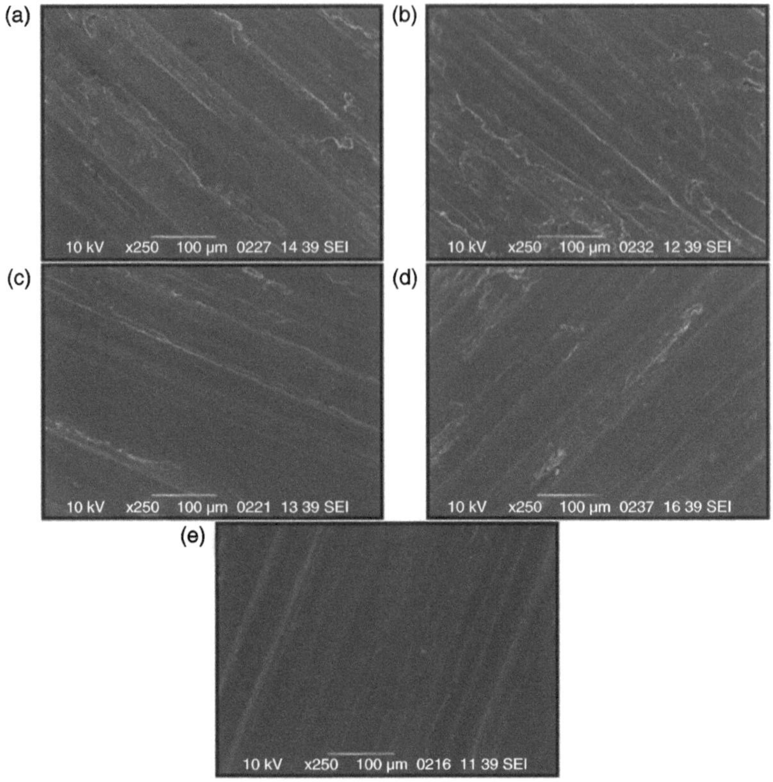

Figure 8.37 **SEM micrograph of wear surface of specimens with: (a) 0 wt.% ZrB_2; (b) 2.5 wt.% ZrB_2; (c) 5.0 wt.% ZrB_2; (d) 7.5 wt.% ZrB_2; and (e) 10 wt.% ZrB_2 [92]**

particles creates the interface stress. When the interface stress exceeds the bond strength, the particles are pulled off. The pulled-off particles begin to act as wear particles in the sliding wear course.

8.6 Tool wear rate in friction stir welding and processing

In FSW, the tool plays an important role. The tool material is determined based on the BM. For materials with a low softening temperature such as aluminium, tool steel can be used, whereas metals such as steel, which has a high softening temperature, tungsten carbide and polycrystalline

Table 8.2 **Summary of current friction stir welding tool materials [11]**

Alloy	Thickness		Tool material
	mm	in.	
Aluminum alloys	<12	<0.5	Tool steel, WC-Co
	<26	<1.02	MP159
Magnesium alloys	<6	<0.24	Tool steel, WC
Copper and copper alloys	<50	<2.0	Nickel alloys, PCBN[a] tungsten alloys
	<11	<0.4	Tool steel
Titanium alloys	<6	<0.24	Tungsten alloys
Stainless steels	<6	<0.24	PCBN, tungsten alloys
Low-alloy steel	<10	<0.4	WC, PCBN
Nickel alloys	<6	<0.24	PCBN

[a]PCBN, polycrystalline cubic boron nitride

cubic boron nitrate (PCBN) are used. The tool grade is selected based on basic requirements, such as the tool should be able to withstand high temperature, should have good wear resistant properties, and should have enough strength, etc. Table 8.2 [11] summarizes the most commonly used tool materials in FSW and FSP.

To apply FSW in steel or other high-temperature materials, the difficulty is mainly associated with finding the proper tool material; a material that can withstand the high temperatures that are experienced during the process. Resistance to wear (durability) is one important aspect, especially as many of the intended applications are considered critical, hence there can be no traces of the tool left in the seam.

PCBN is one of the most promising tool materials so far. Owing to its insolubility in iron, nickel and related alloys at high temperatures, PCBN abrasives are used for machining steel.

Boron nitride binds well with metals, due to formation of interlayers of metal borides or nitrides, making the manufacturing of tools easier and more reliable and economical.

Figure 8.39 shows that the tool for FSW from this material, after 260 feet of work in 1018 mild steel, has no visible wear [93], although the tool life for some materials depends on the FSW/FSP conditions (Table 8.3) [94]. However, the PCBN tools have the following limitations:

- high temperatures and pressures required in the manufacturing of PCBN;
- tool costs are very high;
- low fracture toughness;
- tendency to fail during the initial plunge stage;
- maximum weld depths are currently limited to 10 mm for welding of steels and Ti alloys;
- PCBN is brittle and boron from PCBN may get dissolved into base material (BM) to form an undesirable phase.

Materials such as aluminium or magnesium alloys, and aluminium matrix composites (AMCs) are commonly welded by using steel tools. Steel tools have also been used for the joining of dissimilar materials in both lap and butt configurations.

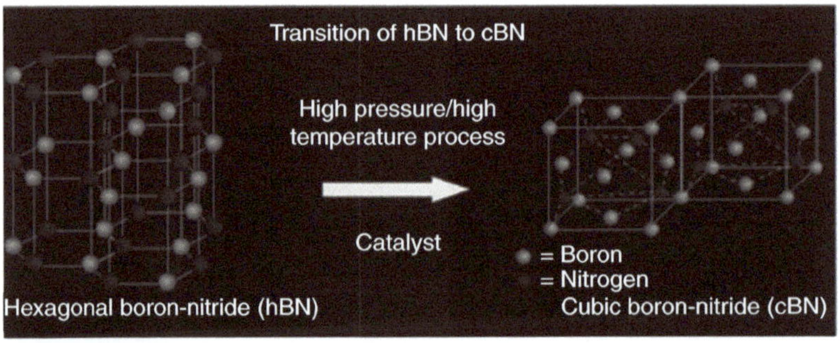

Figure 8.38 Schematic of receiving PCBN from hBN [93]

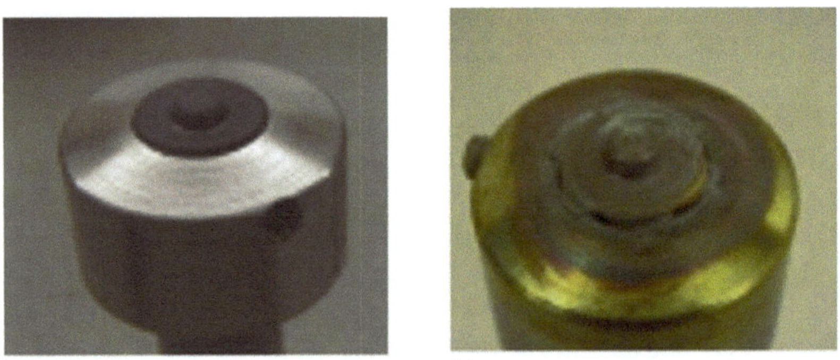

(a) (b)

Figure 8.39 Post-weld analysis of the tool (a) before; and (b) after 200 feet

Table 8.3 Summary of tool life for some materials to be welded [94]

Materials	Tool life, m
6000 series aluminum alloy	up to 1000 m
6061 aluminum alloy	no measurable tool wear
6061 AA+Al_2O_3 MMC	0.64%/cm wear rate (1000 rpm),
	0.42 (1500 rpm), 0.56 (2000 rpm)
Steels	It has been reported that tools were replaced after they were used to produce 1.5–2.0 m of weld of steel
	PCBN tool → 10 m in HSLA-65, 30 m in 304 Stainless steel, 80 m in A36 steel
	Tool replacement usually occurs in response to severe tool wear, and pin and shoulder fracture.

Tool wear during welding of metal matrix composites is greater when compared with welding of soft alloys, due to the presence of hard, abrasive phases in the composites. Total wear was found to increase with rotational speed and decrease at lower traverse speeds, which suggests that process parameters can be adjusted to increase tool life.

The rotation and translation of tool through the workpiece result in its wear. The FSW tool may also deform plastically, due to a reduction in yield strength at elevated temperatures in an environment of high loads. Therefore, FSW tools for welding of high strength materials such as steels are often liquid cooled [95]. When the stresses are higher than the load bearing ability of the tool, failure may occur. Not many detailed studies have been done on the tool wear in FSW, but diffusion and abrasion are the expected wear mechanisms. Reaction of the tool material with its environment, including both the workpiece and the surrounding gases, is also expected to contribute to tool wear. Depending on the results, a particular tool material may be a good choice for one workpiece material, but not so good for another of similar physical properties.

Excessive tool wear changes the tool shape, thereby increasing the probability of defect generation, and possibly degrading the weld quality. The exact wear mechanism depends on the interaction between the workpiece and the tool materials, the selected tool geometry and the welding parameters. For example, in the case of PCBN tools, the wear at low tool rotation rate is mainly caused by adhesive wear (also known as scoring, galling or seizing), while at the higher rotation rates is caused by

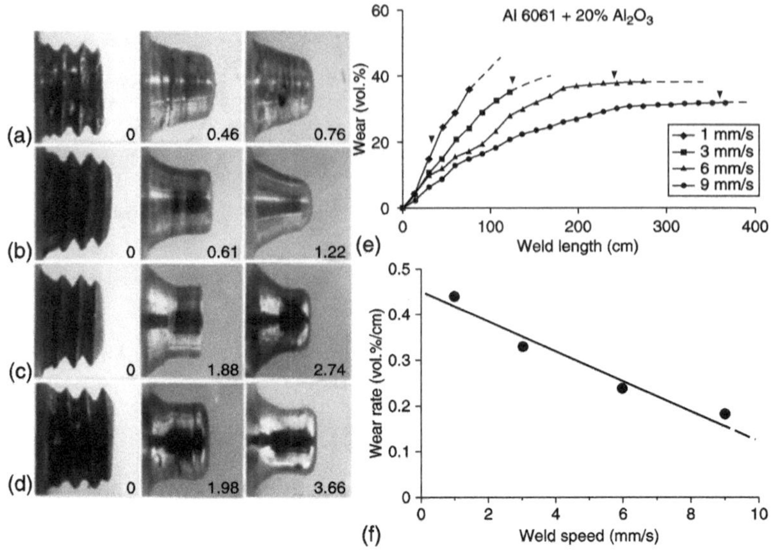

Figure 8.40 **Wear features of probes for Al-MMC at 1000 rev min^{-1} welding speeds at: (a) 1; (b) 3; (c) 6; and (d) 9 mm s^{-1}; (e) probe wear (vol.-%) vs. weld length; and (f) wear rate vs. weld speed [97]**

abrasive wear [11,96,97]. [98] reported on the tool wear for Al–20SiC and Al–20Al_2O_3 particle reinforced composites. The tool used consisted of an AISI oil hardened tool steel initially with screw nib right-hand threads.

Owing to the abrasive particles in the Al-MMCs, the threads of the probe wore away, leading to a slightly curved shaped probe (Figure 8.40). Astoundingly, the self-optimized shape (worn tool) with no threads could result in homogeneous and integral welds without further visible tool wear. These observations suggest that tool consumption can be greatly minimized, even for metal-matrix composites (MMCs), when using the optimized tool shape. Hence to reduce tool wear and extend the tool life, understanding and controlling the material flow associated with the probe profile in the solid state are important. It should be noted that the self-optimized shape does change somewhat with increasing welding speed due to increasing tool wear [99].

For FSSW, severe tool wear has been reported after producing hundreds of welds. For light metals such as Al and Mg alloys, the welding tool is commonly made of tool steel and suffers from little wear. It was reported that no significant wear was found on the steel tool, even after hundreds of thousands of spot welds for Al.

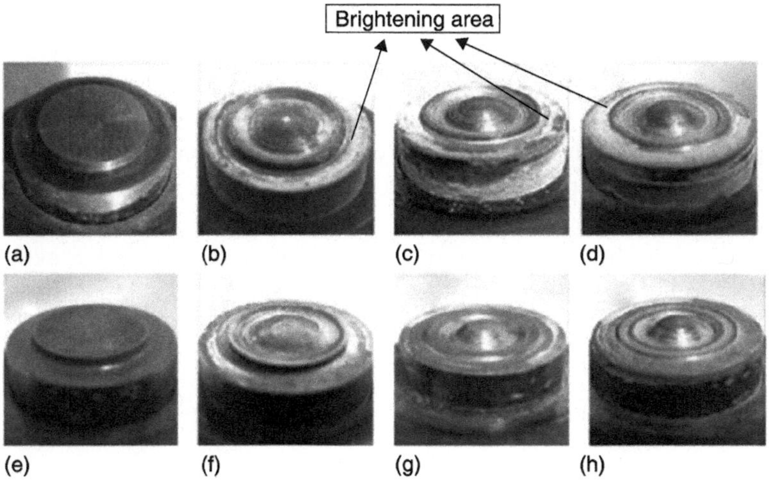

Figure 8.41 External shape of WC–Co tools: (a and e) before welding; (b and f) after 100; (c and g) after 300; and (d and h) after 500 spot welds [100]

However, steel tools are not suitable for high melting point materials, such as Ti, Ni, steels, etc. For these high strength materials, the welding tools are usually made of hard metals, carbides and metal matrix composites with superior thermal and wear resistance at temperatures higher than 1000 °C, such as WC–Co, TiC and PCBN [100]. Figure 8.41 shows the external WC–Co tool shape after some welds [101]. It clearly reveals that the extreme wear for WC–Co tools mainly occurred between the probe centre and the external edge.

Cost-effective and long-life tools are available for the FSW of aluminium and other soft alloys. They are needed but not currently available for the commercial application of FSW to high strength materials. Tool material properties, such as strength, fracture toughness, hardness, thermal conductivity and thermal expansion coefficient, affect the weld quality, tool wear and performance.

Reactivity of tool material with oxygen from the atmosphere and with the workpiece is also an important consideration. PCBN and W-based alloys are important candidate materials for the FSW of high strength materials. High strength, hardness and high temperature stability of PCBN allow much less wear compared with other tools. Low fracture toughness and high cost of PCBN are issues that require attention. W-based alloys, although not as hard and wear resistant, are more affordable options and have been used to weld steels and Ti alloys on a limited scale. There is also an interest in Si_3N_4 as a prospective tool material, because it has produced welds comparable with PCBN tools at a much lower cost. Further developments in FSW tool materials are required to address the problem of high tool cost with low tool life during welding of harder alloys.

Figure 8.42 shows the appearance of the threaded tool after each FSW experiment [102]. It can be seen from the figure that the threads near the pin root almost disappear after the first welding (Figure 8.42(b)). The maximum diameter, i.e. minimum wear, occurs in the upper half of the pin, while the minimum pin diameter, i.e. maximum wear, exists in the lower half near the shoulder (Figures 8.42(b–h)). After the sixth

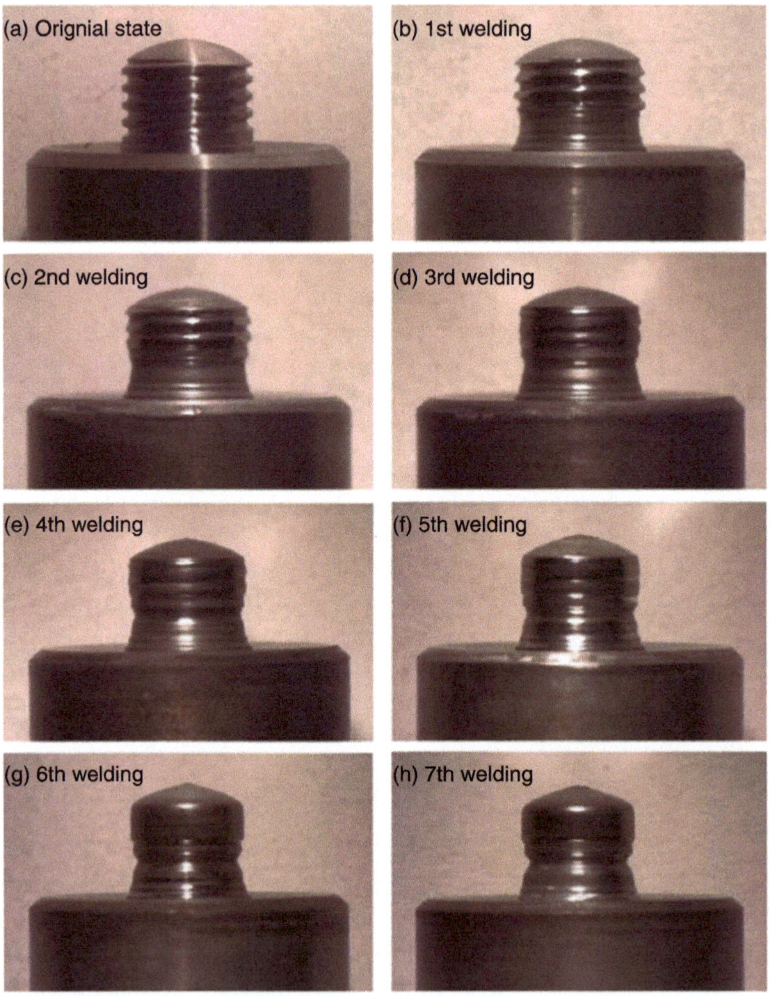

Figure 8.42 Appearance of the threaded tool after each FSW experiment [101]

welding is performed, all the original threads of the pin have been completely worn away (Figures 8.42(g and h)).These results indicate that the tool is clearly consumed or its shape and size are significantly changed. In order to quantitatively assess the tool wear, the percent variation in tool size is used as an evaluation index, i.e. variation = (original size – measured size)/(original size) × 100%.

Different from the variation in pin length, the variations in pin diameter are very remarkable (Figure 8.43). The pin diameter significantly decreases with increasing travel distance, especially when the welding speed is relatively low. For example, after an initial welding is performed at a welding speed of 25 mm/min, the pin diameter has at most decreased by 11%. After the seventh welding is performed, 27% of the pin diameter at the maximum-wear location has disappeared (Figure 8.43(a)). This indicates that the radial wear of the pin is appreciable and is related to the welding process parameters. It should be pointed out that the variations in pin diameter are not the same for different locations of the pin, although they all increase with increasing travel distance. With respect to the three typical locations of 0.6, 1.5 and 3.2 mm distant from the pin root, the moderate, high and low variations in pin diameter take place, respectively (Figure 8.43(b)). This indicates that the radial wear of the pin is uneven at different locations on its surface.

The maximum and minimum wear of the pin are produced at apparently different locations of the pin and such locations change with the travel distance (Figure 8.44(a)). With increasing travel distance, the maximum-wear location gradually reaches a distance of about 1.5 mm from the pin root, while the minimum-wear location steeply approaches a definite distance of about 3.2 mm. This indicates that the maximum wear is finally produced at a location of about one-third pin length from the pin root.

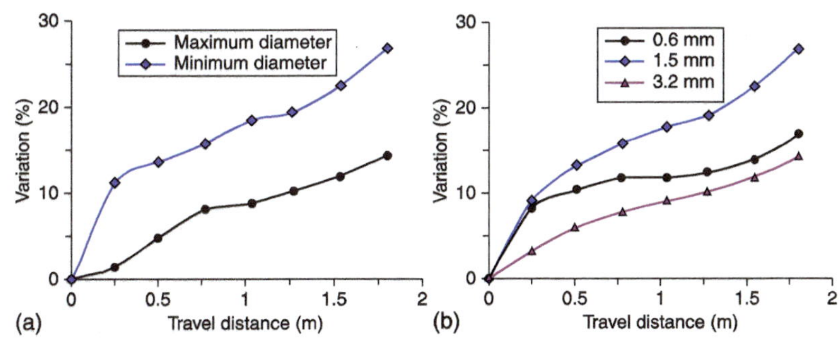

Figure 8.43 Variations in pin diameter: (a) the extreme-wear locations; and (b) typical locations [101]

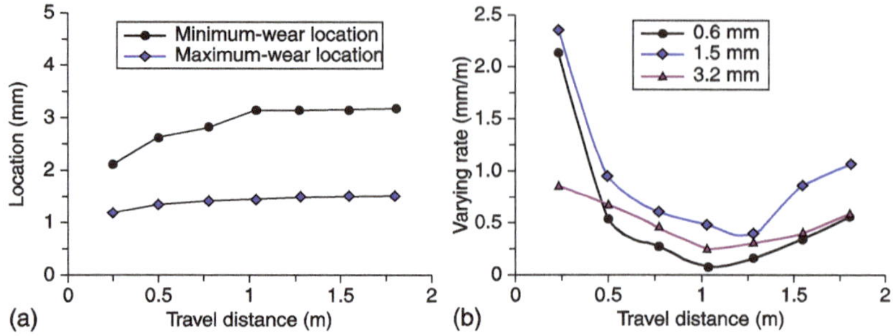

Figure 8.44 **(a) Extreme-wear location of the pin; and (b) varying rate of the pin diameter [101]**

Radial wear rate of the pin is defined as the varying quantity of the pin diameter per unit travel distance of the pin, i.e. wear rate = varying quantity/travel distance. As shown in Figure 8.44(b), the radial wear rate of the pin is changed with the travel distance. When the travel distance is smaller than 1.27 m, the wear rate decreases with an increase in travel distance. However, when the travel distance is greater than 1.27 m, the wear rate tends to change in the opposite direction.

The wear rates are different at different locations of the pin. With regard to the three typical locations of 0.6, 1.5 and 3.2 mm distant from the pin root, the low, high and moderate wear rates are produced, respectively, but this trend is not included in the initial welding. The wear rate is related to the welding sequence. The maximum wear rate is always produced in the initial welding, regardless of any locations of the pin. In fact, the welding speed has a decisive effect on the wear rate. The lower the welding speed, the higher the wear rate.

8.7 References

1. Fernandez, G.J. and Murr, M.L.E. (2004), 'Characterization of tool wear and weld optimization in the friction stir welding of cast aluminium 359 + 20% SiC metal-matrix composite', *Mater. Charact.*, 52: 65–75.
2. Mishra, R.S., Ma, Z.Y. and Charit, I. (2003), 'Friction stir processing: A novel technique for fabrication of surface composite', *Mater. Sci. Eng. A*, 341: 307–10.
3. Ni, D.R., Wang, D., Feng, A.H., Yao, G. and Ma, Z.Y. (2009), 'Enhancing the high-cycle fatigue strength of Mg-9Al-1Zn casting by friction stir processing', *Scr. Mater.*, 61: 568–71.
4. McNelley, T.R., Swaminathan, S. and Su, J.Q. (2008), 'Recrystallization mechanisms during friction stir welding/processing of aluminium alloys', *Scr. Mater.*, 58: 349–54.

5. Ma, Z.Y., Sharma, S.R. and Mishra, R.S. (2006), 'Effect of multiple-pass friction stir processing on microstructure and tensile properties of a cast aluminium-silicon alloy', *Scr. Mater.*, 54: 1623–6.
6. Thomas, W.M., Nicholas, E.D., Needham, J.C., Murch, M.G., Templesmith, P., and Dawes, C.J. (1991), GB Patent 9125978.8, December.
7. ASM Handbook (1994), *Friction, Lubrication and Wear Technology*, vol. 18, Materials Park, Ohio: ASM International.
8. Mishra, R.S. and Ma, Z.Y. (2005), 'Friction stir welding and processing', *Materials Science and Engineering R: Reports*, 50(1–2): 1–78.
9. Thomas, W.M., Johnson, K.I. and Wiesner, C.S. (2003), 'Friction stir welding-recent developments in tool and process technologies', *Advanced Engineering Materials*, 5(7): 485–90.
10. Threadgill, P.L. (1999), 'Friction stir welding-state of the art', *TWI Report*, 678.
11. Mishra, R.S. and Mahoney, M.W. (2007), 'Friction stir welding and processing', *ASM International*, 1–333.
12. Basavakumar, K.G., Mukunda, P.G. and Chakraborty, M. (2007), 'Influence of grain refinement and modification on dry sliding wear behaviour of Al-7Si and Al-7Si-2.5Cu cast alloys', *Journal of Materials Processing Technology*, 186(1–3): 236–45.
13. Basavakumar, K.G., Mukunda, P.G. and Chakraborty, M. (2008), 'Influence of grain refinement and modification on microstructure and mechanical properties of Al-7Si and Al-7Si-2.5Cu cast alloys', *Materials Characterization*, 59(3): 283–9.
14. Prasada, A.K., Das, K., Murty, B.S. and Chakraborty, M. (2004), 'Effect of grain refinement on wear properties of Al and Al-7Si alloy', *Wear*, 257(1–2): 148–53.
15. Prasada, A.K., Das, K., Murty, B.S. and Chakrabort, M. (2008), 'Microstructure and the wear mechanism of grain-refined aluminium during dry sliding against steel disk', *Wear*, 264(7–8): 638–47.
16. Chandrashekharaiah, T.M. and Kori, S.A. (2009), 'Effect of grain refinement and modification on the dry sliding wear behaviour of eutectic Al-Si alloys', *Tribology International*, 42(1): 59–65.
17. Collier, M., Steel, R., Nelson, T., Sorensen, C. and Packer, S. (2003), 'Grade development of polycrystalline cubic boron nitride for friction stir processing of ferrous alloys', *Proceedings of the Fourth International Conference on Friction stir Welding*, 14–16 May.
18. Soleymani, S.I., Abdollah-Zadeh, A. and Alidokht, S.F. (2011), 'Improvement in tribological properties of surface layer of an al alloy by friction stir processing', *JSEMAT*, 1: 9514–16100.
19. Simar, A. (2006), 'A multi-scale multi-physics investigation of aluminium friction stir welds', PhD Thesis, Universite Catholique de Louvain, 1–257.
20. Frigaard, O. (1999), 'A process model for friction stir welding of age hardening aluminium alloys', PhD Thesis, The Norwegian University of Science and Technology (NTNU).
21. Zahedul, M., Khandkar, H. and Khan, J.A. (2001), 'Thermal modeling of overlap friction stir welding for Al-alloys', *J Mater. Process. Manuf. Sci.*, 10: 91–105.

22. Mishra, R.S., Mahoney, M.W., McFadden, S.X., Mara, N.A. and Mukherjee, A.K. (2000), 'High strain rate superplasticity in a friction stir processed 7075 al alloy', *Scr. Mater.*, 42: 163–8.
23. Mishra, R.S., Ma, Z.Y. and Charit, T. (2003), 'Friction stir processing: A novel technique for fabrication of surface composite', *Mater. Sci. Eng. A*, 341: 307–10.
24. Morisada, Y., Fujii, H., Nagaoka, T. and Fukusumi, M. (2006), 'Effect of friction stir processing with SiC particles on microstructure and hardness of AZ31', *Mater. Sci. Eng. A*, 433: 50–4.
25. Johannes, L.B., Yoweii, L.L., Sosa, E., Arepaiii, S. and Mishra, R.S. (2006), 'Survivability of single-walled carbon nanotubes during friction stir processing', *Nanotechnology*, 17: 3081–4.
26. Asadi, P., Besharati-Givi, M.K., Abrinia, K., Taherishargh, M. and Salekrostam, R. (2011), 'Effects of SiC particle size and process parameters on the microstructure and hardness of AZ91/SiC composite layer fabricated by FSP', *JMEPEG*, 20: 1554–62.
27. Lee, C.J., Huang, J.C. and Hsieh, P.J. (2006), 'Mg based nano-composites fabricated by friction stir processing', *Scr. Mater.*, 54: 1415–20.
28. Mahmoud, E.R.T., Tkeuchi, K. and Takahashi, M. (2008), 'Fabrication of SiC particle reinforced composite on aluminium surface by friction stir processing', *Sci. Technol. Weld. Join.*, 13: 607–18.
29. Mahmoud, E.R.T., Tkeuchi, K., Shibayanagi, T. and Takahashi, M. (2009), 'Effect of friction stir processing tool probe on fabrication of SiC particle reinforced composite on aluminium surface', *Sci. Technol. Weld. Join.*, 14: 713–25.
30. Smith, C.B., Friction stir processing for Efficient Manufacturing. Final Report DE-EE0003465, 2011. available from: *http://www.osti.gov/bridge*
31. Santella, M.L., Engstrom, T., Storjohann, D. and Pan, T.Y. (2005), 'Effects of friction stir processing on mechanical properties of the cast aluminium alloys A319 and A356', *Scripta Mater.*, 53(2): 201–6.
32. Stachowiak, G.W. (2005), *Wear: Materials, Mechanisms and Practice*, UK: John Wiley & Sons Ltd.
33. Singh, J., Lal, H. and Bala, N. (3013), 'Investigation on the wear behaviour of friction stir processed magnesium-based AZ91 alloy', *Int. J. Mech. Eng. & Rob. Res.* 2(1): 271–4.
34. Lee, W.B., Lee, C.Y., Kim, M.K. Yoon, J.Y., Kim, Y.K. et al. (2006), 'Microstructures and wear property of friction stir welded AZ91 Mg/SiC particle reinforced composite', *Compos. Sci. Tech.*, 66: 1513–20.
35. Ni, D.R., Wang, D., Feng, A.H., Yao, G. and Ma, Z.Y. (2009), 'Enhancing the high-cycle fatigue strength of Mg-9Al-1Zn casting by friction stir processing', *Scripta. Mater.*, 61: 568–71.
36. Santella, M.L., Feng, Z., Degen, C. and Pan, T.Y. (2005), 'Friction stir processing to improve the surface properties of aluminium and magnesium castings', *Materials Science & Technology 2005 Conference*, Pittsburgh, PA: ASM International.
37. ASM Special Handbook (1999). *Magnesium and Magnesium Alloys*, Pittsburgh, PA: ASM International, 106–18.

38. Ma, Z.Y., Sharma, S.R. and Mishra, R.S. (2006), 'Effect of friction stir processing on the microstructure of cast A356 aluminum', *Materials Science and Engineering*, 433: 269–78.
39. Sharma, S.R. and Mishra, R.S. (2008), 'Fatigue crack growth behavior of friction stir processed aluminum alloy', *Scripta. Mater.*, 59: 395–98.
40. Chang, C.I., Lee, C.J. and Huang, J.C. (2004), 'Relationship between grain size and Zener-Holloman parameter during friction stir processing in AZ31 Mg alloys', *Scripta. Mater.*, 51: 509–14.
41. Arora, H.S., Singh, H.S. and Dhindaw, B.K. (2011), 'Observations on microstructural changes in a Mg-based AE42 alloy subjected to friction stir processing', *Metall. Mater. Trans.*, B43: 92.
42. Zhang, X.N., Geng, L. and Wang, G.S. (2006), 'Fabrication of Al-based hybrid composites reinforced with SiC whiskers and SiC nanoparticles by squeeze casting', *J. Mater. Process Technol.*, 176: 146–51.
43. Arora, H.S., Singh, H.S. and Dhindaw, B.K. (2013), 'Wear behaviour of a Mg alloy subjected to friction stir processing', *Wear*, 303: 65–76.
44. Qu, J., Blau, P.J., Watkins, T.R., Kavin, O.B. and Kulkami, N.S. (2005), 'Friction and wear of titanium alloys sliding against metal, polymer and ceramic counter-faces', *Wear*, 258: 1348–56.
45. Budinski, K.G. (1991), 'Tribological properties of titanium alloys', *Wear*, 151: 203–17.
46. Molinari, A., Straffelini, T.B. and Bacci, T. (1997), 'Dry sliding wear mechanisms of the Ti6Al4V alloy', *Wear*, 208: 105–12.
47. Dong, H. and Bell, T. (1999), 'Tribological behaviour of alumina sliding against Ti6Al4V in unlubricated contact', *Wear*, 225–9: 874–84.
48. Evans, A.G. and Marshall, D.B. (1980), 'Fundamentals of friction and wear of materials', *ASM*, 439.
49. Rice, S.L., Nowotny, H. and Wayne, S.F. (1981), 'Characteristics of metallic subsurface zones in sliding and impact wear', *ASME Proc. Wear Mater.*, 47–52.
50. Schmidt, R.F. and Schmidt, D.G. (1993), 'Selection and application of copper alloy castings', *ASM handbook II*, 3446–557.
51. Davim, J.P. (2000), 'An experimental study of the tribological behaviour of the brass/steel pair', *Journal of Materials Processing Technology*, 100: 273–7.
52. Cederqvist, L. (2004), 'A weld that lasts for 100,000 years: Friction stir welding of copper canisters', 807: 477–82.
53. Lee, W.B. and Jung, S.B. (2004), 'The joint properties of copper by friction stir welding', *Materials Letters*, 58: 1041–6.
54. Prasad, B.K. (1997), 'Dry sliding wear response of some bearing alloys as influenced by the nature of micro constituents and sliding conditions', *Metallurgical and Materials Transactions A*, 28: 809–15.
55. Bhushan, B. (1996), *Tribology and Mechanics of Magnetic Storage Devices*, 2nd edn, New York: Springer-Verlag.
56. Blau, P.J. (1992), 'Scale effects in sliding friction: an experimental study', in: Singer, I.L. and Pollock, H.M. (eds), *Fundamentals of Friction: Macroscopic and Microscopic Processes*, vol. E220, Dordrecht, Netherlands: Kluwer Academic, 523–34.
57. Bhushan, B. (1999a), *Handbook of Micro/Nanotribology*, 2nd edn, Boca Raton, FL: CRC Press.

58. Bhushan, B. (1999b), *Principle and Applications of Tribology*, New York: John Wiley & Sons, Inc.
59. Nuruzzaman, D.M. and Chowdhury, M.A. (2013), 'Friction coefficient and wear rate of copper and aluminium sliding against mild steel', *Int. Trans. J. Eng. Manag. & App. Sci. & Tech.*, 4(1): 29523–34, 40.
60. Andersson, C.G. and Andrews, R.E. (1999), *Proceedings of the First International Symposium on Friction stir Welding*, Thousand Oaks, CA, June.
61. Andersson, C.G., Andrews, R.E., Dance, B.G.I., Russell, M.J., Olden, E.J. and Sanderson, R.M. (2000), *Proceedings of the Second Symposium on Friction stir Welding*, Gothenburg, Sweden, June.
62. Hautala, T. and Tianien, T. (2003), in: David, S.A., DebRoy, T., Lippold, J.C., Smartt, H.B. and Vitek, J.M. (eds), *Proceedings of the Sixth International Conference on Trends in Welding Research*, Pine Mountain, GA: ASM International, 324.
63. Lee, W.B. and Jung, S.B. (2004), 'The joint properties of copper by friction stir welding', *Mater. Lett.*, 58: 1041–6.
64. Occupational Safety and Health Administration (2001), 'Occupational exposure to hexavalent chromium', *OSHA RIN*: 1218-AB45.
65. Thomas, W.M., Threadgill, P.L. and Nicholas, E.D. (1999), 'Feasibility of friction stir welding steel', *Sci. Tech. Weld. Joining*, 4: 365.
66. Lienert, T.J., Stellwag, W.L, Grimmett, B.B. and Warke, R.M. (2003), 'Friction stir welding studies on mild steel', *Weld. J.*, 82(1): 1s–9s..
67. Posada, M., Deloach, J., Reynolds, A.P. et al. (2001), in: Jata, K.V., Mahoney, M.W., Mishra, R.S., Semiatin, S.L. and Filed, D.P. (eds), *Friction stir Welding and Processing*, Warrendale, PA: TMS, 159.
68. Park, S.H.C., Sato, Y.S., Kokawa, H., Okamoto, K., Hirano, S. and Inagaki, M. (2003), 'Rapid formation of the sigma phase in 304 stainless steel during friction stir welding', *Scripta Mater.*, 49: 1175–80.
69. Johnson, R. and Threadgill, P.L. (2003), in: David, S.A., DebRoy, T., Lippold, J.C., Smartt, H.B. and Vitek, J.M. (eds), *Proceedings of the Sixth International Conference on Trends in Welding Research*, Pine Mountain, GA: ASM International, 88–92.
70. Posada, M., Deloach, J., Reynolds, A.P. and Halpin, J.P. (2003), in: David, S.A., DebRoy, T., Lippold, J.C., Smartt, H.B. and Vitek, J.M. (eds), *Proceedings of the Sixth International Conference on Trends in Welding Research*, Pine Mountain, GA: ASM International, 307–12.
71. Li, Y., Trillo, E.A. and Murr, L.E. (2000), 'Friction stir welding of aluminum alloy', *J. Mater. Sci. Lett.*, 19: 1047–51.
72. Ouyang, J.H. and Kovacevic, R. (2002), 'Material flow and microstructure in the friction stir butt welds of the same and dissimilar aluminum alloys', *J. Mater. Eng. Perform.*, 11: 51–63.
73. Kazi, S.H. and Murr, L.E. (2001), in: Jata, K.V., Mahoney, M.W., Mishra, R.S., Semiatin, S.L. and Filed D.P. (eds), *Friction stir Welding and Processing*, Warrendale, PA: TMS, 139.
74. Murr, L.E., Li, Y., Flores, R.D., Trillo, E.A. and McClure, J.C. (1998), 'Intercalation vortices and related microstructural features in the friction stir welding of dissimilar metals', *Mater. Res. Innovat.*, 2: 150–63.

75. Larsson, H., Karlsson, L., Stoltz, S. and Bergqvist, E.L. (2000), in: *Proceedings of the Second International Symposium on Friction stir Welding*, Gothenburg, Sweden, 26–28 June.
76. Sterling, C.J., Nelson, T.W., Sorensen, C.D., Steel, R.J. and Packer, S.M. (2003), in: Jata, K.V., Mahoney, M.W., Mishra, R.S., Semiatin, S.L. and Lienert, T. (eds), *Friction stir Welding and Processing II*, Warrendale, PA: TMS, 165–71.
77. Lederich, R.J., Baumann, J.A. and Oelgoetz, P.A. (2001), in: Jata, K.V., Mahoney, M.W., Mishra, R.S., Semiatin, S.L. and Filed, D.P. (eds), *Friction stir Welding and Processing*, Warrendale, PA: TMS, 71.
78. Murr, L.E., Li, Y., Trillo, E.A. and McClure, J.C. (2000), 'Fundamental issues and industrial applications of friction stir welding', *Mater. Technol.*, 15: 37–48.
79. Lee, W.B., Yeon, Y.M. and Jung, S.B. (2003), 'The joint properties of dissimilar formed Al alloys by friction stir welding according to the fixed location of materials', *Scripta Mater.*, 49: 423–28.
80. Lim, S., Kim, S., Lee, C.G. and Kim, S. (2004), 'Tensile behavior of friction stir-welded A356-T6/Al 6061-T651 bi-alloy plate', *Metall. Mater. Trans. A*, 35(9): 2837–43.
81. Chen, C.M. and Kovacevic, R. (2004), 'Joining of Al6061 alloy to AISI 1018 steel by combined effects of fusion and solid state welding', *Int. J. Mach. Tool. Manufac.*, 44: 1205–14.
82. Jiang, W.H. and Kovacevic, R. (2004), 'Feasibility study of friction stir welding of 6061-T6 aluminium alloy with AISI 1018 steel', *Proc. Instn Mech. Engrs.*, vol. 218, Part B: *J. Engineering Manufacture*, 1323–31.
83. Shafiei-Zarghani, A., Kashani-Bozorg, S.F. and Zarei-Hanzaki, A. (2009), 'Microstructures and mechanical properties of Al/Al_2O_3 surface nano-composite layer produced by friction stir processing', *Mater. Sci. Eng. A*, 500: 84–91.
84. Mahmoud, E.R.I., Takahashi, M., Shibayanagi, T. and Ikeuchi, K. (2009), 'Fabrication of surface-hybrid MMCs layer on aluminium plate by friction stir processing and its wear characteristics', *Materials Transactions*, 50(70): 1824–31.
85. Shorowordi, K.M., Laoui, T., Haseeb, A.S.M., Ceiisand, J.P. and Froyen, L. (2003), 'Microstructure and interface characteristics of B_4C, SiC and Al_2O_3 reinforced Al matrix composites: A comparative study', *J. Mater. Process. Technol.*, 142: 738–43.
86. Ahiatchi, H., Kooer, T., Candan, E. and Gmenoglu, H. (2006), *Tribol. Int.*, 39: 213–20.
87. Munro, R.G. (1997a), 'Evaluated material properties for a sintered alpha-Al_2O_3', *J. Am. Ceram. Soc.*, 80: 1919–28.
88. Munro, R.G. (1997b), 'Material properties of a sintered alpha-SiC', *J. Phys. Chem. Ref. Data*, 26: 1195–1203.
89. Saravanan, R.A., Lee, J. and Kang, S. (1999), 'Dry sliding wear behavior of A356-15% SiCp composites under controlled atmospheric conditions', *Metal. Mater. Trans. A*, 30: 2523–37.
90. Lee, J., Kang, S. and Han, J. (2008), 'Dry sliding wear of MAO-coated a356/20 vol% SiCp composites in the temperature range 25–180 °C', *Wear*, 264: 75–85.

91. Mo, J.L. and Zhu, M.H. (2008), 'Sliding tribological behavior of AlCrN coating', *Trib. Tnt.*, 41: 1161–8.
92. Dinaharan, I. and Murugan, N. (2011), 'Influence of friction stir welding parameters on sliding wear behaviour of AA6061/0-10 wt.% ZrB_2 *in-situ* composite butt joints', *J. Minerals & Materials Characterization & Engineering*, 10(14): 1359–77.
93. Kumar, P.B. 'Friction stir welding of stainless steel 304', Research Scholar, Department of Mechanical Engineering, Delhi Technological University, Delhi, India, available from: *http://iosrjournals.org/iosr-jmce/papers/vol1-issue2/D0122223.pdf*
94. Park, K. (2009), 'Development and analysis of ultrasonic assisted friction stir welding process', PhD Thesis, University of Michigan.
95. Cam, G. (2011), 'Friction stir welded structural materials: Beyond Al alloys', *Int. Mater. Rev.*, 56(1): 1–48.
96. Zhang, Y.N., Cao, X., Larose, S. and Wanjara, P. (2001), 'Review of tools for friction stir welding and processing', *Canadian Metallurgical Quarterly*, 51(3): 250–61.
97. Collier, M., Steel, R., Nelson, T., Sorensen, C. and Packer, S. (2003), 'Grade development of polycrystalline cubic boron nitride for friction stir processing of ferrous alloys', *Proceedings of the 4th International Conference on Friction stir Welding*', Park City, UT: TWI, May.
98. Prado, R.A., Murr, L.E., Soto, K.F. and McClure, J.C. (2003), 'Self-optimization in tool wear for friction stir welding of Al 6061+20% Al_2O_3 MMC', *Mater. Sci. Eng. A*, A349: 156–65.
99. Prado, R.A., Murr, L.E., Shindo, D.J. and Soto, K.F. (2001), 'Tool wear in the friction stir welding of aluminium alloy 6061+20% Al_2O_3: A preliminary study', *Scr. Mater.*, 45: 75–80.
100. Hovanski, Y., Santella, M.L. and Grant, G.J. (2007), 'Friction stir spot welding of hot-stamped boron steel', *Scr. Mater.*, 57: 873–6.
101. Choi, D.H., Lee, C.Y., Ahn, B.W., Choi, J.H., Yeon, Y.M. et al. (2009), 'Frictional wear evaluation of WC-Co alloy tool in friction stir spot welding of low carbon steel plates', *Int. J. Refract. Met. Hard. Mater.*, 27: 931–6.
102. Liu, H.J., Feng, J.C., Fujii, H. and Nogi, K. (2005), 'Wear characteristics of a WC-Co tool in friction stir welding of AC4A+30vol% SiCp composite', *J. Machine Tools & Manufacture*, 45: 1635–9.

9

Mechanical alloying by friction stir processing

D. Ahmadkhaniha and P. Asadi, University of Tehran, Iran

DOI: 10.1533/9780857094551.387

Abstract: During recent years, friction stir processing (FSP) has become known as a beneficial method for mechanical alloying (MA). High inserted heat and strain during FSP can be applied to produce *in situ* intermetallics and modify the microstructure of alloys by increasing the alloying elements. These fabricated materials by FSP have shown an improvement in mechanical properties, such as hardness and tensile properties. However, there is limited literature about this application of FSP. Therefore, in this chapter, different examples of FSP application as a new route for MA are presented.

Key words: mechanical alloying, friction stir processing, intermetallic, *in situ* composites, mechanical properties, alloy transformation, alloy modification.

9.1 Introduction

Most commercial alloys are produced by casting. However, many alloys are difficult or virtually impossible to prepare by this procedure, including alloys in which the components have widely different melting points or vapour pressures or alloys whose components are not mutually soluble in the liquid state. Incongruently melting intermetallic phases (or 'intermetallic compounds') are also difficult to prepare in a pure state by melting and casting. For these alloys, other alloying techniques such as mechanical alloying (MA), which circumvents many of the limitations of melting and casting, must be employed.

MA is a processing technique that produces alloys starting from mixtures of elemental and/or compounds powders. The loading powder is mixed and grinded with a grinding medium (generally hardened steel or tungsten carbide balls) in a stainless steel container sealed under a protective argon atmosphere (to avoid/minimize oxidation and nitridation during milling) and milling for the desired length of time. Particles are fractured and cold welded during MA. Therefore, refinement of the microstructure and an intimate mixing of the constituents that favours the formation of a metastable structure can be achieved. MA was developed around 1966 by John Benjamin and coworkers at the Paul D. Merica Research Laboratory of the International Nickel Company (INCO), to produce a material that is strengthened by oxide dispersion as well as gamma prime precipitation in a nickel-based super alloy for gas turbine applications [1].

MA has been applied in different material processes to obtain different materials. It can be combined with consolidation techniques to develop bulk materials, such as oxide dispersion strengthened (ODS) materials [2], intermetallics [3], nanomaterials [4], composites [5], ceramics [6], polymers [7], amorphous materials [8], hydrogen storage [9] and extended solid solutions [10].

9.2 Technological applications of mechanical alloying

The most important application of mechanically alloyed (MA) products is formation of ODS alloys, which are difficult to process through conventional ingot metallurgy (IM) methods. INCO production annually consists of about 350 t of commercial ODS alloys mainly based on nickel, iron and aluminium. Table 9.1 shows the compositions and applications of MA-ODS alloys [11].

Apart from the ODS alloys, MA is also commercially used to fabricate about 5 t per year physical vapour deposition (PVD) targets for the electronic industry by Tosoh, USA. Another commercial application of MA products is in the use of Meal, Ready-to-Eat (MRE) heaters, which were extensively applied during the Desert Storm Operation by the US in 1996. These heaters contain MA powders of Mg and Fe, which produce heat on contact with water. It has also been reported [12] that Zoz in

Table 9.1 Typical applications of mechanically alloyed ODS alloys [11]

Alloys	Composition (wt%)									Typical applications
	Ni	Fe	Cr	Al	Ti	Mo	W	Ta	Y_2O_3	
MA754	Balance	–	20	0.3	0.5	–	–	–	0.6	Vanes and bands for aircraft gas turbine engines
MA6000	Balance	–	15	4.5	0.5	2.0	4.0	2.0	1.1	Gas turbine blades and vanes
MA956	–	Balance	20	4.5	0.5	–	–	–	0.5	Vacuum furnace muffles, trays, baskets and in glass-processing industry
MA957	–	Balance	14	–	1.0	0.3	–	–	0.25	Fuel cladding for fast neutron fast breeder reactors, heat exchanger tubes

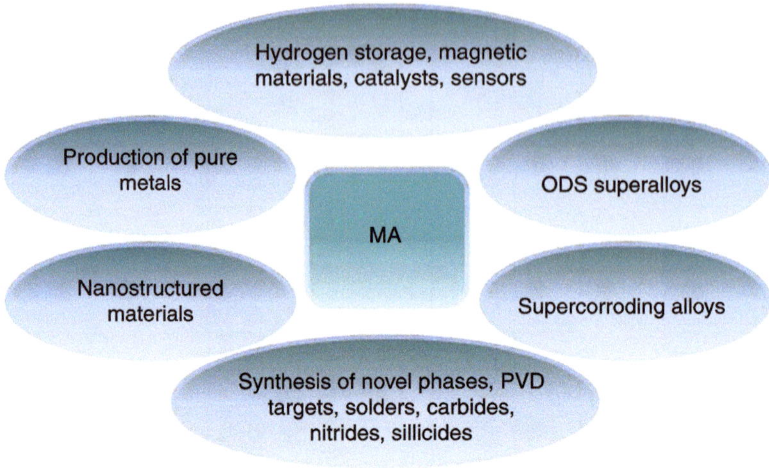

Figure 9.1 Typical current and potential applications of mechanically alloyed products [11]

Germany and Fukuda Metal Foil and Powder in Japan have produced jointly about 600 kg per day (~200 t per year) of MA material for paints and solders. The most important advantage of MA in all these applications is production of a highly homogeneous product without any segregation. Typical current and potential applications of MA are shown in Figure 9.1.

Conventional MA has its limitations. The contamination of powders during MA is a major concern. The small size of the powder particles, availability of large surface area, and formation of new surfaces during milling, all lead to contamination of the powder. In addition, the milling conditions (grinding medium, grinding vessel, time of milling, intensity of milling, etc.) and the atmosphere under which the powder is milled, also contribute to the contamination level [11].

To improve the application of MA, several new versions of the original process have been developed by [13] and the different processes have been designated differently. In the following section, one of them, named the mechanochemical processing (MCP), will be explained.

9.3 Mechanochemical processing

Mechanochemical processing (MCP) is the term applied to powder processing in which the application of mechanical energy helps chemical reactions and phase transformations to take place during milling. In this process, plastic deformation and chemical processes occur almost simultaneously. The first publication of this process dates back to 1892. The mechanochemical reactions can result in the formation of novel materials, reduction/oxidation processes, exchange reactions, decomposition of compounds and phase transformations in both organic and inorganic solids. The materials produced in this way can be applied in several areas, such as hydrogen storage materials, gas absorbers, fertilizers and catalysts. This technique has become important in the general field of MA and milling. In fact, a series of conferences are being regularly held on this aspect [14].

The major difference between varieties of methods is that while MA produces the material in the powder form, most of these new methods deal with bulk materials. Another difference is that these new methods focus on producing nanocrystalline and amorphous materials with new potential applications. A common theme among all of these methods is heavy plastic deformation that is similar to MA in producing amorphous alloys and other non-equilibrium phases. Some of these new methods are briefly described below.

9.4 Methods for producing amorphous alloys and other non-equilibrium phases

9.4.1 Repeated cold rolling

Repeated cold rolling can produce effects very similar to those of MA. In this case, rolling a stack of two metals (bimetal layer) reduces the thickness. Subsequent rolling passes reduce the thickness further (Figure 9.2) [14].

The actual process includes stuffing a mixture of the elemental powders into a stainless steel tube. The tube containing the powder is pressed and then cold rolled (Figure 9.3). The process is repeated several times.

Alternatively, thin foils of the constituent metals can be taken instead of stuffing the powders into a stainless steel tube. We could also press the mixture of powder, break them into smaller pieces, and repeat the process. Figure 9.4 demonstrates schematic views of the repeated pressing of layered thin foils and of powders.

Obtaining supersaturated solid solutions of Ag-70 at% Cu samples, both by MA and by repeated cold rolling methods, suggests that both methods yield similar types of metastable phases [14].

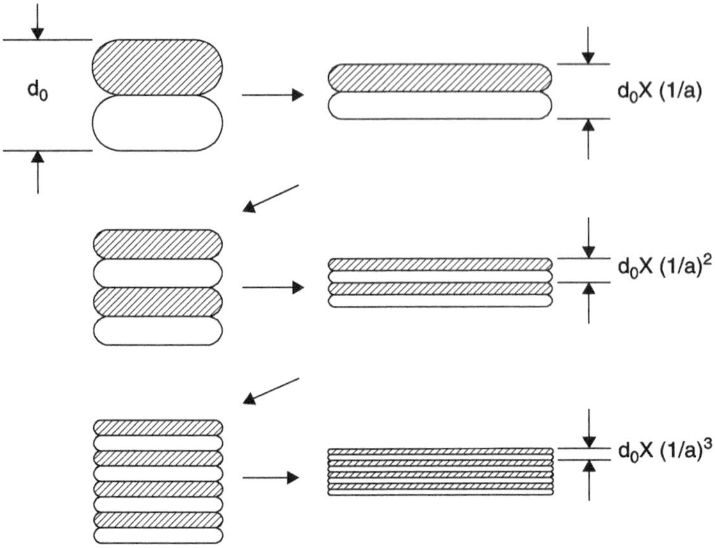

Figure 9.2 Thickness reduction by repeated rolling and folding operations [14]

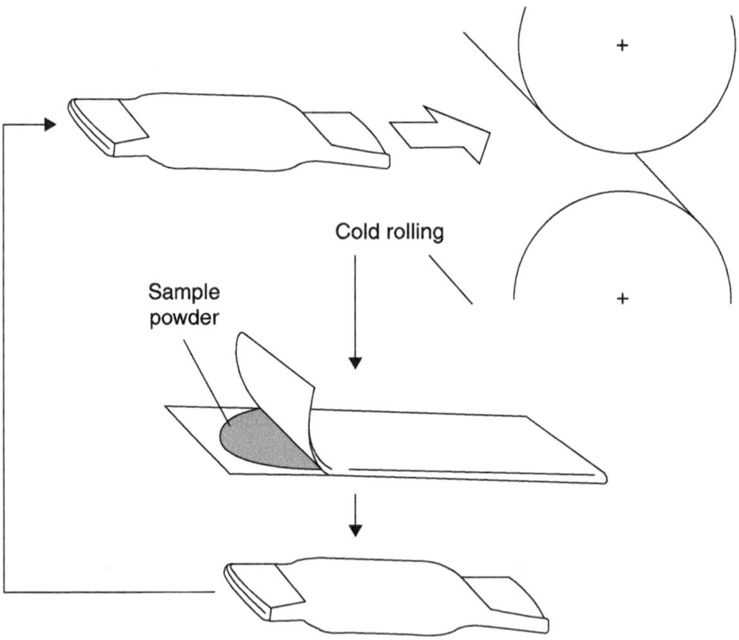

Figure 9.3 Schematic of experimental procedure for repeated cold rolling [14]

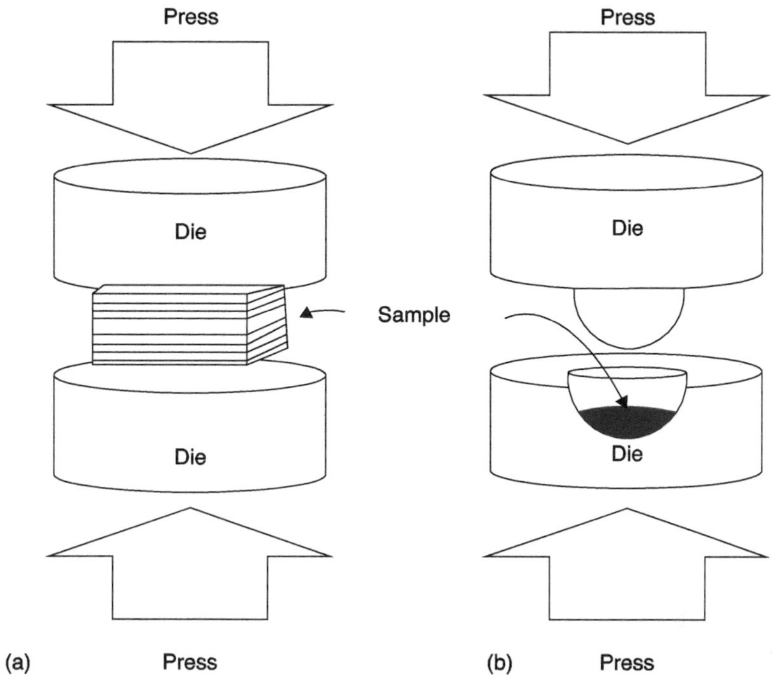

Figure 9.4 Schematic views of the repeated pressing procedure: (a) repeated pressing of layered thin foils; and (b) repeated pressing of powders [14]

9.4.2 Severe plastic deformation

Heavy deformations, for example by cold rolling or drawing or multiple forging, result in significant refinement of microstructures at low temperatures. However, the microstructures produced by these methods are usually substructures of a cellular type having low-angle grain boundaries, while significant improvement in the properties of nanocrystalline materials is due to high-angle type of the grain boundaries.

Severe plastic deformation (SPD), such as torsion straining under high pressure and equal channel angular pressing (Figure 9.5), has been developed to produce ultrafine grained materials in the submicron range, by inducing very large deformations at relatively low temperatures.

In recent years, apart from the methods mentioned above, friction stir processing (FSP) has been used to synthesis and develop new materials by *in situ* formation of intermetallics or alloying. In the following section, different applications of FSP from the MA view will be reported.

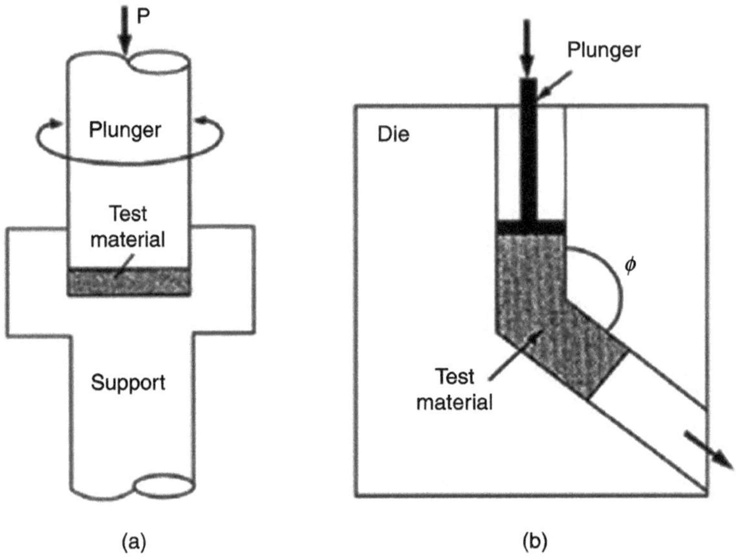

Figure 9.5 Principles of severe plastic deformation by: (a) torsion under high pressure; and (b) equal channel angular pressing [14]

9.5 Applications of friction stir processing as a mechanical alloy treatment

9.5.1 *Producing Al-Ni in situ composite*

Aluminium intermetallic compounds are applied as practical materials, due to their resistance to wear, high hardness and stability at elevated temperatures. However, aluminium intermetallic compounds such as Al_3Ni cannot serve alone as a structural material since they are so brittle. The brittleness can be compensated by embedding them in a ductile matrix material. Thus, with these advantages, they can practically be utilized in a structural material and the ductility of matrix material can nullify their brittle shortcomings. In the past few years, these Al–Al_3Ni composites have been fabricated by directional solidification [15], equal-channel angular pressing [16], electromagnetic separation method [17] and MA [18]. Apart from these methods, FSP can be used to fabricate these intermetallics. For this purpose, pure aluminium plate and pure nickel powder are used as the initial materials. Nickel powders are filled into the holes on the surface of the Al plate. To enhance the Al–Ni reaction, multi-pass FSP with subsequent heat treatment is applied.

Due to SPD and heat input during FSP, Al_3Ni can be produced (Figure 9.6(a)). The existence of nickel peaks after three FSP passes shows that the reaction between Al and Ni is not promoted completely. However, after subsequent heat treatment, nickel peaks will disappear and be replaced by Al_3Ni peaks. More Al_3Ni peaks with increased intensity can be detected from the X-ray diffraction (XRD) pattern of a heat treated specimen. In addition, two obvious Al_3Ni_2 peaks in the XRD curves with $2\theta=44-45$ degrees are seen in the magnified view of the XRD pattern (Figure 9.6(b)).

A scanning electron microscopy (SEM) equipped with energy dispersive spectroscopy (EDS) results (Figure 9.7) show that the phase in the region B is most likely to be Al_3Ni and the phase in the region C is probably Al_3Ni_2. It is suggested that the brightest white phase is Ni, the dark grey matrix is Al, the light grey phase is Al_3Ni_2, and the grey phase, which surround Al_3Ni_2 particle or dispersed in the Al matrix, is Al_3Ni.

One of the advantages of FSP is to produce an *in situ* composite without grain growth due to severe deformation and short time at elevated temperatures, which can have significant impression on the mechanical properties of the materials. Grain refinement and the precipitation hardening effect of the Al_3Ni intermetallics result in the significant increase of the microhardness for the Al–Al_3Ni composites. For example, the microhardness for the Al_3Ni_2 and Al_3Ni intermetallics, which are produced during FSP, are 1283 and 841 HV (Hardness Vickers) respectively, and for Al matrix is 52 HV. In addition, the presence of a large amount of Al_3Ni in the composite results in a significant improvement in the tensile strength (Figure 9.8) [19].

9.5.2 Al-Al_2Cu intermetallic composite

Pure aluminium powder and pure copper powder can be used to produce an Al–Al_2Cu composite via FSP. Cu reacts completely with Al and forms Al_2Cu during FSP. In addition, copper oxides (CuO and Cu_2O), which result from oxidation during sintering in air, disappeared after FSP, which might be a result of the reaction between Al and copper oxides to form Al_2O_3 and Al_2Cu (Figure 9.9) [20.21].

A lower thermal conductivity of sintered Al–Cu powder more compact than that of a solid aluminium alloy can raise the peak temperature

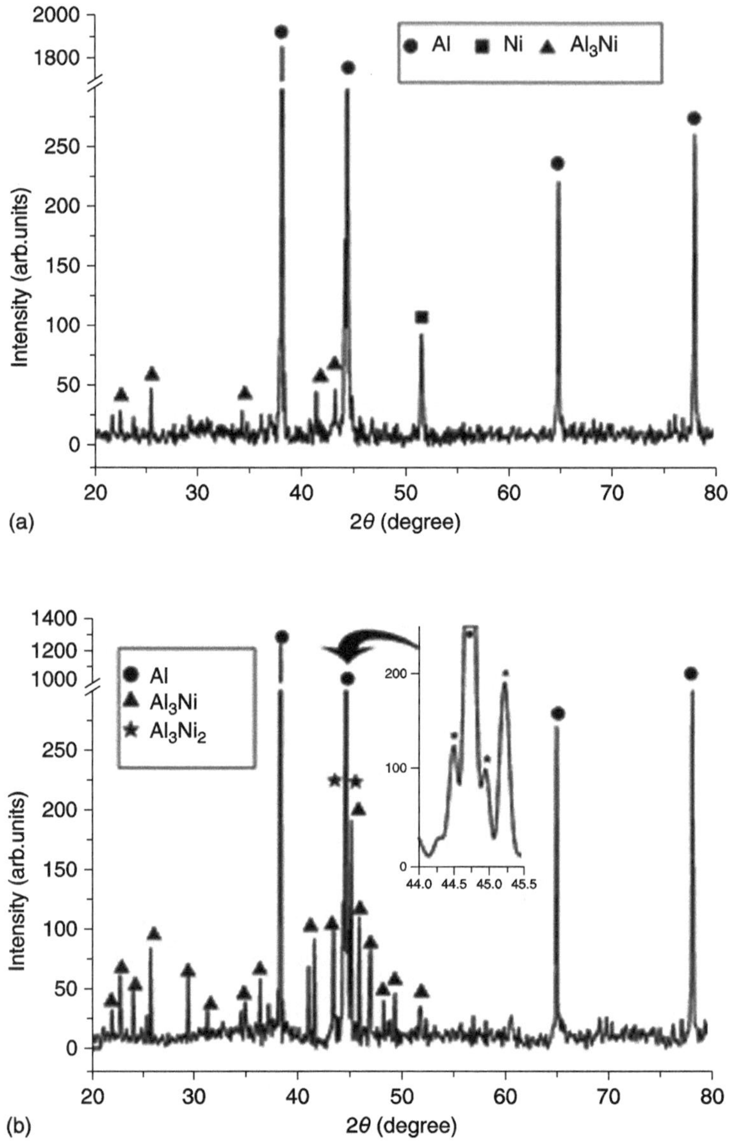

Figure 9.6 XRD patterns of Al-Ni composites for: (a) 3 passes of FSP: and (b) heated at 550 °C for 6 h after FSP [19]

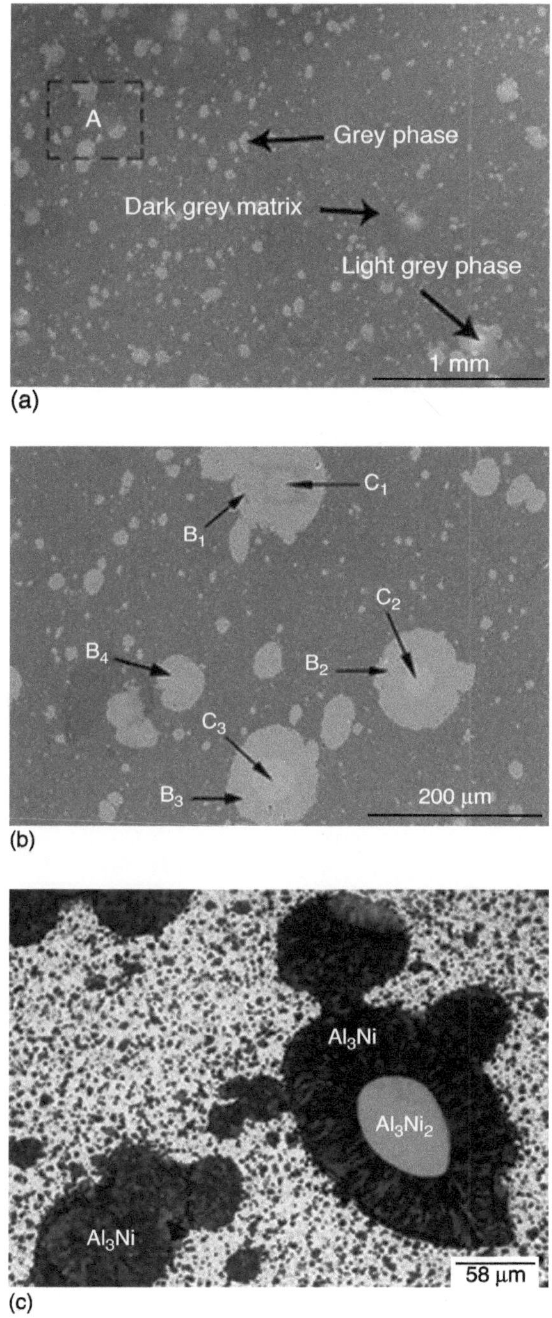

Figure 9.7 SEM micrographs showing the particle dispersion: (a) three-pass FSP with subsequent heat treatment at 550 °C for 6 h; (b) the enlarged micrograph of marked A region in Figure 9.8 (a); and (c) optical microscope image of Al_3Ni grain [19]

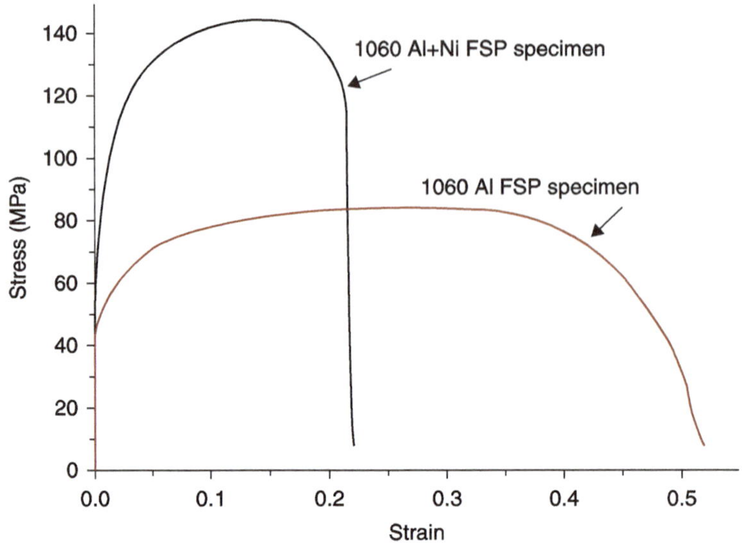

Figure 9.8 Tensile stress-strain curves for FSP specimens machined from Al-Ni composites and 1060Al [19]

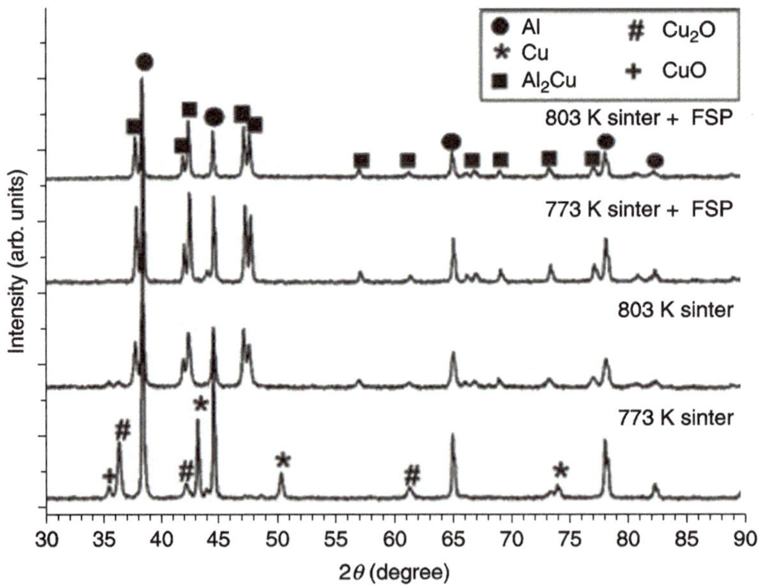

Figure 9.9 X-ray diffraction patterns for as-sintered specimens and specimens after two FSP passes [22]

during FSP to above 663°K. In addition to high temperature, severe deformation, which not only promotes mixing and refining of constituent phases in the sample, but also by increasing the diffusion rate of elements, accelerates the reaction between Al and Cu. The resultant composite possesses an enhanced Young friction stirs modulus (88 ± 8 GPa), satisfied compressive strength (450 MPa yield strength (YS) and 650 MPa ultimate tensile strength (UTS)), with reasonable good compressive ductility (0.15 failure strain).

9.5.3 Al-Ti intermetallic composite

Among aluminide particles (Al_3Ti, $NiAl_3$, $FeAl_3$, etc.) reinforced Al matrix composites (AMCs), *in situ* Al_3Ti/Al composites are even more interesting, since they have a relatively low density (3.4 g cm^{-3}) and high Young friction stir modulus (220 GPa), as well as low diffusivity and solubility of Ti in Al. Hence, Al_3Ti is expected to have a low coarsening rate at elevated temperatures. In previous researches, *in situ* Al_3Ti/Al composites were produced by powder metallurgy (e.g. sintering, hot pressing) and stirring casting [23,24]. However, the size of Al_3Ti as a result of high processing temperatures or coarse raw material powders was large (generally above 10 μm). The coarse Al_3Ti blocks are prone to crack under low stress during tensile tests, resulting in low strength and ductility.

MA fabricates *in situ* Al_3Ti/Al composites with high strength due to the presence of the fine Al_3Ti particles, but the resultant composites are always accompanied by significant ductility loss, due to inhomogeneous microstructures and contaminations introduced by MA. In addition, MA increases the cost and duration of production.

FSP can be used as an alternative process to fabricate these composites with better properties. *In situ* Al_3Ti in the Al matrix can be produced by hot pressing of Al and Ti mixed powders and subsequent FSP. For this purpose, the billets are forged to obtain adequate strength before FSP. Al_3Ti peaks are found in XRD patterns of FSP specimens, while they are not observed in forged ones, which indicate that the reaction between Al and Ti did not take place during forging (Figure 9.10). However, after 4-pass FSP with a rotation rate of 2000 rpm, the Ti peaks still exist, indicating that the Ti did not react with Al completely [25].

The diffusion coefficient of Al in Ti is much higher than that of Ti in Al ($D_{Al} \approx 1.5 \times 10{-}13\ m^2s^{-1}$ and $D_{Ti} \approx 3.2 \times 10{-}17 m^2\ s^{-1}$ at 600 °C). Thus Al is always plentiful during diffusion. In this case, the intermediate intermetallics among Ti_3Al, TiAl, Al_2Ti, and Al_3Ti will be Al_3Ti.

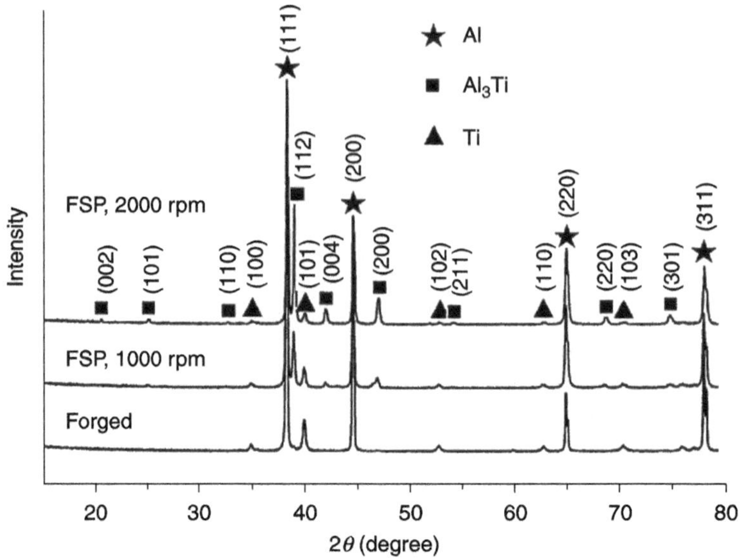

Figure 9.10 XRD pattern of different specimens [26]

The heat release due to the *in situ* chemical reaction during FSP can also contribute to temperature rise. Thus, it is likely that the peak temperature of the stir zone (SZ) might be higher than that reported for FSP/FSW aluminium alloys.

The large plastic deformation of FSP breaks up the oxide film surrounding Al, which causes close contact between Al and Ti. In addition, intense plastic deformation during FSP reduces the diffusion distance and enhances the diffusion rate. In this case, the heat provided by FSP can raise the temperature enough to induce the reaction between Al and Ti. Once the Al_3Ti is formed at the Al–Ti interface, the large plastic strain imposed by the FSP tool can remove it immediately. Therefore, close contact between Al and Ti can be maintained, and the reaction can proceed rapidly at the interface.

In the forged sample, the coarse white Ti particles are uniformly distributed in the Al matrix, and a few voids exist in the matrix, while 4-pass FSP refines the coarse Ti particles and almost eliminates the voids (Figure 9.11).

With increase in the tool rotation rate, the size of the Ti particles decreases as a result of higher strain and fracture force (Figure 9.11). In this case, the reaction between Al and Ti particles can be accelerated. Al_3Ti particles formed in the Al matrix by FSP will be in equiaxed shaped with approximate sizes of 140 nm for the rotation rates of 1000 to 2000 rpm [26].

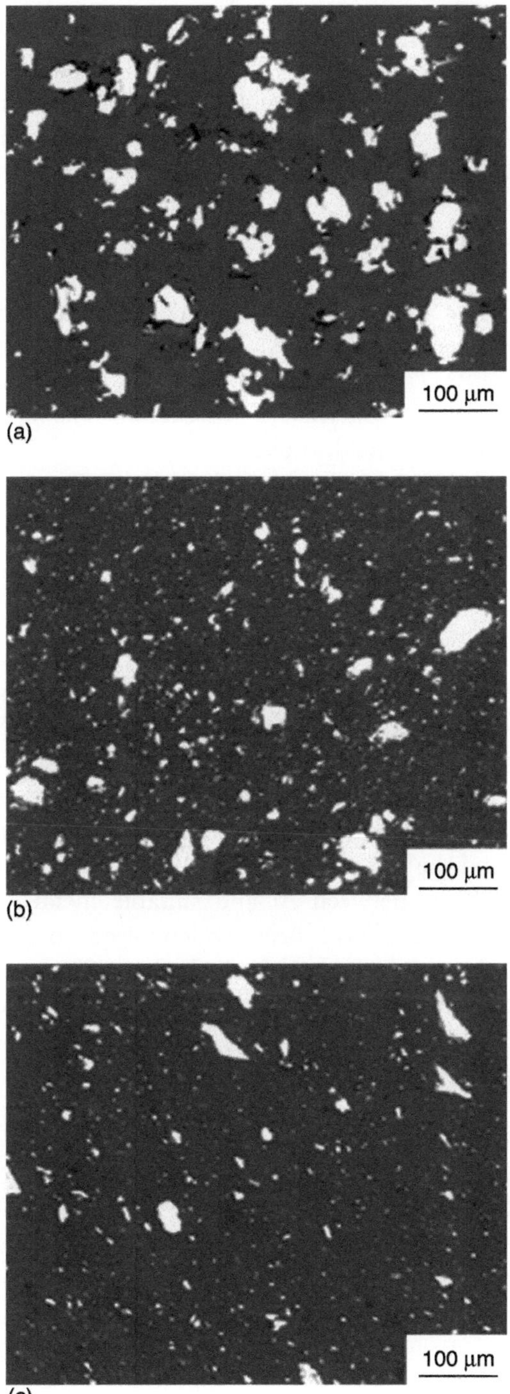

Figure 9.11 SEM micrographs of samples: (a) forged; (b) FSP, 1000 rpm; and (c) FSP, 2000 rpm [26]

Figure 9.12 shows the SEM/backscattered electron image (BEI) of Al–5Ti alloy produced by 4-pass FSP. The micro-sized white particles in Figure 9.12 are pure Ti, which are confirmed using EDS, and the very fine grey particles (<100 nm) uniformly dispersed in the Al matrix are Al_3Ti particles. After FSP, the average size of Ti particles is refined from 40 μm to about 1 to 5 μm. Some of the Ti particles after FSP show an irregular morphology (Figure 9.12(b)). The appearance of irregular morphology at the interface between Ti and Al + Al_3Ti predicts that a liquid phase might be present at the Al–Ti reaction front. The liquid phase is referred to Al melt, as a result of the exothermic Al–Ti reaction. The liquid Al can react rapidly with Ti to form an Al_3Ti phase.

The ultra-fine Al_3Ti particles formed during FSP exert a remarkable strengthening effect on the Al matrix through the Orawan bowing mechanism. Meanwhile, the finer Ti particles, due to possessing higher strength than Al, exert an additional strengthening effect on the Al matrix. Therefore, 4-pass FSP results in a significant improvement in YS and UTS, but a reduction in elongation relative to the forged sample.

A higher tool rotation rate will result in a higher strength, due to the increased refinement of the Ti particles and increased Al_3Ti intermetallics amount. Fracture surfaces of FSP specimens have coarse dimples and some deep pits (Figure 9.13). Large deeps are associated with the pulling out of the Ti particles. Increasing the tool rotation rate decreases the size of the dimples and the pits substantially [26].

Chemical reactions between Al and suitable metal oxides, such as CuO, Fe_2O_3 and TiO_2, have been widely used to fabricate *in situ* aluminium matrix composites (AMCs). Reduction of such oxides by Al results in the formation of another metal and/or aluminium oxide. The

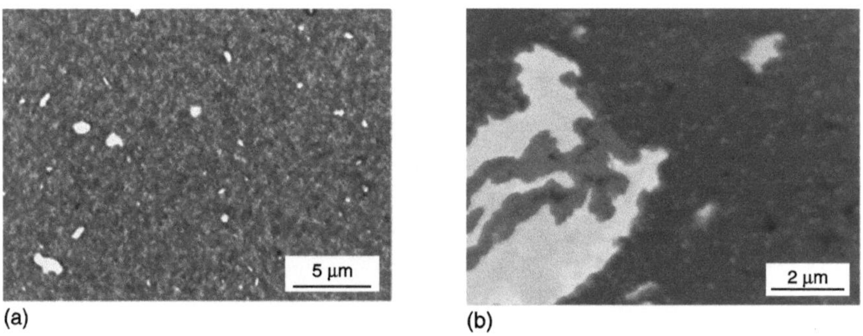

Figure 9.12 SEM/BEI of: (a) Al–5Ti alloy; and (b) Al–10Ti alloy, produced by four-pass FSP. Tool rotation rate was 700 rpm [26,27]

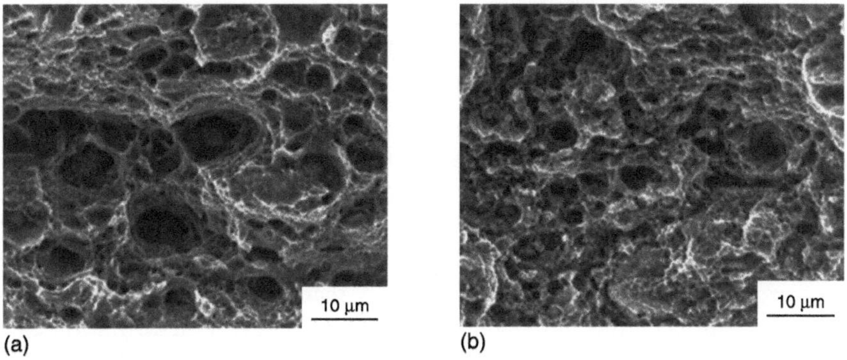

Figure 9.13 Fracture surfaces of FSP samples: (a) 1000 rpm; and (b) 2000 rpm [26]

produced composite has interesting properties since aluminium oxide, especially Al_2O_3, is a good reinforcement for AMCs. In addition, the resulting metal can be a good alloying element for the Al matrix, or produce new intermetallic phases as reinforcement.

These composites can be fabricated by applying 4-pass FSP on the mixture of pure Al powder and rutile TiO_2. Figure 9.14 shows the XRD

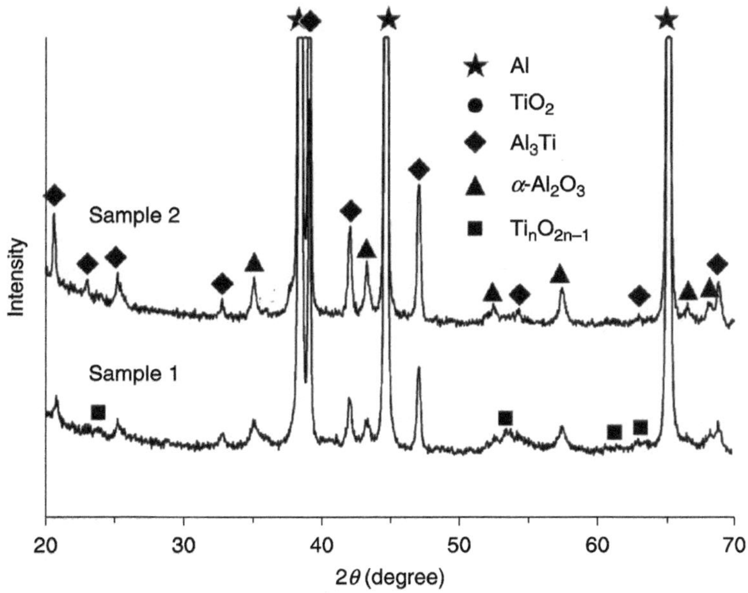

Figure 9.14 XRD patterns of various composites as-FSPed [28]

pattern of the specimen. Furthermore, some weak peaks corresponding to Ti_nO_{2n-1} were detected in one of the FSP specimens.

Formation of these compounds via FSP can be explained by a high density of lattice defects, such as dislocations, which are introduced by SPD during FSP and enhance pipe diffusion. Since lattice defects can be annealed out rapidly at relatively high temperatures, they probably contribute little to thermally-induced diffusion. In addition, severe deformation during FSP, which functions like a mechanical activation, can reduce the activation energy of the reaction and so increase the reaction rate. The mechanical activation during severe deformation can be achieved in different ways, such as:

- When dislocations move along a slip plane, the mechanical energy is transformed into the kinetic energy of the atoms. Therefore, the translational mobility of the atoms increases.
- When the chemical bond is distorted as a result of stress during mechanical treatment, bond length is extended. If the imposed stress is beyond the chemical bonding energy, the solid bonds can be broken and their crystalline structures collapse. Therefore, the molecules or atoms of the reactants become activated and can react with other reactants.
- When the grain size decreases dramatically during severe deformation, the onset temperature of the reaction will be decreased and the reaction enhanced.
- Schematic illustration of the formation of reinforcing particles is shown in Figure 9.15. In the hot-pressed or sintered samples [19], TiO_2 particles aggregate into the initial Al particles. Solid-state diffusion at point contacts controls the reaction. Al_2O_3 can be formed at the contact points. With the proceeding reaction, a TiO_2 particle gradually transforms into one or two Al_2O_3 particles from the contact points, where its size is comparable with the size of TiO_2 (Figure 9.15(a)). In FSP samples, many Al_2O_3 particles nucleate and grow in the Al/TiO_2 interfaces (Figure 9.15(b)) [28].

9.5.3.1 Al-Ti-X *in situ* composite

Three types of *in situ* composites with nominal compositions of Al-16.4wt.%Ti, Al-16.4wt.%Ti-2.5wt.%Cu and Al-16.4wt.%Ti-2.5wt.%Mg, can be fabricated by FSP (denoted as Al-Ti, Al-Ti-Cu and Al-Ti-Mg samples).

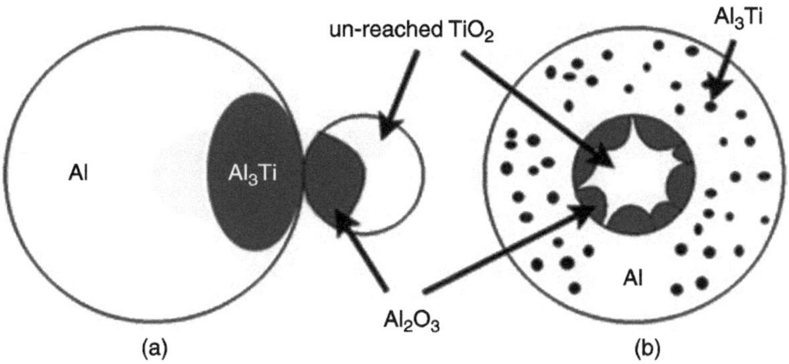

Figure 9.15 **Schematic of the formation mechanism of Al_3Ti and Al_2O_3 during: (a) hot pressing or sintering; and (b) FSP [28]**

For the FSP Al-Ti-Cu sample, Al_2Cu peaks are detected (Figure 9.16). For the Al-Ti-Mg sample, except for Al, Al_3Ti and Ti peaks, some peaks corresponding to $Ti_2Mg_3Al_{18}$ are found, while they are absent in the forged Al-Ti-Mg sample.

From analysis of SEM images (Figure 9.17), the volume fractions of Ti particles in the FSP Al-Ti, Al-Ti-Cu and Al-Ti-Mg samples, are listed in Table 9.2. It can be seen that less Ti particles remained in the FSP Al-Ti-Cu and Al-Ti-Mg samples than in the FSP Al-Ti one [29]. This fact demonstrates that Cu addition enhances the Al-Ti reaction during FSP and results in the formation of more Al_3Ti particles.

In the TEM images of the FSP samples (Figure 9.18), some Al_2Cu particles with an irregular shape are revealed for the Al-Ti-Cu sample, and some fine Al_3Ti particles are often seen in the inner of Al_2Cu particles (Figure 9.18(b)). For the Al-Ti-Mg sample, except for Al_3Ti particles, some particles with fine twin lamellas are observed (Figure 9.18(c)), which are confirmed by selected electron diffraction patterns that are face centred cubic (f.c.c)-structured $Ti_2Mg_3Al_{18}$ (Figure 9.18(d)).

Since the peak temperatures of FSP Al-Ti, Al-Ti-Cu and Al-Ti-Mg samples are 568, 540 and 545 °C, respectively, it can be concluded that Cu or Mg addition has little influence on the peak temperature of FSP.

Three observations from the differential scanning calorimetry (DSC) curves of various forged samples are shown in Figure 9.19. First, all the DSC curves show an evident endothermic peak at about 660 °C attributed to the melting of Al matrix. The onset temperature of the melting of Al for the Al-Ti sample is 637 °C; however, it is reduced to 612 and 623 °C, respectively, for the Al-Ti-Cu and Al-Ti-Mg samples. Second, an

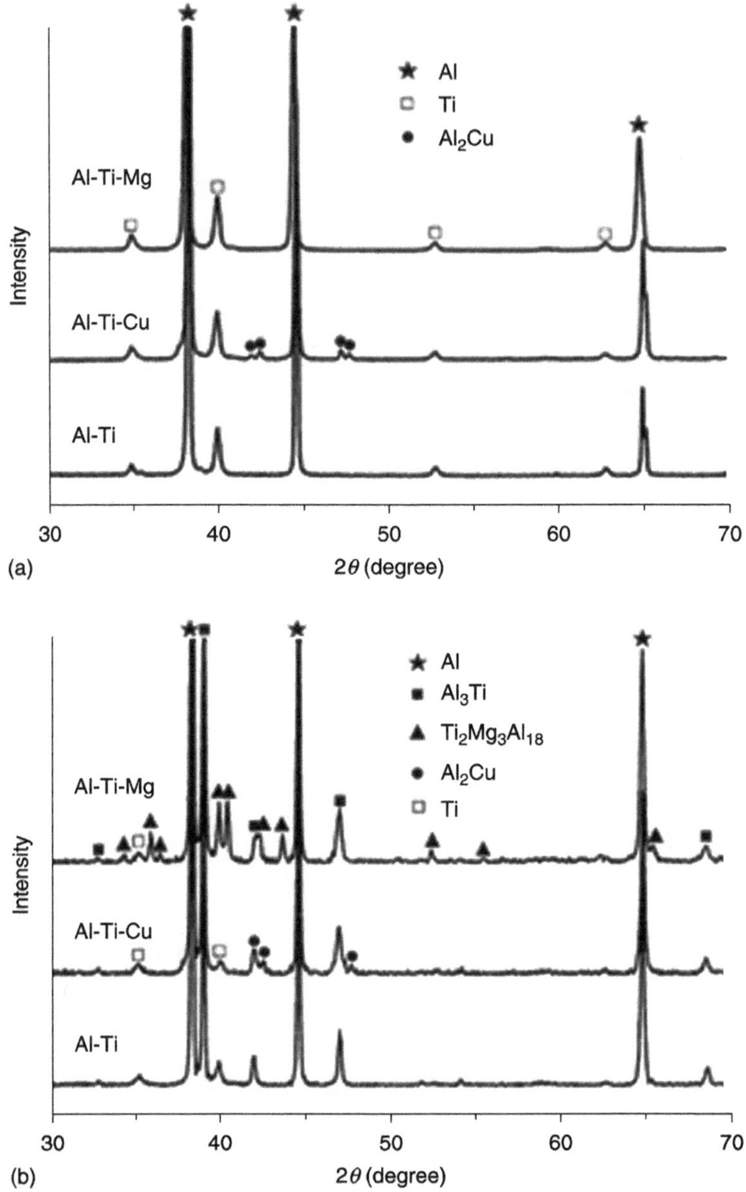

Figure 9.16 XRD patterns of: (a) forged, and (b) FSP samples [29]

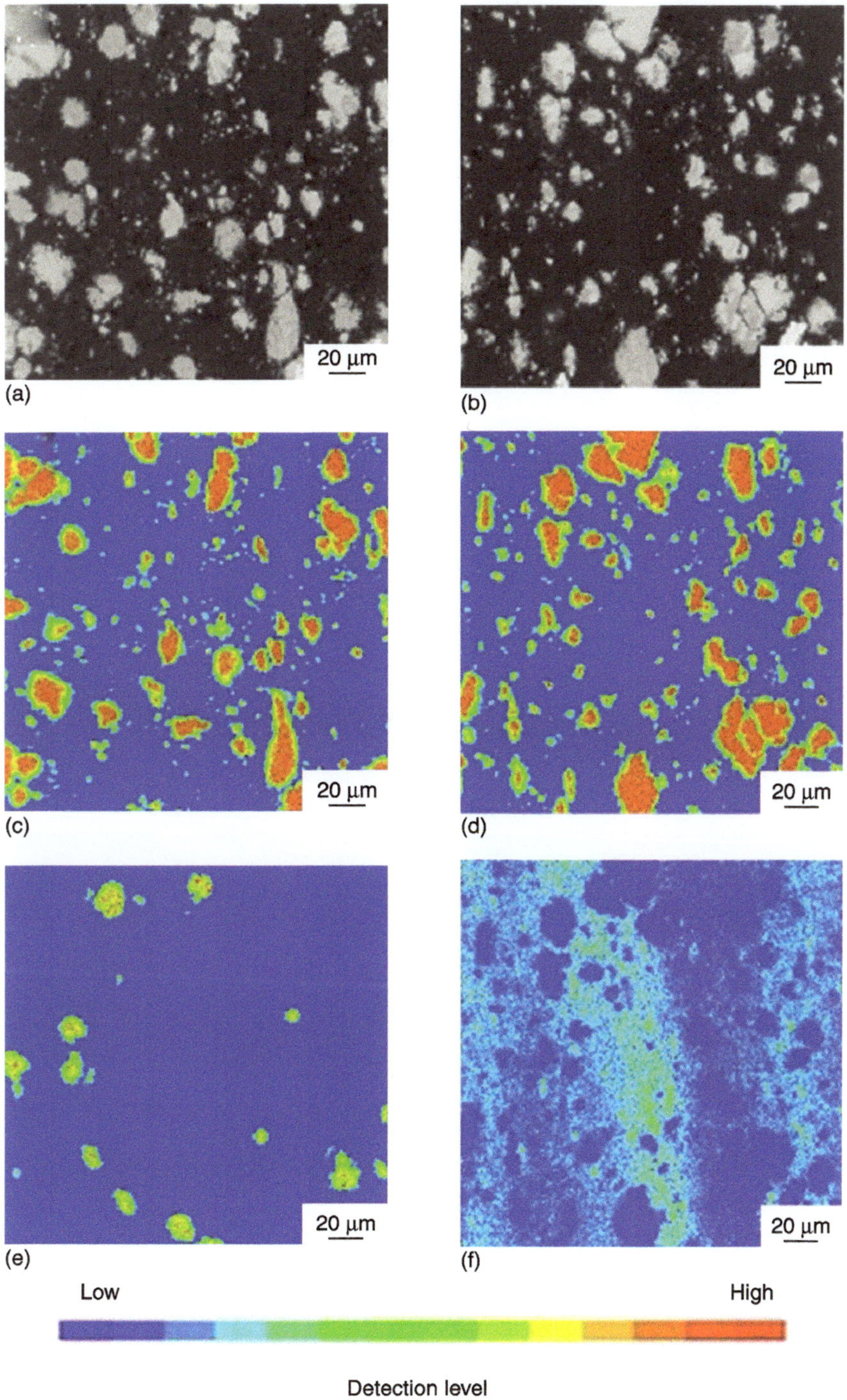

Figure 9.17 SEM/BEI images and the corresponding elemental maps of forged samples: (a) Al-Ti-Cu; (b) Al-Ti-Mg; (c and e) distribution of Ti and Cu in Al-Ti-Cu sample; and (d and f) distribution of Ti and Mg in Al-Ti-Mg sample [29]

Table 9.2 **Volume fractions of untreated Ti and reinforcing particles in FSP samples [29]**

FSP sample	Ti, Vol%[a]	Reinforcing particles, Vol%[b]
Al-Ti	4.6±0.34	29.4±1.2 (Al_3Ti)
Al-Ti-Cu	3.5±0.42	33.3±1.5 (Al_3Ti)
Al-Ti-Mg	3.4±0.23	17.2 (Al_3Ti), 23.8 ($Ti_2Mg_3Al_{18}$)

[a] Experiment values
[b] Calculated values

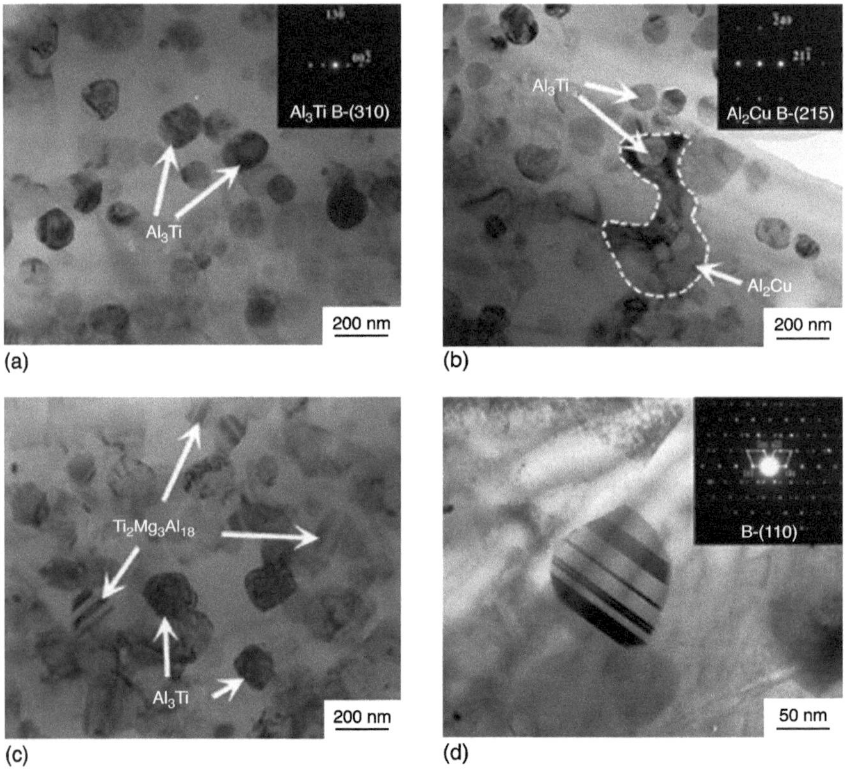

Figure 9.18 **TEM images of FSP samples: (a) Al-Ti; (b) Al-Ti-Cu; (c) Al-Ti-Mg; and (d) $Ti_2Mg_3Al_{18}$ in Al-Ti-Mg sample [29]**

exothermic peak is revealed after the melting of Al for the three samples corresponding to the exothermic reaction resulting from the formation of different intermetallic particles such as Al_3Ti, $Ti_2Mg_3Al_{18}$. Third, a weak endothermic peak at 546 °C was detected in the curve of the Al-Ti-Cu sample, which is the result of the eutectic reaction between Al and Al_2Cu, and no evident peak is seen before the melting of Al matrix in the Al-Ti and Al-Ti-Mg samples.

The results of DSC indicate that the eutectic reaction between Al and Al_2Cu occurs when the temperature reaches 546 °C (Figure 9.19(b)). The reaction formula is

$$Al + Al_2Cu \leftrightarrow \text{liquid} \qquad [9.1]$$

As shown in Figure 9.19(a), the peak temperature near the SZ rises to about 550 °C during FSP, which is more than the temperature of the Al-Al_2Cu eutectic reaction. Thus, it is expected that some liquid phase with eutectic composition (Al 55.9 wt.%, Cu 44.1 wt.%) would appear during FSP. Since, the Al-Ti interdiffusion between Ti and the liquid phase would occur and the diffusion coefficient would be much higher than that at solid-state, the Al-Ti reaction is enhanced and more Al_3Ti forms during FSP in the Al-Ti-Cu sample.

For the forged Al-Ti-Mg sample, due to the low temperature of eutectic reaction in the Al-Mg system, no Mg-containing intermetallics form and Mg is dissolved into the Al matrix, although its distribution is non-uniform (Figures 9.16 and 9.17).

Since Mg is dissolved into the Al matrix in the forged sample, based on the Al-Mg diagram, the liquid phase should appear in the Al solid solution containing 2.5 wt.%Mg at about 620 °C, which is proved by the DSC results in Figure 9.19(b). The peak temperature of the SZ during FSP of the Al-Mg-Ti sample is not as high as this temperature (Figure 9.19(a)). Hence, it is reasonable to expect that no liquid phase would appear during FSP of the Al-Mg-Ti sample. Lower volume fraction of unreacted Ti particles in the FSP Al-Ti-Mg sample than that in the FSP Al-Ti sample (Table 9.2) can be attributed to the formation of a $Ti_2Mg_3Al_{18}$ phase in the Al-Ti-Mg sample.

With respect to the FSP Al-Ti sample, the strength of the FSP Al-Ti-Cu sample increased substantially, while the elongation decreased to 5%. This can be explained by the following factors:

- First, Cu addition enhanced the Al-Ti reaction during FSP, resulting in the formation of more Al_3Ti particles in the Al-Ti-Cu specimen (Table 9.2). This would result in more significant strengthening of the matrix through the Orowan bowing mechanism.
- Second, some part of Cu is held in the Al matrix as a solute, which would exert additional strengthening effect due to solution strengthening mechanism.
- Third, fine Ti particles in the FSP Al-Ti-Cu sample can efficiently bear load during tension. Observing more cracked Ti particles on the

fracture surface of the FSP Al-Ti-Cu sample than that of the FSP Al-Ti sample confirmed this viewpoint (Figure 9.20(c)). Although some Al_2Cu particles precipitated in the FSP Al-Ti-Cu sample, these particles reveal relatively coarse size, similar to that in over-aged Al-Cu alloy, and therefore did not affect the strength of the Al matrix.

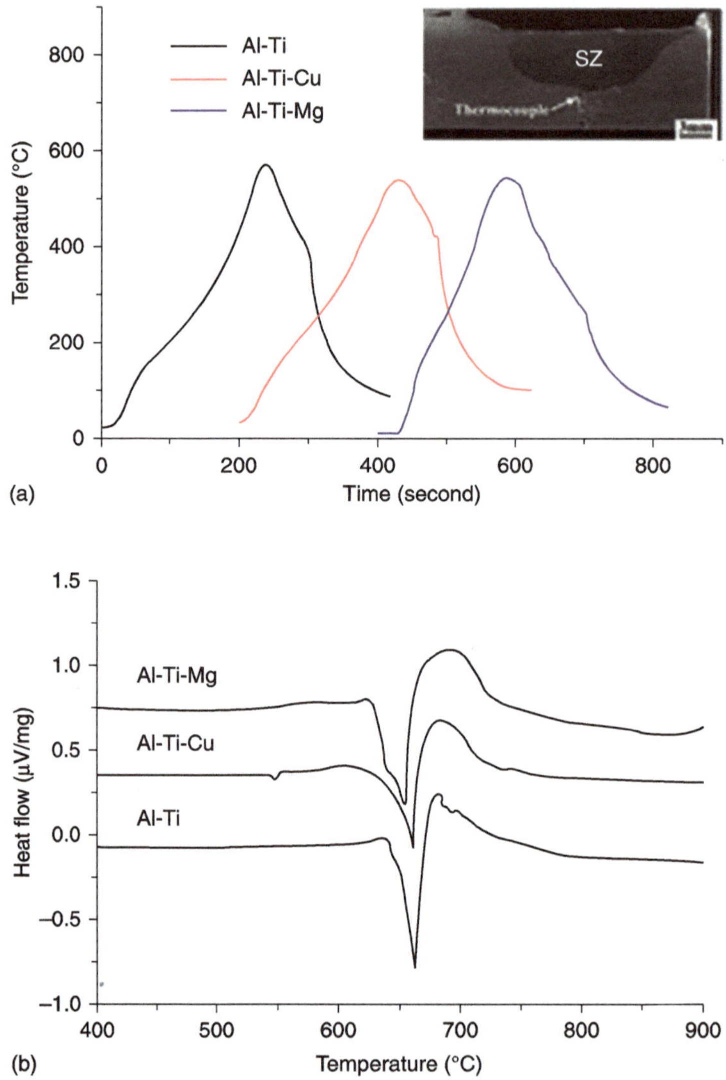

Figure 9.19 (a) Thermal profiles during FSP; and (b) DSC curves of forged samples [29]

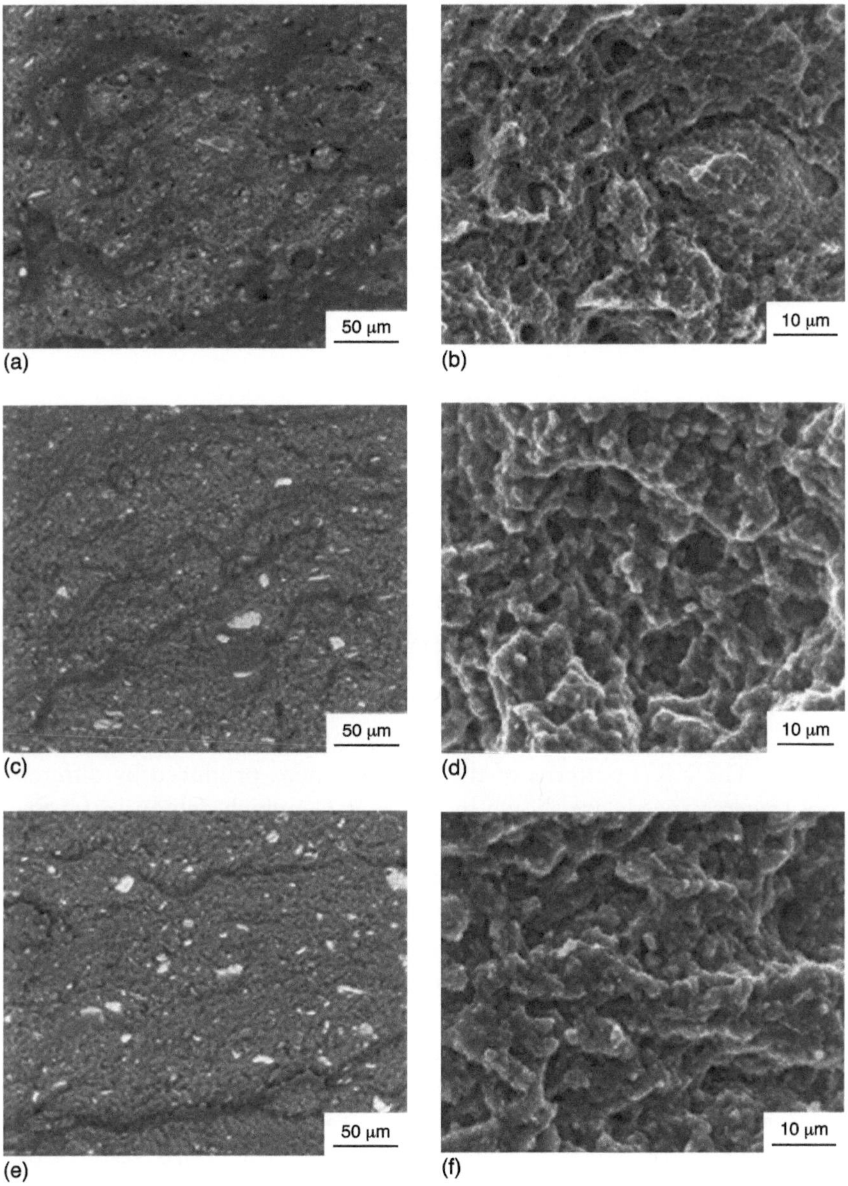

Figure 9.20 SEM images of fracture surfaces of FSP samples: (a) Al-Ti (BEI); (b) Al-Ti (SEI); (c) Al-Ti-Cu (BEI); (d) Al-Ti-Cu (SEI); (e) Al-Ti-Mg (BEI); and (f) Al-Ti-Mg (SEI) [29]

The high volume fraction and the fine size of the reinforcing particles caused the smaller interparticle distance, and resulted in a larger strengthening effect on the matrix. More Ti particles crack during tension when the strengthening of the matrix occurs (Figure 9.2(e)), based on the load transferring mechanism. However, brittle fracture of the FSP Al-Ti-Mg sample (Figure 9.20(f)) resulted due to the small interparticle distance. Consequently, Cu and Mg addition enhanced the strength of the *in situ* composites substantially, due to the introduction of more strengthening modes and more reinforcing particles; however, the elongation reduced dramatically [29].

9.5.4. Al-$Al_{11}Ce_3$-Al_2O_3 intermetallic composite

Al_2O_3 compounds were reported as interesting compounds for enhancing properties of Al matrix. Besides the stability of Al_2O_3 in aluminium, $Al_{11}Ce_3$ compound is also expected to have a low coarsening rate at elevated temperatures, since Ce has low diffusivity and solubility in aluminium. Therefore, Al-$Al_{11}Ce_3$-Al_2O_3 composites are supposed to have good elevated temperature properties.

An Al-$Al_{11}Ce_3$-Al_2O_3 *in situ* composite can be produced by applying FSP to the mixture of pure aluminium powder and pure cerium oxide (CeO_2). The XRD patterns of the FSP specimens produced by different tool traverse speeds are shown in Figure 9.21, which demonstrates that CeO_2 peaks have disappeared after four FSP passes. For FSP specimens with low tool traverse speed, no CeO_2 peak is found, indicating that powders have reacted completely with each other after four FSP passes. However, at higher tool traverse speeds, some of the CeO_2 powders are only partially reduced by Al, which may be attributed to the shorter reaction time and lower reaction temperature [30].

In addition, lower tool traverse speed, as a result of a higher stirring time, causes a finer and more uniform particle size. Most of the fine (submicron-sized) particles are attributed to the *in-situ* reaction during FSP, while those particles with a relatively larger size result from fragmentation of the bulky $Al_{11}Ce_3$ particles formed during sintering. Presence of a large amount of reinforcing particles, $Al_{11}Ce_3$+Al_2O_3, and good interfacial bonding for an effective load transfer in the composite, result in a higher Young friction stir modulus (105 to 109 GPa) than that of aluminium (70 GPa) (Figure 9.22). Lower strength of the specimens produced with higher tool traverse speeds can be attributed to the presence of the lower amount of reinforcing particles in them.

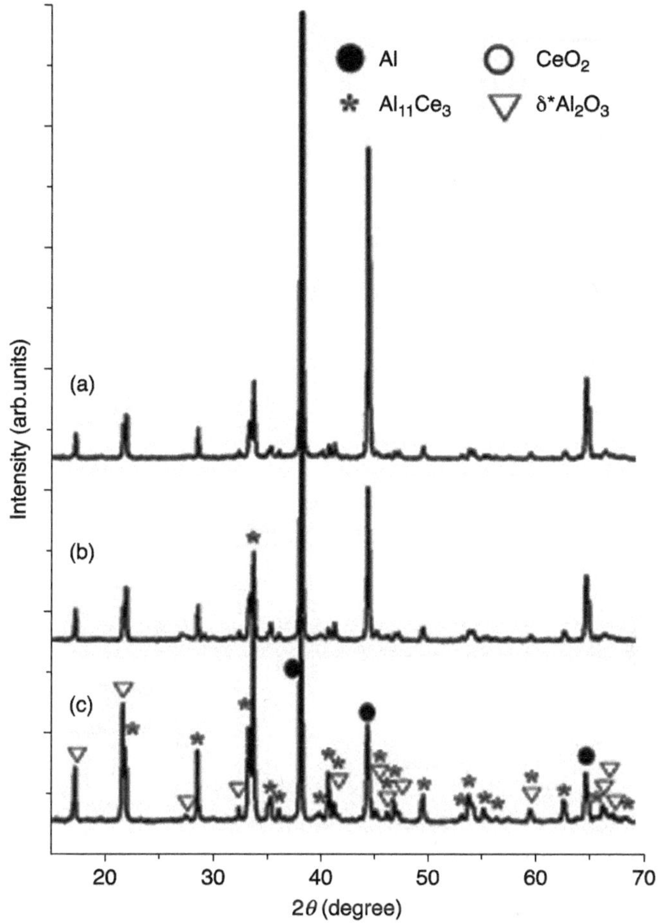

Figure 9.21 **XRD patterns of FSP specimens processed by different tool traversing speeds. Specimens were sintered at 883 K and followed by four FSP passes. The FSP was performed with a tool rotation of 500 rpm and traverse of: (a) 30 mm/min; (b) 85 mm/min; and (c) 120 mm/min [30]**

The temperature resulting from the combination of tool rotational and traverse speeds must be high enough to initiate the reaction; however, it should not be too high. A high temperature not only results in coarser microstructure, but also promotes exothermic Al-CeO_2 reaction, which may cause excessive melting in the material. The presence of a significant

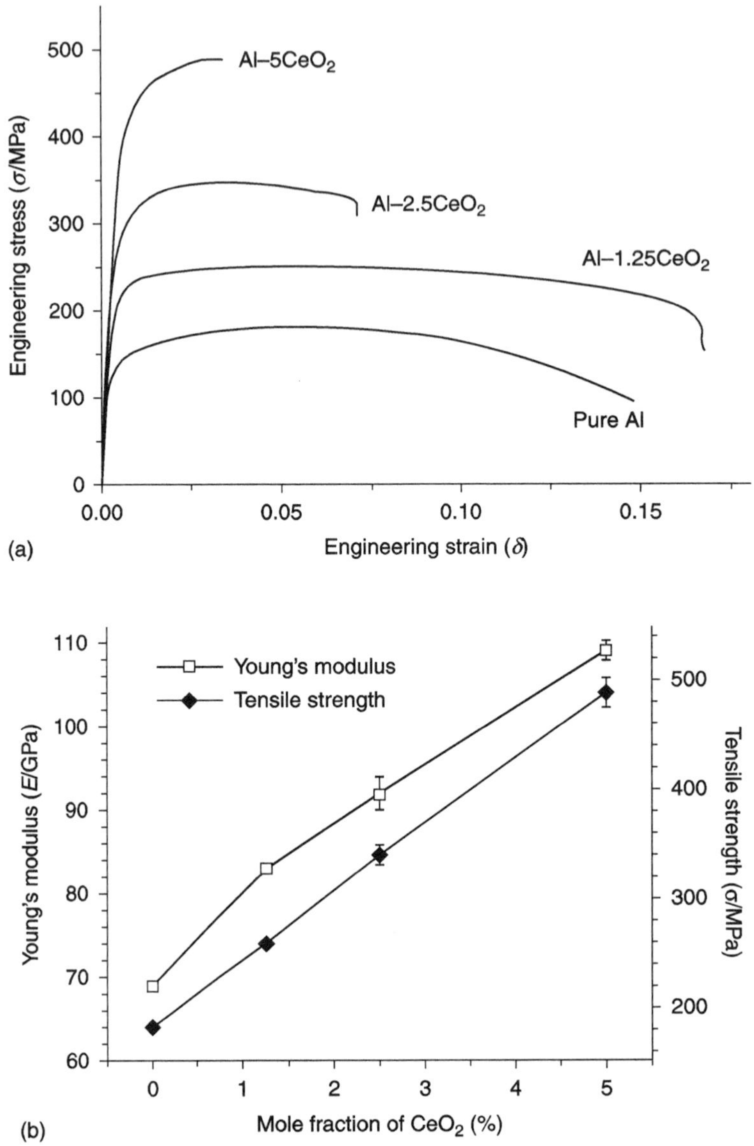

Figure 9.22 (a) Tensile stress-strain curves for FSP specimens with different CeO_2 content; and (b) Young's modulus and tensile strength of the composites increase monotonically with increasing CeO_2 content [30]

amount of liquid in the material during FSP may result in large cavities in the SZ.

The size of the majority of $Al_{11}Ce_3$ particles are in the same order as aluminium grains. Therefore, $Al_{11}Ce_3$ particles are not supposed to interact directly with intragranular dislocations and cannot contribute to the strength via the Orowan bowing mechanism. These particles may contribute to the strength by load sharing.

9.5.5 Al-Fe in situ nanocomposite

Aluminium-iron alloys are applied to high temperature applications due to the dispersion of fine second-phase particles, which are stable at elevated temperatures. Torsion under high pressure, which can produce material only with limited size, was used to fabricate these materials, while a combination of the hot working nature of FSP and the exothermic reaction between aluminium and transition metals can be more beneficial. The FSP can provide the following functions in one step, SPD to mix and refine the constituent phases in the material and hot consolidation to produce a fully dense solid.

The Al–Fe reaction during FSP can result in the production of $Al_{13}Fe_4$ (also denoted as Al_3Fe) with a minor amount of Al_5Fe_2. Three phases with different contrast can be characterized in the BEI images (Figure 9.23). According to EDS analysis, the brightest phase is Fe, the dark grey matrix is Al and the light grey phase, which surround the Fe particle or dispersed in the Al matrix, is the Al–Fe compound. With increasing the passes of FSP, the Fe-containing particles are refined and more uniformly distributed in the Al matrix (Figure 9.23). However, the reaction between Al and Fe is not complete, even after four FSP passes (Figure 9.23(c)), since iron particles still exist in the SZ [31].

Due to the low temperatures and short reaction times in FSP, the Al_5Fe_2 compounds are not abundantly existent in FSP specimens [32]. The heat formation of Al_3Fe is 28.1 kJ/g-atom (at 1173°K) and the heat of fusion of Al at 933°K is 10.8 kJ/g-atom. Therefore, the exothermic reaction can release enough heat to cause eutectic melting of Al and $Al_{13}Fe_4$. The interfacial compounds $Al_{13}Fe_4$ are fragmented as a result of the eutectic melting combined with the large plastic strain introduced by the rotating FSP tool. Then the compounds are rapidly removed and dispersed in the aluminium matrix. Since the heat is rapidly absorbed by the surrounding aluminium matrix, the melt solidifies rapidly.

Rapid cooling and vigorous stirring in FSP result in formation of fine $Al_{13}Fe_4$ particles during eutectic solidification. In addition, a

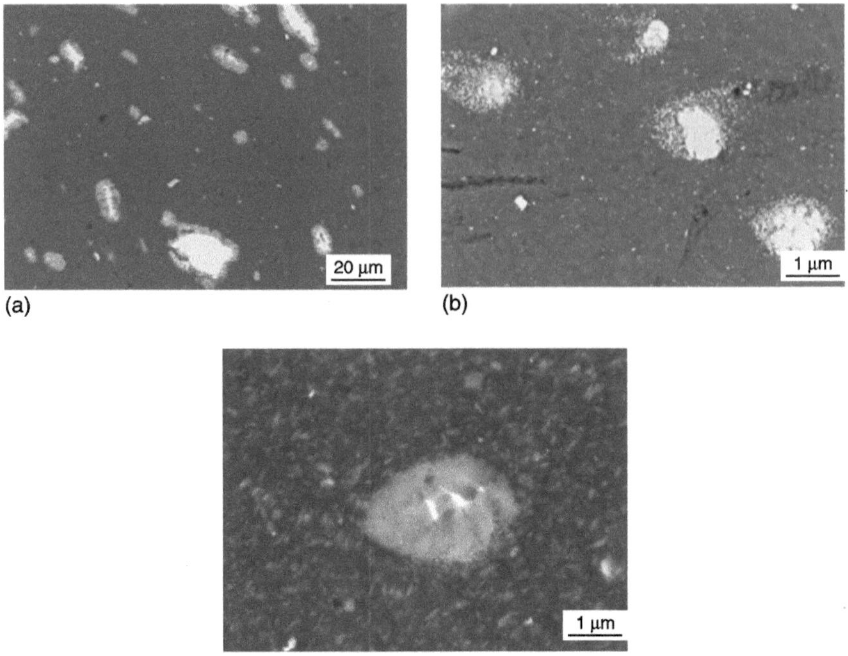

(c)

Figure 9.23 Backscattered electron image (BEI) showing the microstructure of: (a) stir zone; (b) two-pass FSP; and (c) four-pass FSP [31]

supersaturation of Fe in the Al matrix can also form during rapid cooling, which can be followed with precipitation of fine $Al_{13}Fe_4$ particles. The nanometer-sized $Al_{13}Fe_4$ particles, with the elliptical shape shown in Figure 9.24, can be produced by either eutectic solidification or solid state precipitation in FSP. However, some particles of irregular shape and larger size may result from direct fragmentation of the interfacial $Al_{13}Fe_4$.

The presence of local melting at the Al/Fe interface and the large plastic strain imposed in FSP can effectively remove the $Al_{13}Fe_4$ particles, such that direct contact between Al and Fe can be maintained, and consequently the reaction can proceed rapidly at the interface.

Increasing FSP passes enhance the Al–Fe reaction, which results in formation of more $Al_{13}Fe_4$ particles and higher hardness (Figure 9.25). Grain size refinement in the Al matrix and the formation of fine $Al_{13}Fe_4$ particles uniformly distributed in aluminium matrix are the main causes of hardness and modulus increase (Table 9.3). The uniform distribution of fine particles within the aluminium matrix can contribute significantly to Orowan strengthening.

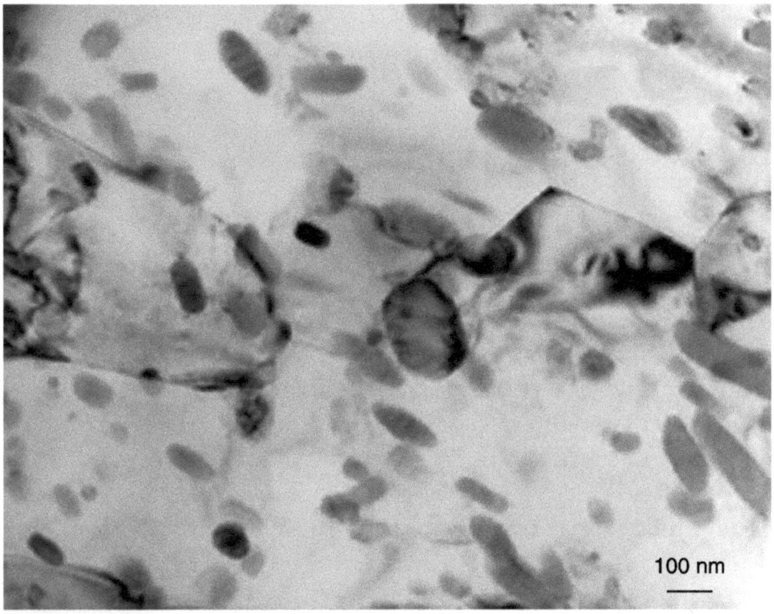

Figure 9.24 The microstructure of FSPed specimens revealed by TEM. Fine $Al_{13}Fe_4$ particles are uniformly distributed in aluminium matrix [31]

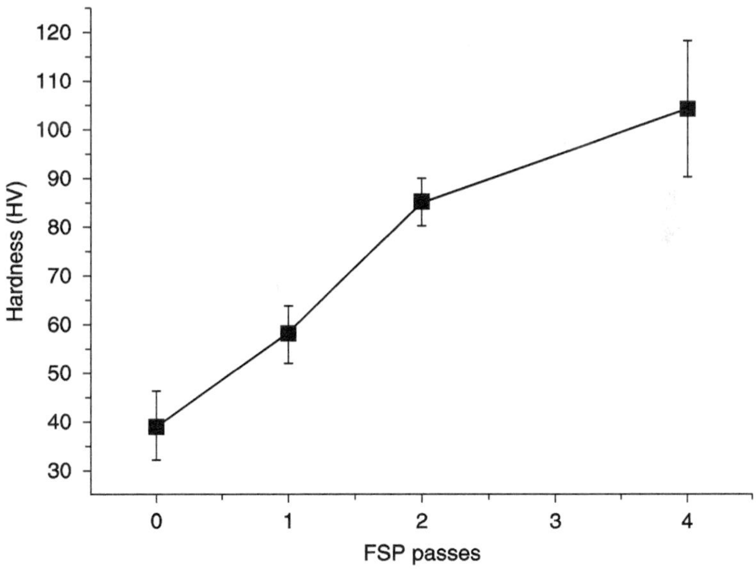

Figure 9.25 The average hardness in the stir zone increases by increasing the FSP passes [31]

9.5.6 Transformation of Al-Si alloy surface to modified hypereutectic alloy

Al–Si alloys can be classified in three groups:

1. hypoeutectic with less than 11% Si;
2. eutectic with 11 to 13% Si; and
3. hypereutectic with higher than 13% Si.

Each of these groups is suitable for particular applications. Hypoeutectic alloys normally contain ductile primary α-Al, which forms and grows dendritically, and a needle-like brittle hard eutectic Si phase. Eutectic alloys are composed of an Al–Si eutectic phase. Hypereutectic alloys are identified with coarse, primary angular Si particles together with a eutectic Si phase. For instance, the LM13 eutectic Al–Si as-cast alloy can be transformed to hypereutectic by inserting silicon powder via FSP.

Figure 9.26(a) shows the microstructure of LM13 base metal (BM), which includes primary α-Al dendrites, needle-like eutectic silicon, porosities (not shown) and some intermetallic phases such as Al_9FeNi, Al_3CuNi and Mg_2Si. Such a microstructure is responsible for the brittle behaviour of this alloy.

According to Figure 9.26(b), FSP can embed the silicon in the matrix, such that the eutectic silicon of the BM and the added silicon cannot be distinguished.

Addition of Si increases the hardness of FSP specimens significantly. Reduction of the hardness in the FSP specimens by increasing the pass

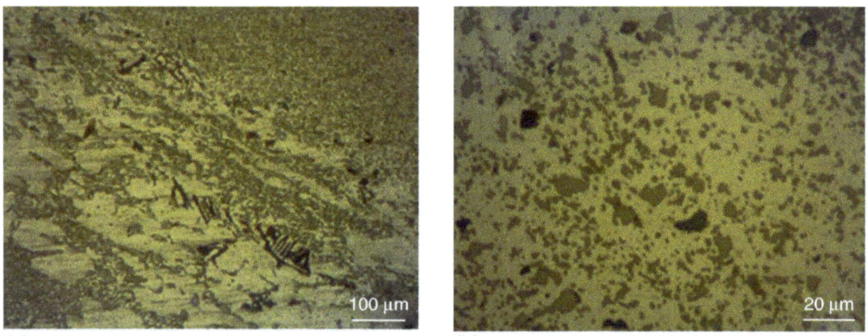

(a) (b)

Figure 9.26 Microstructure of: (a) LM13 base metal; and (b) FSPed specimens [33] number, in spite of decreasing the Si particles size and improvement in particle distribution, can be attributed to dislocation density and precipitate dissolution.

In Al–Si–Mg alloys, the amount of Mg_2Si precipitates in the FSP specimen is lower than that in the as-cast ones. Thus, some Mg_2Si precipitates dissolved in the matrix during FSP, and due to the high rate of cooling, the solutes cannot precipitate again. In addition, multi-passes of FSP result in more dissolution of precipitates, which can facilitate the reduction in hardness after each pass. Increase in the tool rotational speed results in better Si particle distribution and decreases its size, and consequently improves the tensile strength and elongation. Increasing the traverse speed will not make a noticeable variation in the strength; however, elongation reduces and hardness increases.

Figure 9.27(a) shows the SEM micrographs of fractured surfaces of as-cast LM13 alloy, which reveals the brittle fracture. The fracture surface can be categorized in the two-phase region:

1. fracture in the α-Al with some plastic deformation; and
2. brittle fracture in the eutectic phase.

The needle-like and primary Si phase can be seen in Figure 9.27(a), which exhibits the intercrystalline brittle fracture in the Mg_2Si intermetallic phase. The fracture surface of FSP specimen is shown in Figure 9.27(b). Formation of dimples, demonstrating ductile fracture, can be clearly

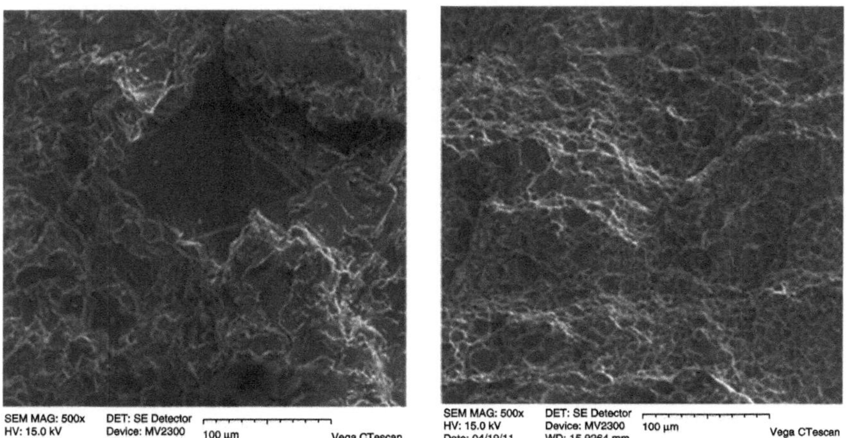

(a) (b)

Figure 9.27 SEM micrographs of fracture surface of: (a) as-cast LM13 Al alloy; and (b) FSPed specimen [33] observed on the fracture surfaces of the FSP specimens. In Al–Si casting alloys, an increase in the Si volume fraction will result in higher brittleness. However, in FSP specimens, in spite of adding 10% volume fraction silicon, the brittle fracture mode perfectly converts to the ductile mode. Grain refinement, disruption of needle-like Si particles and uniform redistribution of equiaxed particles are the main reasons for variation of the fracture mode of FSP specimens compared to the as-cast specimen.

By increasing the pass number of FSP, both tensile strength and elongation can be enhanced. This behaviour can mostly be attributed to the decrease in Si particle size accompanied by better particle distribution. Reduction in particle size increases the required crack nucleating stress, which postpones the fracture of the specimen and improves the tensile strength and elongation.

Finally, due to the presence of dendrite branches, eutectic phase, silicon size and morphology, intermetallic morphology and porosities in the as-cast alloy, the tensile properties are lower than that of the FSP ones. In the FSP specimen without introduced Si powder, the elongation is the highest. This reveals that although the addition of Si powder increases the strength, it decreases the ductility [33].

9.5.7 Fabrication of MgAlZn intermetallic alloys

Commercial cast or wrought type Mg–Al–Zn alloys, such as the AZ series, have been widely used in automobiles and electronic appliances. However, the hardness is typically low (in the range of 50–100 HV), and they suffer from insufficient creep resistance. A lot of effort has been made to improve the strength, hardness and creep properties of commercial magnesium alloys by adding rare earth elements or ceramic reinforcements. An alternative approach is by mixing a much higher portion (10–30 wt.%) of the existing solute elements Al and Zn through FSP.

FSP is applied in mixing elemental thin sheets of Mg, Al and Zn in various portions (Figure 9.28), resulting in hard intermetallic alloys with excess Vickers hardness of 350. The $Mg_3Al_2Zn_3$ phase is formed at approximately 360 °C during FSP, which along with some other nano-size binary or ternary phases, results in a greater hardness.

After two or three passes of FSP, the microstructure includes a fine-grained darker Mg (measuring 0.2–2 μm) surrounded by lighter Al and nearly white Zn (Figure 9.29). The low melting Zn (Tm ~ 420 °C) seems to have melted during FSP and is squeezed into a thin layer surrounding the Mg grains. A few large particles with distinct contrast in Figure 9.30,

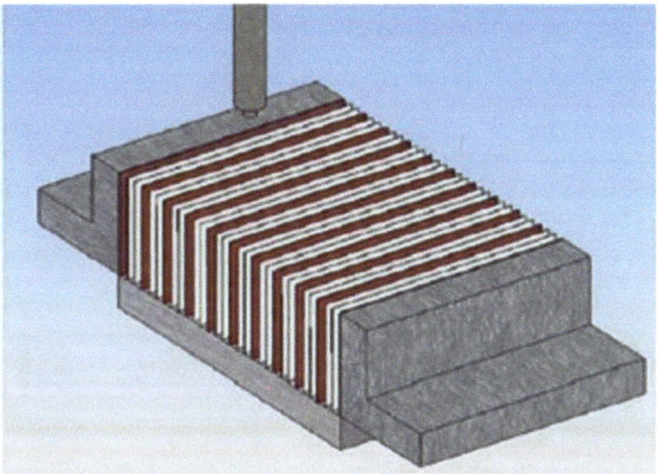

Figure 9.28 Schematic of the stacked AZ31 Mg alloy sheets (1 mm thick in dark contrast) with pure Al and Zn foils (0.2 mm and 0.125 mm thick in lighter contrast) [34]

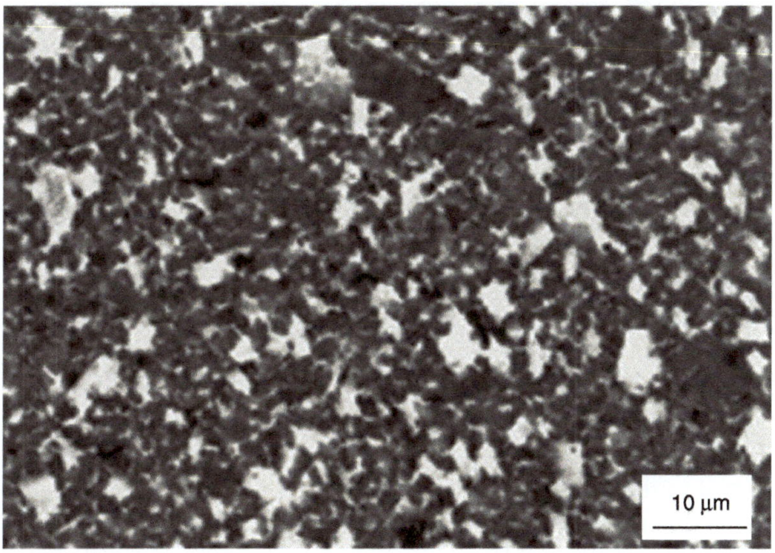

Figure 9.29 SEM/BEI micrograph showing the phase dispersion in the $Mg_{80}Al_{10}Zn_{10}$ system after three passes. The white contrast Zn appears to have melted and coats the Mg/Al phases [34]

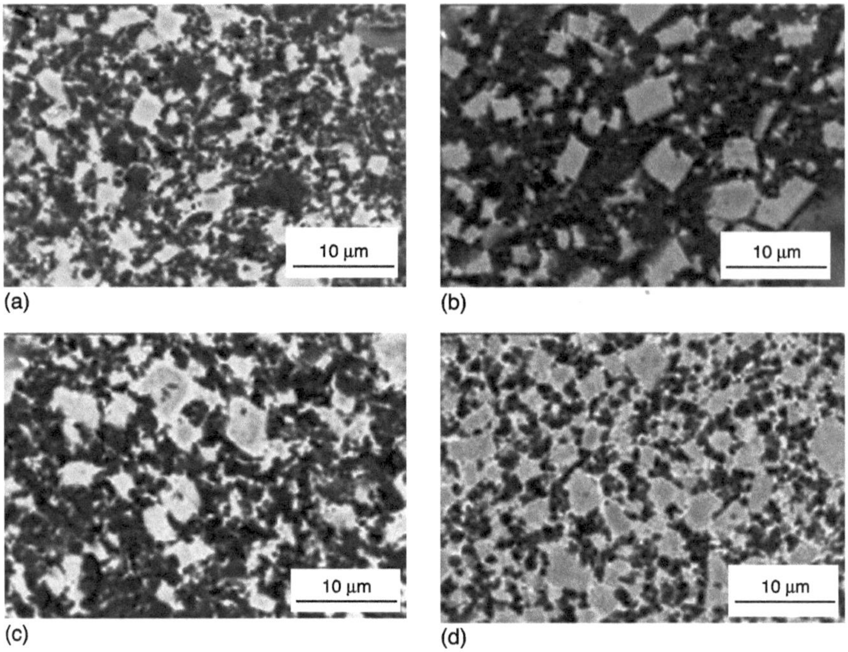

Figure 9.30 SEM/BEI micrographs: (a and b) the $Mg_{70}Al_{15}Zn_{15}$ system after two and three passes; and (c and d) the $Mg_{60}Al_{20}Zn_{20}$ system after two and three passes [34]

which are typically around 5 μm in size, are determined as $Mg_3Al_2Zn_3$. For higher Al and Zn content, this phase emerged even earlier after the first pass of FSP. With further FSP passes, this phase gradually become the dominant one (Figure 9.30).

With higher cooling rate using liquid N_2, the microstructures tend to be more uniformly distributed and even smaller $Mg_3Al_2Zn_3$ phases can be induced.

The typical hardness measurement results along the transverse cross-section are depicted in Figure 9.31. Increasing the FSP passes will enhance the HV values and reduce the variation of HV inside the NZ, which can be related to the formation of more intermetallic phases. Meanwhile, the hardness is seen to gradually increase with increasing Al and Zn. With liquid N_2 cooling, a slightly greater hardness than the same alloy system without liquid N_2 achieved.

Multi-passes and high fractions of Al and Zn elements cause apparent grain refinement, proved by the broadening of diffraction peaks and also SEM results [34].

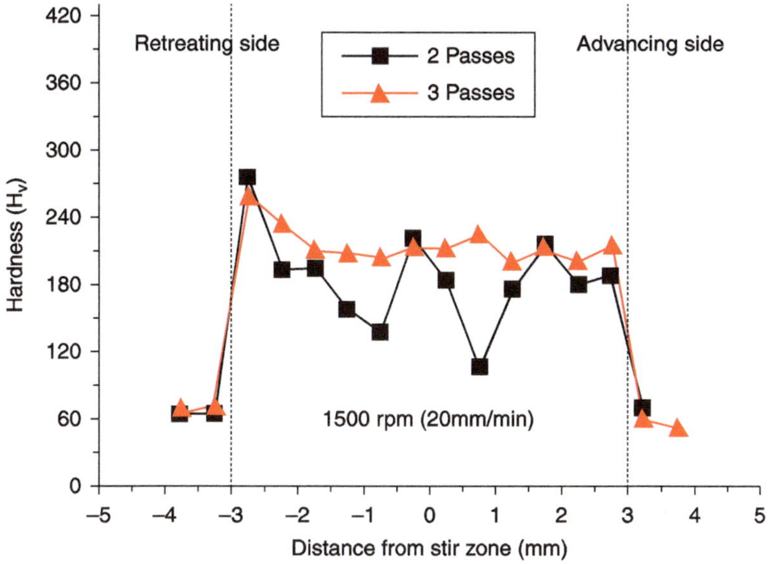

Figure 9.31 The variation of HV along the transverse cross-sectional plane of the $Mg_{60}Al_{20}Zn_{20}$ system [34]

9.6 References

1. Suñol, J.J and Fort, J. (2007), *Materials Developed by Mechanical Alloying and Melt Spinning*, New York: Marcel Dekker.
2. Tang, C.F. (2007), 'Spark plasma sintering of ODS cobalt base alloy powder by mechanical alloying', *Rare Metal Materials and Engineering*, 36: 1461–4.
3. Crivello, J.C. (2007), 'Limits of the Mg–Al γ-phase range by ball-milling', *Intermetallics*, 15: 1432–7.
4. Chawla, V. (2007), 'State of the art: Applications of mechanically alloyed nanomaterials: A review', *Materials and Manufacturing Processes*, 22: 469–73.
5. Cherdyntsev, V.V. (2007), 'Structure and properties of mechanically alloyed composite materials Al/Al-Cu-Fe quasicrystal', *Physics of Metals and Metallography*, 104: 497–504.
6. Krasnowski, M. (2007), 'Nanocrystalline FeAl-TiN composite obtained by hot-pressing consolidation of reactively milled powers', *Scripta Materiala*, 57: 553–6.
7. Karttunen, M. (2005), 'Electrically conductive composite powders and compounds produced with solid state synthesis', *Materials and Manufacturing Processes*, 20: 887–92.
8. Sharma, S. (2007), 'Mechanical crystallization of Fe-based amorphous alloys', *Journal of Applied Physics*, 102: 083544.

9. Lee, E.Y. (2007), 'Synthesis of composite Mg-Ni-Ca metal hydride by mechanical alloying', *Journal of Alloys and Compounds*, 446: 129–33.
10. Ueda, Y. (1996), Magnetoresistance and magnetism in Fe-Cu alloys produced by electroposition and mechanical alloying methods', *Materials Science and Engineering A*, 217: 371–5.
11. Suryanarayana, C., Ivanov, E. and Boldyrev, V.V. (2001), 'The science and technology of mechanical alloying', *Materials Science and Engineering A*, 304–6: 151–8.
12. Zoz, H., Ernst, D., Reichardt, R., Mizutani, T., Nishida, M. and Okouchi, H. (1998), 'Mechanical alloying in a production scale using cycle operation: Part II', *Metallurgy*, 52: 521–7.
13. Benjamin, J.S. (1970), 'Dispersion strengthened superalloy by mechanical alloying', *Metall. Transaction*, 1: 2943–51.
14. Suryanarayana, C. (2004), *Mechanical Alloying and Milling*, New York: Marcel Dekker.
15. Uan, J.Y., Chen, L.H. and Lui, T.S. (2001), 'On the extension microstructural evolution of Al-Al_3Ni *in situ* composite', *Acta Materiala*, 49: 313–20.
16. Zhang, Z.G., Akiyama, E., Watanabe, Y., Katada, Y. and. Tsuzaki, K. (2007), 'Effect of a-Al/A13Ni microstructure on the corrosion behavior of Al–5.4 wt.% Ni alloy fabricated by equal-channel angular pressing', *Corrosion Science*, 49: 2962–72.
17. Song, C.J., Xu, Z.M. and Li, J.G. (2007), '*In-situ* Al/AL_3Ni functionally graded materials by electromagnetic separation method', *Materials Science and Engineering A*, 445–6: 148–54.
18. Gonzalez, G., Sagarzazu, A., Bonyuet, D., Angelo, L.D. and Villalba, R. (2009), 'Solid state amorphisation in binary systems prepared by mechanical alloying', *Journal of Alloys and Compounds*, 483: 289–97.
19. Huanga, C., Xing, L. and Liming, K. (2010), 'Al-Ni intermetallic composites produced *in situ* by friction stir processing', *Journal of Alloys and Compounds*, 503: 494–9.
20. Hsu, C.J., Kao, P.W. and Ho, N.J. (2005), 'Ultrafine-grained Al–Al_2Cu composite produced *in situ* by friction stir processing', *Scripta Materialia*, 53: 341–5.
21. Sort, J., Zhilayer, A., Zielinska, M., Noguès, J., Suriñach, S., Thibault, J. and Barò, M.D. (2003), 'Microstructural effects and large microhardness in cobalt processed by high pressure torsion consolidation of ball milled powders', *Acta Materialia*, 51(20): 6385–93.
22. Greil, P. (2002), 'Advanced engineering ceramics'. *Advance Materials*, 14: 709.
23. Chianeh, V.A., Hosseini, H.R.M. and Nofar, M. (2009), 'Micro structural features and mechanical properties of Al–A_3lTi composite fabricated by *in-situ* powder metallurgy route', *Journal of Alloys and Compounds*, 473: 127–32.
24. Das, K. and Narnaware, L.K. (2009), 'A study of microstructure and tribological behaviour of Al–4.5% Cu/Al_3lTi composites', *Materials Characterizations*, 60: 808–16.
25. Mishin, Y. and Herzig, C.H.R. (2000), 'Diffusion in the Ti-Al system', *Acta Materiala*, 48: 589–623.

26. Zhang, Q., Xiao, B.L., Wang, D. and Ma, Z.Y. (2011), 'Formation mechanism of *in situ* Al_3lTi in Al matrix during hot pressing and subsequent friction stir processing', *Materials Chemistry and Physics*, 130: 1109–17.
27. Hsu, C.J., Kao, P.W. and Ho, N.J. (2007), 'Intermetallic-reinforced aluminum matrix composites produced *in situ* by friction stir processing', *Materials Letters*, 61: 1315–18.
28. Zhang, Q., Xiao, B.L., Wang, W.G. and Ma, Z.Y. (2012), 'Reactive mechanism and mechanical properties of *in situ* composites fabricated from an Al–TiO_2 system by friction stir processing', *Materialia*, 60: 7090–103.
29. Zhang, Q., Xiao, B.L. and Ma, Z.Y. (2013), '*In situ* formation of various intermetallic particles in AleTieX(Cu,Mg) systems during friction stir processing', *Intermetallics*, 40: 36–44.
30. Chen, C.F., Kao, P.W., Chang, L.W. and Ho, N.J. (2010), 'Effect of processing parameters on microstructure and mechanical properties of an Al-$Al_{11}Ce_3$-Al_2O_3 *in-situ* composite produced by friction stir processing', *Metallurgical and Materials Transactions A*, 41: 513–22.
31. Lee, I.S. and Kao, P.W. (2008), 'Microstructure and mechanical properties of Al–Fe *in situ* nanocomposites produced by friction stir processing', *Intermetallics*, 16: 1104–8.
32. Lee, J.M., Kang, S.B., Sato, T., Tezuka, H. and Kamio, A. (2003), 'Evolution of iron aluminide in Al/Fe *in situ* composites fabricated by plasma synthesis method', *Materials Science and Engineering A*, 362: 257–63.
33. Tutunchilar, S., Besharati Givi, M.K., Haghpanahi, M. and Asadi, P. (2012), 'Eutectic Al-Si piston alloy surface transformed to modified hypereutectic alloy via FSO', *Materials Science and Engineering A*, 534: 557–67.
34. Chuang, C.H., Huang, J.C. and Hsieh, P.J. (2005), 'Using friction stir processing to fabricate MgAlZn intermetallic alloys', *Scripta Materialia*, 53: 1455–60.

Friction stir welding/processing as a repair welding

P. Salami, University of Khaje nasireldin toosi, Iran and T. Khandani, P. Asadi and M.K. Besharati Givi University of Tehran, Iran

DOI: 10.1533/9780857094551.427

Abstract: Some metal alloys are difficult to weld and some types of defects are formed during (fusion) welding and casting, including porosity and solidification cracking, which have significant influence on mechanical properties of the joints. Therefore, the repair welding process of defect-existent joints and products attracts much attention. FSW, as a solid state joining technique that has been demonstrated in a variety of metals, such as aluminum, copper, titanium and steel, now can be used as a repair technique for welding joints and casting products, as well as worn or corroded surfaces, in order to modify the microstructure or remove the defects. This chapter presents an overview of repair-welding use of FSW/FSP.

Key words: Friction stir welding/processing, repair welding, microstructural modifications, defect removal.

10.1 Introduction

Repair welding is an essential process in fabrication of some structures. Properly repaired structures may have equivalent static strength, ductility, fracture toughness and fatigue strength levels as the base structures. Repair welding can be carried out as a logical procedure that ensures the production of a usable and safe component, or it can be approached haphazardly. The latter approach results in poor-quality workmanship and can lead to failed parts, large warranty claims and dissatisfied customers [1]. The logical sequence of successful repair welding is as follows:

- *Welding procedure*: the welding procedure must be accessible to the use of the welders. It must include the process that is used and specific information concerning the welding joint technique required.
- *Welding equipment*: sufficient welding equipment should be supplied, then delays will not occur. Standby equipment may also be required. This not only comprises welding equipment, but also includes sufficient holders, grinders, wire feeders if required, cables, etc.
- *Materials*: sufficient materials must be accessible to the entire job. This includes the filler metals, which should be stored properly for use on the repair. It also includes materials such as insert pieces, reinforcing pieces, etc. Materials also include fuel for maintaining preheat and interpass temperature, shielding gases if used, and fuel for engine powered welding machines.
- *Alignment markers*: before weld making, alignment markers are occasionally employed. These can be nothing more than centre punch marks made across the joint at various locations.
- *Welding Sequences*: the welding sequence should be well described in the welding procedure.
- *Safety*: ultimately, safety cannot be overlooked throughout the welding operation. For instance, when fuel gases are used for preheating, etc., ventilation must be provided
- *Weld quality*: the quality of the weld must be constantly checked. The final weld should be smooth and there should be no notches and reinforcing. In order to use it, it should flowed smoothly throughout the existing structure. In fact, grinding should be done to maintain smooth flowing contours.

Repair welding can be in one of three general categories:

1. repair of weld defects;
2. repair of failed parts;
3. repair of worn parts.

10.2 Welding processes

All of the common welding processes can be applied for repair welding based on the particular application.

10.2.1 Gas metal arc welding

One of the serious problems in welding of aluminium alloys is development of hydrogen gas porosity during solidification of the weld. Fatigue strength and tensile strength of welded joints is reduced by inclusion of porosity. The other serious problem in aluminium welding is hot tearing solidification cracking. This occurs when the thermal stresses that build up during freezing, exceed the strength of the solidifying weld metal. Common methods that are used for increasing the tendency of solidification cracking are listed as:

- altering the weld composition;
- close process control; and
- control of grain structure within the fusion zone (FZ).

Liquation cracking tendency within the heat affected zone (HAZ) is a function of heat input, in fact it is the reason of local melting and presence of specific phases of grain boundaries. The aluminium repair process is complex method in comparison with steel, because the range of materials that can be welded is narrower, and also control of welding variables is constantly needed. Moreover, welded components which have failed during service life are often repaired to extend life. Weld repaired specimens have lower life compared to weld specimens, because of greater heat input that can weaken the material to an even greater extent, as well as increase the size and number of defects in the material. One of the main requirements of the aluminium welding technique is reduction of heat passed to the weld metal. Pulse gas metal arc welding (GMAW-P) has been applied to reduce the heat input. This mode is characterized by pulsing the current between low-level background current and high-level peak current. It generally operates in one drop per pulse, and in this way stability of the arc is provided. It also produces less distortions and fumes. Improved deposition character, like improved wetting, is useful for overcoming viscosity at the joint (which is found in reinforced aluminium alloys), is found in GMAW-P when compared to conventional GMAW [2].

10.2.2 Gas tungsten arc welding

Gas tungsten arc welding (GTAW) or tungsten inert gas (TIG) is commonly applied to repair tools and dies, especially components that are made of aluminium and magnesium [3]. The porosity effect on fatigue strength GTAW repaired D357 aluminium alloy has been reported. The repaired alloy underwent a post-weld heat treatment (PWHT) involving solution and artificial ageing. In comparison to the as-cast specimens, the weld-repaired specimens represent a much wider distribution of fatigue cycles. About 20% of the welded specimens had achieved 400 000 fatigue cycles, and about 20% of them had less than 200 000 fatigue cycles.

The remaining welded specimens were approximately distributed between 250 000 and 350 000 fatigue cycles. In contrast, all of the as-cast specimens had more than 250 000 fatigue failure cycles, and about 45% of them had less than 400 000 fatigue cycles. The number of failure cycles of the welded specimens demonstrated a much wider distribution band than those of the as-cast specimens. Crack-like shrinkage porosity was identified as the predominant factor leading to the inconsistent fatigue strength of the welded specimens. Spherical pores also contributed to decreasing the fatigue life of welded specimens, although it is not as significant as the crack-like pores [2].

10.2.3 Laser welding

Owing to occurrence of various types of defects such as oxide films, cracks and cavities, magnesium alloys have not usually been welded to date, except for some repaired structures. The wider application of magnesium alloys needs reliable welding processes. TIG and metal inert gas (MIG) processes are usually employed to remove and repair the casting defects for magnesium alloys. However, low welding speeds, large HAZ and FZ, high shrinkages, variations in microstructures and properties, evaporative loss of alloying elements, high residual stress and distortion of arc-welded joints have caused attention focused on laser welding because of the low and precise heat input, small HAZ, deep and narrow FZ, low residual stress and weldment distortion and high welding speed.

Two main types of industrial lasers are CO_2 and Nd:YAG, which have been used to investigate the weldability of magnesium alloys. Crack-free laser welded joints with low porosity and good surface quality can be obtained for some magnesium alloys, in particular wrought material, using appropriate laser processing parameters. In comparison with electron beam welded joints, the laser-welded specimens can tolerate higher load amplitudes for a given number of cycles to failure, due to lower notch sensitivity for the laser-welded specimens [4].

A recent application of laser forming with a far-reaching industrial significance concerns laser deposition and repair of single-crystal high-pressure high-temperature gas turbine blades. The process combines the advantages of a near net-shape manufacturing with the close control of the solidification microstructure in laser forming. Lately, the feasibility of this proposition has been called epitaxial. Laser metal forming has shown that it may be useful in repair of cracked or worn parts of single crystal turbine blades and so could extend the engine life. A careful study of the solidification microstructure under different laser forming regimes has proposed the processing windows for generation and repair of single crystal super alloy turbines by laser deposition. If the deposited or cladded layer is polycrystalline, electron back scatter diffraction (EBSD) and mechanical properties of the layer correlated with orientation of the grains can determine the orientation of the epitaxial grown grains [5].

10.2.4 Micro-welding

GTAW or an equivalent process is commonly used in the repair of turbine blades and vanes. But unfortunately, these components are often sensitive to HAZ cracking during the weld repair process. It can be a major problem, especially in cast alloys because of their coarse-grain size, and where the (Al + Ti) content is in excess of 3 to 4%. In low stress non-rotating components such as vanes, vacuum brazing is also applied. Micro-welding has the potential to deposit small amounts of filler at low heat input levels with minimum HAZ. Thus, it is an attractive process for depositing a quality weld. Micro-welding process is used to evaluate the effects of process parameters on deposition rate, void content, and crack density with filler alloys Inconel 625, Rene 41, Nimonic 105 and Inconel 738. Moreover, this process is proposed as a suitable process for repairing Ni-base super alloys with (Al + Ti) contents covering a range of 0.8 to 6.8 wt.%, especially when alloys are extremely susceptible to HAZ micro-fissuring [6].

10.2.5 Friction stir welding/processing

Many marine propellers, which are used by the US Navy, are made using nickel aluminium bronze, a Cu–9Al–5Ni–4Fe alloy, due to suitable resistance by sea water corrosion, and it is combined with moderate strength, ease of casting, low magnetic permeability and excellent resistance to shock. Several issues in casting a nickel aluminium bronze propeller are important. Owing to the variable thicknesses in propellers from the blade tip to the blade root and base, variable cooling rates are encountered leading to variability in microstructure. Overall slow cooling rates result in coarse microstructures with reduced mechanical properties. Furthermore, substantial porosity is encountered in as-cast nickel aluminium bronze. A layer of refined, homogeneous microstructure can be created by friction stir processing (FSP) of the propeller surface, with the elimination of porosity, over the entire exterior of the propeller or only in a selected area of the propeller. Thus, the surface mechanical properties, and thereby service performance of the propeller will improve. In addition, this is a solid state process and the workpieces never experience temperatures in excess of their liquids temperatures, and since no filler material is used, the chemical composition of the weld is determined only by the base metal(s).

There were several samples of nickel aluminium bronze in an experimental work of FSP to investigate microstructural properties. The first sample was processed with a single-pass FSP using a tool rotational speed of 800 rpm (revolutions per minute) and a tool traverse speed of 4 ipm. The second sample was processed with multiple FSP passes using the same rpm/ipm combination in a rectangular spiral pattern (Figure 10.1), by working from the inside to the outside of the plate and

Figure 10.1 Example of a rectangular spiral pattern used for FSP of multiple pass plates [7]

keeping the advancing side (AS) of the tool to the outside of the plate. The third sample was processed with a single FSP pass using a tool rotational speed of 1200 rpm and a tool traverse speed of 2 ipm and finally, the fourth sample came from a plate that was processed with multiple FSP passes using a tool rotational speed of 1200 rpm and a traverse speed of 2 ipm. As with the second sample, the fourth was done in a rectangular spiral pattern by working from the inside to the outside of the plate and keeping the AS of the tool to the outside of the plate.

Homogeneous and refined microstructure has been realized on the macro-scale, optical montages. However, in both single pass samples inhomogeneity was described as apparent on the retreating side (RS) of the SZ. The application of successive passes on the RS removed these inhomogeneities. Moreover, it is obvious that the approximate volume fraction of β transformation products seems to be smaller in the multipass sample than that of the single pass sample, as it is evident by less dark etching, which is shown in Figure 10.2(a) rather than in Figure 10.2(b). In addition, the size of α grains were conspicuously coarser in the multipass sample than that of the single pass sample, as it is obvious in Figures 10.3(a and b).

It is announced that a single-pass FSP significantly refines and homogenizes the microstructure, while multipass lead to further homogenization. It is noted that the SZ in samples, which underwent a single-pass FSP, experienced higher temperatures than that in the multiple

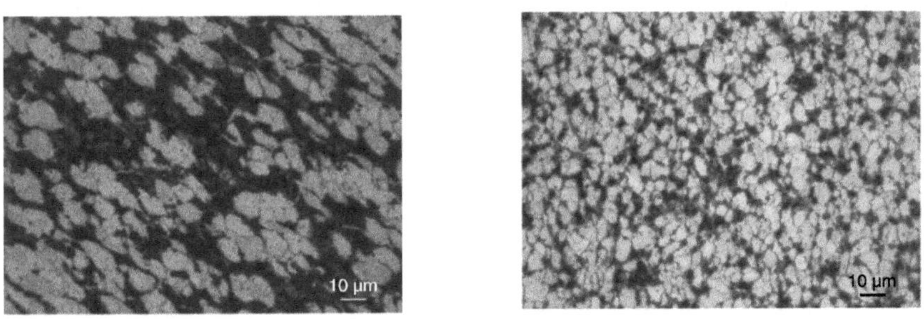

(a) (b)

Figure 10.2 Optical micrographs from 0.5 mm below surface of sample: (a) 800/4 single pass; and (b) 800/4 multipass [7]

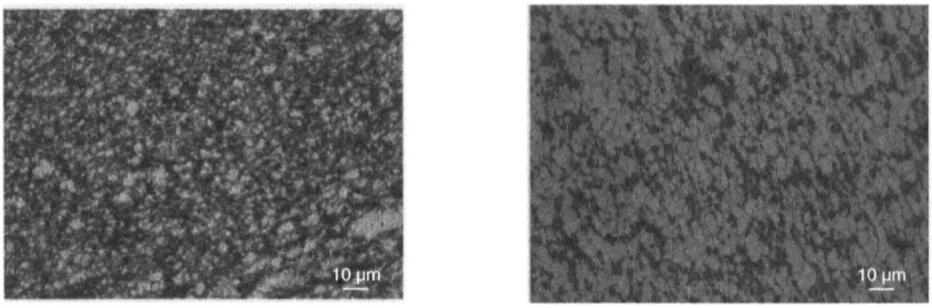

(a) (b)

Figure 10.3 **Optical micrographs from 5.3 mm below surface of sample: (a) 800/4 single pass; and (b) 800/4 multipass [7]**

pass samples. The required temperature of multiple-pass samples show less apparent dependence on processing conditions and location of the SZ, while the temperature attained on single-pass samples varied more with position and conditions [7].

The size and morphology of the microstructure in the material having undergone FSP with multiple passes is more homogeneous, though it is also somewhat coarser. The added strain from the multiple passes accounts for the homogeneity, while the increased dwell time at high temperature accounts for the larger grain size [7].

Generally, commercial transport aircraft are made up of a built-up metallic structure where the skin-to-stringer, skin-to-clip and clip-to-frame joints are mainly riveted. Such joints have been the subject of extensive research for many years, especially in terms of multiple site and widespread fatigue damage. Lately, FSW was identified as the 'key technology' for fuselage and wing manufacturing and repairing by leading aircraft manufacturers.

FSW can be applied to repair voids in an aluminium alloy. The method of repairing a circular void, surface and subsurface longitudinal voids and an in-process exit hole in a weld, includes steps of fabricating filler material of the same composition or compatible with parent material into a plug form to be fitted into the void, positioning the plug in the void and FSW over and through the plug (Figure 10.4) [8].

A repairing method, cracking in a component utilizing FSW, was announced by the Boeing Company for the first time. The cited method includes preparing a surrounding surface of the crack for repairing and

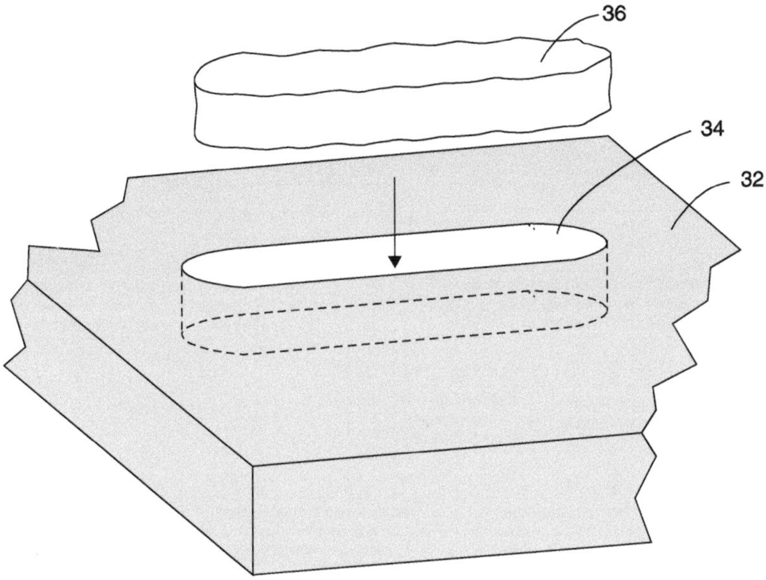

Figure 10.4 Schematic illustrating a method for repairing a void [8]

welding a first portion of the component on a first side of the crack to a second portion of the component on a second side of the crack to form a fused crack area. A logic flow diagram of this method is presented in Figure 10.5. For additional strength, the method may also include a patch over the fused crack area.

There are several advantages over existing crack repairing techniques, one of which is to provide a welding process of repairing cracks that does not melt material of the component, thereby minimal distortion, residual stress and alteration to chemical and physical properties of the component will be provided. Also, a new repairing crack method is provided by the present invention, which is more durable and longer lasting than traditional repair techniques. The service life of the component is increased over that of even the BM by using FSW and the application of the patch. On the basis of the component with the patch, the service life can be potentially increased to more than double. A FSW component of the present method plus the applied patch is able to withstand increased flight cycles, therefore enhancing the productive life of the component [9].

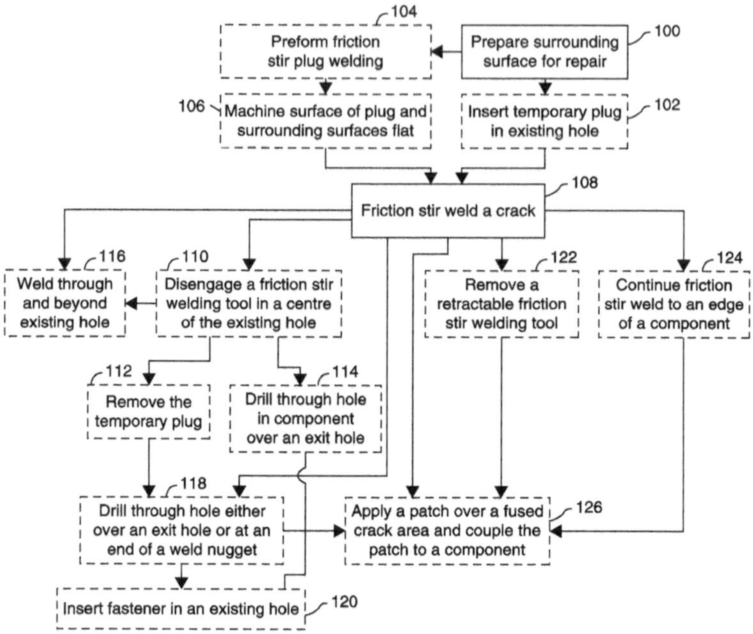

Figure 10.5 Logic flow diagram illustrating the method of repairing the crack [9]

Rapid injection of molten metal into a mould under high pressure produces the aluminium die casting alloys. These are mass produced, and they have dense and fine grain surface. Nevertheless, a casting defect and an abnormal structure, such as a cold flake, is formed in the BM caused by destruction of solidified layers in a metal sleeve. Mechanical properties of the BM are substantially degraded by these defects. According to this perspective, the mechanical properties of aluminium die casting alloys can be improved by FSP.

Four-millimeter thick plates of ADC12 aluminium die casting alloys were employed in a survey. The shape of the FSW tool was like a columnar with a screw probe. The diameter of the tool shoulder was 15 mm. The diameter of the tool probe was 5 mm, and its length was 3.9 mm. Tool plunge down force was 14.2 kN, due to the processing parameters. In addition, tool rotation speed was 1250 rpm, and the plate travel speed was 500 mm/min. Figure 10.6 demonstrates the bead appearance and X-ray radiography of the multipass FSP sample. Obviously for a wide area inside the sample, a smooth surface was obtained, and no inner defect, such as a cavity, was detected on the basis of X-ray radiography.

The hardness profile of a cross-section of the multipass FSP sample is a uniform 110 HV, which is about 20 HV higher than the BM (Figure 10.7). The hardness increase in the multipass FSP part is a result of deleting the casting defect, fine dispersion of Si particles, and grain refinement.

The tensile properties of the ADC12 BM and multipass FSP specimens are shown in Figure 10.8. The tensile strengths of the multipass FSP

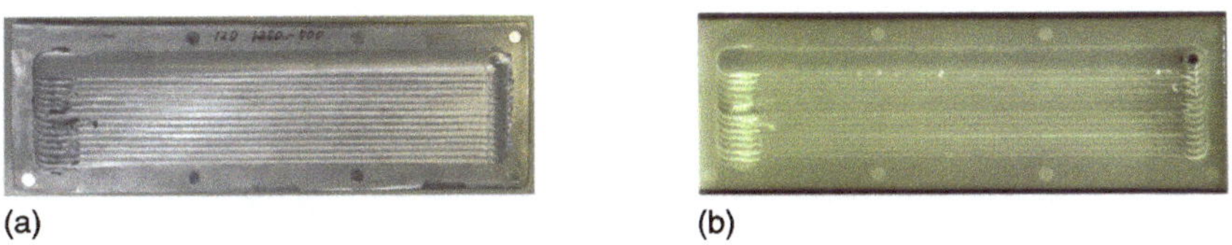

Figure 10.6 (a) Surface appearance; and (b) X-ray radiography of the multipass FSP sample [10]

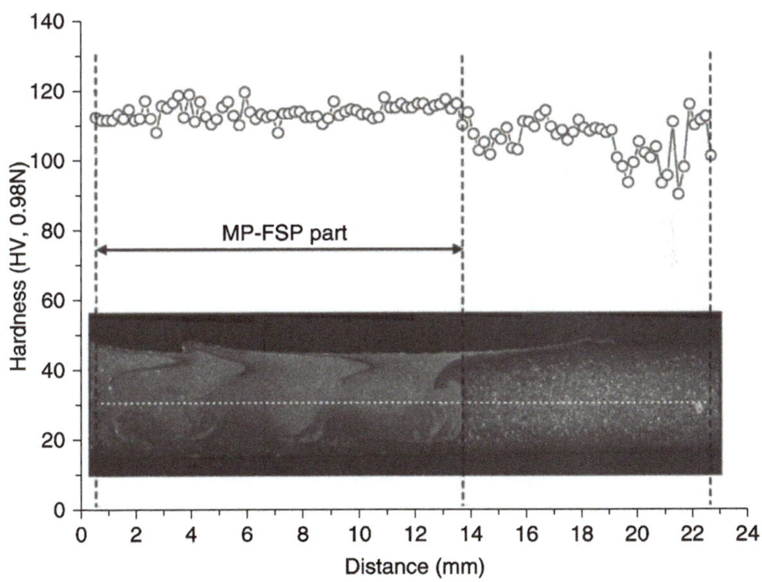

Figure 10.7 Hardness profile of cross-section in the multipass FSP sample [10]

specimens are much higher than the BM. The average tensile strength is 330 MPa, which is increased in the parallel direction around 1.7 times over the BM of 190 MPa, although the strength is noticeably scattered in the BM, the same as in the hardness profile. The factors which increase the tensile strength for the multipass FSP samples are listed as follows:

- elimination of the casting defects (i.e. porosity and cold flake);
- uniform dispersion of the finer Si particles;
- grain refinement of aluminium matrix.

The appearance and scanning electron microscope (SEM) images for the typical fracture surface of the BM and multipass FSP sample are shown in Figure 10.9. In the BM, a coarse fracture structure is obviously detected, which has cold flake as a partly brittle fracture. However, after the multipass FSP and when the cold flakes were hardly seen, a fine ductile fracture emerged. According to this result, the cold flake must substantially affect the tensile properties [10].

Some distinctive opportunities to repair worn or cracked components in aged aircrafts are FSW and FSP presented. From remanufacturing perspective, repairing of ruddervator fitting for the KC–135 re-fueling boom is considered as a case study (Figure 10.10(a)). This structural

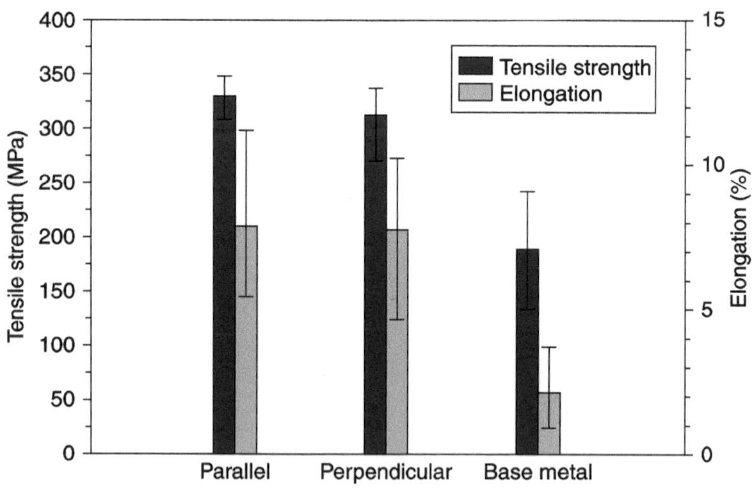

Figure 10.8 Tensile test results of the base metal and multipass FSP samples for different directions [10]

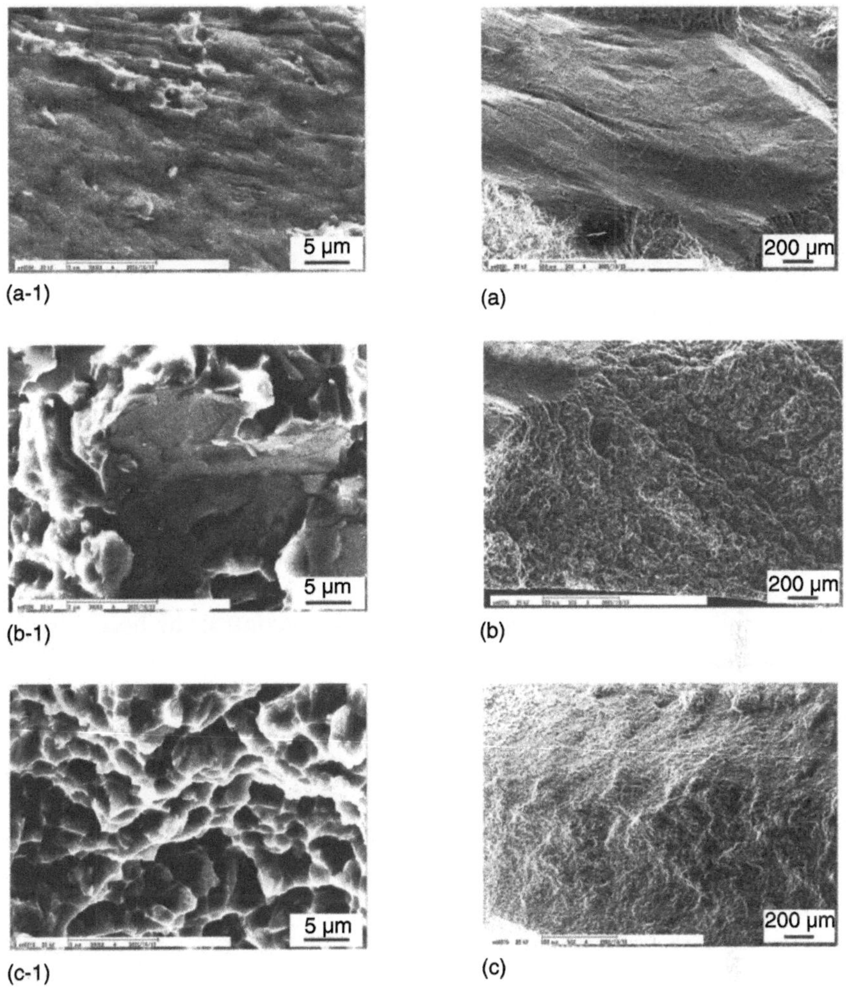

Figure 10.9 **Fracture surface of tensile test specimens in the base metal and the multipass FSP sample: (a-1), (b-1) and (c-1) high magnification of (a), (b) and (c)**

component is responsible for connecting the re-fueling boom to the rudder shaft. The component has two ears, each with a bearing race. When the bearings are installed, usually punch and hammer are used to deform the part and stake the bearing in place. Regrettably, these stake marks are known as stress concentration points, and form fatigue cracking when placed into service. Figure 10.10(b) shows a picture of the typical stake marks in the ruddervator fitting.

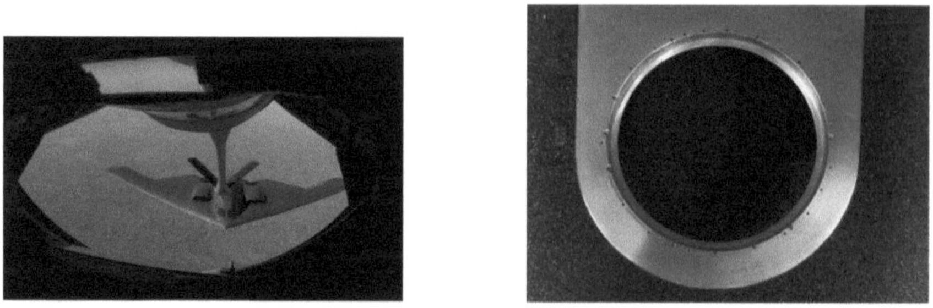

(a) (b)

Figure 10.10 (a) KC-135 refuelling boom [11]; and (b) typical stake damage to a ruddervator ear [12]

The study of defect processing 'stake-marks' were simulated with holes drilled into a piece of 0.125" 7075-T6 sheet stock. These 0.065" diameter holes are similar in diameter to the defects found on the defective parts, and were made to be 50%, 100% and 150% of the observed defect depth (0.030", 0.060" and 0.090", respectively). A vertical mill with digital readout was applied to control the hole's depth and spacing.

The base sheet was aligned in a jig and clamped in place with a 0.040" 7075-T6 sheet on top in a lap-weld configuration. All parts are welded with the same set-up and programme of welding because of the fixture. Two sets of welding parameters were selected from earlier weld schedule development: 600 rpm and 11 ipm yielding a Pseudo Heat Index (PHI) of 2.27, Parts 4, 5 and 6 used 800 rpm and 9 ipm yielding a PHI of 7.57. Weld runs were randomized, and both sets of parameters used a 2-degree lead angle and heel plunge of –0.005". All welds were made with an adjustable pin tool with a scrolled shoulder and tapered, tri-flute, threaded pin profile (Figure 10.11). Welds were centred over the defects 0.5" from the edge of the part and were 8.2 inches in length.

The pathfinder experiment employed two surrogate parts, as in the geometry to the Ruddervator fittings, and two plugs machined from 7075-T753 barstock (Figure 10.12). Four defects, 0.065" in diameter and 0.090" deep were installed in the four quadrants of each bore. Defects were installed approximately 0.060" on centre, radially from the edge of the bore.

Thermocouples were mounted in the surrogate parts at three locations, in order to capture comparative information between 2- and 4-pass welds and to compare the thermal conditions between the

Figure 10.11 Pin tool and shoulder [12]

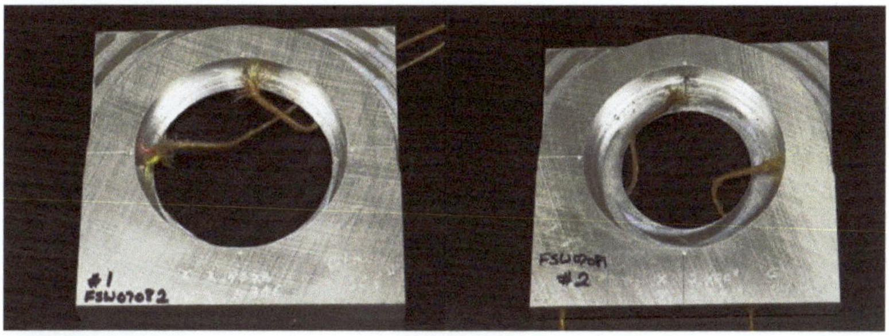

Figure 10.12 Surrogate plugs [12]

Figure 10.13 Installation of plugs and thermocouples [12]

surrogate part and the prototype. Thermocouples were placed in the quadrants of the surrogate parts where the bore meets the bearing stop. By using a manual hydraulic press, plugs were pressed into the bores (Figure 10.13).

The only clamping holding part in position was a single 5/8" bolt through the centre of the plug (Figure 10.14(a)). In order to test the weld programme, a 4-pass test-weld was made on a piece of 0.25", 6061-T6 aluminium fixed in a similar manner to the surrogate parts. A 2-pass weld (Figure 10.14b) was successfully completed on the first surrogate part. Nevertheless, Figure 10.15 shows a lack of consolidation (LOC)

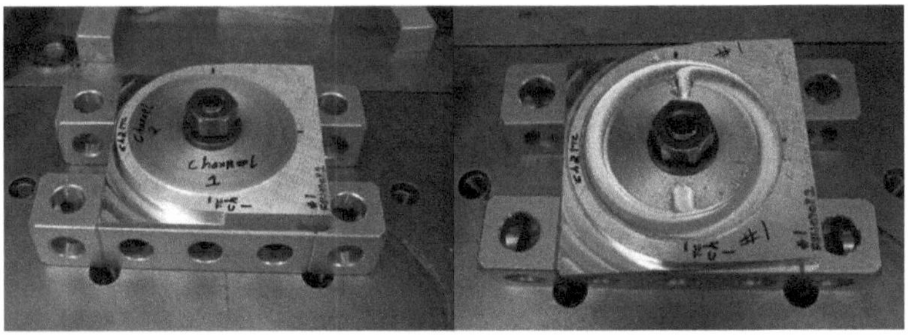

(a) (b)

Figure 10.14 Surrogate 1 (2-pass weld) [12]

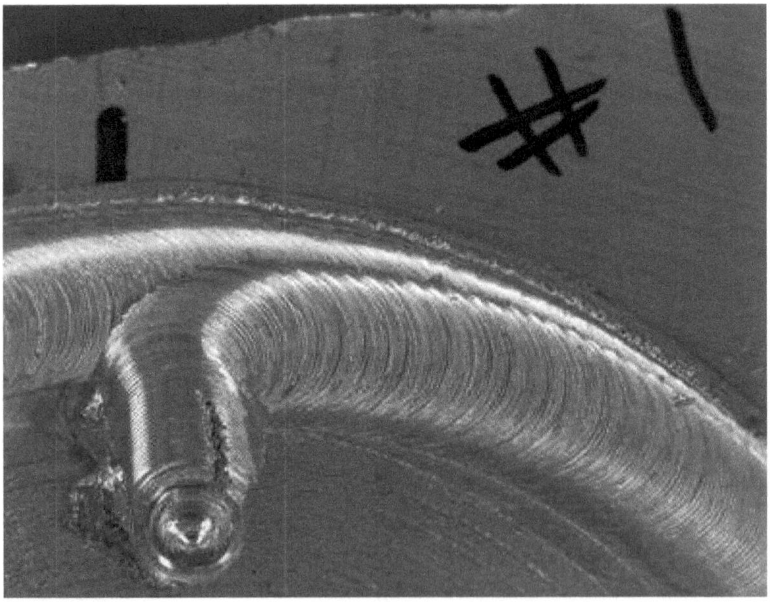

Figure 10.15 Lack of fill on exit radius [12]

Figure 10.16 Surrogate 2 (4-pass weld) [12]

defect formed through the exit radius. A 4-pass weld (Figure 10.16) was successfully completed on the second surrogate part. Through the exit radius, the travel speed was reduced, which reduced the LOC defect, whilst not entirely eliminating it.

In order to remove the plugs from the bearing bores, parts were re-machined and radiography inspection was used to determine presence of volumetric defects (Figures 10.17 and 10.18).

The study of prototype production was done on actual ruddervator fittings. In order to remove the paint from the working surfaces of the ruddervator fittings, a pneumatic abrasive wheel was employed to expose the bare aluminium. The bearing race was then submerged in a NaOH solution (10% by mass) prior to conducting a dye-penetrant inspection. A coordinate measuring machine measured the ruddervator fittings to determine the parallelism of the two-part face to be welded. Measurements determined that Parts 1 and 4 were parallel to within 0.005" and 0.0025", respectively, so these features were acceptable to use as data.

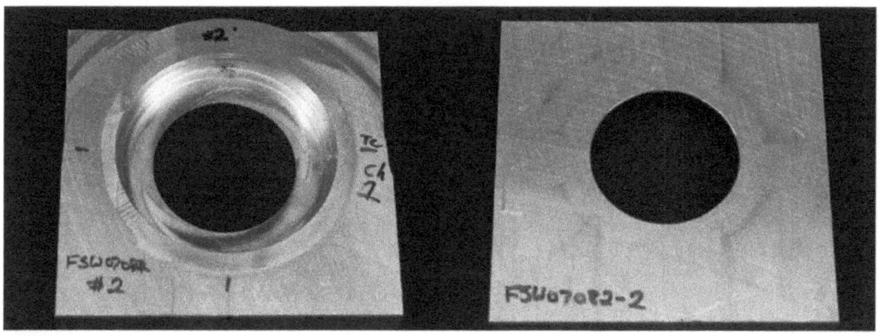

Figure 10.17 Surrogate 2 post-machining [12]

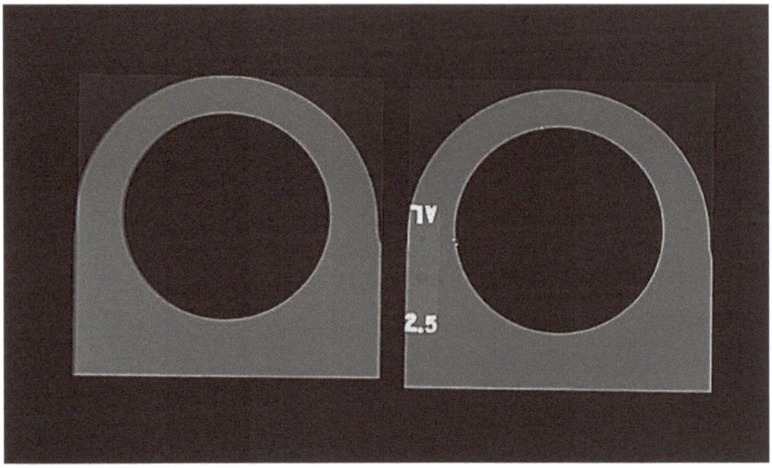

Figure 10.18 X-ray inspection of surrogate parts [12]

Figure 10.19 shows that the prototype ruddervator fitting was cleaned with a solvent wipe and the machined plugs were pressed into place. Standard tooling was used which was similar to the set-up used in welding of the surrogate parts in the previous section. This was vital to avoid differences in the thermal and mechanical conditions that are applied in the development of the repair process to aid portability between surrogate and prototype production. Whilst the acquisition system was collecting thermal data, welds were made by using the developed parameters in the previous section (800 rpm, 8 ipm). Welds were made counter-clockwise about the centre of the plug, with two inner passes followed by two outer

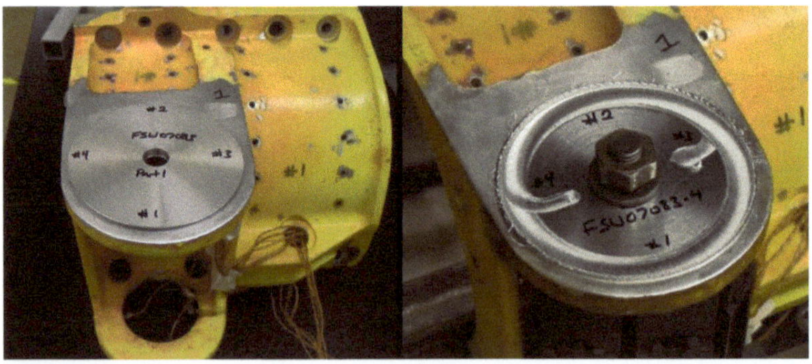

Figure 10.19 Dry-fit of repair plug (left) and post-weld repair (right) [12]

passes. In the following welding, the resultant plugs were machined from the ruddervator fittings, and the repaired part was radiographically inspected, which disclosed that no volumetric defects remained in the part.

By using the developed procedures and information in the defect processing study and the surrogate parts welding, the prototype production sections were processed with some confidence in the quality of the repair. A LOC indication was still present, even with the increased radii employed for the exit procedure. The stake defects were observed to be successfully processed from the part once the plug was machined from the bore (Figure 10.20). The plug machined out of the ruddervator part; however, slight concentricity issues exist between the re-machined bore and the pre-welded bore. No remaining volumetric defects were detected by the radiographic inspection.

To sum up, it is reported that the repair demonstration on the ruddervator fittings was successfully done. Radiographic inspection displayed that the defects were completely processed from the parts. The parameter development welds demonstrating a defect size of 150% could be processed with a single pass. When the defects have been removed, the heat input to the part alters the material temper of the aluminium part, in the weld nugget and in the HAZ; there could be a problem of holding the bearing in place, as long as the ruddervator fitting is a structural component and may require a harder/stronger material. Nevertheless, the fatigue defects resulted in the scrapping of the parts that may not be as critical in the HAZ of the FSW for the general causes. Softer materials are less susceptible to fatigue failure. In order to solve this issue, development work is warranted. By applying friction stir technology, volumetric defects and fatigue cracking have been successfully processed out of the ruddervator fittings with little difficulty. This demonstration suggests that the repair of structural aircraft components is feasible via FSP [13].

Figure 10.20 Pre- and post-repair of ruddervator staking damage [12]

In the power generation industry, FSW technique has been used for repair. A full size canister welding machine has been designed and constructed. This equipment has been used to develop FSW tool technology and optimization of weld parameters. Welds in 50 mm thick copper have been produced in 120-degree segments of approximately 1 m length (Figure 10.21). Recent studies have focused on the development of tool forms and tool materials [14].

Cast iron has great material features, such as abrasion resistance, corrosion resistance, machinability, vibration absorptivity, and accordingly, it has been used in several industrial fields, such as automobile parts, industrial machine parts and machine tool parts. FSP was applied to the surface hardening of cast irons. In order to investigate the validity of this method, flake graphite cast iron (FC300) and nodular graphite cast iron (FCD700) were used. The matrices of the FC300 and FC700 cast irons are perlite, and consequently the Vickers hardness of the nodular graphite cast iron is 200 to 230 HV, whilst for the flake graphite cast iron, it is 170 to 210 HV. A cylindrical tool with a 25 mm diameter was applied, and the tilt angle was 3 degrees.

The travelling speed was varied between 50 and 150 mm/min, in order to control the heat input, whenever the rotation speed of the tool was constant at 900 rpm. The heat input noticeably influences the formation

Figure 10.21 Friction stir welded segment in 50 mm thick copper [14]

of the defects. As long as the heat input is insufficient, a defect is formed during the early stage by scooping out the material. This formation is because the cast iron does not soften due to the insufficient heat input, and then a flash is formed by a cutting-like phenomenon. In addition, when the heat input is excessively high, the surface material is peeled off as if the material is melted because it softens extremely. Figure 10.22 depicts the Vickers hardness distribution in the depth direction at the

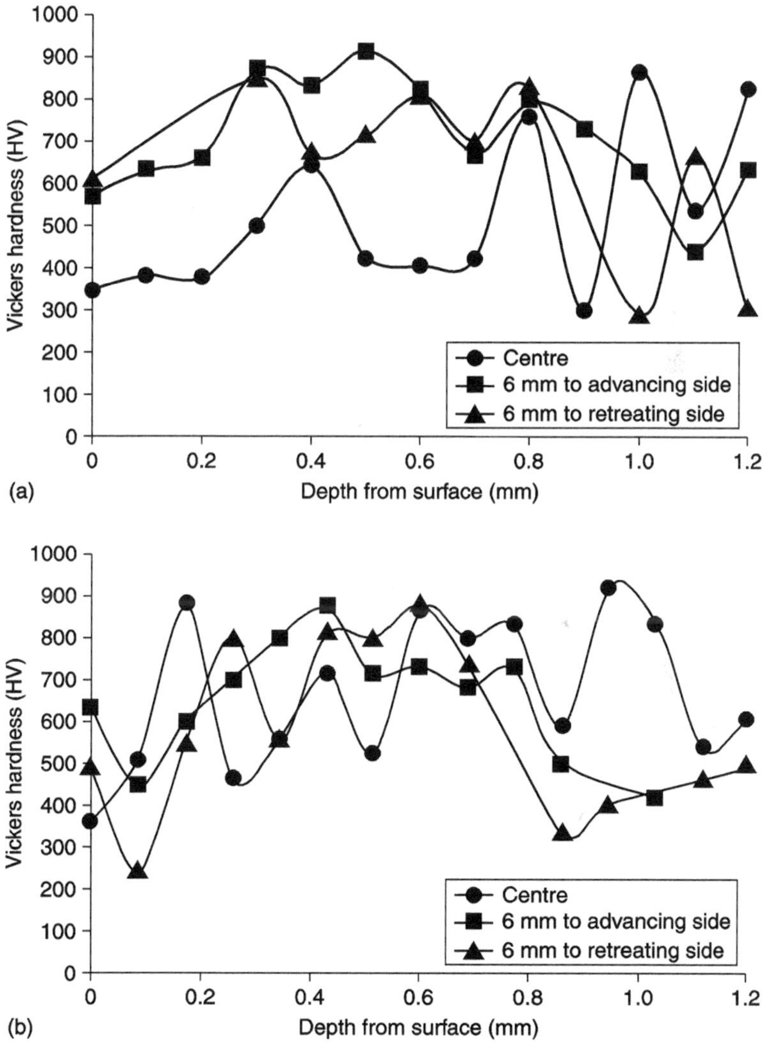

Figure 10.22 Vickers hardness distribution of FSP: (a) FCD300; and (b) FCD700 [15]

central part and at 6 mm from the centre on the AS and RS on a cross-section vertical to the welding direction after the FSP. The average hardness of approximately 700 HV for the matrix is obtained for both the flake graphite and the spheroidal graphite cast iron. This is because this process forms an extremely fine martensite phase. It is considered that a very fine martensite structure is formed, because the FSP generates the heat very locally, and also a very high cooling is constantly obtained. Moreover, the hardness changes depending on the size of the martensite, which can be governed by the process conditions, such as the tool travelling speed and the load [15].

Furthermore, the effect of repairing the engine blocks by FSW was evaluated. Grey cast iron contains an abundance of graphite flakes that act as a lubricant; indeed it does not allow enough frictional heat to enter the welding process. The use of a steel cover plate to interact with graphitic microstructures to allow more frictional heat was evaluated. In order to have the cast iron heated uniformly, a heating anvil is placed on the welding table and heated up to approximately 200 °C before welding. The steel cover plate is then centred and clamped down on top of the cast iron plate. The welding head is then positioned for the purpose of obtaining a zero and a maximum distance for the weld. The weld proceeds at 0.25 ipm and at 1000 rpm. The FSW specimen at rotational speed of 100 rpm and traverse speed of 0.25 ipm and different forge force is shown at Figure 10.23. A 9 inch weld, welded with 2700, 3000 and 3300 lbf forge force, appeared to be defect free and an overall acceptable weld. Although, when the steel cover plate was removed, cracks were noticed between the nugget and parent-material interface in the 3000 and 2700 lbf forge welds. Looking at the microstructures there was the evidence of a poor parent [15].

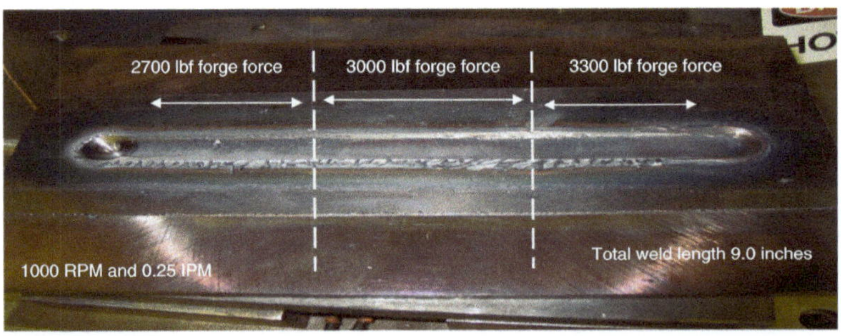

Figure 10.23 The FSW specimen at different forge force [15]

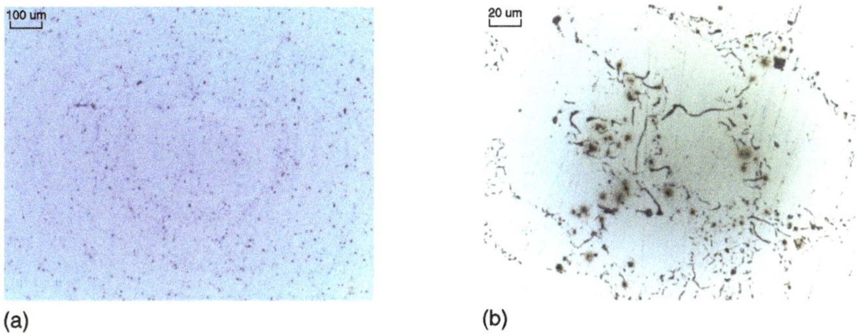

Figure 10.24 Graphite flakes in: (a) centre of weld; and (b) parent material [6]

In Figure 10.24, the welds show how the graphite flakes in the centre of the nugget are extremely fine, while in the parent material they are still long and relatively large.

Only the 3300 lbs weld was able to be tested because of a poor interface. All tensile samples failed on the RS. As it is obvious from Figure 10.25, the 3300 weld failed due to graphite segregation in the thermo-mechanically affected zone (TMAZ).

A method was created to repair a void in an aluminium alloy, especially a void resulting from an exit hole left from FSW. This method consists of machining the void to provide a tapered bore through a parent material. According to the present invention, grooves and ridges are provided on a sidewall of the tapered bore. As presented in Figure 10.26, a consumable tapered plug with an included angle less than, or equal to, the tapered bore is fitted to the bore, preferably with the tip of the plug extending beyond the bore depth. The tapered plug is preferably attached at its

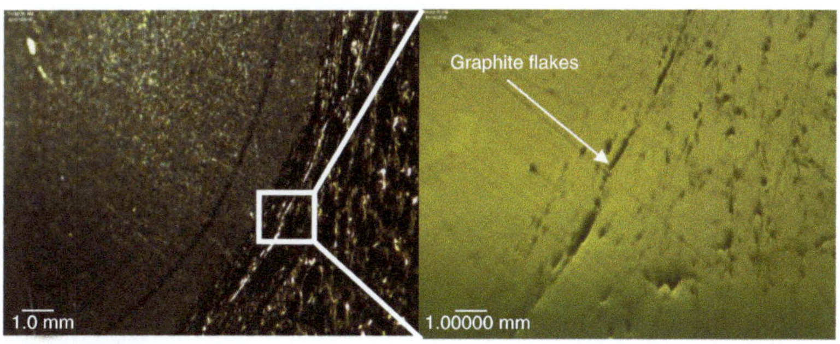

Figure 10.25 Graphite segregation in TMAZ [17]

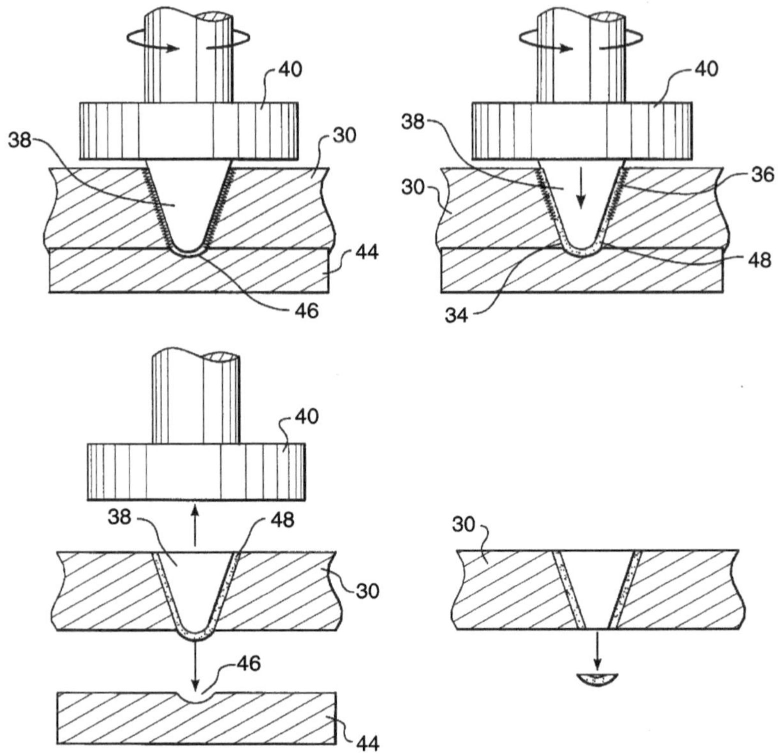

Figure 10.26 A schematic illustrating procedure of the invention [17]

proximal end to a larger non-consumable part, which serves as a heat sink. The heat sink serves to remove excessive heat generated during rotation of the plug. Otherwise, external cooling, such as a cooling fluid, can be used to remove excess heat.

A support bracket is temporarily attached to the aluminium alloy adjacent to the tapered bore. The support bracket includes a cavity that is machined, so that the tip of the plug is fitted to the cavity. As the consumable tapered plug is rotated inside the tapered bore under an applied load, the plug and the parent material in proximity to their interface become plasticized and commingled to form a bonding interface, due to the normal component of the applied load, frictional heat and hydrodynamic force. In order to further enhance local deformation and commingling of the plug material and the parent material, the plug material crushed and plasticized the ridges on the surface of the tapered bore. Simultaneous with hardening of the plasticized material at the interface, the heat sink and the support bracket can be trimmed away, leaving the tapered bore completely bonded

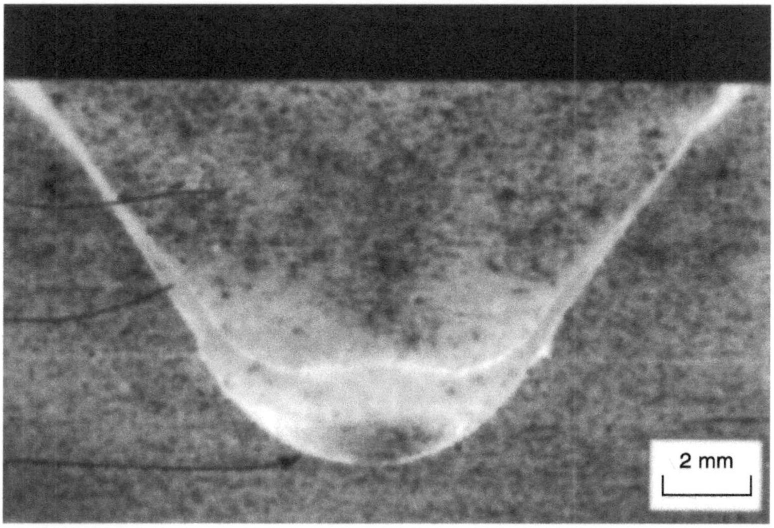

Figure 10.27 An optical micrograph showing an overall cross-sectional detailed view of a filled void [17]

with the plug material at the interface consisting of refined and recrystallized fine metal. Figure 10.27 shows the optical micrograph of a void filled with the consumable pin [16].

AISI type 316 stainless steel and its modified grades such as 316L are used as structural material in nuclear power plants for the construction of water storage tanks. This alloy is chosen on the basis of its excellent high temperature tensile and creep-fatigue strengths in combination with good fracture toughness. While nuclear plants grow older around the world, the incidences of stress corrosion cracking (SCC) problems are increasing. Therefore, the SCC repair is becoming a crucial need to expend the service life of stainless steel water storage tanks that have experienced SCC cracks at the external surface. The FSP technique has basically provided a high integrity smooth repair of shallow surface defects. The rotation speeds of 4, 5 and 6 rad/s and welding speeds of 16, 20 and 24 mm/min are the processing parameters. Also, the working loads of the tool are 25, 30 and 35 KN.

The tool is fabricated from polycrystalline cubic boron nitride (PCBN), and includes a convex shoulder having a diameter of 22 mm and a tapered pin. Investigation results indicate that the area with sigma phase precipitation has direct ratio relations with working loads and rotation speeds, while it has an indirect ratio with welding speeds in current experimental situations. For the purpose of obtaining the illustrative diagram of this relationship,

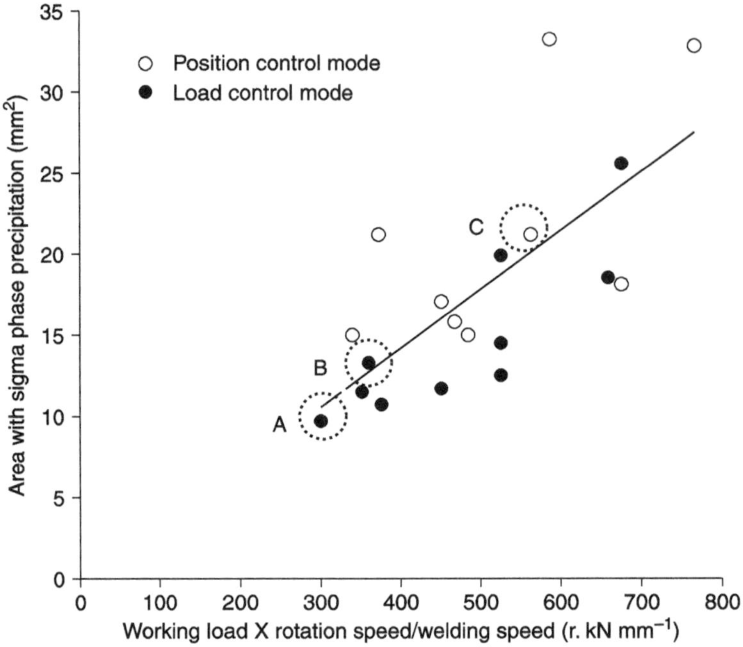

Figure 10.28 The area with sigma phase precipitation as a function of heat input in both control modes [18]

the expression of rotation speed multiplying load divided by welding speed is taken as x-axis in Figure 10.28. Transmission electron microscopy (TEM) in FSP of 316L austenite stainless steel confirmed the sigma phase precipitation. From the standpoint of heat input, the amount of sigma phase precipitation has direct ratio relations with heat input in current experimental conditions. As the heat input is higher, the amount of sigma phase formation is larger. The formation of sigma phase produces Cr reduction in banded structure, which results in the largely degraded corrosion resistance in this region. Low heat input (through the lower rotational speeds and higher travelling speeds) contributes to restrain sigma phase precipitation during FSP of 316L stainless steel [17].

As distinguished from fusion welding techniques, metals are joined below their melting point by FSW. Consequently, the whole thermal history of the solid-stating FSW process contributes for the purpose of avoiding the metallurgical weld-defects, such as hot-cracks and blowholes. However, different types of FSW defects (i.e. tunneling cavity, void, porosity, surface groove, flash, 'S-curve', etc.) could emerge a as result of some inappropriate process parameters. However, the overlapping FSW technique can always repair the FS-weld defects, in order to form a second-pass FS-weld on the original weld. Weld defects of the single pass FSW of aluminium 2219-T6 were repaired by overlapping FSW technique [18]. The typical FS-weld defects in the top region of the weld are shown in Figures 10.29(a and b), and a schematic diagram of the repair FSW process and the specific shape of the FS-tool, employed for the 20 mm-thick defective weld, are shown in Figure 10.29(c).

Through a series of initial experiments with the aim of repairing the defective welds by using the overlapping FSW technique, the following appropriate process parameters were selected to produce repair welds on the original defective welds: v = 120 mm/min, n = 450 r/min, d = 1.0 mm and $\alpha = 2.5°$. The results proved that the former weld defects in the top region of the original defective weld were entirely repaired after the stated overlapping FSW technique. In addition, a dense distribution of the large-sized particles and the sub-micron precipitates were detected. The results of energy dispersive spectroscopy (EDAX) showed that the particles or precipitates were mainly Al_2Cu phase, and there were also a few impurity phase particles. The comparison showed that the large-sized Al_2Cu particles in the repair weld were commonly larger in size than those of the original weld. This was the result of the second thermal cycle and heat-input provided by the overlapping repair FSW (Figures 10.30(a–e)).

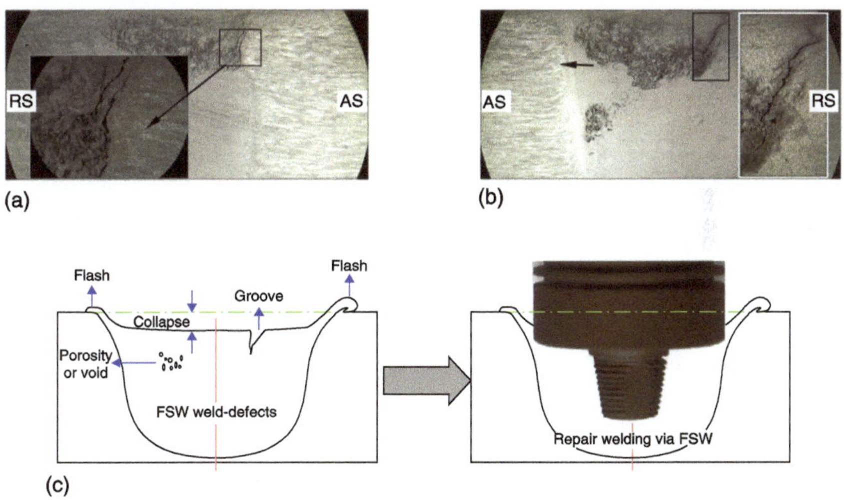

Figure 10.29 (a and b) Appearance of FS-weld defects in the top regions of original welds; and (c) schematic diagram of the repair FSW process and the specific shape of the FS tool [17]

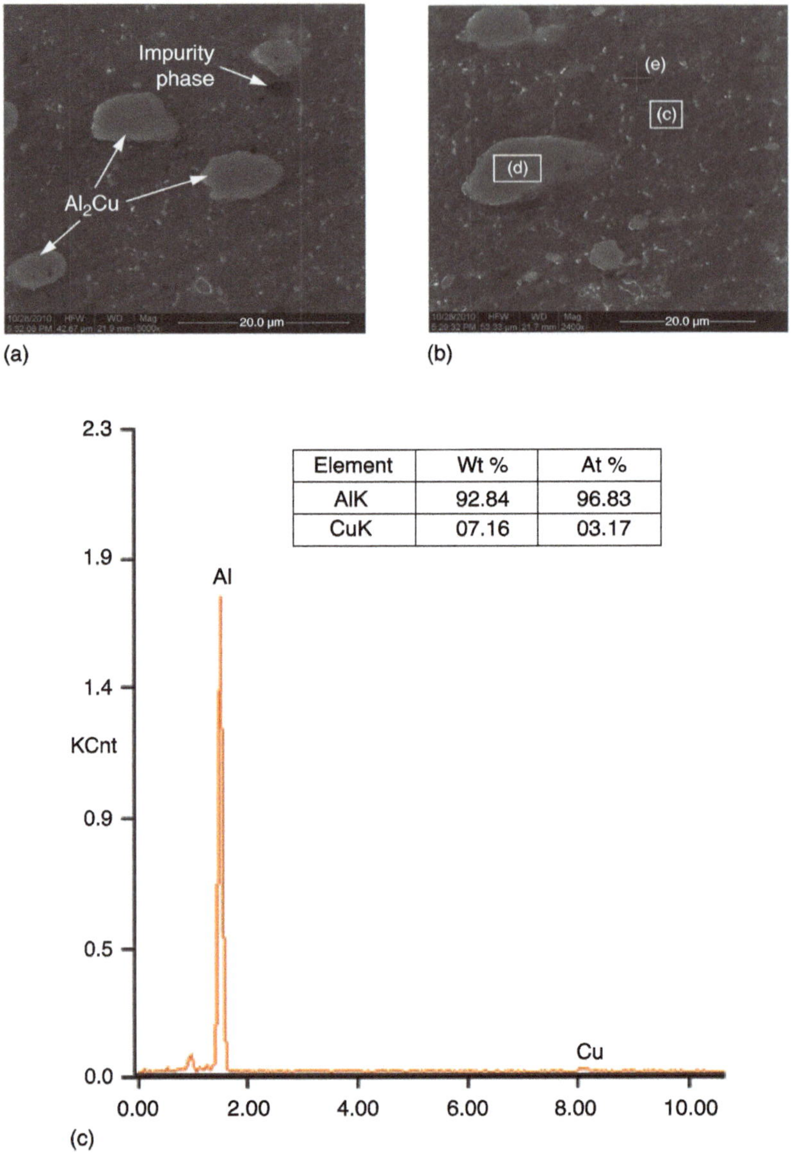

Element	Wt %	At %
AlK	92.84	96.83
CuK	07.16	03.17

Figure 10.30 SEM images of normal-sized particles in: (a) original FSW; and (b) repair FSW; and (c and e) their EDAX testing results [17]

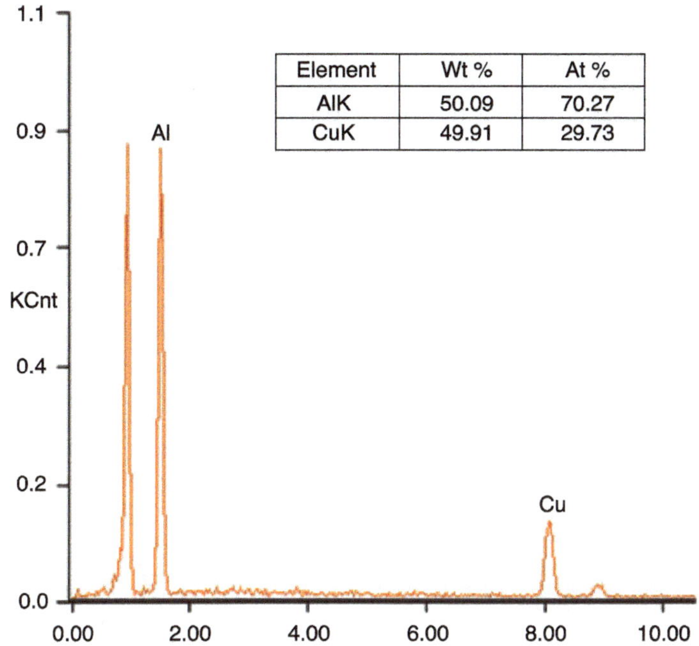

Element	Wt %	At %
AlK	50.09	70.27
CuK	49.91	29.73

(d)

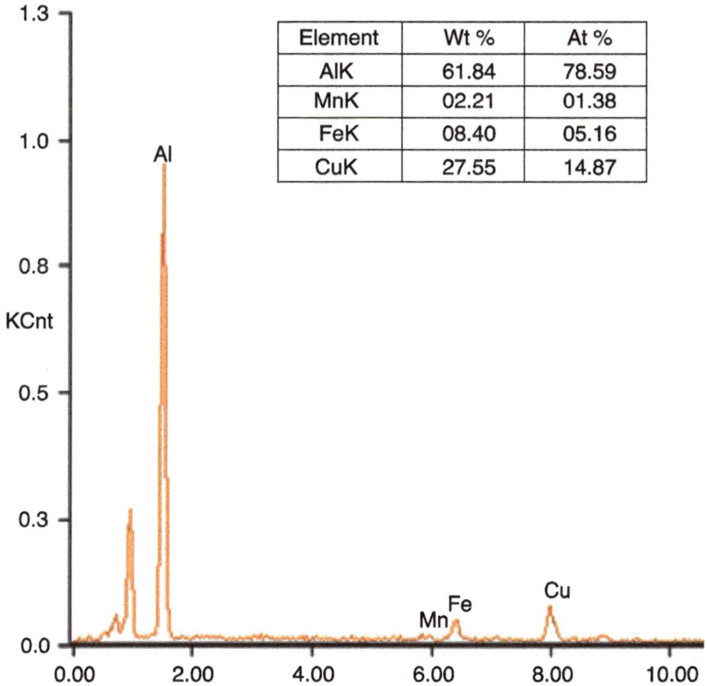

Element	Wt %	At %
AlK	61.84	78.59
MnK	02.21	01.38
FeK	08.40	05.16
CuK	27.55	14.87

(e)

Figure 10.30 Continued

Furthermore, results indicated that abnormal coarsening of Al_2Cu particles occurred in the overlapping repair FSW of AL2219-T6 thick plates without any PWHT process [18].

10.3 References

1. ASM Handbook (1992), *Welding, Brazing, and Soldering*, vol. 6, ASM International.
2. Li, L., Liu, Z. and Snow, M. (2006), 'Effect of defects on fatigue strength of gtaw repaired cast aluminum alloy', *Welding Research*, November: 1–6.
3. Cary, H.B., Scott, P.D. and Helzer, C. (2005), *Modern Welding Technology*, Charlottesville, Virginia: University of Virginia.
4. Cao, X. et al. (2006), 'A review of laser welding techniques for magnesium alloys', *Journal of Materials Processing Technology*, 171(2): 188–204.
5. Gäumann, M. et al. (2001), 'Single-crystal laser deposition of superalloys: Processing-microstructure maps', *Acta Materialia*, 49(6): 1051–62.
6. Durocher, J. and Richards, N.L. (2011), 'Evaluation of the low heat input process for weld repair of nickel-base superalloys', *Journal of Materials Engineering and Performance*, 20(7): 1294–303.
7. Nelson, E.A. (2009), *Microstructural Effects of Multiple Passes during Friction Stir Processing of Nickel Aluminum Bronze*, December, Monterey: Naval Postgraduate School.
8. Rosen, C.D., Litwinski, E. and Valdez, J.M. (1998), *Friction stir Welding Process to Repair Voids in Aluminum Alloys*, Seal Beach: The Boeing Company.
9. Rajesh Talwar, R.P. (2006), *Method of Repairing a Crack in a Component utilizing Friction stir Welding*, in: *Grant*, T.B. Company Editor.
10. Nakata, K. et al. (2006), 'Improvement of mechanical properties of aluminum die casting alloy by multi-pass friction stir processing', *Materials Science and Engineering A*, 437(2): 274–80.
10. *Refueling boom of a KC–135R*. Available from: http://www.defense.gov/photos/newsphoto.aspx?newsphotoid=9912.
12. Tweedy, B.M.A, Hrabe, R.H. and Hutto, J.M. (2009), 'Ageing weapons systems repair using friction stir welding', in: *Friction stir Welding and Processing V*. TMS, The Minerals, Metals & Materials Society.
13. Tweedy, B.M.A., William, A., Robert H. and Hutto, J.M. (2009), 'Ageing weapons systems repair using friction stir welding', *Proceedings of the 5th Symposium on Friction stir welding*.
14. *Novel Joining Techniques for Repair in the Power Generation Industry* (2002), available from: http://www.twi.co.uk/technical-knowledge/published-papers/novel-joining-techniques-for-repair-in-the-power-generation-industry-june–2002/
15. Fujii, H.Y., Toshifumi, K., Shoji, K. and Kiyoshi, N. (2009) 'Surface hardening of two cast irons by friction stir processing', *Journal of Physics: Conference Series*, 154(1): http://iopscience.iop.org/1742–6596/165/1/012013/

16. Mahoney, M.W. et al. (1998), *Method to Repair Voids in Aluminum Alloys*, The Boeing Company.
17. Chen, Y.C. et al. (2012), 'Banded structure and its distribution in friction stir processing of 316L austenitic stainless steel', *Journal of Nuclear Materials*, 420(1–3): 497–500.
18. Li, B. and Shen, Y. (2011), 'The investigation of abnormal particle-coarsening phenomena in friction stir repair weld of 2219-T6 aluminum alloy', *Materials & Design*, 32(7): 3796–802.

Force and torque in friction stir welding

P. Shahi and M. Barmouz, Tarbiat Modares University, Iran and P. Asadi, University of Tehran, Iran

DOI: 10.1533/9780857094551.459

Abstract: Forces in friction stir welding, as well as torque and power, induced by welding input parameters, including tool shape, material and process parameters, play a key role in welding. Forces and torque in FSW can be considered as online indicators of weld quality, in which their values during processing should be in a safe range to obtain defect-free welds. While processing for a given set of tool and material characteristics, changes in process parameters result in force variation in traverse and axial directions, as well as torque around the tool axis. These values can be monitored and controlled by welding parameters adjustment, in order to avoid defective welds. Hence, forces and torque measurements and understanding their relationship with welding parameters are highly important in successful FSW, which are the focus of this chapter. Presented empirical and analytical models could also be successfully employed to predict welding forces and torque values used for instrument (e.g. fixture) design.

Key words: friction stir welding, forces, torque.

11.1 Introduction

A fundamental understanding of the main parameters effect is highly crucial for the successful FSW process [1]. The independent input parameters in this process include tool pin and shoulder diameter, tool design, workpiece and tool material, tool rotational speed, welding speed, plunge depth, etc. To obtain a defect-free weld, the employed forces and torque should be in a safe domain, beyond which the welds could be defective. These forces, in three coordinate directions and torque around the *z*-axis, are functions of the aforementioned parameters. Hence, the forces and torque relationships with the process parameters should be understood, which is the aim of this chapter.

The FSW process can be classified as the pre-welding and welding stages. In the pre-welding stage, a non-consumable tool moves downward in the *z*-direction and the pin and shoulder penetrates into the plates (plunging stage). The tool rotates for a while at the same position (e.g. 10 s) to soften the material by friction heating (dwelling time). Then the tool starts to travel in a longitudinal *x*-direction (welding direction), stirring the material and joining the plates [2].

The FSW process is complex, since the weld quality depends on many parameters. The main independent input parameters in this process are classified as tool, workpiece and process parameters. Tool parameters include shoulder diameter and pin profile design and diameter. The workpiece material is also of great importance, as the materials which could be welded by this method are limited. In addition, the tool rotational speed, welding speed and plunge depth are the main process parameters. These parameters significantly influence the required forces in the welding process and the quality of the fabricated components. To obtain a defect-free weld, these parameters should be selected properly for any type of workpiece.

The dependent parameters in FSW are the forces and torque. The tool in the FSW process, due to its rotational and linear movement, experiences forces in a three coordinate axis:

- F_z: the z force, which is also named as the downward, forging, thrust or vertical force;
- F_x and F_y: the in-plane forces. F_x is also referred as the longitudinal/welding force.

Moreover, regarding the rotational movement, the torsional torque is exerted to the tool. The F_Z, F_y and torque applied on the tool are schematically presented in Figure 11.1.

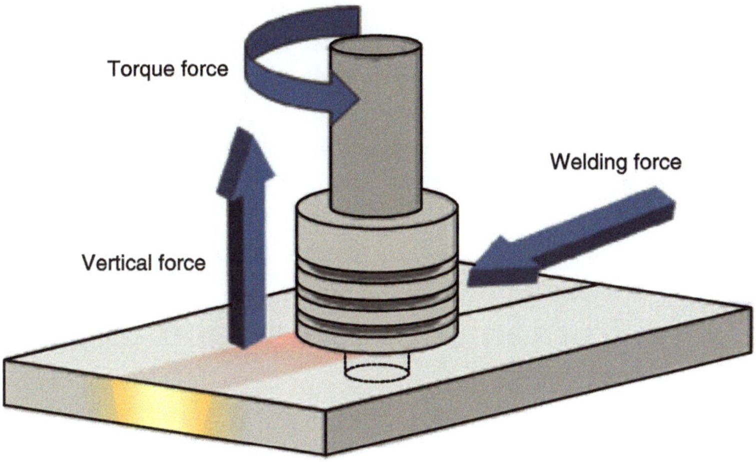

Figure 11.1 FSW applied forces and torque on the tool [3]

11.2 Importance of force and torque measurement and prediction

In the FSW process, the tool, due to its rotational and linear movements, exerts vertical and horizontal forces to the welding plates, besides its thermal impact. These forces can lead to tool wear and deformation of the welding plates and fixtures. Hence, predicting, monitoring and controlling the FSW forces are highly important, having many advantages such as:

- optimized tool design;
- tool life prediction;
- optimized fixture design;
- prediction of the required clamp forces;
- higher quality of the welded joints;
- higher strength of the joints;
- selecting a suitable welding machine.

The effective parameters on the weld quality include material, tool rotational and linear speeds, tool and shoulder design, etc., which makes it difficult to drive a comprehensive model. However, some empirical and analytical models are developed to predict the forces and torque, which can be used to optimize the design as well as the process. The presented empirical models are based on experimental investigation via monitoring the forces during the welding progress. Hence, several characteristics can be identified in the force-time and torque-time plots along with the maximum values.

11.3 Instruments for forces and torque measurements

For force and torque measurements, various methods are proposed. The axial, in-plane forces and torque can be measured directly or indirectly by means of strain gauges, load cells, dynamometers and output signals of driver motors. The measurement methods can be classified as direct or in-direct types. Using strain gauges, load cells and dynamometers are types of direct measurement instruments, while monitoring current or power of the driver motors during the FSW process is indirect measurement method. Here some employed methods are described briefly.

Using strain gauges and loads is one of the efficient methods in this regard. As the tool moves in the welding direction, bending forces are imparted on the tool and positive and negative normal stresses are experienced by the tool. Since the tool rotates, the stresses at an arbitrary point alternatively change in a sine wave form. Strain gauges make possible the use of measurements of the translational (F_x) and lateral (F_y) forces, simultaneously. Four strain gauges can be mounted on the tool surface, 180 degrees apart in a Wheatstone bridge configuration (Figure 11.2). The strains change alternatively from tension to compression and the output force diagram data will be in sine wave form, which is the vector sum of F_x and F_y.

It should be noted that the output sine form plot presents the magnitude of F_x and F_y vector sum. When the strain gauges are along the x-axis, the output value in the sine wave form plot corresponds to F_X. In a similar way, as the strain gauges become collinear with the y-axis, the output value corresponds to F_Y. Hence, finding the position of the strain gauges will be helpful in finding the X and Y components of the total force. This is possible by using the Hall Effect sensor and two attached magnets (Figure 11.3). Here the LVDT was employed to monitor and control the plunge depth of the tool pin.

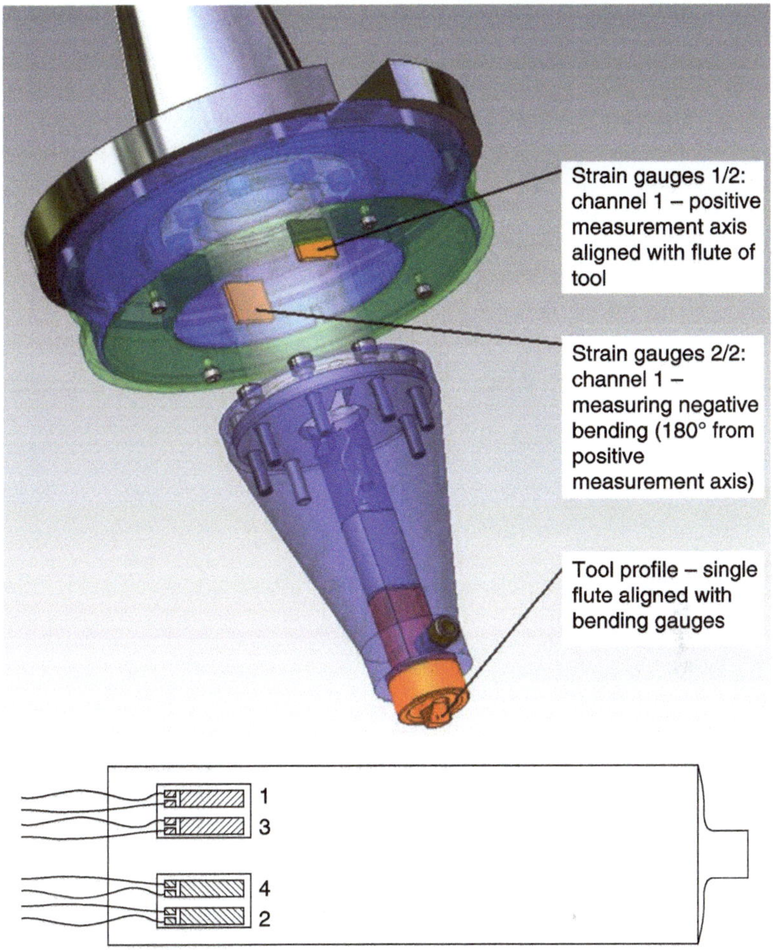

Figure 11.2 Schematic of mounted strain gauges on the tool [4,5]

Two magnets are also attached on the tool holder surface, which are 90 degrees apart. As the magnet passes in front of the sensor, a square shape pulse will be produced in the output. The sensor and magnets should be positioned precisely, in which the end tail of each pulse takes place while the stain gauges are along the x- or y-axes. Using this approach and post-data processing, the forces in the x- and y-directions can be extracted from the total force plot. A typical plot of total force and generated pulses using the Hall Effect sensor with the extracted X and Y forces are presented in Figure 11.4.

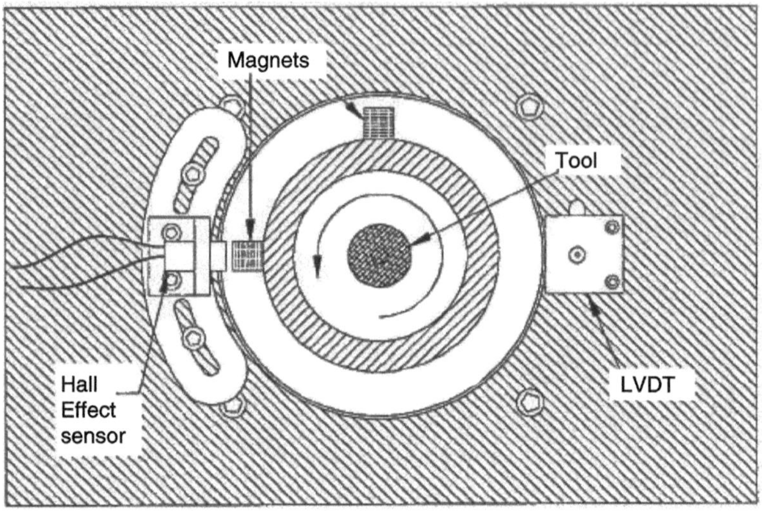

Figure 11.3 Hall Effect sensor location [5]

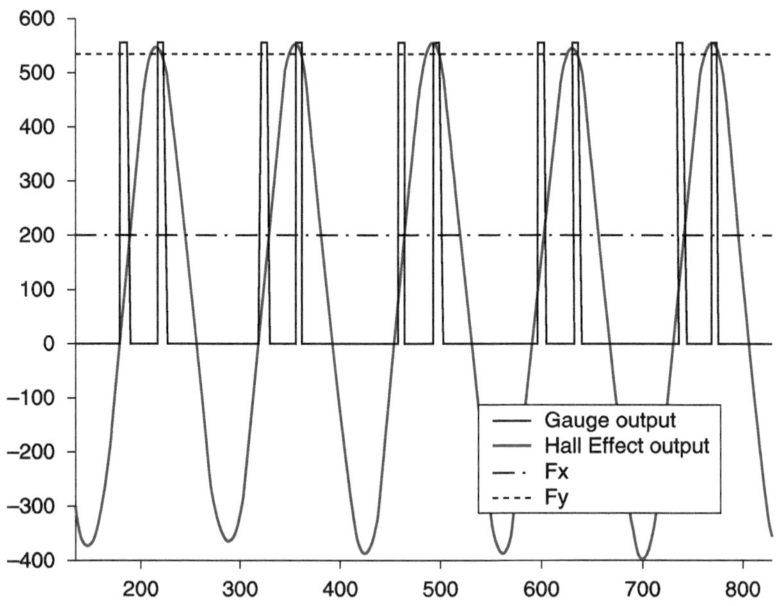

Figure 11.4 Separation of F_x and F_y from the total force curve [5]

The X (translational) and Y (traverse) components of force have nearly the same magnitude, which are both 7 to 80 times less than F_Z (axial force) [6]. To sense and record the *z*-axis force, it is possible to use the load cell. The position of the load cell is shown clearly in Figure 11.5. The schematic of the whole system, including the position of the strain gauges and load cells, are shown in Figure 11.6.

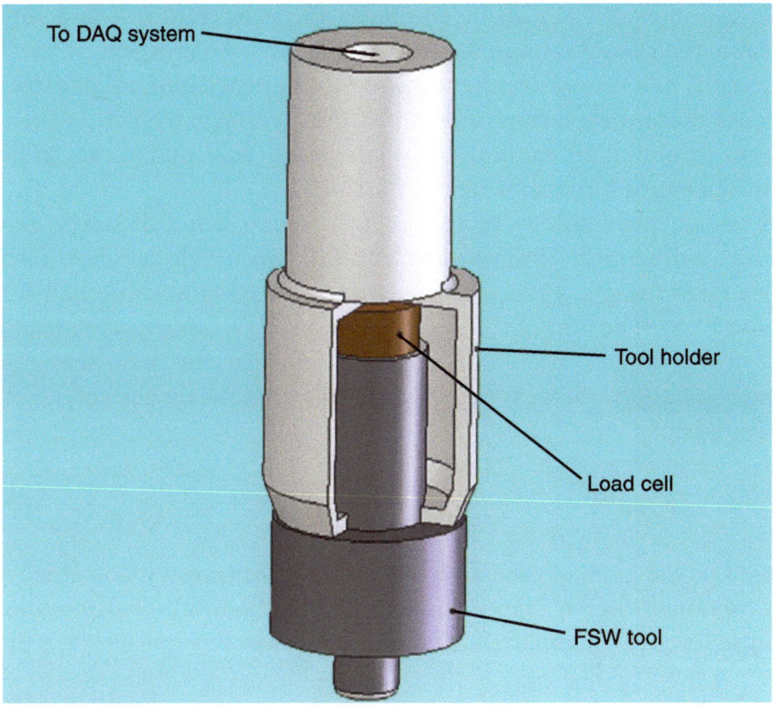

Figure 11.5 Schematic of axial force load cell location and the entire tool [7]

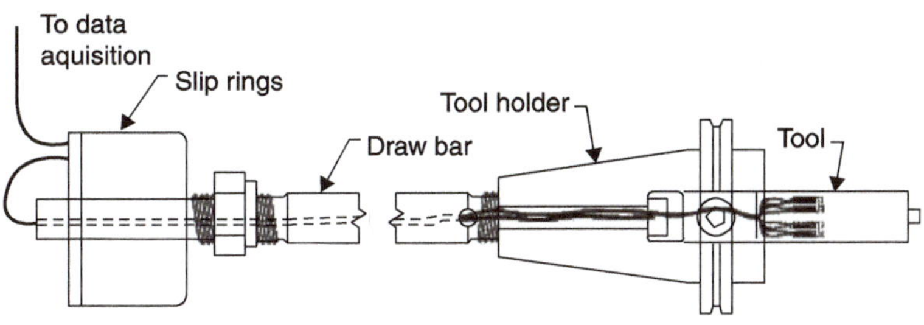

Figure 11.6 The whole measurement system showing the strain gauges and load cell locations [5]

To measure the torque, similar to bending gauges, four torsional strain gauges can be placed at the tool base 180 degrees apart. As an alternative, the strain gauges and load cells can be embedded on the workpiece, by which the forces on the tool are indirectly being measured by this method. For example, a typically designed FSW system equipped with two load cells to measure both longitudinal and axial forces are shown in Figure 11.7.

Dynamometers can also be employed to measure torque and forces; however, their usage is limited due to higher costs. An inexpensive method to quantify the torque and forces is to monitor the electrical parameters of driver motors. By means of recorded electrical signals, such as current, power, torque and calibration, these data will make the indirect measurement of forces and torque possible.

Here the methodology of torque and force for a ball screw driven machine is described. The linear motion of the worktable in the x-direction is generated using a servo motor, planetary gear reducer and ball screw (Figure 11.8). The output torque of the x-axis servo motor is the result of two acting forces: frictional force (f: between the ball screw and worktable) and traverse force (F). The total output torque (M_f) can be computed as

$$M_f = \frac{C}{2\pi\mu}.(f + F) \qquad [11.1]$$

where C is the pitch of the ball screw and μ is the transmission efficiency. The frictional force (f) can be assumed to be the total torque, while no traverse force (F) is imparted. The total torque in this case is given by

$$M_{f0} = \frac{C}{2\pi\mu}f \qquad [11.2]$$

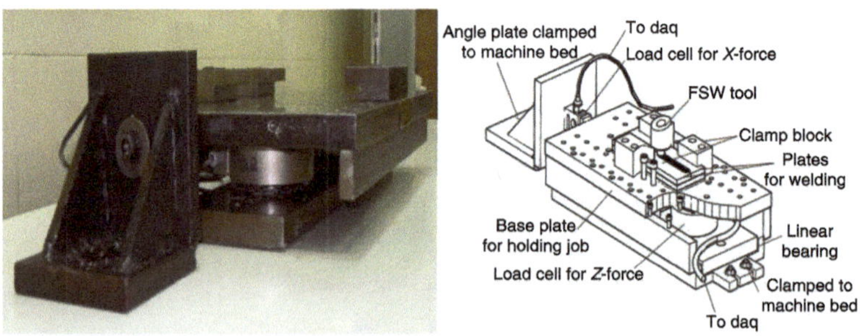

Figure 11.7 Designed instrument equipped with two load cells to measure F_x and F_y [8]

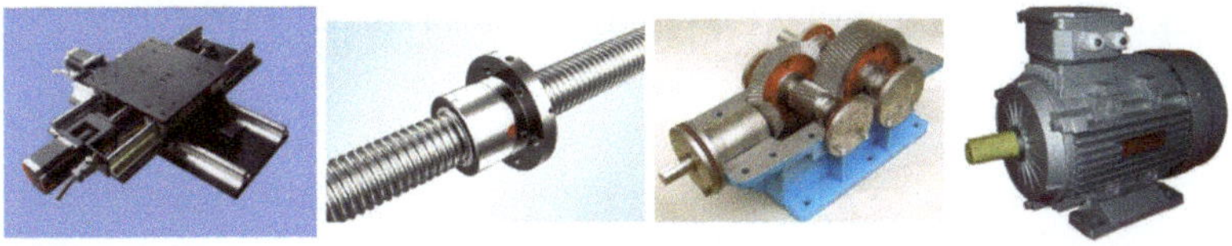

(a) (b) (c) (d)

Figure 11.8 *x*-axis driven system: (a) worktable; (b) ball screw; (c) gear reducer system; and (d) servo motor [9]

Hence, the traverse force finally can be computed as

$$F = \frac{2\pi\mu(M_f - M_{f0})}{C} \qquad [11.3]$$

The tool torque makes it possible to measure via the spindle motor output. While a conveyor system and spindle motor were used for *z*-axis rotation (Figure 11.9), the tool torque (M_T) is given by

$$M_T = (TRC).M_0 \qquad [11.4]$$

where the TRC is the transfer ratio of the conveyor and M_0 is the recorded spindle motor torque. To obtain the axial force value, the torque of the *z*-axis servo motor can be analyzed, which is responsible for the tool *z*-axis linear movement.

(a) (b)

Figure 11.9 Spindle system: (a) spindle motor; and (b) conveyor [9]

11.4 Force variation in welding progress

During welding, the forces can be measured and recorded by means of the instruments, as described above. Hence, the force and torque vs. time makes it possible to capture, monitor and analyze, to gain an insight into the whole welding procedure. Typical force-time plots for the joining of AA2219 alloys are presented in Figure 11.10. The two pre-welding and welding phases can be clearly identified in these plots. The pre-welding stage is prior welded, while the tool moves downwards, plunging the workpiece (Figure 11.11). Then, in the welding stage, the tool moves in the welding direction as it rotates, softening and welding the plates. In the pre-welding stage, the two observed surges are representative of the tool pin and shoulder contact with the workpiece as the tool moves in an axial direction. However, the forces in the welding stage have less variation and seem to be constant in the welding distance.

In addition, the figure indicates the higher values for the downward force (z-direction force) compared with the x-direction force. The lowest level between forces belongs to the y-direction force (Figure 11.12). From the figure, different trends for F_x, F_y and F_z plots are noticeable. The in-plane forces (forces in the x- and y-directions) show a quasi-static

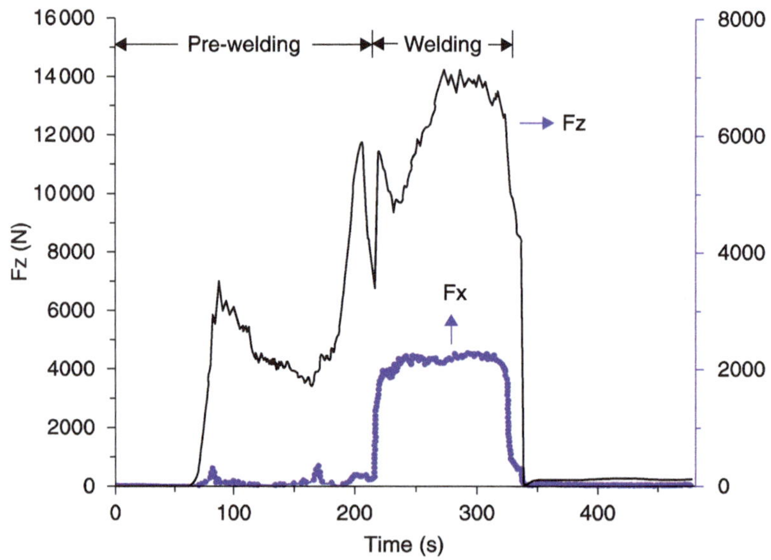

Figure 11.10 F_x and F_y plots in welding progress [2]

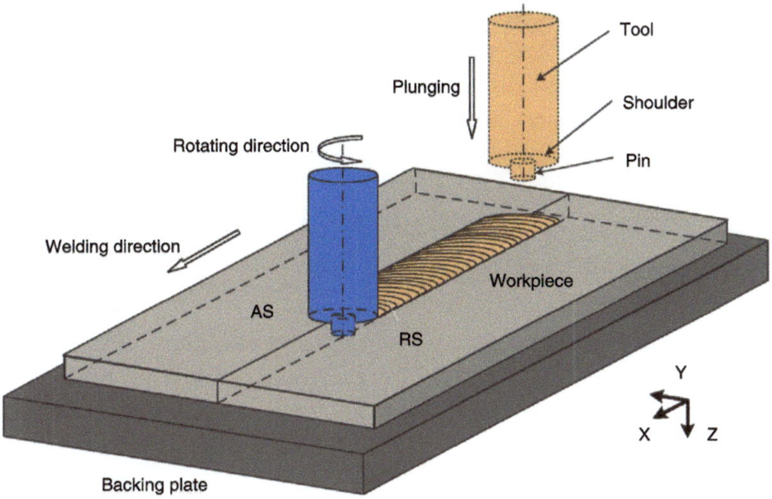

Figure 11.11 Schematic of tool movements during FSW [9]

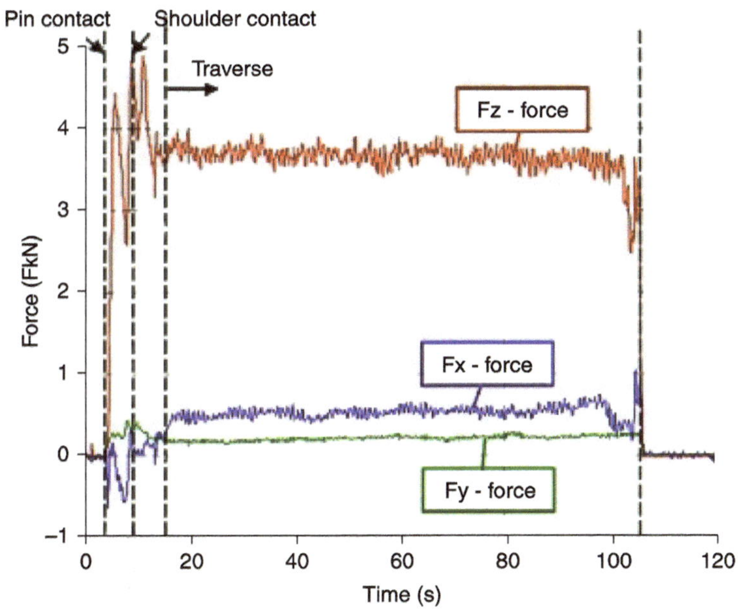

Figure 11.12 Variation of the forces in *x*-, *y*- and *z*-directions during FSW [11]

behaviour by the welding time. For the force in the *y*-direction, only a slight decay near the peak value can be noticed, while for the axial force (*z*-direction), the significant decrease by around 30% near its initial maximum value is observed [10].

11.5 Effect of friction stir welding parameters on forces

Welding speed, rotational speed, and shoulder and pin diameters, as the main input parameters, affects the torque, thrust and longitudinal forces. Here the effect of these parameters on F_x, F_z and T was investigated comprehensively.

11.5.1 Effect of processing parameters on F_x

In Figure 11.13, the effect of the processing parameters on the F_x is presented. Increasing the welding speed and shoulder and pin diameter

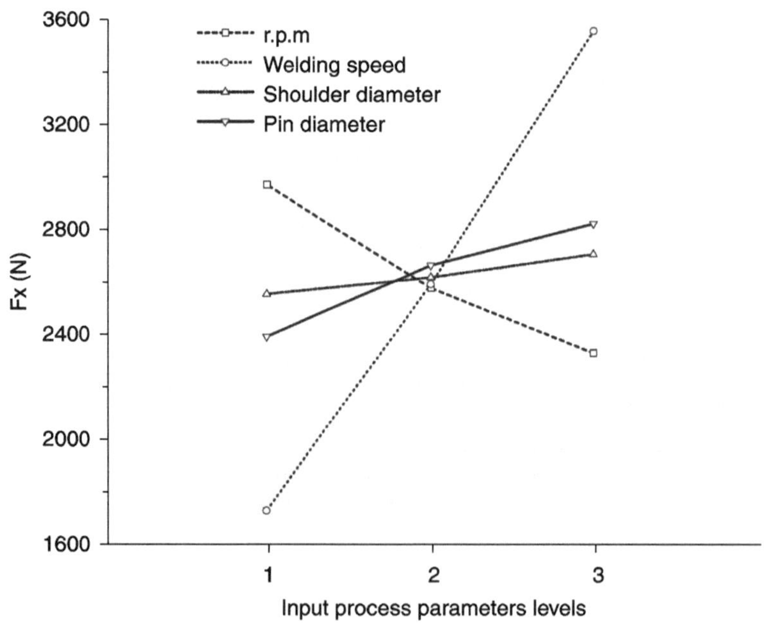

Figure 11.13 Effect of process parameters on F_x [2]

all result in an increase of F_x. Among these parameters, the effect of welding speed is more significant on F_x. However, the X force is inversely proportional to the tool rotational speed [2].

The effect of shoulder diameter on F_x seems to be complicated, since the decrease in the X force by the shoulder diameter increase was also reported [8]. The interaction effect of the tool shoulder diameter and its rotational speed was also found to be significant (Figure 11.14).

The increase of F_x with the pin diameter is contrary to other research, which indicates no noticeable trend with pin diameter (Figure 11.15). This is unexpected; however, the linear regressions show a very slight decrease of the X force by the pin diameter.

The increase of F_x by welding speed is clear, since more material should be plasticized per time unit. Moreover, as the pin diameter increases, the higher contact surface of the tool is subjected to the material, hence higher values for the F_x is expected. The opposite relation of the tool rotational speed with F_x can be attributed to the higher amounts of generated frictional heat. This leads to the decrease in flow stress of the material (aimed to be softened) and decrease in the X force as a consequence.

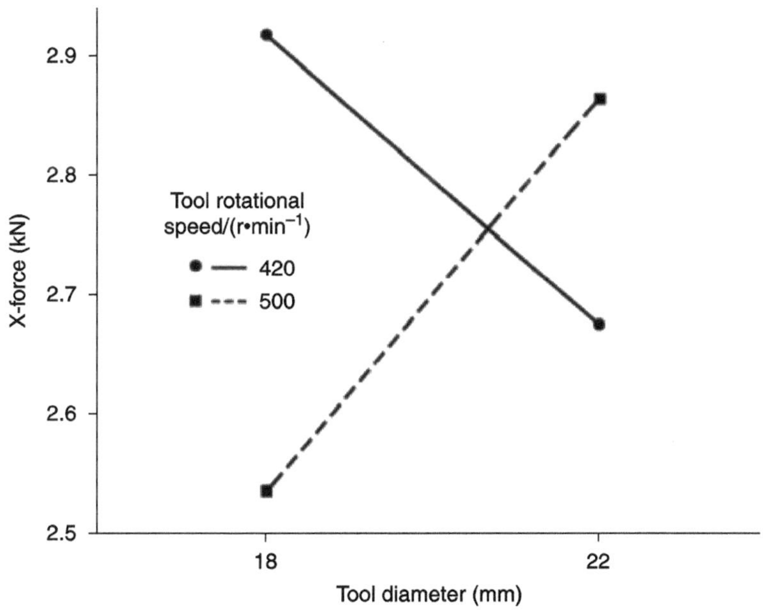

Figure 11.14 Interaction effect of tool diameter and rotational speed [8]

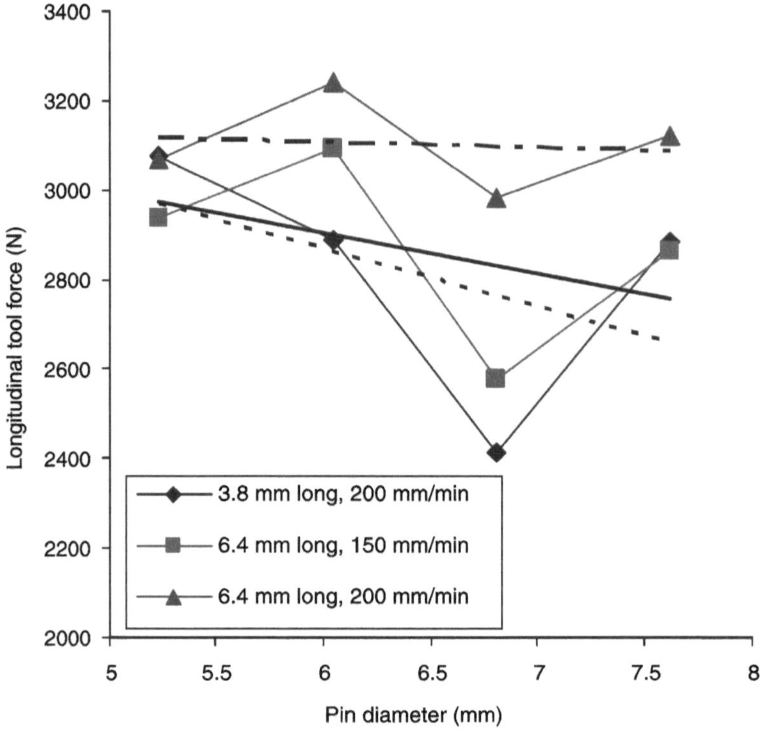

Figure 11.15 Variation of X force with pin diameter [12]

To explain the shoulder diameter effect, the main components of F_x should be considered. First, the shear (drag) force is experienced by the tool pin due to the material resistance against the tool to advance. Second, the frictional force is exerted to the tool shoulder. The increase in shoulder diameter leads to an increase in frictional heating and decrease in friction coefficient, which tends to decrease the F_x. However, the drag force will be increased by the shoulder diameter increase. Hence, these two contradicting components govern the variation of the X force by the tool shoulder diameter.

The effect of pin length on the longitudinal (X) force is presented in Figure 11.16. From the figure, it can generally be stated that the higher pin lengths result in higher longitudinal forces. This is expected since the tools experience higher material resistance against deformation. However, in some tool lengths, the decrease in the X force is noticeable, in which a physical cause for this drop has not yet been identified [12].

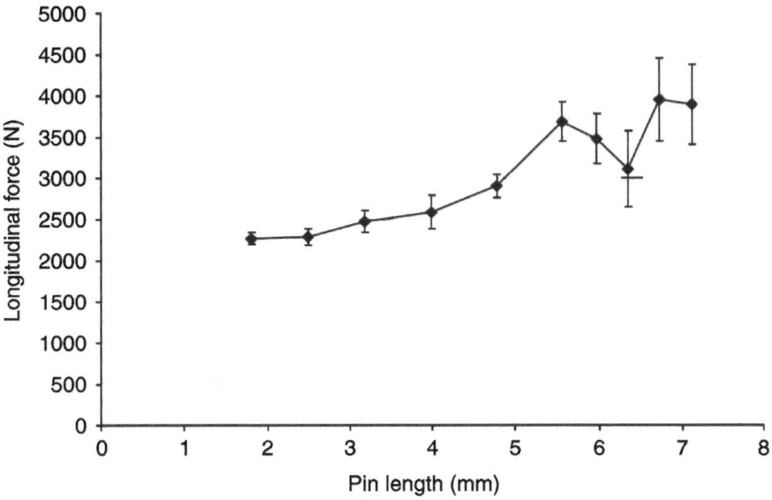

Figure 11.16 **Longitudinal force variation with pin length [12]**

11.5.2 Effect of welding parameters on F_x

The axial force plays a key role in the FSW process, as defect-free weld fabrication is highly dependent on this parameter. The forging action during welding takes place by this force and it is highly effective in the weld quality. The downward force is also influenced by welding speed, rotational speed, tool shoulder and pin diameter. The effect of all these parameters on the F_z (axial force) is presented in Figure 11.17.

F_z increases with welding speed, since more material should be stirred in per unit time (as also explained for F_x). However, the influence of welding speed is less, compared to its effect on F_x, due to the indirect relationship. The increase in tool rotational speed gives rise to a higher temperature at the shoulder contact surface, material softening and decrease in F_z [13]. The shoulder diameter affects the F_z strongly by a power of 2, according to the simplified equation:

$$F_z - \pi.D^2 \cdot \frac{\sigma_y}{4} \tag{11.5}$$

This also confirms the significant effect of the shoulder diameter on F_z (Figure 11.17).

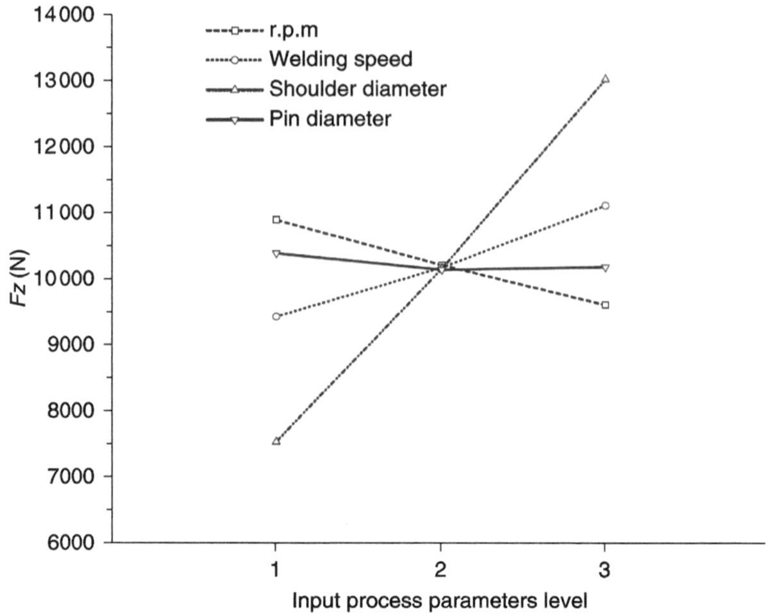

Figure 11.17 Effects of processing parameters on F_z [2]

11.5.3 Effect of friction stir welding parameters on torque

The proposed analytical model for torque prediction can be focused on better understand of the process parameters effects. The model is based on the exposed shear stress on the tool workpiece contact area. The shear stress value can be computed by

$$\tau_t = [(1-\delta)\tau + \delta\mu_f P_N] \quad [11.6]$$

The first and second terms are the sticking and sliding shear stresses, respectively. Here, τ is the shear yield stress and P_N is the axial force. δ and μ are the spatially variable fractional slip and coefficient of friction (COF), respectively, and can be computed using the formulas below:

$$\delta = 0.31 \times exp\left(\frac{\omega r}{1.87}\right) - 0.026 \quad [11.7]$$

and

$$\mu = 0.5 \times exp\ (-\delta\omega r) \quad [11.8]$$

The shear yield stress is temperature dependent and given by [14]

$$\tau = \frac{Y}{\sqrt{3}} \qquad [11.9]$$

where Y is the tensile yield stress. The total required torque, M, consists of two fractions: sticking and sliding torque named as M_T and M_L, respectively. The sticking torque is the resistance of the plasticized material against the tool rotation. The sliding and sticking torque can be computed as

$$M_T = \oint r_A \times (1-\delta)\tau \times d_A \qquad [11.10]$$

and

$$M_L = \oint r_A \times \delta\mu P \times d_A \qquad [11.11]$$

where r_A is the radial distance of the infinitesimal area (dA) from the tool axis.

The sticking torque, as well as plastic flow behaviour, depends on the temperature because of the heat generated during the process. The generated heat per unit area, S, of the tool and workpiece contact surface area is given by [15]

$$S = [(1-\delta)\mu\tau + \delta\mu_f P_N]\ (\omega r - U_1\ sin[\![(\theta)]\!] \qquad [11.12]$$

where U_1 is the welding velocity and the term $(\omega r - U_1 \sin\theta)$ represents the relative velocity of a point on the tool. θ is the angle with the welding direction, if the point location is considered in the polar coordinate system (Figure 11.18). The term U_1 sine θ is somewhat negligible compared to ωr. Hence, regarding the above formula, the generated heat (S) strongly depends on the tool rotational speed (ω) and the shoulder diameter. As a result, the shoulder diameter and tool rotational speed have a significantly effect on the sticking torque.

The effect of tool diameter on the total torque and sliding and sticking components, individually, were computed and are presented in Figure 11.10. The sliding torque, M_L, increases exponentially by tool shoulder, while the decrease in sticking torque, M_T, is clearly observed for high shoulder diameters. However, the decrease in sticking torque is not high enough and so the total torque increased.

As the shoulder diameter increases, the higher surface area results in higher torque regarding the presented formula. However, the decrease in sticking torque is due to the higher heat generation at the welding zone and a subsequent decrease in the material flow stress. The temperature and flow stress in the FSW process using various shoulder diameters are presented in Figure 11.20.

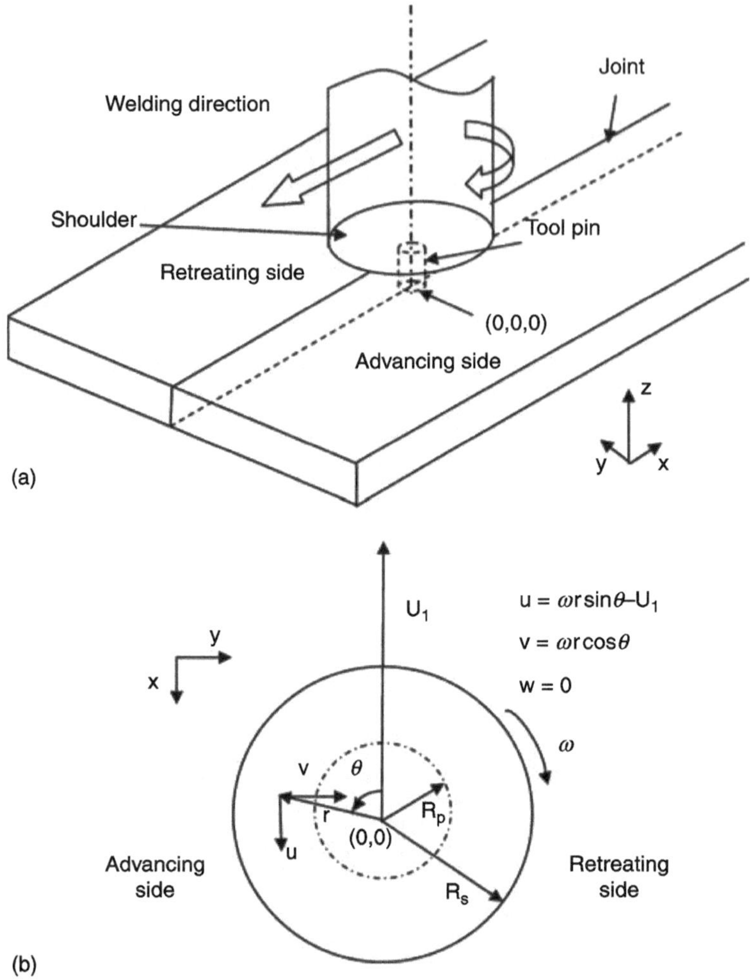

Figure 11.18 (a) Schematic of FSW process; and (b) top view of the tool showing velocity for a typical point at the tool work piece interface [15]

From the figure, it can be implied that for the large shoulder diameters the material in the welding zone deforms at high temperatures and low flow stresses and vice versa. From experimental measurement, effects of tool rotational speed and shoulder diameter, for a typical welding of AA6061 alloy, are both depicted in Figure 11.21. From the figure it can be implied that the required torque for welding increases by shoulder diameter, while it is inversely proportional with the tool rotational speed. Figure 11.22 shows the effect of tool plunge depth on the experimentally

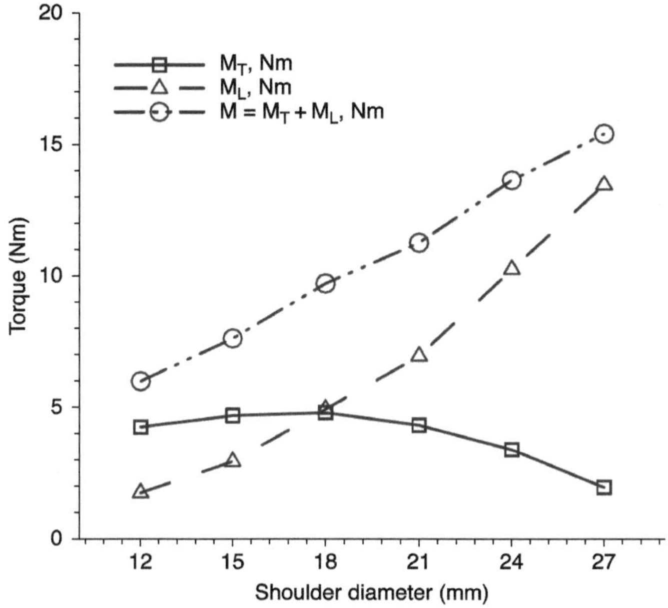

Figure 11.19 Effect of shoulder diameter on the sliding, sticking and total required torque [16]

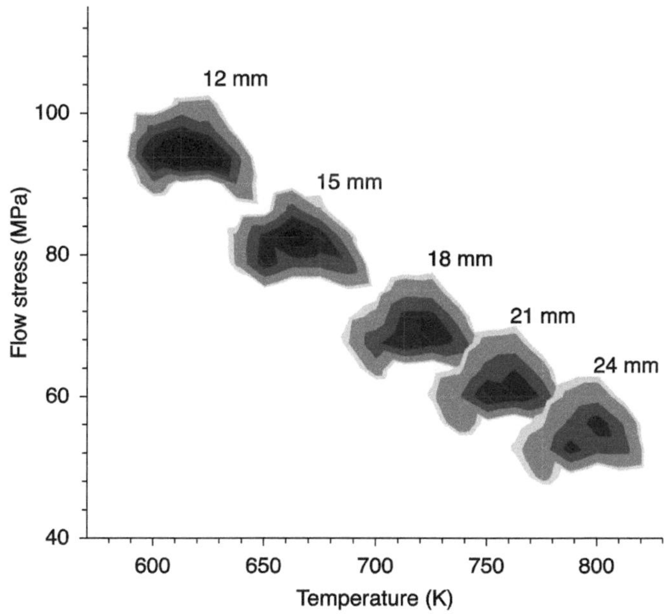

Figure 11.20 Flow stress and temperature during welding of AA60601 with various shoulder diameters [16]

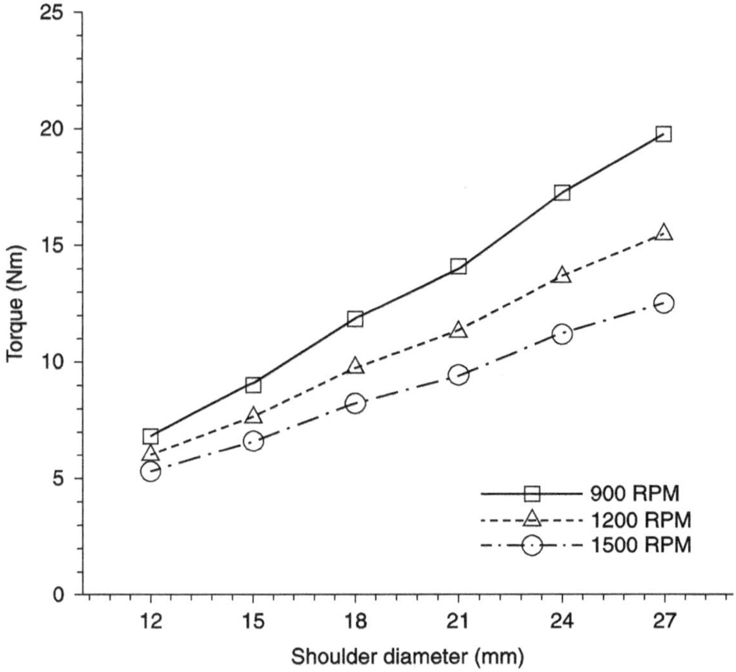

Figure 11.21 Required torque for the FSW process of AA 6061 for various shoulder diameters and rotational speeds [16]

Figure 11.22 Effect of plunge depth on the required axial force and torque in FSW process of 5xxx alloy (tool rotational speed = 1300 rpm, welding speed = 90 mm/min) [17]

measured axial force and torque. From the figure, the slight effect of this parameter can be concluded. These results are recorded in defect-free welding stages, while the force and torque variations in defective welding are noticeable.

11.6 Force and torque variations as representative of weld quality

During the welding process, the changes in welding parameters result in force and torque variations. To fabricate an ideal joint, the main input parameters should be selected appropriately. Ranges for the main input parameters can be identified, named as the process window, where beyond these ranges the production of a defect free joint is not achievable. Typically for the AA6061-T651 alloy, the appropriate process input regions are presented in Figure 11.23 beside the inappropriate range of parameters.

Regarding the conditions under which the experiments were conducted, the ranges of 1300 to 13600, 80–120 and 1.85–2, respectively for the tool rotational speed, welding speed and plunge depth, were found to be the appropriate process settings resulting in defect-free welds. However, beyond this domain, the defective welds are noticeable. In this study, the

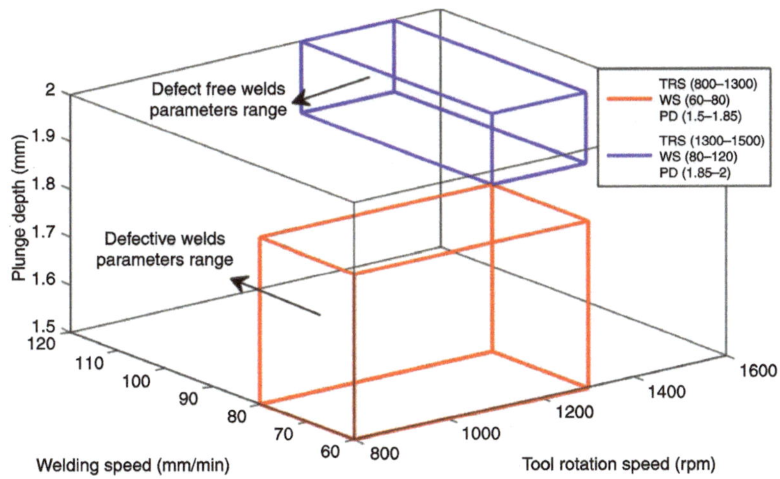

Figure 11.23 Domain of FSW input parameters resulting in defective and defect-free joints [17]

tool rotational speed varies between 800 and 1600 rpm and defect-free welds are manufactured for rotational speeds higher than 1300 rpm. If the tool rotational speed is too high for the welding speed, the surface defects will reappear (Figure 11.24). So, for a specific welding speed, the higher limits for the tool rotational speed should be taken into account to create defect-free welds. Since the rotational and traverse speeds are interrelated, the speed ration (ω/v) can be beneficially used as an indicative parameter in this regard [18].

To find the appropriate set of welding parameters in the fabrication of defect-free joints, trial and error experiments are time-consuming and costly. This will be more problematic when FSW of different base materials are considered. Hence, proposing a criterion which represents the weld quality during welding is beneficial.

In a successful solid-state FSW process, the forces and torque values should be sufficient to stir the plasticized material properly. As discussed earlier, the forces and torque are also influenced by the tool, workpiece and process parameters. As a consequence, the improper combination of parameters will result in insufficient plastic deformation, stirring action and defective welds. For a given set of parameters, the variation of forces and torque could be monitored as well as the weld quality. This will help to find a relationship between force and torque variation in defective and defect-free welds fabrication.

For a typical welding process of AA6061-T651 alloy, the axial (z-direction) force variation with tool rotational speed is depicted in Figure 11.25. The inspection of welded plates shows defective welds

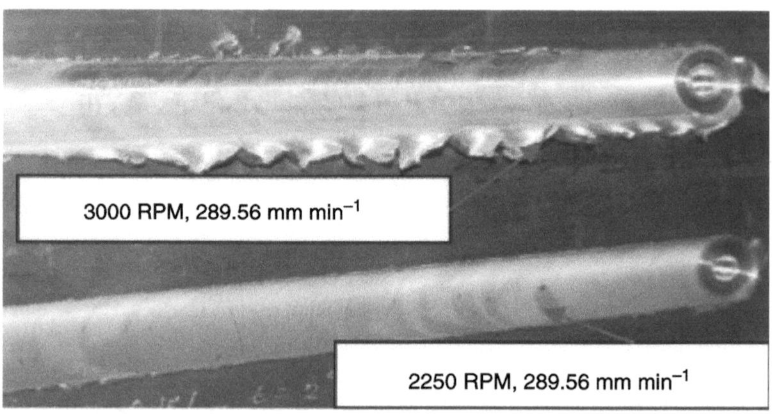

Figure 11.24 Surface deformation for high tool rotational speeds [13]

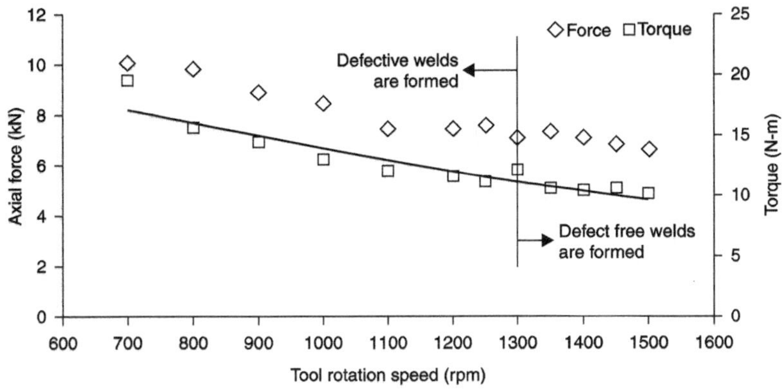

Figure 11.25 **Axial force and torque variation with tool rotational speed (tool shoulder diameter = 18 mm, plunge depth = 1.9 mm and welding speed = 90 mm/min) [12]**

formation below a critical value of tool rotational speed (here 1300 rpm), in which above this value the fabricated welds are defect-free.

Welding speed is another important process parameter and force-welding speed variations are depicted in Figure 11.26. In this case, a critical value for welding speed (here 80 mm/min) can also be identified, in which above this value defect-free welds fabrication is not possible. From the figure, opposite the tool rotational speed, the axial force increases with the welding speed.

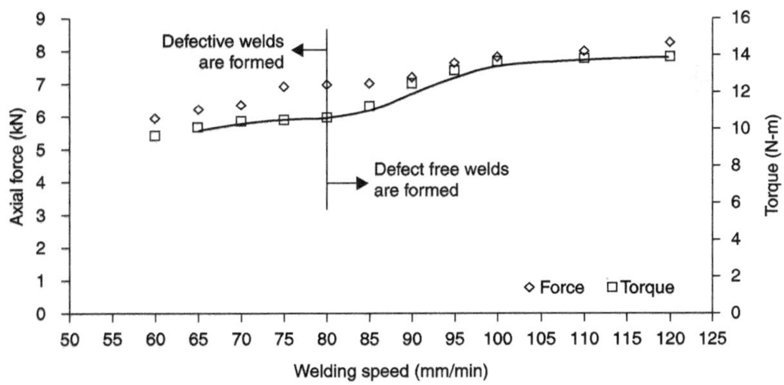

Figure 11.26 **Axial force and torque variation with welding speed (tool shoulder diameter = 18 mm, plunge depth = 1.9 mm and tool rotational speed = 1350 rpm) [12]**

If the force and torque variation with welding speed and tool rotational speed were monitored and plotted in the defect-free domains, slight changes could be observed (Figures 11.27 and 11.28). This proposes a method to online controlling of the process in fabricating successful welds.

A similar trend was also observed in the axial force-plunge depth and torque-plunge depth plots. Hence, it can be implied that the force and torque variation in the defect-free domains are slight, stated as [17]

$$\left(\frac{\partial \tau}{\partial p}\right)_{defective} > \left(\frac{\partial \tau}{\partial p}\right)_{defect\text{-}free} \quad [11.13]$$

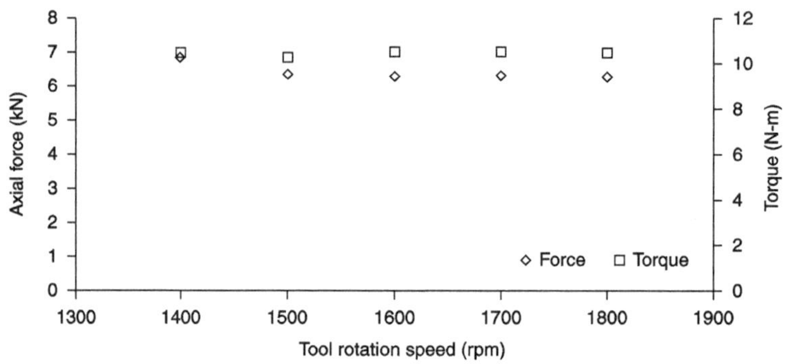

Figure 11.27 Axial force and torque variation with welding speed at defect-free region (tool shoulder diameter = 18 mm, plunge depth = 1.9 mm and welding speed = 90 mm/min) [12]

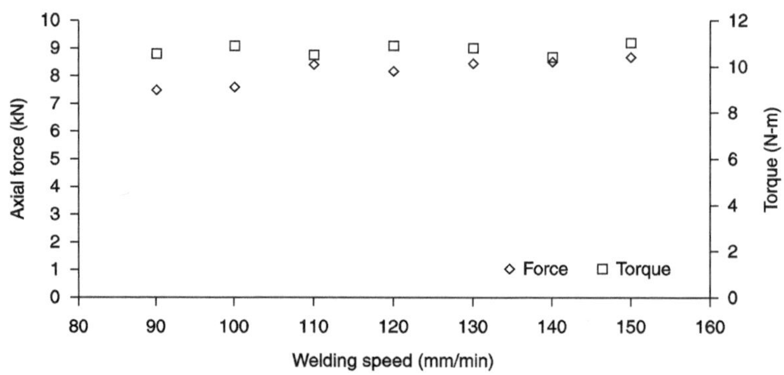

Figure 11.28 Axial force and torque variation with tool rotational speed at defect-free region (tool shoulder diameter = 18 mm, plunge depth = 1.9 mm and tool rotational speed = 1300 rpm) [12]

where p is the process parameter (tool rotational speed, welding speed or plunge depth). This criterion is independent from the material and was also proved for AA5xxx base material. As a result, to select an appropriate set of process parameters, the constant values of axial force and torque can be considered as indicators aiming to fabricate defect-free welds.

Figures 11.29 and 11.30 show the magnitude of force and torque vs. time, for defective and defect-free welds, respectively. For the defective

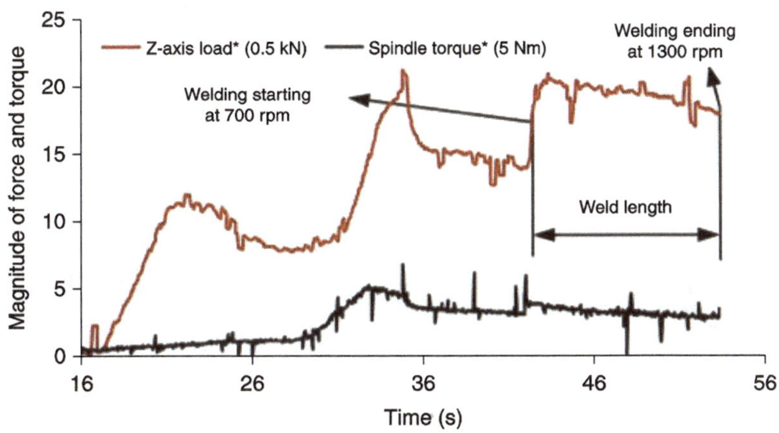

Figure 11.29 Axial force and torque variation vs. time for the defective weld [12]

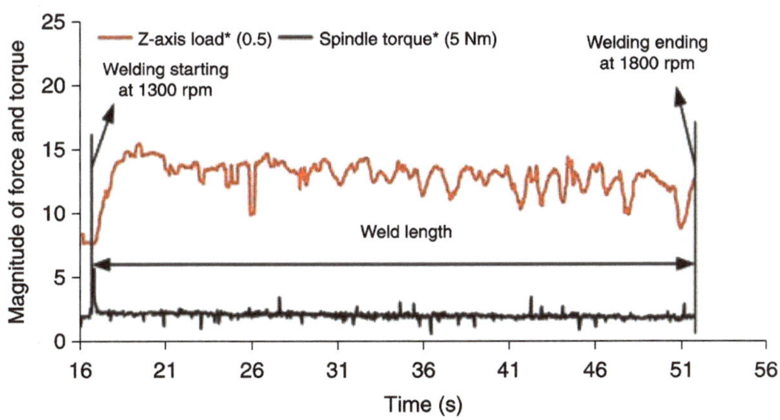

Figure 11.30 Axial force and torque vs. time for the defect-free weld [12]

welds, which are fabricated at low tool rotational speeds, the force-time and torque-time plots show sever fluctuation. However, the force and torque variations are uniform over the welding time for the defect-free welds formed at higher tool rotational speeds. Hence, the force and torque monitoring over time can also be employed as an indicator of weld quality control.

11.7 Analytical models

During the FSW process, as the rotating tool moves in the welding direction, the torsional and bending moments are exerted on the tool. Hence, the tool experiences normal and shear stresses besides the thermal effects of the process. The maximum shear stress can be computed and compared with the tool shear strength to find the safety factor of the tool for the given set of material and process parameters. The stress calculation for the pin is of great importance, since the failure of the FSW process occurs due to tool pin failure [19]. Tools that are designed with low safety factors are the subject of static or fatigue failures. The design safety factor should be considered as high enough to avoid severe tool deformation, which disrupts the welding progress and influences the weld quality. Another important outcome of the stress calculation is to drive a model for tool force prediction, which will be the focus of this section.

To gain an insight into the exerted stresses, a schematic of distributed force on the cylindrical tool and its cross-section are shown in Figure 11.31(a), where q(z) is the applied force on the infinitesimal length (dz) of the pin length. The s-s cross-section of the tool pin is also shown in Figure 11.31(b).

For a point A, at Z_1 distance from the pin root, the bending moment can be calculated as

$$M_y = \int_{z_1}^{L} z.q(z)\,dz \qquad [11.14]$$

Hence, the normal stress can be computed using the formula [20]:

$$\sigma_B = \frac{M_y x}{I_{yy}} = \frac{M_y (r\cos[\theta])}{\frac{\pi r^4}{4}} = \frac{4\cos\square\theta}{\pi r^3}\int_{z_1}^{L} z\, q(z)\,dz \qquad [11.15]$$

where I_{yy}, r, x and θ are the second moment of area, pin radius, normal distance of the point A from y-axis and the angle in polar coordinate,

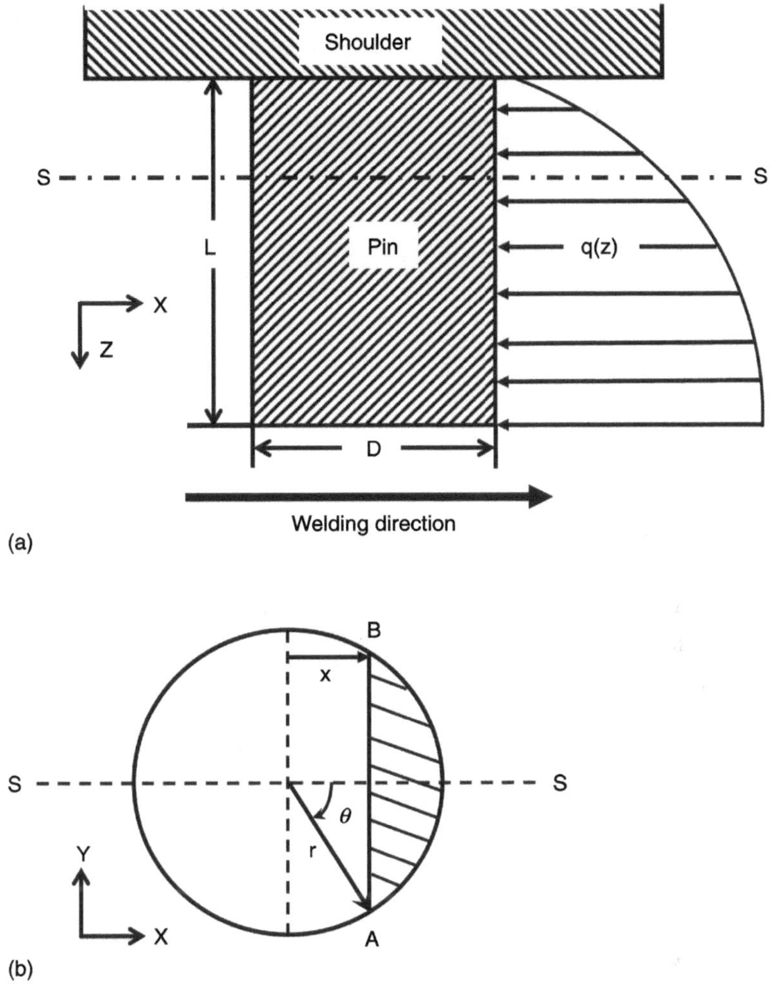

Figure 11.31 **Schematic of tool and cross-section exposed to distributed forces [19]**

respectively. On the other hand, the torsional shear stress can be calculated using the formula [20]:

$$\tau_T = \frac{M_T r}{J_{zz}} = \frac{M_T r}{\frac{\pi r^4}{2}} \qquad [11.16]$$

where M_T is the sticking component of the total required torque in welding, as discussed earlier. The direction of the torsional shear stress at different positions of the tool cross-section is shown in Figure 11.32(a).

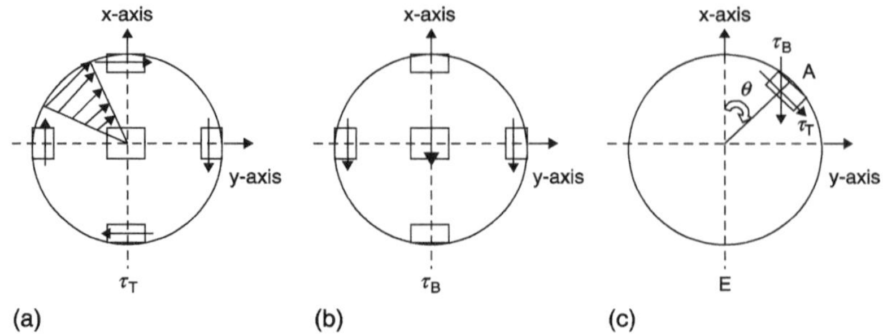

Figure 11.32 Direction of torsional and bending shear stresses in different positions and at Point A [19]

Another component of shear stress, because of exposed bending moments, should be taken into account as the pin length to diameter is small (< 20). This shear stress component, τ_B, is maximum at the neutral axis and directions at all positions are the same (Figure 11.32(b)). For an arbitrary point A, the direction of τ_T and τ_A presented in Figure 11.32(c) should be combined to obtain the total shear stress at this point.

For point A, located at Z_1 distance from the pin root, the shear force *(V)* and first moment of area *(Q)* of the hatched area in Figure 11.30(b) can be computed as

$$V = \int_{z_1}^{L} q(z)\, dz \qquad [11.17]$$

and

$$Q = \int_{x}^{r} g.x\, dx \qquad [11.18]$$

where g is the chord height of infinitesimal area of dA located in the hatched area. Thus, g is equal to $2\sqrt{r^2 - x^2}$ or $2r \sin\theta$ in the Cartesian or polar coordinate system, respectively. Consequently, the first moment of area *(Q)* can be simplified and rewritten as

$$Q = \int_{x}^{0} g\left(-g\frac{dg}{4}\right) = \int_{0}^{g} \frac{g^2}{4} dg = \frac{g^2}{12} = \frac{\left(2\sqrt{r^2 - x^2}\right)^2}{12} = \frac{2}{3}(r \sin\theta)^2 \qquad [11.19]$$

Replacing the Q from above in the shear stress formula, the bending shear stress can be computed as [20]

$$\tau_B = \frac{VQ}{I_{yy}g} = \frac{4}{3}\frac{\sin^2\theta}{\pi r^2}\int_{z_1}^{L} q(z)\,dz \qquad [11.20]$$

Therefore, the principal stresses of σ_1 and σ_2 and maximums shear stress, representative of the exposed mechanical stresses magnitude, can be calculated as [20]

$$\sigma_{1.2} = \frac{\sigma_B}{2} \pm \sqrt{\left(\frac{\sigma_B}{2}\right)^2 + (\tau_B + \tau_T \sin[\theta)]^2 + (\tau_T \cos\theta)^2} \qquad [11.21]$$

and

$$\tau_{max} = \frac{\sigma_1 - \sigma_2}{2} = \sqrt{\left([\frac{\sigma_B}{2}\right)]^2 + (\tau_B + \tau_T \sin[\theta)]^2 + (\tau_T \cos[\theta)]^2} \qquad [11.22]$$

These stresses could be applied in the design of the FSW tool, as well as other instruments. As mentioned above, the total required torque in the FSW process can be computed as

$$M = M_T + M_L \qquad [11.23]$$

M_T and ML fractions can be computed using [21]

$$M_T = \oint r_A \times (1-\delta) r \times d_A \qquad [11.24]$$

and

$$M_L = \oint r_A \times \delta\mu P \times d_A \qquad [11.25]$$

where δ and μ are the fraction of slip at the interface and COF, respectively. Then, the total torque can be also computed from total shear stress as [22]

$$T = \oint r \times (\tau_t dA) = \tau_t \int_0^{R_s} 2\pi r^2\, dr = \frac{2\pi R_s{}^2 \tau_t}{3} \qquad [11.26]$$

where the τ_t is the total shear stress and R_s is the shoulder diameter. In addition, the spindle required power is computed as

$$P = \oint \{(1-\delta)\tau + \delta\mu_f P_N\}\omega r_A dA \qquad [11.27]$$

Moreover, the tool force in the welding (x) direction results from two components, the force experienced by the tool shoulder *(F_s)* and tool pin *(F_P)*. Hence, the total force can be written as

$$F = F_s + F_P \qquad [11.28]$$

The shoulder and pin forces can be computed by using [21]

$$F_S = \oint \delta \times \mu P \times dA \qquad [11.29]$$

and

$$F_p = \oint \sigma \times dA \qquad [11.30]$$

In these equations, σ is the yield strength of the base material considering the temperature of the workpiece and dA is the projected arc of the tool pin in the *y-z*-plane. These analytical equations can be employed to calculate the force values for a given set of input parameters. Typical examples are given here to compare the results calculated from models with the experimental measured values. Measured traverse force at different tool rotational speeds for FSW of AA2524 are depicted in Figure 11.33, besides the computed values from presented models.

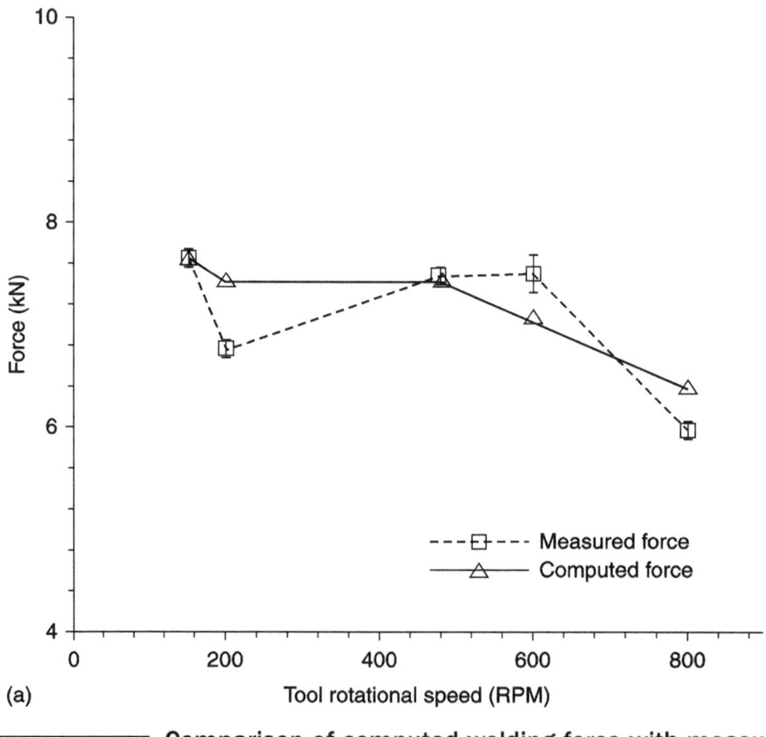

Figure 11.33 Comparison of computed welding force with measured values at different tool rotational speeds during FSW of AA2524 (welding speed = 2.11 mm/s) [19,23]

The calculated and experimentally measured welding force values as a function of pin length are also presented in Figures 11.34 and 11.35. Considering the error bars, there is good agreement between the proposed analytical model and experimental values.

Since the tool failure occurs at the pin due to the exposed stresses, it is crucial to calculate the pin force. This term, as a fraction of total tool force, can be computed directly using the proposed model. However, when comparing with the experimented values, it should be noticed that the pin force measurement from experiments is not easy. Regression-based approaches were proposed to find the pin force fraction of the total tool force. The total longitudinal force for a typical FSW process is shown in Figure 11.36. From the figure, it is clear that the force values approach a limit for lower pin length (envisaged parameter in this case).

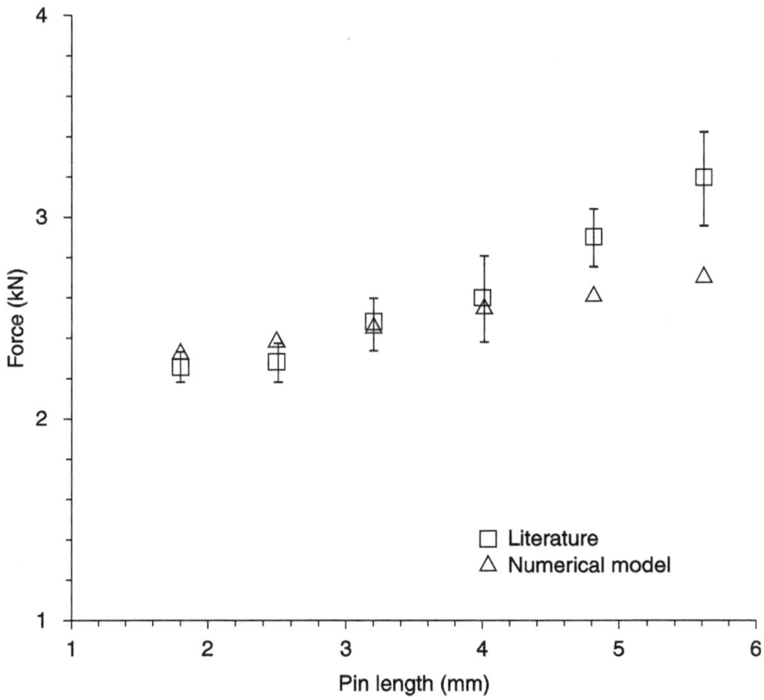

Figure 11.34 **Comparison of computed welding force with measured values at different pin lengths during FSW of AA2524 (welding speed = 3.33 mm/s, tool rotational speed = 650 rpm, pin diameter = 7.6 mm) [12,19]**

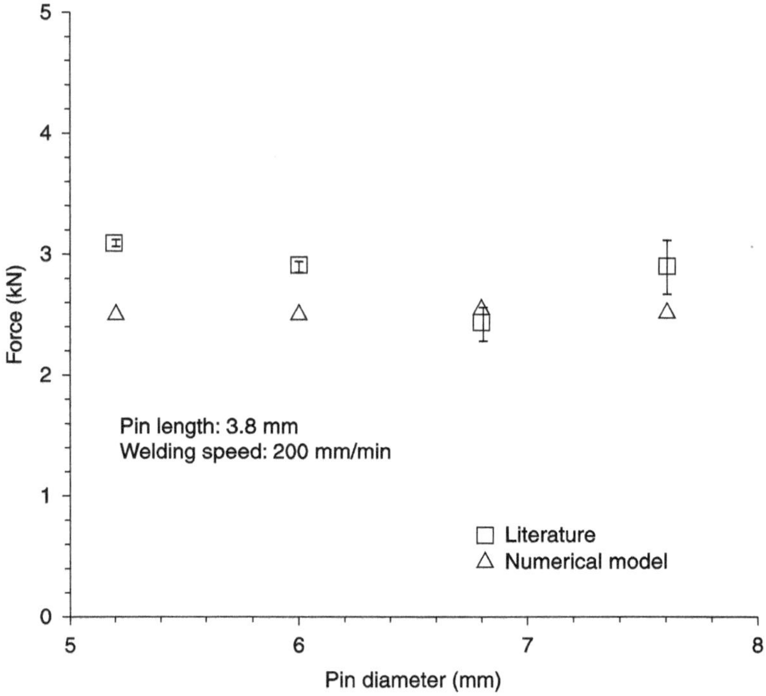

Figure 11.35 **Comparison of computed welding force with measured values at different pin diameters during FSW of AA2524 (welding speed = 3.33 mm/s, tool rotational speed = 650 rpm, pin length = 3.8 mm) [14]**

If the shoulder force is assumed to be independent of the tool length, the limiting constant force, as the tool length tends to lower values, can be attributed to the shoulder force. Hence, subtraction of this constant shoulder force gives rise to the tool pin force, which is shown in Figure 11.37 for different pin lengths.

Using the proposed method, the experimental values are compared with the calculated values by the model (Figure 11.38).

Differences are noticeable between the estimated values from the experiments and calculated values from the proposed model. To explain this, it should be considered that here the pin forces were not directly measured. Hence, the assumption of the regression based model may be the source of observed errors.

To validate the proposed analytical model for torque calculation, comparison with the measured experimental values is shown in

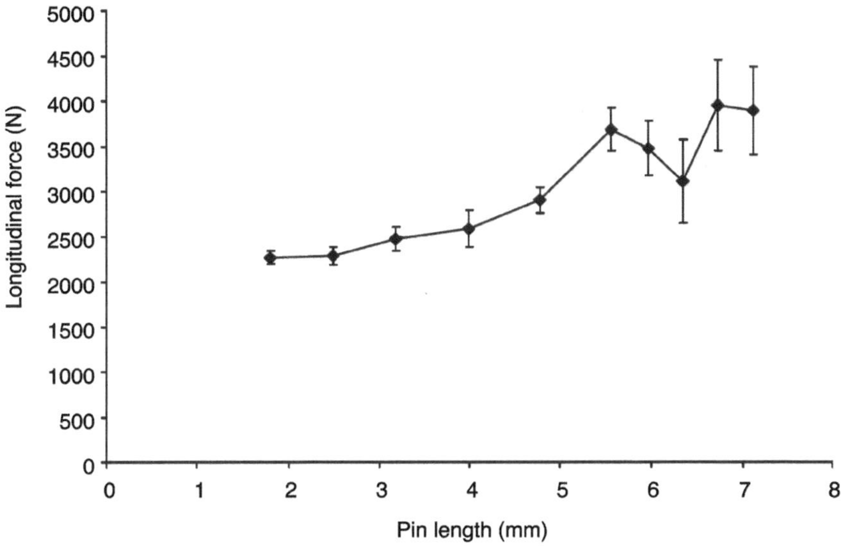

Figure 11.36 Longitudinal force vs. pin length during FSW of 6061-TS alloy [12]

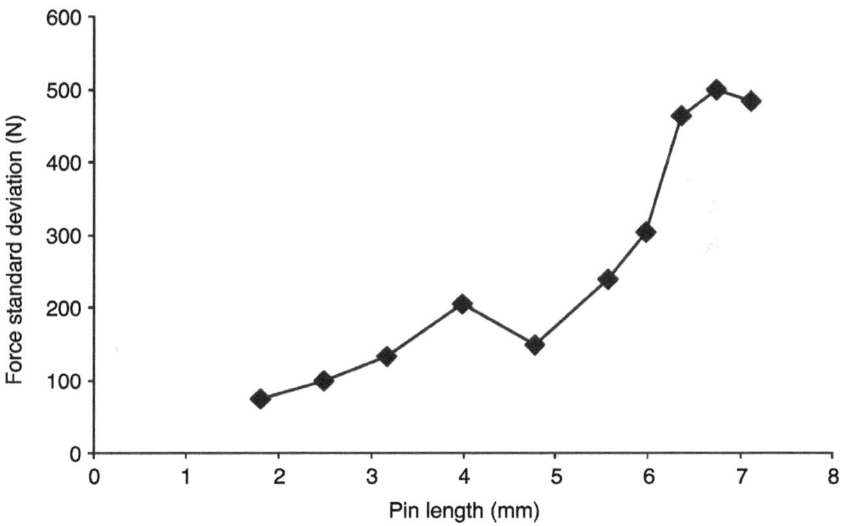

Figure 11.37 Calculated pin force vs. pin length [12]

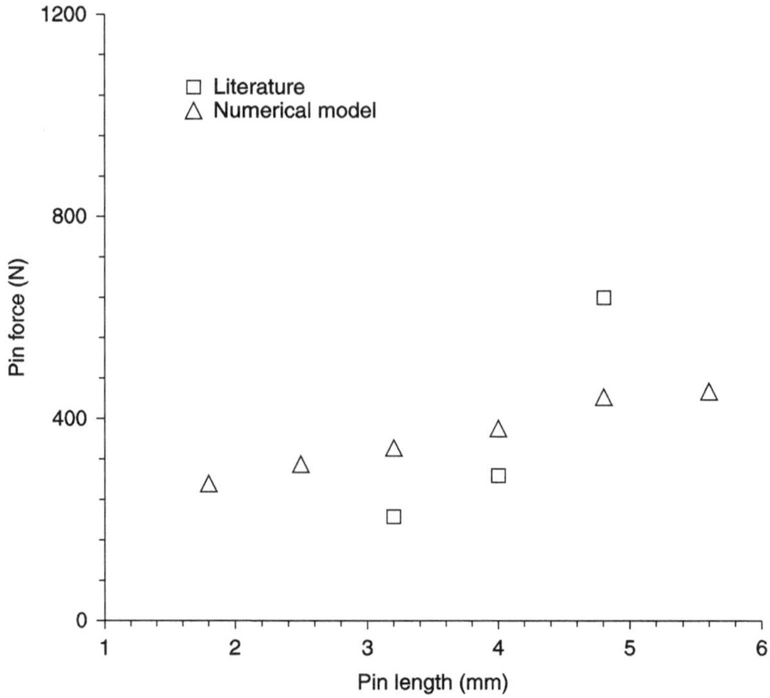

Figure 11.38 **Comparison of regression based estimated pin forces with computed ones for different pin lengths during FSW of AA6061 alloy (welding velocity = 3.33 mm/s, pin diameter = 7.6 mm, rotational speed = 650 rpm) [19]**

Figure 11.39. Here, the shoulder diameter and rotational speed were the variables for the welding of AA7075-T6 alloy.

The experimental and calculated torque values at different tool rotational speeds for the FSW process of two base materials are presented in Figure 11.40, which are in close agreement.

The comparison between computed and corresponding measured power values for the welding of AA7075-T6 alloy are also presented in Figure 11.41. The shoulder diameter and tool rotational speed are the variables while other parameters remain constant. According to Figure 11.41, there is a close agreement between proposed models and measured values.

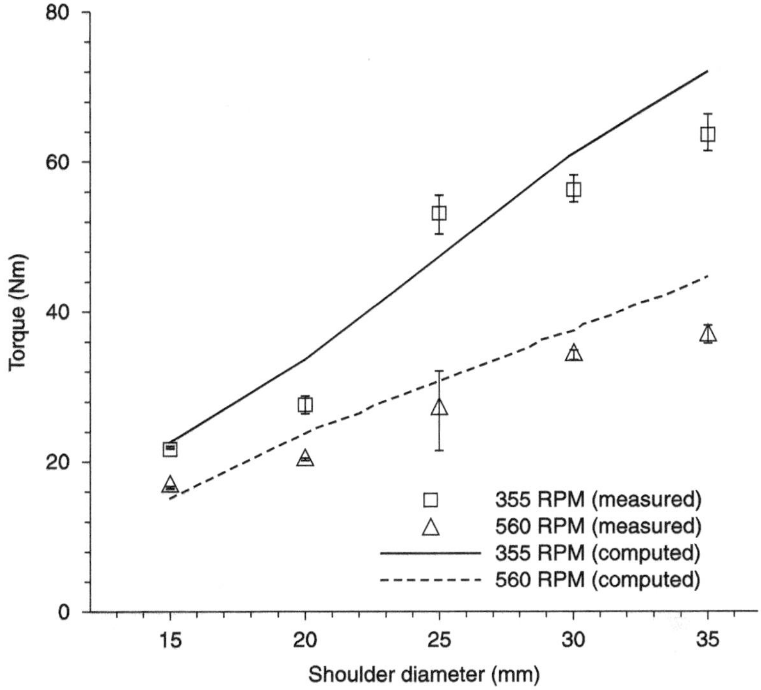

Figure 11.39 Comparison between torque obtained from experiments and calculated from proposed analytical models [21]

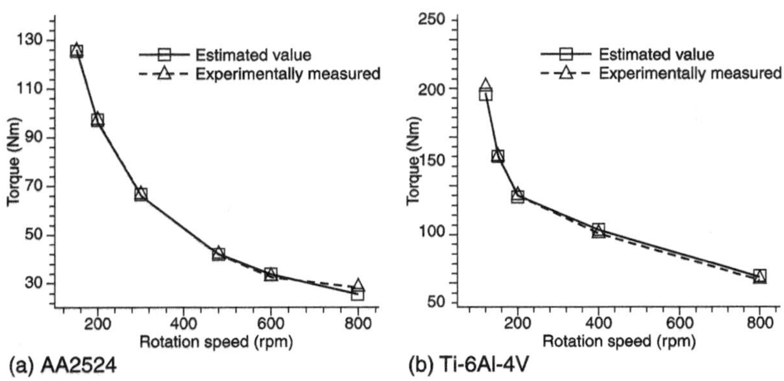

Figure 11.40 Torque variations with rotational speed obtained from experiments and analytical model for: (a) AA2524; and (b) Ti-6Al-4V [24]

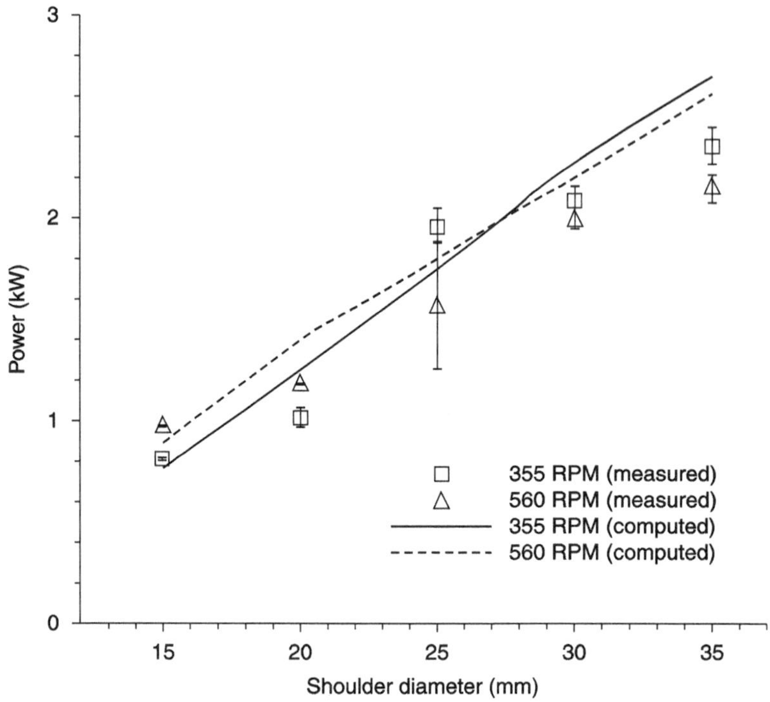

Figure 11.41 Comparison of measured and computed values of required power [21]

11.8 Empirical models

To develop an empirical model, first the force was recorded during welding by the above-mentioned instruments. These plots show fluctuations over time in both pre-welding and welding phases, due to the welding procedure and mechanism. Then the average value of the force in the welding phase was obtained.

For the FSW process of aluminum alloy 2219, the empirical models below are developed and presented to calculate forces in the *x*- and *z*-directions [2]:

$$F_x = 478.2\ (N)^{-0.468}\ (S)^{0.646}\ (D)^{0.078}\ (d)^{0.523} \qquad [11.31]$$

and

$$F_z = 9.68\ (N)^{-0.312}\ (S)^{0.146}\ (D)^{2.72}\ (d)^{-0.121} \qquad [11.32]$$

where N, S, D and d stand for the rotational speed, welding speed, shoulder diameter and tool pin diameter, respectively.

Regarding the models, the X force is proportional to welding speed, shoulder and welding diameter and inversely proportional to tool rotational speed. The most effective parameter is the welding speed associated with highest power in the model among the other parameters. The model for the Z force indicates the significant effect of the tool shoulder diameter compared to other parameters. Moreover, from the model, the opposite effect of tool rotational speed and pin diameter, unlike the welding speed and shoulder diameter, can be inferred.

In another research study, the following models were presented for the welding of AA5083. In these models, the pin diameter effect was found to be negligible and so excluded [8]:

$$F_x = 2748 + 219\,d + 146\,DN + 326\,v \qquad [11.33]$$

and

$$F_z = 17412 + 2426\,D - 551\,N + 531\,v \qquad [11.34]$$

where d, D, N and v stand for the pin diameter, shoulder diameter, rotational and welding speeds, respectively. The highest coefficient of welding speed and shoulder diameter in respective F_x and F_z empirical models also indicate these parameters to have significant effects.

The welding force consists of two fractions, shoulder force and pin force. The pin force plays a key role in the tool design, hence its estimation is highly significant. From experimental results, it was found that the pin force shows quadratic behaviour with the pin diameter. However, the decrease in higher pin lengths was also noticed. Hence, for lower pin lengths, empirical models were developed to predict the pin force. The pin force obtained by subtraction of the shoulder force from the total tool force was measured from the experiments. The shoulder force is assumed to be the limit approach of the tool force at the lower pin lengths. For the conditions in which the experiments were done (FSW of 6061-T6 alloy, pin diameter = 7.6 mm, tool rotational speed = 650 rpm, welding speed = 200 mm/min), the pin force vs. pin length and fitted curve are depicted in Figure 11.42.

The following empirical model is proposed for the tool pin force estimation, where L_0 is the minimum in the fit force curve and A is the constant of the regression. In this case, $F_{shoulder}$, L_0 and A was chosen as 2260 N, 1.8 mm and 86.3 MPa, respectively [12]:

$$F_p = F_{tool} - F_{shoulder} = \begin{cases} 0, & L \le L_0 \\ A\,[(L - L_0)]^2, & L > L_0 \end{cases} \qquad [11.35]$$

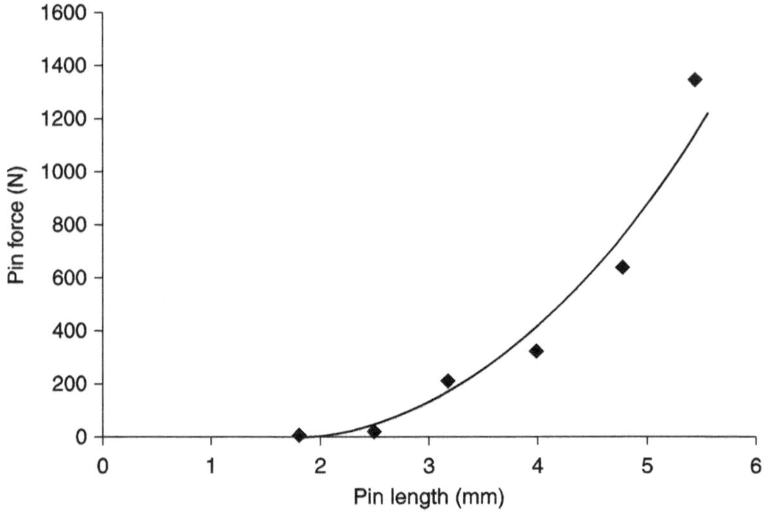

Figure 11.42 Pin force variation with pin length and the fitted curve [12]

For the same experimental conditions, to predict the pin force for various tool diameters, the following linear model is also presented [12]:

$$F_{tool} = F_0 - B.r_p \quad [11.36]$$

where F_0 and B are the constant values presented in Table 11.1 [12]. The coefficient of r_p, B, is negative, indicating the opposite effect of tool pin diameter on the pin force.

Table 11.1 Linear fitting coefficients for different pin diameters [9]

Pin diameter, mm	Feed rate, mm/min	F_0, N	B, N/mm
3.8	200	3180	–12
6.4	150	3450	–91
6.4	200	3660	–131

11.9 References

1. Mishra, R.S. and Ma, Z. (2005), 'Friction stir welding and processing', *Materials Science and Engineering R: Reports*, 50(1): 1–78.
2. Arora, K.S., Pandey, S., Schaper, M. and Kumar, R. (2010), 'Effect of process parameters on friction stir welding of aluminum alloy 2219-T87', *The International Journal of Advanced Manufacturing Technology*, 50(9–12): 941–52.
3. Trimble, D., Monaghan, J. and O'Donnell, G. (2012), 'Force generation during friction stir welding of AA2024-T3', *CIRP Annals-Manufacturing Technology*, 61(1): 9–12.
4. Hattingh, D., Blignault, C., Van Niekerk, T. and James, M. (2008), 'Characterization of the influences of FSW tool geometry on welding forces and weld tensile strength using an instrumented tool', *Journal of Materials Processing Technology*, 203(1): 46–57.
5. Mitchel, J.E., Cook, G.E. and Strauss, A.M. (2003), 'Force sensing in friction stir welding', *Proceedings of the 6th Internatinal Trends in Welding Research*. Pine Mountain.
6. Crawford, R., Cook, G., Strauss, A.M., Hartman, D. and Stremler, M. (2006), 'Experimental defect analysis and force prediction simulation of high weld pitch friction stir welding', *Science and Technology of Welding & Joining*, 11(6): 657–65.
7. Mandal, S., Rice, J. and Elmustafa, A. (2008), 'Experimental and numerical investigation of the plunge stage in friction stir welding', *Journal of Materials Processing Technology*, 203(1): 411–19.
8. Kumar, R., Singh, K. and Pandey, S. (2012), 'Process forces and heat input as function of process parameters in AA5083 friction stir welds', *Transactions of Nonferrous Metals Society of China*, 22(2): 288–98.
9. Su, H., Wu, C., Pittner, A. and Rethmeier, M. (2013), 'Simultaneous measurement of tool torque, traverse force and axial force in friction stir welding', *Journal of Manufacturing Processes*.
10. Lombard, H., Hattingh, D., Steuwer, A. and James, M. (2008), 'Optimising FSW process parameters to minimise defects and maximise fatigue life in 5083-H321 aluminium alloy', *Engineering Fracture Mechanics*, 75(3): 341–54
11. Vukčević, P.D.M., Savićević, P.D.S, Janjić, P.D.M. and Šibalić, P.D.N. 'Measurement in friction stir welding process',
12. Sorensen, C.D. and Stahl, A.L. (2007), 'Experimental measurements of load distributions on friction stir weld pin tools', *Metallurgical and Materials Transactions B*, 38(3): 451–9.
13. Longhurst, W.R. (2009), *Force Control of Friction Stir Welding*, Vanderbilt University.
14. Nandan, R., Roy, G. and DebRoy, T. (2006), 'Numerical simulation of three-dimensional heat transfer and plastic flow during friction stir welding', *Metallurgical and Materials Transactons A*, 37(4): 1247–59.
15. Nandan, R., Roy, G., Lienet, T. and DebRoy, T. (2007), 'Three-dimensional heat and material flow during friction stir welding of mild steel', *Acta Materialia*, 55(3): 883–95.

16. Arora, A., De, A. and DebRoy, T. (2011), 'Toward optimum friction stir welding tool shoulder diameter', *Scripta Materialia*, 64(1): 9–12.
17. Ramulu, P.J., Narayanan, R.G., Kailas, S.V. and Reddy, J. (2013), 'Internal defect and process parameter analysis during friction stir welding of Al 6061 sheets', *The International Journal of Advanced Manufacturing Technology*, 1–14.
18. Leitão, C., Louro, R. and Rodrigues, D. (2012), 'Using torque sensitivity analysis in accessing friction stir welding/processing conditions', *Journal of Materials Processing Technology*, 212(10): 2051–7.
19. Arora, A., Mehta, M., De, A. and DebRoy, T. (1990), 'Load bearing capacity of tool pin during friction stir welding', *The International Journal of Advanced Manufacturing Technology*, 61(9–12): 911–20.
20. Popov, E.P. et al. (1990), *Engineering Mechanics of Solids*, vol. 2, Englewood Cliffs, NJ: Prentice Hall.
21. Mehta, M., Atora, A. De, A. and DebRoy, T. (2011), 'Tool geometry for friction stir welding: Optimum shoulder diameter', *Metallurgical and Materials Transactons A*, 42(9): 2716–22.
22. Arora, A., Nandan, R., Reynolds, A. and DebRoy, T. (2009), 'Torque, power requirement and stir zone geometry in friction stir welding through modeling and experiments', *Scripta Materialia*, 60(1): 13–16.
23. Boyles, J. (2008), *Process Development for Property Evaluation of Friction stir Welded Titanium Alloys*, University of South Carolina: Columbia.
24. Arora, A., DebRoy, T. and Bhadeshia, H. (2011), Back-of-the-envelope calculations in friction stir welding: Velocities, peak temperature, torque and hardness', *Acta Materialia*, 59(5): 2020–8.

Simulation of friction stir welding and processing

P. Asadi, University of Tehran, Iran and M. Akbari and H. Karimi-Nemch, Iran University of Science and Technology, Iran

DOI: 10.1533/9780857094551.499

Abstract: FSW/FSP results in intense plastic deformation and temperature increase within and around the stir zone. FSW is based on strong couplings of thermo-mechanical phenomena. An understanding of mechanical and thermal processes during FSW/FSP is needed for optimizing process parameters and controlling microstructure and properties of welds. Still an under-developed process, its numerical simulation is expected to aid a better understanding of the physics, by observing the influence of the input parameters on obtained joints, and optimizing the overall process for a large range of tools, processing conditions and materials. In this chapter, understanding the simulation of FSW and important issues related to the topic provide an overall view of the subject. First, a detail finite element formulation and computational aspects of FSW such as FEM software, material and friction model are discussed. Then, some important results of finite element modelling of FSW such as temperature, welding force, strain distribution, residual stress, material flow and grain size are presented.

Key words: finite element methods, thermo-mechanical simulation, material flow, strain distribution.

12.1 Introduction

FSW and FSP are based on a strong combination of thermal and mechanical phenomena, which results in severe plastic deformation and temperature rise around and inside the stir zone (SZ). It brings about very complex material movements and large shear forces. An understanding of mechanical and thermal phenomena during FSW/FSP is needed for process parameters optimization and controlling microstructure and properties of welds. Being a recent and under-developed process, its numerical simulation is expected to aid a better understanding of its nature. In fact, finite element (FE) simulation of FSW/FSP to investigate of the influence of the processing parameters on the obtained joints and how to optimize them. However, the development of a satisfactory numerical model requires considerable effort.

The numerical modelling of FSW poses a challenge due to high strains, strain rates and temperatures involved in the process, which leads to an intricate problem involving non-linear material behaviour. In addition, the simultaneous presence of three phenomena, mechanical heat generation, heat transfer and mass transfer, make it more complicated.

This chapter provides an understanding of the simulation of FSW and important issues related to the topic, to provide an overall view of the subject, which is elaborated on in individual sections.

First, the physical description of the process with the aim of explicit description of the heat generation model and interaction between thermal and mechanical model is addressed. A detailed finite element formulation and computational aspects are also included. Then the numerical models utilized to investigate different aspects of the FSW/FSP, such as temperature field, welding forces, strain distribution, residual stress, material flow, grain size and simulation of dissimilar material, are described.

12.2 Finite element method software

In recent years, a number of researchers have attempted to develop numerical codes for FSW. Many kinds of commercial codes, such as ABAQUS [1–3], ANSYS [4,5], DEFORM [6–9], FIDAP [10], FLUENT [11,12], LSDYNA [13] and FORGE3 [14], as well as user-defined codes [15], have been utilized for simulation of the FSW process. In these researches, two basic approaches were employed:

1. a solid mechanics explicit finite element method (FEM) with an updated Lagrangian description; and
2. the computational fluid dynamics (CFD) finite volume method (FVM) with the Eulerian description.

In the Lagrangian description model, the domain is divided into material nodal points and elements. These points and elements move and deform during the analysis. In the Eulerian description model, the domain is divided into spatial nodal points and elements, which remain stationary in space during the analysis. However, material in this model is allowed to flow across element boundaries in a rigid mesh. Field variables, including temperature, velocity and strain rate, can be calculated at these stationary points and elements.

Each software has its own distinct ability for FSW simulation. For instance, DEFORM-3D commercial code has the ability to predict material flow, grain size and hardness. Based on pervious works by different authors on FSW modelling, the ability of software to simulate of FSW is shown in Table 12.1.

12.2.1 Material model

The selection of an appropriate constitutive law to reflect the relation of flow stress with temperature, plastic strain and strain rate is essential for modelling the FSW process. For this purpose, different material laws were used to describe the relationship between the flow stress strain rate and temperature by different authors (Table 12.2). A rigid-viscoplastic, whose flow stress depends on temperature and strain rate, was modelled with different constitutive equations for FSW simulation. Also, the temperature and strain rate-dependent elastic–plastic Johnson–Cook law were utilized by [6–9, 16].

12.2.2 Friction model

Generally, two friction models have been employed for modelling of FSW by different authors:

1. shear friction model; and
2. Coulomb friction model.

Table 12.3 summarizes friction coefficients and models that were used by these various authors.

Table 12.1 Software ability for FSW simulation

Commercials code	Material model	Material laboratory	Temperature calculation	Strain calculation	Residual stress calculation	Material flow	Grain size calculation	Hardness calculation
Deform 3D	Visco-plastic	✓	✓	✓	–	✓	✓	✓
Abaqus	Elastic-plastic	–	✓	✓	✓	✓	Not reported	Not reported
Ansys	Visco-plastic	–	✓	✓	✓	Not reported	Not reported	Not reported
Forge3	Visco-plastic	✓	✓	✓	Not reported	Not reported	Not reported	Not reported

Table 12.2 Material laws used by different authors

Reference	Material	Material model	Flow stress equation
[6]	AZ91 magnesium	rigid-viscoplastic	$\dot{\bar{\varepsilon}} = A[Sinh(\alpha\bar{\sigma})]^n e^{-\frac{\Delta H}{RT_{abs}}}$
[17]	AA2139-T8	rigid-viscoplastic	$\sigma = KT^A(\dot{\varepsilon})^B(\bar{\varepsilon})^C$
[8]	AA6082-T6	rigid-viscoplastic	$\sigma = KT^A(\dot{\varepsilon})^B(\bar{\varepsilon})^C$
[18]	AA 6061	rigid-viscoplastic	$\sigma = Ae^{BT}\dot{\bar{\varepsilon}}^C e^{(D/\bar{\varepsilon})}$
[2]	AA2024-T3	elastic–plastic	$\bar{\sigma} = [A + B(\varepsilon^{-pl})^n][1 + c ln(\frac{\dot{\bar{\varepsilon}}^{pl}}{\dot{\varepsilon}_0})](1 - \hat{\theta}^m)$
[9]	AA5083	rigid-viscoplastic	$\dot{\bar{\varepsilon}} = A[Sinh(\alpha\bar{\sigma})]^n e^{-\frac{\Delta H}{RT_{abs}}}$
[19]	AA6082-T6	rigid-viscoplastic	Tabulated values

Table 12.3 Friction model used by different authors

Reference	Material	Friction model	Friction coefficient
[7]	Ti6Al4V	shear	0.31
[6]	AZ91	shear	0.4
[20]	Al–Si	shear	0.5
[9]	AA5083	shear	0.46
[21]	AA5083	Coulomb	–
[16]	AA2024	Coulomb	0.3

Shear friction model

The constant shear friction model was used by many researchers [6–9,17] and the frictional force in this model is calculated by

$$f = mk \quad [12.1]$$

where f denotes the frictional stress, k is the shear yield stress and m is the friction factor. The friction coefficient between tool and workpiece is an input parameter in the FE model, used to formulate heat generation. Selection of an appropriate friction factor is important, since it has a considerable impact on heat generation. If the friction factor is too small, heat generation will be too low to deform the materials and consequently to weld them; and the other hand, too large friction factors may result in an unreasonable thermal field.

Some methods were used by different authors in order to determine the friction factor [20]. The friction factor was approximated by calculating the axial forces for three different friction factors, 0.4, 0.5 and 0.6 in FE modelling. By comparison of simulation results and experimental measurements (Figure 12.1), a friction factor of 0.5 was the best choice.

In an alternative method [7], in which the shear friction factor was optimized through an inverse approach, initially the temperature was measured at various process parameters via experimental methods and then a function representing the calculated quadratic error between the measured and predicted temperature profiles was built and minimized (Figure 12.2). This quadratic error function is obtained by changing both process parameters and friction factor value.

Coulomb friction model

Besides the shear friction model, the Coulomb friction model is also utilized by some researchers. The frictional force in the Coulomb law model is defined by

$$f_s = \mu p \qquad [12.2]$$

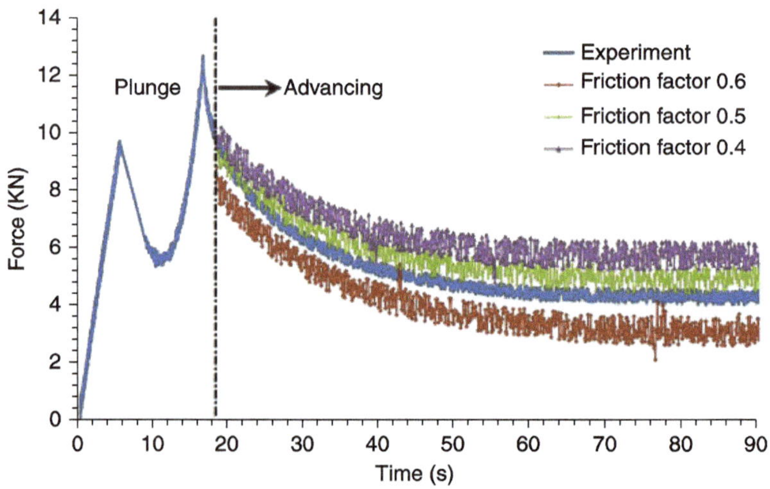

Figure 12.1 Simulation and experimental axial force [20]

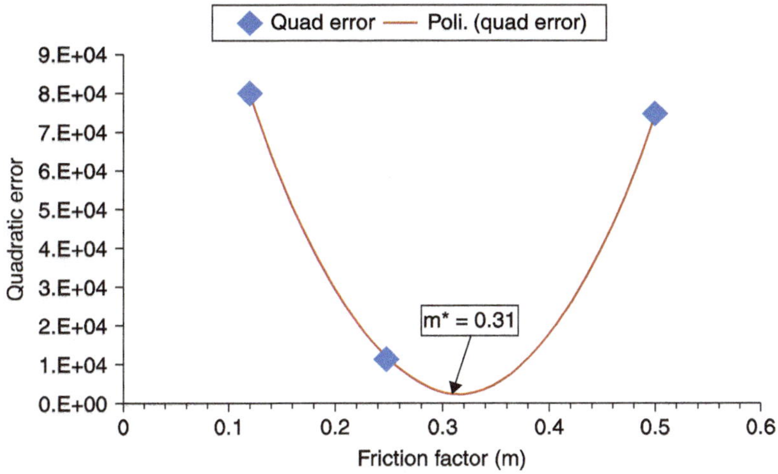

Figure 12.2 Quadratic error function for temperature distributions vs. friction factor value m [7]

where *fs* is the frictional stress, *p* is the interface pressure between two bodies and μ is the friction coefficient.

12.3 Finite element method result

12.3.1 Temperature distribution

FSW results in severe plastic deformation around the rotating tool and intense friction between tool and workpiece. Both of these factors lead to the temperature increase within and around the SZ, which has dramatic effects on the microstructure of the welds, such as grain size, grain boundary characteristics, coarsening or dissolution of precipitates, and resultant mechanical properties of the welds. Consequently, it is important to predict temperature distribution precisely during FSW. It should be mentioned that owing to temperature, measurements within the SZ are very difficult due to the intense plastic deformation produced by the rotation and translation of the tool temperature distribution around weld zone, which is usually determined numerically.

The predicted temperature achieved by the FE model was compared to the experimental results in many studies. It was observed that the predicted temperature matches well with the experimental one. Comparison of simulation and experimental results in temperature history, reported for aluminum [8] and magnesium [6], are illustrated in Figures 12.3 and 12.4. It is observed that there is an acceptable agreement between experimental and simulated temperature history curves.

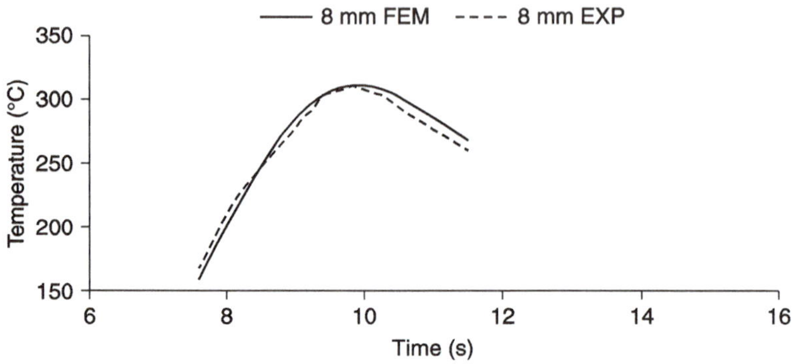

Figure 12.3 Temperature history derived from simulation and experiment for 6082-T6 aluminum alloy [8]

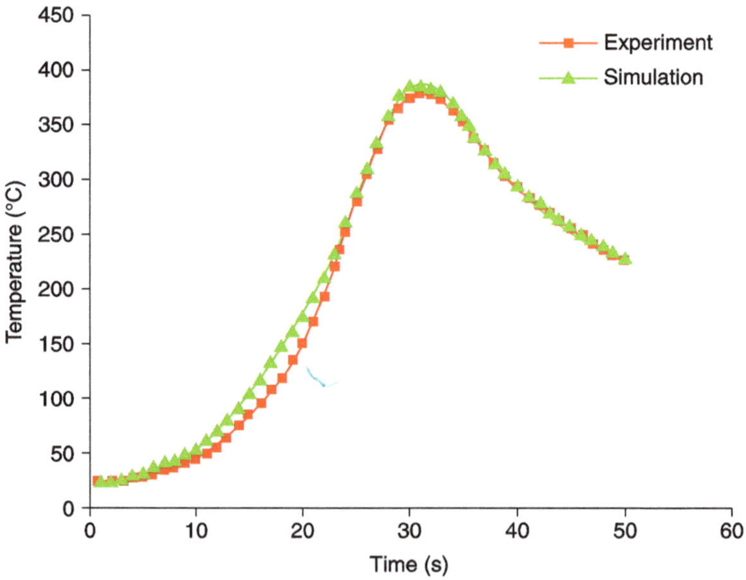

Figure 12.4 Comparison of simulation and experimental temperature history for AZ91 magnesium alloy [6]

The temperature profiles around the weld zone at a traverse speed of 30 mm/min and a rotational speed of 1400 rpm for FSW welding of AA5083 is shown in Figure 12.5. As can be seen, temperature distribution along the cross-section is approximately symmetrical. This can be clarified by considering actual revolving and traverse velocity used in industrial applications of FSP/FSW. As a pertinent case in a FSW process with a revolving speed of 1400 rpm and a tool pin diameter of 3 mm, the peripheral pin velocity due to rotation is about 13 194 mm/min at the pin surface, much higher than the traverse speed that is in the range of 50 to 120 mm/min. This difference would be even greater at a generic point of the shoulder surface, where most of the frictional heat is generated. Furthermore, frictional heat that supplies most of the heat converted to the welding is directly proportional to the relative velocity between tool and workpiece. Since the relative velocity due to the revolving speed is much higher than the traverse speed, the heat generation dominated by the pin rotation leads to an approximately symmetrical temperature distribution about the weld line.

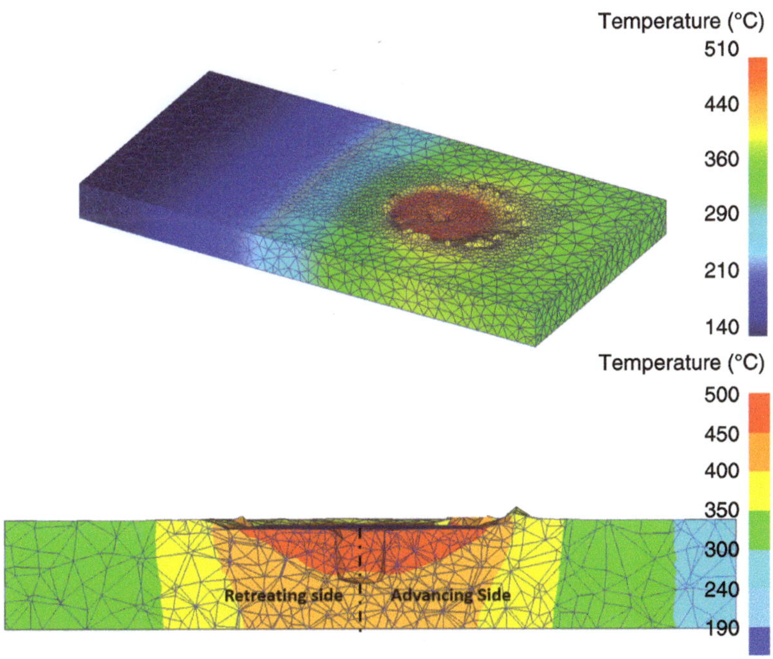

Figure 12.5 Temperature profiles around the weld zone

It is interesting to note that the temperature on the advancing side (AS) is somewhat higher than that on the retreating side (RS). This is due to the fact that the most severe plastic deformation occurs at the AS, where the tool rotation and material flow directions are opposite to each other (Figure 12.6).

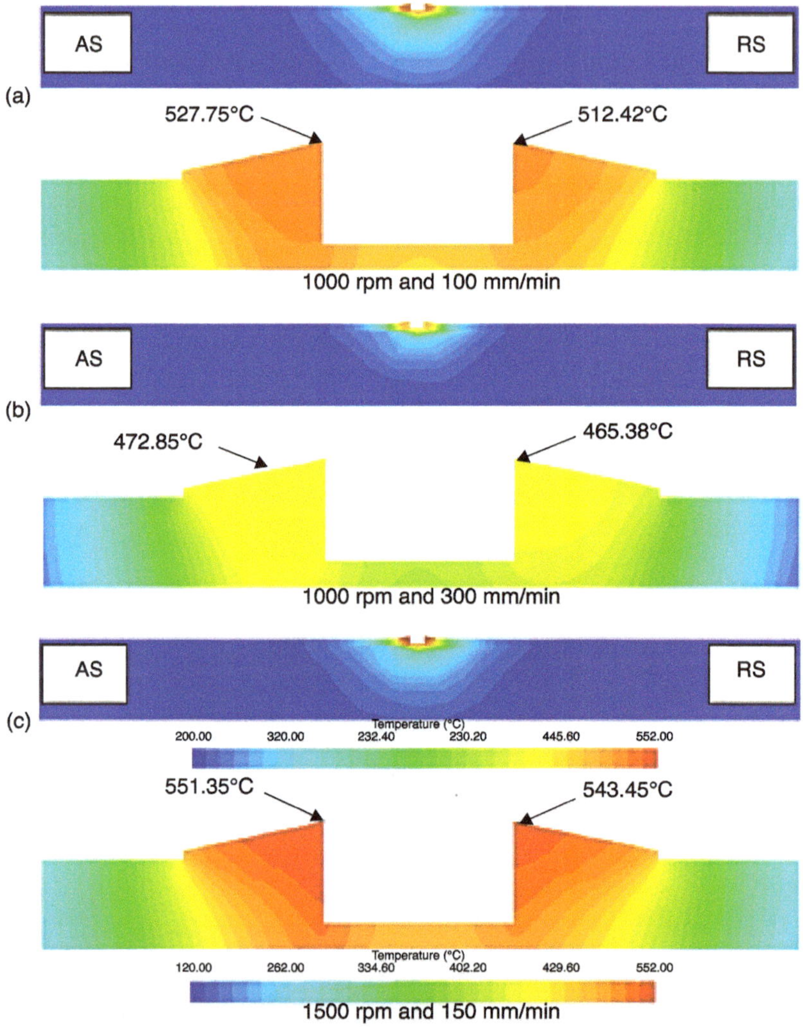

Figure 12.6 Temperature profiles at the cross-section containing the middle of the tool for the welding conditions: (a) 1000 rpm and 100 mm/min; (b) 1000 rpm and 300 mm/min; and (c) 1500 rpm and 150 mm/min (in °C)

In Figure 12.7, a magnified view of FSW tool temperature contours for a typical numerical simulation of FSW of AA 7050-T7451 thick plates is presented. In this illustration, tool rotational and welding speeds are 11.7 rev/s and 1.05 mm/s, respectively [10]. It is supposed that the tool region above the plates has a convection heat loss to the ambient temperature. As expected, maximum predicted temperatures are placed at the rim of the shoulder, where maximum peripheral or tangential velocities are found. However, minimum predicted temperatures are positioned at the pin tip where less deformation takes place.

An investigation [17] studied the effect of pin geometries (cylindrical and conical) on temperature distribution and the heat affected zone (HAZ). All the pins had the same geometries: height of 2.8 mm, fillet radius at the pin–shoulder interface of 0.5 mm and different values of the bottom pin diameters (namely d_{pin}) and of the taper angle (Figure 12.8).

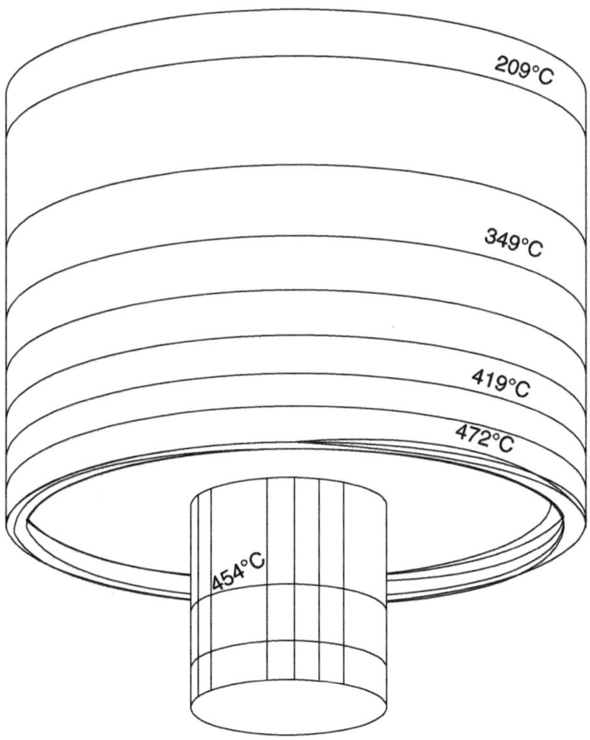

Figure 12.7 Predicted isotherms (in °C) on an enlarged view of the stir welding tool [10]

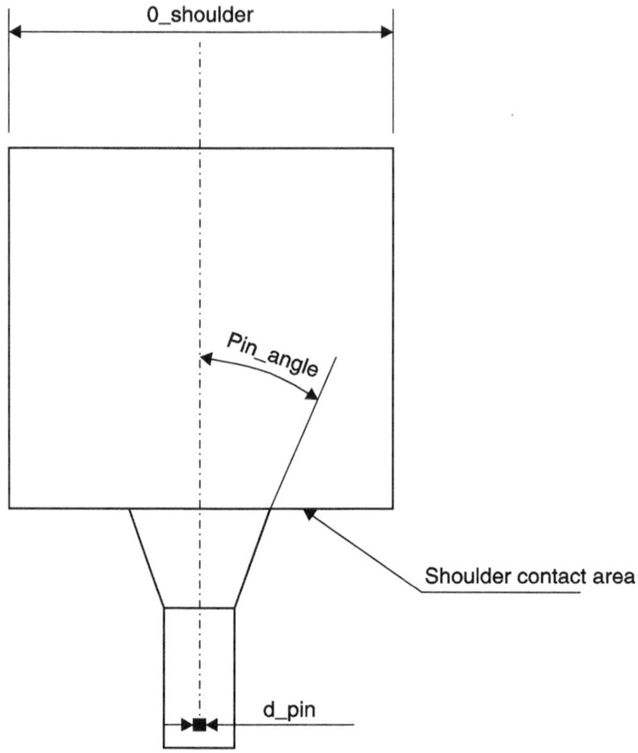

Figure 12.8 Tool design showing geometric parameters [17]

The temperature contours around the weld zone for different pin taper angles are shown in Figure 12.9. It can be seen that the HAZ enlarges by increase in pin taper angle. The enlargement of HAZ can be mainly attributed to the increase in frictional heat due to the large pin angle, which increases the welding pin/workpiece contact area, and to the significant increase in deformation energy [17].

It is observed that the temperature varied much less at a large pin angle as compared to a small pin angle. For example, when the pin angle changes from 0 to 10 degrees and from 10 to 20 degrees, the temperature increases by more than 18 °C, while by changing the pin angle from 30 to 40 degrees, the temperature increases only by about 2 °C. This indicates that the effect of the pin angle on the heat generation during the FSW process is only considerable for the pin angle below a certain value. The same relationship is also observed for a maximum temperature at the sheet–pin interface. When the pin angle is above 30 degrees, there is almost no change in the maximum temperature.

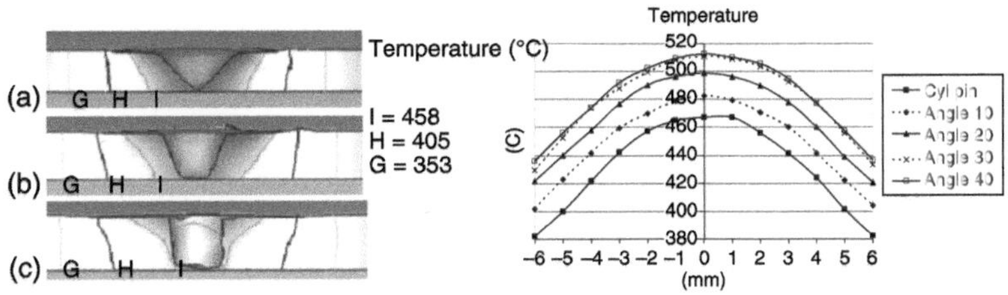

Figure 12.9 **Temperature distributions along the *x-z* section in an FSW joint as a function of pin angle α [17]**

12.3.2 Welding force and torque

A substantial insight into the FSW process can be provided in terms of machine and tool limitations and design by force measurement. The forces and torques acting on the tool are important for several reasons. First, a larger torque relates to a greater power requirement for the FSW process. Second, tool deformation and wear are enhanced by increase of load on the tool, which leads to a greater processing cost due to decreased lifetime of the tool. Third, tool wear may lead to contamination of the weld and deterioration of the joint properties [22].

A typical tool used in FSW processing experiences axial, longitudinal and lateral forces due to viscous and inertial effects (Figure 12.10) [23]. As the tool rotates inside the workpiece, it is faced with an axial force that tends to lift the tool. This axial force is countered by the applied axial force through the tool shoulder. It is necessary to maintain the position of the tool at or below the material surface, so a downwards force is needed. Some FSW machines operate by controlling the load, but in many cases the vertical position of the tool is controlled and thus the load will vary during welding. The pin thrust is determined as expected, primarily by the material being processed (57%). The traverse rate contributes 18%. Other contributing parameters are grain direction, rpm and dwell time. The unexplained factors are only 15%. It is believed that the traverse rate causes a downward and then upward material flow, which pushes against the pin. There is no significant interaction of the factors.

The longitudinal forces on the FSW tool result from the linear motion of the tool through the material. The longitudinal acting force is parallel to the tool motion and positive in the traverse direction. Since

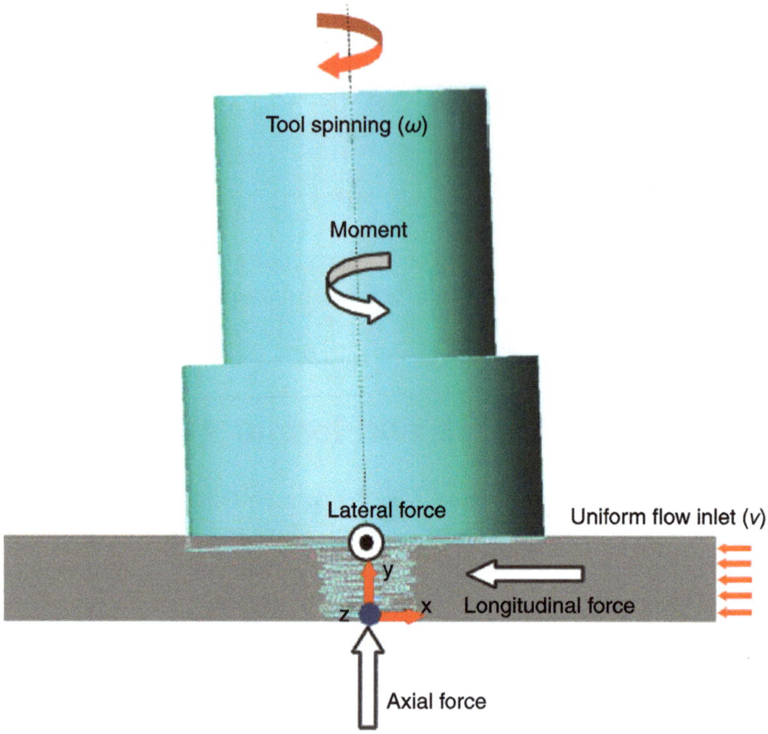

Figure 12.10 Presumed loads applied to the FSW tool [23]

resistance of the material to the motion of the tool leads to this force, it might be expected that it will decrease by the increase of material temperature around the tool. The lack of parameters, except for the two affecting the traverse, was surprising. Only the material (78%) and grain direction (2%) contributed, and the unexplained factors were 20%. This result indicates that by knowing the material, the X force necessary in machine design can be determined.

Combination of tool rotation and its linear motion results in an asymmetrical flow field around the tool, leading also to a lateral force on the tool, which is perpendicular to the direction of linear motion due to the Magnus effect [23].

In some publications [9,17], accurate numerical modelling (using DEFORM-3D) of the forces generated during FSW is reported.

Transverse force calculation in the tool pin in the 3-D material flow and temperature field model is [24]:

$$F_p = \int_0^L \int_0^{2\pi} \sigma(r) d\,\theta dL \qquad [12.3]$$

In this equation, σ is the temperature compensated yield strength of the deforming material around the tool, r is the average pin radius and L is the length of the pin.

The generated tool forces and the torque that were recorded during testing are shown in Figure 12.11 [25]. As the workpiece comes in contact with first the pin, and then the shoulder during the initial plunge, the forces acting on the tool vary substantially due to the combination of work hardening (under axial compression and shear) along with softening (as a result of heat generation) [23,24]. Both the vertical force and torque reach peak values during the plunging of the tool into the workpiece when the shoulder reaches the target depth. After the plunge, as the tool traverses some distance into the workpiece, the forces on the tool reach a steady state at a value that is generally lower than the peak forces during plunging. Hence, the tool can be vulnerable and susceptible to failure during the plunge stage.

Numerical analysis along experimental investigation of FSW process of AA6061 with a threaded tool pin for calculation of axial, longitudinal and lateral forces acted on the pin and the shoulder shows that axial forces increased with increasing rotational speed and decreasing tool travel speed. This finding is demonstrated both experimentally and by FEA. However, the computed increase in rotational speed and decrease in

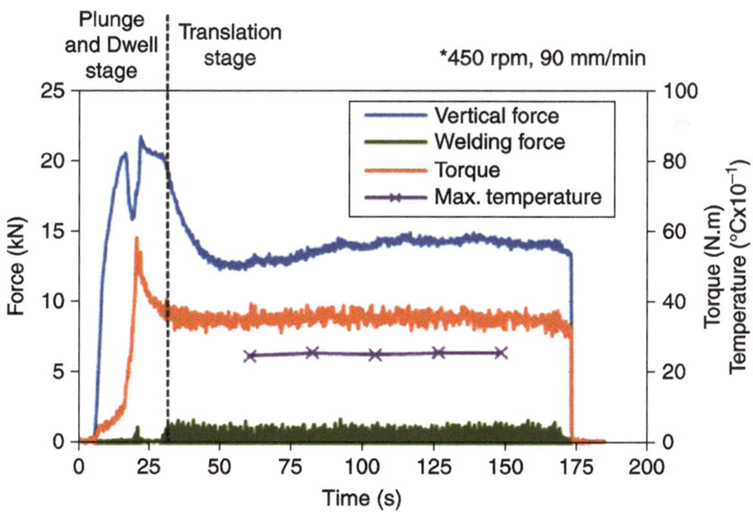

Figure 12.11 Experimental force and temperature generation [25]

tool travel speed resulted in a decrease in the calculated longitudinal forces on both the tool pin and the tool shoulder. As increasing the rotational speed leads to higher heat generation rate and accordingly lower flow stress, the tool will experience a lower longitudinal force. The effect of travel speed on the longitudinal force was also ascribed to the variation in the dynamic pressure distribution along the welding line. It was demonstrated that variation of rotational speed compared with the travel speed has much more influence on both the lateral and the axial forces. The axial, longitudinal and lateral forces acting on the tool shoulder were found to be much larger than the corresponding forces on the tool pin. The calculated moments were high at low rotational and high travel speeds. In Table 12.4 [22], the effects of travel and rotational speeds on the forces acted on the welding tool are summarized. It should be noted that the sign ± in a cell signifies that an increase in the column parameter results in an increase/decrease in the corresponding row parameter, and approximately denotes weak or no effect.

The investigation of variation of forces and moment acted on the FSW tool with the changing of the welding parameters (rotational and traverse speed) [23] are illustrated in Figure 12.12. As can be seen in

Table 12.4 **Effect of travel speed and rotational speed on moment and forces**

		Travel speed	Rotational speed
On pin	Longitudinal force	+	–
	Axial force	~	+
	Lateral force	~	+
	Moment about tool axis	~	–
On shoulder	Longitudinal force	+	–
	Axial force	–	+
	Lateral force	~	+
	Moment about tool axis	~	–
Total	Longitudinal force	+	–
	Axial force	–	+
	Lateral force	~	+
	Moment about tool axis	+	–

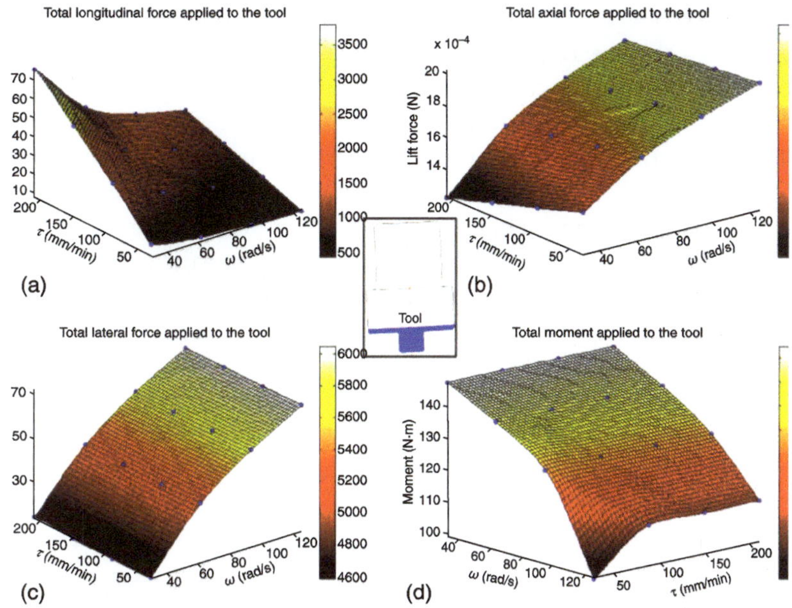

Figure 12.12 **Numerical results of the total loads applied to the FSW tool with change of the welding parameters [23]**

Figure 12.12(a), by increasing the angular velocity, the magnitude of the longitudinal force decreases. Adversely, increasing traverse velocity increases the magnitude of the longitudinal force. The magnitude of longitudinal force would be at its maximum value when the tool angular velocity is at its minimum (31.4 rad/s), and the traverse velocity of the tool is maximum (210 mm/min) and vice versa. Figure 12.12 also infers that in order to obtain high speed welding with lower resistive force on the FSW tool, it is imperative to increase both the angular and traverse velocities of the process.

Figure 12.12 also depicts the variation of total axial force by changing the welding parameters. The minimum axial force is obtained when the angular and traverse velocities reach their minimum and maximum values, respectively. Consequently, increasing the angular and decreasing the traverse velocity enlarges the amount of the total axial force applied to the FSW tool. In Figure 12.12(c), the amount of the total lateral force is shown. As can be seen, the lateral force is increased significantly by increasing the angular velocity. Nevertheless, the traverse velocity of the FSW tool has a negligible effect on the value of the lateral force.

Also, the effect of shoulder and pin diameter on welding force is investigated. Figure 12.13 shows the welding forces for various pin and shoulder diameters. As can be inferred from this figure, the larger the pin diameter, the more the welding forces are developed. This is attributed to not providing sufficient heat in order for material softening of the workpiece along with increasing the moving material mass, which increases the resistance acting on the tool as it advances along the weld line. Adversely, increasing the shoulder diameter has the opposite effect. The reason why this happens is ascribed to an increase in material softening of the workpiece, due to increased heat generation [9].

Figure 12.14 depicts the forces acting on the welding tool for various translational speeds, which are predicted numerically. All forces were found to increase significantly with increasing translational speeds. This is attributed to a reduction in thermal softening of the workpiece, due to reduced heat generation, which consequently increases the resistance acting on the tool as it traverses along the joint line. Adversely, increasing the rotational speed has the opposite effect, due to an increase in heat generation [25].

12.3.3 Strain distribution

As thermal history and plastic deformation of the weld region have dramatic effects on the microstructure of the weld, better knowledge

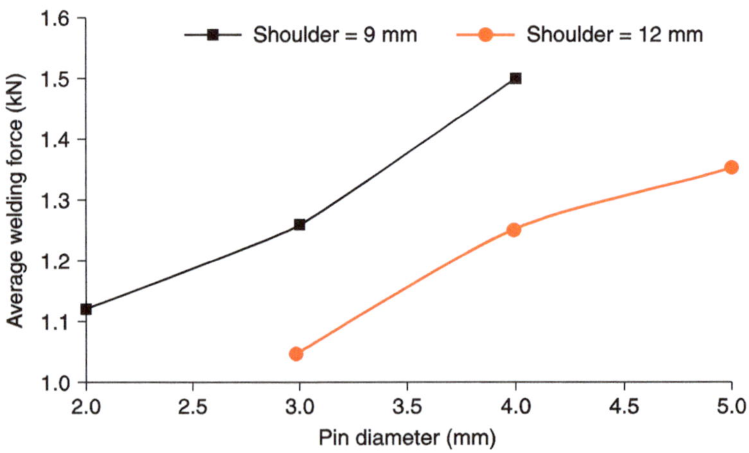

Figure 12.13 The numerical welding forces for various pin and shoulder diameters [9]

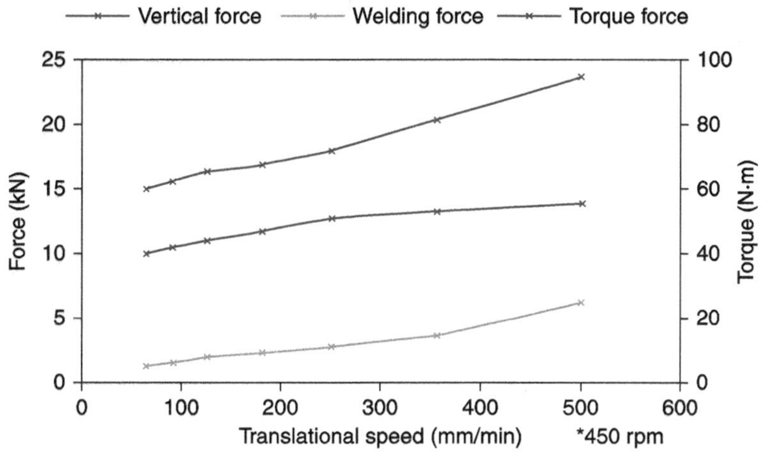

Figure 12.14 FEM predicted forces for various translational speeds [25]

about the effects of process parameters on the strain distribution in the joint during the FSP/FSW is very helpful for optimal design and process control. In FSW, the strain distributions are not symmetrical about the weld line and maximum strains are on the material AS where a 'positive' combination of the tool feed rate and of the peripheral tool velocity is observed. This implies that the combination of rotating and advancing speeds of the tool have dramatic impact on material flow and/or deformation. Figures 12.15 and 12.16 illustrate the strain distribution and the contours of effective strain in the weld zone, respectively. From both figures, it is obvious that the AS undergoes higher effective strain compared with the RS.

This finding is also evident in Figure 12.17(a), which shows the contours of effective strain in FSP of AZ91. This finding results from this fact that at the AS compared to RS, more shearing occurs. To investigate how effectively strain is distributed in the cross-section of the FSP sample behind the tool pin, the strain contour in this area is plotted in Figure 12.17(b). It can be deduced from this figure that the effective strain in the cross-section increases from the bottom to the top surface. As the rotation of the tool shoulder hastens the material flow on the top surface, plastic deformation near the top is higher than that near the bottom. This explains why the shape of the SZ becomes conical [6].

Evaluation of the effect of pin and shoulder diameter variation on strain during FSW is investigated and Figure 12.18 shows the strain

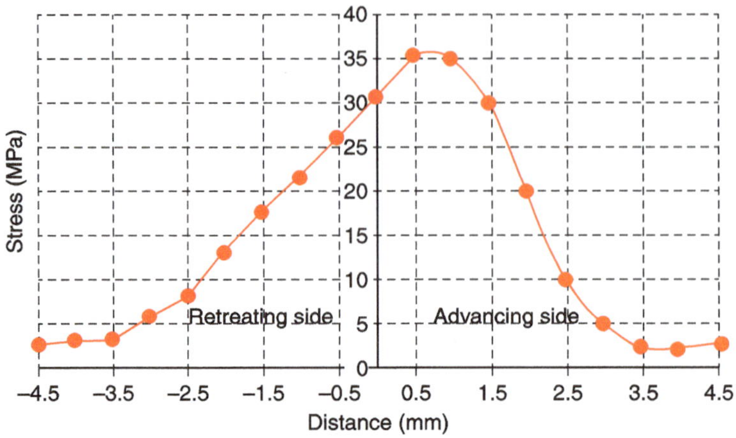

Figure 12.15 Effective strain distributions along a cross-section in an FSP joint as a function of pin angle α [17]

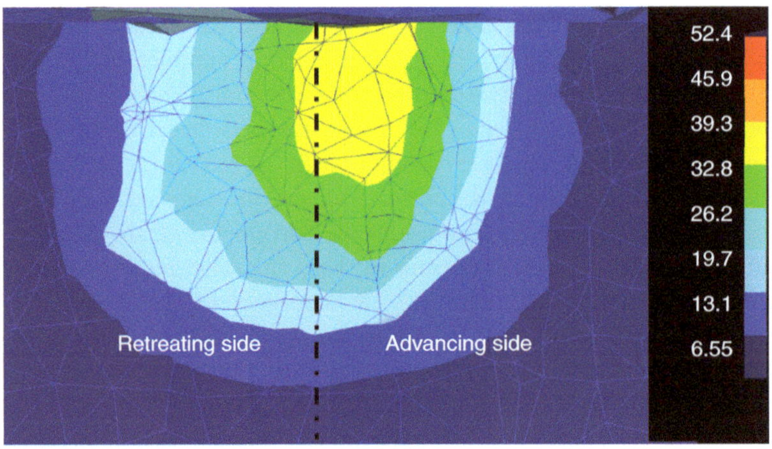

Figure 12.16 Contour of effective strain in a cross-section of a friction stir processed specimen

profile in the transverse section of the weld zone for different pin and shoulder diameters. As shown in the figure, the strain profiles are asymmetrical about the weld line. It can also be inferred that with increasing pin diameter, maximum strain increased from 30 to 45 mm/mm. With comparing Figures 12.18(a and b), we can conclude that

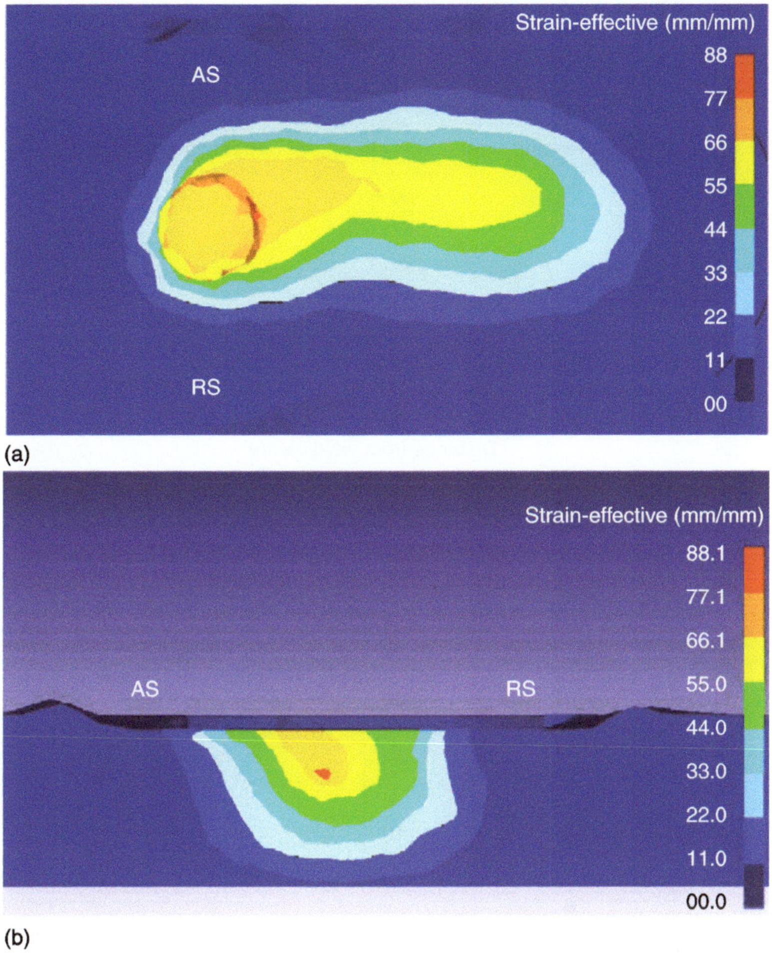

Figure 12.17 **Contour of effective strain in: (a) perspective view; and (b) cross-section of friction stir processed specimen [6]**

the shoulder diameter has a trivial impact on maximum strain. At the same time, with increasing shoulder diameter, the plastically deformed area increases, which is due to extra heat generation and material softening [9].

12.3.4 Residual stress

The existence of a high value of residual stress exerts a significant effect on post-weld mechanical properties, particularly fatigue properties.

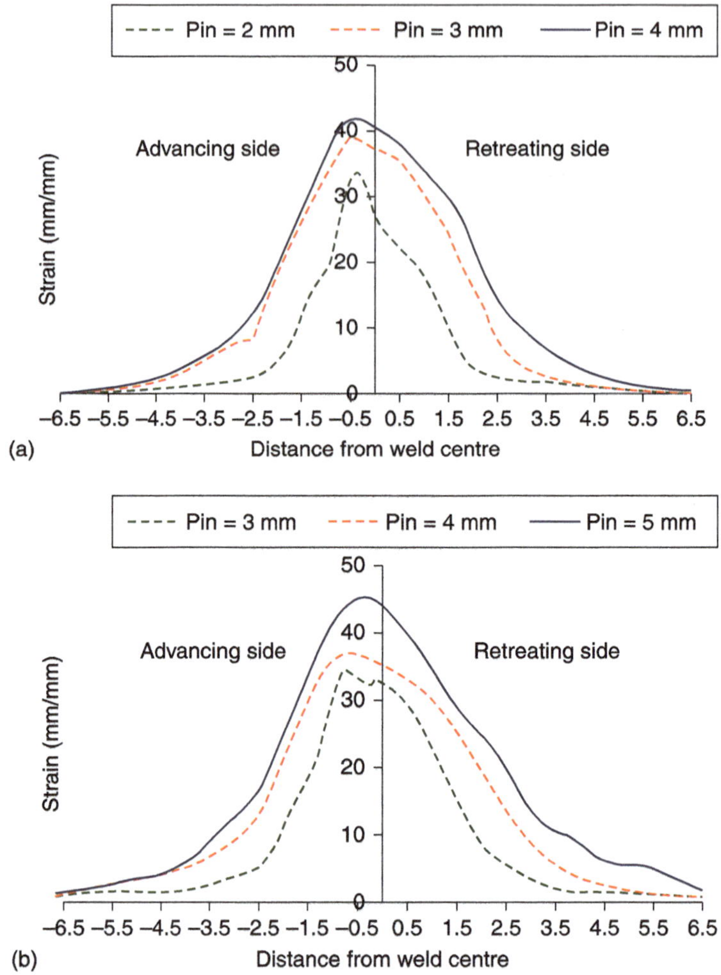

Figure 12.18 Strain profile in the transverse section for: (a) shoulder diameter of 9 mm; and (b) shoulder diameter of 12 mm [9]

Therefore, it is of paramount importance to investigate the residual stress distribution in the FSW welds.

In fusion welding, owing to concurrent application of heat and associated constraint, a complex combination of thermal and mechanical stresses develop in the workpiece. In fusion welding, these stresses, which are called residual stresses, usually approach the yield strength of the base material. Nonetheless, in FSW process, these stresses are low due to the fact that FSW is a low-temperature solid-state welding. However, rigid fixtures utilized in FSW exert a much higher constraint on the joint plates and consequently a higher value of residual stresses. These constraint stresses are rooted in impeding the contraction of the weld.

To study the residual stresses developed during FSW, a sequentially coupled thermal-mechanical FEM is utilized and thermal analysis is performed first to generate temperature histories for the welding process. The results are then utilized as thermal input to the mechanical analysis for the prediction of residual stresses and strains that are generated owing to the temperature changes within the constrained metal plates. In research that investigates numerically the residual stresses in FSW joints, some considerations have developed. First, the local mechanical action of the tool, and in particular of the tool pin, is not taken into account. Merely the thermal flux is considered, so that a kind of macro-effect of the process on the material is investigated. Moreover, the adopted thermal models, which describe the heat flux due to the tool action, are always axisymmetrical and therefore the effect of the asymmetrical material flow in FSW is not considered.

Most of these limitations were overcome by a modified FEM model [26], in which the actual thermal flow is achieved. Such thermal histories, at each node of the model, are passed to a further elasto-plastic FEM model. Thus, the residual stress state is obtained for the investigated butt joint. In this way, the final stress state, occurring in the specimen after the releasing of the clamping constraints, is highlighted in terms of longitudinal stress, i.e. in the direction of the welding, transverse stress and normal one. Figures 12.19 and 12.20 illustrate the contours of longitudinal and transverse residual stresses. As expected, due to the asymmetrical nature of material flow, both longitudinal and transverse residual stresses are distributed asymmetrically. In Figure 12.21, which shows the normal residual stresses, the asymmetrical nature of stresses also is evident.

More information can be acquired by observing the trends of the residual stresses in the transverse section of the joint. Figure 12.22 shows the average residual stress in the middle of the traverse section. Referring to this figure, the normal stresses are almost negligible, while longitudinal stresses reach slightly larger values than transverse components. Moreover, as far as longitudinal stresses are concerned, larger values are reached in the AS of the joint with respect to the RS, close to the tool shoulder edge [27].

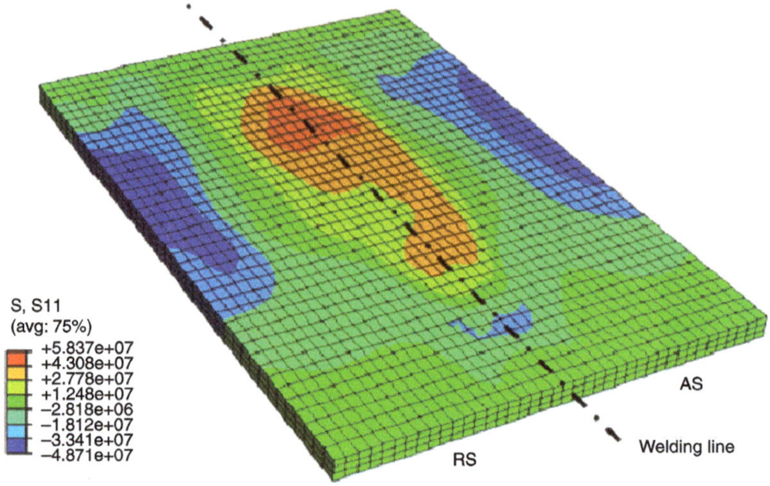

Figure 12.19 Strain profile in the transverse section for: (a) shoulder diameter of 9 m; and (b) shoulder diameter of 12 mm [26]

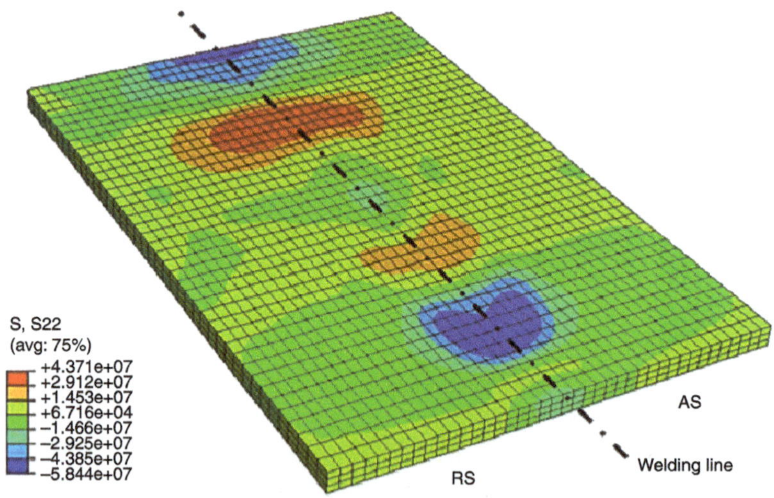

Figure 12.20 Transverse residual stresses in MPa; this residual stress component is responsible for the out-of-plane distortion of the welded plate [26]

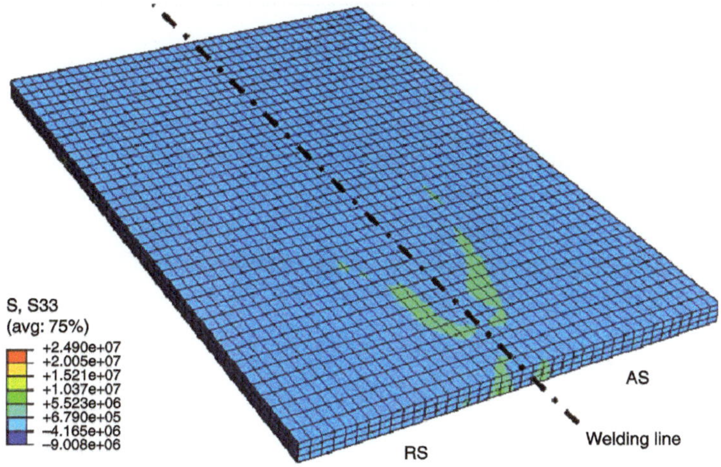

Figure 12.21 Normal residual stresses in MPa; the through-thickness variation of the residual stress can be neglected [26]

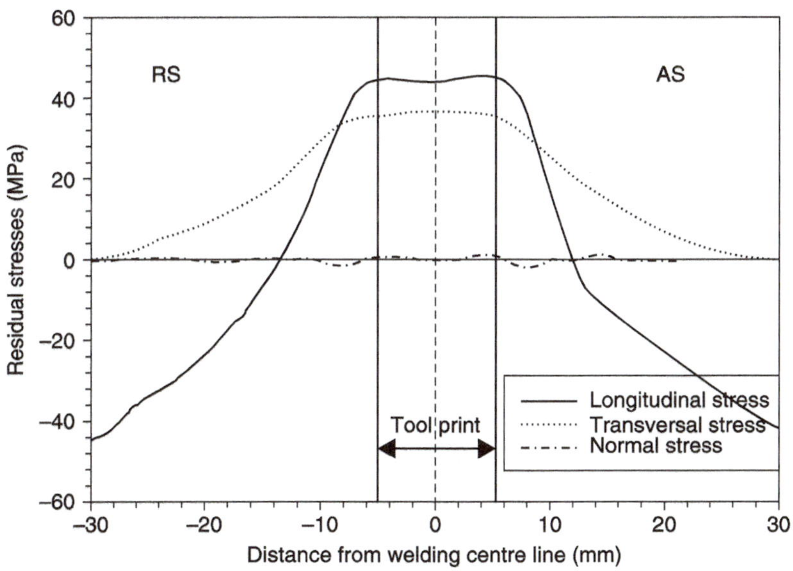

Figure 12.22 Average residual stresses in the transverse section in the middle of the joint; the longitudinal stress is higher than the transversal one while the normal stress is close to zero [26]

12.3.5 Material flow

The material flow during FSW is intricate and depends on various parameters, including tool geometry, process parameters and the material to be welded. Material flow characteristics have considerable effects on weld quality and mechanical properties. Therefore, it is of paramount importance to investigate the material flow characteristics for optimal tool design and so obtain high structural efficiency welds. This has led to numerous investigations on material flow behaviour during FSW. A number of computational methods including FEA have also been used to model the material flow [3].

In a study, material flow is investigated by dividing the workpiece into three sections (top surface, middle surface and bottom surface). It is found that the RSs and ASs are different in material flow characteristics. The material particles at the RS do not enter into the AS and those that are on the top surface are rotated behind by the tool but do not enter into the SZ. The material particles at the AS, which are rotated to the RS on the top surface, pile up at the border of the SZ at the RS, which can lead to the formation of weld flash [3].

The material flows in the middle of the welding plate thickness can be observed in Figures 12.23. In comparison with the flow pattern on the top surface, it can be concluded that the material particles in the middle of the welding plate can enter into the region under the shoulder, but the ones on the top surface cannot enter the region under the shoulder. The material particles at the RS are directly pushed back by the welding tool, but the ones at the AS can be rotated with the tool for several revolutions and then some of them slough off from the rotating pin. Experimental investigation also proved this type of material flow.

The flow pattern on the bottom surface is shown in Figure 12.24. Bottom surface and middle of the welding plate show different flow patterns compared with top surface. The rotating zone is greatly decreased in the middle of the welding plate thickness and on the bottom surface. This is the reason why the SZ has an inverse trapezoid cross-section.

Investigations on the effect of tool rotational and traverse speed on material flow show that the stirring effect increases as the rotational speed increases. On the other hand, extreme increase in the tool rotational speed can also lead to the formation of weld flash and consequently may have an adverse effect on the weld quality.

The material flow is also investigated in FSP of LM13 eutectic Al–Si as-cast alloy [20]. The dimensions of plates were 100 mm in length, 50 mm in width and 7 mm in thickness, and in order to recognize the material flow,

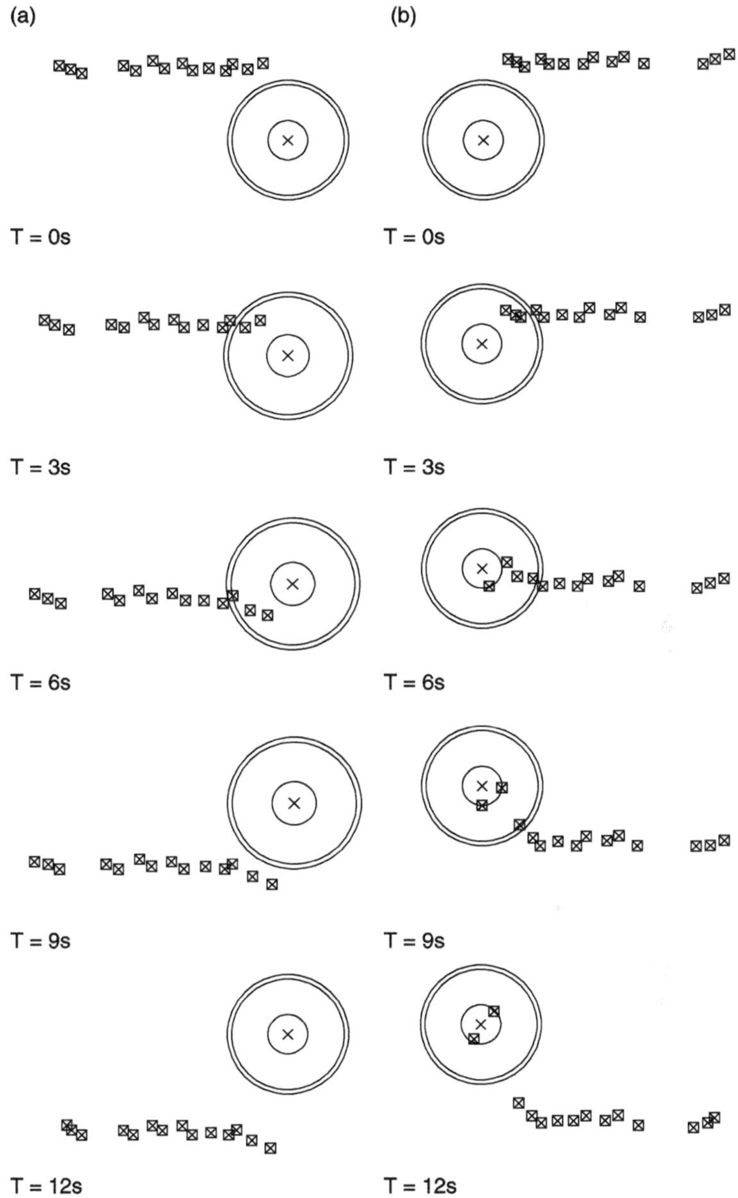

Figure 12.23 Material flow at: (a) retreating side; and (b) advancing side in the middle of the welding plate thickness when v=2.363 mm/s and ω=240 rpm [3]

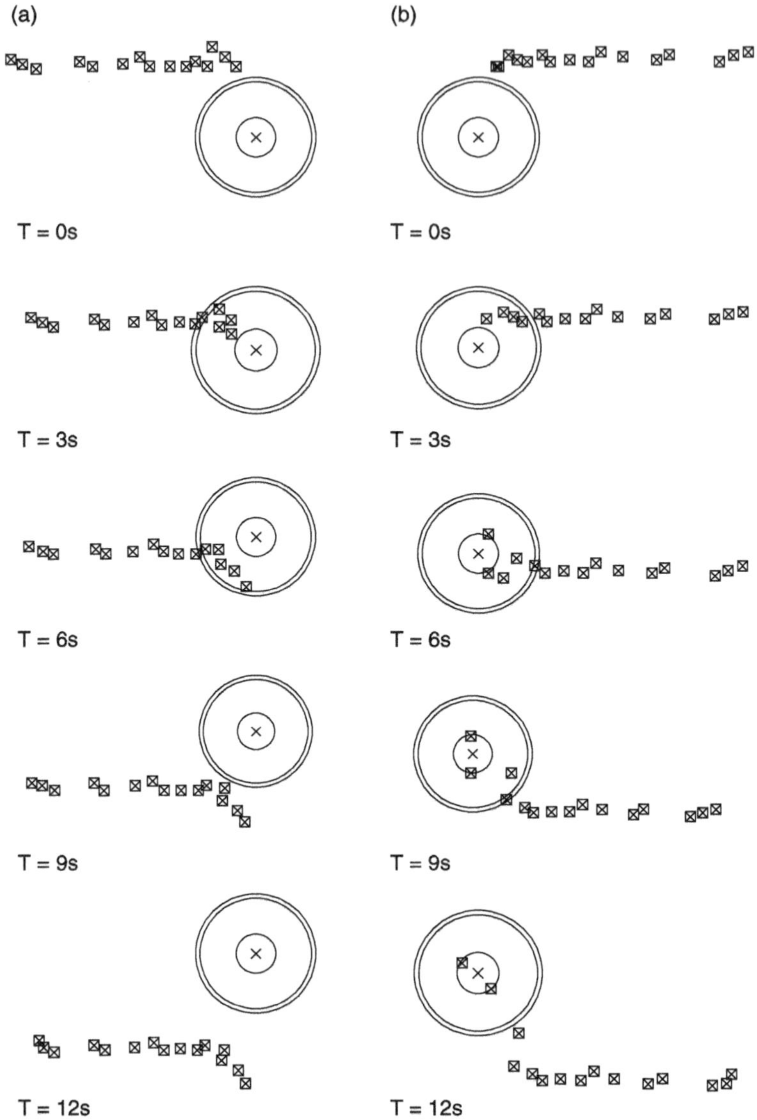

Figure 12.24 **Material flow at: (a) retreating side; and (b) advancing side on the bottom surface when v = 2.363 mm/s and ω = 240 rpm [3]**

graphite powder (with 5 mm particle size) was embedded in a groove with the depth and width of 3 and 1 mm, respectively. The shoulder and pin diameters and pin height were 18, 6 and 4 mm, respectively. The tool rotational and the traverse speeds were also 900 rpm and 50 mm/min. For simulation they used DEFORM-3D software and a constant shear friction of 0.5. Figure 12.25 shows the initial position of points selected to study material flow pattern in the SZ. Points P1–P8 are located 0.5 mm from each other in the weld centre. Similarly, points P9–P16 and P17–P24 are located at the ASs and RSs with the distance of 2.5 mm from the centre line.

Figure 12.26 also shows the material flow at the weld centre. Figures 12.26(a) illustrates the initial position of points located along the centre line and Figures 12.26(b–e) track the aforementioned points during the process. As can be seen, the points near the bottom of the SZ (P7 and P8) have a semicircular movement from the leading side towards the trailing side. However, the part of the material that is close to the top surface, in spite of the movement of the tool pin behind, is stretched towards the AS. Such behaviour is clearly expected, due to the frictional shear force induced by the tool shoulder. An interesting observation is that the final position of the points, except point P1, is similar to the SZ shape. Point P1 is an exception, because it is stuck to the shoulder and rotates with the shoulder for an unknown period of time.

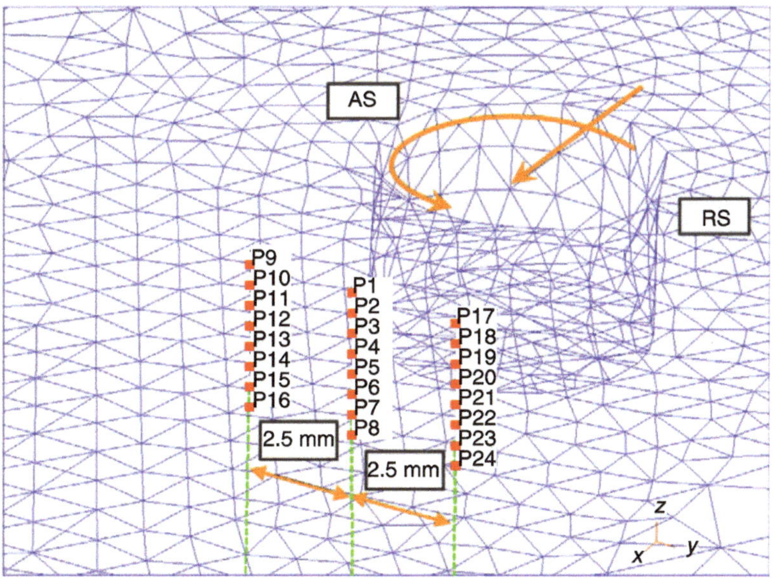

Figure 12.25 Initial positions of points that have been selected for material flow in the stir zone [20]

Figure 12.27 indicates the plastic strain applied to the points P1–P8. The plastic strain for the different points is about 30. Moreover, the amount of effective plastic strain decreases upon moving to the SZ root. The effective plastic strain of point P1, which is stuck to the shoulder, is considerably higher than that of the other points, which can be attributed to the accumulation of the effective strain during several successive rotations around the pin.

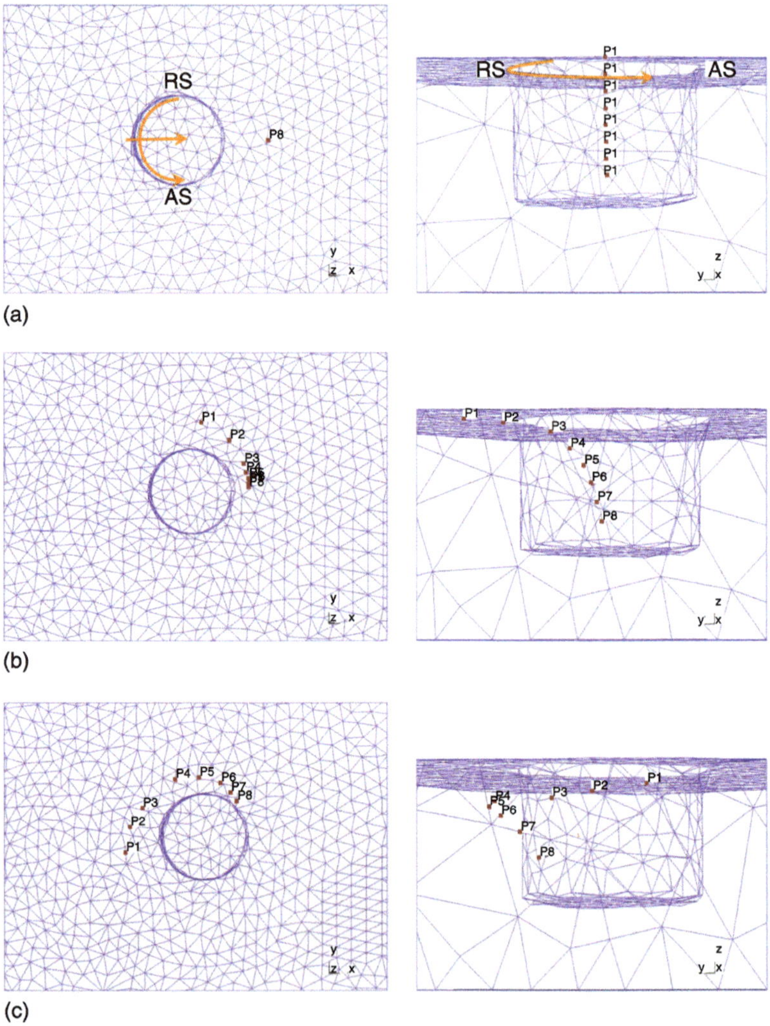

Figure 12.26 Centre line points flow pattern [20]

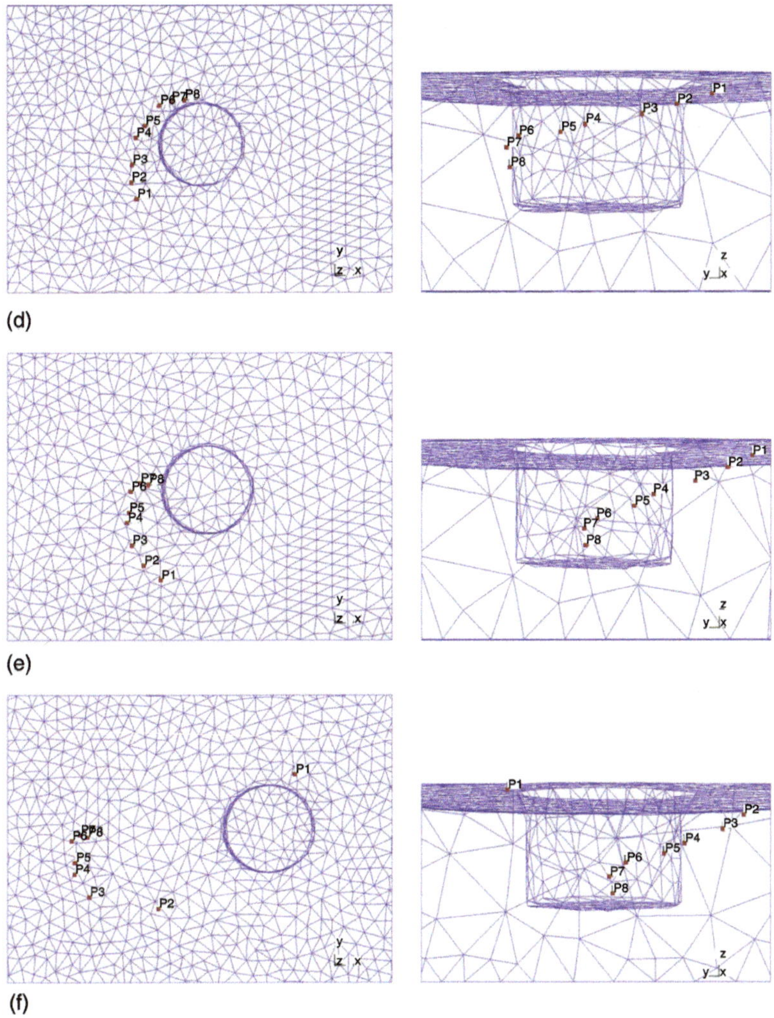

Figure 12.26 Continued

Furthermore, Figures 12.28(a and b) indicate the velocity of points P1–P8 and velocity contour in cross-section. As expected, velocity decreases upon increasing the distance from the top surface. Figure 12.28(b) also reveals that the velocity at the RS is higher than that at the AS. X-ray transmission examination also confirms these results.

Many researchers have discussed frictional boundary conditions, but still it is not clear whether the friction is sticky or slippery, or something

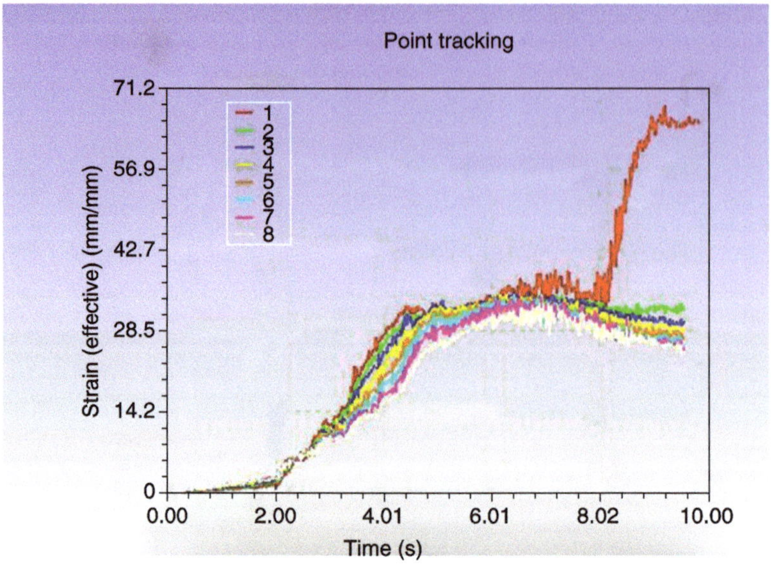

Figure 12.27 Applied plastic strain to the points P1–P8 [20]

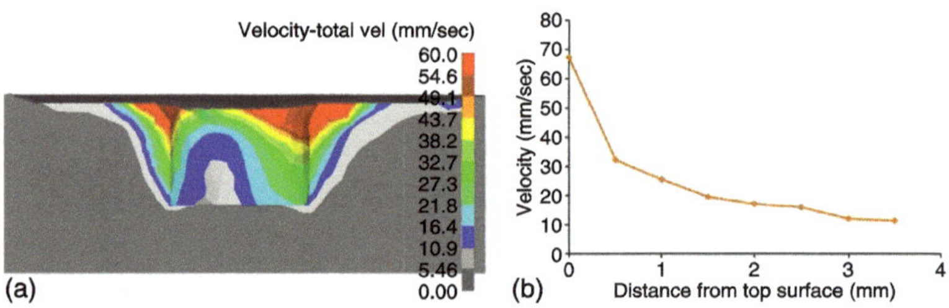

Figure 12.28 Material velocity: (a) velocity contour in the cross-section of the weld; and (b) velocity diagram of points P1–P8 [20]

in between. At the AS near the trailing side, the pin velocity has its own minimum value and due to the pressure reduction, the material barely flows into this region. This is the main reason for formation of tunnelling cavity and thus the cavity is mostly formed at the AS near the trailing side.

The probability of tunnelling cavity formation increases by increasing the traverse speed. Figure 12.29(a) demonstrates the cavity that formed in the higher traverse speed (200 mm/min). As shown in Figure 12.29(b), a slot was created at the AS of the trailing side. This slot and tunnelling cavity is also visible in the experiment (Figure 12.30).

Flow patterns of points P9–P16 and P17–P26 are also shown in Figures 12.31 and 12.32, respectively. As shown in Figure 12.31, points initially located at the AS revolve around the pin and finally come back to the AS. However, this behaviour is not observed for the points located at the RS and they are scattered over the AS and RS after passing the tool. Moreover, the points near the SZ root remain at the RS and upon decreasing the distance from the top surface, material stretching towards the AS diminishes. In other words, the part of material near the tool shoulder at the AS, RS and centre line moves towards the AS. Most of the SZ materials are finally located in the AS after passing the tool. This phenomenon (Figure 12.33) can explicate the powder agglomeration at the AS near the top surface in the case of FSP with powder. This was obviously seen in the experiment. Figure 12.34 reveals the graphite agglomeration occurring at the AS near the top surface. The agglomeration of embedded powder or marker at the AS has been reported in several works.

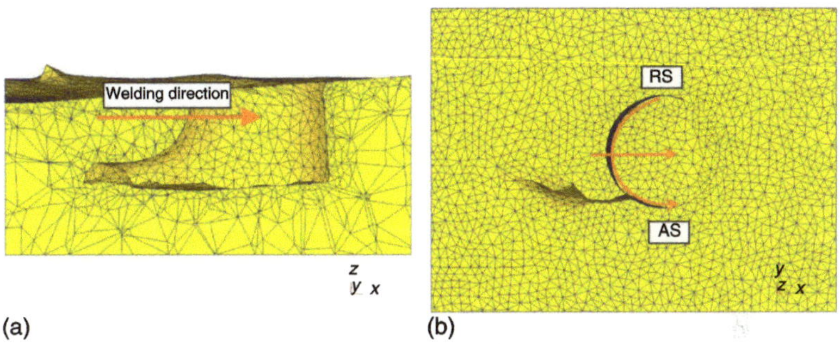

Figure 12.29 (a) Tunnelling cavity formation behind the pin; and (b) slot formation at advancing side behind the pin [20]

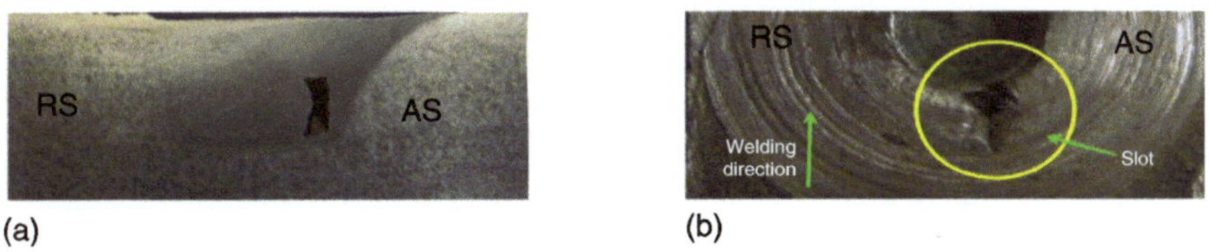

Figure 12.30 (a) Tunnelling cavity formation behind the pin; and (b) slot formation at advancing side behind the pin [20]

Also, the shape of the SZ can be predicted through using the point tracking option of DEFOM-3D software [20]. Thus, in order to achieve a better understanding of material flow during the process, more points were picked inside the SZ before tool passing (Figure 12.35(a)). The final position of these points after tool passing is shown in Figure 12.35(b). The shape of SZ predicted by simulation was projected in Figure 12.34 (SZ shape resulted from the experiment) and their outcome is shown in Figure 12.36.

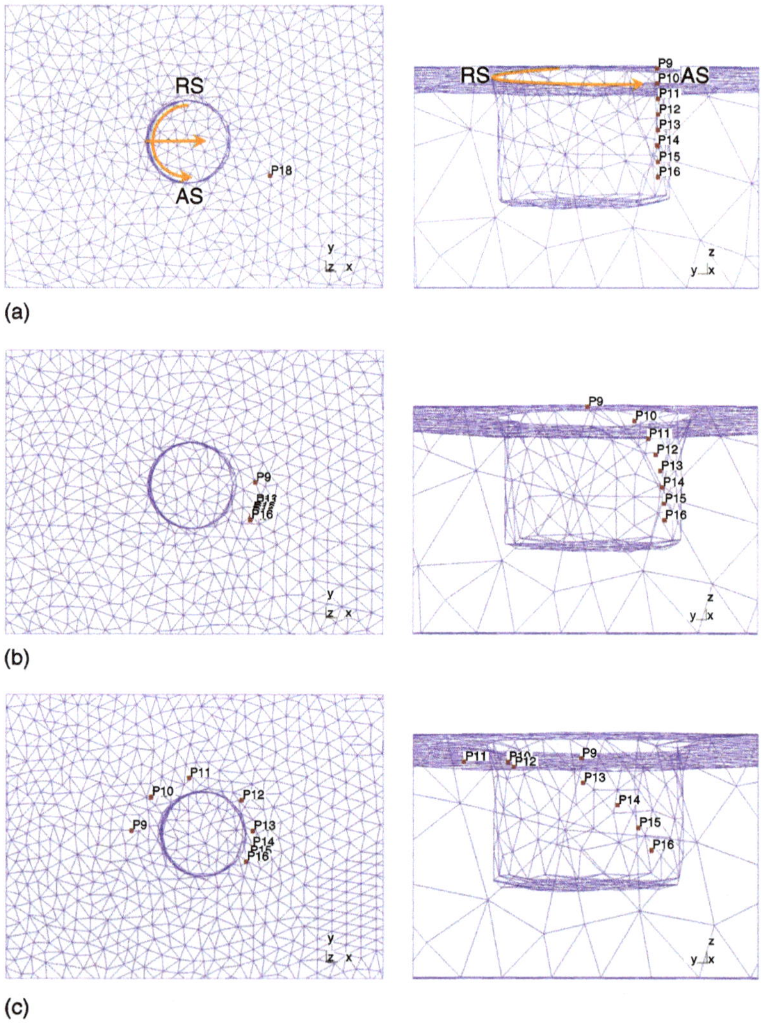

Figure 12.31 Advancing side points flow pattern pin [20]

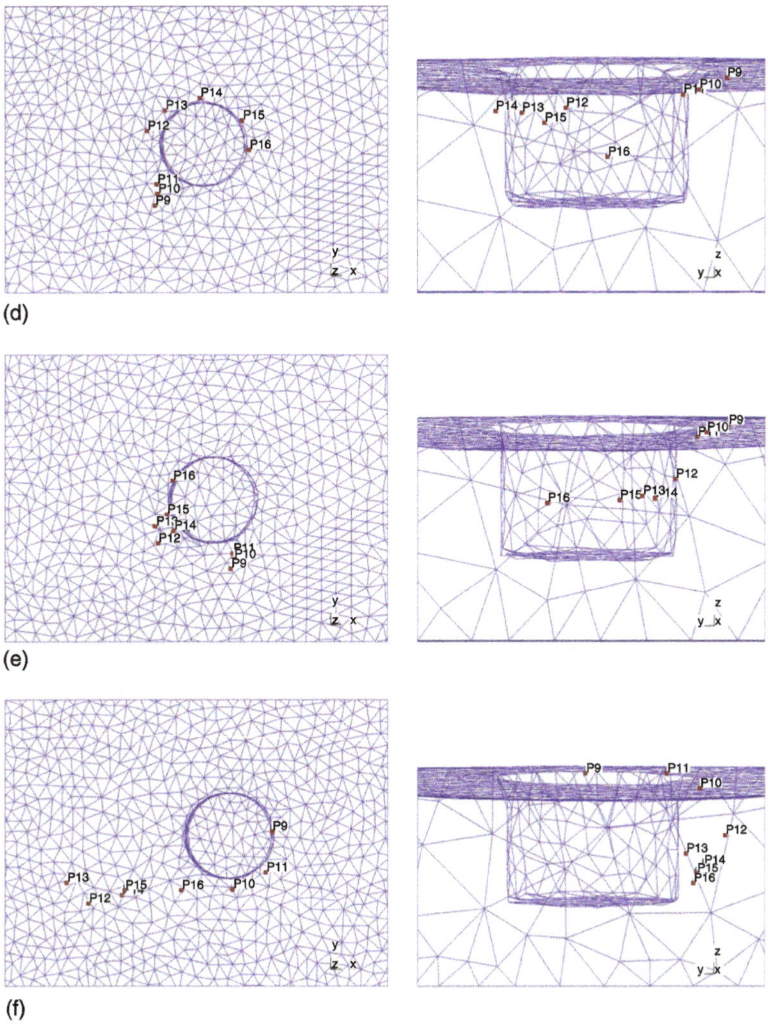

Figure 12.31 Continued

Figure 12.37 opens new perspectives towards a better understanding of how materials move into the SZ. According to Figure 12.37(a), nine points at the RS (P1–P9) and ten points at the AS (P10–P19) were picked. Figures 12.37(b and c) represent the final positions of the aforementioned points. As can be seen from the figures, the points at AS remain in contact with the tool shoulder and rotate several times during the process. Nevertheless, the points at RS never revolve around the tool and are scattered over the SZ. Thus, the main part of material inside the marked zone probably comes from the zone located at a certain distance from the tool pin and near the top surface at the RS. Furthermore, the material at the AS travels a longer path compared to the material at the RS and the SZ centre. Therefore, shifting the tool towards the RS can lead to a better powder distribution during the process. Figure 12.38 shows the graphite powder distribution in the FSP specimen with a 2 mm shifted tool towards

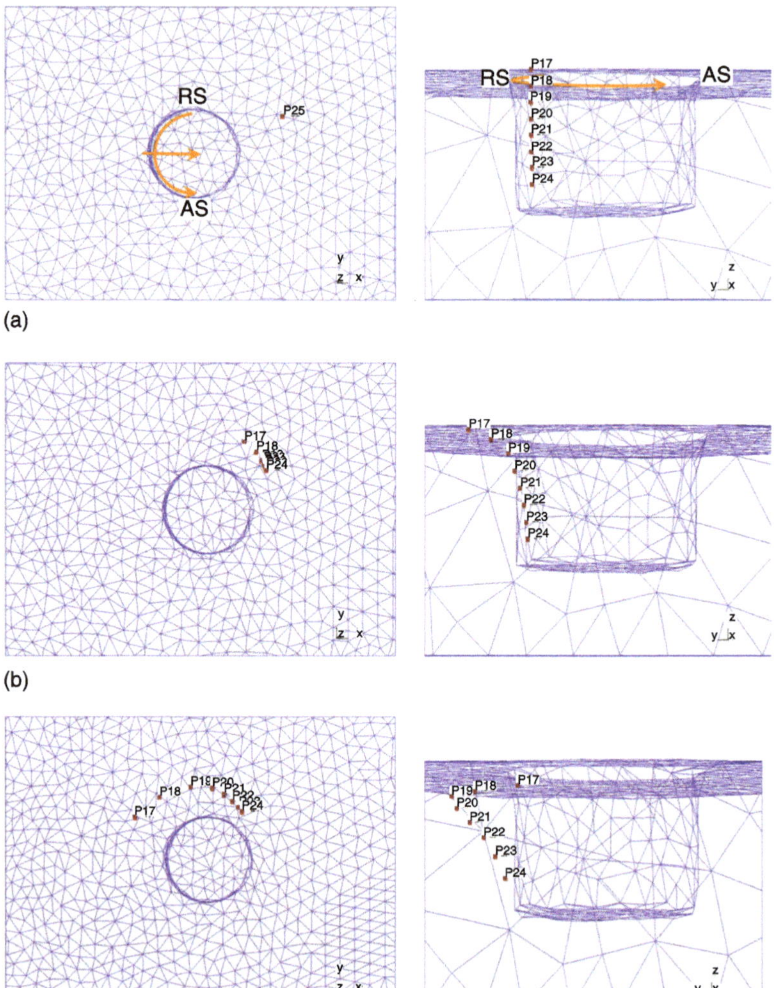

(a)

(b)

(c)

Figure 12.32 Retreating side points flow pattern pin [20]

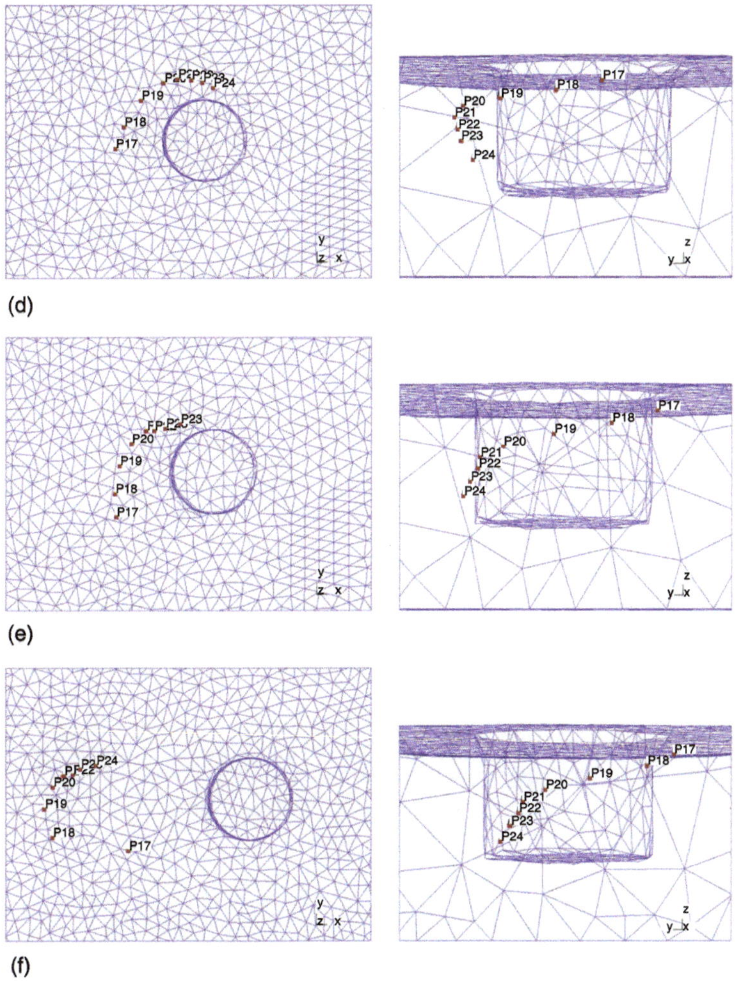

Figure 12.32 Continued

the RS. As can be seen, shifting the tool towards the RS considerably leads to a better powder distribution.

12.3.6 Grain size

The size of DRX grains in the SZ is a function of various parameters, including tool geometry, composition and temperature of the workpiece, and amount of vertical pressure.

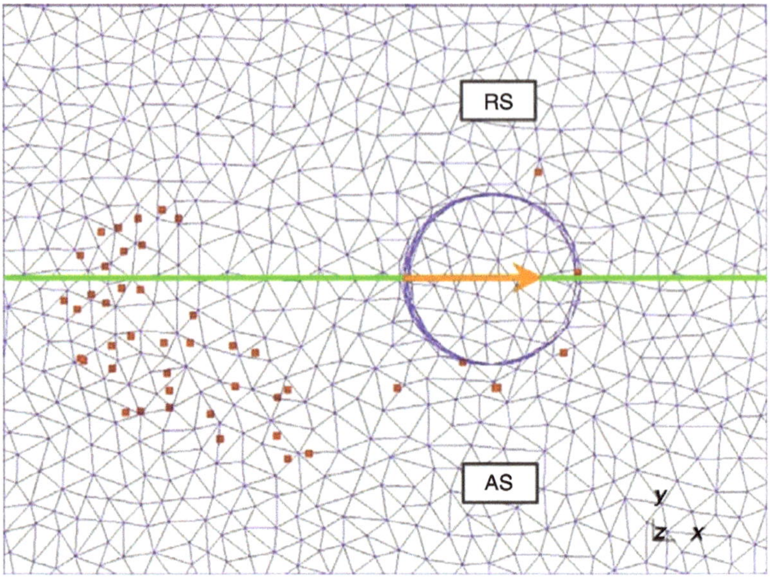

Figure 12.33 Aggregation of points at the AS in final position [20]

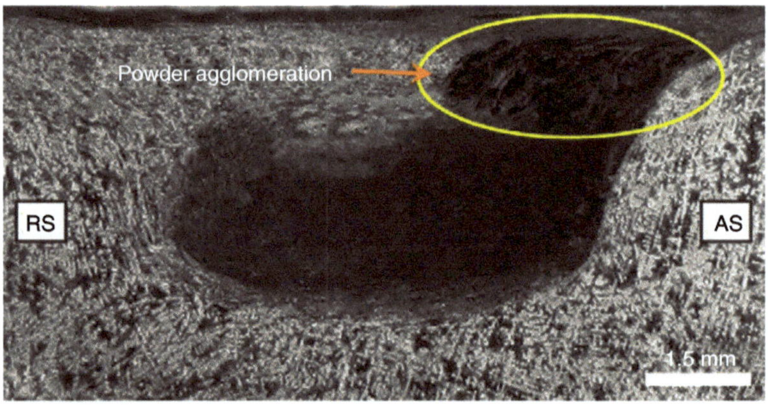

Figure 12.34 Cross-section macro-image of FSPed specimen [20]

Gao et al. (2013), by taking into account three internal state variables, implemented a physically-based model into the commercial FE package DEFORM-3D to describe the microstructure evolution during FSW. The merit of their model was the ability to predict the

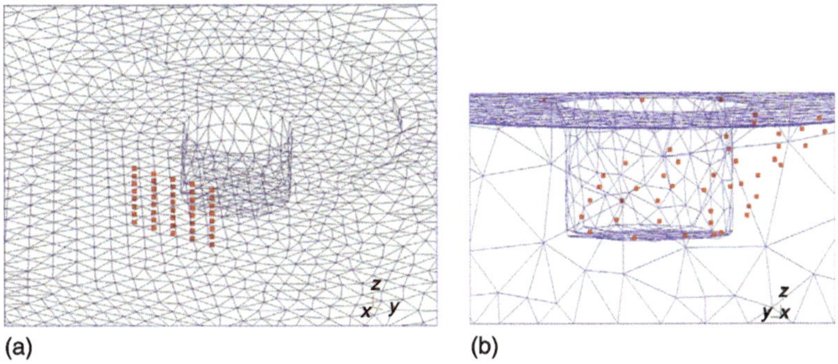

Figure 12.35 (a) Initial position of points picked for studying the SZ shape; and (b) final position of points [20]

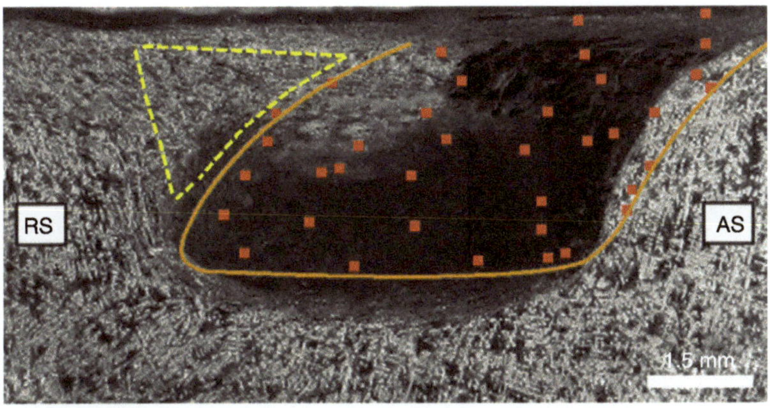

Figure 12.36 Comparison of simulated and experimental SZ shape [20]

dislocation density, grain size, temperature, effective strain and strain rate during FSW.

Based on their study, geometrical dynamic recrystallization (GDRX) occurs only if the effective strain exceeds the critical effective strain, which is estimated to be 2.0 mm/mm. Figure 12.39 illustrates the distribution of the grain size after FSW. As can be seen, the predicted grain size in the SZ at the end of the process is around 5.8 μm compared to a measured value of 4.9 μm. Due to low heat input in areas outside the shoulder, grain size in these areas remains approximately unchanged [19].

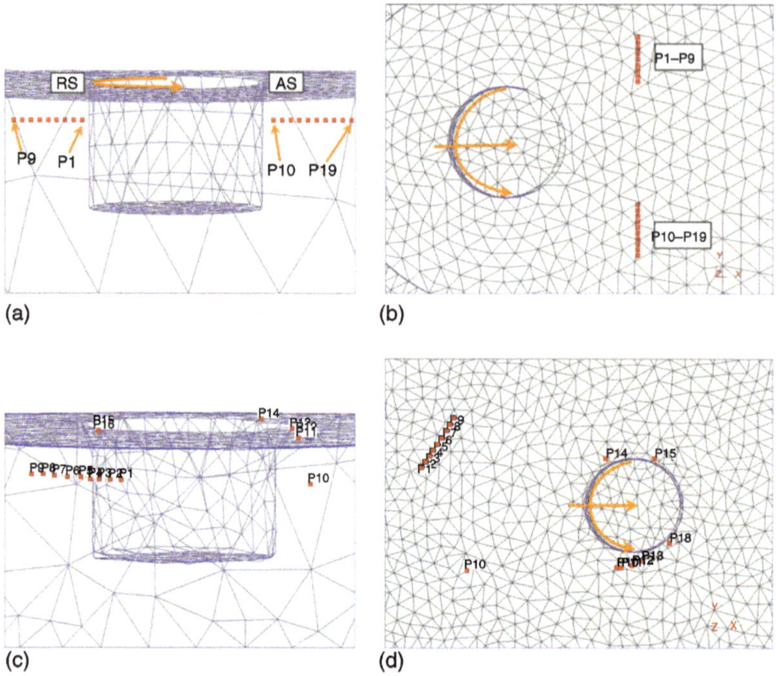

Figure 12.37 (a) and (b) Initial positions of points picked for specification of initial location of points which finally are located in the dashed line; and (c and d) final position of points [20]

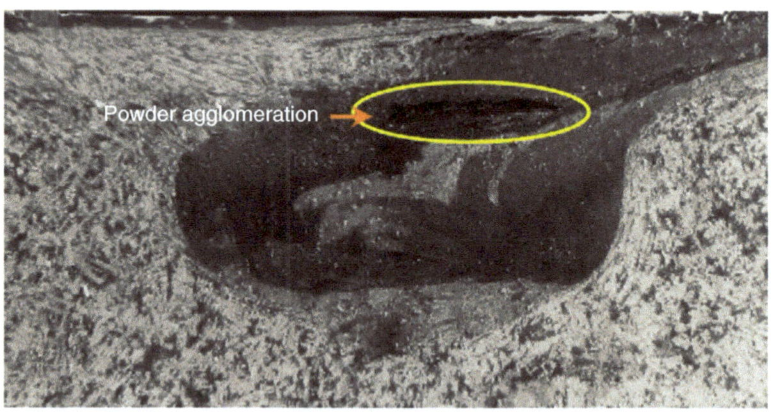

Figure 12.38 Cross-section macro-image of FSPed specimen with 2 mm tool shifting towards the RS [20]

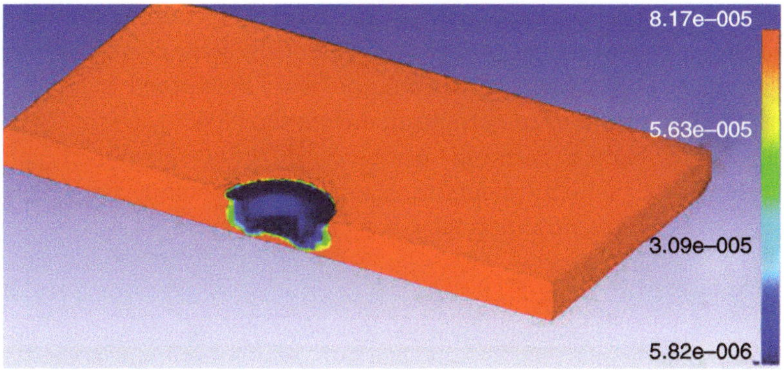

Figure 12.39 FE simulation: distribution of the grain size after FSSW [19]

12.3.7 Simulation of dissimilar material joints by FSW

In recent years, researchers proposed a 3-D continuum-based FEM to highlight FSW mechanisms and to show local values of the field variables, which was based on a 'single block' approach. Just a single blank was considered in these analyses, and the interaction of the tool with the workpiece material, both in its plunging and advancing stages of the process, was investigated. It should be noted that no actual bonding between different blanks was modelled.

Considering the utilized single block approach in numerical simulation of the FSW process, different materials cannot be modelled. Hence, simulation of FSW in dissimilar alloys is not a simple task and is very challenging. However, recent development of a numerical 'trick' paved the way to simulate FSW of dissimilar materials. In an investigation of [28], by maintaining the 'single block' approach and then considering a single workpiece in the model, the two different materials were modelled as two different phases of a single material.

In order to take into account the different thermo-mechanical behaviour of two different materials in the single continuum body, a numerical trick was utilized on the basis of the possibilities offered by the used DEFORM-3D code.

Initially, considering one material (AA2024-T4), the joint was modelled in the preprocessor. Afterwards, the material was allowed to fully transform from AA2024-T4 to AA7075-73511. This transformation was done in the area of the single block model of the joint corresponding to the stringer. Then phase transformation was interrupted to avoid the occurrence of further transformation during the FSW.

The code allows carrying out the 'diffusion simplified' phase transformation, which calculates the material volume fraction that undergoes the transformation as

$$\zeta_j = 1 - exp[A(\frac{T - T_s}{T_e - T_s})^D] \qquad [12.4]$$

in which ζ_j is the material volume fraction, which experiences the phase transformation in the *j* element, and *Ts* and *Te* are the temperatures at which the transformation starts and ends, respectively. Finally, *A* and *D* are two coefficients, which govern the transformation and should be properly determined.

12.4 References

1. Khandkar, M.Z.H. et al. (2006), 'Predicting residual thermal stresses in friction stir welded metals', *Journal of Materials Processing Technology*, 174(1–3): 195–203.
2. Mandal, S., Rice, J. and Emustafa, A.A. (2008), 'Experimental and numerical investigation of the plunge stage in friction stir welding', *Journal of Materials Processing Technology*, 203(1–3): 411–19.
3. Zhang, Z. and Zhang, H.W. (2009), 'Numerical studies on controlling process parameters in friction stir welding', *Journal of Materials Processing Technology*, 209(1): 241–70.
4. Soundararajan, V., Zekovic, S. and Kovacevic, R. (2005), 'Thermo-mechanical model with adaptive boundary conditions for friction stir welding of Al 6061', *International Journal of Machine Tools and Manufacture*, 45(14): 1577–87.
5. Chen, C.M. and Kovacevic, R. (2006), 'Parametric finite element analysis of stress evolution during friction stir welding', *Proceedings of the Institution of Mechanical Engineers, Part B: Journal of Engineering Manufacture*, 220(8): 1359–71.
6. Asadi, P., Mahdavinejad, R.A. and Tutunchilar, S. (2011), 'Simulation and experimental investigation of FSP of AZ91 magnesium alloy', *Materials Science and Engineering A*, 528(21): 6469–77.
7. Buffa, G., Ducato, A. and Fratini, L. (2013), 'FEM-based prediction of phase transformations during friction stir welding of Ti6Al4V titanium alloy', *Materials Science and Engineering A*, 581(0): 56–65.
8. Buffa, G. et al. (2006a), 'A continuum based fem model for friction-stir welding: Model development', *Materials Science and Engineering A*, 419(1–2): 389–96.

9. Shojaeefard, M.H., Khakhali, A. and Akbari, M. (2013), 'Investigation of friction stir welding tool parameters using FEM and neural network', *Proceedings of the Institution of Mechanical Engineers, Part I: Journal of Materials: Design and Applications*, **In press.**
10. Ulysse, P. (2002), 'Three-dimensional modelling of the friction stir-welding process', *International Journal of Machine Tools and Manufacture*, 42(14): 1549–57.
11. Colegrove, P.A. and Shercliff, H.R. (2005), '3-Dimensional CFD modelling of flow round a threaded friction stir welding tool profile', *Journal of Materials Processing Technology*, 169(2): 320–7.
12. Chen, G-Q. et al. (2013), 'Computational fluid dynamics studies on heat generation during friction stir welding of aluminum alloy', *Computational Materials Science*, 79(0): 540–6.
13. Aramayo G. et al. (2002), *Modelling of Friction stir Welding Process for Fusion Energy Applications*, Oak Ridge National Laboratory.
14. Fourment, L. et al. (2004), 'Numerical simulation for the friction stir welding process using both Lagrangian and arbitrary Lagrangian Eulerian formulations', *Proceedings of the 5th International Friction stir Welding Symposium*, TWI, France.
15. Heurtier, P. et al. (2006), 'Mechanical and thermal modelling of friction stir welding', *Journal of Materials Processing Technology*, 171(3): 348–57.
16. Mandal, D., Pal, S.K. and Saha, P. (2007), 'Modeling of electrical discharge machining process using back propagation neural and multi-objective optimization using non-dominating sorting genetic algorithm-II', *Journal of Materials Processing Technology*, 186(1–3): 154–62.
17. Buffa, G. et al. (2006b), 'Design of the friction stir welding tool using the continuum based FEM model', *Materials Science and Engineering A*, 419(1–2): 381–8.
18. Assidi, M. et al. (2010), 'Friction model for friction stir welding process simulation: Calibrations from welding experiments', *International Journal of Machine Tools and Manufacture*, 50(2): 143–55.
19. Gao, Z. et al. (2013), 'FE modelling of microstructure evolution during friction stir spot welding in AA6082-T6', *Welding in the World*, 57(6): 895–902.
20. Tutunchilar, S. et al. (2012), 'Simulation of material flow in friction stir processing of a cast Al–Si alloy', *Materials & Design*, 40(0): 415–26.
21. Grujicic, M. et al. (2010), 'Modelling of AA5083 material-microstructure evolution during butt friction stir welding', *Journal of Materials Engineering and Performance*, 19(5): 672–84.
22. Rai, R. et al. (2011), 'Review: Friction stir welding tools', *Science and Technology of Welding &* Joining, 16(4): 325–42.
23. Atharifar, H., Lin, D. and Kovacegic, R. (2009), 'Numerical and experimental investigations on the loads carried by the tool during friction stir welding', *Journal of Materials Engineering and Performance*, 18(4): 339–50.
24. Nandan, R. et al. (2006), 'Numerical modelling of 3-D plastic flow and heat transfer during friction stir welding of stainless steel', *Science and Technology of Welding & Joining*, 11(5): 526–37.

25. Trimble, D., Monaghan, J. and O'Donnell, G.E. (2012), 'Force generation during friction stir welding of AA2024-T3', *CIRP Annals – Manufacturing Technology*, 61(1): 9–12.
26. Fratini, L., Macaluso, G. and Pasta, S. (2009), 'Residual stresses and FCP prediction in FSW through a continuous FE model', *Journal of Materials Processing Technology*, 209(15–16): 5465–74.
27. Ge, Y.Z. et al. (2006), 'Limited weld residual stress measurements in fatigue crack propagation: Part I. Complete field representation through least-squares finite-element smoothing', *Fatigue & Fracture of Engineering Materials & Structures*, 29(7): 524–36.
28. Fratini, L., Buffa, G. and Monaco, L.L. (2010), 'Improved FE model for simulation of friction stir welding of different materials', *Science and Technology of Welding and Joining*, 15(3): 199–207.

13

Artificial neural network and optimization

M. Akbari, Iran University of Science and Technology, Iran,
P. Asadi and M.K. Besharati Givi, University of Tehran, Iran and
G. Khodabandehlouie, Islamic Azad University, Iran

DOI: 10.1533/9780857094551.543

Abstract: The artificial neural network (ANN) can be used as a model for friction stir welding (FSW), predicting the correlation between FSW parameters and weld properties. ANN for predicting properties of welded plates by FSW is explained and pervious works in this field are reviewed. Furthermore, some examples about applications of ANN in FSW are presented. The development of sound joints between materials is of great importance in FSW. One of the most challenging problems is choosing appropriate welding parameters in order to produce sound joint. Traditionally, a time-consuming trial and error development was carried out to determine welding parameters. Additionally, an optimized welding parameters combination was not achieved, because welds can be fabricated with very different ideal welding parameters. Different optimization techniques can be employed to determine the optimized output parameters by specifying the relation between the input and output variables. Applications of optimization methods in FSW are explained, and basic principles of these methods, such as Taguchi, genetic optimization and multi-objective optimization methods, are described

Key words: artificial neural network, optimization, FSW, Taguchi, genetic algorithm, multi-objective optimization.

13.1 Introduction

Friction stir welding/friction stir processing (FSW/FSP) involves complex material movement and plastic deformation. As shown in Figure 13.1, various parameters influence material flow pattern and temperature distribution, and consequently affect the microstructural and mechanical properties of the material. It is well-known that the main challenge for manufacturers is selecting the parameters leading to production of an excellent weld joint. Conventionally, determining the weld input parameters for new welding products in order to produce a proper weld joint that will pass the demanded requirements is a time-consuming trial. This process may contain error development efforts that finally lead to parameters chosen by operator-dependent skills.

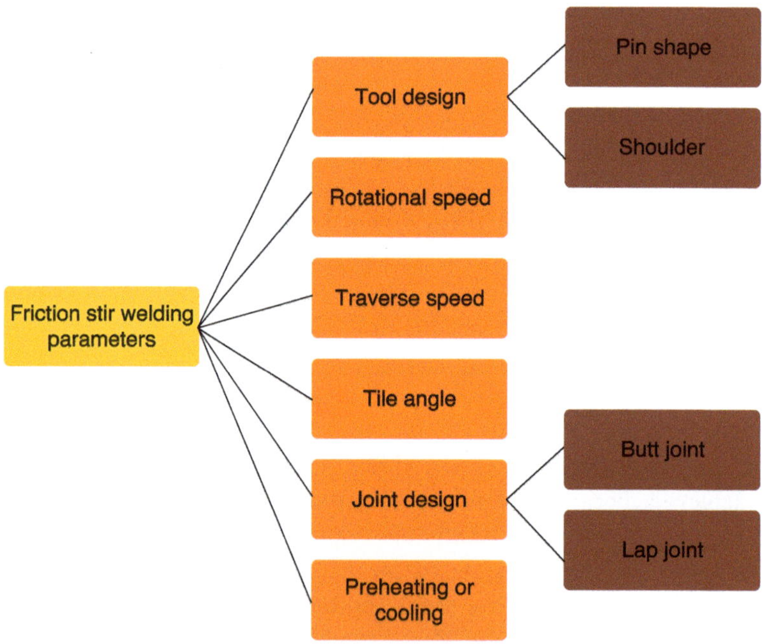

Figure 13.1 Friction stir welding parameters

In addition, a combination of optimized welding parameters has not been achieved, because welds can be fabricated with many different ideal welding parameters. To prevent this problem, different optimization and modelling techniques can be used to determine the optimized output parameters by specifying the relation between the input and output variables. For predicting and specifying the relation between the input and output variables of the welding parameters accurately without consuming time, materials and labour effort, various modelling methods are available, and one such method is the artificial neural network (ANN). An ANN development can accurately predict welding parameters, whilst various methods can be used for optimizing welding parameters. The methods introduced in this chapter are the Taguchi method and multi-objective optimization.

ANNs are used to solve a wide variety of problems in science and engineering, particularly in areas where conventional modelling methods fail. A well-trained ANN can be employed as a prediction model for a particular application. The ability of prediction in an ANN results from the training experimental data, and then validation by independent data. An ANN has the ability to relearn and so improve its performance as new data become available. Also, it can accommodate multiple input variables to predict multiple output variables. The ability to learn the system, modelled without former knowledge about the process relationships, makes ANN different from conventional modelling methods. A well-trained ANN is usually much faster in prediction than formal simulation software or mathematical models, since no time-consuming iterative calculations are required to solve differential equations by numerical approaches. However, the selection of a proper neural network topology is essential from model accuracy and simplicity points of view.

Optimization is, indeed, finding one or more possible solution related to extreme values of one or more objectives. The necessity to find such optimized solutions for a problem is raised mainly from the extreme goal of either minimum achievable cost or maximum achievable quality, etc. On account of such extreme characteristics of optimal solutions, optimization approaches have attracted attention in practice, specifically in engineering and scientific experiments.

Single-objective optimization is when an optimization problem includes just one objective function, while an optimization problem involving more than one objective function is referred as multi-objective optimization. Because multi-objective optimization consists of multiple objectives, it can be stated that the single-objective optimization is a very simplified model of multi-objective optimization.

13.2 Artificial neural network

13.2.1 What is a neural network

ANN is an approach inspired by the brain's structure and attempts at simulating its processing capabilities. A neural network is an immense parallel-distributed processor composed of simple processing units that has a natural tendency to store practical knowledge and make it available for use when needed. ANN is like a brain in two aspects [1]:

1. The knowledge is achieved by the network via a learning process.
2. Interneuron connection strengths, known as synaptic weights, are used for storing the achieved knowledge.

ANN operates like a 'black box' model, and thus there is no need to acquire detailed information about its system. The relationship between the input and output parameters is learned by the system through observing the formerly recorded data. The use of neural networks offers the beneficial properties and capabilities shown in Figure 13.2.

13.2.2 Brief historical review

The history of ANN involves several creative individuals working in different fields; many of them have strived for decades to develop a concept that we are now using. ANN research has experienced three periods of extensive activity. The first peak in the 1940s was due to McCulloch and Pitt's pioneering work, which showed that networks of artificial neurons could calculate any mathematical or logical function [2]. The second spurt occurred in the 1960s, with Rosenblatt's perceptron

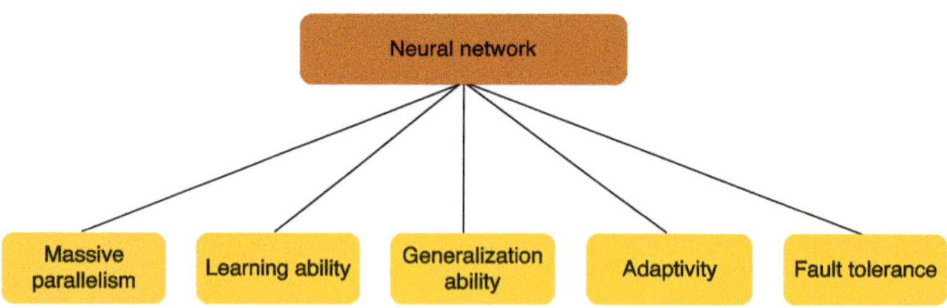

Figure 13.2 Capabilities and benefits of a neural network

convergence theorem [3] and in 1972 showing the limitations of a simple perceptron were shown [4]. The results weakened the interest of most researchers, particularly those in the community of computer science. This appeasement in research of neural networks continued for almost 20 years. Eventually, in the early 1980s, ANNs received substantial renewed enthusiasm. The major developments behind this revival involve Hopfield's energy approach [5] in 1982 and the back-propagation learning algorithm for multilayer perceptrons proposed first by Werbos, and re-invented several times, and finally famed by Rumelhart et al. [6] in 1986.

13.2.3 Biological neural networks

ANN is indeed an effort to model the information processing ability of nervous systems, therefore the important properties of biological neural networks should be considered from the information processing point of view. In the next section, those properties of brain function that have motivated the development of ANN will be briefly described.

The brain involves a large number of neurons, which have three principal components:

1. the dendrites;
2. the cell body; and
3. the axon [7].

The dendrites are a tree-shaped receptive network of nerve fibres carrying the electrical signals into the cell body. The cell body efficiently sums and thresholds the received signals. The axon is a single long fibre that carries the signal from the cell body out to other neurons. The point of contact between an axon of one cell and a dendrite of another cell is called a synapse. The function of the neural network is established by the arrangement of neurons and the strengths of the individual synapses, determined by complex chemical process [8]. Figure 13.3 is a schematic illustration of two biological neurons.

Some of the neural structure is defined at birth; however, since new connections are made and others waste away, other parts of the neural structure are developed through learning. Neural structures change continually throughout life. These later changes are primarily composed of strengthening or weakening of synaptic junctions.

ANN is not as complicated as the brain; however, biological and artificial neural networks do resemble each other in two aspects. First,

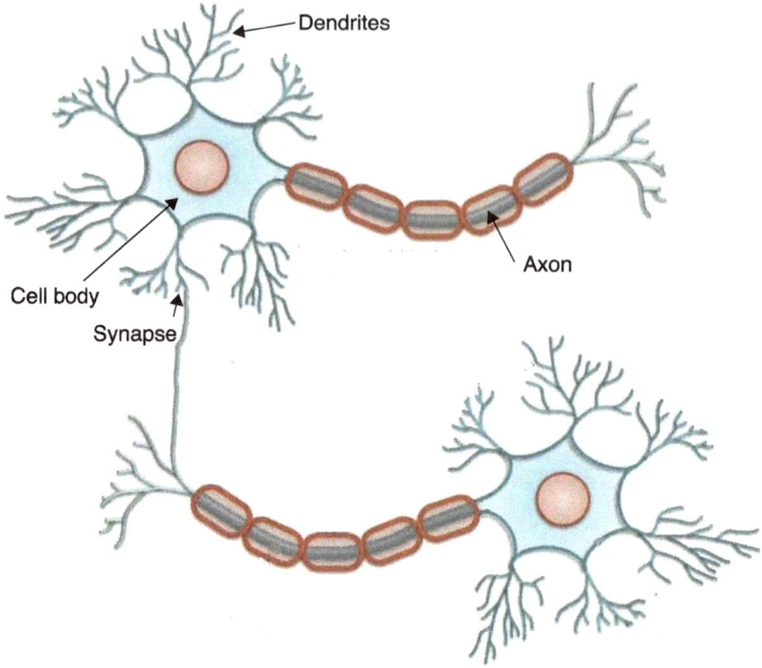

Figure 13.3 Schematic drawing of a biological neuron

the building blocks of both networks are simple computational devices that are highly interconnected. Second, the networks function is determined by the connections between neurons.

13.2.4 Computational models of neurons

A large number of highly interconnected processing elements, known as neurons, are organized in a neural network. An elementary neuron with single inputs is shown in Figure 13.4. Furthermore, an artificial neuron primarily composes of weight, bias and activation functions. The scalar p, as an input, is weighted with an appropriate w. The sum of the weighted input and the bias forms the input to the transfer function f, which produces the scalar neuron output a [8].

The neuron output is calculated as

$$a = f(wx + b) \quad [13.1]$$

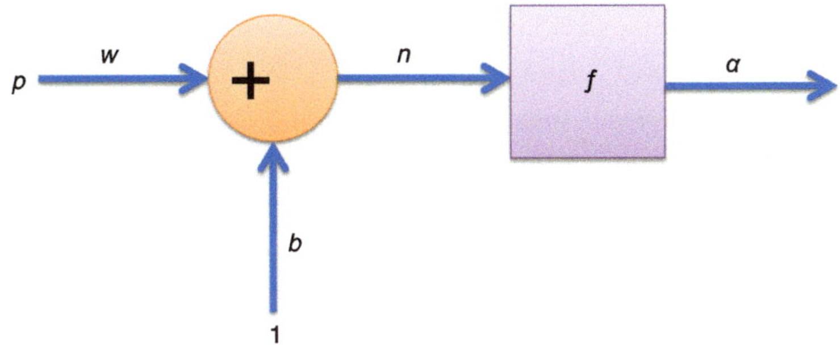

Figure 13.4 A single input neuron

Note that w and b are both flexible scalar parameters of the neuron. Generally, neurons input/output relationships can meet some specific goal, because the transfer function is chosen by the designer and then the parameter w and b will be adjusted by some learning role.

13.2.5 Transfer function

The transfer function can be linear or nonlinear. Solving some specification of the problem by neurons needs the choice of a particular transfer. The most commonly used function is discussed below:

The hard limit transfer function

The hard limit transfer function (Figure 13.5) sets the output of the neuron to 0 if the function argument is less than 0, or 1 if its argument is greater than or equal to 0.

Linear transfer function

The output of linear transfer function is equal to its input $(a=n)$ (Figure 13.6).

Log sigmoid transfer function

The log sigmoid transfer function is shown in Figure 13.7. This function generates outputs between 0 and 1, as the neurons net input goes from negative to positive infinity according to

$$a = \frac{1}{1+e^{-n}} \qquad [13.2]$$

The log sigmoid transfer function is a commonly used multi-layer neural network that is trained by the back propagation algorithm [8].

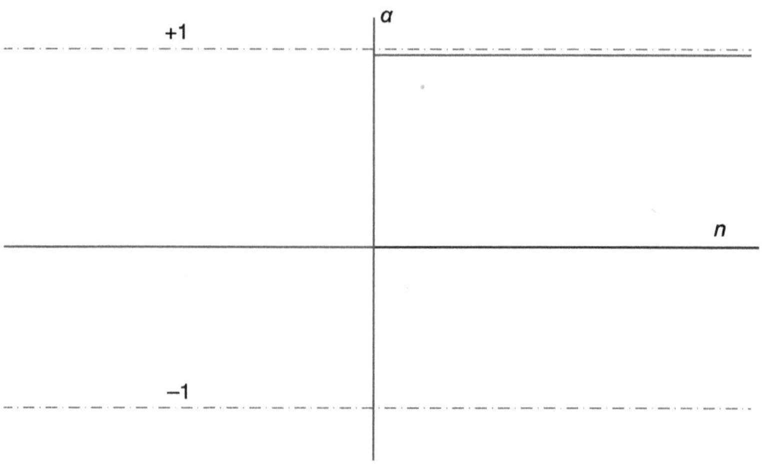

Figure 13.5 Hard limit transfer function

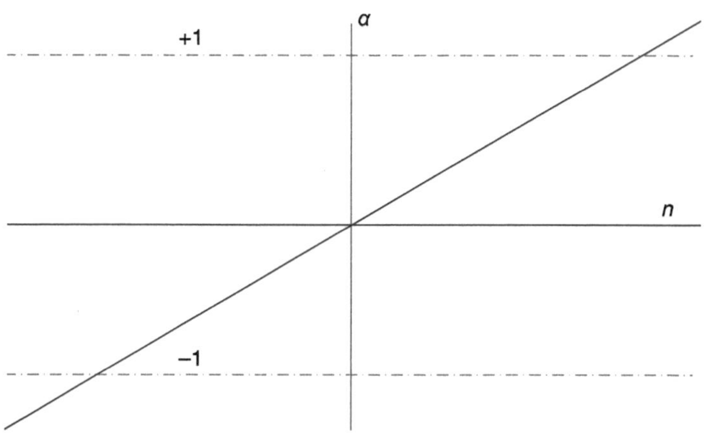

Figure 13.6 Linear transfer function

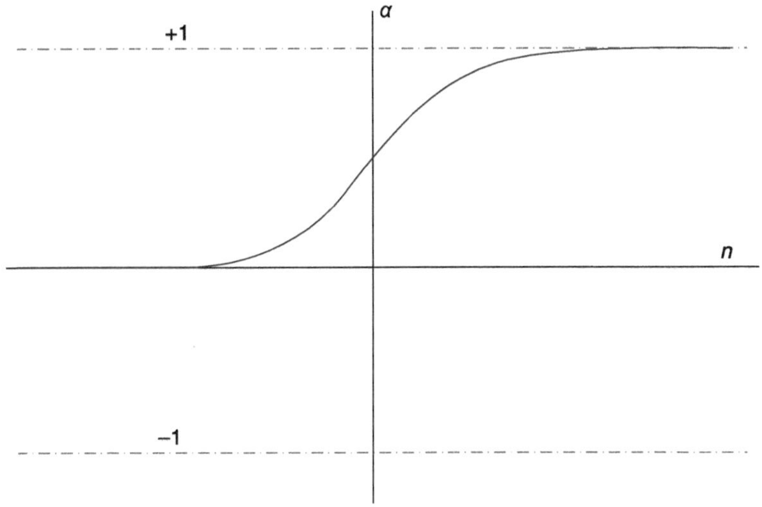

Figure 13.7 Log sigmoid transfer function

Hyperbolic tangent sigmoid transfer function

The hyperbolic tangent transfer function is shown in Figure 13.8. This transfer function takes the input and condenses the output into the range of (–1, 1) according to

$$a = \frac{e^n - e^{-n} 1}{e^n + e^{-n}} \qquad [13.3]$$

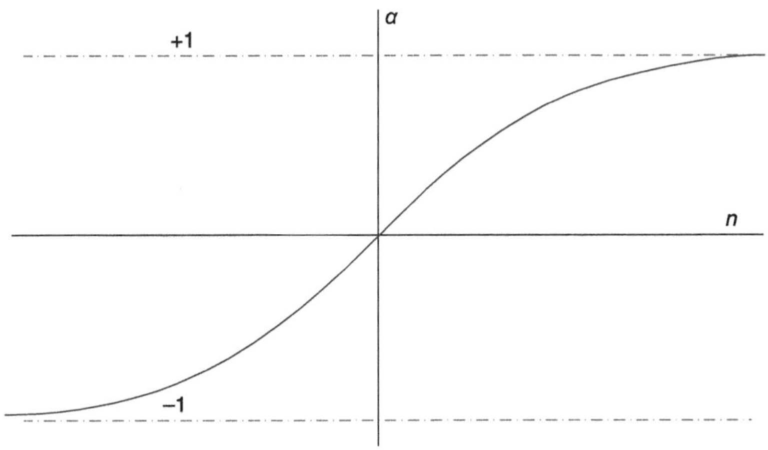

Figure 13.8 Hyperbolic tangent sigmoid function

13.2.6 Network architectures

This section underlines the pattern of connections between the units and the propagation of data. ANNs can be categorized into two groups according to this pattern of connection (Figure 13.9) [9]:

- *feed-forward networks*: in which graphs have no loops (Figure 13.10(a)), and
- *recurrent (or feedback) networks*: in which loops occur due to feedback connections (Figure 13.10(b)).

Appropriate learning algorithms are required for different network architectures. The next section provides an overview of the learning processes.

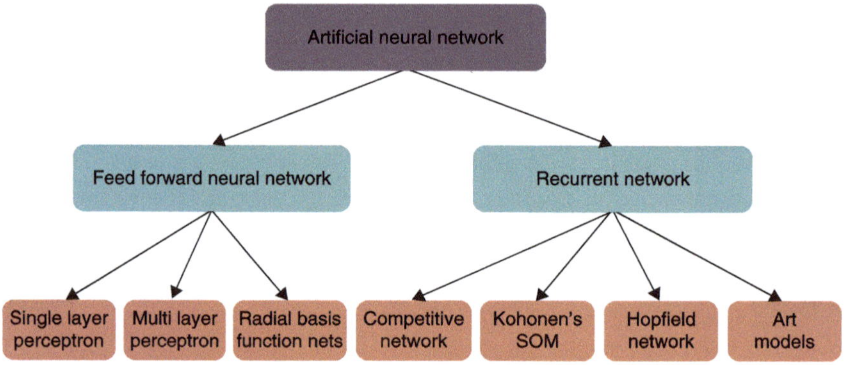

Figure 13.9 Taxonomy of feed-forward and recurrent network architectures

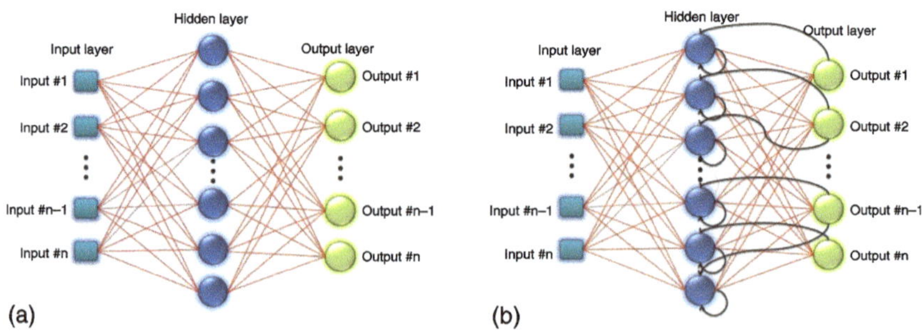

Figure 13.10 Schematic of: (a) feed-forward neural network; and (b) recurrent network

13.2.7 Learning

The ability of the network to learn from its environment and to improve its performance through learning is a principal characteristic for a neural network. The improvement in performance is a time-consuming process in accordance with some prescribed measure. A neural network learns about its environment through an interactive process of adjustments applied to its synaptic weights and bias levels. Ideally, every interaction of the learning process expands the environmental knowledge of the network.

To justify the definition of 'learning' in a precise way, many activities can be used. Recognizing that our specific interest is in neural networks, we use a definition of learning that is adapted from [10]:

> Learning is a process by which the free parameters of a neural network are adapted through a process of stimulation by the environment in which the network is embedded. The type of learning is determined by the manner in which the parameter changes take place.

In order to understand or design a learning process, at first you should know what information is accessible to the network; this information derived from, having a model of the environment, in which a neural network operates, is essential. We refer to this model as a learning paradigm. Second, you must understand how network weights are updated, and which *learning rules* control the updating process. A *learning algorithm* refers to a procedure in which learning rules are used for regulating the weights.

Paradigm learning

There are two main learning paradigms: supervised and unsupervised, as follows [11]

1. *Supervised learning or associative learning*: in which the network is trained by providing it with input and matching output patterns. An external trainer or system, which contains the network, can provide these input-output pairs.
2. *Unsupervised learning or self-organization*: in which an (output) unit is trained to respond to clusters of patterns within the input. In this paradigm, the significant characteristics of the input population are

statistically discovered by the system. Unlike the supervised learning paradigm, not only is there no *a priori* set of classified patterns, but also the system must develop its own representation of the input stimuli.

Learning rule

In this section, we discuss four fundamental learning rules:

1. error-correction learning;
2. Boltzmann learning;
3. Hebbian learning; and
4. competitive learning.

Error-correction learning

In the supervised learning paradigm, for each input pattern a desired output is given to the network. Over the learning process, the actual output y is generated by the network may not equal the desired output d. The fundamental principle of error-correction learning rules decreases this error gradually by using the error signal $(d\text{-}y)$ to modify the connection weights. The back-propagation learning algorithm is also based on the error-correction principle [7].

Boltzmann learning

Boltzmann machines are symmetrical recurrent networks consisting of binary units (+1 for 'on' and −1 for 'off). Symmetry indicates that the weight on the connection from unit i to unit j is precisely equal to the weight on the connection from unit j to unit i ($W_{ij} = W_{ji}$). A subset of the neurons, called 'visible', communicate with the environment, while the rest, called 'hidden', do not. Each neuron is a stochastic unit that generates an output (or state) based on the Boltzmann distribution of statistical mechanics. Boltzmann machines operate in two modes: clamped, in which visible neurons are clamped onto particular states determined by the environment; and free-running, in which both visible and hidden neurons are allowed to operate freely. Boltzmann learning is a stochastic learning rule derived from information-theoretic and thermodynamic principles. The purpose of Boltzmann learning is to adjust the connection weights so that the states of visible units satisfy a particular desired probability distribution [12].

Hebbian rule

Hebb's postulation of learning is the oldest and most famous of learning rules. Hebb based the rule on the following observation from neurobiological experiments. If neurons on both sides of a synapse are activated simultaneously and repeatedly, the strength of the synapses is substantially increased [7].

Competitive learning rules

In competitive learning, as its name implies, the output neurons of a neural network compete among themselves to become active (fired). In fact, the significant difference between competitive learning and Hebbian learning is in the number of active neurons at any one time. Whereas neural network based on Hebbian learning, several output neurons may be active simultaneously in competitive learning, only a single output neuron is active at any one time. According to this feature, competitive learning is highly suitable for discovering statistically salient features, which makes it useful for classification of input patterns.

13.2.8 Multilayer perceptron

The most popular class of multilayer feed-forward ANNs is the multilayer perceptron, with one or more layers between the input and output layer. The multilayer perceptron network is most commonly used with the back-propagation algorithm. Multiple layers of neurons with nonlinear transfer functions allow the network to learn nonlinear and linear relationships between input and output vectors.

The back-propagation algorithm is composed of two passes through different layers of the network. The first is the forward pass, in which the input vector is applied to the network, and the network is trained with the back-propagation learning algorithm layer by layer. In this pass, the synaptic weights are assumed to be constant. The second one is the backward pass in which, unlike the forward pass, parameters of the network are inconstant and modified. First the error, the discrepancy between the output of the network and the expected output, is carried out through back-propagation to update the synaptic weights. The weights are continuously updated after each piece of data is processed. The input vector is presented to the network and the process continues until the output of the network comes closer to the desired output [13].

13.2.9 Pervious works

The field of neural networks is immense and interdisciplinary. Speed of calculation, capability of learning from examples and simplicity are the advantages of ANNs. These characteristics enable ANNs to be used in materials science. The ability of ANNs for modelling and predicting FSW has been investigated by several researchers. [14] led and tested multilayer feed-forward networks were led and tested based on a the back propagation algorithm, with results from a phenomenological model to generate tool durability maps that show the ratio of the shear strength of the tool material to the maximum shear stress on the tool pin for various combinations of welding variables (Figure 13.11). Their maps

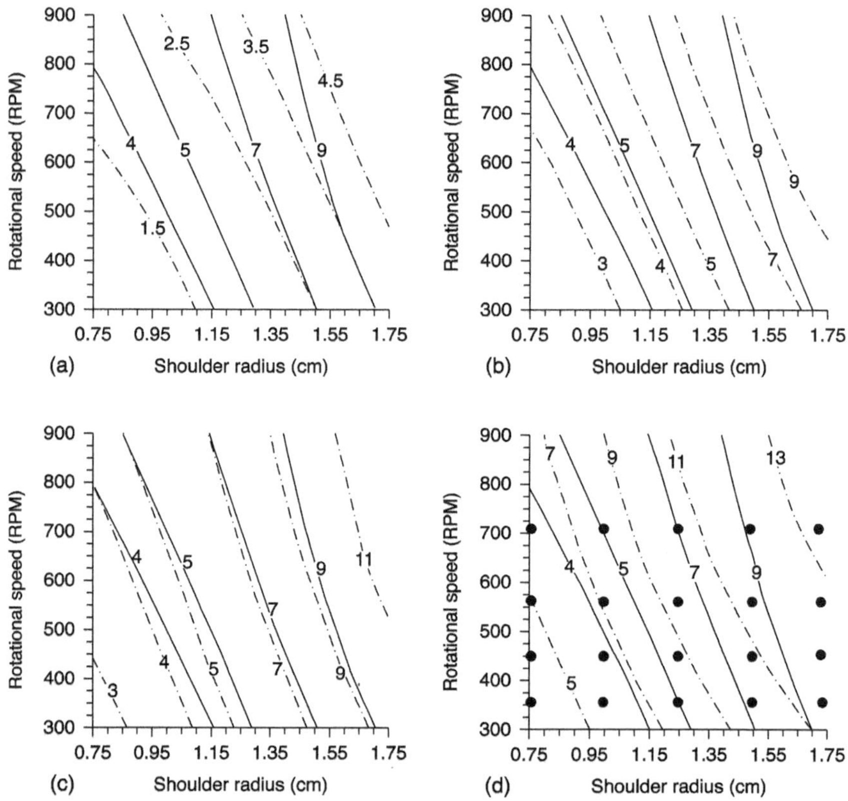

Figure 13.11 Tool durability index contours computed using the ANN model for the FSW of AA7075 as function of tool shoulder radius and tool rotational speed for various welding variables [13]

demonstrate how the thicker plates and faster welding speeds adversely affect tool durability, and how that can be optimized.

A set of five neural networks were developed, tested and made available for calculating the peak temperature, torque, traverse force and bending and equivalent stresses on the tool pin for the FSW of an aluminum alloy [15]. Therefore, according to their claim, the models can be used even beyond the range of training with predictable levels of uncertainty.

Multilayer feed-forward networks were properly trained and linked [16] based on back-propagation to a three-dimensional (3-D) finite element model (FEM) model for the FSW of Ti–6Al–4V titanium alloy, to predict both the microhardness values and the microstructure of the welded butt joints with the varying of the main process parameters. Furthermore, a good agreement between experimental values and calculated results were found. Figure 13.12 demonstrates a satisfying agreement between their experimental and numerical result.

A neural network was linked to an FEM of the FSW process to predict the average grain size values [17]. The comparison between their measured average grain sizes and calculated ones is shown in Figure 13.13.

An ANN model was developed to simulate the correlation between the FSW parameters and mechanical properties [18]. Performance of the ANN model was excellent, and also the model was applied to predict the ultimate tensile strength and hardness of butt joint of AA7075–AA5083 as functions of weld and rotational speeds. The trial-and-error approach was used to choose the optimal network architecture. On the basis of the analysis, the optimal architecture of the proposed ANN was 2–6–2 NN

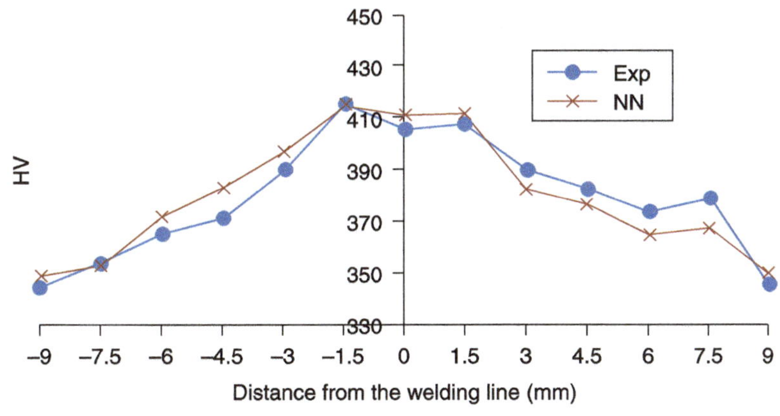

Figure 13.12 Comparison between experimental and numerical [16]

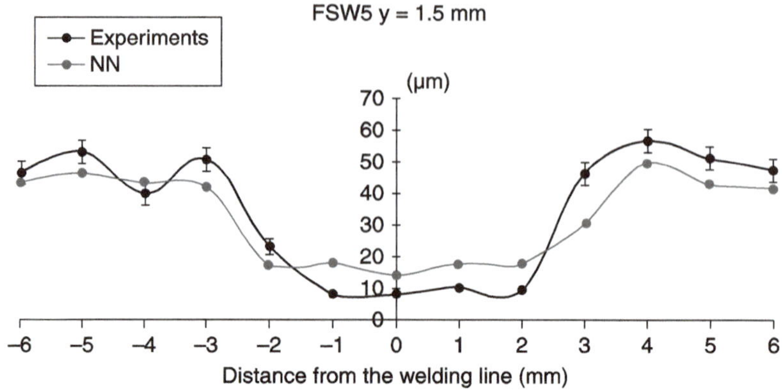

Figure 13.13 Comparison between measured and calculated average grain size [17]

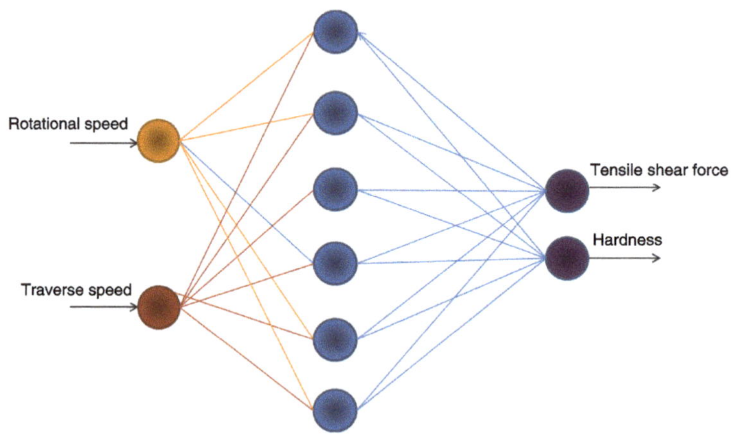

Figure 13.14 Three-layer neural network [18]

(Figure 13.14), where both activation functions in the hidden layer and output layer were 'logsig'.

A relation between FSP parameters and grain size and hardness of nanocomposite using ANN was published [19]. In this ANN modelling, the inputs included traverse speed, rotational speed and region types, and the outputs were hardness and grain size. According to this investigation, the model can be used to predict hardness and grain size as functions of rotational, traverse speeds and region types. Figures 13.15(a and b) show their experimental and predicted data at the training stage. Obviously, the data predicted by ANN are compared with the experimental results.

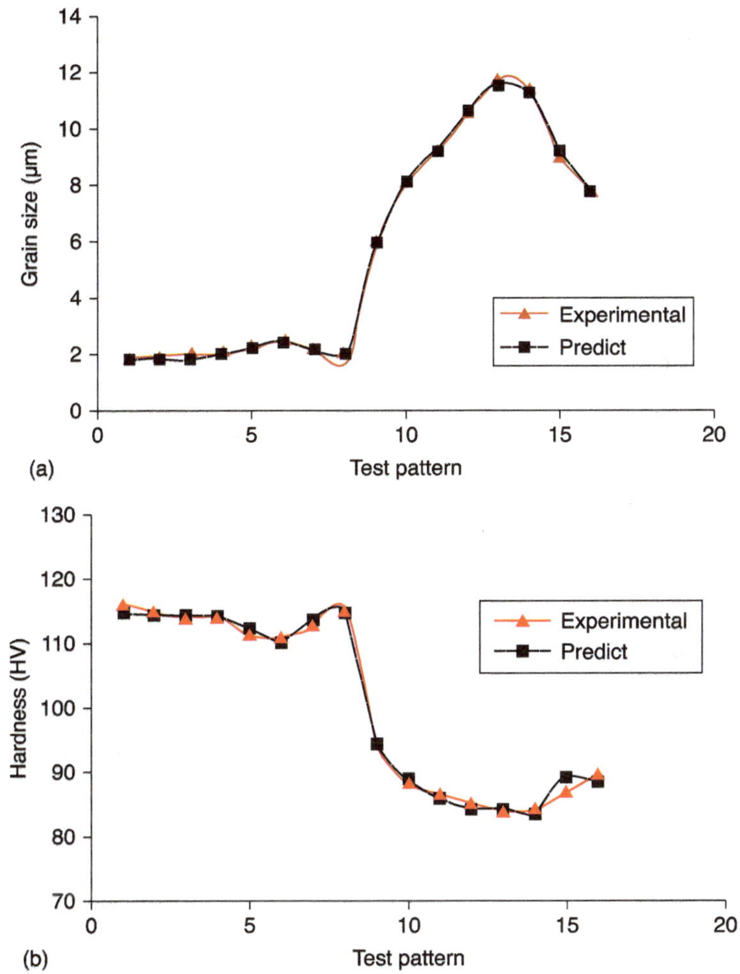

Figure 13.15 **Experimental and predicted: (a) grain size; and (b) hardness at the training stage of ANN model [19]**

An ANN model was developed to simulate the correlation between the friction stir lap welding (FSLW) parameters and mechanical properties [20]. Later, a sensitivity analysis was performed to investigate the effect of each input parameter on the output in terms of magnitude and direction. Four methods were compared, namely the 'PaD' method, the 'Weights' method, the 'Profile' method and the 'backward stepwise' method, which can give the relative contribution and/or the contribution profile of the input factors.

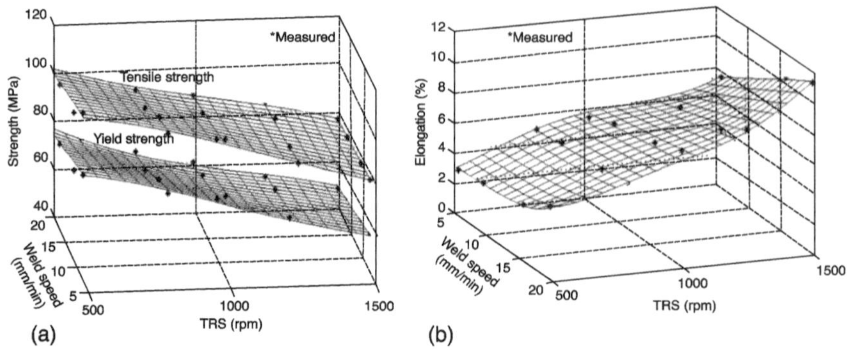

Figure 13.16 Variation of predicted and measured hardness of: (a) HAZ; and (b) weld metal with TRS and weld speed [21]

An ANN model was developed for the analysis and simulation of the correlation between FSW parameters of aluminum plates and mechanical properties [21]. In the proposed model, input parameters consist of weld speed and tool rotation speed (TRS), and outputs include property parameters, namely tensile strength, yield strength, elongation, hardness of weld metal and hardness of the heat affected zone (HAZ). Good performance of the ANN model (Figure 13.16) was achieved, and based on this investigation, the model can be used to compute the mechanical properties of welded Al plates as functions of weld speed and TRS.

13.2.10 Application of an artificial neural network in friction stir welding

In this section, an ANN model is developed to simulate the correlation between the FSW parameters and mechanical properties of a butt joint of AA7075–AA5083, as function of weld and rotational speeds. The summary of experimental results are presented in Table 13.1. As shown in the table, the inputs are the rotational speed and the traverse speed, and the outputs are tensile strength and hardness.

This example has used back-propagation algorithm for feed-forward neural network. Moreover, inputs and outputs have been normalized in the range of 0 to 1. In all phases of the model development, including training and testing of the network, MATLAB software was applied. One of the most crucial tasks in ANN studies is choosing the optimal network architecture, which is associated with the activation function and number of neurons in the hidden layer. In this example, the trial-and-error

Table 13.1 Experimental results

No.	Inputs		Outputs	
	Traverse speed	Rotational speed	Tensile strength	Hardness
1	20	1400	267	92.3
2	20	1200	232	94.2
3	20	900	216	93.7
4	20	700	192	96.7
5	20	565	198	97.9
6	36	1400	221	90.23
7	36	1200	201	92.1
8	36	900	192	94.1
9	36	700	180	95.25
10	36	565	177	97.3
11	63	1400	195	89
12	63	1200	180	90.1
13	63	900	173	91
14	63	700	165	90.9
15	63	565	161	95.2
16	96	1400	195	87.21
17	96	1200	180	88.23
18	96	900	150	90.1
19	96	700	165	91.67
20	96	565	142	92.89

approach is used to generate the optimal network architecture. Generally, different activation functions and number of neurons were tried to obtain the optimal architecture of the network. The performance of each network was checked by Mean Relative Error (*MRE*), as shown in Eq. 13.4:

$$MRE = \frac{1}{P}\sum_{1}^{P}\left(\frac{|experimental\ value - predicted\ value|}{experimental\ value}\times 100\right) \qquad [13.4]$$

The target is to reach the minimum quantity of *MRE*, which leads to obtaining the network with the best generalization; therefore, *MRE* values were calculated for many different network models. Based on this analysis, the optimal architecture of the ANN was constructed as

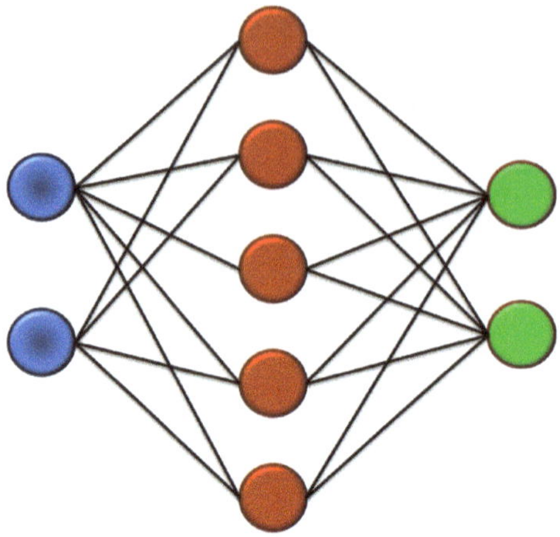

Figure 13.17 Three-layer neural network in the present example [18]

2–6–2 NN, and both activation functions in the hidden layer and the output layer were logsig. A schematic diagram of this three-layer neural network is shown in Figure 13.17.

Linear regression analyses were employed to calculate the correlation coefficient (R) between the experimental and predicted values. The correlation coefficients for ultimate tensile strength and hardness at the training and testing phases were 0.9999, 0.9916, 0.9799 and 0.9891, respectively. Obviously, the neural network prediction of FSLW parameters follows the experimental results very closely, and also the developed ANN can accurately predict the ultimate tensile strength and hardness. It is evident that the predicted data by the ANN is in good agreement with the experimental results, otherwise it would be possible to predict the ultimate tensile strength and hardness without carrying out the experiment.

The hardness and ultimate tensile strength can be computed by Eqs 13.5 and 13.6, which are derived from the ANN:

$$\textit{Ultimate tensile strength} = \left(\frac{1}{1+e^{-(-1.447\times F_1+0.390\times F_2-0.497\times F_3+0.055\times F_4+0.415\times F_5-0.331\times F_6+2.14)}}\right) \quad [13.5]$$

and

$$\textit{Hardness} = \left(\frac{1}{1+e^{(-(-0.198\times F_1+0.169\times F_2-0.0762\times F_3+0.303\times F_4+0.281\times F_5+0.080\times F_6+1.86))}}\right) \quad [13.6]$$

where $F_{(i=1, 2, 3, ..., 6)}$ can be calculated as

$$F_i = \frac{1}{1+exp^{-U_i}} \qquad [13.7]$$

where U_1 to U_6 may be determined as

$$U_i = C_{1i} \times R + C_{2i} \times T + C_{3i} \qquad [13.8]$$

where constants, C_{ji}, are given in Table 13.2 for LM algorithm with 6 neurons. T and R are traverse and rotational speeds, respectively. For normalization, T and R values are divided by 1600 and 100, respectively. For outputs, ultimate tensile strength and hardness values need to be multiplied by 300 and 110, respectively.

Table 13.2 Weights and biases between input layer and hidden layer for Eqs 13.5 and 13.6

	$U_i = C_{1i} \times R + C_{2i} \times T + C_{3i}$		
i			
1	50.53	-58.73	34.21
2	-30.54	23.086	1.76
3	-45.85	-13.57	54.87
4	2.95	62.84	-48.98
5	10.32	-51.71	25.50
6	13.17	-233.73	107.30

13.3 Optimization

In the process of adjusting the inputs or characteristics of a device, a mathematical processor experiment to find the minimum or maximum output result is called optimization (Figure 13.18). The input is made up

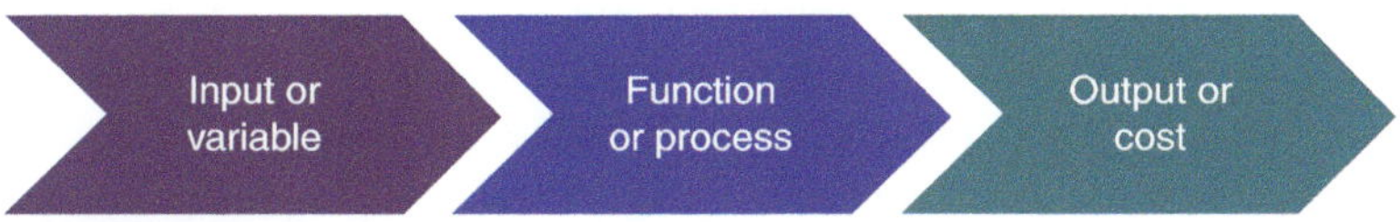

Figure 13.18 Diagram of a function or process that is to be optimized

of variables. The process or function is known as the cost function, objective function or fitness function, thus the output is the cost or fitness. If the process is an experiment, the variables are physical inputs to the experiment [1].

13.3.1 Taguchi method

Taguchi techniques were developed in 2012 [22]; more recently these techniques have been utilized widely in engineering analysis to optimize the performance characteristics within the combination of design parameters. The Taguchi technique is also a powerful tool for designing high-quality systems. It introduces an integrated, simple and systemic approach, which simply and efficiently finds the best set of designs for quality, performance and computational cost.

To sum up, the parameter design of the Taguchi method includes the following steps:

1. determination of the quality characteristic, which should be optimized;
2. determination of the number of levels for the design parameters;
3. designing the matrix experiment and defining the data analysis procedure;
4. conduct the matrix experiment;
5. analysis of the experimental results by the S/N and ANOVA analyses
6. selection of the optimal levels of design parameters;
7. predicting the performance at these levels; and
8. verification of the optimal design parameters through the confirmation experiment [22].

Design of experiment (DOE)

The technique of laying out the conditions (design) of experiments, involving multiple factors, was first proposed by Sir R.A. Fisher in the 1920s. He has been highly influential in the field of DOEs. A full factorial design evaluates simultaneously all possible combinations for a given set of factors. Since most industrial experiments usually consist of a large number of factors, a full factorial design results in a significant number of experiments [23]. For instance, in an experiment involving seven two-level factors, the total number of combinations will be $128(2^7)$. As a result, only a small set from all the possibilities will be selected, to reduce the number of experiments to a practical level.

The method of selecting a limited number of experiments is known as a partial factorial experiment. Although this shortcut method is well known, there is no general guideline for its application or the analysis of the results, which are obtained by performing the experiments. Taguchi's approach complements these two important weaknesses. First, he clearly defined a set of orthogonal arrays (OAs), each of which can be used in many experimental situations. Second, he devised a standard method for analyzing the results. The synthesis of standard experimental design techniques and the analysis methods in his approach form consistency and reproducibility, which rarely can be found in any other statistical method [23,24].

It is essential to comprehend how a set of factors result in all possible combinations before discussing how Taguchi decreases the number of experiments. Suppose that we are concerned about one factor A (say pressure). If we were to study its effect on a product at two levels (say *100* and *200 kPa*):

$$\text{Factor } A\text{: Level } 1 = A_1 \ (100\,kPa) \qquad \text{Level } 2 = A_2 \ (200\,kPa) \qquad [13.9]$$

As we cited in the previous section, with three factors, each at two levels, there are 2^3 (eight) possible experiments. If A, B and C represent these factors, the eight experiments can be expressed as [23]

$$A_1B_1C_1,\ A_1B_1C_2,\ A_1B_2C_1,\ A_1B_2C_2,\ A_2B_1C_1,\ A_2B_1C_2,\ A_2B_2C_1,\ A_2B_2C_2. \qquad [13.10]$$

By using the above general principle, the total number of possible experiments for different numbers of factors at two or three levels and the corresponding suggested Taguchi number of experiments is shown in Table 13.3.

Table 13.3 **Comparison of factorial design and Taguchi design**

Factors	Level	Factorial Design	Taguchi
2	2	$4(2^2)$	4
3	2	$8(2^3)$	4
4	2	$16(2^4)$	8
7	2	$128(2^7)$	8
11	2	$2048(2^{11})$	12
4	3	$81(3^4)$	9

Consider a factorial experiment of seven 2-level factors with 128 possible combinations. The factors are represented by the letters *A*, *B*, *C*, *D*, *E*, *F* and *G* (Table 13.4). The subscripts 1 and 2 refer to the value of each factor at level 1 or 2, respectively. Cells T-l to T-8 indicated the 8 trial numbers defined by Taguchi's partial factorial OA for this experiment.

Taguchi has established OAs to describe a large number of experimental situations. OA designs are often used in design experiments with multiple-level factors. The symbolic designation for these arrays carries the key information on the size of the experiment. The array of Table 13.4 is designated as L-8 or L8, where the number 8 indicates that 8 trials are required. The next lower size of the OA is L4. An L4 experiment requires 4 trial runs. The array handles up to 3 factors at 2 levels each. To fit a situation with factors between 4 and 7, all at 2 levels, an L8 will be used. For situations that demand a larger number of factors, such as higher levels as well as mixed levels, other OAs are available [23]. Predominantly, the selection of OA depends on these items in order of priority:

1. the number of factors and interactions of interest;
2. the number of levels for the factors of interest;
3. the desired experimental resolution or cost limitations.

Signal-to-noise ratio

The signal-to-noise (S/N) ratio approach is used to measure the quality characteristic deviating from the desired value by the Taguchi technique.

Table 13.4 **Experiment design using an L8 array**

Trial number	Column						
	A	B	C	D	E	F	G
T-1	1	1	1	1	1	1	1
T-2	1	1	1	2	2	2	2
T-3	1	2	2	1	1	2	2
T-4	1	2	2	2	2	1	1
T-5	2	1	2	1	2	1	2
T-6	2	1	2	2	1	2	1
T-7	2	2	1	1	2	2	1
T-8	2	2	1	2	1	1	2

This approach is also used instead of the average value to convert the experimental results into a value for the evaluation characteristic in the optimum parameter analysis. The S/N ratio is quoted in *dBi* units and can be defined as

$$\eta = -10.log(M.S.D.) \qquad [13.11]$$

where *M.S.D.* is the mean-square deviation for the output characteristic. The *S/N* ratio characteristics can be divided into three stages: the smaller-the-better, the nominal-the-better and the higher-the better, when the quality characteristics are continuous for engineering analysis [25]. For smaller-the-better:

$$M.S.D. = \frac{I}{N}\sum_{i=1}^{n}(y_i)^2 \qquad [13.12]$$

For nominal-the-better:

$$M.S.D. = \frac{I}{N}\sum_{i=1}^{n}(y_i - m)^2 \qquad [13.13]$$

For higher-the-better:

$$M.S.D. = \frac{I}{N}\sum_{i=1}^{n}(\frac{1}{y_i})^2 \qquad [13.14]$$

where y_i is the results of experiments, observations or quality characteristics such as length, weight or surface finish, *M is the* target value of result and *N* is the number of repetitions.

Determination of the optimum

As a general principle, the optimum performance will be calculated by using

$$Y_{opt} = Y_m + \sum_{i=1}^{q}(Y_i + y_m) \qquad [13.15]$$

where Y_m is the total mean of all results, Y_i is the mean result at the optimal level and q is the number of the main design parameters that significantly affect the performance characteristics.

Analysis of variance (ANOVA)

ANOVA is a collection of statistical models used to determine whether there is a statistical association between an output and one or more inputs. It should be noted that ANOVA is different from regression analysis, which is a statistical technique used to form a predictive model; however, ANOVA is a general technique that can be used to provide a statistical test of whether the means among two or more groups are equal or not, based on the assumption that in each group the mean of the outputs is normally distributed with the same variance.

Application of the Taguchi in determining friction stir welding parameters

Many researchers have used Taghuchi's OA experiments widely to find the optimum levels of FSW parameters. Process parameters were optimized for FSW of dissimilar aluminum alloys (AA2024-T6 and AA6351-T6) by using the Taguchi method [26]. The effects of process parameters, rotational speed, traverse speed and axial force were evaluated on tensile strength. Optimized welding conditions also were estimated for maximize tensile strength due to improving the productivity and weld quality. Ultimately, the results were verified by conducting the confirmation tests at identified optimum conditions. In addition, the effect of FSP process variables, such as rotational speed, traverse speed and tool tilt angle on the tensile strength of magnesium alloy AZ31, was investigated by [27]. The experiments were conducted according to the Taguchi parametric design L9 at several combinations of process parameters. The results illustrate that the rotational speed is the most important factor, followed by the traverse speed and tool tilt angle for maximizing the tensile strength of FSP magnesium alloy. [28] The simultaneous effect of the major FSP parameters was studied on the mechanical behaviour of a magnesium-based AE42 alloy. Taguchi's experimental design approach was utilized to determine the optimized set of investigated FSP parameters for processing the AE42 alloy.

Cooling temperatures during FSP, FSP tool rotational speed and numbers of FSP passes were found as the most influential FSP parameters. DOE was applied to determine the most important factors that influence ultimate tensile strength (UTS) of AA6061/SiC nanocomposites produced by FSP [29]. The optimum of process parameters was determined by the Taguchi method. Taguchi's approach of parameter design as a statistical DOE technique was utilized to set the optimal welding parameters [30]. The experiments were arranged by using Taguchi's L9 OA.

The most influential significant parameter and percentage contribution of each parameter was identified on tensile strength of FSW AA 7075-T6 aluminum joints by conducting the minimum number of experiments using Taguchi's OA. S/N ratio and ANOVA were employed to find the contribution of the main welding parameters such as tool rotational speed, weld traversing speed and plunge depth on the tensile strength of the welded joint. The investigation detected that both of the tool profiles and the tool rotational speed have more influence on tensile strength than weld traversing speed (welding speed) and plunge depth, and also the tool with a conical profile results in better joint efficiency than the tool with a square profile.

The optimum FSW process operating conditions were found for two plates of aluminum alloy AA5083 welded in a butt joint using a grey-based Taguchi method [32]. The FSW of Al-Mg alloy studied optimization of process parameters using the Taguchi method [33].

Example of using Taguchi's method to optimize friction stir welding

In this example, Taguchi L9 orthogonal DOEs are used to optimize FSP parameters of AA5083/CuZn34 alloys. The rotational speed, tool tilt angle and traverse speed were the parameters taken into consideration. Because we are considering three levels and three factors, L_9 OA is used in this example. The values, their symbols and levels are listed in Table 13.5.

Signal-to-noise ratio

The purpose of this example is to maximize tensile shear force through optimum process parameters in FSW. In order to determine better quality characteristics, the higher value is employed in this study. The tensile strength of the FSW joint values is studied to consider the effects of FSW process parameters. The experimental data are converted into the S/N

Table 13.5 Process parameter values and their levels

Level	A Rotational speed (rpm)	B Tool tilt angle $\theta°$	C Traverse speed (mm/min)
Level 1	900	0.5	6.5
Level 2	1100	1.5	12
Level 3	1400	2.5	25

ratio. The calculated S/N ratio values are tabulated in Table 13.6. The average mean and S/N ratio values of all levels are also calculated and listed in Tables 13.7 and 13.8. The larger S/N ratio corresponds to the better-quality characteristics. The optimal level setting is $A_2B_2C_1$, according to mean and S/N ratio values.

Analysis of variance (ANOVA)

The aim of the ANOVA in this example is to evaluate the importance of process parameters on tensile shear force. It gives a clear picture as to

Table 13.6 Experimental layer: L9 orthogonal array, mean value and SN ratio value

No.	A Rotational speed	B Tilt angle	C Traverse speed	Mean tensile shear force (N)	SN(ratio)
1	1	1	1	4058	72.16624
2	1	2	2	4471	73.00809
3	1	3	3	3615	71.16217
4	2	1	2	4095	72.24508
5	2	2	3	4057	72.1641
6	2	3	1	4520	73.10277
7	3	1	3	2627	68.3892
8	3	2	1	4380	72.82948
9	3	3	2	2960	69.42583

Table 13.7 Response table for means

Level	A Rotational speed	B Tilt angle	C Traverse speed
1	4048	3593	**4319**
2	**4224**	**4303**	3842
3	3322	3698	3433
Delta	902	709	886
Rank	1	3	2

The optimal setting parameters based on mean $A_2B_2C_1$.

Table 13.8 Response table for signal-to-noise ratios

Level	A Rotational speed	B Tilt angle	C Traverse speed
1	72.11	70.93	**72.7**
2	**72.5**	**72.67**	71.56
3	70.21	71.23	70.57
Delta	2.29	1.73	2.13
Rank	1	3	2

The optimal setting parameters based on mean $A_2B_2C_1$.

how far the process parameters affect the response and the level of significance of the factor considered. The results from ANOVA are shown in Table 13.9. Statistically, an F-test is used to indicate which design parameter has a significant effect on the quality characteristic. Generally, when $F > 4$, it means that the change of the design parameter significantly influences the quality characteristic. Based on our investigations, rotational speed is a highly significant factor, and plays a dominant role in affecting the tensile strength of the weld.

Determination of the maximum tensile shear force

According to the experiments, the optimum level setting is $A_2B_2C_1$. The average values of the factors at their levels are taken from Table 13.5 and the predicted value of the response is

$$Tensile\ shear\ force = A_2 + B_2 + C_1 - 2T = 4224 + 4303 + 4319 - 2 \times 3864 = 5117\ N \qquad [13.16]$$

Table 13.9 Analysis of variance

Source	Sum Sq.	d.f	F	Prob > F
Rotational Speed	1370579.9	2	7.97	0.1115
Tilt angle	1180714.9	2	6.87	0.1271
Traverse speed	879397.6	2	5.11	0.1636
Error	171984.2	2		
Total	3602667.6	8		

where A_3 is the average mean value of rotational speed at the third level, B_3 is the average mean value of transverse speed at the third level, and C_1 is the average mean value of tile angle at the first level.

13.3.2 Genetic algorithm

The genetic algorithm (GA) is used to generate useful solutions to optimize and search problems according to the principles of genetics and natural selection. In a GA, a population of many individuals is evolved under specified selection rules to a state that maximizes the 'fitness' (i.e. minimizes the cost function). The term 'genetic algorithm', which nowadays is universally identified as the abbreviated GA, was first used by John Holland [34]. His book, *Adaptation in Natural and Artificial Systems*, was instrumental in creating what is now a flourishing field of research and application that goes much wider than the original GA [34].

A binary genetic algorithm

The start and end of GA are like any other optimization algorithm; it begins with definition of optimization variables, the cost function and the cost, and ends by testing for convergence. However, in between, this algorithm is quite different. A path through the components of the GA is shown as a flow chart in Figure 13.19.

Selecting the variables and the cost function

An output is generated from a set of input variables (a chromosome) by a cost function. The cost function can be a mathematical function, an experiment or a game. The object is to modify the output in some desirable fashion by finding the appropriate values for the input variables. In the GA literature, the fitness term is extensively used to designate the output of the objective function.

At the beginning, the GA defines a chromosome or an array of variable values that should be optimized. If the chromosome has N_{var} variables (an N_{var}-dimensional optimization problem) given by $p_1, p_2, \ldots, p_{var}$, then the chromosome is written as an N_{var} element row vector [1]:

$$Chromosome = [p_1, p_2, p_3 \ldots, p_{var}] \quad [13.17]$$

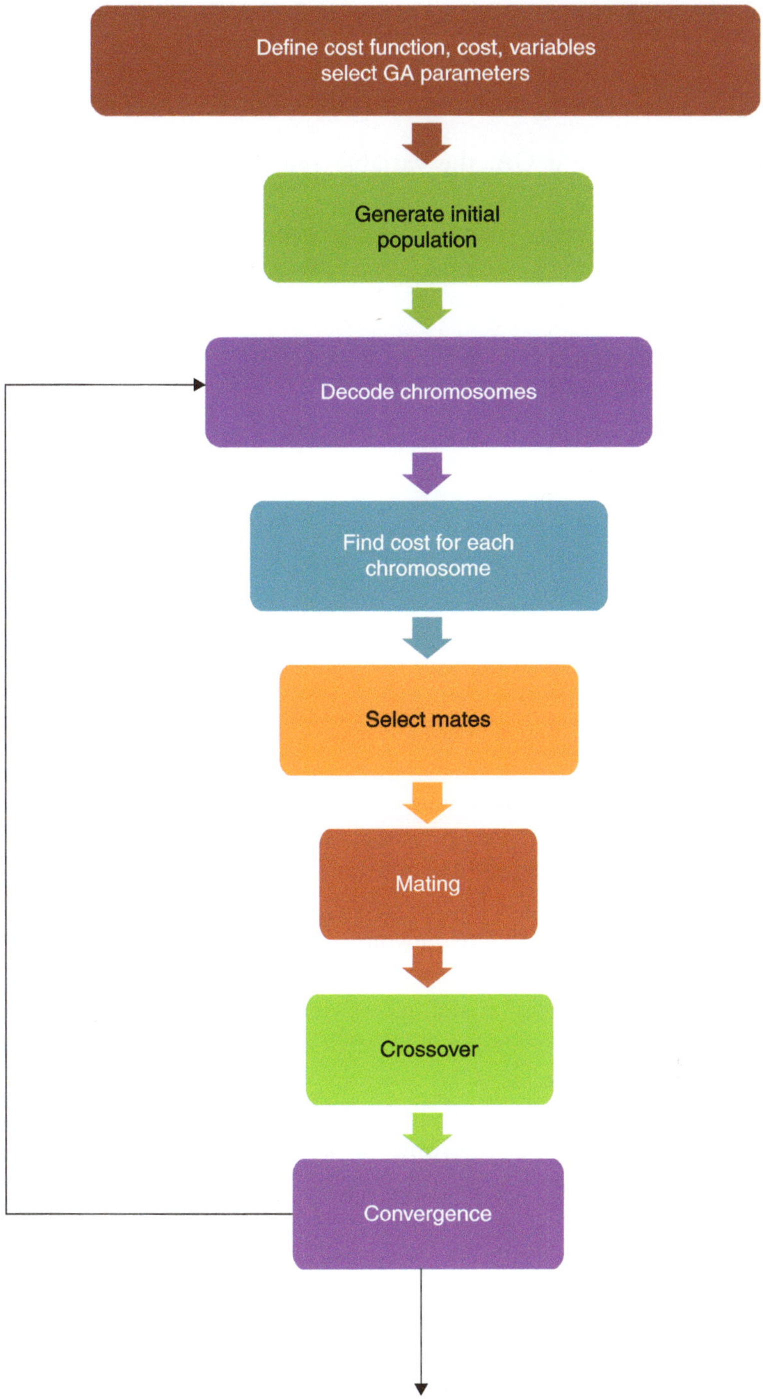

Figure 13.19 Flow chart of a binary GA

Variable encoding and decoding

Before you can use a GA, the variable values must be converted into binary. The binary GA works with bits. The variable x has a value represented by a string of bits that is N_{gene} long. An example of a binary encoded chromosome that has N_{var} variables, each encoded with $N_{gene} = 10$ bits, is

$$choromosome = [1110110010, 1010011010, \ldots, 1010001011] \quad [13.18]$$

The population

The GA starts with a group of chromosomes known as the population. The population has N_{pop} chromosomes, and it is a $N_{pop} \times N_{bit}$ matrix that is filled with random ones and zeros. Next, variables are passed to the cost function for evaluation.

Selection

The selection term refers to survival of the fittest, which translates into discarding the chromosomes with the highest cost. First, the N_{pop} costs and associated chromosomes are ranked from lowest cost to highest cost. Then only the best are selected to continue, while the rest are deleted. The selection rate, *Xrate*, is the fraction of N_{pop} that survives for the next step of mating. The number of chromosomes that are kept in each generation is N_{keep}.

Now its time to play matchmaker. In order to produce two new offspring, two chromosomes are selected from the mating pool of N_{keep} chromosomes. Pairing takes place in the mating population until N_{pop}- N_{keep} are born to replace the discarded chromosomes. As shown in Figure 13.20, pairing chromosomes in a GA can be as interesting and varied as pairing in an animal species.

Crossover

The crossover is a process that divides a string into segments; these segments are swapped with the segments corresponding to another string. This process generates two new strings, which are different from those that produced them. A crossover point is randomly selected between the first and last bits of the parents' chromosomes. Then each parent

Figure 13.20 Selection methods

is cut and recombined with a piece of the other. Many crossover techniques exist.

One-point crossover

A single crossover point on both parents' strings is selected. All data beyond that point in both strings is exchanged between the two parents, and the result is the offspring (Figure 13.21).

Two-point crossover

Two-point crossover calls for two points to be selected on the parents' strings. Everything between the two points is exchanged between the parents, rendering two offspring (Figure 13.22).

Cut and splice

Another crossover variant is the 'cut and splice' approach. This approach changes the length of the offspring strings. The reason for this difference

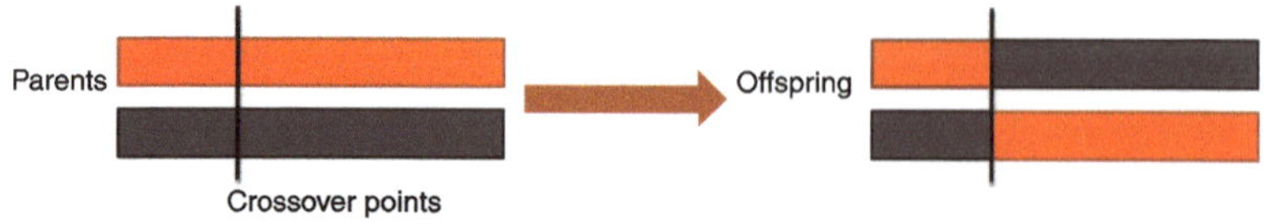

Figure 13.21 One-point crossover

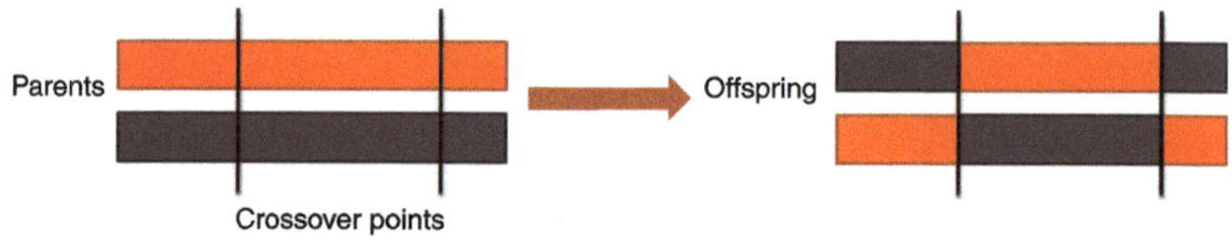

Figure 13.22 Two-point crossover

is that each parent string has a separate choice of crossover point (Figure 13.23).

Uniform crossover

The uniform crossover uses a fixed mixing ratio between two parents; unlike one- and two-point crossovers, the uniform crossover enables the parent chromosomes to contribute to the gene level rather than to the segment level (Figure 13.24).

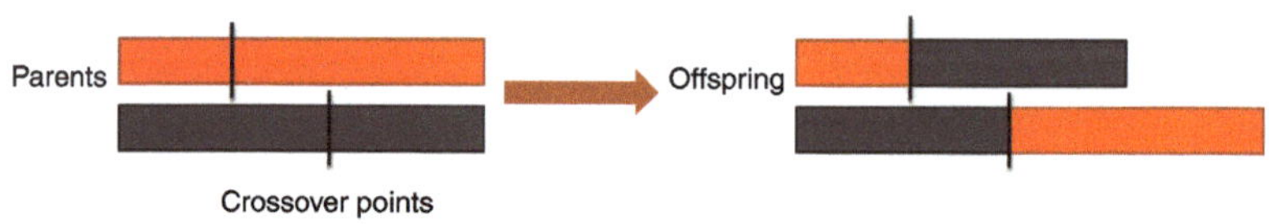

Figure 13.23 Cut and splice crossover

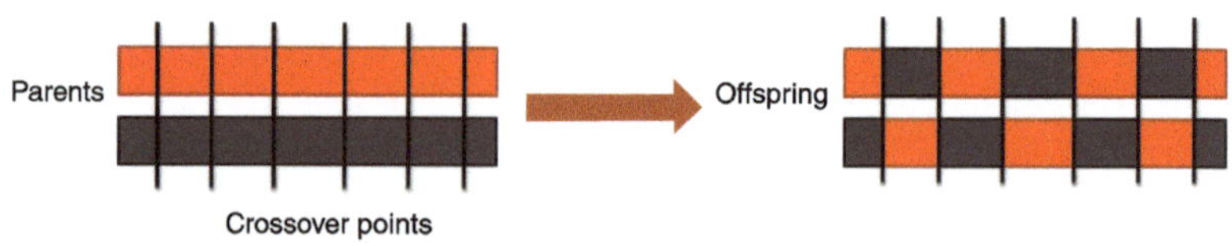

Figure 13.24 Uniform crossover

Mutations

A certain percentage of the bits in the list of chromosomes can be altered by random mutation. Mutation is the second way that a GA explores a cost surface. It can introduce traits not in the original population and keeps the GA from converging too fast before sampling the entire cost surface. A single point mutation changes a 1 to a 0, and vice versa. Mutation points are randomly selected from the $N_{po}N$ total number of bits in the population matrix. The algorithms freedom, for searching outside the current region of variable space, can be increased by increasing the number of mutations.

Convergence

After the mutations take place, the costs associated with the offspring and mutated chromosomes are calculated, which is an iterative process. The number of generations that evolve depends on whether an acceptable solution is reached or a set number of iterations are exceeded. After a while, all the chromosomes and associated costs would become the same if it were not for mutations. At this point the algorithm should be stopped.

The continuous genetic algorithm

The continuous GA is very similar to the binary GA. The main difference is that variables are no longer represented by bits of zero and one, but instead by floating-point numbers over whatever range is deemed appropriate.

Previous work

GA can be found in various research fields for parameter optimization. Genetically optimized neural network systems (GONNS) were developed to simulate the intelligent decision-making capability of humans [35]. GONNS was proposed for modelling and optimizing the FSW process. The GONNS use the BP-type ANNs to model the relationship between the inputs and the outputs of the system under consideration. Separate ANNs were used for estimation of each output. After the training of the ANNs, GA was used to find optimal solutions. One GA was used for searching the optimal TRS and feed rate by using five ANNs representing the FSW operation. The block diagram of the GONNS for this problem is presented in Figure 13.25. The proposed GONNS program was

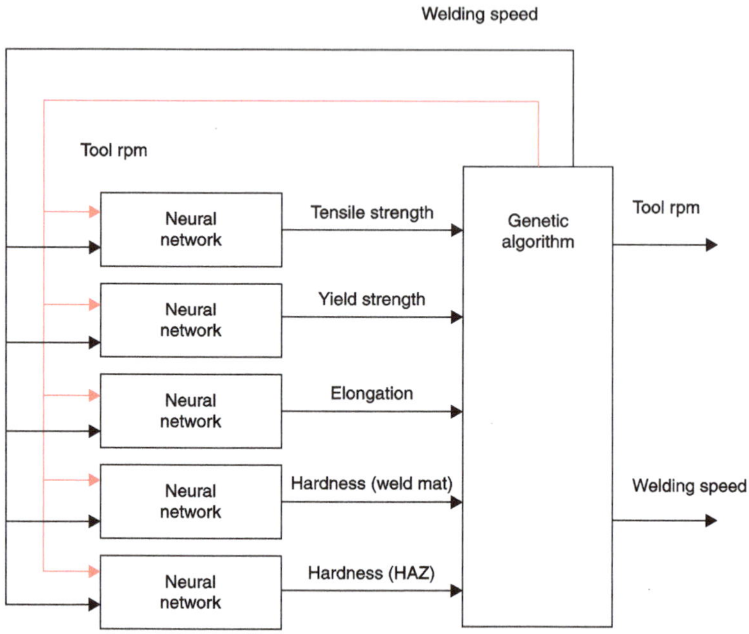

Figure 13.25 Genetically optimized neural network structure [35]

developed to minimize or maximize one of the outputs, while the other outputs are kept within the desired range.

An approach was presented for determination of the best cutting parameters leading to minimum surface roughness in end milling mould surfaces of an Ortez part used in biomedical applications by coupling neural network and GA [36]. A feed-forward neural network model is developed exploiting experimental measurements from the surfaces in the mould cavity. In their proposed method, GA coupled with a neural network was applied to find optimum cutting parameters leading to minimum surface roughness without any constraint.

Soft computing techniques were applied to optimize the powder metallurgy processing of pure iron [37]. An ANN was trained to predict the stress resulting from a given trend in strain and sintering temperature, and a GA was created to predict the optimum sintering temperature for a desired stress–strain behaviour.

Combining ANN and GA (ANN/GA) methods was proposed to optimize the injection moulding process [38]. They developed a BP neural network model to map the complex non-linear relationship between process conditions and quality indexes of the injection moulded parts, and a GA is used in the process conditions optimization with the fitness function based on an ANN model. Their results indicate that the combining ANN/GA method is an effective tool for the process optimization of injection molding.

ANNs and GA were employed to optimize laser-cutting parameters for QFN packages [39]. An ANN was employed to build a predicting model for cutting Quad Flat Non-lead (QFN) packages by using a Diode Pumped Solid State Laser System. The predicting model includes three input variables of the current, the frequency and the cutting speed, and six cutting qualities of depths of the cutting line, widths of HAZ and cutting line for epoxy and for copper-compounded epoxy (Figure 13.26).

The results show that the ANN model has the predicting ability to estimate the laser-cutting qualities of QFN packages. Eventually, a GA is employed to find the optimal cutting parameters that lead to the least HAZ width and fast cutting speed with complete cutting. The investigation claimed that GA is helpful to determine the ideal laser-cutting parameters, in order to meet the desired cutting qualities and to avoid unnecessary adjustments in the subsequent cutting process.

Synthesis of ANN and GA was applied to optimize chemical composition for satisfying mechanical properties for TC11 titanium alloy based on the large amount of experimental data [41]. Chemical elements were chosen as input parameters of the ANN model, and the output parameters were mechanical properties. The fitness function for GA was obtained from a trained ANN model. Their results strongly show that the proposed optimization model offers an optimal chemical composition for TC11 titanium alloy, which implies that it is a novel and effective approach for optimizing materials chemical composition.

Example of using GA method to optimize FSW

In order to investigate the optimal FSW parameters, the obtained neural network models in the previous sections are now deployed in an optimization procedure. In this section, GA is used to optimize the process conditions, and the fitness function, which is used in the process conditions optimization of FSW, is based on the ANN model. The outline of the combining ANN/GA optimization algorithm is given in Figure 13.27).

The objective of this example is to optimize ultimate tensile strength with respect to the design variables, namely *rotational speed (R)* and

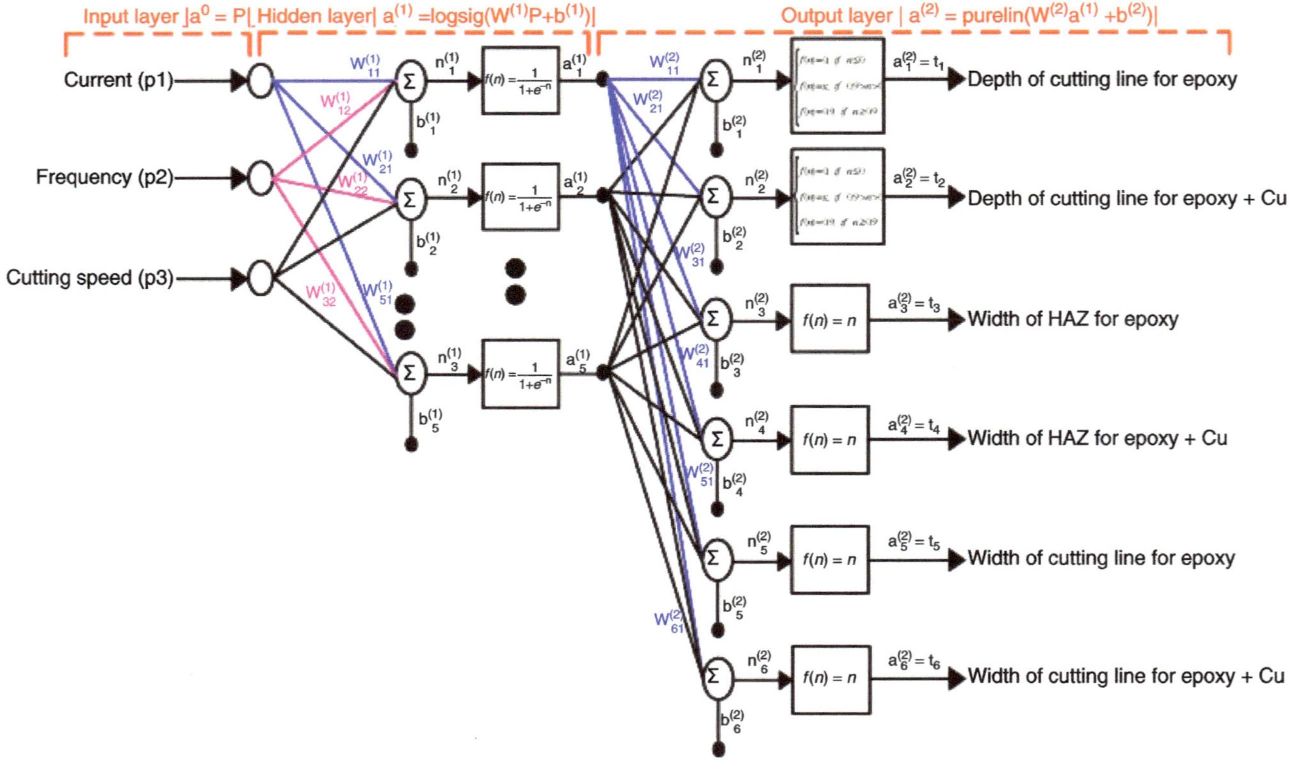

Figure 13.26 The ANN structure for predicting six laser-cutting qualities of a QFN package [39]

Figure 13.27 Flowchart of combining ANN/GA optimization

traverse speed (T). The optimization problem was formulated in the following form:

$$\begin{cases} Max\ Ultimate\ Tensile\ Strength\ (T,R) \\ 565 \leq Rotational\ Speed \leq 1400 \\ 20 \leq Traverse\ Speed \leq 90 \end{cases} \qquad [13.19]$$

The defined problem is optimized by real-coded GA. The population size is 50, and the maximum number of generations is 100. The evolution of the generations for FSW is shown in Figure 13.28. The optimized process conditions are traverse speed of 565 mm/min, rotational speed of 1400, and the neural network prediction of ultimate tensile strength under the optimized process conditions is 266.45.

13.3.3 Multi-objective optimization

Multi-objective optimization, which is also known as multi-criteria optimization or vector optimization, has been defined as mathematical optimization problems, which finds a vector of decision variables that satisfies constraints and gives acceptable values to all objective functions. In these problems, there is more than one objective or cost functions (a vector of objectives) to be optimized (minimized or maximized) simultaneously. These objectives often are in conflict with each other, so an improvement in one of them will deteriorate another. Therefore, there is no single optimal solution with respect to all the objective functions. Instead, there is a set of optimal solutions, known as Pareto optimal solutions or Pareto front [18,40].

Over the years, an important number of techniques have been produced by a substantial number of operational researchers to deal with multi-objective optimization problem. However, it was relatively new until research realized the potential of evolutionary algorithms (EA) in this area [42].

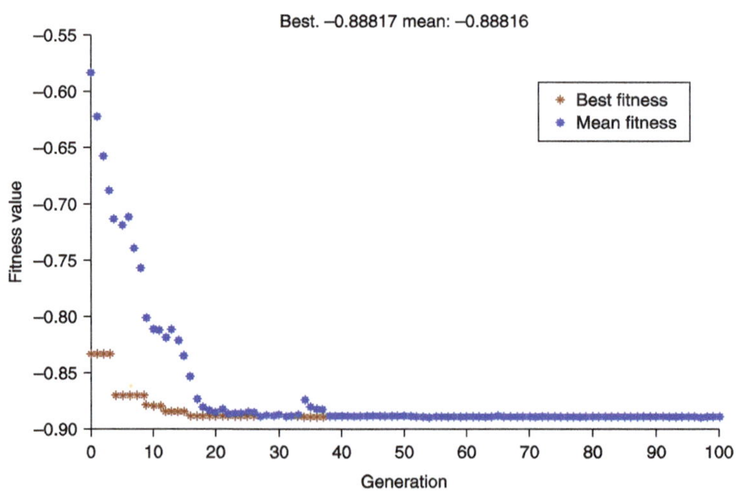

Figure 13.28 Evolution of generations of FSW

Basic concepts

A multi-objective optimization problem has a number of objective functions, which should be minimized or maximized. Also a single-objective optimization problem usually has a number of constraints that any feasible solution (including the optimal solution) must satisfy. In the following, we state the multi-objective optimization problem (MOOP) in its general form [43]:

Find the vector $X^* = [x_1^*, x_2^*, \ldots, x_n^*]^T$ to optimize:

$$F(X) = [f_1(X), f_2(X), \ldots, f_k(X)]^T, \qquad [13.20]$$

subject to m inequality constraints

$$g_i(X) \leq 0, \qquad i = 1 \text{ to } m \qquad [13.21]$$

and p equality constraints

$$h_j(X) = 0, \quad j = 1 \text{ to } p \qquad [13.22]$$

where $X^* \in \Re^n$ is the vector of decision or design variables, and $F(X) \in \Re^k$ is the vector of objective functions, both of which must be either minimized or maximized.

One of the noticeable differences between single-objective and multi-objective optimization is that in multi-objective optimization the objective functions, in addition to the usual decision variable space, constitute a multi-dimensional space. This additional space is called the objective space. For each solution to x in the decision variable space, there is a point in the objective space. The mapping takes place between an n-dimensional solution vector and an m-dimensional objective vector. Figure 13.29 shows decision and objective spaces, and also the mapping between them is visible [44].

The concept of domination is used in most multi-objective optimization algorithms. In these algorithms, two solutions are compared on the basis of whether one dominates the other solution or not. The following definition covers mixed problems with minimization of some objective functions and maximization of the rest of them.

Definition of Pareto dominance

A vector $U = [u_1, u_2, \ldots, u_k] \in \Re^k$ is said to dominate $V = [v_1, v_2, \ldots, v_k] \in \Re^k$ (denoted by $U < V$) if and only if U is partially less than V $\forall i \in \{1,2, \ldots, k\}, u_i \leq v_i \wedge \exists j \in \{1,2, \ldots, k\}: u_j < v_j$. In other words, there is at least one u_j that is smaller than v_j, whilst the remaining u_s are either smaller or equal to the corresponding v_s [45].

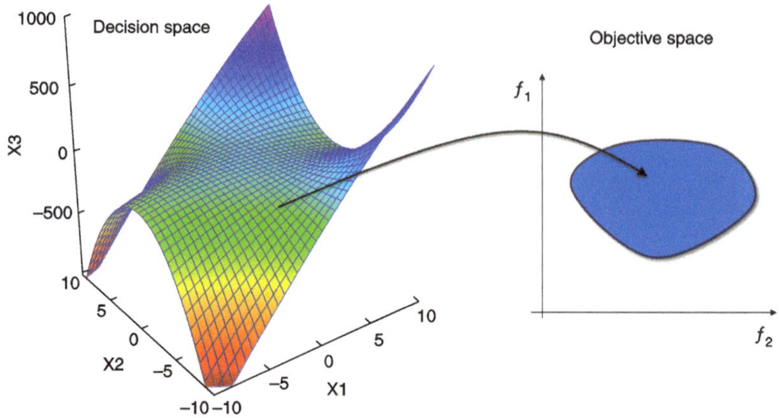

Figure 13.29 Decision and objective space

Consider a two-objective optimization problem with three different solutions, which is illustrated in the objective space (Figure 13.30). Assume that the objective function 1 and objective function 2 need to be maximized. Also three solutions with different objective function values are shown in this figure. While both objective functions are important to us, it is usually difficult to find the best solution with respect to both objectives. Nevertheless, the above definition of domination can be used to decide which solution is better among any two given solutions in terms of both objectives. For instance, if solutions 1 and 2 are compared, we observe that solution 1 is better than solution 2 in the first objective function, and solution 2 is better than solution 1 in the second objective function 2. Thus, solution 1 is non-dominated by solution 2. Whilst, if comparing solutions 2 and 3 m we can perceive that solution 3 is better than solution 2 in the first objective, and solution 3 is also better than solution 2 in the second objective. Thus, both of the above conditions for domination are also satisfied, and we may write that solution 3 dominates solution 1.

For a given finite set of solutions, all possible pair-wise comparisons can be performed, and it can be found which solution dominates which, and which solutions are non-dominated with respect to each other. In the end, we expect to have a set of solutions, any two of which do not dominate each other. This set is called the *non-dominated set* for the given set of solutions.

Definition of non-dominated set

Among a set of solutions *P*, the non-dominated set of solutions pi are those that are not dominated by any member of the set *P*.

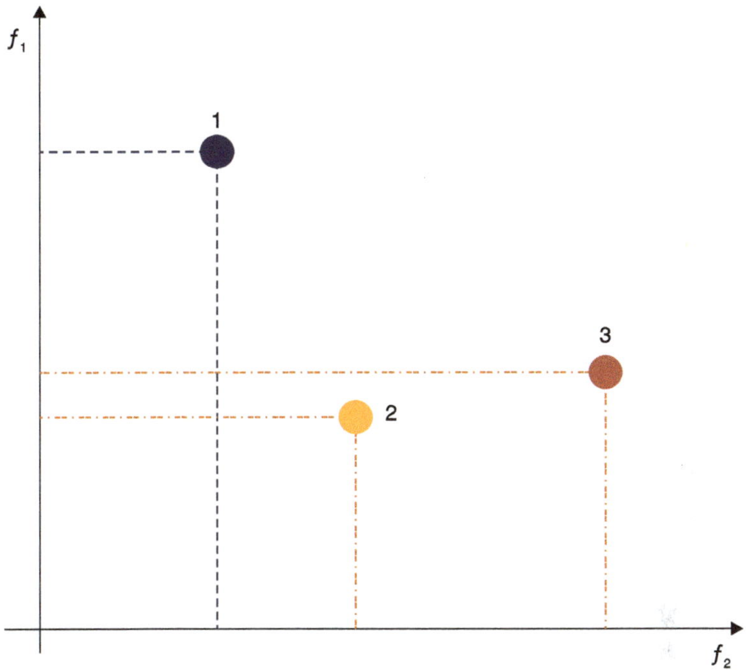

Figure 13.30 A population of three solutions

When the set P is the entire search space, or P-S, the resulting non-dominated set is called the *Pareto-optimal set*. Figure 13.31 marks the Pareto-optimal set with continuous curves for four different scenarios with two objectives. Each objective can be minimized or maximized. In Figure 13.31(a), the task is to minimize both objectives f_1 and f_2. The solid curve marks the Pareto-optimal solution set. If f_1 is to be minimized and f_2 is to be maximized for a problem having the same search space, the resulting Pareto-optimal set is different (Figure 13.31(b)). Here, the Pareto-optimal set is a union of two disconnected Pareto-optimal regions. Likewise, the Pareto-optimal sets for two other cases (maximizing f_1, minimizing f_2) and (maximizing f_1, maximizing f_2) are also shown in Figures 13.31(c and d), respectively. In any case, the Pareto-optimal set always consists of solutions from a particular edge of the feasible search region.

Definition of a Pareto set

The non-dominated set of the entire feasible search space S is the globally Pareto-optimal set.

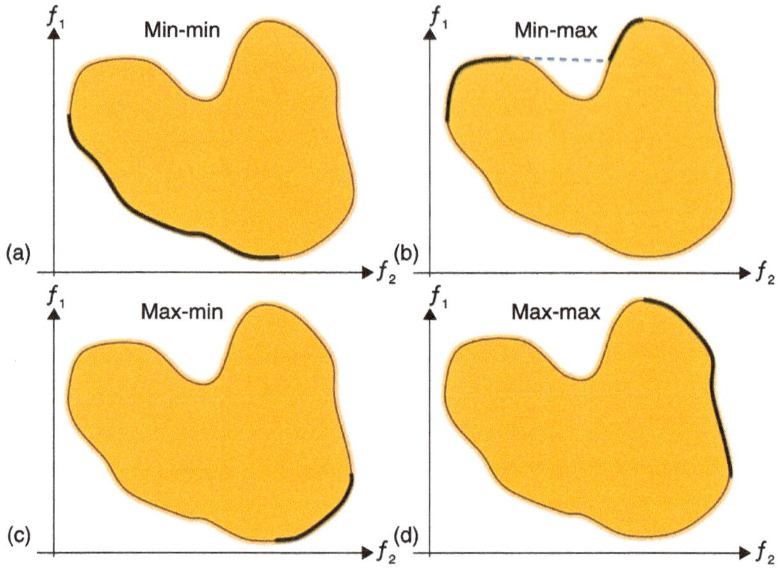

Figure 13.31 Pareto optimal solutions

The concept of Pareto front or set of optimal solutions in the space of objective functions in multi-objective optimization problems (MOOPs) stands for a set of solutions that are non-dominated to each other but are superior to the rest of solutions in the search space. It means that it is not possible to find a single solution that is superior to all other solutions with respect to all objectives, so that changing the vector of design variables in such a Pareto front consisting of these non-dominated solutions could not improve all objectives simultaneously. Therefore, such a change will worsen at least one objective. Thus, each solution of the Pareto set includes at least one objective inferior to another solution in that Pareto set, although both are superior to others in the rest of search space.

Classical methods

We have loosely shown all search and optimization algorithms as classical methods, primarily to distinguish them from evolutionary methods [44]. Classical optimization algorithms use a single solution update in every iteration, and that mainly use a deterministic transition rule. Classical multi-objective optimization methods have been used for at least the past four decades. During this period, many algorithms have also been

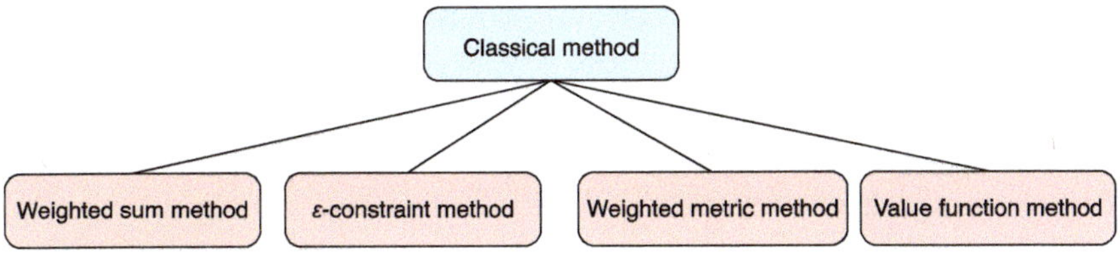

Figure 13.32 **Classical methods**

proposed, some of which are shown in Figure 13.32. Most of these algorithms convert the MOOP problem into a single objective optimization problem by using some user-defined procedures.

Classical optimization methods are divided into two distinct groups: direct methods and gradient-based methods [46]. In direct search methods, only the objective function and the constraint values are used to guide the search strategy, whereas gradient-based methods use the first- and/or second-order derivatives of the objective function and/or constraints to guide the search process. The direct search methods are slow, and require many function evaluations for convergence as long as derivative information is not used. For the same reason, these can also be applied to many problems without a major change in the algorithm. However, gradient-based methods quickly converge near an optimal solution, but are not efficient in non-differentiable or discontinuous problems. Furthermore, some common difficulties with most classical direct and gradient-based techniques are mentioned as follows:

- The convergence to an optimal solution depends on the chosen initial solution.
- Most algorithms tend to get *stuck* to a suboptimal solution.
- An efficient algorithm in solving one optimization problem may not necessarily be efficient in solving a different one.
- Algorithms cannot efficiently handle the problems with a discrete search space.

Evolutionary algorithms

Evolutionary algorithms (EAs) use mechanisms inspired by biological evolutionary principles to constitute search and optimization procedures. EAs are different from classical search and optimization procedures in a variety of respects. EAs especially solve multi-objective optimization problems, because these algorithms simultaneously consider sets of possible solutions. Therefore, several members of the Pareto optimal set can be found in a single run, instead of performing a series of separate runs as in the case of traditional mathematical programming technique. Furthermore, EAs are less delicate to the shape or continuity of the Pareto front, whilst these two issues are a main concern for the classical method.

Historical

Rosenberg reported the early use of evolutionary search in 1960s. After that, different EAs for MOPs were introduced. Among the proposed methods, the most important ones are the vector evaluated GA (VEGA) [47], GA (MOGA) [48], non-dominated sorting GA (NSGA) [49], the strength Pareto evolutionary algorithm (SPEA) [50], the Pareto archived evolution strategy (PAES) [51] and the Niched-Pareto GA (NPGA) [52].

Structures of various MOEAs

A wide variety of MOEAs have been proposed in the last few years. Nevertheless, providing a comprehensive survey of them is not the purpose of this section as long as these reviews are available elsewhere. Actually, this section provides a brief discussion of a relatively small set of MOEAs that are representative of the conducted research in this area.

Vector evaluated genetic algorithm

The first multi-objective GA was implemented to find a set of non-dominated solutions by [47,53]. Subsequently, Schaffer compared his proposed GA with an adaptive random search technique, and obtained a better performance by his algorithm. This GA was called the *vector evaluated GA*, or VEGA. The name was chosen appropriately for multi-objective optimization, because this GA evaluated an objective vector (instead of a scalar objective function) with each element of the vector representing each objective function. The simplest possible GA is VEGA. In fact VEGA is a straightforward expansion of a single-objective GA for multi-objective optimization. As a number of objectives (say M) should be satisfied, Schaffer considered the possibility of dividing the GA population at every generation into M equal sub-populations randomly. Each sub-population is assigned a fitness according to a different objective function. This method provides the possibility of using each of the M objective functions to evaluate some members in the population.

Multiple-objective genetic algorithm

A multi-objective GA (called MOGA) was introduced for the first time in 1993 [48]. The MOGA used the non-dominated classification of a GA population. A MOGA explicitly emphasizes non-dominated solutions and simultaneously maintains diversity in the non-dominated solutions.

The MOGA differs from a standard tripartite GA in how fitness is assigned to each solution in the population. The rest aspects of the algorithm (the stochastic universal selection (SUS), a single-point crossover, and a bit-wise mutation) are the same as the classical GA.

Non-dominated sorting genetic algorithm (NSGA)

In 1994, Goldberg's idea of using the non-dominated sorting concept in gas was directly implemented [49]. This method is based on several layers of classified individuals. Before selection is carried out, the population is ranked on the basis of non-domination: all non-dominated individuals constitute one category with a dummy fitness value, which is directly proportional to the population size for providing an equal reproductive potential for these individuals. These classified individuals are shared with their dummy fitness values to maintain the population diversity. Afterwards this group of classified individuals is ignored, and another layer of non-dominated individuals is considered. Since all individuals in the population are classified, the process continues. Stochastic remainder proportionate selection is adopted for this technique. The individuals in the first front always get more copies than the rest of the population until they have the maximum fitness value. An offshoot of this approach, the NSGA-II, uses elitism and a crowded comparison operator that ranks the population on the basis of both Pareto dominance and region density. This crowded comparison operator makes the NSGA-II substantially faster than its predecessor, while providing very good results.

Niched-Pareto genetic algorithm

A multi-objective GA has been proposed according to the non-domination concept [52]. The selection operator of the proposed method is different from the previous methods. The niched Pareto genetic algorithm (NPGA) uses the binary tournament selection, unlike the proportionate selection method used in VEGAs, NSGAs and MOGAs. Their choice of tournament selection over proportionate selection is motivated by the theoretical studies on the selection operators used in single-objective GAs. The study has shown that the tournament selection has better growth and convergence properties in comparison to proportionate selection. However, there is one difficulty when a tournament selection (or any other non-proportionate selection) is applied to MOGAs, which uses the sharing function approach.

Pareto archived evolution strategy (PAES)

The Pareto Archived Evolution Strategy (PAES) was designed and implemented in 1999 [51]. The concept of proposed approach is straightforward [54]. PAES consists of a (1 + 1) evolution strategy (i.e. a single parent that generates a single offspring) in combination with a historical archive, which records some of the non-dominated solutions previously found. This archive is used as a reference set against which each mutated individual is compared. This is similar to the tournament competitions. PAES also uses a novel approach to retain diversity. This approach consists of a crowding procedure that divides objective space in a recursive manner. Each solution is placed in a certain grid location according to the values of its objectives. A map of such a grid is maintained, which indicates the number of solutions that reside in each grid location. No extra parameters are required as long as the procedure is adaptive. Moreover, the computational complexity of this procedure is lower than the traditional niching methods.

Previous works

Mechanical properties of FSW AA7075/AA5083 butt joints were modelled and optimized by applying neural network and particle swarm algorithms [28]. The neural network models are used in a multi-objective optimization procedure to gain optimal ultimate tensile strength and hardness. In this study, ultimate tensile strength and Al average hardness are two conflicting objectives, which should be simultaneously optimized with respect to the design variables, namely *rotational speed (R)* and *traverse speed (T)*.

A group method of data handling (GMDH) type neural network and EAs for modelling the effects of intake valve-timing and engine speed of a spark-ignition engine on both developed engine torque and fuel consumption using some experimentally obtained training and test data [43]. Using such obtained polynomial neural network models, a multi-objective EA (NSGA-II) is used for Pareto-based optimization of the variable valve-timing engine considering two conflicting objectives such as torque and fuel consumption. The optimal Pareto front of two objectives (Fc) and (T) is shown in Figure 13.33.

The complex electrical discharge machining (EDM) process was attempted to model and optimize using soft computing techniques [55]. The ANN with a back propagation algorithm is used to model the process. As the output parameters are conflicting in nature, there is no single

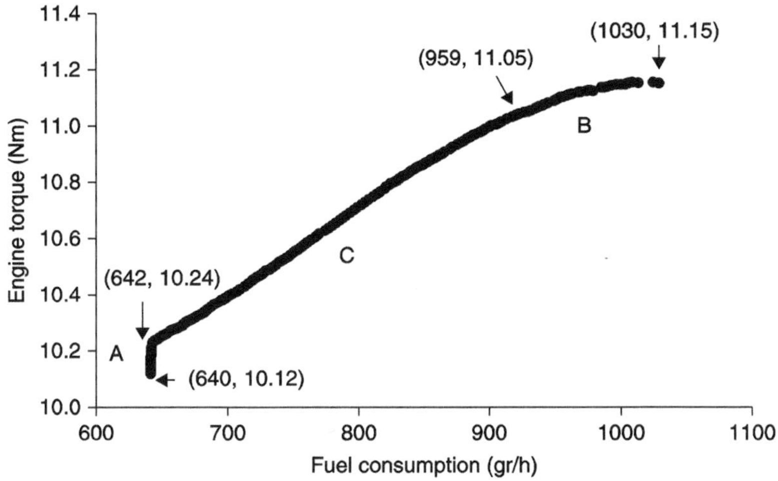

Figure 13.33 Optimal Pareto front of conflicting objective functions engine torque and fuel economy based on GMDH type model of experimental data [43]

combination of cutting parameters, which provides the best machining performance. A multi-objective optimization method and non-dominating sorting genetic algorithm-II is used to optimize the process. The objectives of this study are maximizing the MRR and minimizing the tool wear. Their non-dominated solution set is shown in Figure 13.34.

A multi-objective optimization study of the gas cyclone separator was performed in 2012 [56]. Two radial basis function neural networks (RBFNNs) were developed and employed to model the pressure drop and the cut-off diameter for cyclone separators, for the purpose of accurately predicting the complex nonlinear relationships between the performance parameters (pressure drop and cut-off diameter) and the geometrical dimensions. They used two trained ANNs were used as two objective functions to obtain new optimal ratios for minimum pressure drop and minimum cut-off diameter using the MOGA optimization technique.

Multi-objective optimization of cyclone separators at three phases was performed in 2011 [57]. At the first phase, pressure drop and the cut-point in a set of cyclone separators are numerically investigated using CFD techniques. At the second phase, two meta-models based on the evolved group method of data handling (GMDH) type neural networks are obtained for modelling cyclone parameters. Finally, using obtained polynomial neural networks, MOGAs are used for Pareto-based

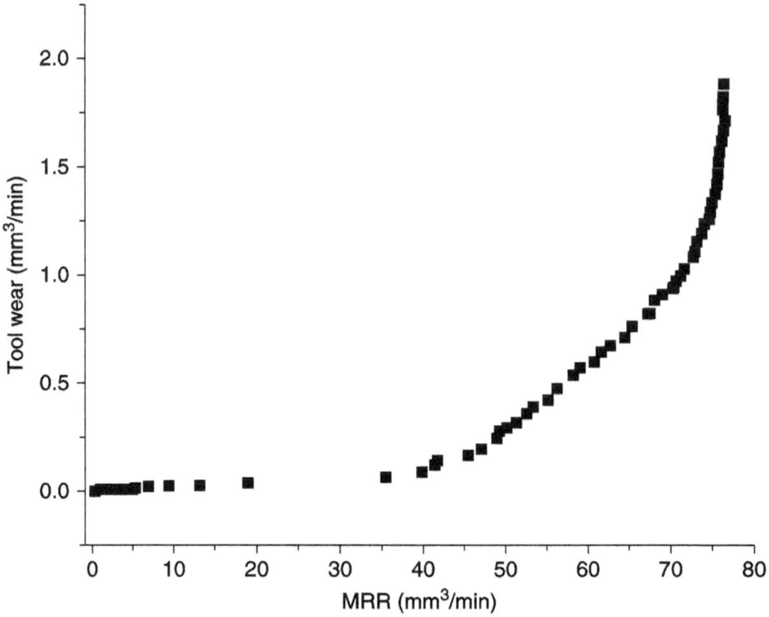

Figure 13.34 Pareto optimal set [10]

optimization of cyclone separators considering two conflicting (Figure 13.35).

Application of multi objective optimization in FSW- an example

In order to handle this multi-objective optimization problem, a hybrid multi-objective EA is proposed. This proposed method consists of two steps: generation of a Pareto set by NSGA-II, and using TOPSIS to obtain the best compromise solution from the Pareto set.

Multi-objective optimization of friction stir parameter using NSGA-II method

The neural network models described above were used in a multi-objective optimization procedure for the purpose of obtaining optimal ultimate tensile strength and hardness. Two conflicting objectives in this case study were ultimate tensile strength and average hardness, to be simultaneously optimized with respect to the design variables, namely *rotational speed (R)* and *traverse speed (T)*. The optimization problem was formulated in the following form:

$$\begin{cases} \max \textit{Hardness}\ (T,R) \\ \max \textit{Tensile shear force}\ (T,R) \\ 565 \le T \le 1600 \\ 11 \le \textit{Tensile shear force} \le 90 \end{cases} \qquad [13.23]$$

The corresponding Pareto front of two objectives, hardness and traverse speed, is shown in Figure 13.36. This figure clearly indicates that choosing appropriate values for rotational speed and traverse speed for gaining a better value of one objective would worsen the value of another objective.

TOPSIS to determine the best compromise solution

The given alternatives of the Pareto solutions, obtained by NSGA II, are ranked by TOPSIS. The basic concept of TOPSIS determines the positive

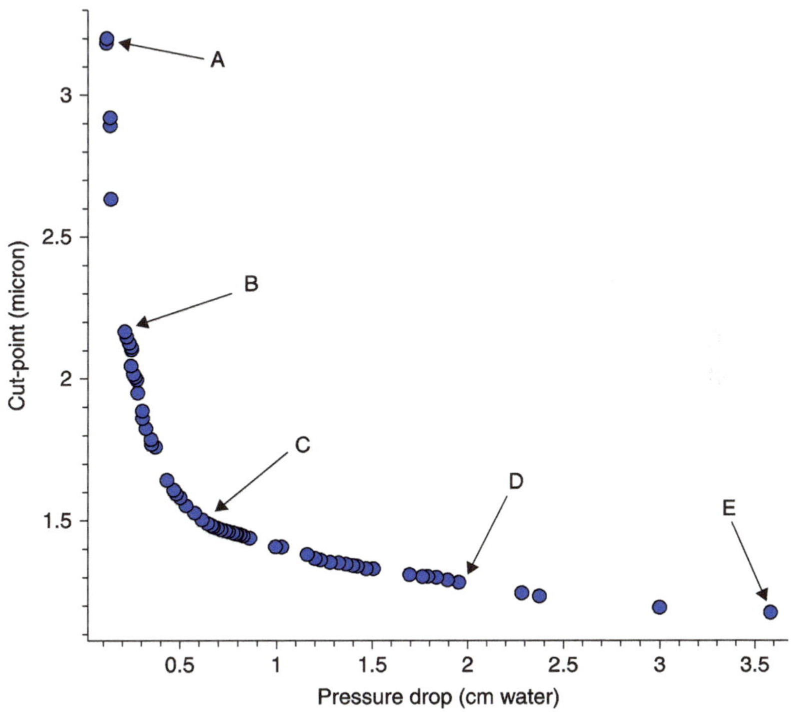

Figure 13.35 Multi-objective Pareto result for cyclone separators [57]

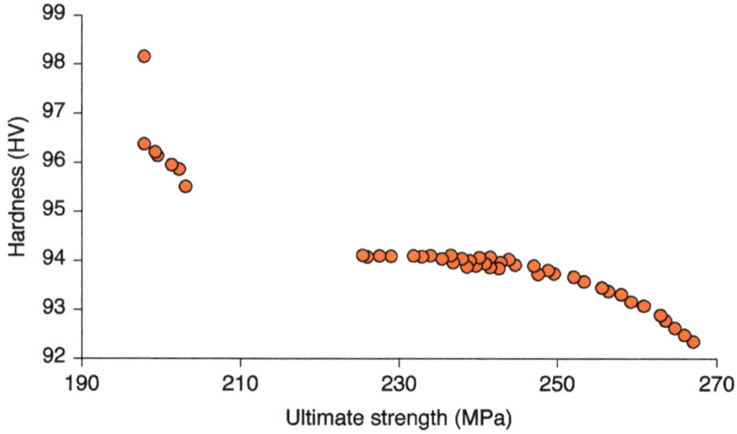

Figure 13.36 Optimization result (Pareto front)

ideal solution (S^+) as well as the negative ideal solution (S^-). TOPSIS, on the basis of chosen alternative, should have the shortest distance from S^+ and the longest distance from S^-. The alternative is chosen from the Pareto set according to the decision-maker's objective weights. The process of TOPSIS to determine the best compromise solution is carried out as follows [58]:

a) Input S, where the element *j*th, s_{ij} is the objective value of the *i*th alternative (i.e. S is composed of the Pareto solutions).

b) Calculate normalized rating (vector normalization is used) $\hat{s}_{ij}$..

c) Develop a set of importance weights (*W*) for each of the criteria. The basis for these weights can be anything, but usually is the expert's reflection of relative importance.

$$\sum_{j=1}^{2} w_j = 1 \qquad [13.24]$$

and

$$\hat{s}_{ij} = w_j \times \tilde{s}_{ij} \qquad [13.25]$$

d) Determine *S*+ and *S*– as follow:

$$S^+ = \{(max\ \hat{s}_{ij}\ |j \in J_1),\ (min\ \hat{s}_{ij}\ |j \in J_2),\ i = 1, 2, \ldots, n\} \qquad [13.26]$$

and

$$S^- = \{(min\ s_{ij}^{\wedge} | j \in J_1), (max\ s_{ij}^{\wedge} | j \in J_2), i = 1, 2, \ldots, n\} \qquad [13.27]$$

where J_1 is a set of benefit attributes and J_2 is a set of cost attributes.

e) Develop a distance measure over each criterion to both ideal (D^+) and nadir (D^-):

$$D_i^+ = \sqrt{\sum_j (s_{ij}^{\wedge} - s_j^+)^2} \qquad [13.28]$$

and

$$D_i^- = \sqrt{\sum_j (s_{ij}^{\wedge} - s_j^-)^2} \qquad [13.29]$$

f) Calculate relative closeness D_i for each Pareto solution according to following equation:

$$D_i - \frac{D^-}{D^- + D^+} \qquad [13.30]$$

g) Choose the best compromise solution whose relative closeness *Di* is the closest to 1.

After acquiring the Pareto set, one compromise solution must be determined to implement the assignment by a decision-maker. Assuming an equal weight for each objective, the solution having the highest TOPSIS score is the compromise solution. In this experiment under W = (0.5, 0.5), the best compromise solution has a rotational speed of 1182.11 and traverse speed of 11 mm/min.

13.4 References

1. Haupt, R.L. and Haupt, S.E. (2004), *Practical Genetic Algorithms*, John Wiley.
2. McCulloch, W. and Pitts, W. (1943), 'A logical calculus of the ideas immanent in nervous activity', *The Bulletin of Mathematical Biophysics*, 5(4): 115–33.
3. Rosenblatt, F. (1962), *Principles of Neurodynamics: Perceptrons and the Theory of Brain Mechanisms*, Spartan Books.
4. Minsky, M.L. and Papert, S. (1972), *Perceptrons: An Introduction to Computational Geometry*, MIT Press.
5. Hopfield, J.J. (1982). 'Neural networks and physical systems with emergent collective computational abilities', *Proceedings of the National Academy of Sciences*, 79(8): 2554–8.

6. Rumelhart, D.E., MacClelland, J.L. and S.D.P.R G. University of California (1986), 'Parallel Distributed processing: Explorations in the microstructure of cognition', *Psychological and Biological Models*, vol. 2, MIT Press.
7. Haykin, S.S. (1999), *Neural Networks: A Comprehensive Foundation*, Prentice Hall.
8. Hagan, M.T., Demuth, H.B. and Beale, M.H. (2002), *Neural Network Design*, Campus Publishing Service, Colorado University Bookstore, University of Colorado at Boulder.
9. Anthony, M. and Bartlett, P.L. (2009), *Neural Network Learning: Theoretical Foundations*, Cambridge: Cambridge University Press.
10. Mendel, J.M. and Fu, K.S. (1970), *Adaptive, Learning, and Pattern Recognition Systems: Theory and Applications*, Academic Press.
11. Kröse, B. and van der Smagt, P. (1996), *An Introduction to Neural Networks*, University of Amsterdam.
12. Anderson, J.A. and Rosenfeld, E. (1989), *Neurocomputing Foundations of Research*, MIT.
13. Muthukrishnan, N. and Davim, J.P. (2009), 'Optimization of machining parameters of Al/SiC-MMC with ANOVA and ANN analysis', *Journal of Materials Processing Technology*, 209(1): 225–32.
14. Debroy, T., De, A., Bhadeshia, H.K.D.H., Manvatkar, V.D. and Arora, A. (2012), 'Tool durability maps for friction stir welding of an aluminium alloy', *Proceedings of the Royal Society A: Mathematical, Physical and Engineering Sciences*, 468(2147): 3552–70.
15. Manvatkar, V.D., Arora, A., De, A. and DebRoy, T. (2012), 'Neural network models of peak temperature, torque, traverse force, bending stress and maximum shear stress during friction stir welding', *Science and Technology of Welding and Joining*, 17(6): 460–6.
16. Buffa, G., Fratini, L. and Micari, F. (2012), 'Mechanical and microstructural properties prediction by artificial neural networks in FSW processes of dual phase titanium alloys', *Journal of Manufacturing Processes*, 14(3): 289–96.
17. Fratini, L., Buffa, G. and Palmeri, D. (2009), 'Using a neural network for predicting the average grain size in friction stir welding processes', *Computers and Structures*, 87(17–18): 1166–74.
18. Shojaeefard, M.H., Behnagh, R.A., Akbari, M., Givi, M.K.B. and Farhani, F. (2013), 'Modelling and Pareto optimization of mechanical properties of friction stir welded AA7075/AA5083 butt joints using neural network and particle swarm algorithm', *Materials & Design*, 44(0): 190–8.
19. Asadi, P., Givi, M.K.B., Rastgoo, A., Akbari, M., Zakeri, V. and Rasouli, S. (2012), 'Predicting the grain size and hardness of AZ91/SiC nanocomposite by artificial neural networks', *The International Journal of Advanced Manufacturing Technology*, 63(9–12): 1095–107.
20. Shojaeefard, M.H., Akbari, M., Tahani, M. and Farhani, F. (2013), 'Sensitivity analysis of the artificial neural network outputs in friction stir lap joining of aluminum to brass', *Advances in Materials Science and Engineering*, 2013: 7.
21. Okuyucu, H., Kurt, A. and Arcaklioglu, E. (2007), 'Artificial neural network application to the friction stir welding of aluminum plates', *Materials & Design*, 28(1): 78–84.

22. Koilraj, M., Sundareswaran, V., Vijayan, S. and Koteswara Rao, S.R. (2012), 'Friction stir welding of dissimilar aluminum alloys AA2219 to AA5083 – Optimization of process parameters using Taguchi technique', *Materials & Design*, 42(0): 1–7.
23. Roy, R.K. (2010), *A Primer on the Taguchi Method*, Society of Manufacturing Engineers.
24. Ross, P.J. (1996), *Taguchi Techniques for Quality Engineering: Loss Function, Orthogonal Experiments, Parameter and Tolerance Design*, McGraw-Hill Professional Publishing.
25. Wu, Y. and Wu, A. (2000), *Taguchi Methods for Robust Design*, New York: The American Society of Mechanical Engineers.
26. Murali Krishna, P., Ramanaiah, N. and Prasada Rao, K. (2013), 'Optimization of process parameters for friction stir welding of dissimilar Aluminum alloys (AA2024-T6 and AA6351-T6) by using Taguchi method', *International Journal of Industrial Engineering Computations*, 4(1): 71–80.
27. Venkateswarlu, G., Davidson, M.J. and Tagore, G.R.N. (2012), 'Taguchi optimisation of friction stir processing parameters to achieve maximum tensile strength of Mg AZ31B Alloy', *Transactions of the Indian Institute of Metals*, 65(5): 491–6.
28. Arora, H.S., Singh, H. and Dhindaw, B.K. (2012), 'Parametric study of friction stir processing of magnesium-based AE42 alloy', *Journal of Materials Engineering and Performance*, 21(11): 2328–39.
29. Salehi, M., Saadatmand, M. and Aghazadeh Mohandesi, J. (2012), 'Optimization of process parameters for producing AA6061/SiC nanocomposites by friction stir processing', *Transactions of Nonferrous Metals Society of China (English Edition)*, 22(5): 1055–63.
30. Bozkurt, Y. (2012), 'The optimization of friction stir welding process parameters to achieve maximum tensile strength in polyethylene sheets', *Materials and Design*, 35: 440–5.
31. Suresha, C.N., Rajaprakash, B.M. and Upadhya, S. (2011), 'A study of the effect of tool pin profiles on tensile strength of welded joints produced using friction stir welding process', *Materials and Manufacturing Processes*, 26(9): 1111–16.
32. Chien, C.H., W. Lin, W.B. and Chen, T. (2011), 'Optimal FSW process parameters for aluminum alloys AA5083', *Journal of the Chinese Institute of Engineers, Transactions of the Chinese Institute of Engineers, Series A/ Chung-kuo Kung Cheng Hsuch Kan*, 34(1): 99–105.
33. Vijayan, S., Raju, R., Subbaiah, K., Sridhar, N. and Rao, S.R.K. (2010), 'Friction stir welding of AL-MG alloy optimization of process parameters using Taguchi method', *Experimental Techniques*, 34(5): 37–44.
34. Holland, J.H. (1975), *Adaptation in Natural and Artificial Systems: An Introductory Analysis with Applications to Biology, Control, and Artificial Intelligence*, University of Michigan Press.
35. Tansel, I., Demetgul, M., Okuyucu, H. and Yapici, A. (2010), 'Optimizations of friction stir welding of aluminum alloy by using genetically optimized neural network', *The International Journal of Advanced Manufacturing Technology*, 48(1–4): 95–101.

36. Oktem, H., Erzurumlu, T. and Erzincanli, F. (2006), 'Prediction of minimum surface roughness in end milling mold parts using neural network and genetic algorithm', *Materials & Design*, 27(9): 735–44.
37. Reihanian, M., Asadullahpour, S.R., Hajarpour, S. and Gheisari, K. (2011), 'Application of neural network and genetic algorithm to powder metallurgy of pure iron', *Materials & Design*, 32(6): 3183–8.
38. Shen, C., Wang, L. and Li, Q. (2007), 'Optimization of injection molding process parameters using combination of artificial neural network and genetic algorithm method', *Journal of Materials Processing Technology*, 183(2–3): 412–18.
39. Tsai, M.-J., Li, C.-H. and Chen C.-C. (2008), 'Optimal laser-cutting parameters for QFN packages by utilizing artificial neural networks and genetic algorithm', *Journal of Materials Processing Technology*, 208(1–3): 270–83.
40. Amanifard, N., Nariman-Zadeh, N., Borji, M., Khalkhali, A. and Habibdoust, A. (2008), 'Modelling and Pareto optimization of heat transfer and flow coefficients in microchannels using GMDH type neural networks and genetic algorithms', *Energy Conversion and Management*, 49(2): 311–25.
41. Sun, Y., Zeng, W.D., Han, T.F., Ma, X. and Zhao, Y.Q. (2011), 'Optimization of chemical composition for TC11 titanium alloy based on artificial neural network and genetic algorithm', *Computational Materials Science*, 50(3): 1064–9.
42. Deb, K., Pratap, A., Agarwal, S. and Meyarivan, T. (2002). 'A fast and elitist multiobjective genetic algorithm: NSGA-II', *IEEE Transactions on Evolutionary Computation*, 6(2): 182–97.
43. Atashkari, K., Nariman-Zadeh, N., Gölcü, M., Khalkhali, A. and Jamali, A. (2007), 'Modelling and multi-objective optimization of a variable valve-timing spark-ignition engine using polynomial neural networks and evolutionary algorithms', *Energy Conversion and Management*, 48(3): 1029–41.
44. Deb, K. (2001), *Multi-Objective Optimization using Evolutionary Algorithms*, Wiley.
45. Coello, C.A.C., Lamont, G.B. and van Veldhuizen, D.A. (2007), *Evolutionary Algorithms for Solving Multi-Objective Problems*, Springer.
46. Deb, K. and Beyer, H.-G. (2001), 'Self-Adaptive Genetic Algorithms with Simulated Binary Crossover', *Evol. Comput.*, 9(2): 197–221.
47. Schaffer, J.D. (1985), 'Multiple objective optimization with vector evaluated genetic algorithms', *Proceedings of the 1st International Conference on Genetic Algorithms*, L. Erlbaum Associates Inc., 93–100.
48. Fonseca, C.M. and Fleming, P.J. (1993), 'Genetic algorithms for multi-objective optimization: formulation, discussion and generalization', *Proceedings of the 5th International Conference on Genetic Algorithms*, San Francisco, CA, Morgan Kaufmann Publishers Inc.
49. Srinivas, N. and Deb, K. (1994), 'Muiltiobjective optimization using nondominated sorting in genetic algorithms', *Evolutionary Computation*, 2(3): 221–48.
50. Zitzler, E. and Thiele, L. (1999), 'Multiobjective evolutionary algorithms: A comparative case study and the strength Pareto apporach', *Proceedings of the IEEE Transactions on Evoluationary Computation*, 3(4): 257–71.

51. Knowles, J. and Corne, D. (1999), 'The Pareto archived evolution strategy: A new baseline algorithm for pareto multi-objective optimization', *Proceedings of the Congress on Evolutionary Computation*, IEEE Press.
52. Horn, J., Nafpliotis, N. and Goldberg, D.E. (1994), 'A niched Pareto genetic algorithm for multiobjective optimization: Evolutionary computation', *Proceedings of the First IEEE Conference on Computational Intelligence.* IEEE Press.
53. Schaffer, J.D. and Grefenstette, J.J. (1985). 'Multi-objective learning via genetic algorithms', *Proceedings of the 9th International Joint Conference on Artificial Intelligence*, vol. 1, Los Angeles, CA: Morgan Kaufmann Publishers Inc.: 593–5.
54. Coello, C.A. and Christiansen, A.D. (2000), 'Multiobjective optimization of trusses using genetic algorithms', *Computers & Structures*, 75(6): 647–60.
55. Mandal, S., Rice, J. and Elmustafa, A.A. (2008), 'Experimental and numerical investigation of the plunge stage in friction stir welding', *Journal of Materials Processing Technology*, 203(1–3): 411–19.
56. Elsayed, K. and Lacor, C. (2012), 'Modeling and Pareto optimization of gas cyclone separator performance using RBF type artificial neural networks and genetic algorithms', *Powder Technology*, 217(0): 84–99.
57. Safikhani, H., Hajilo, A. and Ranjbar, M.A. (2011). 'Modeling and multi-objective optimization of cyclone separators using CFD and genetic algorithms', *Computers & Chemical Engineering*. 35(6): 1064–71.
58. Lin, Y.-K. and Yeh, C.-T. (2012), 'Multi-objective optimization for stochastic computer networks using NSGA-II and TOPSIS', *European Journal of Operational Research*, 218(3): 735–46.

14

Friction stir welding/processing of polymeric materials

M. Barmouz and P. Shahi, Tarbiat Modares University, Iran and P. Asadi, University of Tehran, Iran

DOI: 10.1533/9780857094551.601

Abstract: Friction stir welding and processing (FSW/FSP) are relatively new solid-state joining and processing techniques. In particular, they have been used to join and process high-strength aerospace aluminum alloys and other metallic alloys that are hard to weld by conventional fusion welding. Recently, FSW/FSP were developed for joining and processing of polymeric materials. In this chapter, the current understanding and development of the FSW and FSP of polymers and polymer composites are addressed. Particular emphasis has been given to mechanisms responsible for the formation of welds and composites and the effects of FSW/FSP parameters on resultant microstructure and final mechanical properties and appearances.

Key words: friction stir welding, friction stir processing, polymer composites, microstructure, mechanical properties.

14.1 Introduction

Prior to the early 1920s, chemists doubted the existence of molecules having molecular weights greater than a few thousand. This limited view was challenged by Hermann Staudinger, a German chemist with experience in studying natural compounds such as rubber and cellulose. In contrast to the prevailing rationalization of these substances as aggregates of small molecules, Staudinger proposed that they were made up of macromolecules composed of 10 000 or more atoms. He formulated a polymeric structure for rubber, based on a repeating isoprene unit (referred to as a monomer). For his contributions to chemistry, Staudinger received a Nobel Prize in 1953. The terms 'polymer' and 'monomer' are derived from the Greek roots, poly (many), mono (one) and meros (part) [1–3].

Synthetic polymers are produced by chemical reactions, termed 'polymerizations'. Polymerizations occur in various forms, far too many to explore here, but such reactions consist of the repetitive chemical bonding of individual molecules, or monomers. Assorted combinations of heat, pressure and catalysis alter the chemical bonds that hold monomers together, causing them to bond with one another. Most often, they do so in a linear fashion, creating chains of monomers called polymers [2,4].

Some polymerizations join entire monomers together, whereas others join only portions of monomers and create 'leftover' materials or by-products. Co-polymers can be formed using two or more different monomers. Two or more polymers can be combined to produce an alloy or blend that displays characteristics of each component [2].

As an example, let us consider the common plastic polyethylene, which is found in such items as grocery bags, toys and bottles. The monomer ethylene is composed of two carbon atoms, each bonded to two hydrogen atoms and sharing a double bond with one another. Polyethylene consists of a chain of single-bonded carbon atoms, each still carrying its two hydrogen atoms. Figure 14.1 shows the structure of linear polyethylene.

One way to produce polyethylene is by what is called 'free radical polymerization'. As in other polymerizations, the process has three stages, initiation, propagation and termination. First, we need to add a catalyst to our supply of ethylene. A common catalyst is benzoyl peroxide, which when heated has the habit of splitting into two fragments, each with one unpaired electron, or a free radical. These fragments are known as initiator fragments [1,2,5,6]. The unpaired electron naturally seeks

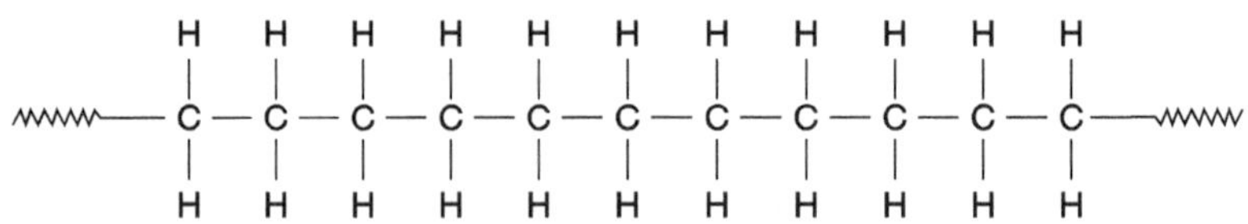

Figure 14.1 Linear polyethylene [4]

Figure 14.2 Synthetic macromolecular architectures: (a) linear; (b) branches; and (c) cross-linked [4]

another and finds a convenient target in the double bond between the carbon atoms in the ethylene molecule. Taking an electron from the carbon bond, the initiator fragment bonds itself to one of the monomer's carbon atoms.

The radical is now happy, but this initiating reaction creates another free radical associated with the ethylene molecule's other carbon atom. The new radical also seeks a partner. And so ethylene monomers begin attaching themselves in a chain, creating new radicals each time and lengthening the chain, a stage called propagation [5]. Growing chains may also attach themselves to one another. Most commonly, chains join end to end, but sometimes they join end to backbone, forming branched polyethylene molecules (Figure 14.2). Eventually, free radical polymerization stops due to termination reactions. For example, instead of stealing an electron from double-bonded carbons or a nearby propagating chain, the carbon atom with the free radical sometimes steals an entire hydrogen atom from another chain end. The polymer end, robbed of its hydrogen, easily forms a double bond with its adjacent carbon atom, and polymerization stops [4,6].

Because every part of the ethylene monomer is included in the finished polymer, the free radical polymerization of polyethylene is referred to as an addition polymerization; the ethylene molecules are simply added together. However, polymerizations that use only portions of a monomer are known as condensation polymerizations. The monomers that condense with each other must contain at least two reactive groups in order to form a chain. Condensation reactions result in 'condensed' polymers that have less total mass than the monomers used to create them and the by-products, or 'condensates,' combined [7].

For example, poly(ethyleneterepthalate), a polyester known as PET, commonly found in soda bottles, forms from a reaction of two monomers, ethylene glycol and terephthoyl chloride. At the reaction's end, an atom of hydrogen and an atom of chlorine are left out of each PET molecular junction, resulting in a by-product of hydrogen chloride (HCl) gas.

14.2 Properties of macromolecules

A comparison of the properties of polyethylene (both low-density polyethylene (LDPE) and high density polyethylene (HDPE)) with the natural polymers rubber and cellulose is instructive. Synthetic HDPE macromolecules have masses ranging from 10^5 to 10^6 amu (LDPE molecules are more than a hundred times smaller). Rubber and cellulose molecules have similar mass ranges, but fewer monomer units because of the monomer's larger size. The physical properties of these three polymeric substances differ from each other and of course from their monomers [8].

HDPE is a rigid translucent solid which softens on heating above 100 °C, and can be fashioned into various forms including films. It is not as easily stretched and deformed as is LDPE. HDPE is insoluble in water and most organic solvents, although some swelling may occur on immersion in the latter. HDPE is an excellent electrical insulator.

LDPE is a soft translucent solid, which deforms badly above 75 °C. Films made from LDPE stretch easily and are commonly used for wrapping. LDPE is insoluble in water, but softens and swells on exposure to hydrocarbon solvents. Both LDPE and HDPE become brittle at very low temperatures (below –80 °C). Ethylene, the common monomer for these polymers, is a low boiling (–104 °C) gas [7].

Natural (latex) rubber is an opaque, soft, easily deformable solid that becomes sticky when heated (> 60 °C) and brittle when cooled below –50 °C. It swells to more than double its size in non-polar organic solvents such as toluene, eventually dissolving, but is impermeable to water. The C5H8 monomer isoprene is a volatile liquid (b.p. 34 °C) [8].

Pure cellulose, in the form of cotton, is a soft flexible fibre, essentially unchanged by variations in temperatures ranging from –70 to 80 °C. Cotton absorbs water readily, but is unaffected by immersion in toluene or most other organic solvents. Cellulose fibres may be bent and twisted, but do not stretch much before breaking. The monomer of cellulose is $C_6H_{12}O_6$ aldohexose D-glucose. Glucose is a water soluble solid, melting below 150 °C [4,8].

To account for the differences noted here, we need to consider the nature of the aggregate macromolecular structure or morphology of each substance. Because polymer molecules are so large, they generally pack together in a non-uniform fashion, with ordered or crystalline-like regions mixed together with disordered or amorphous domains. In some cases the entire solid may be amorphous, composed entirely of coiled and tangled macromolecular chains. Crystallinity occurs when linear polymer chains are structurally orientated in a uniform three-dimensional (3-D) matrix.

Increased crystallinity is associated with an increase in rigidity, tensile strength and opacity (due to light scattering). Amorphous polymers are usually less rigid, weaker and more easily deformed. They are often transparent [8,9]. Three factors that influence the degree of crystallinity are:

1. chain length;
2. chain branching;
3. interchain bonding.

The importance of the first two factors is nicely illustrated by the differences between LDPE and HDPE. As noted earlier, HDPE is composed of very long unbranched hydrocarbon chains. These pack together easily in crystalline domains that alternate with amorphous segments, and the resulting material, while relatively strong and stiff, retains a degree of flexibility. In contrast, LDPE is composed of smaller and more highly branched chains, which do not easily adopt crystalline structures. This material is therefore softer, weaker, less dense and more easily deformed than HDPE. As a rule, mechanical properties such as ductility, tensile strength and hardness rise, and eventually level off with increasing chain length [10].

The nature of cellulose supports the above analysis and demonstrates the importance of the third factor. To begin with, cellulose chains easily adopt a stable rod-like conformation. These molecules align themselves side by side into fibres that are stabilized by inter-chain hydrogen bonding between the three hydroxyl groups on each monomer unit. Consequently, crystallinity is high and the cellulose molecules do not move or slip relative to each other. The high concentration of hydroxyl groups also accounts for the facile absorption of water that is characteristic of cotton.

Natural rubber is a completely amorphous polymer. Unfortunately, the potentially useful properties of raw latex rubber are limited by temperature dependence; however, these properties can be modified by chemical change. The Cis-double bonds in the hydrocarbon chain provide planar segments that stiffen, but do not straighten the chain. If these rigid segments are completely removed by hydrogenation (H_2&Pt catalyst),

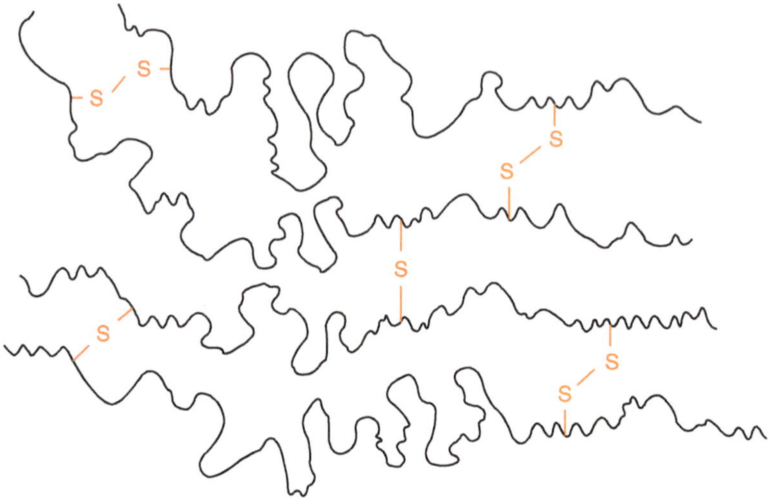

Figure 14.3 Unstretched rubber [10]

the chains lose all containments and the product becomes a low melting paraffin-like semi-solid of little value. If instead the chains of rubber molecules are slightly cross-linked by sulfur atoms, a process called vulcanization discovered by Charles Goodyear in 1839, the desirable elastomeric properties of rubber are substantially improved. At 2 to 3% cross-linking, a useful soft rubber, which no longer suffers stickiness and brittleness problems on heating and cooling, is obtained. At 25 to 35% cross-linking, a rigid hard rubber product is formed (Figure 14.3). The more highly-ordered chains in the stretched conformation are entropically unstable and return to their original coiled state when allowed to relax [10].

On heating or cooling, most polymers undergo thermal transitions that provide insight into their morphology. These are defined as the melt transition, Tm, and the glass transition, Tg.

- T_m is the temperature at which crystalline domains lose their structure, or melt. As crystallinity increases, so does T_m.
- T_g is the temperature below which amorphous domains lose the structural mobility of the polymer chains and become rigid glasses.

T_g often depends on the history of the sample, particularly previous heat treatment, mechanical manipulation and annealing. It is sometimes interpreted as the temperature above which significant portions of

Table 14.1 List of most important polymers with their melting and glass transition temperatures

Polymer	LDPE	HDPE	PP	PVC	PS	PAN	PTFE	PMMA	Rubber
T_m (°C)	110	130	175	180	175	>200	330	180	30
T_g (°C)	110	100	–10	80	90	95	–110	105	–70

polymer chains are able to slide past each other in response to an applied force. The introduction of a relatively large and stiff substituent (such as benzene rings) will interfere with this chain movement, thus increasing T_g. The introduction of small molecular compounds called plasticizers into the polymer matrix increases the inter-chain spacing, allowing chain movement at lower temperatures, with a resulting decrease in T_g. The out-gassing of plasticizers used to modify interior plastic components of automobiles produces the 'new-car smell' to which we are accustomed.

T_m and T_g values for some common addition polymers are listed in Table 14.1. Note that cellulose has neither a T_m nor a T_g.

Rubber is a member of an important group of polymers called elastomers, amorphous polymers that have the ability to stretch and then return to their original shape at temperatures above T_g. This property is important in applications such as gaskets and O-rings, so the development of synthetic elastomers that can function under harsh or demanding conditions remains a practical goal. At temperatures below T_g, elastomers become rigid glassy solids and lose all elasticity. A tragic example of this caused the space shuttle Challenger disaster. The heat and chemical resistant O-rings used to seal sections of the solid booster rockets had an unfortunately high T_g near 0 °C. The unexpectedly low temperatures on the morning of the launch were below this T_g, allowing hot rocket gases to escape the seals.

Recognition that polymeric macromolecules make up many important natural materials was followed by the creation of synthetic analogues having a variety of properties. Indeed applications of these materials, as fibres, flexible films, adhesives, resistant paints and tough but light solids, have transformed modern society [4,8,10].

14.3 Plastic welding

Plastic welding is the process of creating a molecular bond between two compatible thermoplastics. Welding offers superior strength, and often drastically reduced cycle times, to mechanical joining (snap fits, screws) and chemical bonding (adhesives). There are three main steps to any weld, pressing, heating and cooling. The application of pressure, which is often used throughout both the heating and cooling stages, is used to retain the parts in the proper orientation and to improve melt flow across the interface. The purpose of the heating stage is to allow intermolecular diffusion from one part to the other across the faying surface (melt mixing). Cooling is necessary to solidify the newly-formed bond; the execution of this stage can have a significant effect on weld strength.

There are several possible methods of plastic welding: Ultrasonic, Vibration, Spin, Hot Plate, Laser/Infrared, Radio Frequency and Implant are the most common. These plastic welding processes are primarily differentiated by their heating methods. The application of pressure and allowances for cooling are mechanical considerations that may vary from machine to machine within the general process category. Thus the methods for joining plastic components can be divided into three major categories, mechanical joining, adhesive bonding and welding (Figure 14.4)

Welding processes are often categorized and identified by the heating method that is used. All processes can be divided into the two general categories of internal heating and external heating (Figure 14.5)

Historically, pressure has been applied for plastic welding through the use of pneumatic presses. Recently, however, servo motors have been employed for at least a few of the common processes. Pneumatic welders

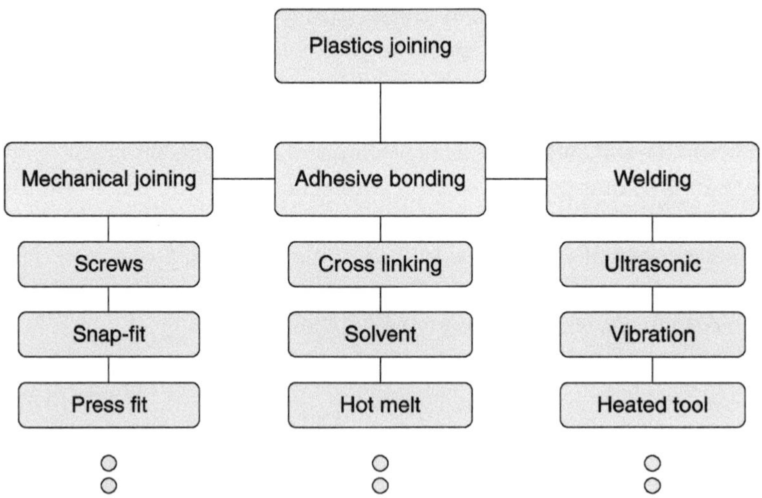

Figure 14.4 Possible plastic joining techniques [11]

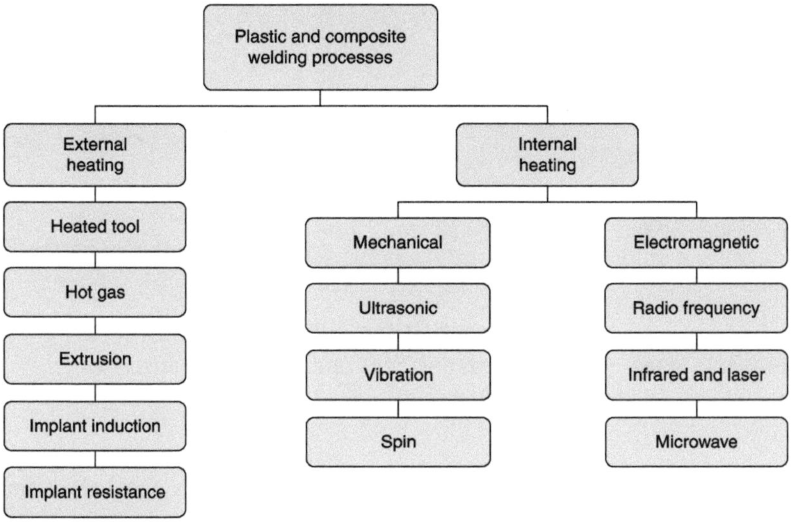

Figure 14.5 Classification of welding techniques [11]

are economical and well-suited to most simple applications. The precision of servo motion, however, offers greater control and precision, which is desirable for more difficult applications or when the equipment is used for a wide variety of applications.

14.3.1 Heating

It is crucial to the plastic welding process to form a melt layer at the faying surface to allow intermolecular diffusion for formation of a molecular bond. In the solid state, polymer chains will not flow. Therefore, the joint surface on both of the parts must be melted to allow the plastic molecules to diffuse across the interface and bond with molecules of the other part. The hotter the melt, the more molecular movement achieved, and a weld can be made in a shorter cycle time. Amorphous polymers must be heated to above their glass transition temperature, while semi-crystalline polymers must be heated to above their melting temperature.

14.3.2 Ultrasonic

The ultrasonic process works through the translation of the AC current into mechanical vibrations using piezo-electric ceramics. The vibrations

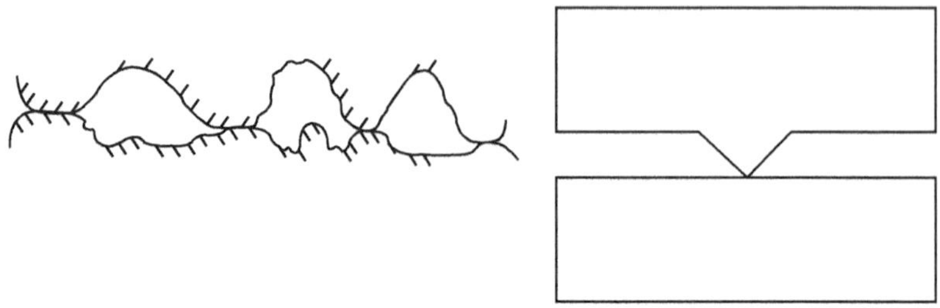
(a) (b)

Figure 14.6 Ultrasonic heating through cyclical deformation of: (a) surface asperities; and (b) man-made energy directors [11]

are then passed through the part to the joint, at which point they cause stress and strain in the contact area between the two parts, which leads to localized heating and melting of the polymer. Ultrasonic welding typically has weld times from one-tenth to two seconds and Dukane's ultrasonic welding equipment can handle parts from approximately one-eighth inch diameter at high frequencies up to 12 inches square at low frequencies (Figure 14.6).

14.3.3 Vibration

The vibration process works through either linear or orbital motion of one part relative to the other. This results in surface friction that leads to heat generation and the formation of a melt layer at the joint. Vibration welding typically has weld times from one to five seconds and Dukane's vibration welders can handle parts from approximately two inch diameter up to five feet long (Figure 14.7).

14.3.4 Spin

The spin process works through rotational motion of one part relative to the other. This results in surface friction that leads to heat generation and the formation of a melt layer at the joint. Spin welding typically has weld times from one-half to five seconds and Dukane's spin welding equipment can handle parts from approximately one-half inch diameter using high speed motors up to nine inches diameter using high torque motors (Figure 14.8).

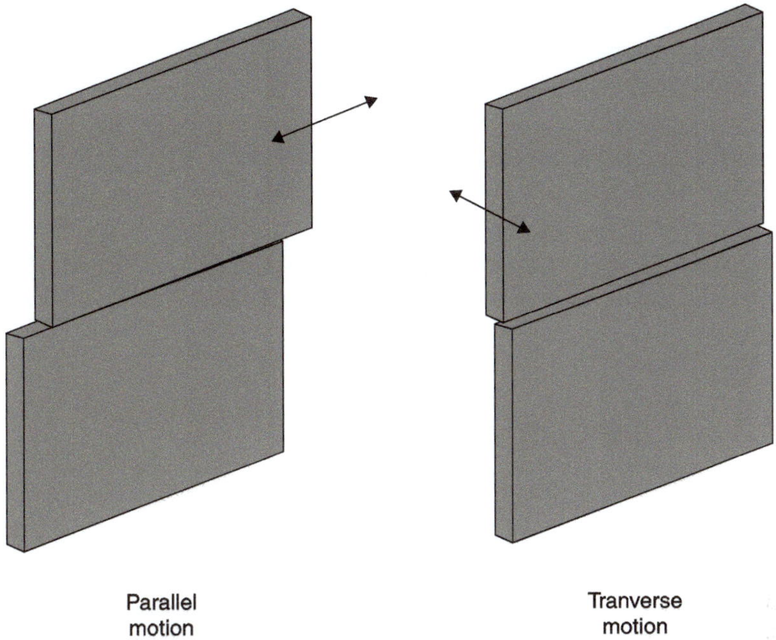

Figure 14.7 Example of parallel and traverse vibration used for plastic welding [11]

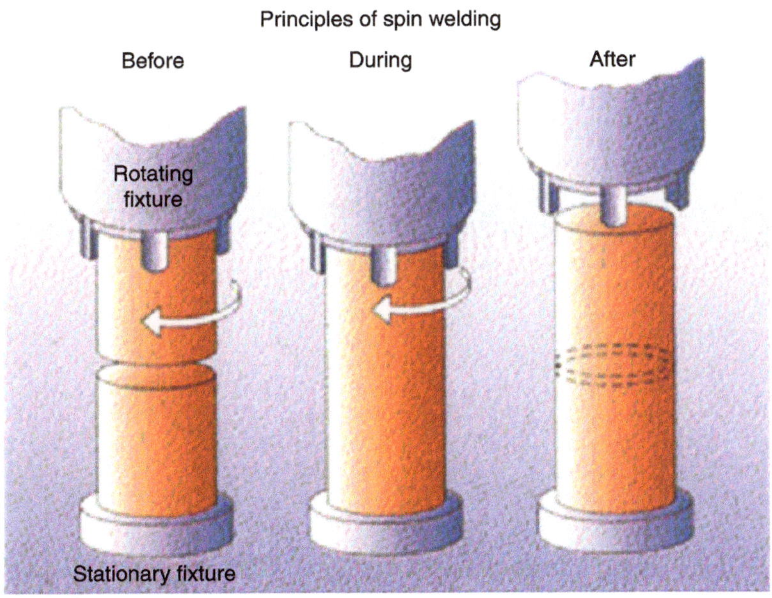

Figure 14.8 Schematic of spin welding of plastics [10]

14.3.5 Hot plate

In the hot plate process, the faying surfaces of the two parts are heated through conduction, convection and/or radiation from an actual heated plate. The two parts are either pressed against the hot plate, or held next to it for some period of time, then the hot plate is removed and the parts are pressed together to form the weld. Hot welding typically has weld times from 10 to 20 seconds and Dukane's hot plate welding equipment can handle parts from approximately one inch diameter up to five feet long (Figure 14.9).

14.3.6 Laser/infrared

There are two means of using laser/infrared energy to create a weld. One way is to heat the joining surface of both parts to create a melt layer, then press the parts together to form a weld. The second way is to use one part that is transparent to the laser/infrared wave length, and another that absorbs it. This way, the laser infrared beam can be passed through the transparent, or transmitting, part to the absorbing part which will then heat. The molten polymer of the absorbing part then heats the surface of the transmitting part. Laser/infrared welding typically has weld times from three to five seconds and Dukane's laser welding equipment can

Figure 14.9 Photograph of saddle point for pipeline installation using hot-plate welding [11]

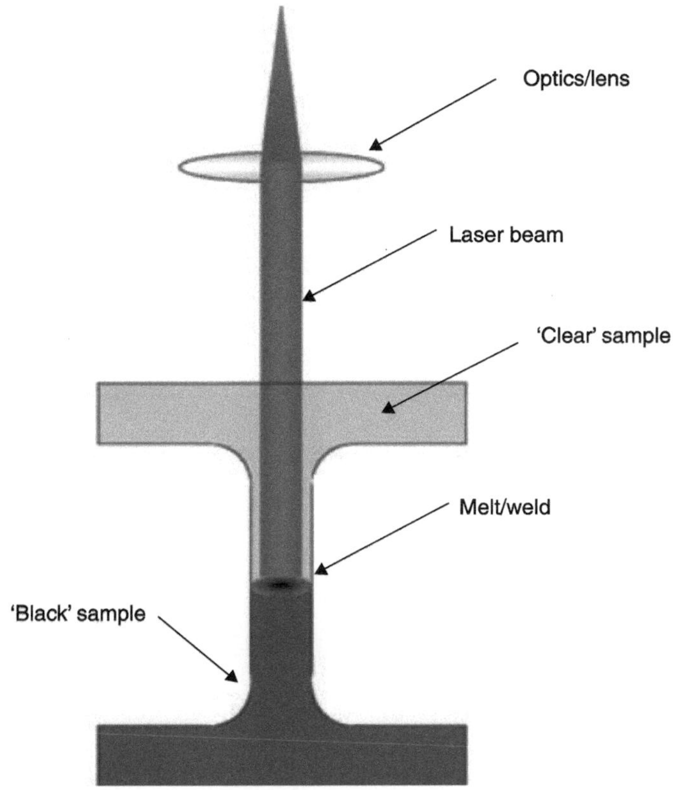

Figure 14.10 Laser welding of plastics [11]

handle parts from approximately ten-thousandths inches thick to one-quarter inches thick. Figure 14.10 shows the schematic of plastic laser welding.

14.3.7 Radio frequency

Radio frequency welding only works with thermoplastics that have a high dielectric constant. In this process, an electric field is generated near the joint, the direction of which is alternated approximately 27 million times a second. Every time the electric field changes, the dipole molecules of the polymer attempt to flip around to realign themselves to the new field orientation. Because this movement is impaired by inertia and friction, the molecules do not flip in unison. This constant relative

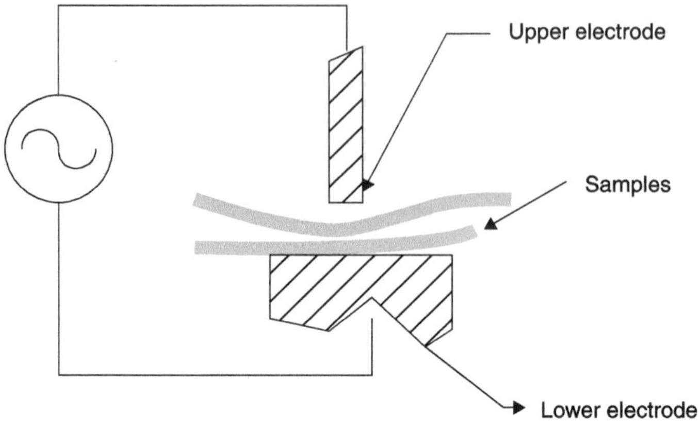

Figure 14.11 Schematic of the infrared welding [11]

movement of the molecules results in intermolecular friction, which results in heat and, subsequently, melting. Radio frequency welding typically has weld cycle times from two to five seconds and can handle parts from approximately one-thousandth inches thick to fifty-thousandths inches thick (Figure 14.11).

14.3.8 Implant

Implant welding relies on heat generated by a foreign substance implanted into the weld joint. There are two methods of generating this heat, resistance and induction. In resistance implant welding, an electrical current is passed through a conductive implant (typically a wire) that is located in the weld joint. The implant then heats due to electric resistance, which causes heating and melting of the surrounding polymer. In induction implant welding, a gasket (typically made of the same polymer to be welded filled with conductive or ferromagnetic material) is placed at the joint. The parts are then placed in an electromagnetic field and the gasket is heated though induction. As the polymer in the gasket heats and melts, the plastic of the parts nearby is heated through conduction and convection, creating a weld between the parts and the gasket. Implant welding typically has weld cycle times from one-half to one minute and can handle parts from approximately one inch diameter up to eight feet long.

14.4 Friction stir welding of polymeric materials

As the request for rapid and high efficiency technologies in polymer engineering are growing in scientific laboratories, the joining techniques for polymers have drawn considerable attention to themselves. There are several methods for polymer joining, which can be categorized as irradiation (e.g. laser) based, heat conduction (e.g. heat-mirror, hot gas) based and friction (e.g. ultrasonic) based methods, which were explained in the previous section.

Friction stir processing (FSP) is a solid-state joining technology, which has been commonly used for metals throughout the last decade. In this method, the non-consumable tool, composed from the two basic parts containing pin and shoulder, is used to join the components. The shoulder is responsible for creation of friction on the surface of the joining parts and the pin plunges into the zone containing a fraction of parts to bind them. The heat produced by this method softens the zone and moves the material from the advanced side (AS) to the retreating side (RS) of the tool, leading to joining of the compounds. Hence, the solid-state nature of the process makes this method practicable and suitable for plastics.

14.4.1 Friction stir welding of polyethylene sheets

In 2005, the mechanical performance of friction stir welded (FSW) polyethylene (PE), which underwent double passes, was investigated [12]. The PE plate used for this study was 5 mm thick. The tool shoulder and pin diameter were 16 and 5 mm, respectively, and the length of the pin was chosen to be 2.8 mm. The parameters that have been considered as variants had rotational speeds (600, 800 and 1000 rpm), welding speeds (12.5, 25, 40 and 60 mm) and tool angles (0 and 1 degrees). Because of the low quality of the one-pass FSW due to the root defect, the double-pass FSW was applied to eliminate this shortage. Table 14.2 shows the tensile results of the FSW specimens with rotational speeds of 1000 rpm containing the tensile strength and types of fracture locations. The tensile yield strength of the reference PE was 20 MPa. As can be demonstrated from Table 14.1, the specimens FSW with the tool angle of 1 degree have better mechanical strength rather than those FSW by tilt angle of 0 degrees. Also the joints performed by a 1-degree tool angle

Table 14.2 The results of tensile test [12]

Sample code	Tool angle	Welding speed (mm/min)	Tensile strength (MPa)	Type of fracture location
A	0	12.5	16.46	Type c
B	0	25	19.35	Type b and type c
C	0	40	15.81	Type a
D	1	12.5	20.45	Type e
E	1	25	19.30	Type b and type d
F	1	40	17.70	Type b and type d

show no voids in the stir zone (SZ). Figure 14.12 shows schematic images of the different processed zones in the FSW polyethylene.

Figure 14.13 shows the different fracture zones of the FSW polyethylene, which contains five locations. When considering Table 14.1 and Figure 14.2,

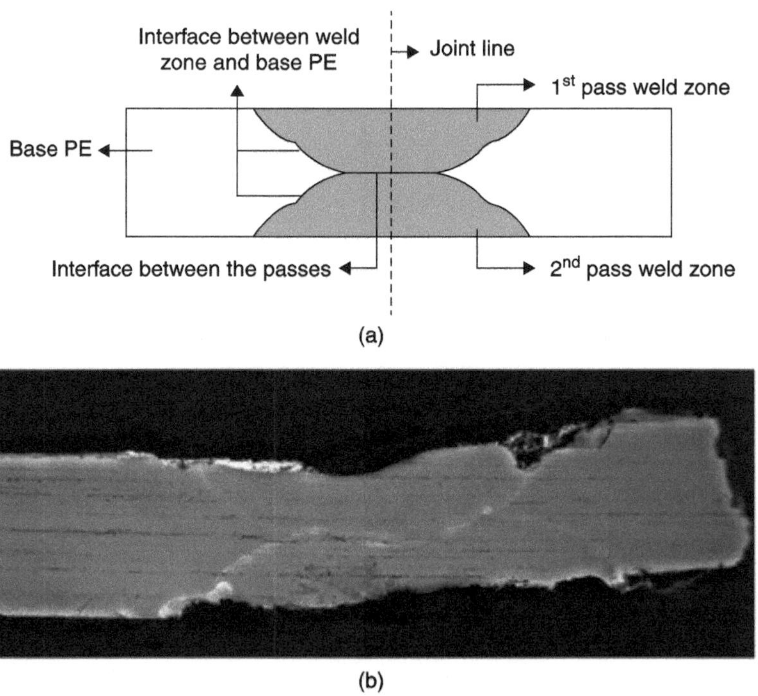

Figure 14.12 The welded zone for double passes FSW: (a) schematic illustration; and (b) photograph [12]

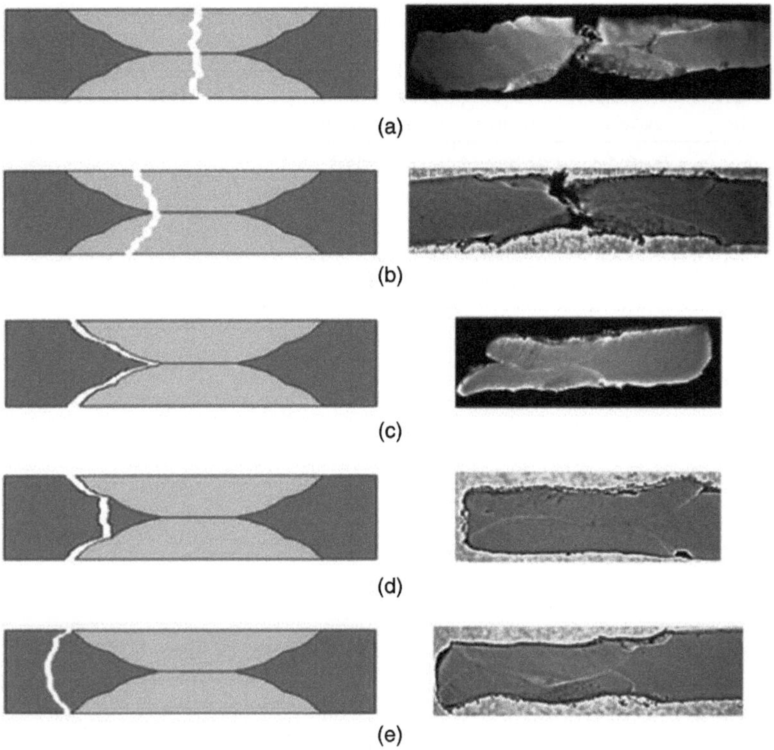

Figure 14.13 **Different fracture zone of FSWed specimens [12]**

it can be concluded that the fracture from the welded zone has the higher strength.

Table 14.3 presents the bending results of the above-mentioned specimens for two types of the specimens (type 1: 1-pass FSW, type 2: double-pass FSW). The results of crack existence, failure displacement, bending angle and yield stress are also presented in this table. Formation of the crack after bending is in type d (Figure 14.13). Bending test results show that the tool angle of 1 degree is better than 0 degrees. Crack formation in types A, B, C and D coded specimens have occurred in the samples. Also the results presented in Table 14.3 suggest a tendency for crack formation to occur in bending of double-pass FSW specimens.

Also the FSW of HDPE was investigated [13], in which the effects of the pin diameter, pin rotational speed, feed rate and temperature on the weld quality were evaluated. They found that these parameters can significantly affect the properties of the HDPE joints. Figure 14.14 shows

Table 14.3 The results of bending test [12]

Sample code	Bending type	Crack formation	Failure displacement	Yield displacement	Bending angle
A	Type 1	Yes	11.6	11.51	18.4
A	Type 2	Yes	15	12.71	23.2
B	Type 1	No	–	19.77	–
B	Type 2	Yes	30	18.01	40.6
C	Type 1	No	–	17.10	–
C	Type 2	Yes	32	17.71	42.4
D	Type 1	Yes	14.4	14.33	22.3
D	Type 2	Yes	18	16.57	27.2
E	Type 1	No	–	17.69	–
E	Type 2	No	–	16.92	–
F	Type 1	No	–	16.86	–
F	Type 2	No	–	18.02	–

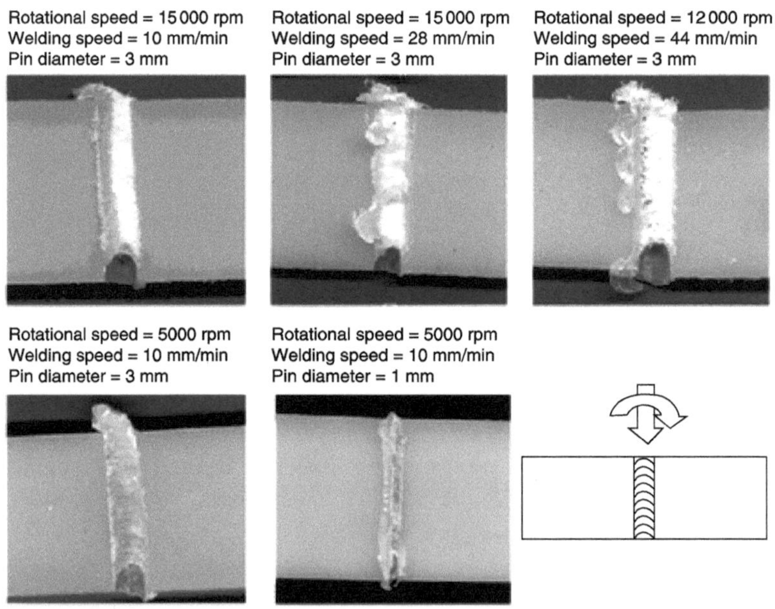

Figure 14.14 Appearance of the welds [13]

the different images of the FSW specimens in various conditions. As can be seen in this figure, all the specimens have a hole in the end part and material being pulling out at the beginning of the joint. The high value of traverse speed leads to low quality of weld due to the inability of the material to be replaced behind the pin. Good results were obtained under the conditions where the rotational speed was the highest (5000 rpm) and the traverse speed was the lowest (10 mm/min).

Several experiments show that the mechanical properties are very different between specimens. Figure 14.15 shows the comparison of the mechanical properties of a welded specimen with the reference HDPE. As can be seen, the elongation of the FSW specimen shows a remarkable reduction but a strength value similar to the reference HDPE was achieved.

The DSC test was performed to evaluate the melting points of the different zones of the welded parts (Figure 14.16). The results show the melting point reduction in the SZ owing to the low crystallinity of this zone.

Figures 14.17(a and b) show the tensile results of the welded specimens with a tool diameter of 1 and 3 mm. As can be seen from the figure, the good quality is obtained by the 1 mm pin diameter at the rotational speed of around 5000 rpm. In the case of traverse speeds, the higher one (28 mm/min) has better strength in comparison with the 10 mm/min one. It should be concluded that in this case the lower pin diameter and rotational speed in conjunction with higher traverse speed is the optimum condition for these experiments.

Figure 14.17(c) shows the previous results for the heated pin, which was performed to achieve the better quality in welding. The results

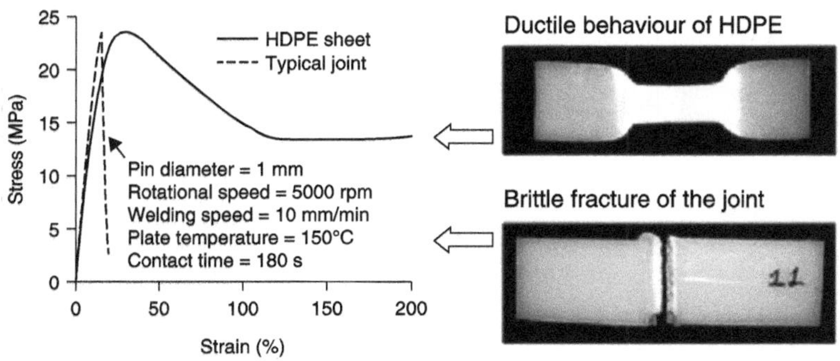

Figure 14.15 Typical tensile test on base material and FSW joint [13]

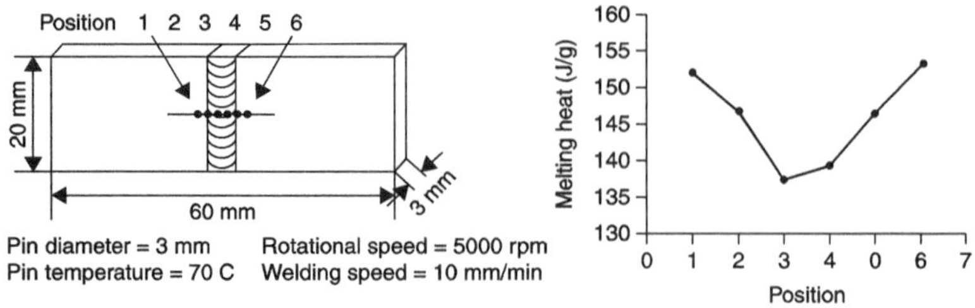

Figure 14.16 DSC test result [13]

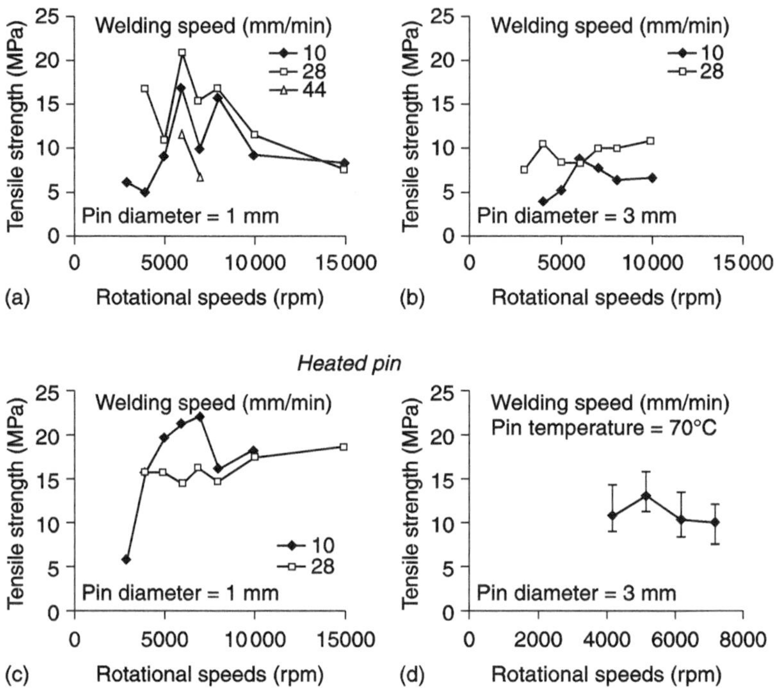

Figure 14.17 Process maps [13]

indicated that the heating of the pin had good effect on the strength of the joints for the 1 mm pin diameter. It was also clear that in the case of the heated pin, the lower traverse speed (10 mm/min) gave better strength to the welded zone.

Optimization of the polyethylene parameters, including the rotational and traverse speeds and tool plunged surface, was performed in other

Table 14.4 Level for operating parameters for the FSW process [14]

Factor	Label	Low level (-)	Centre point (0)	High level (+)
Rotational speed (rpm)	N	653	1280	1700
Feed speed (mm/min)	F	16	29	44
Tool plunged surface (mm^2)	S	377	452	528

research [14]. Table 14.4 shows the levels for operation parameters for the FSW process.

This experimental strategy was based on selection of an appropriate orthogonal array (for 2^3 factor design), which depends on the total degree of freedom. Table 14.4 represents the experimental design for the FSW process. They found that in both the full and reduced model, the rotational speed effect was not tangible at 5% confidence level. Multi regression analysis was performed to predict the longitudinal flow stress from F and S and it was demonstrated that the maximum value for the longitudinal flow stress is 26 MPa, which was obtained from F and S fixed at around 24 mm/min and 401 mm^2, respectively.

The other important experiment in the field of polyethylene welding via FSP was spot welding, in which the tool has three main motions (Figure 14.18), including plunging, stirring and retracting.

First, the tool plunges into the coupled specimens and subsequently the stirring action begins and finally the tool retracts to its initial position. This trend creates the high heat input into the softened welding zone, accompanying the high shear stress, which leads to material flow around the tool and joint plates. The welded zone is schematically illustrated in Figure 14.19, where two main points within it can be specified, including

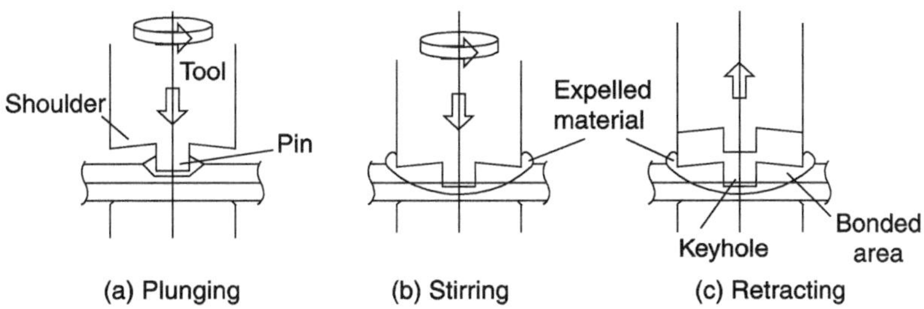

Figure 14.18 Three-phase of friction stir welding process: (a) plunging; (b) stirring; and (c) retracting [15]

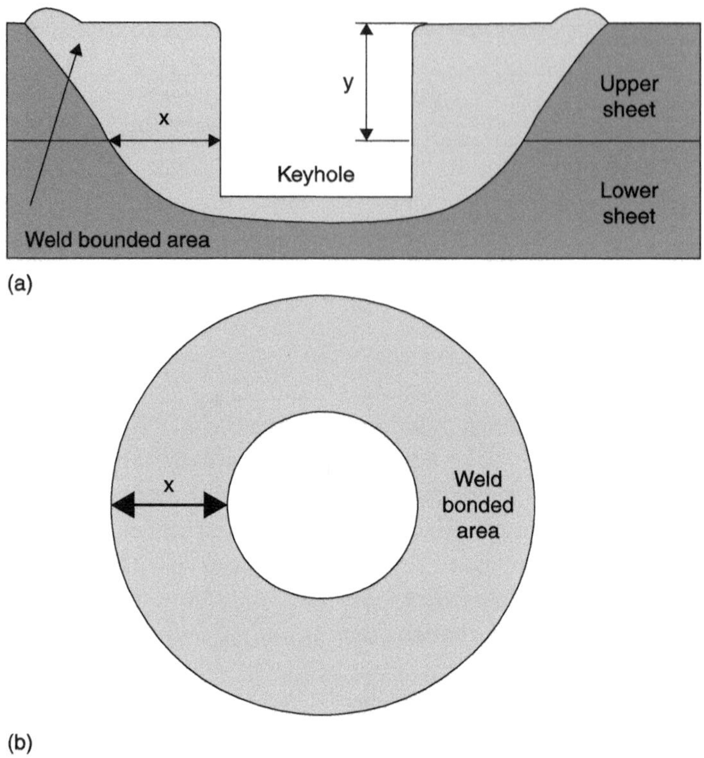

Figure 14.19 (a) Schematic illustration of the cross-section of a friction stir spot weld; and (b) geometry of the weld bonded area. x: nugget thickness; y: thickness of the upper sheet [15]

the thickness of the weld nugget (x) and thickness of the upper plat (y). The sizes of these two points strongly affect the strength of the joints.

Friction stir spot welding (FSSW) of the polyethylene plates with a thickness of 4 mm was investigated (Figure 14.20) [15]. The tool geometry is shown in Figure 14.21.

Table 14.5 contains the parameters that were varied in this experiment.

Figure 14.22 shows the temperature rise with an increase in the dwell time under constant experimental conditions of 1100 rpm rotational speed, 3.3 mm/s rate of tool plunge and 6.00 mm tool plunge depth. As seen, the maximum temperature was 142 °C at 34 s.

The geometry of the cross-section depends upon the waiting time after dwell time. The results show that the welded specimens, in which the tool was extracted immediately, did not show the keyhole (Figure 14.23(a)),

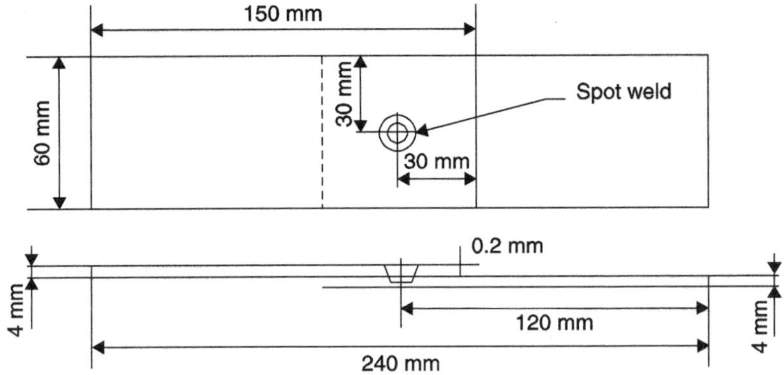

Figure 14.20 Configuration of the lap-share test specimen [15]

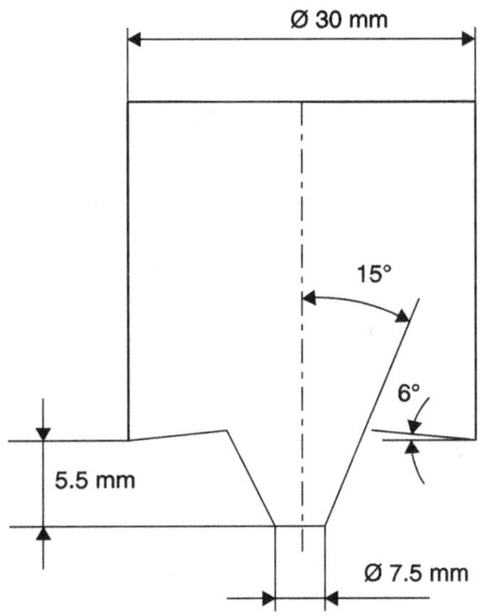

Figure 14.21 Magnified cross-sectional view of the tool [15]

while in the specimens with a waiting time of 30 s after dwell time, the keyhole was revealed (Figure 14.23(b)).

Figure 14.24 shows the results obtained for the different dwell times with constant welding parameters of 1120 rpm rotational speed, 3.3 mm/s plunge rate and plunge depth of 6 mm. The results indicated that by increasing the dwell time until 30 s, the joint strength improved and after that it remained constant.

Table 14.5 Welding parameters and their ranges [15]

Parameters	Units	Ranges
Tool rotational speed	Round per minute (rpm)	280–1400
Tool rotational rate	Millimetre	3.3–20.8
Tool plunge depth	Millimetre (mm)	5.5–7.0
Dwell time	Seconds (s)	8–90

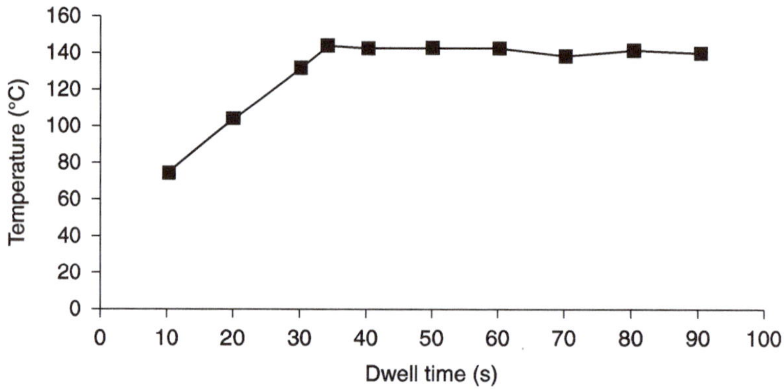

Figure 14.22 The effect of dwell time on weld zone temperature [15]

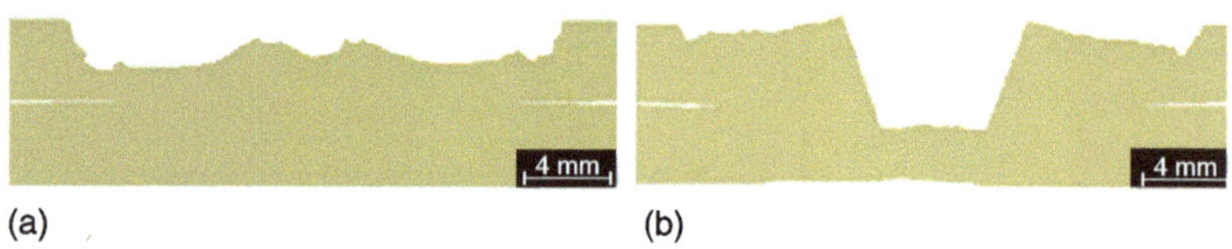

Figure 14.23 The effect of tool retracting delay on weld cross-section: (a) no delay; and (b) 30 s delay [15]

Figure 14.25 shows the diagram of the strength of the different failure mechanisms occurring in the welded specimens. As is clear from the figure, the specimens with pull nugget failure mechanisms have a remarkably higher strength in comparison with the others.

Figures 14.26 and 14.27 show the lap-shear gravity force at different rotational speeds and tool plunge depths. As shown in these images, an increase in the rotational speeds leads to improvement in the mechanical strength of the joint until 710 rpm and slight diminution was observed

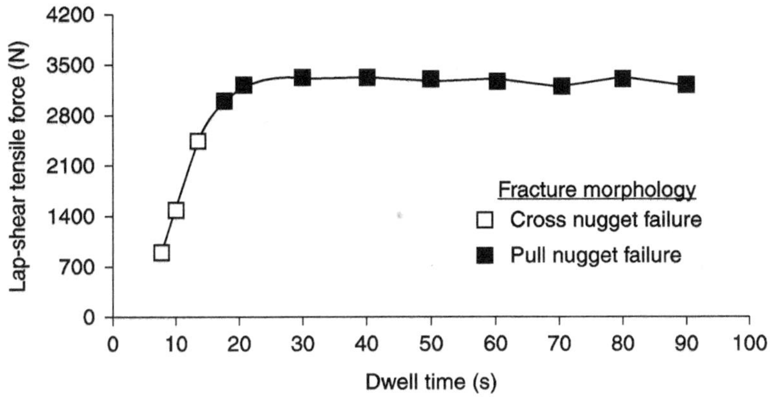

Figure 14.24 The effect of dwell time on lap-share tensile fracture [15]

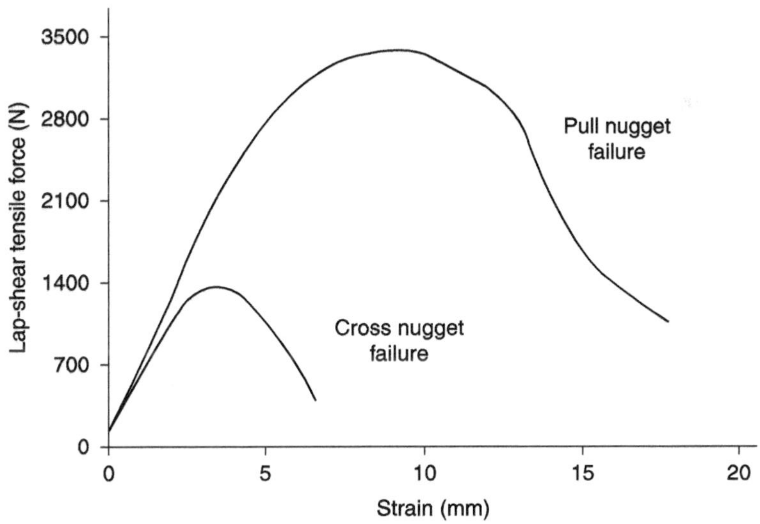

Figure 14.25 Load vs. displacement curves of the specimens which have different morphology [15]

after that, which can be attributed to the higher heat input in the SZ. Several experiments show the considerable effect of the tool plunging depth on the strength of the joints. It is found that plunge depth has an interesting effect on the joint strength, in which a rise in the plunge depth results in high strength in the welded zone until 5.7 mm and the reduction of the joint quality is seen at further plunge depths.

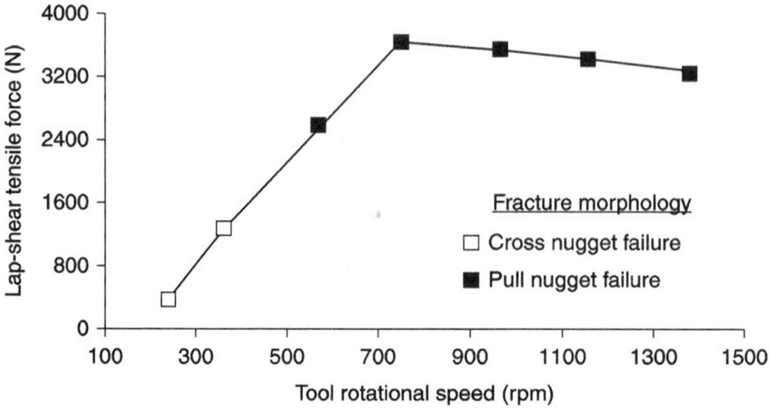

Figure 14.26 The effect of tool rotational speed on lap-share facture load [15]

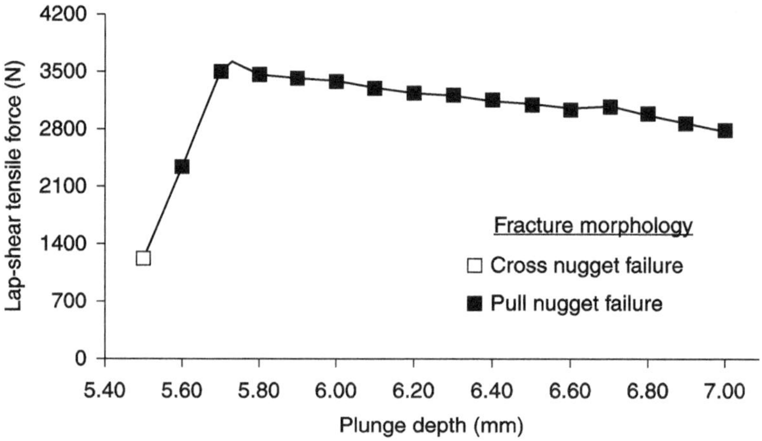

Figure 14.27 The effect of tool plunge on lap-shear tensile fracture load [15]

Figure 14.28 contains schematics of the plunge depth effect on the cross-section of the joints. It is worth noting that the above-mentioned results were obtained at rotational speed and dwell time of 1120 rpm and 50 s, respectively.

It is clear from Figure 14.28 that the thickness of the upper plate reduces by increase the plunge depth. Also the effect of the plunge rate on the strength of the joints was investigated by [15], who show that this

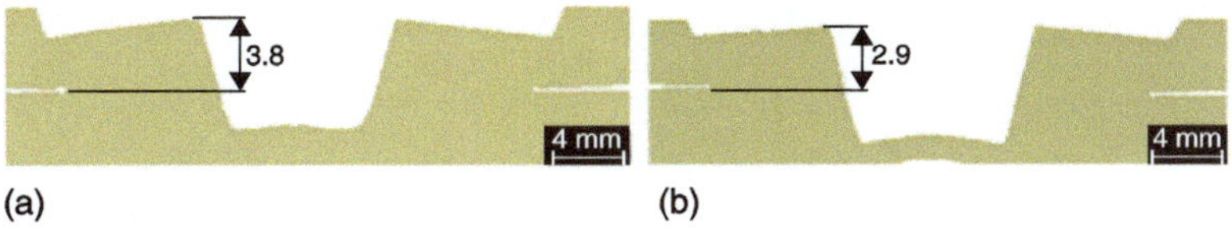

Figure 14.28 The effect of plunge rate on joint cross-section: (a) 5.7 mm plunge depth; and (b) 6.7 mm plunge depth [15]

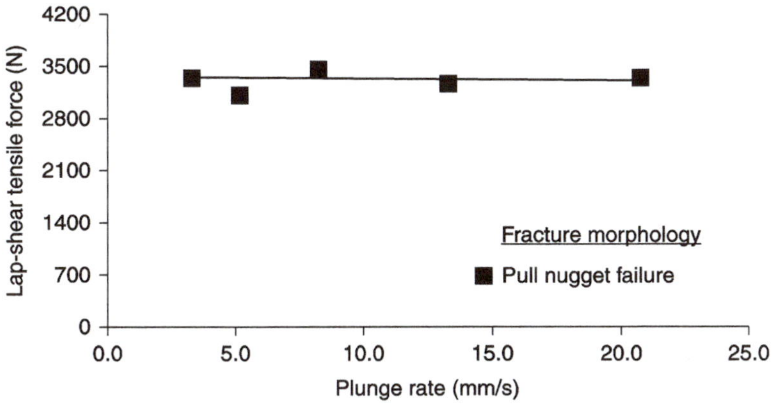

Figure 14.29 The effect of tool plunge rate on lap-share tensile fracture load [15]

factor has no notable effects. Figure 14.29 shows these investigation results for different plunge depths.

The effects of tool geometry and processing parameters on the static strength and macrostructure of the FSSW polyethylene sheets were also evaluated experimentally in 2012 [16]. Regarding the works done by these researchers, the results show a notable improvement in joint quality. Figure 14.30 shows different zones formed during the FSSW, which contain the four critical areas, known as the heat affected zone, pin profile, shoulder profile and keyhole.

The schematic of the tool used for FSW is presented in Figure 14.31. Six different tool pin profiles, straight cylindrical, tapered cylindrical, threaded cylindrical, triangular, square and hexagonal, were used to fabricate different quality of joints. Figure 14.32 indicates all these geometries.

All the tools had a similar pin depth of 5.5 mm and pin size of 7.5 mm and the tapered ones had an angle of 1 degree. Pin size in cylindrical tools refers to the bottom diameter of the pin, and the diameter of the

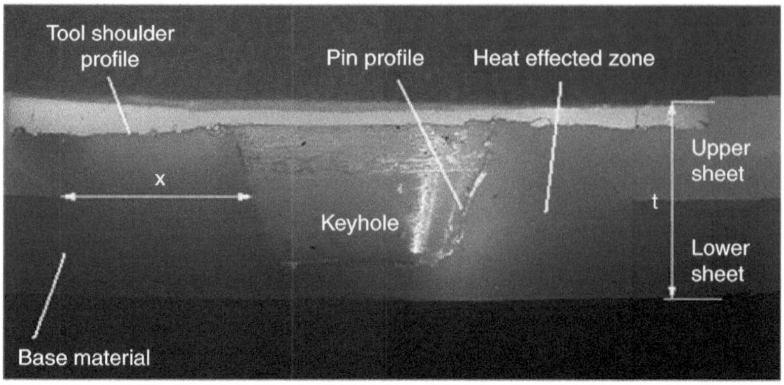

Figure 14.30 Macrostructure of FSW joint. x: nugget thickness; t: material thickness [16]

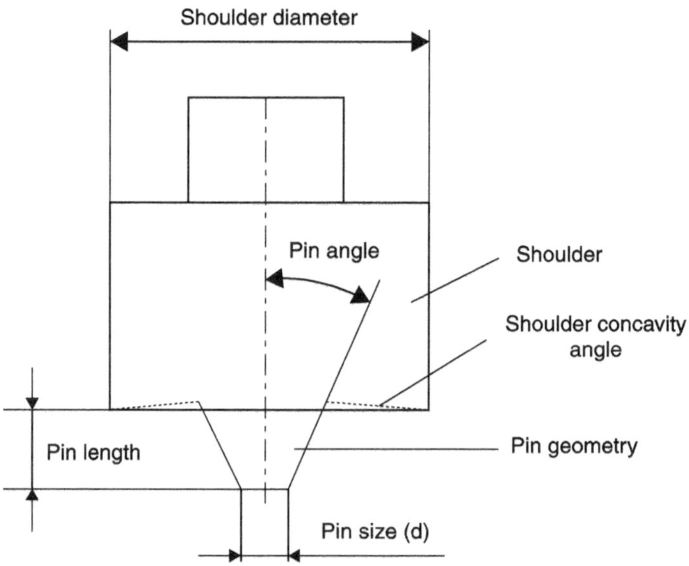

Figure 14.31 Friction stir spot welding tool design showing geometric parameters [16]

cross-sectional area in the triangular, square and hexagonal represents pin size. Table 14.6 presents the tool dimensions and their change ranges.

Table 14.7 contains the welding parameters and their ranges used for evaluation of the processing parameters effect on joints quality.

The cross-section of the joints is highly dependent on the delay time of the tool after the end of the dwell time, and when the tool was retracted

Figure 14.32 FSSW tool profile and pin size (d): (a) straight cylindrical; (b) tapered cylindrical; (c) threaded cylindrical; (d) square; (e) triangular; and (f) hexagonal [16]

Table 14.6 Tool geometry parameters and their ranges [16]

Dimensions	Unit	Range
Pin diameter	Millimetre (mm)	5.0–11.25
Pin angles	degrees	0–25
Pin length	Millimetre (mm)	4–7
Shoulder diameter	Millimetre (mm)	15–35
Shoulder concavity angle	degrees	0–12
Tool geometry	6 types	7.5

Table 14.7 Welding parameters and their ranges [16]

Ranges	Units	Parameters
560–1120	Round per minute (rpm)	Tool rotational speed
8–90	Seconds (s)	Dwell time
0–60	Seconds (s)	Delay befor retracting
0.2–1.2	Millimeter (mm)	Plunge depth

immediately, the keyhole did not appear. Figure 14.33(a) shows the absence of the keyhole in this specimen. However, the keyhole was seen when the tool had a 30 s delay time after the dwell time Figure 14.33(b). Also, as can be seen in Figure 14.33(c), the joint with a delay time of 60 s after dwell time contains the keyhole.

The results presented in Figure 14.34 show that the delay time has significant effect on the strength of the joint. It can be seen that increasing

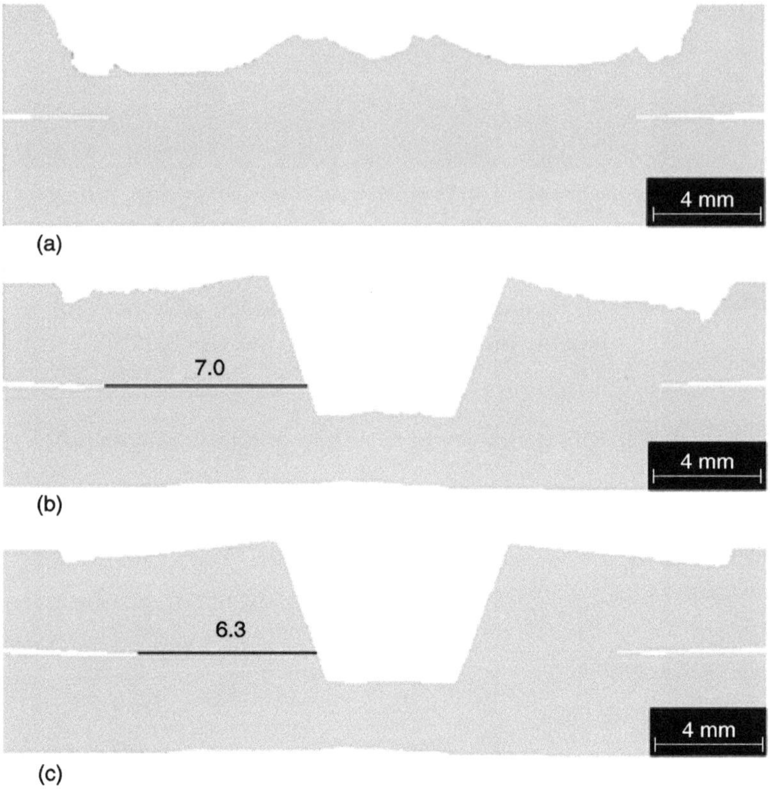

Figure 14.33 Effect of tool retracting delay on weld cross section: (a) no delay; (b) 30 s delay; and (c) 60 s delay [16]

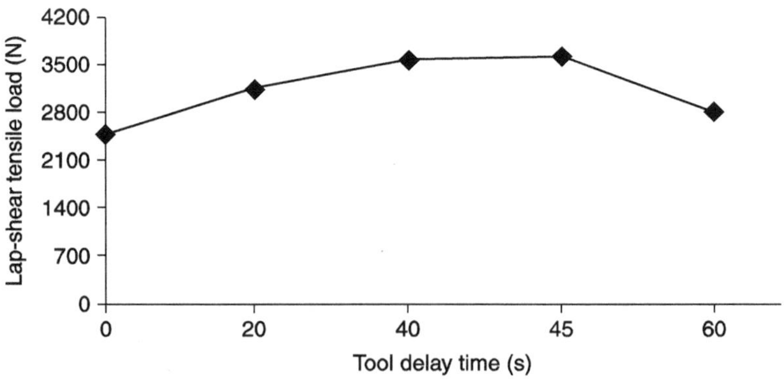

Figure 14.34 Effect of tool staying time in the weld zone after the stirring phase on the weld strength [16]

the delay time by 45 s improves the joints strength, and the joint without delay time shows the lower strength.

Figure 14.35 shows the effect of the tool pin profile and rotational speeds on the weld strength. The results present the idea that in all the rotational speeds the tapered tool geometry has the highest strength and the cylindrical one has the lowest strength.

Figure 14.36 shows the effect of the tool pin profile on the FSSW zone, in which the keyhole sizes significantly depend on the pin profile. The nugget thickness for tapered, angle tapered and treated tool is 6.1, 7.0 and 6.4 mm, respectively.

It is also shown that the tapered pin formed the thicker nugget and a larger weld bond area than the straight one.

Figure 14.37 shows the force induced during FSSW for different pin profiles. The force during the FSSW is changing, which in the higher force belongs to the end of the plunging tool stage. As can be seen, the cylindrical pin, tapered cylindrical, threaded cylindrical and square have forces of 3850, 4775, 3120 and 2160 N, respectively. This could be attributed to the higher heat input for the square pin, which leads to the lower force.

It is interesting to note that the thickness of the nugget zone (NZ) increases with increase in dwell time (Figure 14.38), in which the 13 s dwell time has a thinner nugget compared to the one with 50 s dwell time.

It is also found that the tool's concave angle has a significant effect on the strength of the joint. The higher concave angle in the tool leads to better strength of the joint and this started from 4.5 degrees. This is

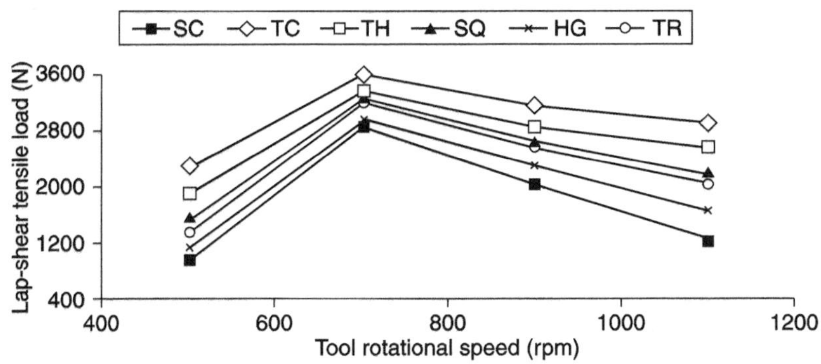

Figure 14.35 Effect of tool profile and rotational speed on the weld strength [16]

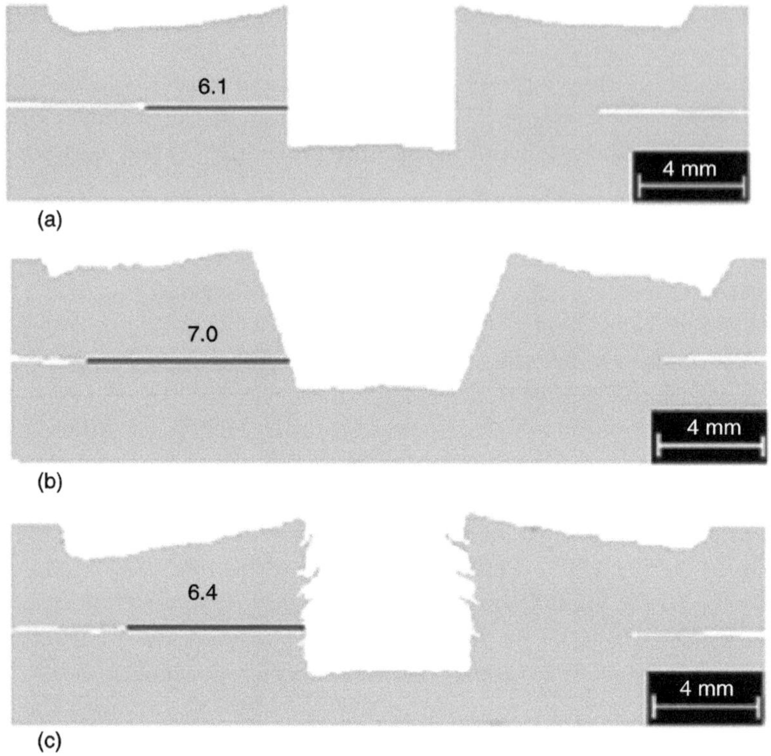

Figure 14.36 Effect of pin angle on nugget formation: (a) straight cylindrical pin; (b) 15 degree pin angle tapered cylindrical pin; and (c) threaded cylindrical pin [16]

shown in Figure 14.39 and the effect of the concave angle on weld nugget formation is shown in Figure 14.40. It is clear that the nugget thickness increases with concave angle increase.

Figure 14.41 shows the effect of the shoulder diameter on joint strength, indicating that the higher shoulder diameter leads to a higher strength of the joints, which could be due to the higher input of heat and consequently higher opportunity of the materials joints.

14.4.2 Friction stir welding of ABS

Figure 14.42 presents the material flow pattern in the AS and RS of the tool. As is indicated, the material from the RS of the tool does not rotate with the tool and material from the AS forms around pin materials.

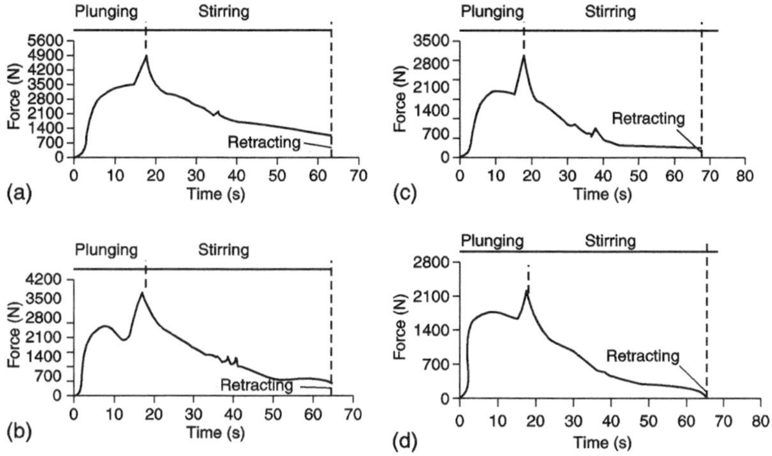

Figure 14.37 Welding force of: (a) the tapered cylindrical pin; (b) the straight pin; (c) the threaded pin; and (d) the square pin [16]

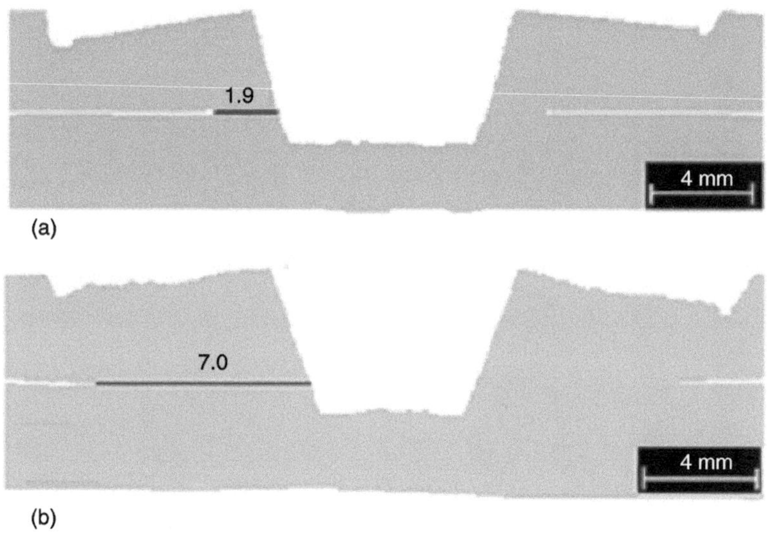

Figure 14.38 Effect of the dwell time on the joint cross-section: (a) 13 s dwell time; and (b) 50 s dwell time [16]

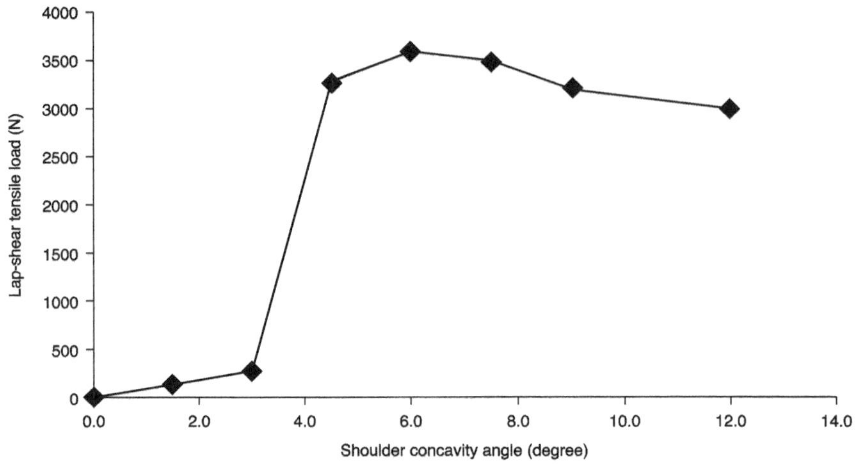

Figure 14.39 Effect of shoulder concavity angle on weld strength [16]

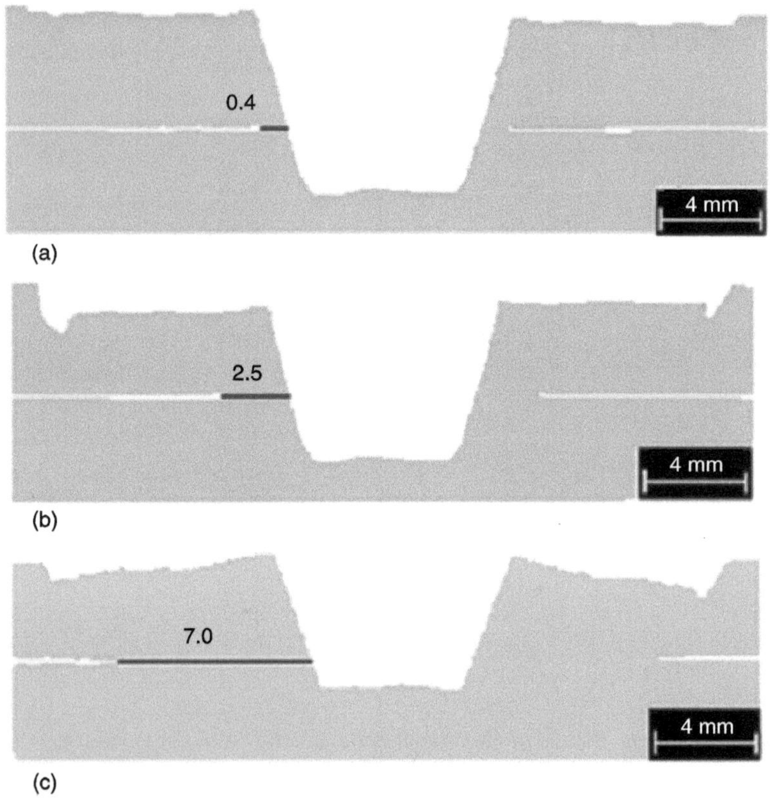

Figure 14.40 Effect of the shoulder concavity angle on the joint cross-section: (a) 0 degree shoulder concavity angle; and (b) 3 degree shoulder concavity angle; and (c) 6 degree shoulder concavity angle [5]

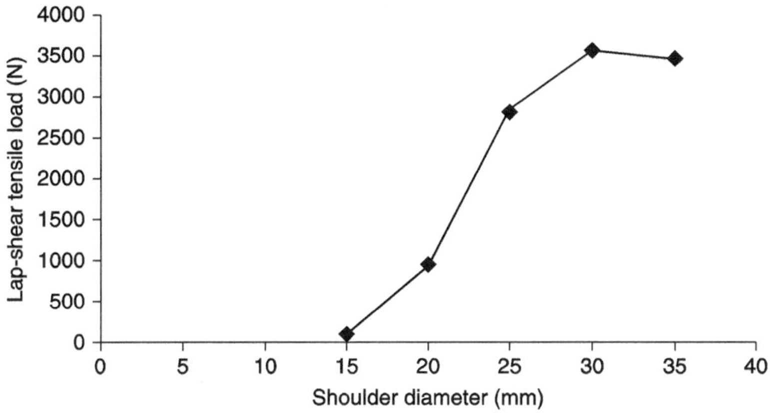

Figure 14.41 Effect of tool shoulder diameter on weld strength

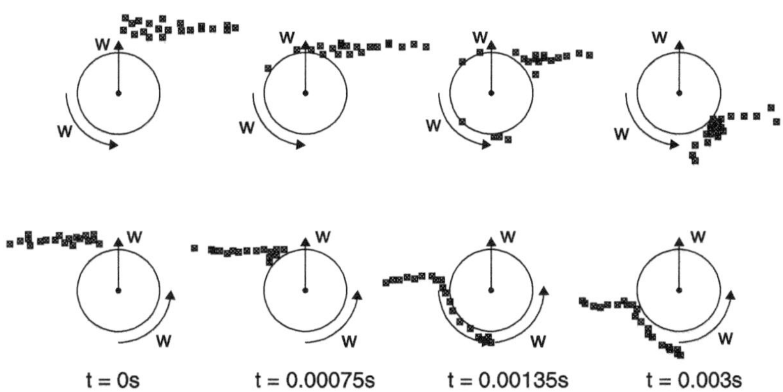

Figure 14.42 Different types of flow at advancing (above) and retreating side (below) of the weld [17]

FSW of the ABS (butadiene styrene) polymer is one of the common commodity polymers with wide applications, in which lots of experiments was carried out [17]. Table 14.8 shows some important properties of this material. They use the hot shoe method for eliminating the voids and poor mixing, and also to achieve high strength of joints.

An improved method for FSW, the hot shoe method, resulted in a high-quality joint surface and simplification of the fusion process. In this method, the tool shoulder is the shoe, and does not rotate with the tool mounted on the bearing. The shoe completely covers the weld zone and does not allow the material to flow out of the SZ. The shoe stirs and is heated by an external source.

Table 14.8 **Mechanical properties of Aida plastic ABS sheets [17]**

Parameters	Tensile strength	Tensile elongation	Flexural strength	Rockwell hardness	Heat distortion temperature	Density
Unit	MPa	%	Mpa	–	°C	g/cm^3
Standard	D-638	D-638	D-790	D-785	D-648	D-792-
Condition	23 °C	–	23 °C	R Scale	1.82 MPa	
Value	54	25	72	112	91	1.04

It was found that the unthreaded tool leads to poor volume retention and excessive machine stress. Figure 14.43 shows a schematic of the tool and rotating pin. A radial ball bearing is used for separating the rotating movement of pin from the fixed hot shoe.

The schematic structure of the test containing the hot shoe, the rotating pin, the ball bearing and thermo-couple system is shown in Figure 14.44. The assembled milling machine with test equipment is presented in Figure 14.45.

Three levels of the welding parameters were selected, based on the processing parameters of weld, including the rotational speed, travel speed and shoe temperature. Table 14.9 shows the above-mentioned levels in this research.

In order to determine mechanical properties, three parameters at three levels full factorial DOE were used to show effects of the operating parameters and an ANOVA study was performed to state the importance of the process parameters. The results show that the very high rotational speeds lead to degradation of the polymer as a result of a higher amount of heat generation by the frictional contact of the shoulder. Figure 14.46 shows the image of the degraded polymer joints at high rotational speed.

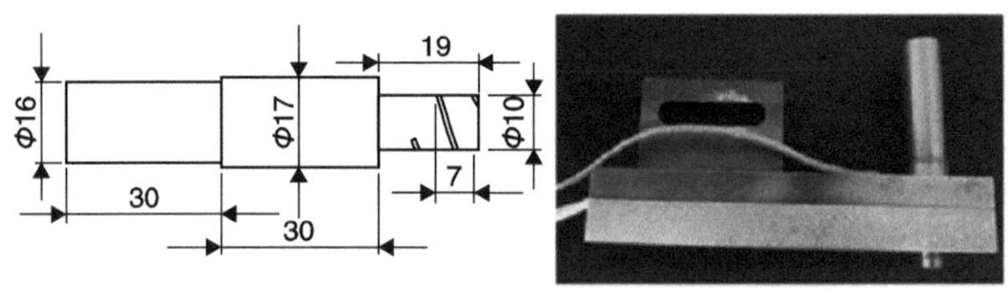

Figure 14.43 **Schematic of the rotating pin and photograph of the final tool design [17]**

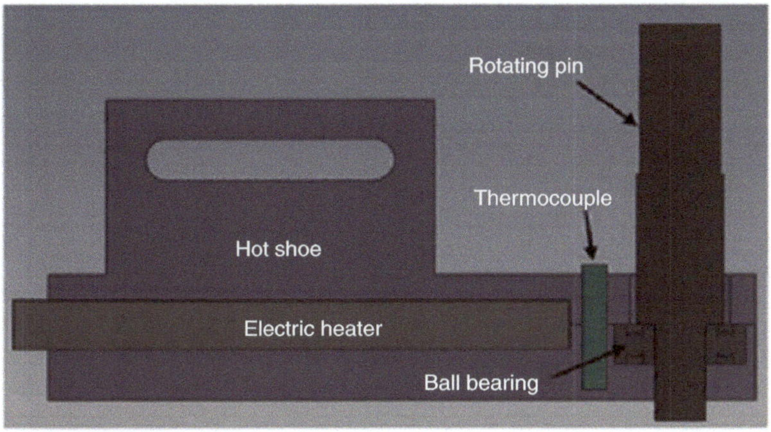

Figure 14.44 Schematic of the hot tool design [17]

Figure 14.45 Welding equipment installed on the milling machine [17]

Table 14.9 Processing parameters [17]

Parameter	Low level	Medium level	High level
Rotational speed (rpm)	800	1250	1600
Travel speed (mm/min)	20	40	80
Temperature (°C)	50	80	100

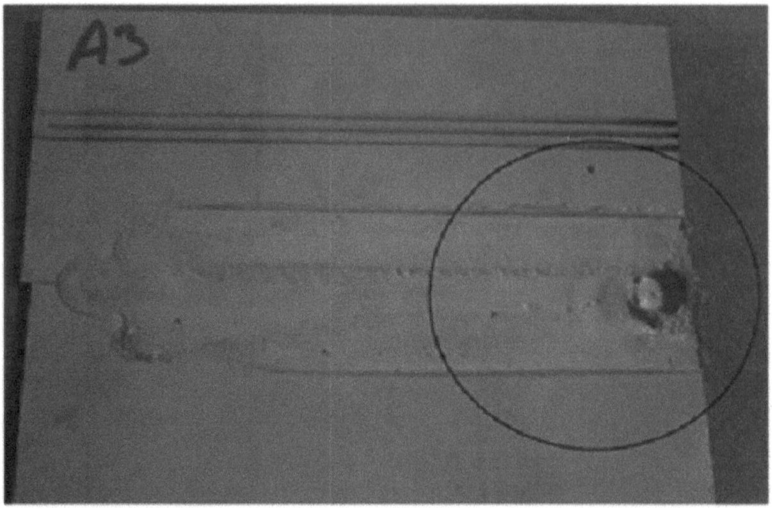

Figure 14.46 Burning and surface peeling in: rotational speed = 1600 rpm; temperature = 50 °C; travel speed = 20 mm/min [17]

As shown in Figure 14.47, the higher amount of traverse speed also leads to poor tensile properties and formation of voids and cracks in the welded zone. However, the lower amount of traverse speed produces an undesirable heat rise in the welded zone, which enhances the probability of polymer degradation. As a result, it is found that the traverse speed of 20 mm/min is the optimum amount.

It was also found that in all the tensile specimens, fracture occurred in the RS of the tool, which is due to the lack of material in this zone.

It is observed from the many experiments done by the authors that the processing conditions are of high importance for the tensile strength. Some of the critical parameters that control the strength of the weld are listed below.

Rotational speed

As shown in previous research, the higher spindle rotational speed leads to more stirring in the weld zone, which results in better mixing of welded parts. Also, increase in rotational speed clearly leads to higher heat input and subsequently higher material flow. Regarding the above-mentioned explanation, the higher rotational speed increases the strength of the

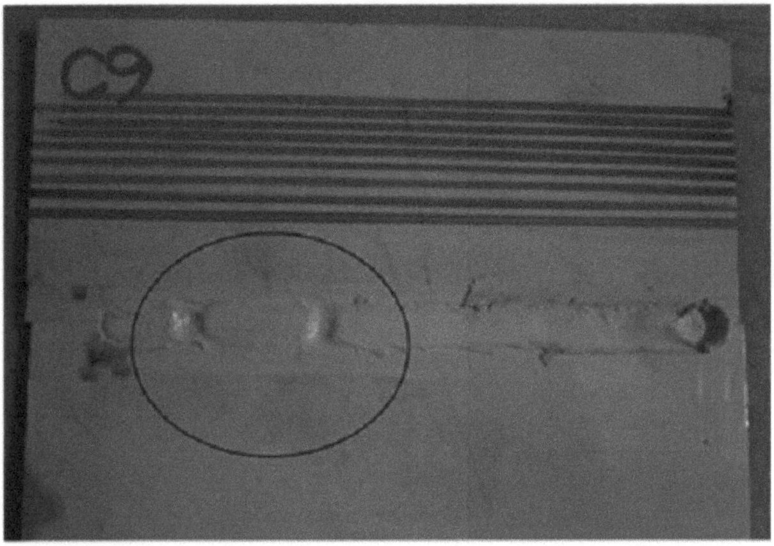

Figure 14.47 Voids and deformation in: travel speed = 80 mm/min; rotational speed = 1600 rpm; temperature = 100 °C [17]

weld (Figure 14.48). It should be noted that the very high rotational speed degrades the polymer and decreases the strength of the joint.

Shoe temperature

Figure 14.49 shows the effect of temperature on the tensile strength of the joint. As was seen through the experiments with the shoe temperature, the higher temperature generation leads to better mixing of the welding component and improves the strength of the weld.

Travel speed

A change in the travel speed of the process decreases the time of stirring, which is the opportunity for good mixing of the material, but reduces the heat input in the weld zone, resulting in low flowability of the materials. These two important factors play dominant roles in the weld quality and the shortage means the strength of the joints is diminished (Figure 14.50).

Contour plots of tensile strength vs. processing parameters and determination of the optimum area of variables change by response surface method (RSM) analysis are presented in Figures 14.51 to 14.53.

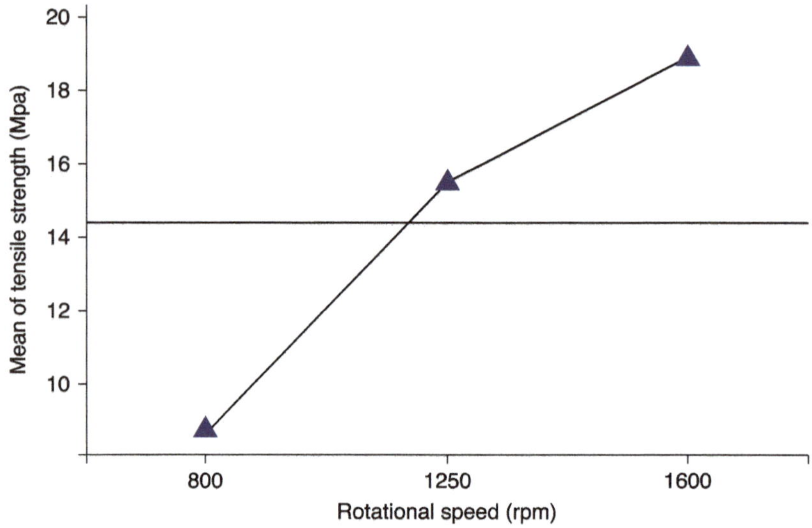

Figure 14.48 Main effects plot for rotational speed and tensile strength [17]

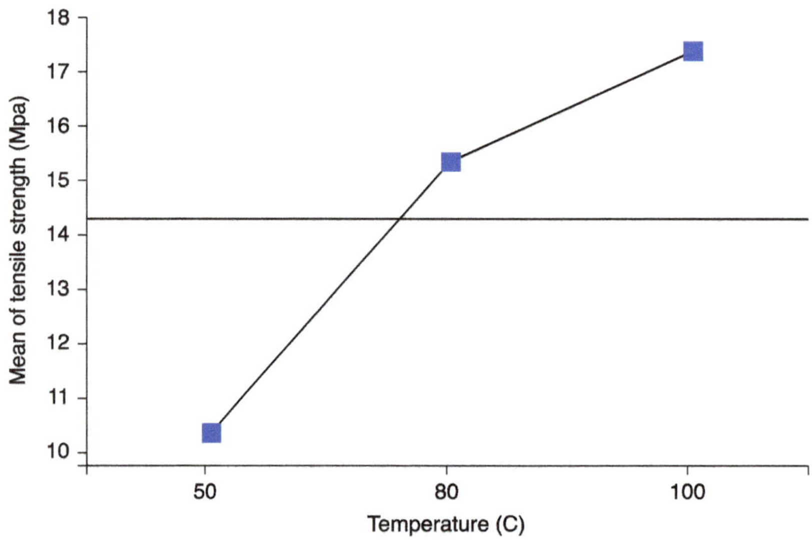

Figure 14.49 Main effects plot for temperature and tensile strength [17]

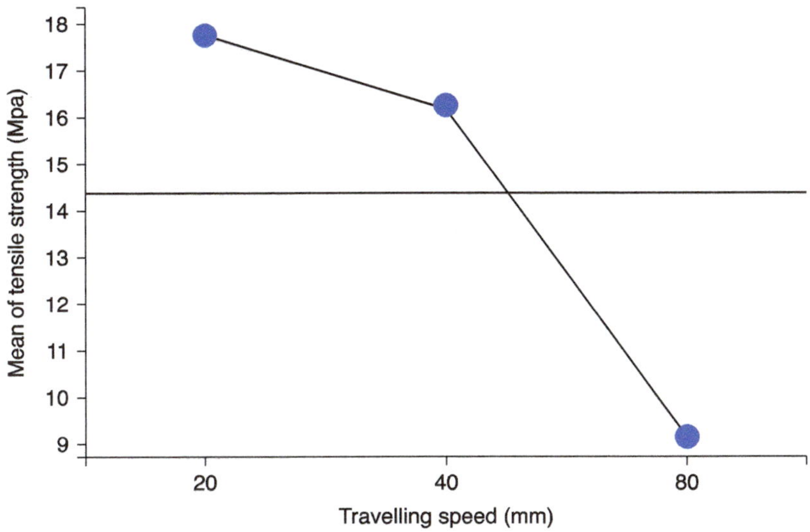

Figure 14.50 Main effects plot for travel speed and tensile strength [17]

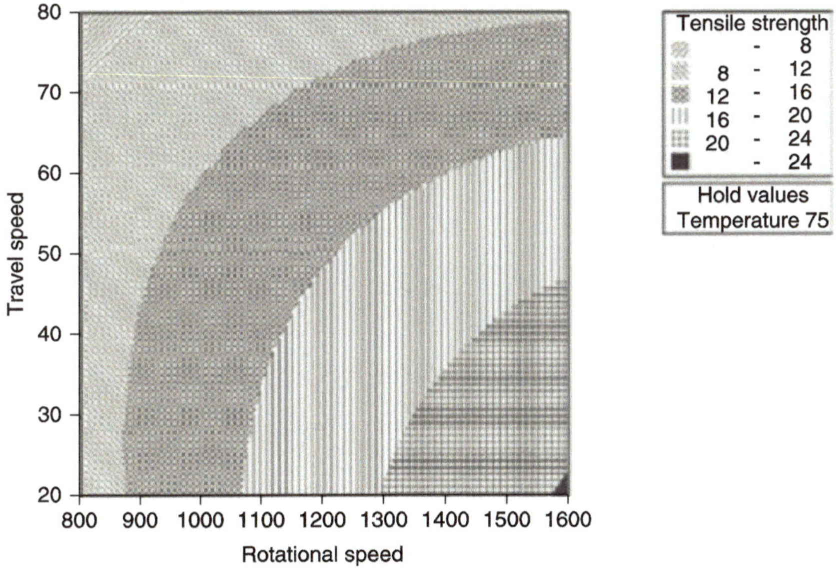

Figure 14.51 Contour plot of tensile strength vs. travel speed; rotational speed [17]

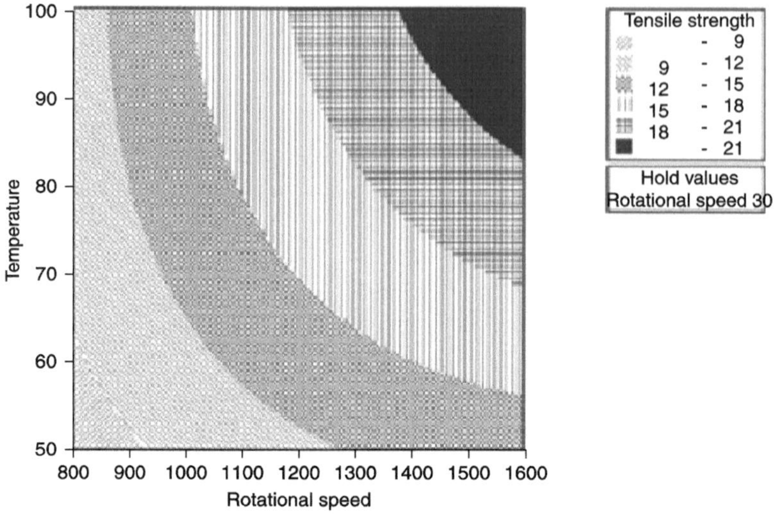

Figure 14.52 Contour plot of tensile strength vs. temperature; rotational speed [17]

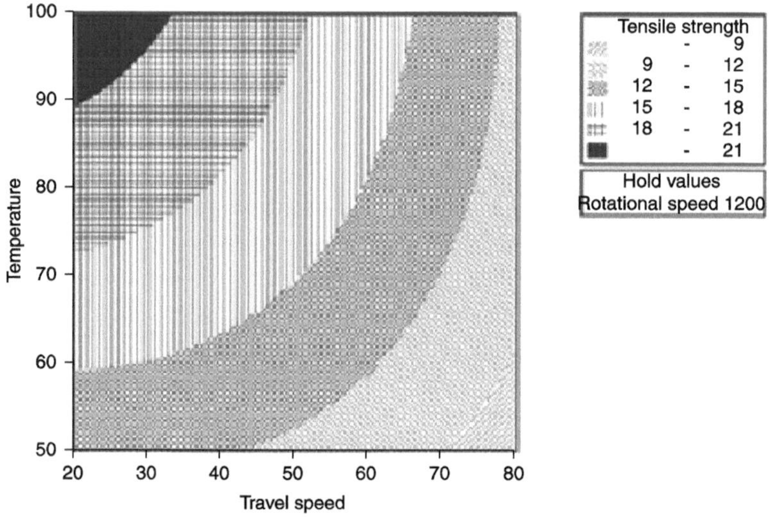

Figure 14.53 Contour plot of tensile strength vs. temperature; travel speed [17]

Table 14.10 presents the tensile strength data under different processing conditions, including the effects of rotational and traverse speeds and shoe temperature. The table shows that the best weld quality (strength of 88.8% of base material (BM)) is obtained under conditions with the

Table 14.10 Results of tensile test [17]

No.	Rotational speed	Travel speed (mm/min)	Temperature (°C)	Tensile strength (MPa)	Relative tensile strength (%)
1	800	20	50	9.79	26.64
2	1250	20	50	12.51	34.03
3	1600	20	50	13.14	36.56
4	800	20	80	10.20	27.76
5	1250	20	80	16.19	44.04
6	1600	20	80	28.98	78.84
7	800	20	100	16.54	44.99
8	1250	20	100	25.47	70.04
9	1600	20	100	32.62	88.76
10	800	40	80	8.01	21.79
11	800	40	50	9.62	26.18
12	800	40	100	14.37	39.09
13	1600	40	80	23.29	63.37
14	1600	40	50	16.19	44.06
15	1600	40	100	23.34	63.51
16	1250	40	80	19.54	53.15
17	1250	40	80	11.18	30.43
18	1250	40	100	19.74	53.71
19	800	80	50	1.99	5.41
20	1250	80	50	7.65	20.82
21	1600	80	50	8.85	24.09
22	800	80	80	6.98	19.00
23	1250	80	80	15.37	41.81
24	1600	80	80	10.31	28.04
25	800	80	100	7.06	19.21
26	1250	80	100	11.55	31.43
27	1600	80	100	12.99	35.35
Base	–	–		36.76	100

highest rotational and shoe temperature, when the lowest traverse speeds are used to join the components. The table also demonstrates that the worse quality joint is formed under conditions when the lowest rotational speed and highest traverse speed are applied.

Figures 14.54 and 14.55 show the micro- and macro-structure of the welded components under the worse and best condition. Figure 15.54(a) belongs to the welded specimen with lowest rotational and highest traverse speed, which contains the heterogeneous structure, whereas Figure 14.54(b) shows the macrostructure of the specimens welded at highest rotational and lowest traverse speeds, representing a uniform and good appearance. Microstructure images of the cross-section of the above-mentioned specimens are presented in Figures 14.55(a and b), indicating a more sound and uniform structure of the latter in comparison with the former.

In similar research, the possibility of dissimilar polymethylmethacrylate and acrylonitrile butadiene styrene sheets, which were FSSW, were

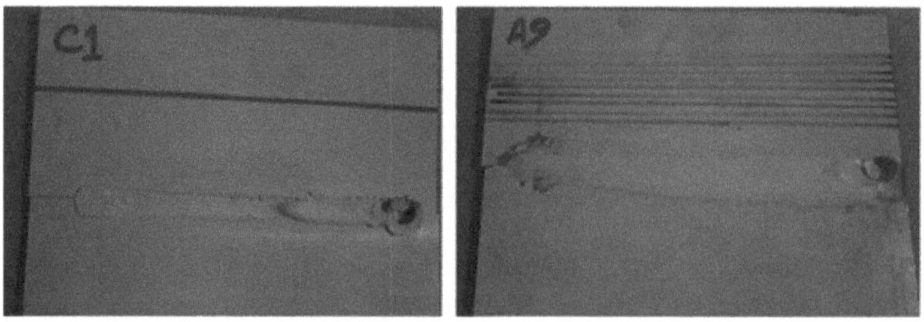

(a) (b)

Figure 14.54 Appearance of friction stir welded samples with: (a) minimum; and (b) maximum tensile strength [17]

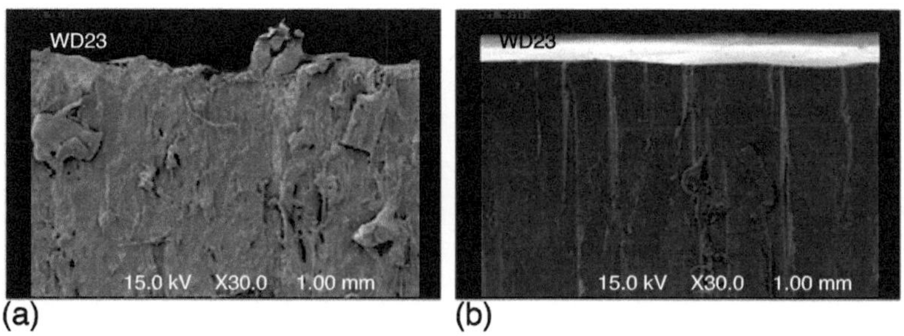

(a) (b)

Figure 14.55 Macrostructure of friction stir welded samples with: (a) minimum; and (b) maximum tensile strength [17] investigated. In this study the authors tried to experiment with the 3^3 (3 parameters in 3 levels) full factorial design to evaluate the effect of the process parameters on the lap-shear strength. The tool rotational speed, tool plunge rate and dwell time were considered as processing parameters. Table 14.11 presents the whole parameters variation in this work.

A tensile test were conducted to determine the lap-shear strength of the joints. Figure 14.56 shows the lap-shear specimens under shear loading conditions.

Based on the fact that the effects of processing parameters on the strength of the lap-shear test are the most important achievement of this joint, the results are reported in Table 14.12. The results are statistically analyzed by minitab V.14 software and the main effect of them on the weld strength was determined.

Figure 14.57 presents the tool rotational speed effect on the lap-shear tensile fracture load. As can be seen, the strength of the joint first increases by an increase in the rotational speed until 800 rpm and then reduces with rotational speed increase. These results are expected based on the previous discussed reasons, which demonstrated that low rotational

Table 14.11 Welding parameters of this [18]

Parameter	Low level	Medium level	High level
Dwell time	10	20	30
Plunge rate (mm/min)	8	16	24
Rotational speed (rpm)	500	800	1250

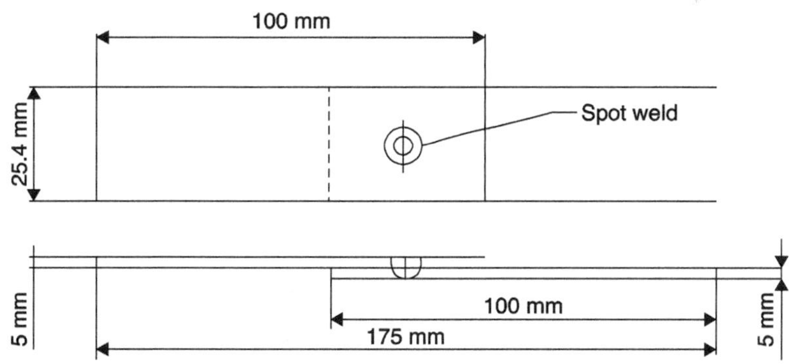

Figure 14.56 Configuration of the lap-shear test specimen [18]

Table 14.12 Lap-shear test results [18]

No.	Rotational speed (rpm)	Tool plunge rate (mm/min)	Dwell time (sec)	Lap-shear strength (N)
1	500	8	10	640.1
2	500	8	20	739.4
3	500	8	30	1005.5
4	500	16	10	479.5
5	500	16	20	529.7
6	500	16	30	755.4
7	500	24	10	279.6
8	500	24	20	480.7
9	500	24	30	474.6
10	800	8	10	683
11	800	8	20	945.4
12	800	8	30	1077.9
13	800	16	10	533.4
14	800	16	20	690.4
15	800	16	30	996.9
16	800	24	10	488
17	800	24	20	703.9
18	800	24	30	800
19	1250	8	10	533.4
20	1250	8	20	840
21	1250	8	30	998.2
22	1250	16	10	334.8
23	1250	16	20	616.8
24	1250	16	30	727.2
25	1250	24	10	436.5
26	1250	24	20	501.5
27	1250	24	30	360.5

speed leads to low material flow, while high rotational speed increases the heat and flowability of the material and consequently strength of the joint. However, it is clear that very high rotational speed leads to excessive heat input, leading to polymer degradation and low strength of joint.

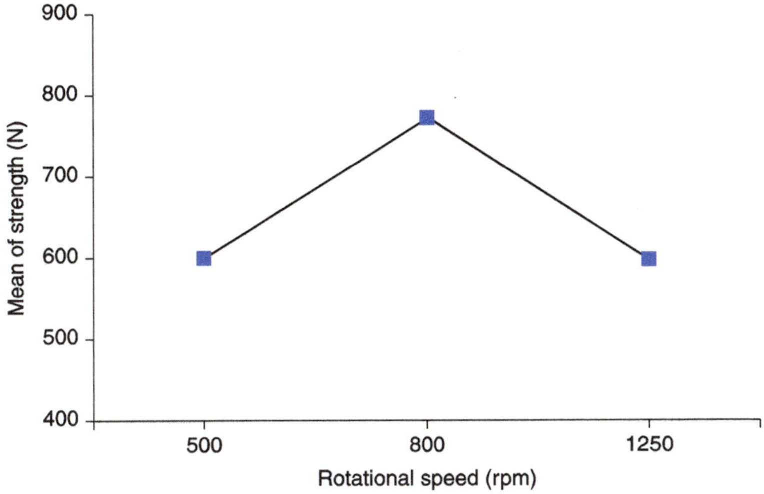

Figure 14.57 Main effects plot of tool rotational speed and lap-shear strength [18]

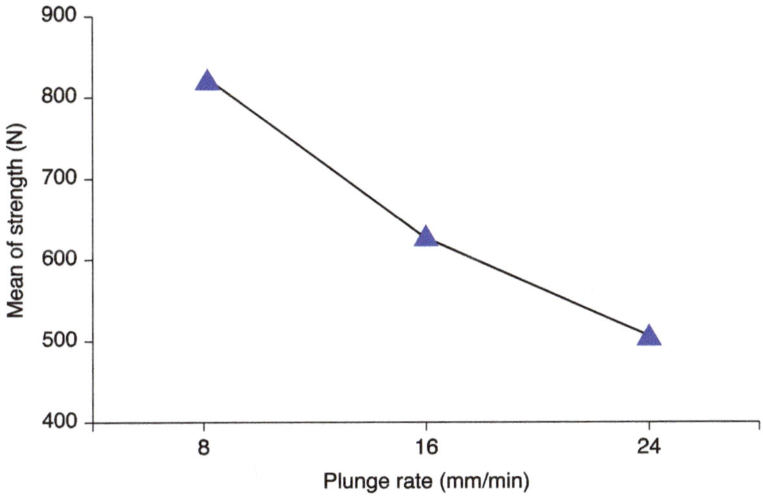

Figure 14.58 Main effects plot of plunge rate and lap-shear strength [18]

The effect of tool plunge rate on lap-shear strength resulting from the tensile shear test is presented in Figure 14.58. The results show that increase in the plunge rate of the tool decreases the heat generation and mixing capability of the components, which leads to low lap-shear strength.

Also, the effect of dwell time has been investigated and the results indicate that increase in dwell time from 10 to 30 seconds improves the strength of the welded parts. The reasons for this phenomenon are similar to earlier discussed parameters, such as rotational speed in which the higher dwell time leads to higher heat input as well as more opportunity to mix the components well. Figure 14.59 shows the results obtained for this evaluation.

To measure the effect of input variables on response variables through the relationship between input and output variables, the ANOVA method was used. The inputs variables for the study under discussion were tool rotational speed, tool plunge rate and dwell time, with the weld strength being considered as an output variable. Table 14.13 presents the results of ANOVA, indicating that the 95% confidence level has a statistically significant effect on weld strength. It is clearly seen from this table that all three input parameters have important effects on the output parameter, because of the larger F values for input variables compared to the F values ($F_{(2,8,0.05)}=4.46$) determined from statistical calculations. Moreover, the intersections F values are found to be smaller than theoretical F values ($F_{(4,8,0.05)}=3.84$), presenting the idea that they confer no significant effect on the weld strength.

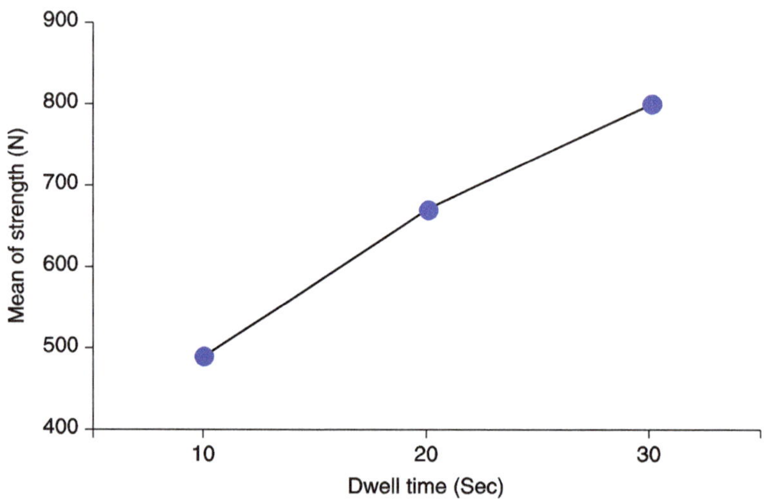

Figure 14.59 Main effects plot of dwell time and lap-shear strength [18]

Table 14.13 ANOVA results of present study [18]

Source	DOF	SS	MS	P	F	$F_{(DOF,8,0.05)}$
Plunge rate	2	487483	243741	0.000	40.92	4.46
Dwell time	2	436266	218133	0.000	36.62	4.46
Rotational speed	2	178539	89269	0.002	14.99	4.46
Rotational speed *plunge rate	4	19503	4876	0.548	0.82	3.84
Rotational speed *Dwell time	4	27262	6815	0.402	1.14	3.84
plunge rate * Dwell time	4	76116	19029	0.076	3.19	3.84
Error	8	47652	5957	–	–	–
Total	26	1272820	–	–	–	–

14.4.3 *Friction stir welding of polypropylene and its fibre reinforced composites*

FSW of the polymer-based composites, which contains 20% carbon fibre in polypropylene base, was reported [18]. Moreover, they investigated the effect of the tool pin profile on weld quality. Table 14.14 shows the different properties of the tools used for this research.

The FSW parameters were considered constant in all tool geometries, where they contained a rotational speed of 1000 rpm, traverse speed of 16 mm/min and tilt angle of 1 degree. Figure 14.60 presents the various tool pin profiles applied for this work.

Table 14.14 Different pin profiles of four friction stir tools [19]

No. tool	Description of the pin	Length of the pin (mm)	Diameter of the pin (mm)	Diameter of the shoulder (mm)
1	Threaded cylindrical	7.7	4	12
2	Threaded cylindrical-conical	7.7	4	12
3	Simple cylindrical-conical	7.7	4	12
4	Threaded conical	7.7	4.12	

Figure 14.60 FSW tools with different pins and shoulders [19]

The shear tensile test specimens were prepared, based on the ASTM D5868, and are shown in Figure 14.61.

Through this research, the effect of tool pin profile on the surface appearance of the welded zone was investigated. As discussed previously, the shoulder and pin have some duties during weld generation, which can be explained as follows: the pin plunges into the zone and stirs the area, which leads to material flow together with partial heat generation as a result of material frictional contact. The shoulder has a frictional contact with the surface of the plates, which are welded and this results in a large

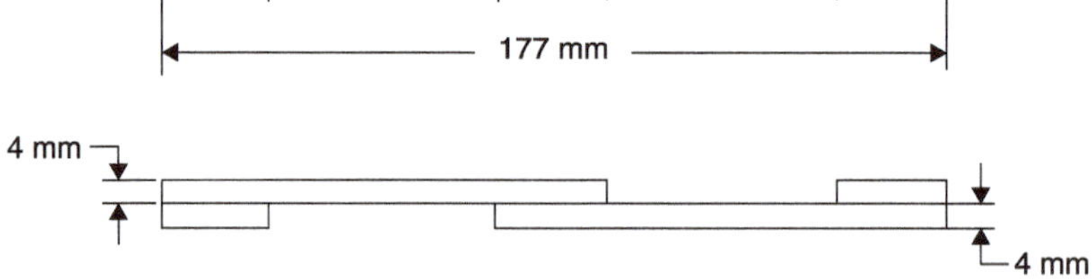

Figure 14.61 Tensile shear strength test specimen [19]

amount of heat input to the welding zone. Figure 14.62 shows the defects appearing as a result of using a pin length shorter than the optimum value.

Based on many previous reports, the pin length should be 0.2 to 0.3 mm shorter than the weld thickness. Figure 14.63 also shows the surface quality produced in the welded zone by using the different tool pin profiles listed above. The results show that the threaded cylindrical-conical tool makes the best surface quality, which could be because of higher contact surface compared to the others, which leads to higher heat input and material flow in the welding zone. It is worth mentioning that the simple cylindrical tool pin geometry resulted in the worse surface quality, which may be due to its low ability to create material flow and heat input.

The effect of the tool pin profile on tensile shear strength is also worth considering. Figure 14.64 shows the results of the tensile shear strength for the welds performed by different pin profiles.

The threaded cylindrical-conical tool pin has a higher shear tensile strength in comparison with the others. This phenomenon was expected because of the previous report on surface quality. This geometry due to the threaded surface has a higher ability to challenge the material and flow around the pin, which also generates higher heat in the welding zone. These specific properties make the joint tighter as a result of good materials combination.

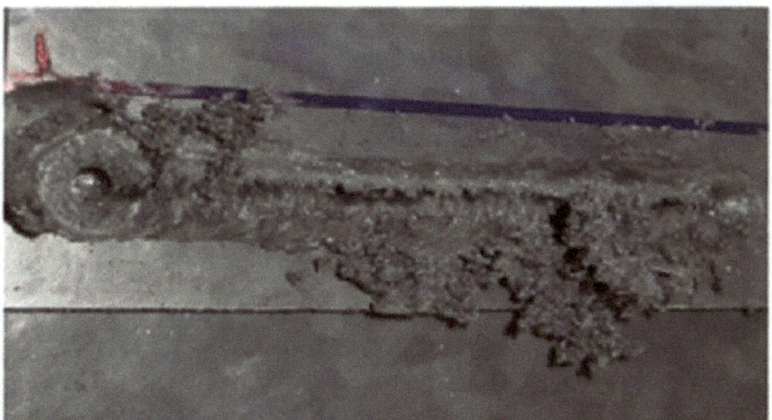

Figure 14.62 Formation of flash defect while welding [19]

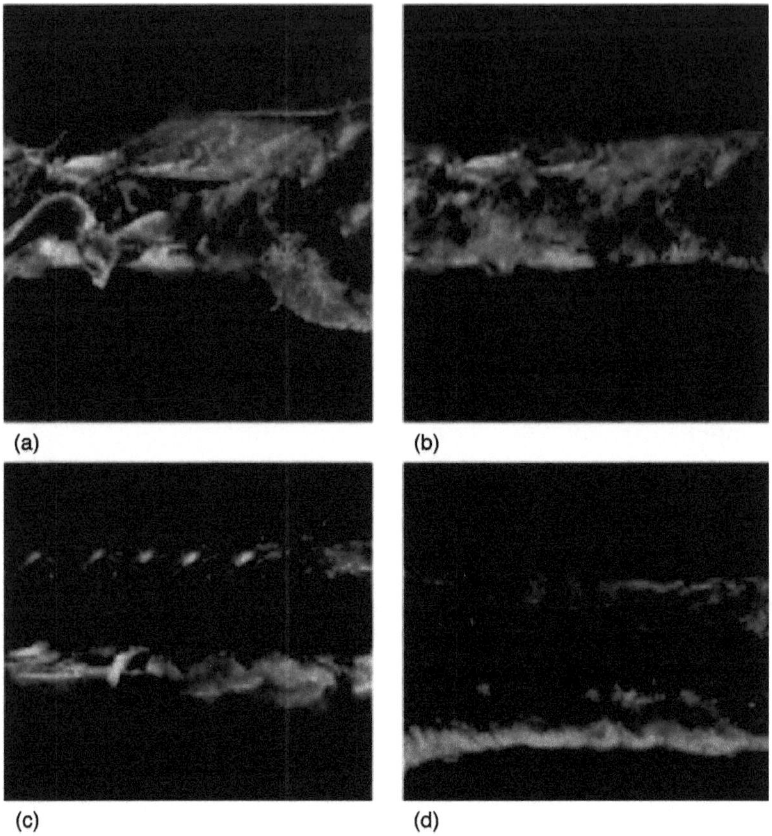

Figure 14.63 Effect of tool pin profile on weld surface appearance: (a) tool 1; (b) tool 2; (c) tool 3; and (d) tool 4 [19]

Microstructural analysis of the FSW zone of polypropylene was recently accomplished [19]. They used a non-heated smoothing shoe made of polytetrafluoroethylene instead of a rotational shoulder (Figure 14.65).

This experiment was performed under processing conditions with a cutting depth of 9.6 mm, a feeding rate of 60 mm/min and two rotational speeds of 2000 and 3000 rpm. The efficiency factors of the joints were the tensile strength of welded parts as compared to the tensile strength of the base metal (BM). The efficiency factor for two different rotational speeds was calculated as $54.9 \pm 5.7\%$ at 2000 rpm and $86 \pm 2.6\%$ at 3000 rpm, compared to the 25 MPa tensile strength of the BM.

Figure 14.66 shows the spherulitic structures formed in the cross-section of the welded zone. It is clear that spherulites are dispersed

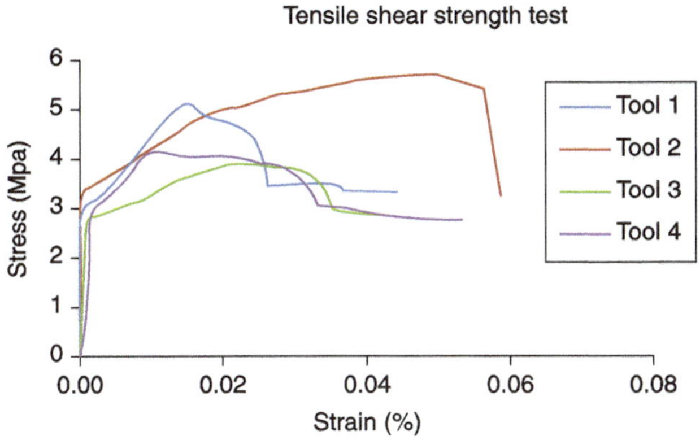

Figure 14.64 Results of tensile shear strength test [19]

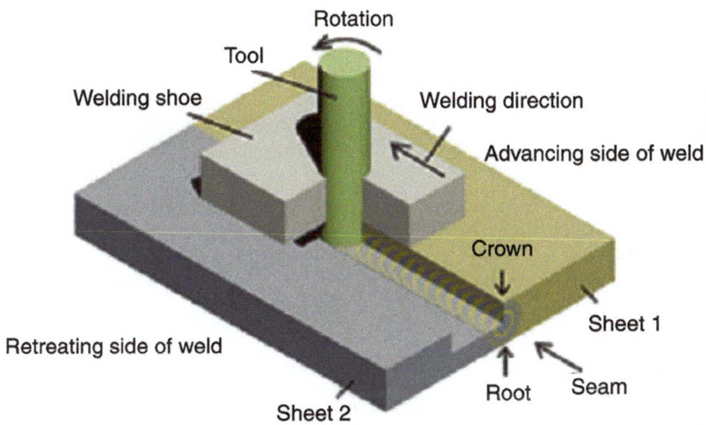

Figure 14.65 Schematic of the principle of friction stir welding for polymers [20]

homogeneously over the cross-section of the seam (Figure 14.66(a)). The spherulitic structures of the BMs (Figure 14.66(b)) are very similar to the seam. However, the spherulites of the BMs are up to 10 to 25 μm, whereas the spherulites of the seam are smaller (10–20 μm), which could be because of the 'memory effect', in the way that the heating by FSW could not erase the prehistory of the extruded sheet. The residual self-nuclei lead to crystallization yielding spherulites with reduced mean diameter in comparison with the base sheet.

Because of negligible differences between the super molecular structures in the central zone of the weld, the study concentrated on a sample taken

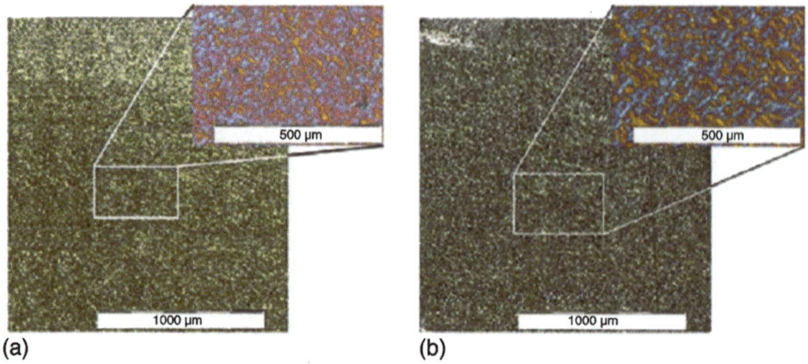

Figure 14.66 Optical micrograph with polarizer: (a) from the centre of the seam; and (b) from the base material. Note: inset images were made by additional use of λ-plate [20]

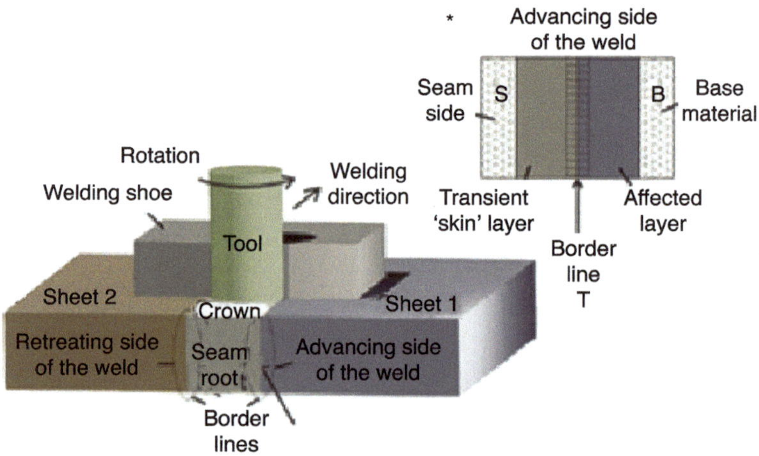

Figure 14.67 Build-up of the seam produced by FSW. Note: the blue region in the inset represents the heat affected zone (HAZ) [20]

from the transition zone (on the border between the BM and the seam (Figure 14.67))

In this research there are no significant differences between the RS and AS of the tool, as the main aim of the investigation was evaluation of the crystalline structure in the immediate neighbourhood of the border. Figure 14.68(a) shows the border in the AS of the specimens prepared at a rotational speed of 2000 rpm. Figures 14.68(b) and (c) show the

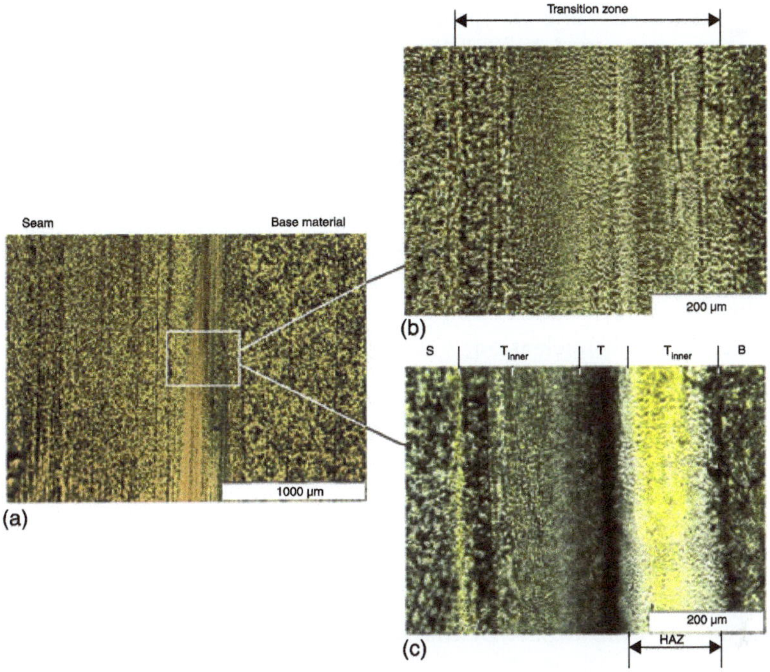

Figure 14.68 **Optical micrographs of the border region between the seam (produced at 2000 rpm) and the base material in transmission mode using crossed polarizers: (a) spherulitic structure of the seam and of the base (bulk) material; and (b) supermolecular structure developed in the border region. Note: (b) and (c) display the transition zone (also referred to as the skin-core structure) of the FSW seam along with the designations introduced. Frames (b) and (c) were taken using a polarizer with the plane diagonal and parallel to the orientated structure of the seam, respectively [20]**

highlighted parts of the transition zone, which reveal the established super-molecular structural layers.

In Figure 14.68(c), from right to left, the following areas can be distinguished:

- B: original spherulitic structure of BM with average size of 24 to 30 µm;

- T_{outer}: sheared zone of distorted spherulitic structure made by the peripheral speed of the welding tool. The high temperature generation in this zone softens the material and smears it along the mantle of the tool;
- T: represents the tool mantle movement line;
- T_{inner}: indicates the flow zone of the seam formed in a similar way to T_{outer}, but on the seam side. In the direction of BM, a severely damaged spherulitic structure can be seen whose formation was conducted by two effects: heat subtraction towards to the bulk, and crystallization affected by self-nucleation;
- S: inside the seam (S) spherulites are formed. Spherulitic crystallization occurred as a result of slow cooling in the middle of the welded zone. However, in the skin, owing to interaction of three main factors including cooling, molecular alignment/relaxation, and more importantly crystallization of the super-molecular structure was formed. After the welding process, the regular spherulitic structures were not revealed in the skin as a consequence of intense heat input, but in the central part of the weld (core), where the crystallization process was performed.

Figure 14.69 shows a SEM micrograph of the seam (Figure 14.69(a)) and of the BM (Figure 14.69(b)). The idea discussed earlier, which pointed out that the size of the spherulites in the seam zone was smaller than the BM, is further supported by these figures. Also it is found that the spherulites in the seam zone are well fragmented, which might be the result of the mixing shearing effect of the tool.

Work was also done on the welding of fibre reinforced polymer composites (polypropylene sheets containing 30 wt% chopped glass

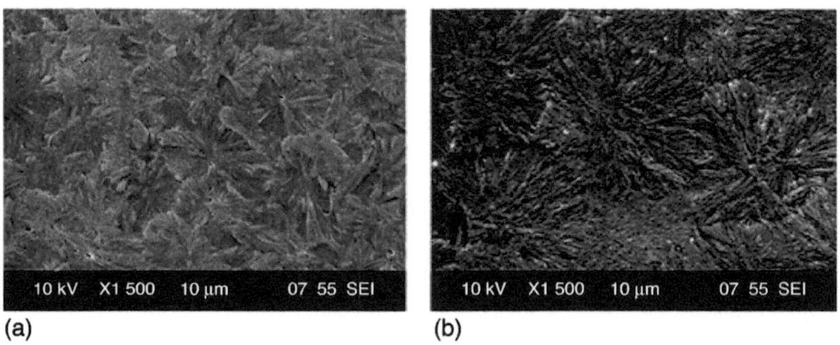

Figure 14.69 SEM micrographs after 4 hours etching: (a) from the seam: and (b) from the base material [20]

fibre), which are the basis of load-bearing structures. The main problem of thermoplastic reinforced polymer joining is lack of continuity in welded parts and dependency of joint strength to matrix-matrix interconnectivity. However, the rotating tool of FSW also has the capability of joining and homogenizing reinforcements. Figure 14.70 shows the uniform intermeshing of the welded parts by FSW.

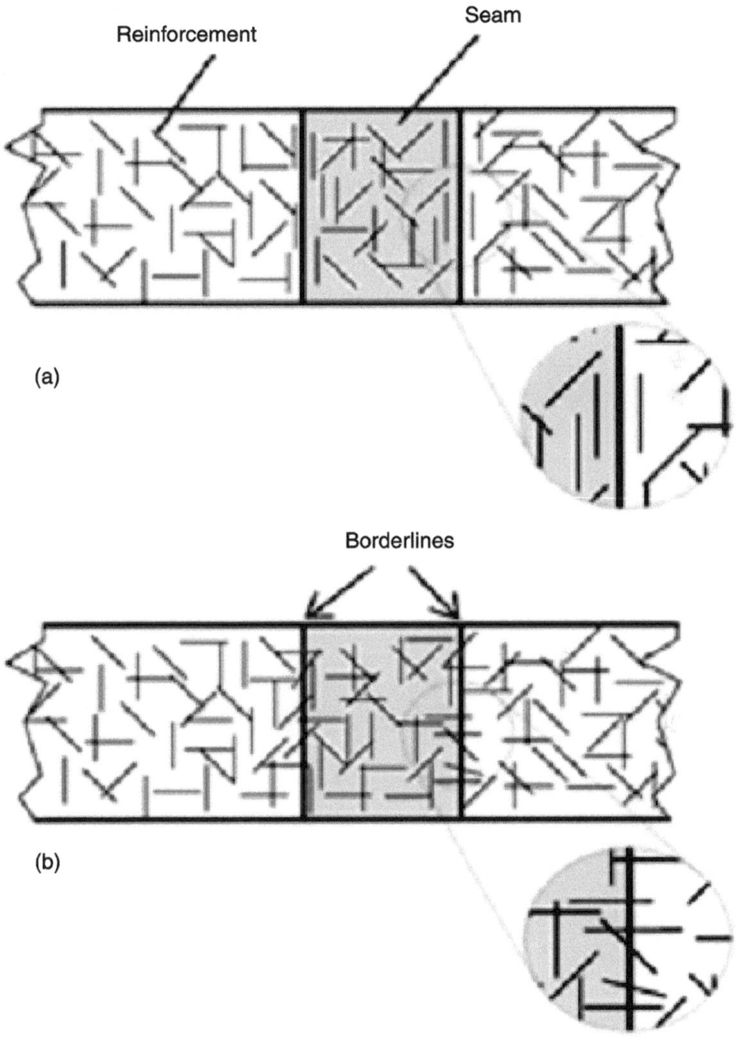

Figure 14.70 Welded seam in fibre reinforced composites with thermoplastic matrix: (a) non-intermeshing fibres; and (b) intermeshing fibre at the interface [21]

The tool, with a milling cutter of 8 mm diameter with 4- or 8-teeth and a 3000 rpm rotational speed was chosen for the experiments. Figure 14.71 shows the relative frequency of the fibre length in the matrix in the BM and welded zone. It is clearly demonstrated that in the latter, the fibre lengths are in the same range, which shows the significant capability of the FSW to reduce the fibre length. This phenomenon can be attributed to the severe shearing stress applied to the welded zone by the rotating tool.

As the strength of the joint performed by the 8-edge tool at 3000 rpm was poor and did not satisfy the design requirements, the author tried to improve the quality of weld by changing the tool from 8 to 4 edges and also to use a rotational speed range of 1000 to 3000 rpm. The results obtained for the different rotational speeds are presented in Figure 14.72, which indicates that the rotational speed of 2100 is the optimum value for giving good strength to the joint. These results originate from the fact that at the higher rotational speed the material can be depredated by high heat input and also at lower rotational speed, the material does not flow well and the fragmentation of the fibres is weak.

Figure 14.73 shows the different fibre length dispersion on the matrix in the seam zone performed by the best rotational speed of 2100 rpm and a 4-edge tool. The results show that distribution of the fibres is still at a high level, but the fibre length average is double that of the sample fabricated by the 4-edge tool. This increase in the distributed fibres leads to better strength of the composite.

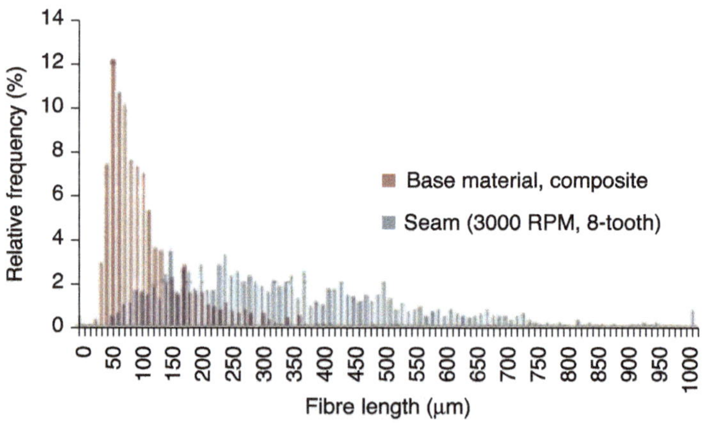

Figure 14.71 Fibre length distribution (frequency) of the glass fibre in the base composite sheet and in the welded seam prepared at 3000 rpm using a milling cutter of 8 mm in diameter with 8-tooth [20]

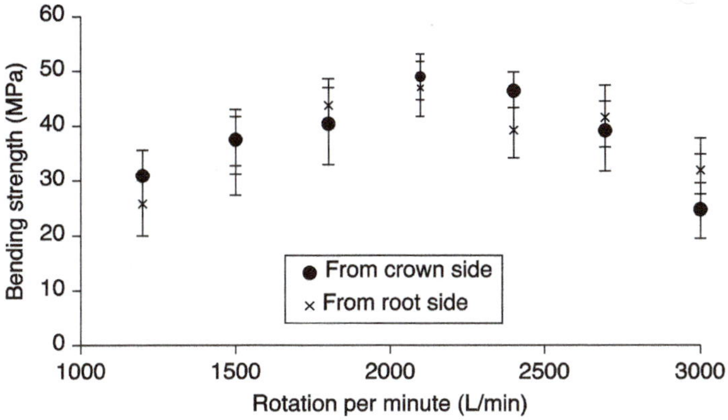

Figure 14.72 Three-point bending strength values of glass fibre reinforced H483F PP matrix composites [21]

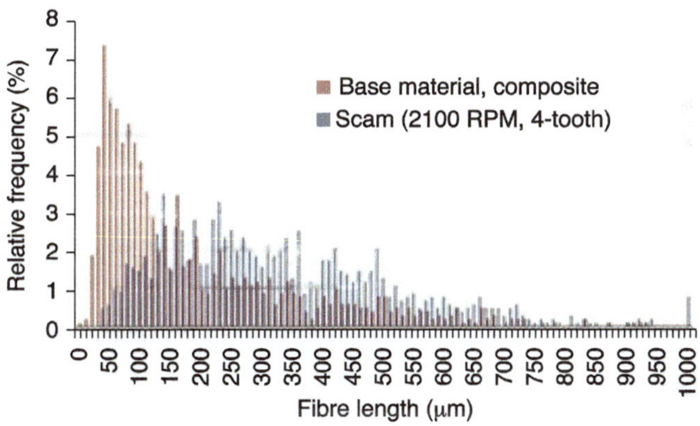

Figure 14.73 Fibre length distribution (frequency) of the glass fibre in the base composite sheet and in the welded seam prepared at 2100 rpm using a milling cutter of 8 mm in diameter with 4-tooth [21]

In the microscopy image of the polished samples, it can be seen that the fibres in the BM are remarkably longer than in the seam zone. Figure 14.74 shows the micrograph image of the seam, border and BM. The image also indicates that the absolute and relative orientation of the fibres in the border and seam are random.

In addition to the optical microscopy image, a scanning electron microscope (SEM) image was also taken of the fractured surface of the welded zone. As shown in Figure 14.75, the broken fibres in the seam or edge of the welded zones are protruding as a result of failure.

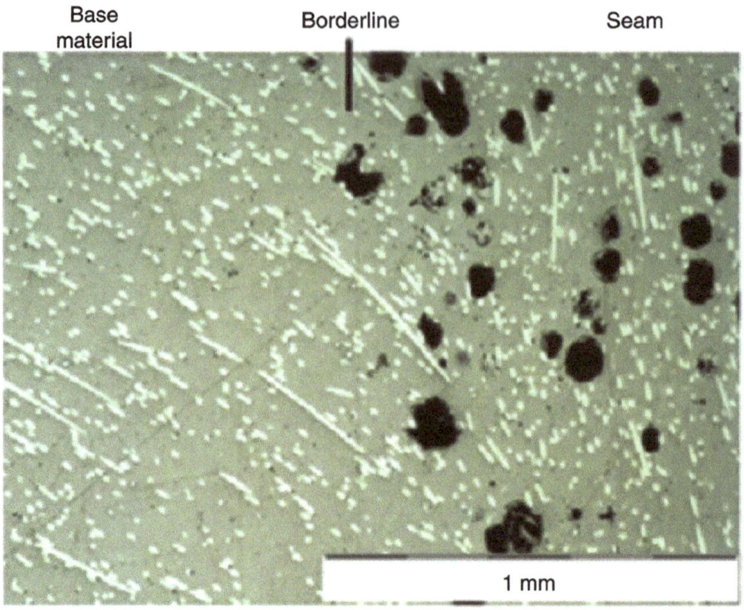

Figure 14.74 Optical micrograph of the polished surface prepared at the borderline of the seam and of the base material [21]

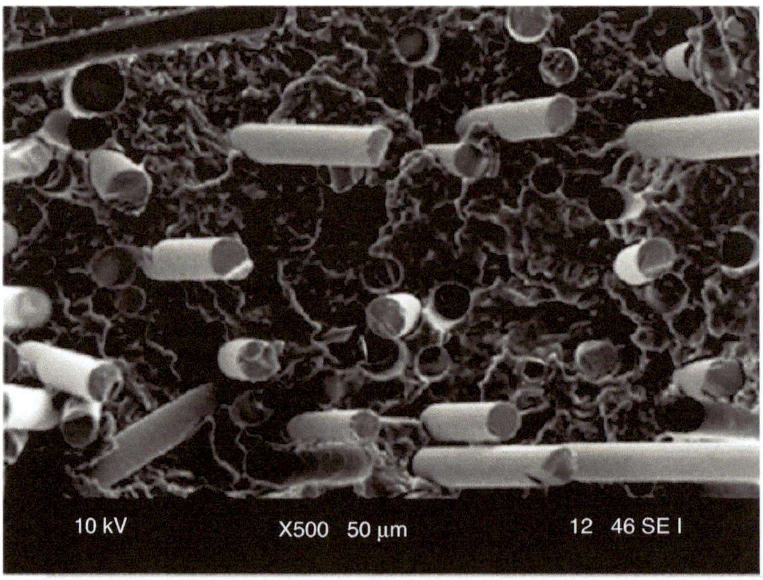

Figure 14.75 SEM micrograph of a fractured surface prepared by FSW on two PP sheets reinforced by 30 wt% glass fibre [21]

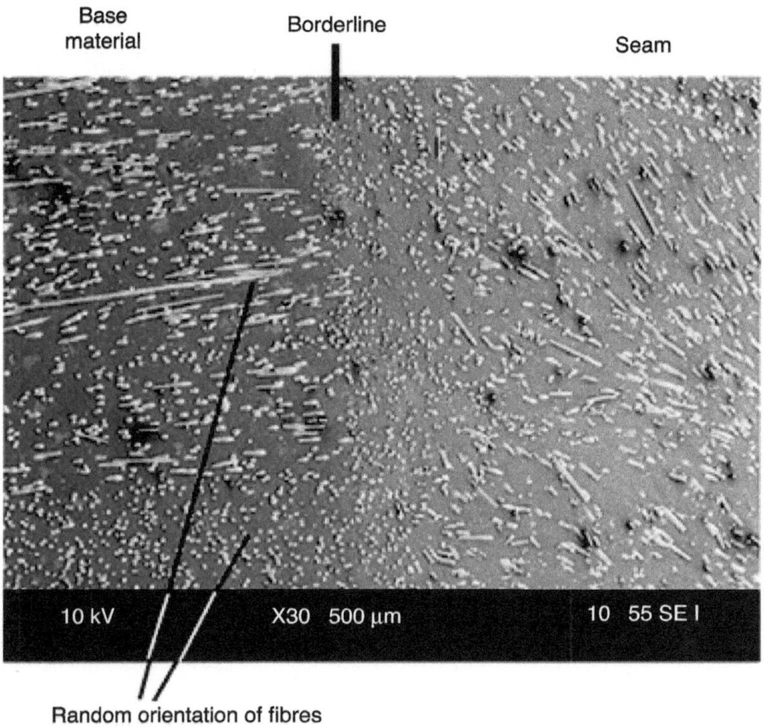

Figure 14.76 SEM micrograph taken at borderline of the seam and the base material [21]

The optical and (SEM) images show that the tool rotational speed could cause intermeshing between the BM and fibres of the seam. Figure 14.76 presents an SEM image of the different areas of the welded components, which further confirm the effect of the rotational speed.

14.5 Composite fabrication via friction stir processing

Although there were many works done on FSW of the polymer and reinforced polymer plates, there was a lack of investigation of FSP. Thus it was planned to experiment with the fabrication of the *in situ* nanocomposite by this method, by reinforcing the HDPE by nanoclay particles [21].

In order to produce HDPE/clay nanocomposites, nanoclay particles were inserted into a groove with dimensions of 1 mm × 1 mm in the centre of samples. When the nanoclay particles had been compressed into the groove, its upper surface was closed with a FSP-shaped tool without the pin, which prevented leakage of the nanoclay particles. The final stage was plunging of the tool with pin inside the plate to produce nanocomposites. The aforementioned stages are presented schematically in Figure 14.77.

The processing tool was tilted to an angle of 1 degree. The FSP tools were made of hot-worked steel with shoulder diameter, square pin diameter and length of 20 mm, 5 mm and 2 mm, respectively. The prepared samples are listed in Table 14.15.

The microhardness properties of the samples were measured on the cross-section of welds and perpendicular to the processing direction using an indenter with a 100 g load for 15 s on a standard Vickers microhardness tester. The standard Vickers microhardness (H) was determined by

$$H = kP/d^2 \tag{14.1}$$

where P is the applied load, k is a geometric factor equal to 1.854 and d is the mean diagonal length of the imprint after removing the indenter. At least 10 imprints were made under each load.

The obtained microhardness values of HDPE/nanocomposites are reported in Figure 14.65 in terms of rotational and traverse speed. It also should be noted that the microhardness value of the parent polymer, which was not processed by FSP, was measured as 44 MPa. According to the results, a 62% increase was observed in the microhardness values

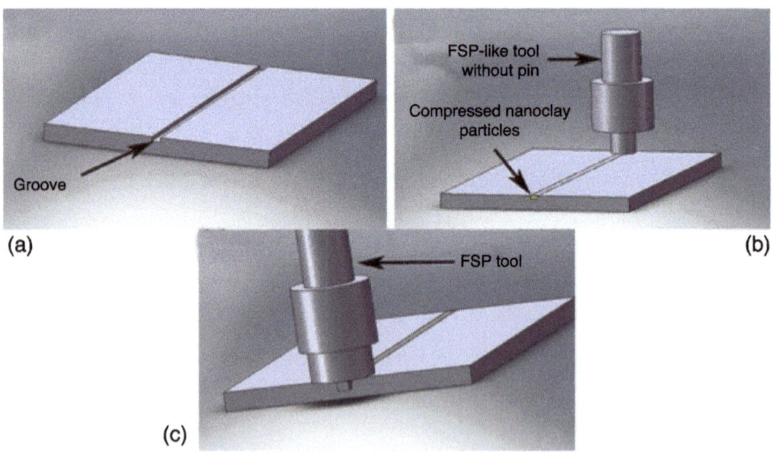

Figure 14.77 Schematic of FSP experiment [22]

Table 14.15 Samples prepared at different processing conditions [22]

Process parameters	Samples					
	T1	T2	T3	R1	R2	R3
Traverse speed (mm/min)	80	120	160	80	80	80
Rotational speed (rpm)	900	900	900	710	900	1120

of the nanocomposites prepared by the FSP method, whereas the sample that was prepared by internal batch mixing showed a 22% enhancement in microhardness values. These results verify the higher efficiency of the FSP method to improve the surface properties of polymers, such as microhardness, rather than by the conventional method of bulk mixing.

The high values of microhardness obtained by the FSP method could be attributed to good dispersion and higher levels of exfoliation of silicate layers, which are known to have high stiffness and hard materials. However, we postulate that the effect of FSP on the crystallization mechanism could be another important factor in determining the microhardness of polymer nanocomposites.

Regarding the role of traverse speed in the dispersion state of nanoclay particles and microhardness of nanocomposites, it should be noted that different opposing factors could be responsible for this behaviour. On the one hand, it is true that at lower traverse speeds, the time that stirring action takes place in the SZ is longer, which leads to an enhanced state of dispersion of nanoclay and microhardness of the system; however, increasing the traverse speed causes the SZ to be compressed, which induces higher amounts of thermo-mechanical stress on the polymer leading to improvement in dispersion state of nanoclay particles. According to the rheological results, it was found that the latter factor is more dominating, leading to an overall enhancement of nanoclay dispersion and consequently microhardness of HDPE nanocomposites.

Figure 14.78 shows the effect of rotational speed on microhardness values. An obvious enhancement in microhardness can be seen by increasing the tool rotational speed. During the FSP, shear forces applied on the polymer macromolecules cause the nanoclay layers to separate, leading to an increase in the ease of penetration of polymer chains into the silicate galleries. Therefore, higher rotational speeds induce higher shear forces on polymer nanocomposites resulting in higher delamination

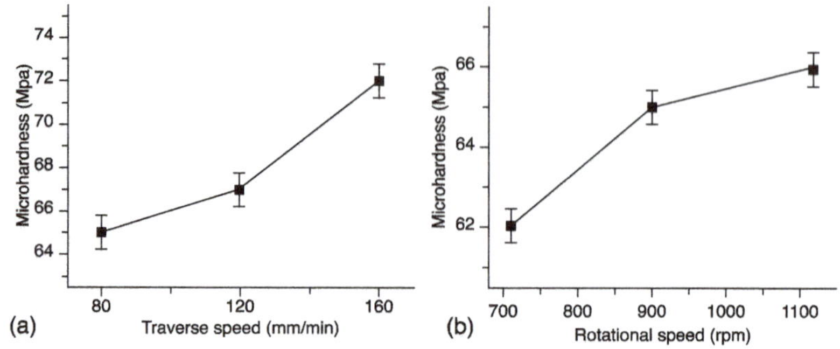

Figure 14.78 **Microhardness of FSPed samples at different processing parameters: (a) traverse speed; and (b) rotational speed [22]**

of silicate layers in the HDPE matrix, which finally increases the microhardness values.

The X-ray patterns of FSP and melt mixed nanocomposites and neat nanoclay are shown in Figure 14.79. Because HDPE is a non-polar

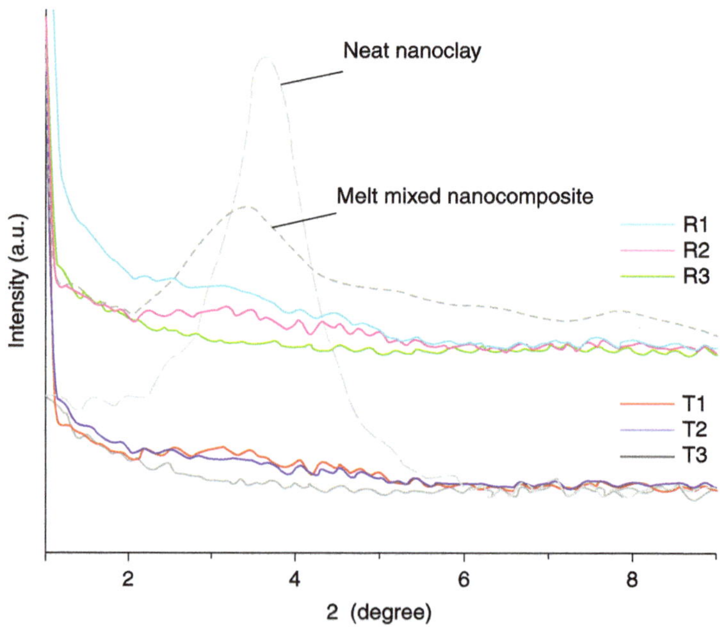

Figure 14.79 **XRD patterns of the melt mixed and FSPed nanocomposites and neat nanoclay [22]**

polymer, it is difficult to prepare HDPE/clay nanocomposites. Generally, melt intercalation has been considered the most desirable method for direct dispersion of clay layers in the molten polymers; however, an exfoliated structure cannot be obtained for HDPE/clay nanocomposites at high nanoclay contents. As can be seen from Figure 14.79, for the neat nanoclay, an intensity maximum occurs at 3.60. For the nanocomposite prepared by the melt mixing method, only a slight shift of the basal peak to lower angles is observed, implying formation of an intercalated structure in this nanocomposite, whereas for nanocomposites FSP at different processing conditions, in addition to a slight shift of the (001) peak to lower angles, significant broadening of the basal peaks is also detected, which indicates an increase in the gallery height of silicate layers and an improved dispersion state of nanoclay particles in the matrix.

Moreover, the absence of the (001) peak for T3 and R3 samples in the low-angle region implies a great deal of delamination of clay layers in nanocomposites, which is impossible via other techniques. However, in the case of the sample prepared at a traverse speed of 120 mm/min (T2), a very broad peak is observed, which suggests that nanoclay layers are largely exfoliated, although there could exist some unexfoliated clay stacks. Thus, it is reasonable to assume that a mixture of intercalated and exfoliated morphologies has been obtained in this sample. X-ray diffraction (XRD) patterns show that FSP could be a very effective method to obtain exfoliated structures in nanocomposites, even those based on non-polar polymers at high nanoclay loadings.

Figure 14.80 shows the transmission electron microscope (TEM) image of HDPE nanocomposite, which is prepared at a rotational speed of 900 rpm and a traverse speed of 120 mm/min (T2). It is observed that a very good level of dispersion of nanoclay layers took place in the matrix and many disordered single platelets can be seen.

Following research into polymer-based composites, investigation of fabrication of PP/Al_2O_3 by imposing FSP was considered [22]. They evaluated the mechanical properties of the nanocomposites fabricated by entering three different volume percentages of Al_2O_3 into the polymer matrix. In this research, an additional heating system was added to the FSP to keep the temperature and cooling rate uniform during the process. The tool designed for this study consisted of a shoe, rotating pin and heater located behind the pin. Figure 14.81 shows the tool geometry and structure.

The processing parameters in this work were applied as follows: tool traverse speed, pin rotational speed and shoulder temperature of

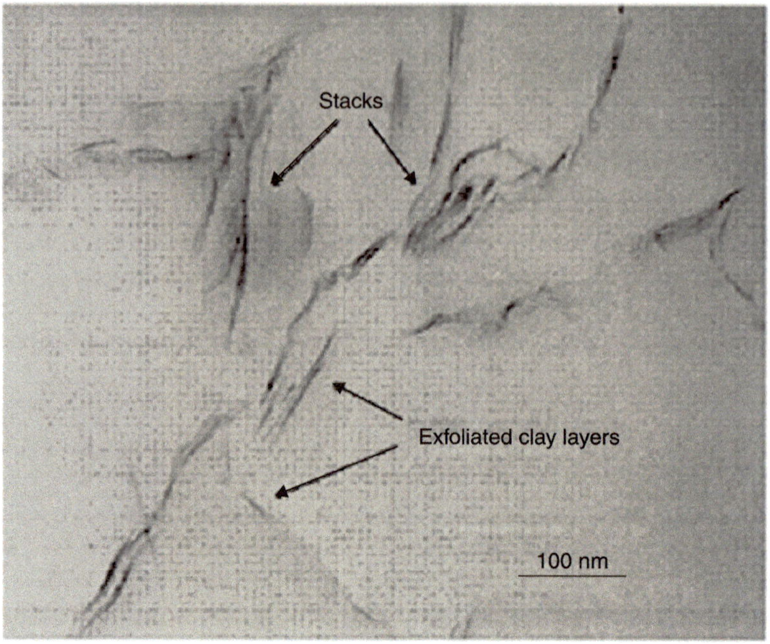

Figure 14.80 TEM image of the same FSPed at traverse speed of 120 mm/min [22]

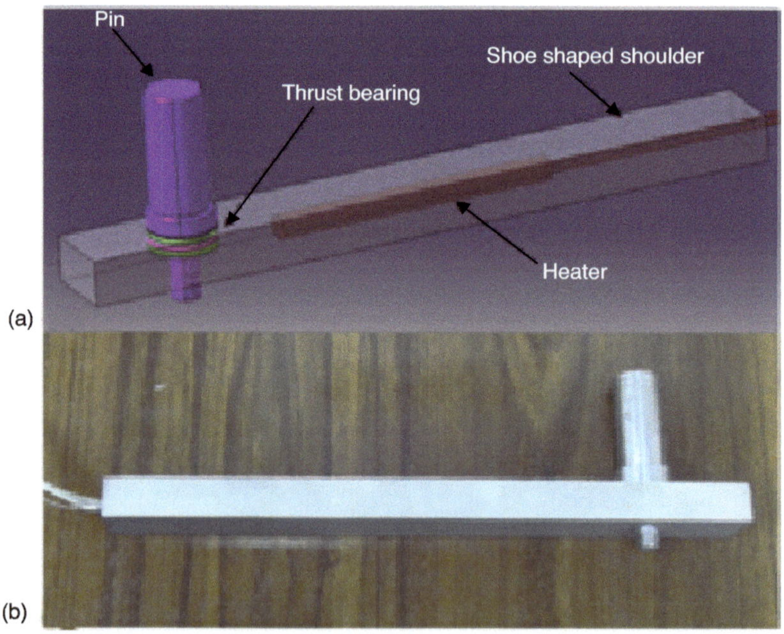

Figure 14.81 Designed tool for FSW of polymer: (a) schematic illustration; and (b) photograph of tool [23]

50 mm/min, 1000 pm and 190 °c, respectively. The Al_2O_3 powder volume percentages were chosen as 5, 10 and 15%, respectively. The hardness measurement was conducted by an indenter with a 100 g load for 15 s on a standard Vickers micro-hardness (H) and was based on the equation:

$$H = KP/d^2 \quad [14.2]$$

where P is applied load, K is geometric factor equal to 1.85 and d is the mean diagonal length of the imprint after removing the indenter. Figure 14.82 shows the microhardness values obtained from the test. It is shown that a composite with 10% filler has 54% enhancement in microhardness values. This could be attributed to the good dispersion of this volume of nano-powder in the polymeric matrix.

Figure 14.83 shows the microstructure of the PP + 10% nano-Al_2O_3, which indicates the high level of fillers distribution in the base. Also it is comprehensible that the composites with higher percentage of nano-powder show a slight reduction in microhardness, which is a result of partial agglomeration of fillers.

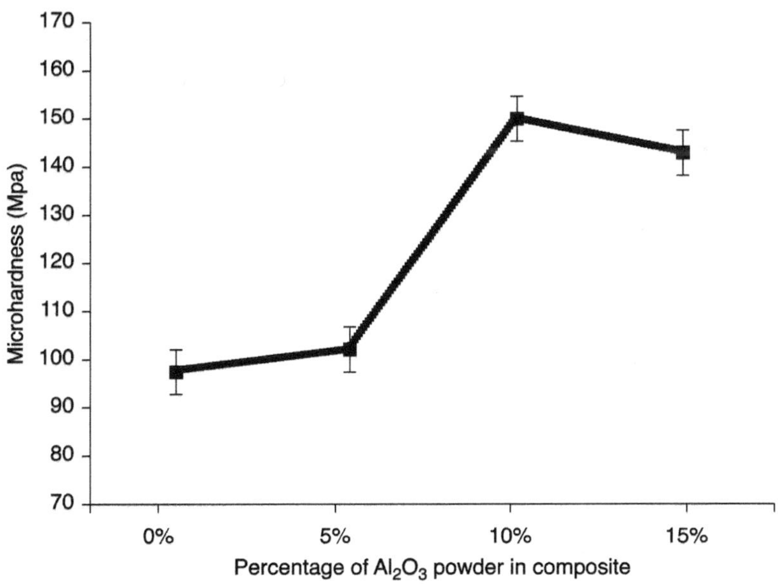

Figure 14.82 Microhardness of fabricated composites in different nano-alumina contents [23]

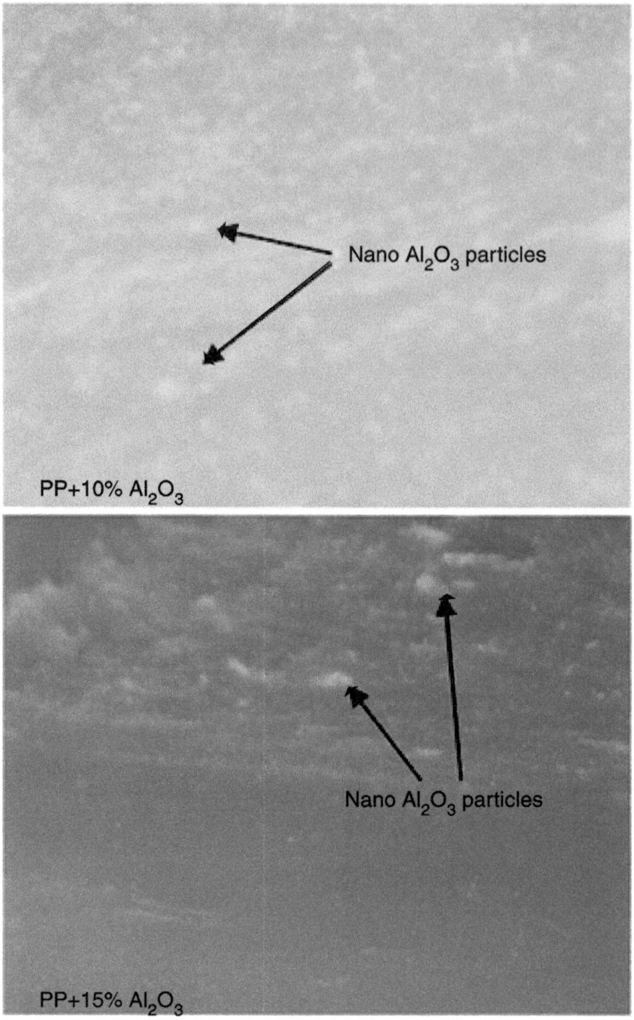

Figure 14.83 Microscopic observation of fabricated nanocomposites [23]

14.6 References

1. Flory, P.J. (1953), *Principles of Polymer Chemistry*, Ithaca, New York: Cornell University Press.
2. Robeson, L.M. (2007), *Polymer Blends: A Comprehensive Review*, Munich: Hanser Gardner.
3. Utracki, L.A. (1995), 'History of commercial polymer alloys and blanks: From a perspective of the patent literature', *Polym. Eng. Sci.*, 35: 1–17.

4. Buchmeise, R.M. (2003), *Polymeric Materials in Organic Synthesis and Catalysis*. Weinheim: Wiley-VCH Verlag GmbH & Co. KGaA.
5. Costello, P.A., Slark, A.T., Sherrington, D.C. et al. (2002), 'Branched methacrylate copolymers from multifunctional monomers: Chemical composition and physical architecture distributions', *Polymer*, 43: 245–54.
6. Findlay, P.H., Leinonen, S.-M., Morrison, J.T. et al. (2000), 'Thin film format for polymer supports synthesis and chemical modification', *J. Mater. Chem.*, 10: 2031–4.
7. Sharma, K.R. (2012), *Polymer Thermodynamics, Blends, Copolymers and Reversible Polymerization*, Boca Raton, FL: CRC Press, Taylor & Francis Group.
8. Braun, D., Cherdron, H., Rehahn, M., Ritter, H. and Voit, B. (2005), *Polymer Synthesis: Theory and Practice: Fundamentals, Methods, Experiments*. Berlin. Heidelberg. New York: Springer.
9. Allen, G. and Bevington, J. (eds) (1989), *Comprehensive Polymer Science*, vol.1, Oxford: Pergamon Press.
10. van Krevelen, D.W. (1990), *Properties of Polymers, their Estimation and Correlation with Chemical Structure*, 3rd edn, Amsterdam: Elsevier.
11. Grewell, D. and Benatar, A. (2007), *Welding of Plastics: Fundamentals and New Developments. Agricultural and Biosystems Engineering*, Ames, IA: Iowa State University.
12. Arici, A. and Sinmaz, T. (2005), 'Effect of double passes of the tool on friction stir welding of polyethylene', *Journal of Materials Science*, 40: 3313–16;
13. Squeo, E.A. and Bruno, G. (2009), 'Friction stir welding of polyethylene sheets', *The Annals of 'Dunărea De Jos*, University Of Galaţi Fascicle V, Technologies in Machine Building.
14. Rezguia, M.-A., Trabelsi, A.-C., Ayadi, M. and Hamrouni, K. (2011), 'Optimization of friction stir welding process of high density polyethylene', *International Journal of Production and Quality Engineering*, 2: 55–61.
15. Bilici, M.-K. and Yükler, A.-I. (2012a), 'Effects of welding parameters on friction stir spot welding of high density polyethylene sheets', *Materials and Design*, 33: 545–50.
16. Bilici, M.-K. and Yükler, A.-I. (2012b), 'Influence of tool geometry and process parameters on macrostructure and static strength in friction stir spot welded polyethylene sheets', *Materials and Design*, 33: 145–52.
17. Bagheri, A., Azdast, T. and Doniavi, A. (2013), 'An experimental study on mechanical properties of friction stir welded ABS sheets', *Materials and Design*, 43: 402–9.
18. Hoseinpour Dashatan, S., Azdast, T., Rash Ahmadi, S. and Bagheri, A. (2013), 'Friction stir spot welding of dissimilar polymethyl methacrylate and acrylonitrile butadiene styrene sheets', *Materials and Design*, 45: 135–41.
19. Kiss, Z. and Czigány, T. (2012), 'Microscopic analysis of the morphology of seams in friction stir welded polypropylene', *eXPRESS Polymer Letters*, 6: 54–62.
20. Czigány, T. and Kiss, Z. (2011), 'Friction stir welding of fibre reinforced polymer COMPO composites', *Proceedings of the 18th International Conference on Composite Materials.*

21. Barmouz, M., Seyfi, J., Besharati Givi, M.K., Hejazi, I. and Davachi, S.M. (2011), 'A novel approach for producing polymer nanocomposites by in-situ dispersion of clay particles via friction stir processing', *Materials Science and Engineering A*, 528: 3003–6.
22. Alyali, S., Mostafapour, A. and Azarsa, E. (2012), 'Fabrication of PP/AL2O3 surface nanocomposites via novel friction stir processing approach', *International Journal of Advances in Engineering & Technology*, 3: 598–605.

15

Friction stir welding applications in industry

A. Amini and P. Asadi, University of Tehran, Iran and P. Zolghadr, Payame Noor University, Tehran, Iran

DOI: 10.1533/9780857094551.671

Abstract: Friction stir welding (FSW) is a solid-state joining technique that was initially applied to aluminum alloys, but nowadays is widely used in many different industrial applications. The process has many advantages in both mechanical properties of the welded material and the energy consumption which directly leads to lower operating costs. FSW can be used for joining many types of materials and also different material combinations, and is also applicable for a wide range of material thicknesses. Being a solid-state welding process, it can be performed in various weld geometries. This chapter provides a relatively complete list of the applications of FSW in different industries and also introduces many other diverse fields of applications in which FSW has proved to be beneficial, but has not yet been commonly used. The last two sections of the chapter are devoted to industrial FSW machines and robotic FSW processes, and the necessary considerations needed when designing or selecting such devices.

Key words: friction stir welding, industrial applications, FSW machines, robotic FSW, weld geometries.

15.1 Introduction

As stated in previous chapters, friction stir welding (FSW) as a solid-state joining technique was initially applied to aluminium alloys, but since then has received worldwide attention and has so many advantages:

- improved mechanical properties proved in fatigue, tensile and bending tests;
- low shrinkage and distortion, even in long welds because of low heat input;
- improved process robustness and repeatable quality, requiring no filler wire nor gas shielding for welding;
- operating in all positions;
- less energy consumption and operating cost advantages;
- less health and environmental problems;
- requiring no grinding, brushing or pickling in mass production;
- producing no fumes, porosity or spatter; and
- having a non-consumable tool, which can typically be used for up to 1000 m of weld length in 6000 series aluminium alloys.

All these features show that FSW can improve existing products and also lead to a number of new product designs not possible previously.

For instance, the Tower Automotive company is one of the industrial plants that have widely utilized FSW for its manufacturing programmes. The company claims that FSW could be a proper substitution for gas metal arc welding (GMAW), because it has an estimated reduced weight of 40%, tensile strength twice that of GMAW in the 6000 series aluminium, and increased fatigue life from 2 up to 20 times of that of GMAW. It requires no consumables and is energy efficient and environmentally friendly [1].

Besides these advantages, like any other process, FSW faces some limitations such as low welding speed, which is usually about 750 mm/min for welding 5 mm thick 6000-series aluminium alloys on commercially available machines. Also, to perform this welding method, workpieces must be rigidly clamped, and thus a backing bar is required. The keyhole at the end of each weld is the other problem to cope with. However, these limitations are being reduced by intensive research and development worldwide.

Considering the limitations mentioned above, the process still benefits from many advantages, which have made it really interesting for different manufacturing industries that are discussed in this chapter.

15.2 Materials and thicknesses

FSW can be used for joining many types of materials and also different material combinations. Most of the efforts done in FSW research and development have been performed to optimize the process for joining aluminium and its alloys. A group of TWI Industrial Members demonstrated that the following aluminium alloys could be successfully friction stir welded to produce high integrity welds within defined parametric tolerances: 2000 series aluminium (Al-Cu), 5000 series aluminium (Al-Mg), 6000 series aluminium (Al-Mg-Si), 7000 series aluminium (Al-Zn) and 8000 series aluminium (Al-Li) [2]. In the case of different alloys, joining aluminium and steel alloys is of significant importance, and recently Honda Company has successfully performed such welding in a vehicle suspension system for mass production [3].

Although parameters for welding of most aluminium alloys have been optimized in a thickness range from 1.6 to 10 mm, FSW can be used for a wide range of thicknesses. For example, in thick aluminium alloy 6082, single-pass butt joints have been made in thicknesses ranging from 1.2 to 50 mm without the need of weld preparation and in thicknesses up to 100 mm, 2-pass welds, one from each side, were performed. Special lap joining tools have also been developed for aluminium alloys with thicknesses from 1.2 to 6.4 mm [2].

15.3 Joint geometries

FSW has been used for the manufacture of butt welds, overlap welds, T-sections, fillets and corner welds. For each of these joint geometries, specific tool designs are required, which are being developed and optimized. The process can also cope with circumferential, annular, non-linear and three-dimensional (3-D) welds. Since gravity has no influence on the solid-phase welding, this method can be used in all positions, that is, horizontal, vertical, overhead and orbital.

Some examples of FSW geometries are presented in Figure 15.1; on the left and on the right are shown different possible categories of geometries suitable for the FSW process.

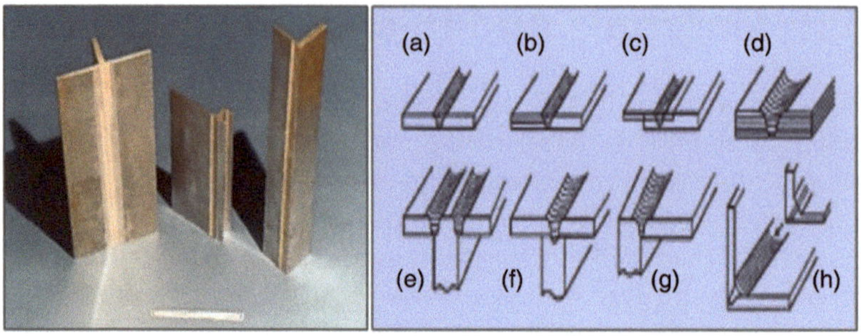

Figure 15.1 Examples of weld geometries (left) and joint geometries (right) suited to FSW: (a) square butt; (b) combined butt and lap; (c) single lap; (d) multiple laps; (e) 3-piece T-butt; (f) 2-piece T-butt; (g) edge butt; and (h) possible corner fillet weld [4]

15.4 Applications

15.4.1 Automotive industry

Two of the largest challenges that the automotive industry has always faced are lightening the vehicles and at the same time improving the safety of passengers. The former aim is important to improve vehicle performance, i.e. acceleration, top speed and fuel consumption. The latter is vital to secure passengers in case of accidents. Considering both of these factors simultaneously is usually difficult to lead to a product. One of the materials that provides enough strength and is also lightweight is aluminium. But the largest problem is that aluminium alloys can rarely be welded with conventional methods. Since the invention of FSW, in 1991, for aluminium alloys, efforts have been done to make this new technology more practical to extend the use of aluminium in vehicles body. This leads to strong structures, light enough to satisfy the performance criteria. Nowadays the use of FSW in car manufacturing has largely widened. Thus in this section we introduce some of the recent applications of this welding method in automotive production.

More and more aluminium is used in transport vehicles to lower dead load. The increasing awareness of environmental issues has also placed pressure on weight reduction in many road applications. As the automotive industry trends towards increased use of aluminium, the FSW process offers many potential benefits for joining aluminium alloys.

FSW in any of its applications has distinct benefits and results in cost reductions and/or other advantages that allow its application to be beneficial.

In general, all aluminium components in a car can be welded by FSW: bumper beams, rear spoilers, crash boxes, suspension systems, rear axles, drive shafts, intake manifolds, stiffening frames, water coolers, engine blocks, cylinder heads, dash boards, rollover beams, pistons, engine and chassis cradles, wheel rims, attachments to hydroformed tubes, tailor welded blanks (TWBs), etc. The application of FSW in large vehicles is even more interesting, as this process could be used in truck bodies, tail lifts for lorries, mobile cranes, armour plated vehicles, fuel tankers, caravans and buses. Motorcycle and bicycle frames are the other potential fields for applying FSW. Although, in all these applications, minor modifications to the structure may be needed in order to make it more suitable for FSW. The advantages of the FSW process are currently being evaluated by several automotive companies and their suppliers, though some other pioneering companies have already utilized this method of welding in their mass production lines.

As can be seen in Figure 15.2, taken at the 2002 Aluminium Fair in Germany at the booth of SAPA, a car is full of application areas for use of aluminium.

Figure 15.2 A car featuring countless application areas of FSW [2,5]

Joining of extrusions for suspension parts

Joining of extrusions with FSW is regarded as an enabling technology to generate larger extrusions profiles, which could not be previously fabricated. For the automotive industry, it is generally accepted that the largest available extrusions are in the range of 200 to 300 mm wide. Larger extrusions can be fabricated, but their cost per mass of material increases with increasing extrusion size. Furthermore, the extrusion tolerances become worse as their size increases, often making them unusable. With FSW, two smaller extrusions can be joined with various weld joint configurations to generate larger extrusions. In this application, the FSW is linear aligned along the extrusions. Once welded, the large extrusion often undergoes a sawing operation, where it is cut into smaller sections. One such example is the suspension component shown in Figures 15.3 and 15.4. This particular part is a suspension link arm, and is cut from several meters long welded sections. For this application, the

Figure 15.3 Automotive link arm fabricated from two extrusions joined by FSW [6]

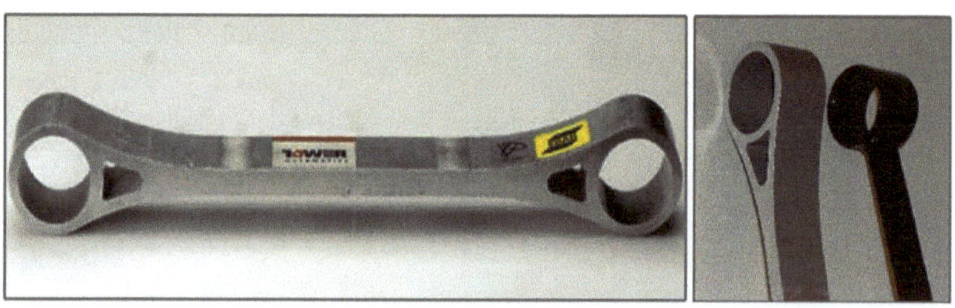

Figure 15.4 Suspension links by Tower Automotive [5], made from three extrusion parts. On the right, the old steel-based design is also shown [1]

Figure 15.5 Ford suspension link at the middle of the welding process [1]

partial penetration butt weld configuration would be the most appropriate [6], though lap penetration, lap fillet or full penetration butt weld configurations can also be used.

Ford Company is the first American automotive manufacturer to adopt FSW technology for its mass production. In Figure 15.5, the Ford suspension link, the first FSW US automotive part, is shown.

Showa Aluminium and Tokai Rubber in Oyama, Japan, joined extruded end-pieces to 20 to 30 mm diameter tubes for the manufacture of suspension arms. The rubber of the suspension arm can be vulcanized before welding, due to the low heat input of the new assembly method.

Joining of dissimilar metals for sub-frame

Honda Motor Company is one of companies that have successfully adopted the FSW technology for its mass production line. In late 2012, the Japanese company announced that it had developed a new technology for the continuous welding of dissimilar metals of steel and aluminium and applied it to the sub-frame, a key component of a vehicle body frame, in a mass-produced vehicle (Figure 15.6). Honda first applied this technology to the North American version of the all-new 2013 Accord, and expanded the application sequentially to other models. This technology generates a new and stable metallic bonding between steel and aluminium, by moving

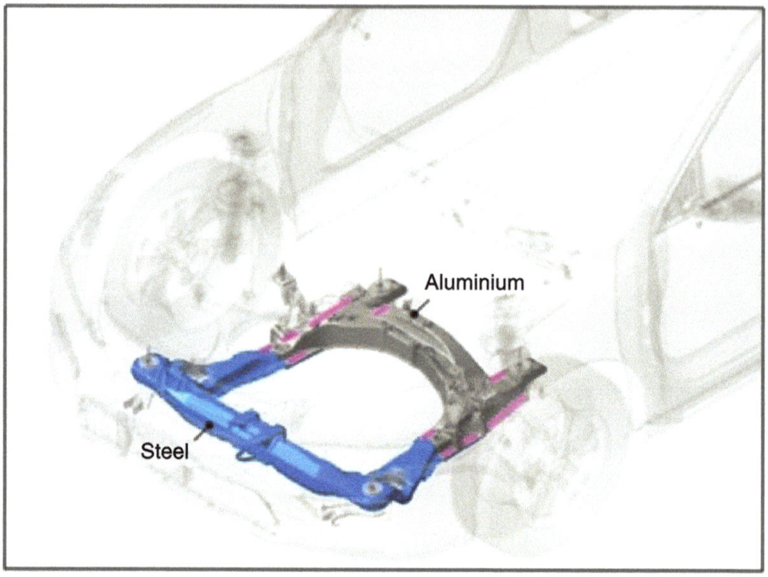

Figure 15.6 **Honda front sub-frame: conceptual diagram of FSW of dissimilar metals [3]**

the rotating tool on the top of the aluminium, which is lapped over the steel with high pressure. As a result, the welding strength becomes equal to or beyond conventional metal inert gas (MIG) welding.

Using this new technology contributes to an improvement in fuel economy by reducing body weight by 25% compared to a conventional steel subframe. In addition, electricity consumption during the welding process is reduced by approximately 50%. It also enables the manufacturer to perform a change in the structure of the subframe and the mounting point of suspension, which increases the rigidity of the mounting point by 20% and also contributes to the vehicle's dynamic performance. Conventionally, FSW requires use of large equipment, but the company developed a FSW continuous welding system applied to a highly versatile industrial robot. This system can also be used for aluminium-to-aluminium welding and thus, the welding system with the same specifications can be used for production of a full-aluminium subframe [3].

General assembly

The general assembly area involves the joining of various components together where the welds performed are relatively short, usually measuring

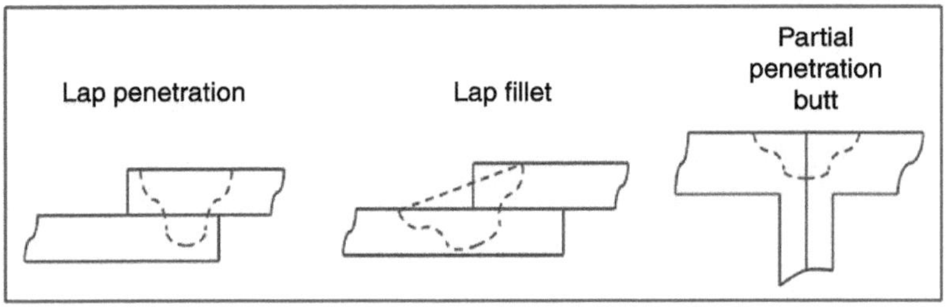

Figure 15.7 Different joint configurations: FSW nugget is indicated by dashed lines [6]

less than a metre, and can have 3-D contours. In high production rates, the joint configuration for this type of application must be robust against probable variations in parts configurations. These variations consist of material thicknesses and other changes in geometry and material properties that cause gap conditions. For this reason, it is likely that joint types for such cases will be limited to lap penetration or lap fillet (lap welds) and partial penetration butt welds [6]. These joint configurations are displayed in Figure 15.7.

As expected, applications of this sort often involve 3-D paths, and thus a machine with at least five degrees of freedom would be necessary. Considering this issue and the cost sensitivity in the automotive industry, it is desirable to utilize industrial robots for this application. However, typical industrial robots lack sufficient stiffness and are questionable for their ability to perform FSW on different components [6].

Tailor welded blanks

Another major potential to apply FSW is the tailor welded blank (TWB) (Figure 15.8). A TWB consists of various sections of flat sheet, sometimes with different thicknesses, which are abutted and then joined together. These joined flat sheets then enter a forming operation to shape the final geometry. The purpose of using a TWB is to optimize material utilization, not only for improved material consumption, but also for reducing the weight of the final formed component. These TWBs often have variations in material thicknesses across the joint line that requires a dissimilar thickness butt weld [6]. Laser and laser hybrid welding have already gained relatively stable positions in joining steels and stainless steels even with different thicknesses, but for aluminium joining, FSW has shown extremely high potential. However, the TWB needs a fully penetration in

Figure 15.8 A tailor welded blank [6]

the butt weld, and production variations are less significant in this application, allowing FSW of TWB to be feasible.

Large vehicles

Use of FSW is design- and material-dependent and the process works best with straight, flat surface welding (Figure 15.9). Thus in large road

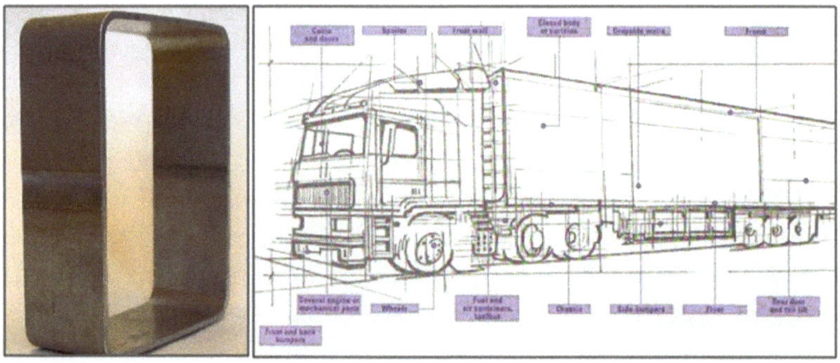

Figure 15.9 A longitudinal FSW weld in a rectangular profile; beams with minimal distortion can be manufactured. Right: a picture by EAA (European Aluminium Association) shows potential application areas for aluminium use in a truck [1]

Figure 15.10 **Aluminium loading bridges, which could easily be made by using FSW [1]**

transport vehicles, the scope for applications is wider and easier to adapt, because of the long, straight or curved welds, as in trailer beams, cabins and doors, spoilers, front walls, closed body or curtains, drop side walls, frames, rear doors, tail lifts, floors, sides, front and rear bumpers, chassis, fuel and air containers, toolboxes, wheels, etc. [1].

As an example, a lorry is full of potential applications for FSW, mainly for the planar straight linear welds with thicknesses typically up to 5 mm. Some samples are shown in Figure 15.10. By replacing conventional arc welding processes with FSW, the straightness of the panels will be magnificently improved and assembly times are shortened to a minimum.

Wheel components

Welding aluminium wheels was one of the earliest applications of FSW in the automotive industry. It was first used for longitudinal welding of an aluminium tube, which was later cut to the correct length and spin-formed to the right shape. Simmons Wheels in Alexandria, Australia, developed a new method of producing a car wheel rim section from a rolled aluminium 6061-O sheet. They described a method of first making a pre-formed cylinder, then cutting it into several individual rim sections, and afterwards post-welding and forming it into the desired rim profile, and finally subjecting this part to heat treatment to the required T6 temper.

For prefabricating rims for lightweight-alloy car wheels, Hayes Lemmerz in Michigan considered chamfering the ends of the originally flat sheets prior to FSW in the weld root area to achieve full penetration. In addition, flats could be formed on the ends to give better contact between the FSW tool shoulder and the workpiece. After performing FSW, the rims of these lightweight wheels can be subjected to a series of flow spinning and roll forming operations to produce a generally uniform thickness wheel rim.

Figure 15.11 FSW used in joining the rim to the hub (Hydro, Norway) [1]

In Norway, an innovative technique of joining two parts of a car wheel by FSW has been invented and successfully demonstrated by Hydro Aluminium in Håvik, for the manufacture of prototype parts (Figure 15.11). Optional design concepts have been developed to either butt or lap welds of cast or forged centre parts to rims that are made from wrought alloys. Branched designs of the centre part are possible by providing two parallel welds per wheel, which run along a cavity to achieve good load transfer and weight reduction. This company has also used FSW in attaching the inner rim to the wheel form.

Engine components

The number of potential applications in the engine of any motor vehicle is incredible. For example, cast aluminium components can successfully be welded with FSW to improve the quality of the cast structure, or to join inserts to the piston (Figure 15.12).

Superplastic forming

FSW is also used in superplastically formed products, as already done by Aston Martin in their Vanguard model (Figure 15.13). This

Figure 15.12 A friction stir processed piston. The metallographic structure was improved with FSW [1,5]

Figure 15.13 Aston Martin Vanguard showing significant use of superplastic forming the hoods [1]

technique often goes hand-in-hand with TWBs. Some applications of door component manufacturing are already available. In superplastic forming, the final shape to products is given with the help of pressurized gas. The idea is to make connecting welds in relatively simple positions and shapes by using FSW before the forming process [1].

15.4.2 Aerospace and aviation

Aerospace

Nowadays, different parts in the aerospace industry are welded by means of FSW. Opportunities include welding skins to spars, ribs and stringers for use in military and civilian aircrafts. This offers significant advantages compared to riveting, such as reduced manufacturing costs and weight savings. In this field of interest, the process could also be used to increase the size of commercially available sheets by welding them before forming. The FSW process can therefore be considered for wings, fuselages, empennages and cryogenic fuel tanks for space vehicles, aircraft fuel tanks, external throw-away tanks for military aircrafts, military and scientific rockets, and also repair of faulty MIG welds. First, we discuss the longitudinal butt welds and circumferential lap welds of aluminium-alloy fuel tanks for space vehicles, which have been friction stir welded and successfully tested (Figure 15.14).

The Space Shuttle's gigantic External Tank, the largest and the only non-reusable element of the spacecraft, was the first commercially successful experience of performing FSW in the aerospace industry. In 1993, NASA challenged Lockheed Martin Laboratories in Baltimore, to develop a high-strength, low-density, lighter-weight replacement for aluminium alloy Al-2219, used on the original Space Shuttle External Tank (Figure 15.15). Lockheed Martin, Reynolds Aluminium and the labs at Marshall Space Flight Center in Huntsville, were successful in developing a new alloy known as Aluminium Lithium, Al-Li-2195, which reduced the weight of the External Tank by 3402 kg. The lithium in the new material made the initial welds of the External Tank far more complex. The repair welds were difficult to make and the joint strength of

Figure 15.14 NASA/Lockheed FSW 2195 for liquid H_2 and liquid O_2 cryotanks [7]

Figure 15.15 Space Shuttle and its gigantic external tank [8]

the External Tank had much lower mechanical properties. So the project looked for alternative welding techniques. Since FSW produces stronger welds, the External Tank project managers decided to use the process on the Super Light Weight Tank, which is made from Al-Li-2195 [8].

The FSW process produced a joint stronger than the fusion arc welded joints that were obtained in the earlier Light Weight Tank program. A significant benefit of FSW is that it has significantly fewer process elements to control. In a fusion weld, there are many process factors that must be controlled, such as purge gas, voltage and amperage, travel speed, shield gas, arc gap and wire feed. However, in FSW there are only three process variables, rotational speed, travel speed and pressure, which are easily controlled. The increase in joint strength combined with the reduction in process variability provides an increased safety margin and high degree of reliability for the External Tank.

One of the early drawbacks to the FSW was the fixed pin, because it limited welding to materials with constant thickness. The Shuttle's External Tank project developed a through-spindle retractable pin tool (RTP) that can retract or expand its pin tip within the material. This allows for changes in thickness such as on the tank's longitudinal barrel. The viability of the technology was demonstrated when NASA's Marshall Center used the RTP to weld a full-scale External Tank. The External Tank implemented FSW on the longitudinal barrel welds on both of the tanks, the liquid oxygen and liquid hydrogen barrels. External Tank-134, launched in January 2005, was the first tank to incorporate the process [8].

On the basis of such successful experiences, nowadays, a much larger number of spacecraft fuel tanks are being produced from difficult-to-weld aluminium alloys. Various FSW procedures for welding end domes into cylindrical high-quality tanks have been developed for cryogenic oxygen storage vessels. It has been proposed to shrink fit an aluminium backing ring into the tank before the last circumferential weld is made. The joint is then simultaneously made between three members, the end dome, the cylindrical shell and the backing ring. The latter may remain in the tank after welding.

FSW is also being considered for producing Ariane-5 motor thrust frames. A study by Fokker Space has shown that FSW can readily be applied to lap joints in aluminium 7075-T7351. Although the tensile strengths measured in this investigation were lower than those obtained with FSW butt joints, they are at an acceptable level to replace bolted lap joints. For unpressurised structures, lap joints offer the significant advantages of generous tolerances at interfaces between components that directly lead to ease of assembly.

In this area, Boeing has been a real pioneer in introducing FSW to industrial manufacturing. In the Delta II and IV programs, FSW has been widely used in manufacturing of rocket fuel tanks, used also on the latest Mars expeditions. For instance, they have applied FSW to the Inter stage Module of a Delta II rocket, which was successfully launched in August 1999. As reported by the company, the production time for a typical tank has been dramatically reduced and cost savings have been gained. Price difference between FSW and riveting is surprisingly large and FSW is only 20% of the costs for riveting. More parts on rockets are constantly being evaluated for future production, not only at Boeing, but almost everywhere related to aerospace or civil aviation industries (Figure 15.16). The potential applications in the commercial and military aircraft industry include carrier beams, floors, whole bodies, wings, etc., as already mentioned.

Figure 15.16 One of two identical vertical machines at the Boeing Company in Decatur, USA. The welding length (height) is 12 metres and the diameter is 5 metres [1]

Civil aviation

The pressure to replace riveting by welding has risen recently. Riveting, in general, is a good process, but it is very time-consuming in practice. Holes for the rivets are typically drilled one by one, thus it is easy to imagine the possible time savings if welding could be used instead. So the main rationale for employing FSW in the manufacture of aerospace components is time and weight savings, which translate directly to cost savings. Reducing weight enables higher speeds and/or reduced fuel consumption. FSW not only eliminates rivets and fasteners, but the need for an overlap sheet configuration. The butt-joint configuration also facilitates joint evaluation and quality assurance, because a homogeneous joint with full penetration eliminates any crack formations. Joints produced in this way have higher strengths than riveted joints and much lower residual stresses than typical fusion welded joints. The main areas of interest are:

- overlapping joints;
- mixed joints (dissimilar alloys);
- welding of anodized plates.

The other advantage of FSW in the aerospace industries is that FSW joints do not suffer from colour changes when anodized, which is a critical issue when compared to conventional arc welding processes, where the matching colour is often gained by sacrificing mechanical strength or corrosion properties [1].

A major breakthrough in the field of civil aviation was surely the approval made by the FAA (Federal Aviation Association), which has approved the FSW process as a joining process for airplanes [1].

Therefore, it is apparent that the FSW process offers tremendous potentials for low-cost fastener less joining of lightweight aluminium airframe structures. For example, The European Aeronautic Defense and Space Company (EADS) also found the high potential and maybe the necessity for joining aluminium alloys by FSW for skin-to-skin fuselage connections of their Airbus 380 production. They presented data showing that the mechanical properties of these welds seemed to approach almost the properties of the parent material. Figure 15.17 shows three research sites of EADS in different parts of the world, all of which are attempting to develop the use of FSW in the aviation and aerospace industry.

It was reported that Eclipse Aviation Corporation decided to use FSW to replace traditional riveting and bonding processes for a then new generation of their business jets (Figure 15.18). This was the first application of FSW in high-volume aviation applications, with the potential to dramatically lower assembly time and cost.

The Phantom Works of the Boeing Company has also been pursuing FSW of thin butt, lap and T-joints and thick butt joints for various aircraft missile and space applications. There is a strong desire for welding these joint configurations on curvilinear paths, and thus enabling welding of complex aircraft parts. Boeing has demonstrated curvilinear FSW of a

Figure 15.17 Research and prototype manufacturing units from left to right: EADS Ottobrunn in Germany, EADS together with Institute de Soudure in France and Alenia Spazio in Italy [1,5]

Figure 15.18 Eclipse, the first commercial aircraft using FSW [1]

complex aircraft landing gear door by using a patented force actuator. They have also successfully demonstrated FSW of sandwich assemblies by welding thin T-joints for a fighter aircraft. Thus the horizon for further utilization of FSW in the aviation industry seems to be wide.

15.4.3 Railway industry

The commercial production of high speed trains made from aluminium extrusions, which could be joined by FSW, has been well spread worldwide. Applications include high speed trains, rolling stock of railways, underground carriages, trams, railway tankers and goods wagons, container bodies, etc.

With rolling stock, such as train cars and coaches, manufacturers are very keen to use FSW in various components of a train car. Hollow profiles and T-stiffener extrusions are used and are currently being friction stir welded in the railway industry for the manufacture of high-speed trains. Innovative joint designs have been proposed to compensate for lateral and angular offset by using the FSW process. For example, since 2001, Alstom LHB has used FSW floor-panels and side walls from Hydro Marine, Norway, in suburban trains. Another pioneer in the train

Figure 15.19 Typical trains in which FSW is used in the floor panels [1,5]

industry has been Hitachi in Japan, who has used friction stir pre-fabricated floor elements in their Shinkansen trains (Figure 15.19). Even in these cases, the profiles and extrusions must be modified suitably for the FSW process. The thickness on the weld and the radii on the corners are the most important parameters of the extrusion panels to make the use of FSW possible in such cases.

The Hitachi company has used an aluminium double-skin frameless structure in its Class-395 mock-up trains to reduce weight of the vehicle, while the body strength is not affected [9]. Such a design, as shown in Figure 15.20 on the left, fits best for modular design, and can also be easily friction stir welded as illustrated in the picture on the right.

The FSW process performed on this product resulted in higher strength, minimal distortion, and higher quality and smoother surface finish. Each

Figure 15.20 Aluminium double-skin frameless structure; possibly FSW process for such a design [9]

Figure 15.21 Class 395 mock-up trains of Hitachi [9]

of 395 trains had to complete 4000 miles of fault-free running prior to acceptance by the operator, which shows the proven quality of the process in practice. Figure 15.21 shows Class-395 mock-up trains in the manufacturing phase and in the finishing sate.

15.4.4 Shipbuilding

Shipbuilding and marine industries are two of the first industrial sectors to have adopted the FSW process for commercial applications. In these industries, FSW is suitable for panels for decks, sides, bulkheads and floors, aluminium extrusions, hulls and superstructures, helicopter landing platforms, offshore accommodation, marine and transport structures, masts and booms in sailing boats, refrigeration plants, etc. (Figure 15.22).

Figure 15.22 Panels for decks, sides, bulkheads and floors joined from several extensions [5]

The first commercial application of FSW concerned the manufacture of hollow aluminium panels for deep freezing of fish on fishing boats. These panels are made from FSW aluminium extrusions at SAPA, Sweden. The minimal distortion and high reproducibility make FSW both technically and economically a very attractive method to produce these stiff panels [10].

New FSW applications in the marine industry have been reported from Japan, where the process is used to produce aluminium honeycomb panels and seawater-resistant panels. The latter are made from five 250 mm wide 5083 aluminium alloy extrusions to make a panel sized 1250 × 5000 mm. They are used for ship cabin walls because of the good flatness of the weld root (Figure 15.23).

Pre-fabricated wide aluminium panels for high-speed ferryboats can also be produced by FSW, which are also commercially available. The panels are made by joining extrusions, which can be produced in standard-size extrusion presses. Compared to fusion welding, there is lower distortion and reduced thermal stresses in components made by means of FSW.

Annually more than 70 km of defect-free FSW have been produced over the last decade at Hydro Marine Aluminium in Haugesund, Norway, mainly for shipbuilding panels (Figure 15.24).

After welding, the panels can be rolled for road transport, as they are stiff only in the longitudinal direction (Figure 15.25).

Using FSW to produce the panels mentioned above makes it possible to construct a large catamaran from building blocks, just like a toy boat. All pieces fit perfectly to each other and dimensional accuracy can be

Figure 15.23 An over-hull structure of Marine Aluminium [5]

Figure 15.24 ESAB SuperStir™ machine at Hydro Marine Aluminium, to weld aluminium extrusions for shipbuilding panels [10]

Figure 15.25 Prefabricated FSW panel for catamaran side-wall, rolled for road transportation (Hydro Marine) [10]

Figure 15.26 A catamaran where lots of friction stir pre-fabricates are used [1]

mastered to a full degree. Thus, one of the earliest examples of a large product where a lot of FSW was used is shown in Figure 15.26; the catamaran made by Kvaerner Fjaellstr and using extruded and FSW profiles produced by Hydro Marine Aluminium. This was actually the kick-start for the industrialization of the process.

Another example in this area is the Littoral Combat Ship, the USS *Freedom*, which was commissioned in 2006 (Figure 15.27). Over 12 miles

Figure 15.27 Littoral Combat Sheet with over 12 miles of FSW [11]

of FSW have been used to fabricate this giant ship. The welded aluminium panels for the superstructure were fabricated by Friction Stir Link, Inc. and delivered to Marinette Marine for final assembly. The use of the solid-state FSW process has resulted in improved strength and fatigue life, reduced distortion, and a robust and repeatable process at a low cost [11].

Due to the low heat input during FSW that results in very low residual stresses, the welded components are perfectly accurate and minimal fit-up work is needed. The resulting savings in both time and money are obvious. The main advantages of FSW in this field can be summarized as:

- industrial production with a high degree of completion;
- extended level of repeatability, ensuring uniform level of performance, quality and narrow tolerances;
- flexible production equipment and capacity allow customer-built solutions with reliability of delivery;
- high level of straightness of the panels ensures easy assembly even at site.

One of the most attractive features of FSW products is that no time-consuming post-weld treatments such as grinding, polishing or straightening are needed. With proper design, the elements are ready to use directly after welding. However, it is important to keep in mind that designs, which are made for MIG or tungsten inert gas (TIG) welding, are not necessarily suitable for FSW. The limiting factor often is the relatively high down-force needed when using FSW. A proper support in the form of backing bar or sometimes design change is often required. Once done, repeatability reaches levels previously not experienced in welding (Figure 15.28).

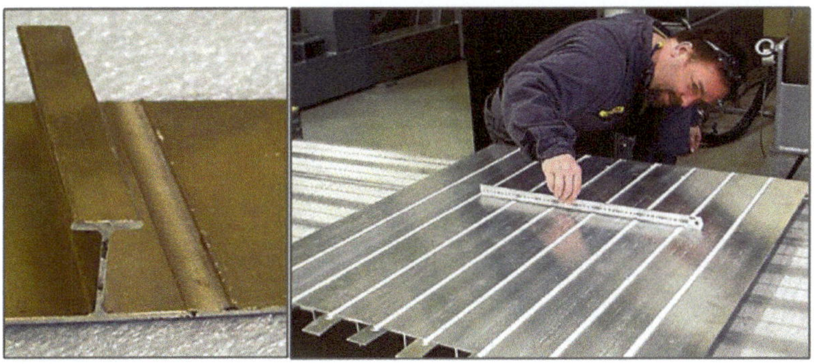

Figure 15.28 Left: high-quality flat panel field after welding at ESAB. Right: instead of using wide profiles, a panel is made from relatively narrow (120 mm) high-quality extrusions [1,5]

When producing large surfaces such as walls or floors, beside the straightness of the panels, the resulting reflections are also an important issue to consider. A lot of time is spent on polishing and making up surfaces, which are architecturally visible. In FSW prefabricated panels, the reflections are merely caused by the surface appearance of the aluminium plates, not by the reflections caused by welding heat input.

15.4.5 Heat and piping

Heat sinks and cooling fins

Other fields of production that FSW has successfully entered in are the piping and heat industries. Using FSW in welding of high pressure vessels and pipelines is still in the exprimental phase, though it is being performed in manufacturing of heat sinks and fins in heat industry.

FSW is ideally suited to manufacture of heat sinks. Instead of large, thick-walled and expensive extrusion profiles, it is possible to produce the heat sink from a number of thin-walled and cost-effective profiles (Figure 15.29). The complete material thickness at the joint is welded, thus better tensile and fatigue strength is achieved and the breakage possibility due to alternating load (temperature or vibration) reduces dramatically. Also according to the high quality of the weld, less finishing processes are needed in comparison to conventional fusion welds. Besides, the modular design of such a process makes it suitable for different geometries and various customer's demands [12].

The cooling block shown on the left in Figure.15.30 is used in a heavy-duty cutting machine. Instead of using one very large extrusion, the new design is to join together two profiles into one. The penetration depth on the weld is approximately 15 mm. Figure 15.30 also shows an alternative solution for attaching cooling fins together to form a cooling element (picture on the right).

Friction stir welding of thick-walled aluminium pressure vessels

Efforts are being made to develop the use of FSW in manufacturing of high-pressure vessels. Presented here is one of the recent experiments done for this purpose. The specific vessel, which has been friction stir welded, is a sphere with an inside diameter of about 1 m and a thickness of 3.81 cm. This basic shape is completed by the addition of cylindrical nozzles at the

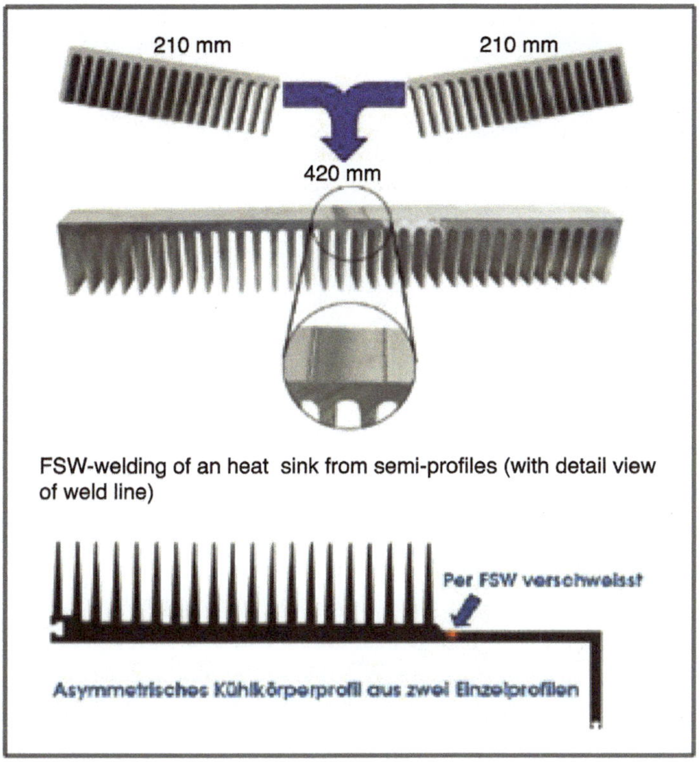

Figure 15.29 Concept of using DSW in heat sink manufacturing [12]

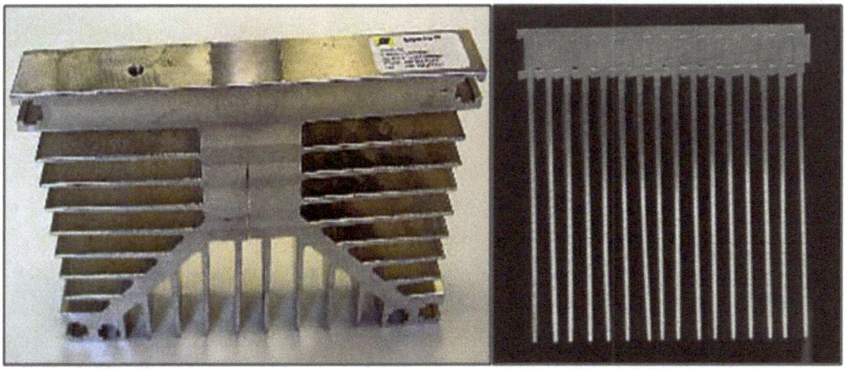

Figure 15.30 Left: cooling block of a heavy-duty cutting machine at ESAB. Right: a potential application of FSW for welding cooling fin packages [1]

Figure 15.31 Alloy 2219 aluminium forgings [13]

top and bottom of the sphere for access to the interior of the vessel. To manufacture the final product, two spherical forgings are joined at their large open ends by a single-pass FSW. The alloy chosen is Alloy 2219, with a nominal composition of 5.8–6.8% Cu, 0.2–0.4% Mn and 0.10–0.25% Zr, with trace amounts of Ti, Si, Fe, Mg and Zn (Figure 15.31) [13].

The challenge in FSW of circumferential parts is providing a backing support for the parts. For single-sided FSW, the forces are in the range of 6000 kg for this alloy and thickness. The force has to be reacted from the back of the weld, which makes assembly and disassembly difficult. The FSW machine used has a rotary device for circumferential welding. This device was used with the SRT machine to weld the vessels. The thicker the material, the greater the torque demand and some modifications of the FSW machine are necessary. Before, Advanced Joining Technologies had performed circumferential welds on 2.5 cm thick rolled plate. The plate was rolled to 102 cm outer diameter and was 122 cm in length.

The vessel halves were tack -welded together using a single-sided FSW tool. The tool created a tack weld that was 1.02 cm deep. Once tack welding was completed, the clamping mechanism was disassembled.

When FSW is completed, a keyhole is present due to tool retraction. Trimming off the keyhole is a big issue in circumferential FSW. A closure method needed to be identified. Examples of closure methods RPT, run-off wedge, friction plug welding and fusion fill. The most viable keyhole closure method for self-reacting FSW is friction plug welding, which maintains the solid-state structure of the FSW (Figure 15.32) [13].

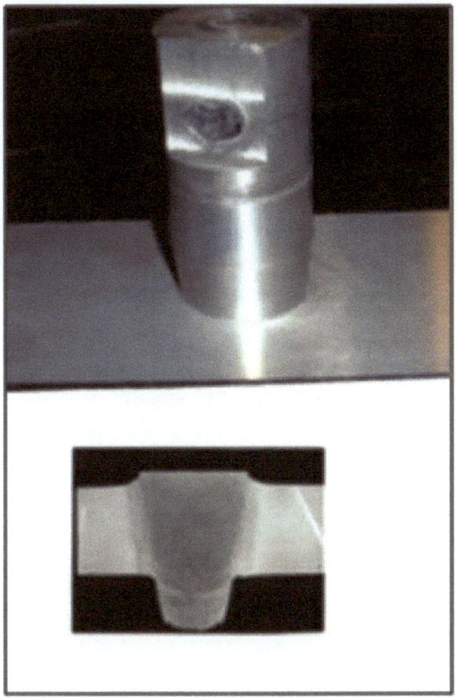

Figure 15.32 A 2219-T6 aluminium plug friction stir welded into 1.5 inch thick 2219-O plate [13]

Friction stir welding in next-generation power plants

Pacific Northwest National Laboratory has reported that the next generation of gains in efficient fuel utilization will require a move to higher system pressures and temperatures. Advanced ultra-supercritical pulverized coal plant designs are calling for some system components to operate at 760C and 5000 psi. This goal requires new materials, especially for performance drivers for heavy section components (i.e. boiler headers and pipes in superheaters and reheaters). Thus, there is a need to develop methods of joining next-generation materials, which result in joints with ideally the same heat strength, creep strength and corrosion/oxidation properties as the base metal, which mostly comes through with FSW joining method [14].

Friction stir welding for small- and large-diameter pipes

Recently the range of applications of FSW in exprimental scale has also extended to small-diameter pipe sections. Here the results of such work is presented to clarify the dimensions of this potential application.

In this study, full penetration FSW is performed on butted sections of aluminium alloy 6061-T6 pipes. These pipe sections are relatively small in diameter (4.2 inches) and relatively thin-walled (0.2 inches). The small radius of curvature distinguishes this weld configuration geometrically from a butted plate configuration and presents unique challenges. These challenges necessitate the use of specialized techniques and specialized equipment. FSW joints of this type are not present widely, although FSW of large-diameter steel pipe sections is being carried out at the research level [15].

In this experiment, an experimental apparatus was designed, shown in the following figures, which rotates the pipe sections beneath a stationary tool axis. The apparatus is mounted on a FSW machine driven by a belt and a computer-controlled motor. The pipe sections are shown, mounted in the welding apparatus (Figure 15.33). The experiment is started from this position. The tool probe position was calibrated, or zeroed, against the surface of the work. The tool rotation

Figure 15.33 Starting point of the FSW of pipes mounted on the apparatus [15]

is started and the probe is plunged into the material until the desired contact is achieved with the shoulder on the cylindrical surface of the work.

The circular nature of the weld path in a butted pipe configuration results in secondary heating at the end of the weld. To complete a full circumferential weld, the tool must cross over the weld initiation site, which remains warm for a small-diameter pipe. This additional heat affects the steady-state portion of the weld and the weld termination [15]. A close-up view of the tool and the butt welded specimens used in the experiment are presented in Figure 15.34.

The weld macro-sections show complete penetration of the weld nugget through the work. Figure 15.36 shows the appearance of lateral cross-sections of welds made at low rates using the 3/16 inch diameter probe tool.

As stated above, according to the less limitations that performing FSW on large-diameter pipes faces, this field is more improved and is much closer to becoming usable in commercial mass production. Figure 15.37 shows a close-up and a complete picture of a typical FSW machine

Figure 15.34 A 5/8 inch diameter scrolled shoulder and 3/16 inch probe diameter tool [15]

Figure 15.35 A still image taken from a video of the experimental weld [15]

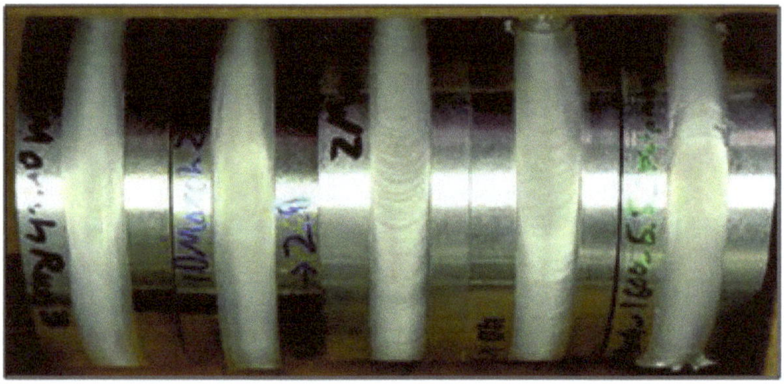

Figure 15.36 The superficial appearance of several pipes welding in the experiment [15]

performing this process on a large-diameter steel pipe. Special attention should be paid to the tool of the mashine, which has turned red because of the high temperatures produced.

Potential winning applications of this method include oil and gas, nuclear and refinery where low-corrosion stainless steel closure welds are required, and drill pipe casing where weld certifications are not a requirement and cost savings are high.

Figure 15.37 **Orbital FSW of a steel pipe [15]**

15.4.6 Other fields of application for Friction stir welding

Polymers

One of greatest potential opportunities for applying the FSW process is in polymers. In recent years, applications of polymers in industry and life have enormously increased. So finding effective techniques to join polymer parts together seems inevitable, which results in higher qualities and also reduced costs. One of the proposed methods for welding polymers is FSW. Studies and experiments so far have demonstrated the high potential of using FSW in joining polymers, although commercial instances of this application are not yet available.

Subwoofer frame

From previous sections, we can conclude that where aluminium exists in the list of the materials of a product, then the FSW has the opportunity to come in. B&O's BeoLab subwoofer is an excellent example of innovative thinking in aluminium construction for loudspeaker frames (Figure 15.38). The cast subwoofer halves are joined together by FSW, which results in no heat-induced deformations in the final product, and a solid structure which is good for acoustics.

Construction industry

The use of portable FSW equipment is possible for aluminium bridges, facade panels made from aluminium, copper and titanium, window

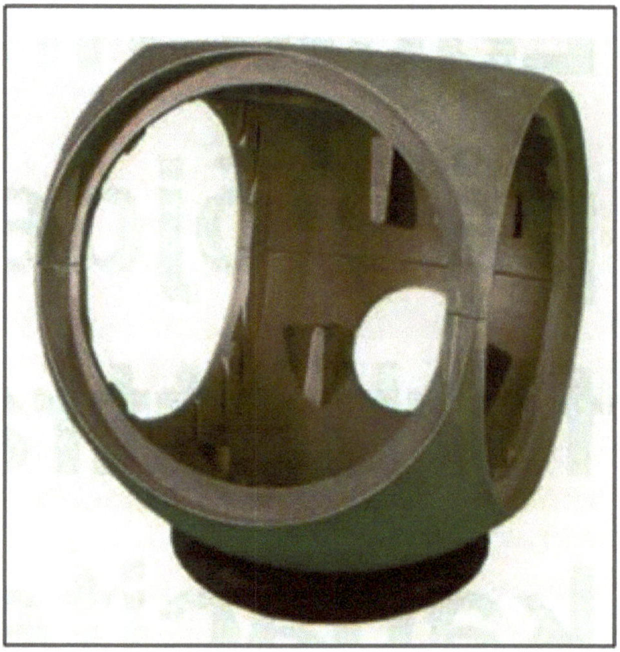

Figure 15.38 B&O loudspeaker frame where two halves are jointed together by means of FSW at Temponik [15]

frames, aluminium pipelines, aluminium reactors for power plants and the chemical industries, heat exchangers and air conditioners, pipe fabrication, etc.

Electrical industry

The electrical industry shows increasing interest in applying FSW for different applications, including electric motor housings, bus bars, electrical connectors, encapsulation of electronics, etc.

Other areas

Further attention on potential applications of FSW shows that this innovative method can also be considered for refrigeration panels, cooking equipment and kitchens, white goods, gas tanks and gas cylinders, connecting of aluminium or copper coils in rolling mills, furniture, etc., all of which strongly demonstrate the future wide range of applications of FSW in most industries.

The content discussed above included a few examples out of hundreds or even thousands of industrial applications of FSW in different fields, which have become available at commercial levels. Up to the present day, the technique of joining has been chosen as an engineering design solution or for process development by 231 organizations in 24 countries. Around half these organizations are end-users and the rest include research organizations, equipment suppliers and academic centres, with the majority based in the USA, Japan, Germany, Norway and Sweden. Table 15.1 summarizes what has been stated before, and also adds some more instances of applications of FSW in industrial activities. Attention should be paid to the fact that the information in this table is limited and typical. To complete this table, we may refer to the preceding paragraph, in which a short review is made on the applications of this attractive method of welding.

Table 15.1 **Sample list of companies adopting FSW in applications in the last two decades**

Year	Application	Company
1995	Heat exchangers	Marine Aluminium, Norway
1996	Shipbuilding	Marine Aluminium, Norway
1998	Delta II rockets	Boeing, United States
1999	Shipbuilding	SAPA, Sweden
2000	Automotive components	SAPA, Sweden
2000	Laser system housings	General Tool, United States
2001	Motor housings	Hydro Aluminium (Marine Aluminium), Norway
2001	Suburban trains	Alstom, Germany, and Hydro Marine
2001	Automotive components	Showa, Japan
2001	Train bodies	Hitachi, Japan
2002	Automotive components	Tower Automotive, United States
2003	Aircraft Structure	Eclipse, United States
2003	Shipbuilding	Advanced Joining Technologies, United States
2004	Space shuttle external tank	Lockheed Martin, United States
2004	Food trays	RIFTEC, Germany
2006	Shipbuilding	Friction Stir Link, United States
2012	Automotive components	Honda Motor Company, Japan

To list the advanced applications demonstrating the success of FSW, it is best to begin with aerospace industry, because of the natural complications associated with that industry. The process has been applied in this domain by Boeing, Lockheed Martin, BAE Systems and EADS, and specifically for some applications in Eclipse, Boeing, Spirit and Airbus aircraft manufacturing. This method has also been used in many of the world's space launch vehicles, including the Space Shuttle, Delta II and IV, SpaceX Falcon 9 and Ariane.

The other field discussed above is the rail industry, where the process is used in the production of aluminium-bodied rail cars, including Hitachi super-fast trains (Shinkansen), which can reach speeds of 320 km/h. As stated earlier, FSW is also commonly used in many other fields, including specialist car manufacture by companies such as Audi, Ford, Honda, Tower Automotive and Mazda, as well as in the manufacture of large heat sinks and in the building process of ships and other superstructures.

On the other hand, there are numerous new applications being discovered and investigated. For instance, the Apple Inc's endorsement of FSW during its recent product launch presentation and in promotional literature about the design of the new iMac is the latest high-profile industrial example of the worldwide success of this relatively new joining technique.

Considering all these various applications of the method and imagining a further vast horizon of possible future applications utilizing this joining technique, leads us directly to the continuously increasing importance and popularity of FSW amongst industries and manufacturers that are seriously considering applying the method in their production lines.

15.5 Machines for friction stir welding

One of the greatest advantages of FSW is that existing machines can be used to perform it, if only the specific head tool and motor are added to them. The process is also suitable for automation and adaptable for robot use. Figure 15.39 illustrates the FSW site at TWI in the UK.

15.5.1 Machines at TWI

TWI installed an ESAB SuperStir machine FW28, in its friction welding laboratory, which is being used in the EUREKA EuroStir Project and for

Figure 15.39 FSW machine and quality verification at TWI [2]

confidential studies. The machine has a vacuum clamping table and can be used for non-linear joint lines. The specifications of this machine are summarized as:

Sheet thickness	Work envelope	Max. down force	Max. rotational speed
1–25 mm	approx. 5 × 8 × 1 m	approx. 60 kN (6t)	5000 rev/min

A moving gantry machine FW21 was built in 1995. This machine uses a moving gantry with which straight welds up to 2 m long can be manufactured. It was used to prove that welding conditions can be achieved, which guarantee constant weld quality over the full length of long welds:

Sheet thickness	Max. welding speed	Max. sheet size
3–15 mm	up to 1.0 m/min	2 m long, 1.2 m wide

In October 1996, a laboratory machine was built to accommodate large sheets and to weld prototype structures. The modular construction of FW22 enables it to be easily enlarged for specimens with even larger dimensions:

Sheet thickness	Max. welding speed	Max. sheet size	Max. working height
3–15 mm	up to 1.2 m/min	3.4 m long, 4 m wide	1.15 m

For welding thin aluminium sheets, TWI equipped one of its existing machines with an air cooled high speed head, which allows rotation speeds of up to 15 000 rev/min and built machine FW20:

Sheet thickness	Max. welding speed	Max. rotational speed
1.2–12 mm	up to 2.6 m/min, infinitely variable	Up to 15 000 rpm

TWI's small transportable machine, FSW demonstrator FW16, produces annular welds with hexagonal aluminium alloy discs. In recent years, it has been exhibited at fairs in USA, Sweden, Germany and the UK. It can be operated with 110 V or 220 to 240 V and has been used by TWI and its member companies.

TWI also commissioned a prototype machine designed and manufactured by their CRC partners at the Department for Mechanical Engineering of the University of Adelaide, Australia. Portable CRC machine can be carried and aligned by two operators without the use of a crane or other lifting device. It has been used to weld curved sheets under site conditions in a shipyard [2].

15.5.2 ESAB machines

ESAB is the company that has contributed so many powerful and accurate FSW machines to industry, and nearly all fields of industry are using its productions to perform FSW.

One of the machines, installed at Marine Aluminium Aanensen, Norway, is capable of making 16 m long welds (Figure 15.40). It was built by ESAB in Laxå, Sweden and is used for the mass production of panels that are made by joining extruded profiles. This gigantic machine is shown in Figure 15.41.

Several shorter-length machines, some of them with CNC systems of up to five axes, have been built for experimental use and for the production of prototype parts. ESAB also introduced a FSW machine with a welding speed of 750 mm/min in the 5 mm thick aluminium 6000 series at the 14th International Welding Fair in Essen.

The other company is SAPA in Finspång, Sweden, which uses FSW in the high-volume production of aluminium components of automotives. An ESAB SuperStir™ machine is installed, which has two welding heads

Figure 15.40 One of the ESAB FSW machines [5]

Figure 15.41 ESAB SuperStir FSW equipment installed at Marine Aluminium for welding panels [5]

to weld hollow extrusions from both sides simultaneously (Figure 15.42). The machine has a carousel-type loading and unloading station and is automatically loaded by an articulated arm robot. The FSW machine is at the same site as the extrusion press, so the surface quality and tolerances of the extrusions are extremely good prior to welding. The automotive

Figure 15.42 SAPA's ESAB SuperStir™ machine with two welding heads for welding hollow aluminium extrusions from both sides simultaneously [10]

components have very tight tolerances, which would be difficult to achieve with conventional fusion welding methods. Now they are produced by FSW with cycle periods of less than 60 seconds.

In general, utilizing FSW in joining of extrusions requires a custom linear machine, and the length of the weld, the required travel speeds to make this process cost competitive, and the associated process forces are the parameters to be controlled. The capital cost of these machines is significant. However, due to the use of lower cost extrusions, a lower net cost is often produced. In this application, travel speed is a critical factor in the net cost of the process, as the largest cost associated with the operation is the capital cost of the machine, since there are generally no consumables with the FSW process, which results in minimal operating cost. An example of a custom production machine, manufactured by ESAB in Laxå, Sweden, is shown in Figure 15.43. This particular machine is used for the production of link arms similar to that illustrated previously.

ESAB SuperStir™ was delivered to Tower Automotive in 2000. The machine is equipped with two separate welding heads for simultaneous welding from top and bottom, in order to obtain symmetrical heat distribution and avoid any root problems. ESAB has also entered the aviation industry; Figure 15.44 shows the first commercial FSW machine, nowadays being used in the Boeing Company. Figures 15.45 and 15.46 show the circumferential welding machine and longitudinal welding machine, respectively.

Figure 15.43 ESAB SuperStir™ at Tower Automotive [1]

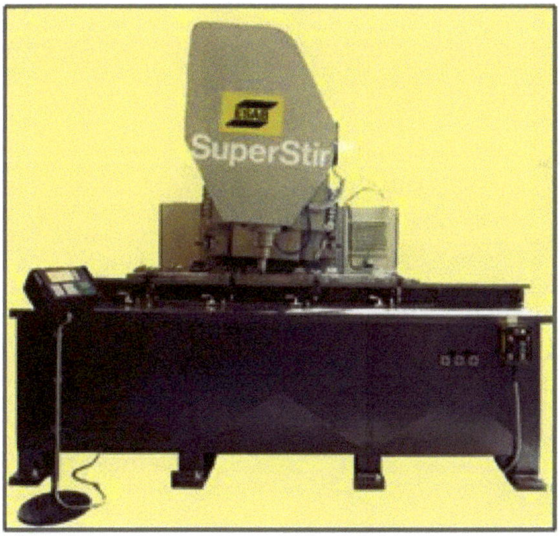

Figure 15.44 SuperStir™ #1: the first FSW machine introduced to the general public in 1997 at the Schweissen and Schneiden Fair, currently used at Boeing laboratories for R&D [1]

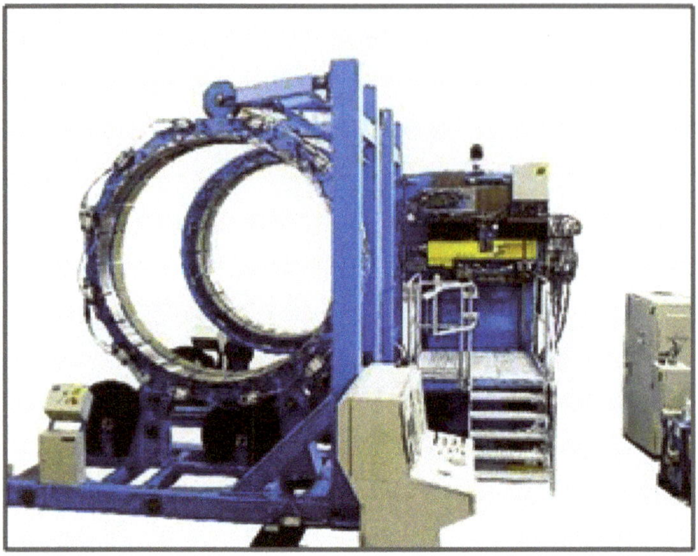

Figure 15.45 Circumferential welding machine [1,5]

Figure 15.46 Longitudinal welding machine for joining of machined honeycomb structured sections. Earlier the segments were joined by MIG or VPPAW (variable polarity plasma-arc welding) [1]

15.5.3 Other machines

Figure 15.47 shows one of the largest multi-axis FSW systems within the USA. The advanced and complicated FSW machine, which is shown in detail in the picture on the right, demonstrates the great ability of this system to perform welds in multidirectional geometries. The strong vertical support of this system allows it to weld thick workpieces as well as conventional plates.

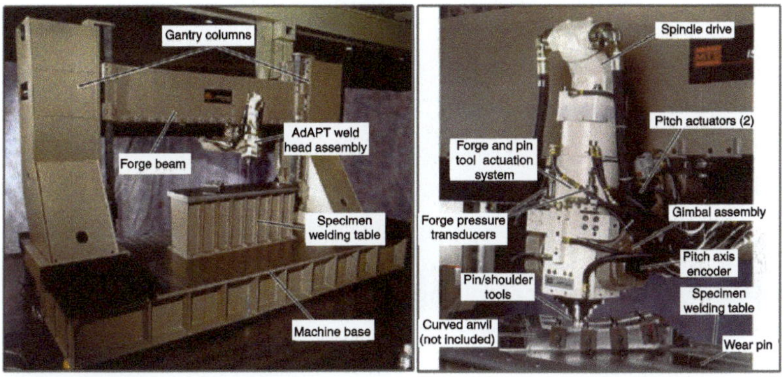

Figure 15.47 One of the largest multi-axis FSW systems within the US, at AMP [16]

Figure 15.48 shows a typical example of a vertical milling machine converted to a FSW machine. With the right tools and simple modifications, the milling machine could be used for FSW. The machine basically consists of a motor, which helps to spin the FSW tool and provide the heat on the specimen due to friction.

15.6 Robotic friction stir welding

The applications of the FSW process, as mentioned in the previous sections, are increasing fast. However, implementing this method on specially designed, high-cost machinery has one major drawback: the process is not flexible, specifically in performing welds in complicated geometries. Any changes in the process parameters require adaption of the hardware, that is, the FSW machine.

Figure 15.48 A vertical milling machine converted to a FSW machine [4]

The use of an industrial robot for FSW could be a major improvement, because many process changes can be intercepted by software adaption. Since the first industrial robots had insufficient stiffness and they could not apply the required down force, so unsuitable for FSW. Nowadays, there are robots with very high payloads, of 1000 kg or more, which are also stiff enough to perform the process.

There are two different robot concepts useful for the FSW process, parallel and serial constructions [17]. Even though a parallel construction is much stiffer and provides good results, it has some major drawbacks. The system becomes more complex, has a very limited workspace and the cost is much higher in comparison to a serial robot with the same payload. Nevertheless, the compliance problems are very detrimental in serial designs; this design is of more interest for applications in the automobile industry.

15.6.1 Robot design

According to the difficulties already mentioned, at present only a few welding companies offer FSW robots. The basic design of this robot is explained briefly as follows.

In general, the sixth axis of the robot frame allows the body to turn around the z-axis. But since the tool of the FSW equipment has the same rotation, the last robot axis becomes redundant and may be removed [17]. Usually this method of welding is associated with high rotational speeds and high torques, therefore a powerful motor is required.

The motor can be either hydraulic or electric. The major advantage of a hydraulic motor is the lower volume-to-power ratio achieved with it. In these experiments, successful welds can only be performed in a very limited workspace in front of the robot, and when the z-axis of the frame does not deviate much from the vertical axis. In the case of an electromotor, for a successful plunge operation, it is found that the motor should be able to provide a torque of at least 60 Nm, which is not feasible with a small motor inside the wrist. However, the motor can be placed outside the wrist. Furthermore, the motor of the sixth axis may be removed, and replaced by a suitable gearbox for the FSW equipment. These two changes result in less compliance issues, lower torque required, and good welding qualities in a predefined workspace.

When it comes to FSW, the increasing forces are no longer an issue, since they are located in the FSW equipment itself and not in the robot joints.

15.6.2 Controlling the system

A robot in general applications moves freely in the air and is subjected to relatively small forces. It uses position control to keep the tool head on the correct path. When it comes to robotic FSW, there are huge forces produced in the joints of the robot, which result in compliance problems. So a new control strategy is implemented, namely 'force control'. This new approach greatly improves the welding quality.

When the real path differs from the planned path using position control, the robot tries to follow the planned path, no matter how difficult it may be. As shown in Figure 15.49, the tool tends to follow the path A to B, but since the surface does not fit with the predefined path, the robot forces increase until the maximum permissible forces for joints are reached. The result is often unstable robotic behaviour and thus a bad welding quality. If the robot has force control in the direction normal to the surface, the tool will be slightly lifted between C and B, on a plate as in Figure 15.49, and the force will be kept stable at a predefined value [17].

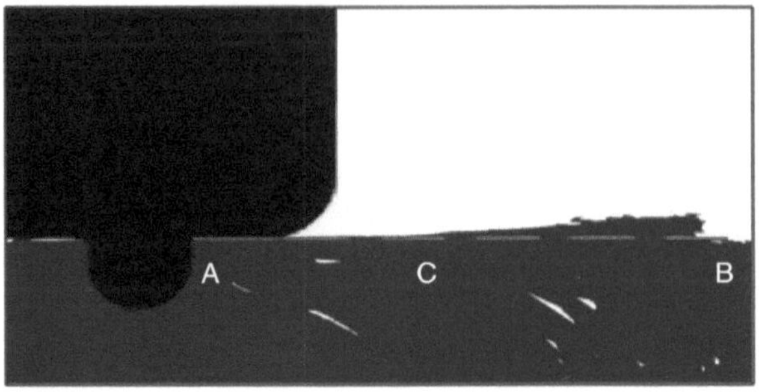

Figure 15.49 Problem with pure position control [17]

15.6.3 Force control methods

Between the end effectors and the last robot axis, a force and a torque sensor are implemented. There are several approaches to handle these measured variables. Generally, they are divided into two categories, namely:

1. *Direct force control*: This includes all the force control strategies that directly generate the actuation signal; computing the force error function commonly using a Proportional, Integral and Derivative (PID) controller. The user can set the desired force value. The main advantage of this system is good results combined with simplicity, because there is no need for a more sophisticated controller if the results of the PID controller are satisfactory (Figure 15.50).

2. *Indirect force control*: This strategy uses an external force loop to generate position commands to an internal loop, constituted by a position-based force controller. The external force loop (with force sensor feedback) generates the position reference for the internal loop (with position encoder feedback), using an admittance relation. This relation is often represented as a second-order mass-spring-damper system. This strategy was introduced for practical reasons, since the current manipulator control systems have built-in position controllers that do not allow direct access to the joint motor torques (Figure 15.51).

The most adopted technique to use force control is the hybrid approach. This means that force control is applied in the z-direction, normal to the surface. The x- and y-directions are controlled by position control. The robot can thus never know where its head is exactly located, since the location in the z-direction is defined by the measured force.

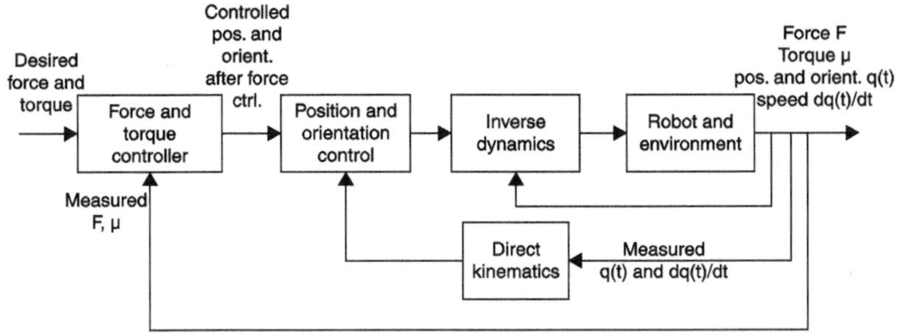

Figure 15.50 Direct force control scheme [17]

When the tool moves to the first welding position, it moves in free air and force control is obviously disabled.

At the beginning of the welding operation, the pin must be plunged into the material. For this operation, force control is necessary for good welding results, so there should be a switch from position control (in free air) to force control (during the plunge). This is a difficult part because as soon as force control is turned on, the pin tends to crash into the material. This problem can be resolved by turning on force control when the pin is very close to the work. By switching the control algorithm right at the moment that the contact between the tool and the workpiece is being made, there are no huge forces produced before the contact is formed and thus no fear of having the tool crashing into the part. After the contact is made, the force control turns on, and gradually increases the force until the proper depth of plunge is achieved [17].

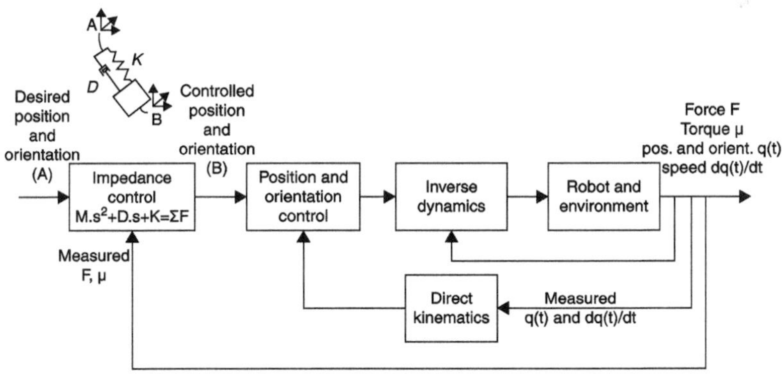

Figure 15.51 Indirect force control scheme [17]

When the tool moves along the path, for example in the x-direction, deviations in the y-direction also arise, due to force differences between the advancing and retreating side of the tool. An extra control element can be implemented, where these deviations are corrected by a linear estimation, based on previous measurements.

15.6.4 Considerations regarding force control

Force control is based on keeping a constant pressure in the z-direction while the tool moves in the x-y plane. During welding, there is heat flow away from the welding point. This implies that the material in the next welding point is warmer and relatively softer. Often the applied force must therefore be reduced along the path. In cases where the workpiece is small, with a smooth surface and a simple welding geometry, it is recommended to use only position control, since it is difficult to define the connection between temperature and required force. It can be useful to investigate the relationship between a measured temperature and the necessary force.

15.6.5 Welding parameters

Compared to traditional welding processes such as Arc and TIG welding, the number of process parameters to be controlled can be remarkably reduced [17]. There is no gas flow, no current and no wire supply and no water flow. The most important parameters for robotic FSW are:

- rotational speed;
- travel speed;
- normal force;
- lateral force;
- tool attitude (or tilt angle);
- shoulder plunge;
- penetration ligament (mostly in butt joints);
- penetration into the bottom member (in lap joints).

In fact, there are only five parameters considered in robotic FSW applications, such as down force, rotational speed, travel speed, plunge depth and tilt angle [17]. It is clear that there are fewer parameters that can cause problems in FSW compared to MIG welding with 13 to 24 parameters, TIG welding with 15 or plasma welding with the impressive number of around 24 parameters. Fewer parameters mean fewer problems, which result in higher repeatability and faster project switching. As mentioned earlier, the lateral forces are not considered in robotic FSW. When the robot welds with lateral force control, it can never have any idea of where its head is located in the workspace.

15.6.6 Path planning

The most frequently applied approach nowadays is offline programming with robot software. For commonly-used applications such as spot welding, there are toolboxes in CAD software to simulate the optimal collision free path. Unfortunately, these toolboxes are not yet available for FSW, so the paths in these applications must be user defined in every single location.

15.6.7 Complex geometries

The main reason to choose a robot instead of the classical FSW machines is the great flexibility obtained by the use of such devices; since in this way the need of changing the whole machine parameters for any new project is eliminated and by applying simple programming changes, we can use the same robot for a wide range of industrial applications, no matter if they are in automotive, shipbuilding, railway, aerospace or other industries. The advantages of applying robotic FSW also include the ability of welding complex paths in a multi-dimensional workspace, instead of just a planar workspace. The current FSW robots are able to weld almost every type of curve in a horizontal plane. Welding in the vertical plane requires some more efforts put into experience, since in such geometries the robot joints are more forced into their maximal deflections.

15.6.8 Instances

The four pictures shown in Figure 15.52 illustrate the conceptual design of a FSW robot to weld TWBs, which are mainly used in automotive manufacturing industries in a wide range of applications. As is clear, such a robot may help the company in many ways; first, because of the wide range of applications that may be considered for it and second, with the perfect quality that FSW contributes to the components and products.

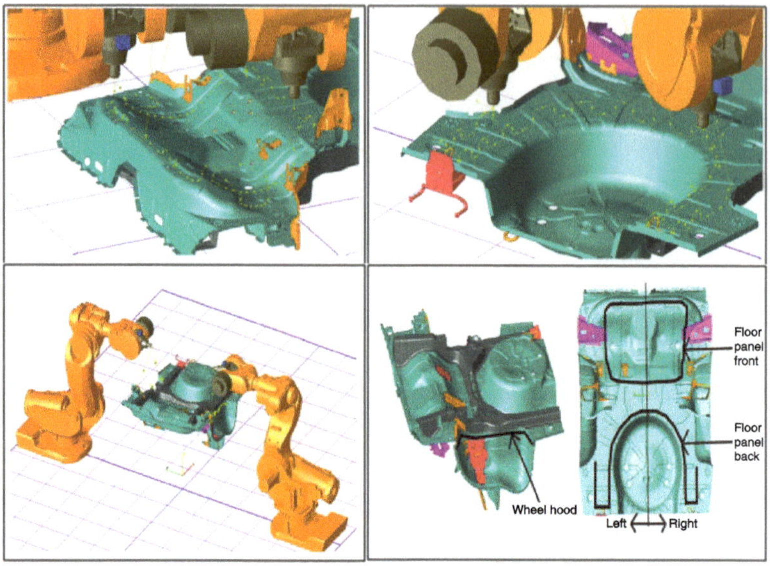

Figure 15.52 **Conceptual design of a FSW robot to weld tailor welded blanks [7]**

Work at Tower Automotive has shown that a low-cost standard industrial robot (ABB IRB 6400) can also perform FSW (Figure 15.53). This robot has been able to make welds in 6061-T6 aluminium in thicknesses up to 4 mm at travel speeds of 1 m/min. With lower material thicknesses, even greater travel speeds have been achieved. The enabling technology for this robot to perform FSW is its open architecture, which allows the development of a force control algorithm, so that the robot can overcome its stiffness limitations. However, for the standard industrial robot to be accepted as a high volume production capable machine, some increase in stiffness will still be necessary.

And finally, Figure 15.54 shows a very high-tech FSW robot, produced by ESAB, which benefits from sufficient stiffness and can perform defect-free 3-D welds in different geometries.

According to what has been stated in this chapter, we may predict a promising future for the applications of FSW in different industries, and in this way, of course, the role of robots for optimizing the mass production lines will be undeniable. All in all, we conclude that special attention should be paid to this innovative welding technique to improve its quality and worthiness further than what has been achieved so far.

Figure 15.53 FSW system integrated to a standard industrial robot [6]

Figure 15.54 ESAB robot for FSW [5]

15.7 References

1. ESAB (2012), *Friction stir Welding – the ESAB Way.*
2. *www.twi.co.uk*
3. 'Honda develops new technology to weld together steel and aluminum and achieves world's first application to the frame of a mass-production vehicle', 6 September 2012, available from: *http://world.honda.com/*
4. Fulufhelo, M. (2009), 'Conceptual design of a friction stir welding machine for joining rails', MSc in Engineering, University of Witwatersrand, Johannesburg.
5. ESAB Technical Handbook (2012), *Friction stir Welding.*
6. Smith, C.B., Crusan, W., Hootman, J.R., Hinrichs, J.F., Heideman, R.J. and Noruk, J.S. (2001), 'Friction stir welding in automotive industry', Tower Automotive, Technology Application.
7. Shah, S. and Tosunoglu, S. (2012), 'Friction stir welding: Current state of the art and future prospects', *Proceedings of the 16th World Multi-Conference on Systemics, Cybernetics and Informatics*, 17–20 July.
8. Space Shuttle Technology Summary (2001), *Friction stir Welding*, NASA.
9. Hitachi Rail Europe Ltd., *Railway Strategies Live 2010*, available from: *www.hitachi-rail.com*
10. Kallee, S.W., Nicholas, E. and Thomas, W.M. (2001) 'Friction stir welding – invention, innovations and applications', *Proceedings of the 8th International Conference on Joints in Aluminium*, TWI Ltd., Munich: 28–30 March.
11. Arbegast, W., Reynolds, T., Mishra, R.S., Nelson, T. and Burford, D. (2011), 'Littoral combat system with improved welding technologies', December, available from: *http://faculty.washington.edu/scottcs/NSF/2012/CFSP.pdf*
12. Euro-Composites, Applications, Friction stir welding, Manufacturing of heat sink extrusion profiles (2012), available from: *www.euro-composites.com*, 2012/08/07, version 3.0.
13. Dalder, E., Pastrnak, J.W., Engel, J., Forrest, R.S., Kokko, E. et al. (2008), 'Friction stir welding of thick-walled aluminium pressure vessels', Lawrence Livermore National Laboratory and Advanced Joining Technologies, Inc., April.
14. Grant, G. and Weil, S. (2011), 'Joining technologies for coal power applications', *DOE-FE Annual Review Meeting of Advanced Research Materials Program*, April.
15. Lammlein, D.H., Gibson, B.T., DeLapp, D.R., Cox, C., Strauss, A.M. and Cook, G.E. (2011), 'The friction stir welding of small-diameter pipe: An experimental and numerical proof of concept for automation and manufacturing', Nashville, TN: Welding Automation Lab, Vanderbilt University.
16. Arbegast, W.J. (2007), *Friction stir Welding and Aircraft Applications*, Part I: University of North Dakota Aerospace Fall Workshop.
17. De Backer, J. and Verheyden, B. (2009), 'Robotic friction stir welding for automotive and aviation applications', MSc, Mechanical Engineering, University West, Department of Technology, Mathematics and Computer Science, May.

Index

FSC
www.fsc.org
MIX
Papier | Fördert gute Waldnutzung
FSC® C083411

Kolibri 360 GmbH
Ferdinand-Jühlke-Straße 7
99095 Erfurt, Deutschland
produktsicherheit@kolibri360.de